Interdisciplinary Applied Mathematics

Volume 16

Interdisciplinary Applied Mathematics

1. *Gutzwiller:* Chaos in Classical and Quantum Mechanics
2. *Wiggins:* Chaotic Transport in Dynamical Systems
3. *Joseph/Renardy:* Fundamentals of Two-Fluid Dynamics:
   Part I: Mathematical Theory and Applications
4. *Joseph/Renardy:* Fundamentals of Two-Fluid Dynamics:
   Part II: Lubricated Transport, Drops and Miscible Liquids
5. *Seydel:* Practical Bifurcation and Stability Analysis:
   From Equilibrium to Chaos
6. *Hornung:* Homogenization and Porous Media
7. *Simo/Hughes:* Computational Inelasticity
8. *Keener/Sneyd:* Mathematical Physiology
9. *Han/Reddy:* Plasticity: Mathematical Theory and Numerical Analysis
10. *Sastry:* Nonlinear Systems: Analysis, Stability, and Control
11. *McCarthy:* Geometric Design of Linkages
12. *Winfree:* The Geometry of Biological Time (Second Edition)
13. *Bleistein/Cohen/Stockwell:* Mathematics of Multidimensional
    Seismic Imaging, Migration, and Inversion
14. *Okubo/Levin:* Diffusion and Ecological Problems: Modern Perspectives
    (Second Edition)
15. *Logan:* Transport Modeling in Hydrogeochemical Systems
16. *Torquato:* Random Heterogeneous Materials: Microstructure and
    Macroscopic Properties
17. *Murray:* An Introduction to Mathematical Biology
18. *Murray:* Mathematical Biology: Spatial Models and Biomedical
    Applications
19. *Kimmel/Axelrod:* Branching Processes in Biology

Salvatore Torquato

Random Heterogeneous Materials

Microstructure and Macroscopic Properties

Springer

Salvatore Torquato
Department of Chemistry
and
Princeton Materials Institute
Princeton University
Princeton, NJ 08544
USA
torquato@electron.princeton.edu

Mathematics Subject Classification(2000): 82Bxx, 78 02, 73-02, 60D05

Library of Congress Cataloging-in-Publication Data
Torquato, S.
Random heterogeneous materials: microstructure and macroscopic properties/Salvatore Torquato.
p.cm.—(Interdisciplinary applied mathematics; v. 16)
Includes bibliographical references and index.
ISBN 0-387-95167-9 (alk. paper)
1. Inhomogeneous materials. 2. Microstructure. I. Title. II. Series.
TA418.9.153 2001
620.1′1—dc21. 2001020203

ISBN 0-387-95167-9
ISBN 978-0-387-95167-6 Printed on acid-free paper.

Printed in the United States of America.

9 8 7 6 5 4 3

springer.com

To My Wife,

KIM

My Daughters,

MICHELLE and LISA

and My Parents,

PALMA and VINCENT

"The fairest thing we can experience is the mysterious. It is the fundamental emotion which stands at the cradle of true art and true science. He who does not know it and can no longer wonder, no longer feel amazement, is as good as dead, a snuffed-out candle."

—Albert Einstein, *Forum and Century* (1930)

"How novel and original must be each new man's view of the universe—for though the world is so old—and so many books have been written—each object appears wholly undescribed to our experience—each field of thought wholly unexplored...."

—Henry David Thoreau, *Journal 4:421* (1852)

Preface

The interdisciplinary subject of random heterogeneous materials has experienced remarkable growth since the publication of the well-known monograph *Statistical Continuum Theories* by Beran (1968). Many of these advances, especially those concerning the statistical characterization of the microstructure and its effect on the physical properties of the material, have not been treated fully in any book. One of the intents of the present book is to fill this gap. This book also distinguishes itself in that it provides a unified rigorous framework to characterize the microstructures and macroscopic properties of the widely diverse types of heterogeneous materials found in nature and synthetic products. Emphasis is placed on providing foundational theoretical methods that can simultaneously yield results of practical utility.

This book treats a wide breadth of topics, but the choice of subjects naturally reflects my own interests. The sheer enormity of the field has prevented me from covering many important topics. I apologize to those colleagues, known and unknown, who may not find enough of their own work cited in the ensuing pages.

This book is intended for graduate students and researchers from various walks of scientific life, including applied mathematicians, physicists, chemists, materials scientists, engineers, geologists, and biologists. In order to reach this broad audience, I have attempted to make the book as self-contained as possible, assuming only a rudimentary knowledge of probability theory, statistical mechanics, advanced calculus, and continuum mechanics. In cases where I have fallen short in this regard, the numerous references provided should satiate the voracious appetite for knowledge of the most curious minded among us. The book contains as many proofs and derivations of key results as can be accommodated within the aforementioned constraints. All of these features and an attempt to avoid technical jargon should make the book accessible to the nonspecialist. Indeed, it is my hope that motivated experimentalists will find this book useful.

The book is divided into two parts. Part I describes basic concepts and recent advances in quantitatively characterizing the microstructure of random heterogeneous materials. Topics covered include the statistical mechanics of many-particle systems, the canonical n-point correlation function, lattice and continuum percolation theory, local volume-fraction fluctuations, computer-simulation methods, image analyses and reconstructions of real materials, and models of microstructures.

Part II treats a wide variety of macroscopic transport, electromagnetic, mechanical, and chemical properties of heterogeneous materials and describes how they are linked to the microstructure of model and real materials. Topics covered include homogenization theory, variational principles and rigorous bounds, phase-interchange relations, exact results, effective-medium approximations, cluster expansions, contrast expansions, and cross-property relations.

A brief description of the topics covered in each chapter of the present book is given towards the end of Chapter 1, which provides the motivation for the book and an overview of its contents. It is unique among the chapters because it is purposely written in the nontechnical style of *Scientific American* in order to introduce the key ideas to the interdisciplinary audience for which the book is intended. The book is an outgrowth of a graduate course that I teach at Princeton University. The course begins with Chapter 1 and then immediately skips to Part II and covers in varying depths Chapters 13–21. I then cover most of the material contained Chapters 2-12 of Part I and subsequently return to Part II to cover Chapters 22 and 23. I follow this sequence because the property/microstructure connection (Part II) provides a motivation for the reasons why we will ultimately need to quantify the microstructure (Part I). However, this sequence certainly does not need to be followed in a course, especially if one is primarily interested in microstructural analysis. Although there is substantial cross referencing between Parts I and II, each part has been designed to be relatively independent of the other. Nonetheless, I believe that Parts I and II make a cohesive unit, and ideally, they should be read together. Those interested in further reading on the theme of Part II of this book are referred to the recent books by Cherkaev (2000) and Milton (2001), which emphasize the general theory of composites.

One of the most enjoyable parts of writing this book is thanking the many people without whose support it would have never been written. The contributions of my collaborators over the years, many of whom are cited in the text, have enriched my scientific experience. I thank George Stell, my former advisor at the State University of New York at Stony Brook, who instilled in me a love for research and introduced me to statistical mechanics and composite media. Hajime Sakai, John Quintanilla, Louis Bouchard, Juan Eroles, Sangil Hyun, Edward Garboczi, Konstantin Markov, Leonid Gibiansky, Tony Roberts, Thomas Truskett, Frank Stillinger, and Leonid Berlyand carefully read various portions of the manuscript and provided valuable criticisms and suggestions. I am deeply indebted to all of them. I gratefully acknowledge fruitful and illuminating discussions with Erhan Cinlar, Thomas Spencer, Marco Avellaneda, and

Robert Kohn on certain aspects of the book. Many thanks are due to Christopher Yeong, Anuraag Kansal, and Juan Eroles, who produced the majority of the numerous figures that have greatly enhanced the book. Thomas Truskett wrote the first version of the Monte Carlo program that appears in one of the appendices. George Dvorak, Nick Martys, and Paul Stutzman each generously supplied a figure from their respective work.

I thank the Department of Energy and Air Force Office of Scientific Research for supporting much of my work. My year-long sabbatical in the School of Mathematics at the Institute for Advanced Study in Princeton, New Jersey as a Guggenheim Fellow during 1999–2000 enabled me to focus my efforts on the final product before you. I am especially grateful to Princeton University for their support and encouragement of my scholarship over the years.

On the home front, I want to express my deep thanks to my wife, Kim, and daughters, Michelle and Lisa, for their love, understanding, perseverance, and patience. Their unfailing support has made this book possible. Finally, I acknowledge the sacrifices made by my parents, Palma and Vincent, that enabled me to pursue my dreams.

The author would be grateful for reports of typographical and other errors to be sent electronically via the following webpage for the book:

http://cherrypit.princeton.edu/book.html,

where an up-to-date errata list will be maintained.

Princeton, New Jersey
June 2001

Salvatore Torquato

Contents

CHAPTER 1

Motivation and Overview

The determination of the transport, electromagnetic, and mechanical properties of heterogeneous materials has a long and venerable history, attracting the attention of some of the luminaries of science, including Maxwell (1873), Rayleigh (1892), and Einstein (1906). In his *Treatise on Electricity and Magnetism*, Maxwell derived an expression for the effective conductivity of a dispersion of spheres that is exact for dilute sphere concentrations. Lord Rayleigh developed a formalism to compute the effective conductivity of regular arrays of spheres that is used to this day. Work on the mechanical properties of heterogeneous materials began with the famous paper by Einstein in which he determined the effective viscosity of a dilute suspension of spheres. Since the early work on the physical properties of heterogeneous materials, there has been an explosion in the literature on this subject because of the rich and challenging fundamental problems it offers and its manifest technological importance.

1.1 What Is a Heterogeneous Material?

In the most general sense, a *heterogeneous material* is one that is composed of domains of different materials (phases), such as a composite, or the same material in different states, such as a polycrystal. This book focuses attention on the many instances in which the "microscopic" length scale (e.g., the average domain size) is much larger than the molecular dimensions (so that the domains possess macroscopic properties) but much smaller than the characteristic length of the macroscopic sample. In such circumstances, the heterogeneous material can be viewed as a continuum on the microscopic scale, subject to classical analysis, and macroscopic or *effective* properties

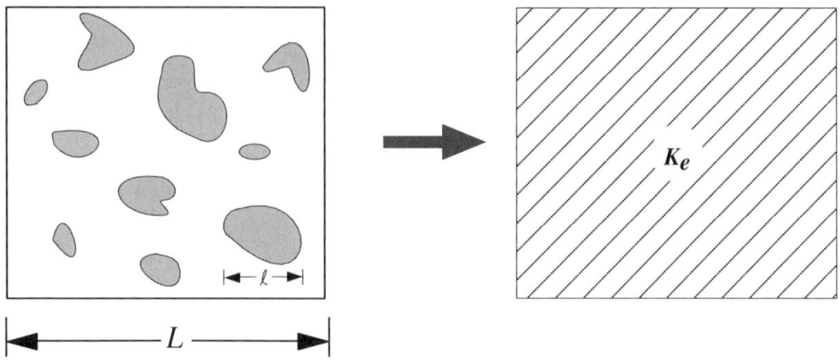

Figure 1.1 Left panel: A schematic of a random two-phase material shown as white and gray regions with general phase properties $\boldsymbol{K}_1$ and $\boldsymbol{K}_2$ and phase volume fractions ϕ_1 and ϕ_2. Here L and ℓ represent the macroscopic and microscopic length scales, respectively. Right panel: When L is much bigger than ℓ, the heterogeneous material can be treated as a homogeneous material with effective property $\boldsymbol{K}_e$.

can be ascribed to it (see Figure 1.1). Such heterogeneous media abound in synthetic products and nature. Synthetic examples include:

- aligned and chopped fiber composites
- particulate composites
- interpenetrating multiphase composites
- cellular solids
- colloids
- gels
- foams
- microemulsions
- block copolymers
- fluidized beds
- concrete

Some examples of natural heterogeneous materials are:

- polycrystals
- soils
- sandstone
- granular media
- Earth's crust
- sea ice
- wood
- bone
- lungs
- blood
- animal and plant tissue
- cell aggregates and tumors

The physical phenomena of interest occur on "microscopic" length scales that span from tens of nanometers in the case of gels to meters in the case of geological media. Structure on this "microscopic" scale is generically referred to as *microstructure* in this book.

In many instances, the microstructures can be characterized only statistically, and therefore are referred to as *random heterogeneous materials*, the chief concern of this book. There is a vast family of random microstructures that are possible, ranging from dispersions with varying degrees of clustering to complex interpenetrating *connected* multiphase media, including *porous media*. A glimpse of the richness of the possible microstructures can be garnered from Figures 1.2 and 1.3, which depict examples of synthetic and natural random heterogeneous materials, respectively.

Beginning from the top, the first example of Figure 1.2 shows a scanning electron micrograph of a colloidal system of hard spheres of two different sizes. The second example is an optical image of the transverse plane of a fiber-reinforced material: ceramic–metal composite (cermet) made of alumina (Al_2O_3) fibers (oriented perpendicular to the plane) in an aluminum matrix. Note the clustering of the fibers. The last example shows a processed optical image of a cermet that is primarily composed of boron carbide (black regions) and aluminum (white regions). Both of these phases are connected across the sample (interpenetrating) even though, from a planar section, it appears that only the black phase is connected. In all of these examples, the microstructure can be characterized only statistically.

Beginning from the top, the first example of Figure 1.3 shows a planar section through a Fontainebleau sandstone obtained via X-ray microtomography. As we will see, this imaging technique enables one to obtain full three-dimensional renderings of the microstructure (see Figure 12.14), revealing that the void or pore phase (white region) is actually connected across the sample. The second example shows a scanning electron micrograph of the porous cellular structure of cancellous bone. The third example shows an image of red blood cells, one of a number of different particles contained in the liquid suspension of blood.

1.2 Effective Properties and Applications

We will consider four different classes of problems as summarized in Table 1.1 on page 7. We will focus mainly on the following four steady-state (time-independent) effective properties associated with these classes:

1. Effective conductivity tensor, $\boldsymbol{\sigma}_e$
2. Effective stiffness (elastic) tensor, $\boldsymbol{C}_e$
3. Mean survival time, τ
4. Fluid permeability tensor, $\boldsymbol{k}$

In each case, the phase properties and phase volume fractions (fractions of the total volume occupied by the phases) are taken to be given information. Depending on

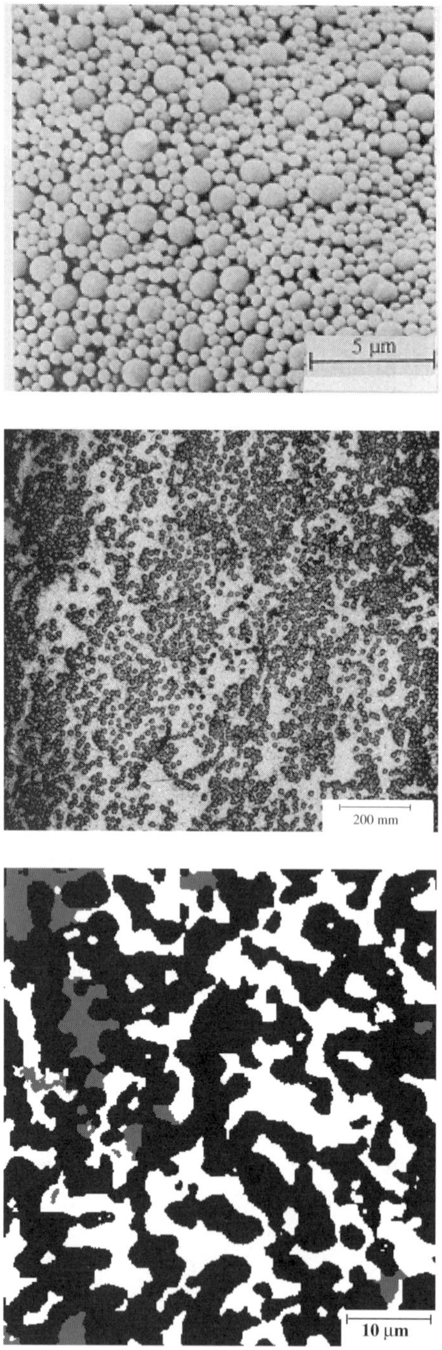

Figure 1.2 Synthetic random heterogeneous materials. From top to bottom: Colloidal system of hard spheres of two different sizes (Thies-Weesie 1995), fiber-reinforced cermet (courtesy of G. Dvorak), and an interpenetrating three-phase cermet composed of boron carbide (black regions), aluminum (white regions), and another ceramic phase (gray regions) (Torquato et al. 1999a).

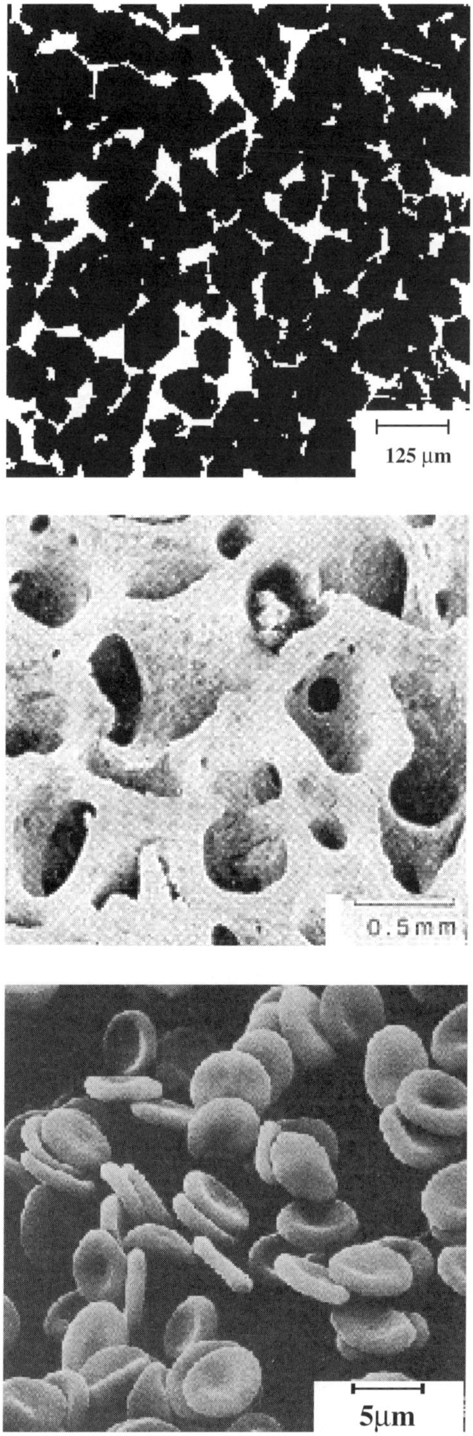

Figure 1.3 Natural random heterogeneous materials. From top to bottom: Fontainebleau sandstone [data taken from Coker et al. (1996)], cellular structure of cancellous bone (Gibson and Ashby 1997), and red blood cells (Alberts et al. 1997).

the physical context, each phase can be either solid, fluid or void. We will also examine certain relaxation times associated with time-dependent transport processes in heterogeneous media.

1.2.1 Conductivity and Analogous Properties

The quantity σ_e represents either the *electrical* or *thermal conductivity tensor*, which are mathematically equivalent properties. It is the proportionality constant between the average of the local electric current (heat flux) and average of the local electric field (temperature gradient) in the composite. This *averaged* relation is Ohm's law or Fourier's law (for the composite) in the electrical or thermal problems, respectively. More generally, for reasons of mathematical analogy, the determination of the effective conductivity translates immediately into equivalent results for the effective dielectric constant, magnetic permeability, or diffusion coefficient (see Chapter 13). Therefore, we refer to all of these problems as class A problems as described in Table 1.1, adapted after a similar table of Batchelor (1974). Of course, each local field within this class will depend on the local phase properties (as depicted in Figure 1.1) and hence generally will be different from one another. Moreover, whereas the electrical conductivity, thermal conductivity, and diffusion coefficient are transport (nonequilbrium) properties, the dielectric constant and magnetic permeability are equilibrium properties. Observe that the determination of the effective diffusion coefficient of a medium in which one phase is impermeable to mass transport is actually just a special limit of the conductivity problem, namely, the limit in which one of the phases has zero conductivity (see Chapter 13).

A key macroscopic parameter characterizing the electrical/thermal characteristics of a heterogeneous material is the effective electrical/thermal conductivity (Beran 1968, Batchelor 1974, Bergman 1978, Hashin 1983, Milton 1984, Torquato 1987). Knowledge of σ_e is of importance in a host of applications. Electrical applications include composites used as insulators for coatings or electrical components and oil drilling operations, where electrical conductivity measurements of the brine-saturated rock are used to infer information about the permeability of the pore space. Thermal applications range from composites used for insulation, heat exchangers, and heat sinks for electronic cooling to geophysical problems (e.g., determination of the geothermal temperature gradient). In the case of composites used as microwave resonator materials, capacitors, and insulators, the effective dielectric constant is a critical macroscopic characteristic. Applications involving composites with desirable values of the effective magnetic permeability include motors, generators, transformers, and computer disks. Diffusion of tracer particles in fluid-saturated porous media occurs in many industrial processes, such as chromatography, catalysis and oil recovery, and biological processes such as blood transport and transport in cells or through cell membranes. In these instances, the effective diffusion coefficient is a key parameter.

Table 1.1 The four different classes of steady-state effective media problems considered here. $F \propto K_e \cdot G$, where K_e is the general effective property, G is the average (or applied) generalized gradient or intensity field, and F is the average generalized flux field. Class A and B problems share many common features and hence may be attacked using similar techniques. Class C and D problems are similarly related to one another.

Class	General Effective Property K_e	Average (or Applied) Generalized Intensity G	Average Generalized Flux F
A	Thermal Conductivity	Temperature Gradient	Heat Flux
	Electrical Conductivity	Electric Field	Electric Current
	Dielectric Constant	Electric Field	Electric Displacement
	Magnetic Permeability	Magnetic Field	Magnetic Induction
	Diffusion Coefficient	Concentration Gradient	Mass Flux
B	Elastic Moduli	Strain Field	Stress Field
	Viscosity	Strain Rate Field	Stress Field
C	Survival Time	Species Production Rate	Concentration Field
	NMR Survival Time	NMR Production Rate	Magnetization Density
D	Fluid Permeability	Applied Pressure Gradient	Velocity Field
	Sedimentation Rate	Force	Mobility

1.2.2 Elastic Moduli

The effective stiffness (elastic) tensor C_e is one of the most basic mechanical properties of a heterogeneous material (Watt, Davies and O'Connell 1976, Christensen 1979, Willis 1981, Hashin 1983, Milton 1984, Kohn 1988, Nemat-Nasser and Hori 1993, Torquato 2000a). The quantity C_e is the proportionality constant between the average stress and average strain. This relation is the averaged Hooke's law for the composite. An obvious class of composites in which it is desired to know C_e is one where the material must bear some mechanical load. This can include synthetic materials, such as structural composites used in a myriad of applications, or biological materials, such as bone or tendon. The speed and attenuation of elastic waves in fluid-saturated porous media (a detection procedure used in oil and gas exploration) depend upon, among other parameters, the elastic moduli of the media. We note that the problem of finding the effective shear viscosity of a suspension of particles in a liquid is related to the problem of determining the effective shear modulus of the suspension under special limits (Chapter 13), and hence we term these class B problems as described in Table 1.1. Moreover, under certain situations, the effective stiffness tensor completely specifies the effective thermal expansion characteristics of a heterogeneous material (Chapter 15). Finally, we note that there is a correspondence between the elastic and *viscoelastic* properties of a heterogeneous material (Chapter 15).

1.2.3 Survival Time or Trapping Constant

Physical problems involving simultaneous diffusion and reaction in heterogeneous media abound in the physical and biological sciences (Prager 1963a, Berg 1983, Zwanzig 1990, Torquato 1991a, den Hollander and Weiss 1994, Zhou and Szabo 1996, Portman and Wolynes 1999). Considerable attention in the *chemical physics* community has been devoted to instances in which the heterogeneous medium consists of a pore region in which diffusion (and bulk reaction) occurs and a "trap" region whose interface can absorb the diffusing species via a surface reaction. Examples are found in widely different processes, such as heterogeneous catalysis, fluorescence quenching, cell metabolism, ligand binding in proteins, migration of atoms and defects in solids, and crystal growth, to mention but a few. A key parameter in such processes is the *mean survival time* τ, which gives the average lifetime of the diffusing species before it gets trapped. Often it is useful to introduce its inverse, called the *trapping constant* $\gamma \propto \tau^{-1}$, which is proportional to the trapping rate. Interestingly, nuclear magnetic resonance (NMR) relaxation in porous media yields an NMR survival time that is mathematically equivalent to the aforementioned one typically studied in chemical physics, and therefore we term these class C problems, as described in Table 1.1.

1.2.4 Fluid Permeability

A key macroscopic property for describing slow viscous flow through porous media is the fluid permeability tensor $\boldsymbol{k}$ (Beran 1968, Scheidegger 1974, Batchelor 1974, Dullien 1979, Torquato 1991b, Adler 1992). The quantity $\boldsymbol{k}$ is the proportionality constant between the average fluid velocity and applied pressure gradient in the porous medium. This relation is Darcy's law for the porous medium. The flow of a fluid through a porous medium arises in a variety of technological problems. Examples include the extraction of oil or gas from porous rocks, spread of contaminants in fluid-saturated soils, and separation processes such as in chromatography, filtration, biological membranes, and bioreactors. We observe that the problem of particles sedimenting through a liquid shares some similarities to the problem of determining the fluid permeability of the suspension, and hence we term these class D problems, as indicated in Table 1.1.

1.2.5 Diffusion and Viscous Relaxation Times

Relaxation processes associated with the previous two problems of trapping and flow in porous media are also of interest. Specifically, it is desired to know how the concentration and velocity fields decay in time from initially uniform values. Such time-dependent processes are exactly described by a spectrum of relaxation times (inverse eigenvalues) that are intimately related to the pore-space topology. In the trapping and flow problems, we refer to $T_1, T_2, \ldots$ and $\Theta_1, \Theta_2, \ldots$ as the diffusion and viscous relaxation times, respectively. It will be shown that these relaxation times are related to their steady-state counterparts (τ and $\boldsymbol{k}$) as well as to each other.

1.2.6 Definitions of Effective Properties

Given the phase properties $\boldsymbol{K}_1, \boldsymbol{K}_2, \ldots, \boldsymbol{K}_M$ and phase volume fractions $\phi_1, \phi_2, \ldots, \phi_M$ of a heterogeneous material with M phases, how are its effective properties mathematically defined? It will be shown in Chapter 13 that the effective properties of the heterogeneous material are determined by averages of local fields derived from the appropriate governing continuum-field theories (partial differential equations) for the problem of concern. Specifically, any of the aforementioned effective properties, which we denote generally by $\boldsymbol{K}_e$, is defined by a *linear* relationship between an average of a generalized local *flux* $\boldsymbol{F}$ and an average of a generalized local (or applied) *intensity* $\boldsymbol{G}$, i.e.,

$$\boldsymbol{F} \propto \boldsymbol{K}_e \cdot \boldsymbol{G}. \tag{1.1}$$

For the conduction, elasticity, trapping, and flow problems, the average generalized flux $\boldsymbol{F}$ represents the average local electric current (heat flux), stress, concentration, and velocity fields, respectively, and the average generalized intensity $\boldsymbol{G}$ represents the average local electric field (or temperature gradient), strain, production rate, and applied pressure gradient, respectively. The precise nature of (1.1) is discussed in Chapter 13.

Table 1.1 summarizes the average local (or applied) field quantities that determine the steady-state effective properties for all four problem classes. As already noted, the individual problems within class A are mathematically equivalent to each other; the same is true of the problems within class C. The elasticity and viscosity problems of class B share some similarities but are generally different (Chapter 13). The fluid permeability and sedimentation problems of class D are related but generally different (Chapter 13). Whereas the properties of classes A and B are *scale invariant*, the properties of classes C and D are *scale dependent* (Chapter 13). This classification scheme is made more mathematically precise in Chapter 13.

At first glance, the effective properties of one class appear to share no relationship to the effective properties of the other classes. Indeed, the governing equations are

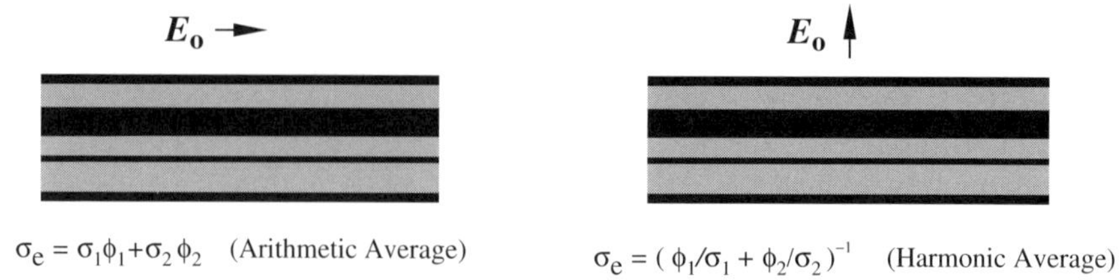

Figure 1.4 A two-phase material consisting of alternating layers of the phases. Formulas (1.2) and (1.3) are exact when the applied field is oriented parallel (left panel) and perpendicular (right panel) to the slabs, respectively.

different from class to class (see Chapter 13 for a complete discussion of this point). Nonetheless, it is shown in Chapter 23 that these apparently different properties can be related to each other via *cross-property* relations.

1.3 Importance of Microstructure

The effective properties of a heterogeneous material depend on the phase properties and microstructural information, including the phase volume fractions, which represent the simplest level of information. It is important to emphasize that the effective properties are generally not simple relations (mixtures rules) involving the phase volume fractions. This suggests that the complex interactions between the phases result in a dependence of the effective properties on nontrivial details of the microstructure.

To illustrate the fact that the effective properties of a random heterogeneous material depend on nontrivial features of the microstructure, we consider two examples. In both instances, we assume that the medium consists of two phases, one with volume fraction ϕ_1 and the other with volume fraction ϕ_2, and so $\phi_1+\phi_2 = 1$. In the first case, it is desired to predict the effective conductivity σ_e of a composite of arbitrary microstructure with phase conductivities σ_1 and σ_2. One might surmise that a reasonable estimate is a simple weighted average of the phase conductivities involving the volume fractions, such as

$$\sigma_e = \sigma_1\phi_1 + \sigma_2\phi_2. \tag{1.2}$$

This arithmetic-average prediction usually grossly overestimates the effective conductivity of isotropic media, especially for widely different phase conductivities. The reason for this discrepancy is that formula (1.2) is exact for the layered composite depicted in the left panel of Figure 1.4 in the direction along the slabs. Thus, because the more conducting phase is always connected across the system along the slab direction, the effective conductivity can be of the order of the more conducting phase. This idealized situation and its close approximants represent a very small subset of possible composite microstructures, and therefore (1.2) can appreciably overestimate the effective conductivities of general composites. On the other hand, one might try to use the

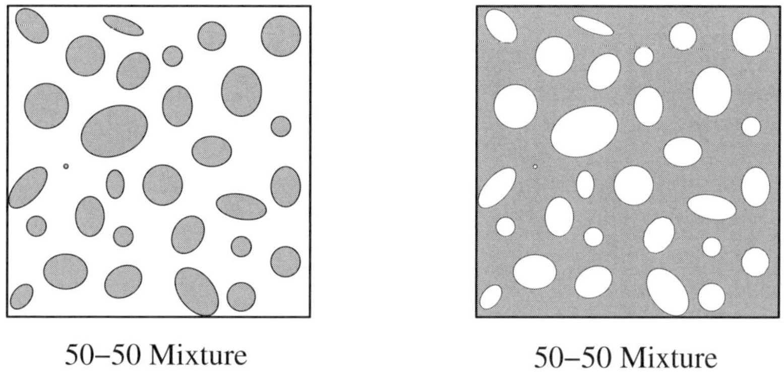

Figure 1.5 Left panel: 50-50 mixture consisting of a disconnected inclusion phase and a connected matrix phase. The gray phase is highly conducting (or stiff) relative to the white phase. Right panel: The same microstructure except the phases are interchanged.

harmonic-average formula

$$\sigma_e = \left(\frac{\phi_1}{\sigma_1} + \frac{\phi_2}{\sigma_2} \right)^{-1} \tag{1.3}$$

to estimate the effective conductivity. This expression, however, typically grossly underestimates the effective conductivity of isotropic media, since it corresponds exactly to the effective conductivity of the layered composite in the direction perpendicular to the slabs (see right panel). It is seen that if one phase is insulating relative to the other, there will be little current perpendicular to the slabs, since the phases are disconnected from one another. In conclusion, estimates based only on incorporating volume-fraction information (i.e., simple mixture rules) cannot capture crucial microstructural features required to estimate accurately the effective conductivity of most composites. Since the conductivity is one of the simplest properties, this last statement applies to all of the other effective properties as well.

In the second example, we consider a 50-50 two-phase system shown in the left panel of Figure 1.5. It consists of a *disconnected* inclusion phase and a *connected* matrix phase. Let the gray "phase" be highly conducting (or stiff) compared to the white "phase." The right panel shows a composite with exactly the same microstructure but with the phases interchanged. Which of the two composites has the higher effective conductivity (or stiffness)? Clearly, the one depicted in the right panel has the higher effective property, since the connected phase here is the more conducting (or stiffer) phase. Thus, even though both composites have the same volume fraction, their effective properties will be dramatically different, implying that the effective properties depend on microstructural information beyond that contained in the volume fractions. Such higher-order microstructural information is the main subject of Part I of this book. We have seen through both examples that connectedness is crucial higher-order information. For this reason, Chapters 9 and 10 are devoted entirely to percolation and clustering in random heterogeneous materials.

To summarize, for a random heterogeneous material consisting of M phases, the general effective property $\boldsymbol{K}_e$ is the following function:

$$\boldsymbol{K}_e = f(\boldsymbol{K}_1, \boldsymbol{K}_2, \ldots, \boldsymbol{K}_M; \phi_1, \phi_2, \ldots, \phi_M; \boldsymbol{\Omega}), \tag{1.4}$$

where $\boldsymbol{\Omega}$ indicates functionals of higher-order microstructural information. The mathematical form that this microstructural information takes is described in the next section.

1.4 Development of a Systematic Theory

In light of the importance of determining the effective properties of heterogeneous media, a vast body of literature has evolved based upon direct measurements (either experimentally or computationally), semiempirical relations, and theoretical techniques. The time and cost to attack this problem by performing measurements on each material sample for all possible phase properties and microstructures are clearly prohibitive. Successful empirical relations tend to be more useful for correlating data rather than predicting them. Inasmuch as the effective property depends not only on the phase properties but is sensitive to the details of the microstructure, it is natural to take the broader approach of predicting the effective property from a knowledge of the microstructure. One can then relate changes in the microstructure quantitatively to changes in the macroscopic property. One of the chief aims of this book is to provide such a *systematic theory* of general random heterogeneous materials.

1.4.1 Microstructural Details

A systematic theory of random heterogeneous materials rests on our ability to describe the "details of the microstructure," by which we mean the phase volume fractions; surface areas of interfaces, orientations, sizes, shapes, and spatial distribution of the phase domains; connectivity of the phases; etc. Quantitatively speaking, we investigate certain n-point correlation functions that statistically describe the microstructure. As will be shown throughout this book, there are a variety of different correlation functions that naturally arise when the averaging process involved in relation (1.1) is rigorously carried out. Roughly speaking, the averaging process results in integrals in which the relevant local fields are weighted with the n-point correlation functions. More precisely, the averages are functionals of the n-point correlation functions.

Many types of correlation functions arise depending on the property and class of microstructures of interest. To give the reader a preview of the concept of a correlation function, we will discuss a specific family of such descriptors that arise in all four problem classes. For simplicity, we consider a two-phase medium that is statistically isotropic (as defined in Chapter 2) and begin with the one-point correlation function

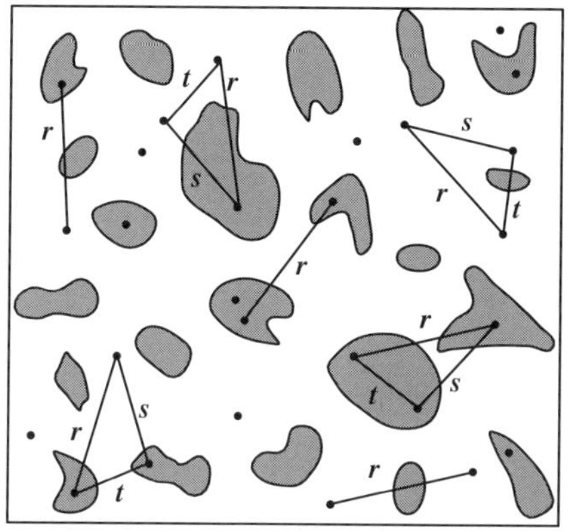

Figure 1.6 A schematic showing attempts at sampling for the correlation functions S_1, S_2, and S_3 from a planar section.

denoted by S_1. Instead of giving here a precise mathematical definition of this quantity, as is done in Chapter 2, we will describe how one would ascertain it from a planar section through the heterogeneous material. The one-point function S_1 is obtained by randomly throwing a single point onto the planar section many times and recording the fraction of times that it lands in one of the phases, say the "white" phase of Figure 1.6. It is clear that if the planar section is big enough and the number of attempts are sufficiently large, S_1 will approach the volume fraction of the white phase. Thus, S_1 is the probability that a single point falls in the white phase. The two-point correlation function $S_2(r)$ is obtained by randomly throwing a line segment of length r into the sample many times and recording the fraction of times that its end points land in the white phase (see Figure 1.6). By performing this experiment for all possible lengths r, one can generate a graph of S_2 as a function of r. Therefore, $S_2(r)$ is the probability that the two end points of a line segment of length r fall in the white phase. Clearly, variations in $S_2(r)$ reflect the extent to which the two points are correlated in the system, and thus $S_2(r)$ contains more information than S_1, which is just a constant. Similarly, $S_3(r,s,t)$ is the probability that the three vertices of a triangle with sides of lengths r, s, and t fall in the white phase. The three-point quantity S_3 embodies more information than S_2. In general, S_n gives the probability that n points with specified positions lie in the white phase.

In Chapters 19 and 20 we demonstrate, using first principles, that the effective properties are indeed generally dependent on an infinite amount of statistical information about the microstructure; this is a direct consequence of the complex field interactions that occur in the heterogeneous material. Of course, for general microstructures, the infinite amount of information can never be ascertained in practice. In light of this limitation, the faint of heart may ask whether one should give up on obtaining structure/property relations? The answer is a definitive *no* for the same reasons that structure/property relations are pursued in any discipline that concerns itself with complex interacting systems, such as materials science, solid and liquid state theory, and statistical physics.

First of all, there are a few special cases in which we do have complete information and hence can compute the effective properties exactly (see Chapters 15, 16, 19, and 20). These examples lend important insight into the salient features that generally determine effective properties. Second, one can develop estimates for the effective properties that incorporate limited microstructural information. Chapter 18 discusses well-known effective-medium approximations that include simple information (volume fractions and shapes). More sophisticated approximations that incorporate three- and four-point information are described in Chapter 21. Third, given partial statistical information on the sample (finite set of correlation functions), one can establish the range of possible values that the effective properties can take, i.e., rigorous upper and lower bounds on the properties. One of the bounds can often yield useful estimates of the effective property even when the other bound diverges from it. Moreover, the study of bounds has important implications for the optimal design of composites. The subject of bounds is treated in Chapters 14, 21, and 22.

It is noteworthy that significant advances have been made recently in the quantitative characterization of the microstructure of random heterogeneous materials both theoretically and experimentally. These breakthroughs, described in Part I, have enabled investigators to compute property estimates (including bounds) that depend upon three- and four-point information for nontrivial models and real materials.

1.4.2 Multidisciplinary Research Area

The study of random heterogeneous materials is a multidisciplinary endeavor that overlaps with various branches of materials science, engineering, physics, applied mathematics, geophysics, and biology, as schematically represented in Figure 1.7. In some cases, the intersections with these disciplines arise because existing models, methods, and results can be applied to study heterogeneous materials and vice versa. In other instances, overlap arises because they share common goals with the study of heterogeneous materials. Moreover, some of the disciplines offer a panoply of intriguing heterogeneous materials that need to be investigated.

One of the central aims of *materials science* is to formulate structure/property relations for single-phase materials (metals, ceramics, and polymers). This formulation is less well developed in the case of composite materials that are composed of combinations of single-phase materials. Because composites can be designed to exhibit the best characteristics of the individual constituents, they are ideally suited in modern technologies that require materials with an unusual combination of properties that cannot be met by conventional single-phase materials. For example, fiber-polymeric composites can be fabricated that have relatively high stiffness, strength, and toughness, and low weight. (The fiber by itself is too brittle, while the polymer alone is too compliant and of low strength.) The ability to tailor composites with a unique spectrum of properties rests fundamentally on a systematic means to relate the effective properties

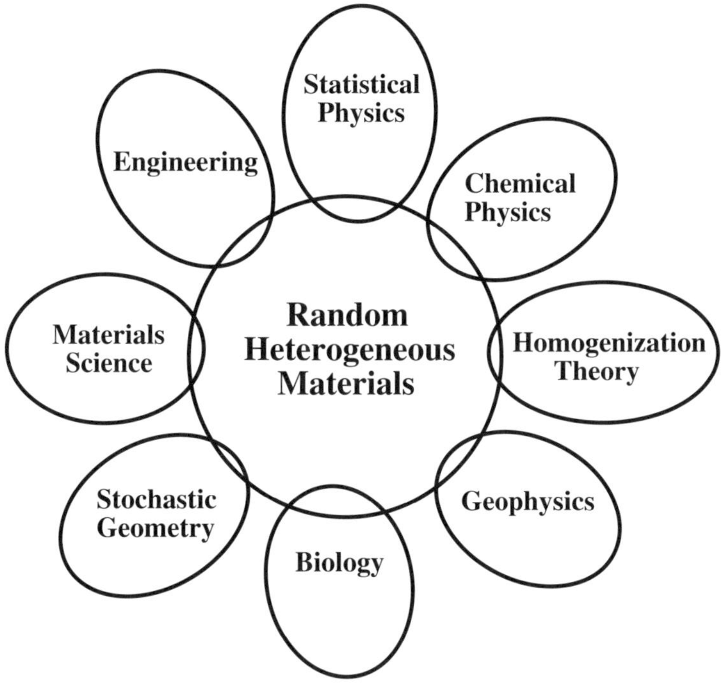

Figure 1.7 The various disciplines that intersect the study of random heterogeneous materials.

to the microstructure, a basic goal of this book. Moreover, the availability of accurate structure/property relations has important implications for improved *materials processing*, since processing controls the microstructure and hence the bulk properties of the heterogeneous material.

Transport, electromagnetic, and mechanical processes that occur in heterogeneous materials are of great importance in *engineering*. In *chemical engineering*, the applications are driven by the petroleum, chemical, electronics, and pharmaceutical industries, and include filtration and separation (flow in porous media), chemical reactor design (thermal properties of packed beds), coatings (polymer dispersions), microelectronic components, inhalation therapy (two-phase aerosols), and drug-delivery systems. In *aerospace and mechanical engineering*, the applications are driven by the defense, space, electronics, transportation, and consumer products industries, and include composites as structural components in aircraft, space vehicles, and automobiles; insulation; heat exchangers; microelectromechanical systems (MEMS); and recreational products (skis and rackets). In *civil engineering*, the applications are driven by the building construction industries, infrastructure, and environmental issues, and include bridges, building materials (concrete and wood), aging of materials (pipes, pressure vessels, exterior of buildings), spread of contaminants in fluid-saturated soils, and soil mechanics. The systematic study of heterogeneous materials in engineering often goes by the names *micromechanics* and *microhydrodynamics*, reflecting concern with primarily solid mechanical properties in the former and fluid mechanical properties in the latter. In this book, we emphasize that such distinctions are unnecessary and indeed are a hindrance, since it will be shown that it is very fruitful to view seemingly disparate phenomena under a unifying light.

The main goal of *statistical mechanics* is to relate the macroscopic properties of a system of many particles (atoms, molecules, spins, etc.) to its microscopic properties, which include the interparticle interactions as well as the spatial statistics of the particles. *Statistical physics* is the broader study of any interacting system of particles, whether it exists at the atomic scale or not. For example, an important research area within statistical physics is *percolation theory*, which seeks to understand connectedness and clustering properties of random media at any length scale (Chapters 9 and 10). In this book we exploit the powerful methods and machinery of statistical mechanics to quantify structure at the larger "microscopic" length scales associated with random heterogeneous materials (see Figure 1.1).

Homogenization theory is an area of applied mathematics that is concerned with the behavior of the partial differential equations that are valid locally within a heterogeneous material in the limit that the ratio of the microscopic to macroscopic length scales tends to zero (Bensoussan, Lions and Papanicolaou 1978, Sanchez-Palencia 1980, Jikov, Kozlov and Olenik 1994). Mathematical questions are the following: What are the homogenized differential equations and how do the solutions converge to this asymptotic limit? A byproduct of the homogenization process is the averaged equation (1.1) that defines the effective property of interest. Chapter 13 is devoted to homogenization theory.

The area of mathematical research that seeks to provide models and methods to characterize random patterns is called *stochastic geometry* (Stoyan, Kendall and Mecke 1995). This subject grew out of the classical area of *geometrical probability* that concerned itself with less general considerations such as the famous Buffon needle question (Chapter 2). *Stereology* is a related area that seeks to recover statistical information on three-dimensional structures from one- and two-dimensional information obtained from linear or planar sections. The contributions of this book concerning the microstructure of heterogeneous materials belong to the domain of stochastic geometry. In particular, we generalize a preponderance of the results of stochastic geometry that have been derived for certain spatially *uncorrelated* models (called *Boolean* models) to a wide class of spatially *correlated* models.

Understanding the effective properties of heterogeneous materials has many applications in *geophysics*. Most earth materials are heterogeneous, frequently on a variety of different length scales. Rocks are aggregates of several different anisotropic minerals that often are characterized by widely varying properties. Determination of the properties of fluid-saturated porous rock is particularly germane to oil and gas exploration. The interpretation of changes in seismic velocities preceding earthquakes and their relation to other precursor phenomena may depend on the effects of cracks on the effective elastic moduli of the medium. Many of the methods and results of this book are of direct relevance in geophysical applications.

Biology is a field that will be playing a larger role in the study of heterogeneous materials in the future. Virtually all biological material systems are composites that are found to have at least one distinct structural level at a variety of length scales. This structural hierarchy is not fractal, i.e., it is neither self-similar nor is the number of levels infinite. Some of these biological materials have superior physical properties [e.g., spider silk is at least five times stronger than steel (Tirrell 1996)]. Thus, biological materials offer fundamental challenges both in terms of microstructure characterization and property predictions. From a practical standpoint, it is desired to employ lessons from biology to produce synthetic composite analogues with a unique spectrum of properties. Finally, we observe that three-dimensional imaging techniques originally developed for biological applications (e.g., confocal microscopy) are now being applied to image inorganic heterogeneous materials.

1.5 Overview of the Book

This book is divided into two parts. Part I deals with the quantitative characterization of the microstructure of heterogeneous materials via theoretical, computer-simulation, and imaging techniques. Emphasis is placed on foundational theoretical methods. Part II treats a wide variety of effective properties of heterogeneous materials and describes how they are linked to the microstructure. This is accomplished using rigorous methods. (Readers primarily interested in property prediction can skip to Part II while referring back to key portions of Part I as indicated.) Whenever possible, theoretical predictions for the effective properties are compared to available experimental and computer-simulation data. The overall goal of the book is to provide a rigorous means of characterizing the microstructure and properties of heterogeneous materials that can simultaneously yield results of practical utility. A unified treatment of both microstructure and properties is emphasized.

1.5.1 Part I

In Chapter 2 the various microstructural functions that are essential in determining the effective properties of random heterogeneous materials are defined. Chapter 3 provides a review of the statistical mechanics of particle systems that is particularly germane to the study of random heterogeneous materials, including sphere packings. In Chapter 4 a unified approach to characterize the microstructure of a large class of media is developed. This is accomplished via a canonical n-point function H_n from which one can derive exact analytical expressions for any microstructural function of interest. Chapters 5, 6, and 7 apply the formalism of Chapter 4 to the case of systems of identical spheres, spheres with a polydispersivity in size, and anisotropic particle systems (including laminates), respectively. In Chapter 8 the methods of Chapter 4 are extended to quantify the microstructure of cell models. Here the random-field approach is also discussed. Chapter 9 reviews the study of *percolation* and *clustering* on a *lattice* and introduces *continuum* percolation. Chapter 10 discusses specific developments in *continuum* percolation theory. Chapter 11 describes a means to study microstructural fluctuations that occur on local length scales. Finally, Chapter 12 discusses computer-simulation techniques (primarily Monte Carlo methods) to quantify microstructure. Moreover, it is shown how to apply the same methods to compute relevant microstructural functions from two- and three-dimensional images of the material. Finally, we describe methods that enable one to reconstruct or construct microstructures from a knowledge of limited microstructural information.

It is noteworthy that the statistical descriptors discussed in Part I are quite general and may also find application in diverse fields where characterization of spatial structure is a vital objective, such as cosmology and ecology. For instance, an important branch of cosmology is concerned with the description and understanding of the spatial distribution of mass and "voids" in galaxies and clusters of galaxies in the universe (Peebles 1993, Saslaw 2000). The study of how spatial patterns arise and are maintained is a major area of research in ecological theory (Pielou 1977, Diggle 1983, Durrett and Levin 1994). It is the opinion of this author that a cross-fertilization of ideas between all of these different fields will be mutually beneficial.

1.5.2 Part II

In Chapter 13 the local governing equations for the relevant field quantities and the method of homogenization leading to the averaged equations for the effective properties are described. The aforementioned classes of steady-state and time-dependent problems are studied. In Chapter 14 minimum energy principles are derived that lead to variational bounds on all of the effective properties in terms of trial fields. Chapter 15 proves and discusses certain phase-interchange relations for the effective conductivity and elastic moduli. Chapter 16 derives and describes some exact results for each of the effective properties. In Chapter 17 we derive the local fields associated with a single spherical or ellipsoidal inclusion in an infinite medium for all problem classes. Chapter 18 presents derivations of popular effective-medium approximations for all four effective properties. In Chapter 19 cluster expansions of the effective properties of dispersions are described. Chapter 20 presents derivations of so-called *strong-contrast* expansions for the effective conductivity and elastic moduli of generally anisotropic media of arbitrary microstructure. In Chapter 21, rigorous bounds on all of the effective properties are derived using the variational principles of Chapter 14 and specific trial fields. Chapter 22 describes the evaluation of the bounds found in Chapter 21 for certain theoretical model microstructures as well as experimental systems using the results of Part I. Finally, cross-property relations between the seemingly different effective properties considered here are discussed and derived in Chapter 23.

1.5.3 Scope

Many of the models, methods, and results reported in this book are obtained for two-phase random heterogeneous materials. The extension to heterogeneous materials with more than two phases is formally straightforward but will be covered less extensively. Such materials include polycrystals, which can be considered to be composites with an infinite number of anisotropic phases in which each phase is defined by the crystallographic orientation of the individual grains.

There are now a variety of computer-simulation techniques that have been developed to evaluate directly the effective properties of realizations of model microstructures (typically with periodic boundary conditions) and of real material microstructures. Such "computer experiments" are invaluable tools in providing benchmarks to test theories and in gaining insight into the nature of the structure/property relation. Some of the theoretical property predictions given in this book will be compared to available simulation data, but a treatment of such numerical methods is beyond the scope of the book. Specific citations to the numerical literature are given in Chapter 22.

Space limitations will not permit us to treat, in any detail, cases in the conduction and elasticity problems in which the multiphase interfaces of the heterogeneous material are characterized by their own properties, i.e., nonideal or imperfect interfaces (Chapter 13). However, the various techniques to obtain effective properties with ideal interfaces (described throughout Part II) may be applied to determine the effective properties with nonideal interfaces. Approaches for nonideal interfaces include approximate methods (Chiew and Glandt 1987, Benveniste 1987, Miloh and Benveniste 1999), exact results for periodic arrays of spheres (Cheng and Torquato 1997a, Cheng and Torquato 1997b), and rigorous bounding techniques (Hashin 1992, Torquato and Rintoul 1995, Lipton and Vernescu 1996, Zoia and Strieder 1997, Lipton 1997).

Heterogeneous materials with nonlinear constitutive relations are not treated. However, it is important to recognize that many of the results and methods in both parts of this book are directly relevant to nonlinear material behavior (Talbot and Willis 1987, Ponte Castaneda and Suquet 1998). In the cases of nonlinear stress–strain or current–electric field laws, it has been shown (Ponte Castaneda and Suquet 1998) that one can obtain estimates of the effective nonlinear behavior based on the behavior of a linear "comparison" material: the subject of this book. Thus, nonlinear behavior involves, at the very least, the same microstructural information as required for the linear material.

The important topic of wave propagation in random media will not be covered. The interested reader is referred to the work of Willis (1981), who discusses variational principles, and of Sheng (1995), who covers a broad range of topics on the theory and physics of wave propagation.

PART I

MICROSTRUCTURE CHARACTERIZATION

CHAPTER 2

Microstructural Descriptors

We have seen that random heterogeneous materials exhibit a remarkably broad spectrum of rich and complex microstructures. Our focus in Part I of this book is to develop a machinery to characterize statistically this broad class of microstructures, i.e., to develop a *statistical, or stochastic, geometry* of heterogeneous materials. How or where does one begin to address this challenging task? The answer, of course, depends on what is the goal of the statistical characterization. Our goal is ultimately the prediction of the macroscopic or effective physical properties of the random heterogeneous material, and thus this determines our starting point. The diverse effective properties that we are concerned with in this book naturally and necessarily lead to a wide variety of microstructural descriptors, generically referred to as *microstructural correlation functions*. As we noted in Chapter 1, such descriptors have applicability in other seemingly disparate fields, such as cosmology (Peebles 1993, Saslaw 2000) and ecology (Pielou 1977, Diggle 1983, Durrett and Levin 1994).

In this chapter we will define and discuss the following microstructural correlation functions, which are fundamental to determining the effective properties of random heterogeneous materials:

- n-point probability functions
- surface correlation functions
- lineal-path function
- chord-length density function
- pore-size functions
- percolation and cluster functions
- nearest-neighbor functions

- point/q-particle correlation functions
- surface-particle function

Whereas the first six types of quantities describe random media of arbitrary microstructure, the last three apply specifically to random particle dispersions. Chapter 4 describes a general formalism to represent and obtain all of these quantities from a canonical correlation function. Chapters 5–8, 10, and 12 deal with the evaluation of these functions for specific models and materials.

2.1 Preliminaries

The use of the term *random heterogeneous material* or simply *random medium* rests on the assumption that any sample of the medium is a realization of a specific random or stochastic process (or random field). An *ensemble* is a collection of all the possible realizations of a random medium generated by a specific stochastic process. We let $(\Omega, \mathcal{F}, \mathcal{P})$ be some fixed *probability space*, where Ω is a sample space (set of "outcomes"), $\mathcal{F}$ is a σ-algebra of subsets of Ω (set of "events") , and $\mathcal{P}$ is a probability measure (a function that assigns probabilities to "events") (Durrett 1996). Let each point $\omega \in \Omega$ correspond to a realization of the random medium that occupies some subset $\mathcal{V}$ of d-dimensional Euclidean space, i.e., $\mathcal{V} \in \Re^d$. The medium is in general statistically characterized by a random variable $\xi(\boldsymbol{x}, t; \omega)$, called the *structure* function, that depends on all values of the position vector $\boldsymbol{x} \in \mathcal{V}$ and on the time t. The time dependence allows for evolving microstructures (e.g., shear flow in a suspension or growth processes in random media).

In this book we will assume that the microstructures are *static* or can be approximated as static, and therefore the structure function $\xi(\boldsymbol{x}; \omega)$ will be taken to be independent of time. For a fixed ω, the structure function may be a continuously varying function of position (e.g., porosity of geologic media or orientation of crystals in a polycrystal), or it may take on discrete values (e.g., fiber composites or colloids). Our primary focus will be on two-phase random media, i.e., cases in which $\xi(\boldsymbol{x}; \omega)$ takes on two different values. However, generalizations to multiphase media with an arbitrary number of discrete phases follow in the obvious way. Some of the results given in this book will apply to multiphase media as well.

Each realization ω of the two-phase random medium occupies the region of space $\mathcal{V} \in \Re^d$ of volume V that is partitioned into two disjoint *random* sets or *phases*: phase 1, a region $\mathcal{V}_1(\omega)$ of volume fraction ϕ_1, and phase 2, a region $\mathcal{V}_2(\omega)$ of volume fraction ϕ_2. Since the *random* sets $\mathcal{V}_1(\omega)$ and $\mathcal{V}_2(\omega)$ are the complements of one another, then $\mathcal{V}_1(\omega) \cup \mathcal{V}_2(\omega) = \mathcal{V}$ and $\mathcal{V}_1(\omega) \cap \mathcal{V}_2(\omega) = \emptyset$. Let $\partial\mathcal{V}(\omega)$ denote the surface or interface between $\mathcal{V}_1(\omega)$ and $\mathcal{V}_2(\omega)$. Figure 2.1 shows a portion of a realization of a two-phase random medium. For a given realization ω, the structure function $\xi(\boldsymbol{x}; \omega)$ is just the *indicator function* $\mathcal{I}^{(i)}(\boldsymbol{x}; \omega)$ for phase i, given for $\boldsymbol{x} \in \mathcal{V}$ by

$$\mathcal{I}^{(i)}(\boldsymbol{x}; \omega) = \begin{cases} 1, & \text{if } \boldsymbol{x} \in \mathcal{V}_i(\omega), \\ 0, & \text{otherwise,} \end{cases} \tag{2.1}$$

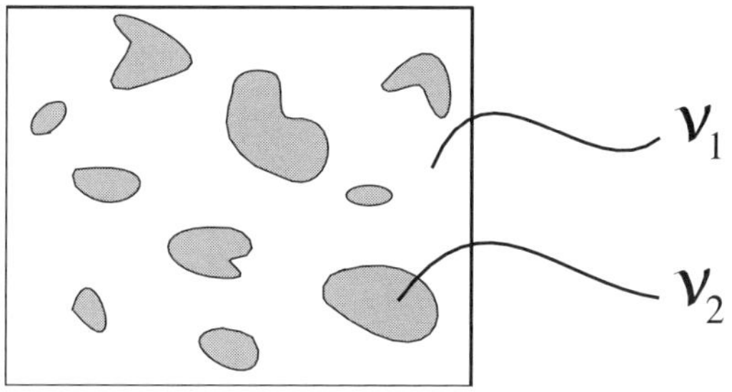

Figure 2.1 A portion of a realization ω of a two-phase random medium, where phase 1 is the white region $\mathcal{V}_1$, phase 2 is the gray region $\mathcal{V}_2$, and $\partial\mathcal{V}$ is the interface between the two regions.

for $i = 1, 2$ with

$$\mathcal{I}^{(1)}(\boldsymbol{x}; \omega) + \mathcal{I}^{(2)}(\boldsymbol{x}; \omega) = 1. \tag{2.2}$$

The random variable $\mathcal{I}^{(i)}(\boldsymbol{x}; \omega)$ is also called the *characteristic* function in the heterogeneous media community, but we will not use this term, since it is usually reserved to mean the Fourier transform of the probability density function in probability theory and stochastic processes. The indicator function $\mathcal{M}(\boldsymbol{x}; \omega)$ for the interface is defined as

$$\mathcal{M}(\boldsymbol{x}; \omega) = |\nabla \mathcal{I}^{(1)}(\boldsymbol{x}; \omega)| = |\nabla \mathcal{I}^{(2)}(\boldsymbol{x}; \omega)| \tag{2.3}$$

and therefore is a *generalized* function (e.g., a function involving Dirac delta functions) that is nonzero when $\boldsymbol{x}$ is on the interface. Depending on the physical context, phase i can be a solid, fluid, or void characterized by some general tensor property. Unless otherwise stated, we will drop ω from the notation and write $\mathcal{I}^{(i)}(\boldsymbol{x})$ for $\mathcal{I}^{(i)}(\boldsymbol{x}; \omega)$ and $\mathcal{M}(\boldsymbol{x})$ for $\mathcal{M}(\boldsymbol{x}; \omega)$.

In what follows we will consider the probabilistic descriptions of these and other random variables. It is assumed that the reader is familiar with the basic notion of a probability distribution of a random variable. The books by Cinlar (1975), Priestley (1981), Vanmarcke (1983), Cressie (1993), and Durrett (1996) cover, in varying depths, fundamental concepts in probability theory and stochastic processes.

2.2 n-Point Probability Functions

2.2.1 Definitions

For fixed $\boldsymbol{x}$, the indicator function $\mathcal{I}^{(i)}(\boldsymbol{x})$ has only two possible values; i.e., for some realizations ω it will be 0 and some other ω it will be 1. Thus, the random variable $\mathcal{I}^{(i)}(\boldsymbol{x})$ does not possess a probability density function (if Dirac delta functions are excluded). The probabilistic description of $\mathcal{I}^{(i)}(\boldsymbol{x})$ is given simply by the probability that $\mathcal{I}^{(i)}(\boldsymbol{x})$ is

1, which we write as

$$\mathcal{P}\left\{\mathcal{I}^{(i)}(\boldsymbol{x})=1\right\}.$$

Given this probability, it follows that

$$\mathcal{P}\left\{\mathcal{I}^{(i)}(\boldsymbol{x})=0\right\}=1-\mathcal{P}\left\{\mathcal{I}^{(i)}(\boldsymbol{x})=1\right\}.$$

A discrete random variable X can equivalently be specified by its *cumulative distribution function* $F(x)$, defined by

$$F(x)\equiv\mathcal{P}\{X\leq x\},$$

which has the properties that it is a nondecreasing, right-continuous function of x with $F(-\infty)=0$ and $F(+\infty)=1$. However, this latter description for the simple binary random variable $\mathcal{I}^{(i)}(\boldsymbol{x})$ is somewhat awkward notationally and will be avoided.

We should note that the expectation (or average) of any function $f[\mathcal{I}^{(i)}(\boldsymbol{x})]$ can be expressed as

$$\left\langle f[\mathcal{I}^{(i)}(\boldsymbol{x})]\right\rangle=\mathcal{P}\left\{\mathcal{I}^{(i)}(\boldsymbol{x})=1\right\}f(1)+\mathcal{P}\left\{\mathcal{I}^{(i)}(\boldsymbol{x})=0\right\}f(0),$$

where angular brackets denote an *ensemble* average, i.e., an average over all realizations ω of the ensemble. In particular, when $f[\mathcal{I}^{(i)}(\boldsymbol{x})]=\mathcal{I}^{(i)}(\boldsymbol{x})$, this expectation relation yields

$$S_1^{(i)}(\boldsymbol{x})\equiv\left\langle\mathcal{I}^{(i)}(\boldsymbol{x})\right\rangle=\mathcal{P}\left\{\mathcal{I}^{(i)}(\boldsymbol{x})=1\right\}. \tag{2.4}$$

Thus, in light of the 0, 1 nature of the indicator function $\mathcal{I}^{(i)}(\boldsymbol{x})$, its expectation is exactly the same as the probability $\mathcal{P}\left\{\mathcal{I}^{(i)}(\boldsymbol{x})=1\right\}$. Accordingly, following Torquato and Stell (1982), we refer to $S_1^{(i)}(\boldsymbol{x})$ as the *one-point probability function* for phase i, since it gives the probability of finding phase i at the position $\boldsymbol{x}$. It is sometimes also referred to as the one-point correlation function for the phase indicator function.

Knowing a realization $\mathcal{V}_i(\omega)$ is the same as knowing $\mathcal{I}^{(i)}(\boldsymbol{x};\omega)$ for all $\boldsymbol{x}$ in $\mathcal{V}$. Therefore, we may regard the random set $\mathcal{V}_i(\omega)$ as the collection of all random variables $\mathcal{I}^{(i)}(\boldsymbol{x})$ for $\boldsymbol{x}\in\mathcal{V}$. Hence, the probability law of $\mathcal{V}_i(\omega)$ is described by the finite-dimensional distributions of the random process $\{\mathcal{I}^{(i)}(\boldsymbol{x}):\boldsymbol{x}\in\mathcal{V}\}$. In other words, the probabilistic description of $\mathcal{V}_i(\omega)$ is given by the joint distribution of $\mathcal{I}^{(i)}(\boldsymbol{x}_1)\mathcal{I}^{(i)}(\boldsymbol{x}_2)\cdots\mathcal{I}^{(i)}(\boldsymbol{x}_n)$ as $n\geq 1$ varies over the integers and $\boldsymbol{x}_1,\boldsymbol{x}_2,\ldots,\boldsymbol{x}_n$ vary over $\mathcal{V}$. Of course, since the $\mathcal{I}^{(i)}(\boldsymbol{x})$ are either 0 or 1, this amounts to specifying the probabilities

$$\mathcal{P}\left\{\mathcal{I}^{(i)}(\boldsymbol{x}_1)=j_1,\mathcal{I}^{(i)}(\boldsymbol{x}_2)=j_2,\ldots,\mathcal{I}^{(i)}(\boldsymbol{x}_n)=j_n\right\}, \tag{2.5}$$

where each j_k is either 0 or 1.

The expectation of the product $\mathcal{I}^{(i)}(\boldsymbol{x}_1)\mathcal{I}^{(i)}(\boldsymbol{x}_2)\cdots\mathcal{I}^{(i)}(\boldsymbol{x}_n)$ is a particularly important average. Following the same line of reasoning leading to (2.4), we get

$$\begin{aligned} S_n^{(i)}(\boldsymbol{x}_1,\boldsymbol{x}_2,\ldots,\boldsymbol{x}_n) &\equiv \left\langle \mathcal{I}^{(i)}(\boldsymbol{x}_1)\mathcal{I}^{(i)}(\boldsymbol{x}_2)\cdots\mathcal{I}^{(i)}(\boldsymbol{x}_n)\right\rangle \\ &= \mathcal{P}\left\{\mathcal{I}^{(i)}(\boldsymbol{x}_1)=1,\mathcal{I}^{(i)}(\boldsymbol{x}_2)=1,\ldots,\mathcal{I}^{(i)}(\boldsymbol{x}_n)=1\right\} \\ &= \text{Probability that } n \text{ points at positions } \boldsymbol{x}_1,\boldsymbol{x}_2,\ldots,\boldsymbol{x}_n \\ &\quad\ \text{are found in phase } i. \end{aligned} \tag{2.6}$$

Following Torquato and Stell (1982), we will refer to $S_n^{(i)}$ as the *n-point probability function* for phase i. Although it is correct to refer to it as an n-point correlation function, we prefer the former term, since it emphasizes its special nature as a probability function. Geometrical probability interpretations of the $S_n^{(i)}$ are given in Section 2.2.3; see also Figure 1.6 for such interpretations of lower-order $S_n^{(i)}$. As we will see, the problem of finding the two-point function $S_2^{(i)}$ bears a close relationship to the classical Buffon needle game of geometrical probability (Kendall and Moran 1962).

The special nature of the indicator function makes it possible to specify the general joint distributions of (2.5) by giving the set of n-point probability functions $S_1^{(i)}, S_2^{(i)}, \ldots, S_n^{(i)}$ for phase i defined by (2.6). This can be seen by noting that

$$\begin{aligned} &\mathcal{P}\left\{\mathcal{I}^{(i)}(\boldsymbol{x}_1)=j_1,\mathcal{I}^{(i)}(\boldsymbol{x}_2)=j_2,\ldots,\mathcal{I}^{(i)}(\boldsymbol{x}_n)=j_n\right\} \\ &\quad= \left\langle \prod_{k\in K}\mathcal{I}^{(i)}(\boldsymbol{x}_k)\prod_{l\in L}[1-\mathcal{I}^{(i)}(\boldsymbol{x}_l)]\right\rangle, \end{aligned} \tag{2.7}$$

where $K=\{k\le n; j_k=1\}$ and $L=\{l\le n; j_l=0\}$, and thus the expectation of the product in (2.7) is computable in terms of the set of n-point probability functions $S_1^{(i)}, S_2^{(i)}, \ldots, S_n^{(i)}$ for phase i.

In particular, one can express the probability $S_n^{(2)}$ of finding n points in phase 2 in terms of the set of phase 1 probabilities $S_1^{(1)}, S_2^{(1)}, \ldots, S_n^{(1)}$. This is easily shown, since

$$\begin{aligned} S_n^{(2)}(\boldsymbol{x}_1,\boldsymbol{x}_2,\ldots,\boldsymbol{x}_n) &= \left\langle \prod_{j=1}^{n}[1-\mathcal{I}^{(1)}(\boldsymbol{x}_j)]\right\rangle \\ &= 1-\sum_{j=1}^{n} S_1^{(1)}(\boldsymbol{x}_j)+\sum_{j<k}^{n} S_2^{(1)}(\boldsymbol{x}_j,\boldsymbol{x}_k) \\ &\quad -\sum_{j<k<l}^{n} S_3^{(1)}(\boldsymbol{x}_j,\boldsymbol{x}_k,\boldsymbol{x}_l)+\cdots+(-1)^n S_n^{(1)}(\boldsymbol{x}_1,\boldsymbol{x}_2,\ldots,\boldsymbol{x}_n). \end{aligned} \tag{2.8}$$

Note that the sth sum in (2.8) contains $n!/[(n-s)!s!]$ terms and carries the factor $(-1)^s$. Indeed, the probability of finding any subset n_1 of the n points in phase 2 and the remaining $n_2=n-n_1$ in phase 1 can be expressed purely in terms of the set of phase 1 probabilities $S_1^{(1)}, S_2^{(1)}, \ldots, S_n^{(1)}$ (or the set of phase 2 probabilities) (Torquato and Stell 1982). For example, the probability $S_2^{(12)}$ of two "dissimilar ends" (i.e., the probability that a point at $\boldsymbol{x}_1$ is in phase 1 and a point at $\boldsymbol{x}_2$ is in phase 2) is given by

$$S_2^{(12)}(\boldsymbol{x}_1,\boldsymbol{x}_2) = \left\langle \mathcal{I}^{(1)}(\boldsymbol{x}_1)[1-\mathcal{I}^{(1)}(\boldsymbol{x}_2)]\right\rangle = S_1^{(1)}(\boldsymbol{x}_1) - S_2^{(1)}(\boldsymbol{x}_1,\boldsymbol{x}_2). \tag{2.9}$$

The n-point probability functions were introduced in the context of determining the effective transport properties of random media by Brown (1955). These statistical descriptors arise in rigorous expressions for the effective transport and mechanical properties of random heterogeneous media, including the following:

- effective conductivity, dielectric constant, magnetic permeability, and diffusion coefficient (Brown 1955, Prager 1963b, Beran 1968, Torquato 1980, Milton 1981a, Phan-Thien and Milton 1982, Torquato 1985a)
- effective elastic moduli (Beran 1968, McCoy 1970, Dederichs and Zeller 1973, Kroner 1977, Willis 1981, Milton 1982, Milton and Phan-Thien 1982, Torquato 1997)
- trapping constant or, equivalently, mean survival time (Prager 1963a, Torquato and Rubinstein 1989)
- fluid permeability (Prager 1961, Weissberg and Prager 1970, Berryman and Milton 1985, Rubinstein and Torquato 1989)

Some general properties of the n-point probability functions have been studied by Frisch and Stillinger (1963) and Torquato and Stell (1982, 1983a). Moreover, lower-order $S_n^{(i)}$ were calculated for various sphere models (Torquato and Stell 1983b, Torquato and Stell 1984, Torquato and Stell 1985a). In Chapters 4–8, we discuss the determination of lower-order $S_n^{(i)}$ for various particle, cell, and random-field models. Chapter 12 describes how to extract such correlation functions from computer simulations and images of real materials.

In what follows we describe some basic properties of the n-point probability functions.

2.2.2 Symmetries and Ergodicity

If the n-point probability function $S_n^{(i)}$ depends generally on the absolute positions $\boldsymbol{x}_1,\boldsymbol{x}_2,\ldots,\boldsymbol{x}_n$, then we say that the medium is *statistically inhomogeneous*. Indeed, even the one-point function $S_1^{(i)}$ can depend on the local position $\boldsymbol{x}_1$ and then can be interpreted as a *position-dependent volume fraction of phase i*. Figure 2.2 depicts two examples of statistically inhomogeneous media.

The medium is strictly spatially stationary or strictly *statistically homogeneous* if the joint probability distributions describing the stochastic process are *translationally invariant*, i.e., invariant under a translation (shift) of the space origin. Thus, the random set $\mathcal{V}_i(\omega)$ generated from the stochastic process $\{\mathcal{I}^{(i)}(\boldsymbol{x}) : \boldsymbol{x} \in \mathcal{V}\}$ is strictly statistically homogeneous, provided that for some constant vector $\boldsymbol{y}$ in $\Re^d$

$$\begin{aligned}
&\mathcal{P}\left\{\mathcal{I}^{(i)}(\boldsymbol{x}_1)=j_1,\mathcal{I}^{(i)}(\boldsymbol{x}_2)=j_2,\ldots,\mathcal{I}^{(i)}(\boldsymbol{x}_n)=j_n\right\}\\
&\quad=\mathcal{P}\left\{\mathcal{I}^{(i)}(\boldsymbol{x}_1+\boldsymbol{y})=j_1,\mathcal{I}^{(i)}(\boldsymbol{x}_2+\boldsymbol{y})=j_2,\ldots,\mathcal{I}^{(i)}(\boldsymbol{x}_n+\boldsymbol{y})=j_n\right\},
\end{aligned}$$

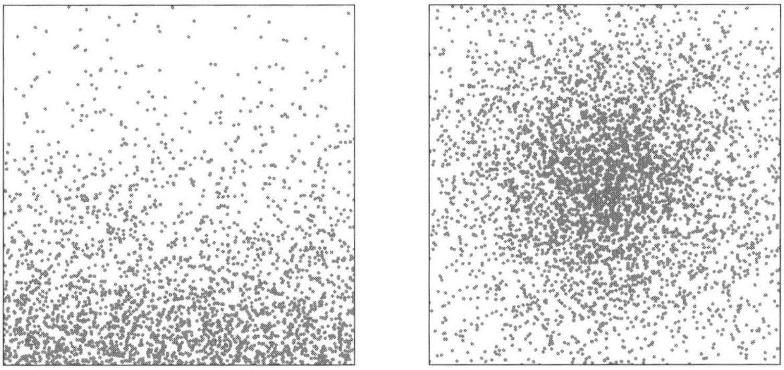

Figure 2.2 Two examples of statistically inhomogeneous media. Left panel: Density of the gray phase decreases in the upward direction. Right panel: Density of the gray phase decreases radially from the center.

for all $n \geq 1$, and $\boldsymbol{x}_1, \boldsymbol{x}_2, \ldots, \boldsymbol{x}_n$ in $\Re^d$, and $j_1, j_2, \ldots, j_n$ in $\{0, 1\}$. (We emphasize that for this statement to have any meaning for $\boldsymbol{y}$ in $\Re^d$, $\mathcal{V}$ must equal $\Re^d$, i.e., the volume V must be infinite.) Equivalently, since such probabilities can be expressed in terms of the n-point probability functions for phase i (see Section 2.1), $\mathcal{V}_i(\omega)$ is strictly statistically homogeneous if and only if

$$S_n^{(i)}(\boldsymbol{x}_1, \boldsymbol{x}_2, \ldots, \boldsymbol{x}_n) = S_n^{(i)}(\boldsymbol{x}_1 + \boldsymbol{y}, \boldsymbol{x}_2 + \boldsymbol{y}, \ldots, \boldsymbol{x}_n + \boldsymbol{y})$$
$$= S_n^{(i)}(\boldsymbol{x}_{12}, \ldots, \boldsymbol{x}_{1n}), \qquad (2.10)$$

for all $n \geq 1$, and $\boldsymbol{x}_1, \boldsymbol{x}_2, \ldots, \boldsymbol{x}_n$ in $\Re^d$, and $\boldsymbol{y}$ in $\Re^d$, where $\boldsymbol{x}_{jk} = \boldsymbol{x}_k - \boldsymbol{x}_j$. We see that for statistically homogeneous media, the n-point probability function depends not on the absolute positions but on their relative displacements. Thus, there is no *preferred origin* in the system, which in relation (2.10) we have chosen to be the point $\boldsymbol{x}_1$. In particular, the one-point probability function is a constant *everywhere*, namely, the volume fraction ϕ_i of phase i, i.e.,

$$S_1^{(i)} = \phi_i. \qquad (2.11)$$

The medium is said to be statistically homogeneous but *anisotropic* if $S_n^{(i)}$ depends on both the orientations and magnitudes of the vectors $\boldsymbol{x}_{12}, \boldsymbol{x}_{13}, \ldots, \boldsymbol{x}_{1n}$ (see Figure 2.3).

When the system is statistically homogeneous, it is meaningful to define volume averages. Roughly speaking, the property of statistical homogeneity states that all regions of space are similar as far as statistical properties of the stochastic process are concerned. This suggests an *ergodic hypothesis*; i.e., the result of averaging over all realizations of the ensemble is equivalent to averaging over the volume for one realization in the infinite-volume limit. Thus, complete probabilistic information can be obtained from a single realization of the infinite medium. The ergodic hypothesis enables us to replace ensemble averaging with volume averaging in the limit that the volume tends

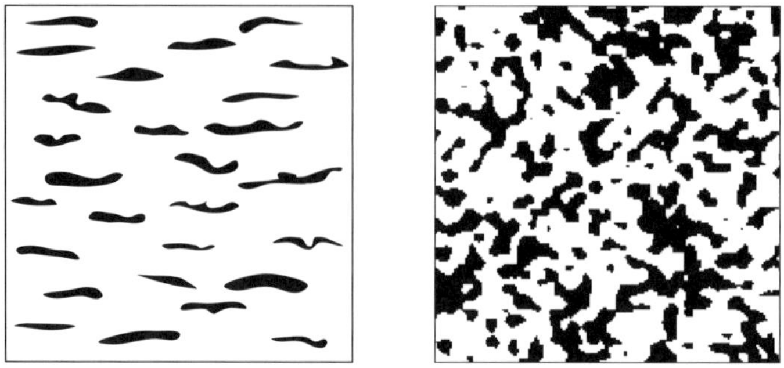

Figure 2.3 Two examples of portions of statistically homogeneous media with black and white phases. Left panel: The layered medium is statistically anisotropic. Right panel: The medium is statistically isotropic.

to infinity, i.e.,

$$S_n^{(i)}(\mathbf{x}_{12},\ldots,\mathbf{x}_{1n}) = \lim_{V\to\infty}\frac{1}{V}\int_{\mathcal{V}} \mathcal{I}^{(i)}(\mathbf{y})\mathcal{I}^{(i)}(\mathbf{y}+\mathbf{x}_{12})\cdots\mathcal{I}^{(i)}(\mathbf{y}+\mathbf{x}_{1n})d\mathbf{y}. \tag{2.12}$$

We will refer to such systems as *ergodic media*.

The medium is said to be strictly *statistically isotropic* if the joint probability distributions describing the stochastic process are *rotationally invariant*, i.e., invariant under rigid-body rotation of the spatial coordinates. For such media, this implies that $S_n^{(i)}$ depends only on the distances $x_{jk} = |\mathbf{x}_{jk}|$, $1 \leq j < k \leq n$ (see Figure 2.3). For example, the two-point function (also known as the *autocorrelation* function) and three-point function have the form

$$S_2^{(i)}(\mathbf{x}_1,\mathbf{x}_2) = S_2^{(i)}(x_{12}), \tag{2.13}$$

$$S_3^{(i)}(\mathbf{x}_1,\mathbf{x}_2,\mathbf{x}_3) = S_3^{(i)}(x_{12},x_{13},x_{23}). \tag{2.14}$$

Relation (2.14) for $S_3^{(i)}$ remains invariant under all permutations of its arguments x_{12}, x_{13}, and x_{23}. Both $S_2^{(i)}$ and $S_3^{(i)}$ can be obtained from any planar cut through a three-dimensional medium when it is isotropic (see Figure 2.4). In practice, this means that the two- and three-point functions can be extracted from cross-sections or two-dimensional images of the isotropic sample (see Figure 1.6), provided that the planar representation is sufficiently large. Moreover, the autocorrelation function $S_2^{(i)}$ can also be found from a *linear* cut through an isotropic medium (see Figure 2.4).

In general, the n-point probability functions for $n \geq 2$ cannot be expressed in terms of lower-order q-point functions, $q < n$. However, in the special case of a medium possessing "phase-inversion" symmetry at $\phi_1 = \phi_2 = 1/2$, it is possible to determine the odd-order probability functions $S_{2m+1}^{(i)}$ from $S_{2m}^{(i)}, S_{2m-1}^{(i)}, \ldots, S_1^{(i)}$. We say that a random medium possesses *phase-inversion symmetry* if the morphology of phase 1 at volume fraction ϕ_1 is statistically identical to that of phase 2 in the system where the volume

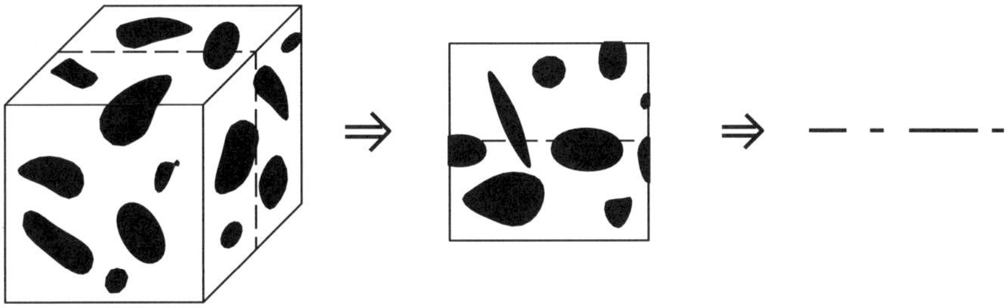

Figure 2.4 Planar and linear cuts through three-dimensional isotropic media. In the infinite-system limit, $S_2^{(i)}$ and $S_3^{(i)}$ can be obtained from a planar cut and $S_2^{(i)}$ can be extracted from a linear cut.

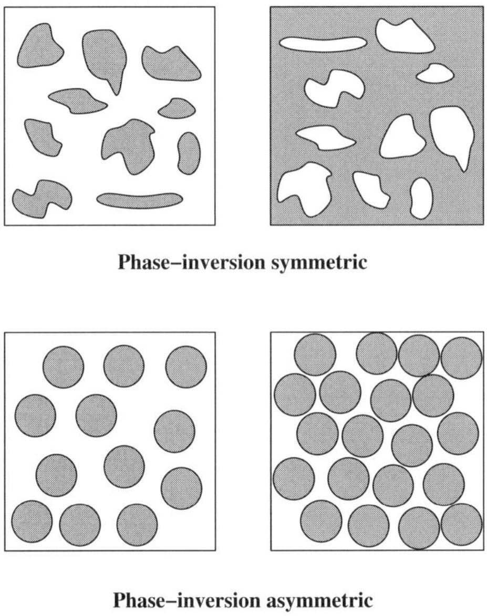

Figure 2.5 Examples of systems possessing phase-inversion symmetry (top) and phase-inversion asymmetry (bottom). In the leftmost and rightmost systems, the volume fractions of phase 1 are ϕ_1 and $1 - \phi_1$, respectively.

fraction of phase 1 is $1 - \phi_1$ (see Figure 2.5) and hence

$$S_n^{(1)}(\boldsymbol{x}^n; \phi_1, \phi_2) = S_n^{(2)}(\boldsymbol{x}^n; \phi_2, \phi_1), \tag{2.15}$$

where $\boldsymbol{x}^n \equiv \{\boldsymbol{x}_1, \boldsymbol{x}_2, \ldots, \boldsymbol{x}_n\}$. The notion of phase-inversion symmetry introduced here quantitatively generalizes the notion of a "symmetric" two-phase material at $\phi_1 = \phi_2 = 1/2$ [discussed by Beran (1968)] to arbitrary volume fractions. Examples of systems with phase-inversion symmetry are *symmetric-cell materials* described in Chapter 8 (see Figures 8.5 and 8.6). To a good approximation, interpenetrating cermets, such as the one depicted in Figure 1.2, can be made to have phase-inversion symmetry. At the point $\phi_1 = \phi_2 = 1/2$, a medium possessing phase-inversion symmetry has the special property that the n-point probability functions for each phase are identical, or in other words, the geometry of one phase is statistically indistinguishable from the other. Thus, from (2.8),

$$2S_{2m+1}^{(2)} = 1 - \sum S_1^{(1)} + \sum S_2^{(1)} - \sum S_3^{(1)} + \cdots + (-1)^{2m} \sum S_{2m}^{(1)} . \tag{2.16}$$

Therefore, for a medium with phase-inversion symmetry at $\phi_1 = \phi_2 = 1/2$, the odd-order probability functions $S_{2m+1}^{(i)}$ can be expressed in terms of all the lower-order probability functions. For example, for such a symmetric medium with $m = 1$, we deduce from (2.16) that

$$S_3^{(i)}(\boldsymbol{x}_1,\boldsymbol{x}_2,\boldsymbol{x}_3) = \frac{1}{2}\left[S_2^{(i)}(\boldsymbol{x}_1,\boldsymbol{x}_2) + S_2^{(i)}(\boldsymbol{x}_1,\boldsymbol{x}_3) + S_2^{(i)}(\boldsymbol{x}_2,\boldsymbol{x}_3) - \frac{1}{2}\right]. \tag{2.17}$$

However, the even-order functions $S_{2m}^{(i)}$ cannot be expressed in terms of the lower-order functions, since the last term in (2.8) is always positive.

It is noteworthy that most random media do not possess phase-inversion symmetry. A common example of a system with such *phase-inversion asymmetry* is a dispersion of particles (see Figure 2.5 and Chapters 3–7).

2.2.3 Geometrical Probability Interpretation

The geometrical-probabilistic significance of the n-point probability function is easily seen for any microstructure. Let $F_n^{(i)}$ be a polyhedron with n vertices located at positions $\boldsymbol{x}_1,\boldsymbol{x}_2,\ldots,\boldsymbol{x}_n$. Then for statistically *inhomogeneous* media, $S_n^{(i)}$ is the probability that all n vertices of $F_n^{(i)}$ with *fixed* positions $\boldsymbol{x}_1,\boldsymbol{x}_2,\ldots,\boldsymbol{x}_n$ lie in $\mathcal{V}_i$. For statistically homogeneous but anisotropic media, $S_n^{(i)}$ is the probability that all n vertices of $F_n^{(i)}$ lie in $\mathcal{V}_i$ when the polyhedron is randomly placed in the volume at fixed orientation i.e., over all translations of the polyhedron. For statistically *isotropic media*, $S_n^{(i)}$ can be interpreted as the probability that all n vertices of $F_n^{(i)}$ lie in $\mathcal{V}_i$ when the polyhedron is randomly placed in the volume, i.e., over all translations and solid-body rotations of the polyhedron.

In light of the above, one can view the determination of $S_n^{(i)}$ as a generalization of the Buffon needle game (Kendall and Moran 1962), in which one tosses a needle of length x onto a grid of equidistant parallel lines separated by a distance $L \geq x$. The probability p that the needle crosses the lines is inversely proportional to π; specifically, $p = 2x/(\pi L)$. One can see that p is closely related to the probability of two dissimilar ends given by (2.9) and thus to the two-point function $S_2^{(i)}$.

For statistically homogeneous media composed of identical spheres of radius R (phase 2) distributed throughout another material (phase 1), we may infer yet another geometrical-probabilistic interpretation of these functions (Torquato and Stell 1982). The function $S_n^{(1)}(\boldsymbol{x}_1,\boldsymbol{x}_2,\ldots,\boldsymbol{x}_n)$ may be interpreted to be the probability that a region $\Omega^{(n)}$, the *union volume* of n spheres of radius R centered at $\boldsymbol{x}_1,\boldsymbol{x}_2,\ldots,\boldsymbol{x}_n$, contains no sphere centers. (Chapters 5 and 6 discuss the evaluation of the n-point probability functions for such models.) A similar interpretation may be inferred for particles of arbitrary shape with a size distribution.

2.2.4 Asymptotic Properties and Bounds

We determine asymptotic properties of and bounds on $S_n^{(i)}$ that apply to any statistically inhomogeneous two-phase random medium.

When any subset of $q+1$ points coincide, so that $\boldsymbol{x}_{i_1} = \boldsymbol{x}_{i_2} = \cdots = \boldsymbol{x}_{i_{q+1}}$, we have

$$\begin{aligned} S_n^{(i)}(\boldsymbol{x}^n) &= S_{n-q}^{(i)}(\boldsymbol{x}_1, \ldots, \boldsymbol{x}_{i_1}, \overline{\boldsymbol{x}_{i_2}, \ldots, \boldsymbol{x}_{i_{q+1}}}, \ldots, \boldsymbol{x}_n) \\ &= \left\langle \mathcal{I}^{(i)}(\boldsymbol{x}_1) \cdots \mathcal{I}^{(i)}(\boldsymbol{x}_{i_1}) \overline{\mathcal{I}^{(i)}(\boldsymbol{x}_{i_2}) \cdots \mathcal{I}^{(i)}(\boldsymbol{x}_{i_{q+1}})} \cdots \mathcal{I}^{(i)}(\boldsymbol{x}_n) \right\rangle, \end{aligned} \tag{2.18}$$

where a bar above a quantity indicates its absence.

Let us now consider partitioning the set $\{\boldsymbol{x}_1, \boldsymbol{x}_2, \ldots, \boldsymbol{x}_n\}$ into L subsets $\{\boldsymbol{x}_1\}$, $\{\boldsymbol{x}_2, \boldsymbol{x}_3\}$, $\{\boldsymbol{x}_4, \boldsymbol{x}_5, \boldsymbol{x}_6\}, \ldots$. Let all of the relative distances between the m elements of these subsets remain bounded, and let F_m^j be the polyhedron with m vertices located at the positions associated with the jth subset. We denote the centroid of F_m^j by R_j. Let R_{jk} be the relative distance between the centroids of F_m^j and F_m^k, where j and k are all possible values such that $1 \le j < k \le L$. A system is said to possess *no long-range order* if the events $R_{jk} \to \infty$ for all i and j are statistically independent, i.e., the n-point function factorizes into L products as follows:

$$\begin{aligned} \lim_{\text{all } R_{jk} \to \infty} & S_n^{(i)}(\boldsymbol{x}_1, \boldsymbol{x}_2, \ldots, \boldsymbol{x}_n) \\ &= \left\langle \mathcal{I}^{(i)}(\boldsymbol{x}_1) \right\rangle \left\langle \mathcal{I}^{(i)}(\boldsymbol{x}_2) \mathcal{I}^{(i)}(\boldsymbol{x}_3) \right\rangle \left\langle \mathcal{I}^{(i)}(\boldsymbol{x}_4) \mathcal{I}^{(i)}(\boldsymbol{x}_5) \mathcal{I}^{(i)}(\boldsymbol{x}_6) \right\rangle \cdots \\ &= S_1^{(i)}(\boldsymbol{x}_1) S_2^{(i)}(\boldsymbol{x}_2, \boldsymbol{x}_3) S_3^{(i)}(\boldsymbol{x}_4, \boldsymbol{x}_5, \boldsymbol{x}_6) \cdots. \end{aligned} \tag{2.19}$$

The above partition, however, is just one of the possible ways to partition the set $\{\boldsymbol{x}_1, \boldsymbol{x}_2, \ldots, \boldsymbol{x}_n\}$. In general, for any partition into sets $\{\gamma\}$, each with $m(\gamma)$ elements, we have in the absence of long-range order that

$$\lim_{\text{all } R_{\alpha\beta} \to \infty} S_n^{(i)}(\boldsymbol{x}_1, \boldsymbol{x}_2, \ldots, \boldsymbol{x}_n) = \prod_{\{\gamma\}} S_{m(\gamma)}^{(i)}(\boldsymbol{x}_1, \boldsymbol{x}_2, \ldots, \boldsymbol{x}_{m(\gamma)}), \tag{2.20}$$

where $R_{\alpha\beta}$ is the distance between the centroids of sets α and β. An example of a system with long-range order, and thus one that does not obey the asymptotic result (2.20), is an infinitely large crystalline (periodic) array of identical spheres.

For concreteness, we apply the aforementioned general asymptotic results for the cases $n = 2$ and $n = 3$ for statistically homogeneous media without long-range order. We have for $n = 2$

$$\lim_{x_{12} \to 0} S_2^{(i)}(\boldsymbol{x}_{12}) = \phi_i, \qquad \lim_{x_{12} \to \infty} S_2^{(i)}(\boldsymbol{x}_{12}) = \phi_i^2, \tag{2.21}$$

and for $n = 3$, under permutations of the distances x_{12}, x_{13}, and x_{23},

$$\lim_{x_{12} \to 0,\, x_{13} \to 0} S_3^{(i)}(\boldsymbol{x}_{12}, \boldsymbol{x}_{13}) = \phi_i, \qquad \lim_{x_{23} \to 0} S_3^{(i)}(\boldsymbol{x}_{12}, \boldsymbol{x}_{13}) = S_2^{(i)}(\boldsymbol{x}_{12}), \tag{2.22}$$

$$\lim_{\substack{x_{13}\to\infty \\ x_{12}\text{ fixed}}} S_3^{(i)}(\boldsymbol{x}_{12},\boldsymbol{x}_{13}) = \phi_i S_2^{(i)}(\boldsymbol{x}_{12}), \qquad \lim_{\text{all } x_{ij}\to\infty} S_3^{(i)}(\boldsymbol{x}_{12},\boldsymbol{x}_{13}) = \phi_i^3. \tag{2.23}$$

Since $0 \le \mathcal{I}^{(i)}(\boldsymbol{x}) \le 1$ for all $\boldsymbol{x}$ in $\mathcal{V}$, we have the elementary bounds

$$0 \le S_n^{(i)}(\boldsymbol{x}^n) \le S_{n-1}^{(i)}(\boldsymbol{x}^{n-1}), \qquad \text{for all } \boldsymbol{x}^n \text{ and } n \ge 2, \tag{2.24}$$

$$0 \le S_1^{(i)}(\boldsymbol{x}_1) \le 1, \qquad \text{for all } \boldsymbol{x}_1. \tag{2.25}$$

The one-point function $S_1^{(i)}(\boldsymbol{x}_1)$ (equal to the volume fraction ϕ_i for homogeneous media) is an upper bound on $S_n^{(i)}(\boldsymbol{x}^n)$ for all $\boldsymbol{x}^n$ and n.

A word on notation is in order here. When possible, we will suppress the superscript in $S_n^{(i)}$ indicating phase i and simply denote the function by

$$S_n(\boldsymbol{x}_1,\boldsymbol{x}_2,\ldots,\boldsymbol{x}_n).$$

In such instances, the phase to which it refers will be specified.

2.2.5 Two-Point Probability Function

As noted earlier, the two-point or autocorrelation function $S_2(\boldsymbol{r}) \equiv S_2^{(1)}(\boldsymbol{r})$ for statistically homogeneous media can be obtained by randomly tossing line segments of length $r \equiv |\boldsymbol{r}|$ with a specified orientation and counting the fraction of times the end points fall in phase 1 (see Figure 2.6). The function $S_2(\boldsymbol{r})$ provides a measure of how the end points of a vector $\boldsymbol{r}$ in phase 1 are correlated. For isotropic media, $S_2(r)$ attains its maximum value of ϕ_1 at $r = 0$ and eventually decays (usually exponentially fast) to its asymptotic value of ϕ_1^2.

The form of $S_2(r)$ provides information about certain *gross* features of the microstructure, as discussed in detail in Chapter 5–7 and 12. For example, two different autocorrelation functions for isotropic particle systems and their associated microstructures are shown in Figure 2.7. In the first case of *nonoverlapping disks* (Section 5.2.1), $S_2(r)$ exhibits oscillations for small r (short-range order) with periodicity roughly equal to the particle diameter D. This is reflective of *spatial correlations* between the particles due to exclusion-volume (hard-core) effects. In the second case of *overlapping disks* (Section 5.1.1), $S_2(r)$ exhibits no short-range order but rather monotonically decays to its asymptotic value at exactly $r = D$. This indicates that particles of characteristic size D are *spatially uncorrelated*. However, the form of S_2 here belies the fact that there are a statistically significant number of clusters in the system that are appreciably larger than D (see Figure 2.7). Quantities that are better able to capture cluster and percolation information are discussed in Section 2.7 and Chapters 9 and 10.

We see that one must be careful in interpreting length scales associated with S_2. To further remark on this point, it is convenient to define, for statistically homogeneous media, the *autocovariance* of phase 1

$$\chi(\boldsymbol{r}) \equiv \left\langle [\mathcal{I}^{(1)}(\boldsymbol{x}) - \phi_1][\mathcal{I}^{(1)}(\boldsymbol{x}+\boldsymbol{r}) - \phi_1] \right\rangle = S_2(\boldsymbol{r}) - \phi_1^2, \tag{2.26}$$

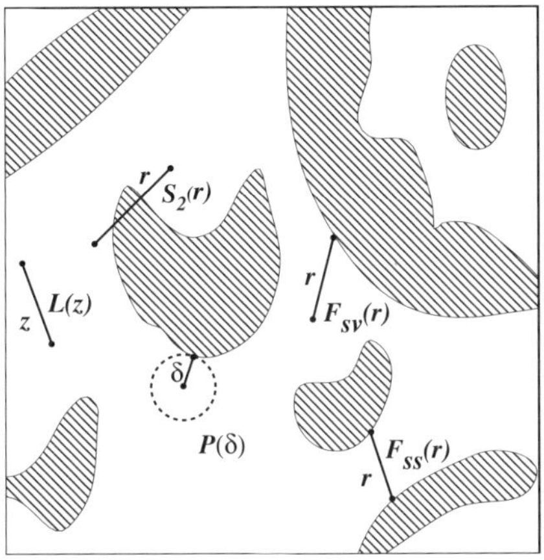

Figure 2.6 A schematic depicting events that contribute to lower-order functions for random media of arbitrary microstructure. Shown are the two-point probability function $S_2 \equiv S_2^{(1)}$ for phase 1 (white region) defined by (2.6) with $n = 2$, surface–void and surface–surface functions F_{sv} and F_{ss} defined by (2.61) and (2.62), lineal-path function $L \equiv L^{(1)}$ defined by (2.66), and the pore-size density function P defined by (2.77).

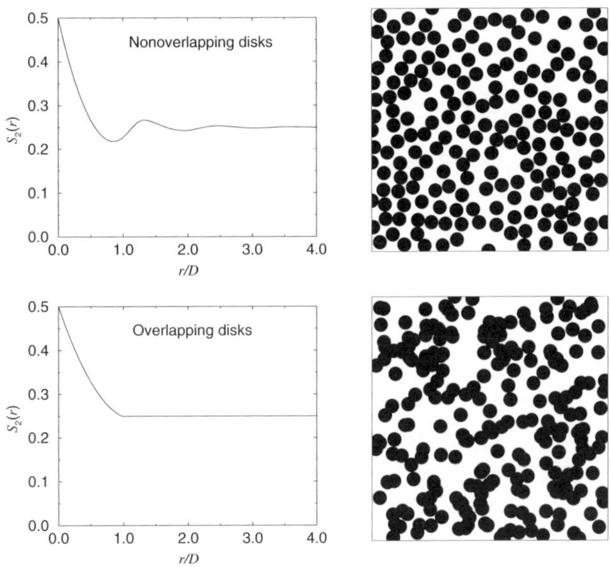

Figure 2.7 The two-point probability function $S_2(r)$ for phase 1 for two different systems at $\phi_1 = \phi_2 = 1/2$: a correlated system of nonoverlapping disks (top) and an uncorrelated system of overlapping disks (bottom). Here D is a disk diameter.

where $\mathcal{I}^{(1)}(\boldsymbol{x})-\phi_1$ is a random variable with zero mean, and $\mathcal{I}^{(1)}$ is the indicator function (2.1) for phase 1. The autocovariance $\chi(\boldsymbol{r})$ has the limiting values $\chi(0) = \phi_1\phi_2$ and $\chi(\infty) = 0$, the latter applying in the absence of long-range order. Moreover, the function $\chi(\boldsymbol{r})$ must be positive semidefinite (nonnegative) in the sense that for any finite number of spatial locations $\boldsymbol{r}_1, \boldsymbol{r}_2, \ldots, \boldsymbol{r}_m$ in $\Re^d$ and arbitrary real numbers $a_1, a_2, \ldots, a_m$,

$$\sum_{i=1}^{m}\sum_{j=1}^{m} a_i a_j \chi(\boldsymbol{r}_i - \boldsymbol{r}_j) \geq 0. \tag{2.27}$$

A variety of length scales associated with S_2 can be defined. One length scale, which we refer to as ℓ_S, is rooted in rigorous considerations:

$$\ell_S = \left\{\int_0^\infty r\chi(r)dr\right\}^{1/2} = \left\{\int_0^\infty r\left[S_2(r) - \phi_1^2\right]dr\right\}^{1/2}. \tag{2.28}$$

This length scale arises in rigorous bounds on the fluid permeability (Prager 1961) and trapping constant (Rubinstein and Torquato 1988) of three-dimensional isotropic random porous media. Since application of (2.8) for any statistically homogeneous medium leads to the result that the autocovariance of phase 1 is equal that of phase 2, i.e.,

$$\chi(\boldsymbol{r}) = S_2^{(1)}(\boldsymbol{r}) - \phi_1^2 = S_2^{(2)}(\boldsymbol{r}) - \phi_2^2, \tag{2.29}$$

it is clear that measures based on the two-point function for the phases are not capable of distinguishing length scales of phase 1 from length scales of phase 2. For example, for isotropic media, the length scale defined by (2.28) for phase 1 is identical to the corresponding one for phase 2.

Debye and Bueche (1949) showed that the two-point probability function $S_2(r)$ of an isotropic porous solid can also be obtained via scattering of radiation. Here phases 1 and 2 are the void and solid phases, respectively. The normalized scattered intensity $i(k)$ at a wave number k for a three-dimensional isotropic porous medium of volume V is proportional to the Fourier transform of the autocovariance $\chi(r)$, i.e.,

$$i(k) = 4\pi V n_o^2 \int_0^\infty \chi(r) r^2 \frac{\sin(kr)}{kr} dr, \tag{2.30}$$

where n_o is the mean density of electrons. To get the real-space two-point function $S_2(r)$ from the scattered intensity $i(k)$, one need only perform the inverse Fourier transform:

$$\chi(r) = S_2(r) - \phi_1^2 = \frac{1}{2\pi^2 V n_o^2}\int_0^\infty i(k) k^2 \frac{\sin(kr)}{kr} dk. \tag{2.31}$$

The accuracy of (2.31) depends on whether the "experimentally bandlimited" scattering curve $i(k)$ approximates sufficiently closely the entire function $i(k)$. The spectral properties of χ will be explored further below.

It has been shown (Guinier and Fournet 1955, Debye, Anderson and Brumberger 1957) that the expansion of the two-point probability function $S_2(r)$ through terms linear in r for any three-dimensional isotropic medium is given by

$$S_2(r) = \phi_1 - \frac{s}{4} r + \mathcal{O}(r^2), \tag{2.32}$$

where s is the *specific surface*, defined to be the *interface area per unit volume*. This formula is valid for any three-dimensional, isotropic two-phase heterogeneous medium, porous or not. The coefficient of the linear term r in (2.32) can be shown (Guinier and Fournet 1955, Debye et al. 1957) to be proportional to the leading term in the asymptotic expansion of the scattering curve $i(k)$ for $k \to \infty$, i.e.,

$$i(k) \sim \frac{2\pi n_o^2 s V}{k^4}, \qquad k \to \infty. \tag{2.33}$$

Thus, given that this asymptotic region can be reached with the value of the wavelength employed, this formula provides a measurement technique to determine the specific surface s of isotropic porous media. Berryman (1987) has shown that formula (2.32) applies to *anisotropic* media as well after angular averaging.

We see that the derivative of $S_2(r)$ at the origin is proportional to the specific surface s for three-dimensional isotropic media. Indeed, for d-dimensional isotropic media, we can extend the arguments of Debye et al. (1957) to obtain for finite s that

$$\left.\frac{dS_2^{(i)}}{dr}\right|_{r=0} = -\frac{\omega_{d-1}}{\omega_d\, d}\, s, \tag{2.34}$$

where $S_2^{(i)}$, more generally, is the two-point probability function for phase i and

$$\omega_d = \frac{\pi^{d/2}}{\Gamma(1 + d/2)} \tag{2.35}$$

is the d-dimensional volume of a sphere of unit radius, with $\omega_0 \equiv 1$. For the first three space dimensions, we have that

$$\left.\frac{dS_2^{(i)}}{dr}\right|_{r=0} = \begin{cases} -s/2\ , & d = 1, \\ -s/\pi\ , & d = 2, \\ -s/4\ , & d = 3. \end{cases} \tag{2.36}$$

Kirste and Porod (1962) examined the next term in the asymptotic expansion of the scattering curve $i(k)$ (proportional to k^{-6}). This was done for a special isotropic medium whose surface separating the void phase from the solid phase could be developed locally in a canonical power series in the local derivatives of the principal radii of curvature R_1, R_2 of the surface (Frisch and Stillinger 1963). The two-point probability function is then given by

$$S_2(r) = \phi_1 - \frac{s}{4} r \left\{ 1 - r^2 \left[\frac{1}{12S} \int K_1 K_2\, dA + \frac{1}{32S} \int (K_1 - K_2)^2\, dA \right] \right\} + \cdots, \tag{2.37}$$

where the integrals are taken over the interface, S is the mean interface area, $K_1 = 1/(2R_1)$, $K_2 = 1/(2R_2)$, and $r < 1/\max(K_1, K_2)$. The first integral in relation (2.37) is related to the topological genus p of the surface by the Gauss–Bonnet integral formula

$$4\int K_1K_2dA = 4\pi(1-p).$$

Notice that there is no quadratic term r^2 in (2.37). Relation (2.37) is valid only for surfaces containing no edges, corners, multiple points, or generally any singular points at which the radii of convergence of the aforementioned canonical expansion of the surface shrink to zero.

Therefore, (2.37) necessarily breaks down for isotropic dispersions of convex impenetrable particles that form interparticle contacts. Indeed, Frisch and Stillinger (1963) showed that for random systems of identical three-dimensional impenetrable spheres of diameter D, $S_2(r)$ is given by

$$S_2(r) = \phi_1 - \frac{s}{4}r + \frac{Z\phi_1}{4}\left(\frac{r}{D}\right)^2 + \mathcal{O}(r^3), \tag{2.38}$$

where Z is the *mean coordination number* defined to be the average number of contacts a given sphere has with its neighbors.

Realizability and Spectral Representation

What are the existence conditions for a valid (i.e., physically realizable) autocorrelation or autocovariance function? In the study of time series (one-dimensional random processes) (Priestley 1981) and the theory of turbulence (Batchelor 1959), it is well known that there are certain nonnegativity conditions involving the spectral representation of the autocovariance $\chi(\boldsymbol{r})$ that must be obeyed. Here we investigate such results for statistically homogeneous two-phase random media in any space dimension d (Torquato 1999). Importantly, we show that these nonnegativity conditions are necessary but *not sufficient* conditions that a valid autocovariance $\chi(\boldsymbol{r})$ of a statistically homogeneous two-phase random medium must meet. We also show that if the random medium is also statistically isotropic, there are d different nonnegativity conditions that one can exploit (Torquato 1999).

Consider an arbitrary stochastically continuous homogeneous process $\{Y(\boldsymbol{x}) : \boldsymbol{x} \in \Re^d\}$ with mean $\mu = \langle Y \rangle$ and autocovariance function

$$\chi(\boldsymbol{r}) = \langle [Y(\boldsymbol{x}) - \mu][Y(\boldsymbol{x}+\boldsymbol{r}) - \mu] \rangle. \tag{2.39}$$

It follows that

$$\chi(0) = \langle Y^2 \rangle - \mu^2 \tag{2.40}$$

and from Schwarz's inequality that

$$|\chi(\boldsymbol{r})| \le \langle Y^2 \rangle - \mu^2. \tag{2.41}$$

We now state the generalization of the Wiener–Khinchtine theorem (Priestley 1981) developed for processes in time to this multidimensional spatial stochastic process (Cressie 1993).

Theorem 2.1 *A necessary and sufficient condition for the existence of an autocovariance function $\chi(\mathbf{r})$ of a general stochastically continuous homogeneous process $\{Y(\mathbf{x}) : \mathbf{x} \in \Re^d\}$ is that it has the spectral (Fourier–Stieltjes) representation*

$$\chi(\mathbf{r}) = \frac{1}{(2\pi)^d} \int e^{i\mathbf{k}\cdot\mathbf{r}} dZ(\mathbf{k}), \tag{2.42}$$

where $Z(\mathbf{k})$ is a nonnegative bounded measure. If $\chi(\mathbf{r})$ is absolutely integrable, i.e.,

$$\int_{\Re^d} |\chi(\mathbf{r})|\, d\mathbf{r} < \infty, \tag{2.43}$$

then $dZ(\mathbf{k})$ can be written as $\tilde{\chi}(\mathbf{k})d\mathbf{k}$ and thus (2.42) becomes the standard Fourier representation

$$\chi(\mathbf{r}) = \frac{1}{(2\pi)^d} \int \tilde{\chi}(\mathbf{k}) e^{i\mathbf{k}\cdot\mathbf{r}} d\mathbf{k}, \tag{2.44}$$

where the spectral function $\tilde{\chi}(\mathbf{k})$ is positive semidefinite, i.e.,

$$\tilde{\chi}(\mathbf{k}) = \int \chi(\mathbf{r}) e^{-i\mathbf{k}\cdot\mathbf{r}}\, d\mathbf{r} \geq 0, \qquad \text{for all } \mathbf{k}. \tag{2.45}$$

Remarks:

1. This theorem may be proved by exploiting a general theorem due to Bochner (1936) that any continuous function $f(\mathbf{r})$ is positive semidefinite in the sense of (2.27) if and only if it has a Fourier–Stieltjes representation with a nonnegative bounded measure. The continuity of $\chi(\mathbf{r})$ follows directly from the requirement that the process $Y(\mathbf{x})$ is stochastically continuous. Thus, Theorem 2.1 may be regarded to be a special case of Bochner's theorem.
2. The quantity $Z/\chi(0)$ is often called the *spectral distribution function*. If $dZ(\mathbf{k}) = \tilde{\chi}(\mathbf{k})d\mathbf{k}$ where $\tilde{\chi}(\mathbf{k}) \geq 0$, then $\tilde{g}(\mathbf{k}) \equiv \tilde{\chi}(\mathbf{k})/\chi(0)$ is referred to as the *spectral density*, since it has the properties of a probability density function, i.e., $\int \tilde{g}(\mathbf{k})d\mathbf{k} = 1$ and $\tilde{g}(\mathbf{k}) \geq 0$.

Although the existence condition of Theorem 2.1 is known in the context of random media (Torquato 1999), it is not commonly known that not all autocovariances can be generated by stochastic processes $\{\mathcal{I}^{(i)}(\mathbf{x}) : \mathbf{x} \in \Re^d\}$ that take only two values, zero or one (Section 2.1). In other words, the class B of autocovariances that comes from the binary stochastic process $\{\mathcal{I}^{(i)}(\mathbf{x}) : \mathbf{x} \in \Re^d\}$ is a subclass of the total class that comes from the general process $\{Y(\mathbf{x}) : \mathbf{x} \in \Re^d\}$ and meets the existence condition of Theorem 2.1. Therefore, the condition of Theorem 2.1 is only necessary but not sufficient for B. An example of a function $\chi(\mathbf{r})$ that meets the requirement of Theorem 2.1 but may not belong to B has been analyzed by Torquato (1999) and is discussed in Section 12.6 [cf. (12.19)].

The task of determining the necessary and sufficient conditions that B must possess is very complex. In the context of stochastic processes in time (one-dimensional processes), it has been shown that autocovariances in B must not only meet the condition of Theorem 2.1 but another condition on "corner-positive" matrices (McMillan

1955, Shepp 1967). Since little is known about corner-positive matrices, this theorem is very difficult to apply in practice. Thus, a meaningful characterization of B remains an open and interesting problem, especially in the context of d-dimensional two-phase random media.

We will not attempt to address the complete characterization of B here but instead will summarize some simple necessary conditions, in addition to Theorem 2.1, that characterize B (Torquato 1999). We have seen that since $S_2^{(i)}(0) = \langle [\mathcal{I}^{(i)}]^2 \rangle = \langle \mathcal{I}^{(i)} \rangle = \phi_i$, the autocovariance at the origin is given by

$$\chi(0) = \phi_1 \phi_2, \qquad \text{for all } \chi(\mathbf{r}) \in B, \tag{2.46}$$

which should be compared to formula (2.40) for general stochastic processes. Application of the inequalities (2.24) to the two-point function $S_2^{(i)}(\mathbf{r})$ for homogeneous media yield the bounds $0 \le S_2^{(i)}(\mathbf{r}) \le \phi_i$, which are a direct consequence of the binary (i.e., zero-one) nature of the process. Combination of these bounds with relations (2.26) and (2.29) give the corresponding bounds that all autocovariances in B must obey:

$$-\min(\phi_1^2, \phi_2^2) \le \chi(\mathbf{r}) \le \phi_1 \phi_2, \qquad \text{for all } \chi(\mathbf{r}) \in B. \tag{2.47}$$

Unlike general stochastic processes for which (2.41) applies, here we have both upper and lower bounds on $\chi(\mathbf{r})$, the lower bound deriving from the *pointwise nonnegativity* of $S_2^{(i)}(\mathbf{r})$. Of course, in the absence of long-range order, $\chi(\infty) = 0$, but this condition is not special to binary processes. Another consequence of the binary nature of the process in the case of isotropic media is that the specific surface s is strictly positive when both phases are present and so (2.34) yields that

$$\left.\frac{dS_2^{(i)}}{dr}\right|_{r=0} = \left.\frac{d\chi}{dr}\right|_{r=0} < 0, \qquad \text{for all } 0 < \phi_i < 1 \quad \text{and} \quad \chi(r) \in B. \tag{2.48}$$

In other words, the slope of $\chi(r)$ at $r = 0$ is strictly negative for nontrivial volume fractions in the range $0 < \phi_i < 1$. Thus, an autocovariance $\chi(r)$ of an isotropic two-phase random medium can neither have a zero nor a positive slope at $r = 0$ when $0 < \phi_i < 1$ (Yeong and Torquato 1998a). Note that when the Fourier transform $\tilde{\chi}(\mathbf{k})$ exists, condition (2.45) implies only that the slope of $\chi(r)$ at $r = 0$ is nonpositive (i.e., negative semidefinite). We recall from an earlier part of this section that S_2 and thus $\chi(r)$ will generally possess not only a linear term r but a quadratic term r^2 for sufficiently small r [cf. (2.38)], although the quadratic term will be zero for a certain subclass of B [cf. (2.37)].

Although the nonnegativity condition of Theorem 2.1 or, equivalently, condition (2.27) is not sufficient to ensure that $\chi(\mathbf{r})$ belongs to B, either condition still provides a stringent test that all physically realizable $\chi(\mathbf{r})$ must meet. Experience shows that the nonnegativity condition coupled with the "binary" conditions (2.46)–(2.48) provide a practical (if not exact) means to test the validity of proposed autocovariances for a wide class of two-phase random media; see Yeong and Torquato (1998a), Cule and Torquato (1999), Torquato (1999), and Section 12.6.

Practically speaking, it is difficult to apply the nonnegativity condition (2.27) in order to test the validity of a proposed $\chi(\mathbf{r})$. For a wide class of statistically homogeneous two-phase media, $\chi(\mathbf{r})$ tends to zero fast enough for the Fourier transform $\hat{\chi}(\mathbf{k})$ to exist [cf. (2.43)]. In such instances, it is much easier to apply the nonnegativity condition (2.45) to test the validity of a proposed $\chi(\mathbf{r})$. In what follows, we will assume that the spectral function $\hat{\chi}(\mathbf{k})$ exists and proceed to show that there are d different nonnegativity conditions that one can exploit if the random medium is also statistically isotropic (Torquato 1999).

The Fourier transform of some absolutely integrable function $f(\mathbf{r})$ in d dimensions is given by

$$\tilde{f}(\mathbf{k}) = \int f(\mathbf{r}) e^{-i\mathbf{k}\cdot\mathbf{r}} d\mathbf{r}, \tag{2.49}$$

and the associated inverse operation is defined by

$$f(\mathbf{r}) = \frac{1}{(2\pi)^d} \int \tilde{f}(\mathbf{k}) e^{i\mathbf{k}\cdot\mathbf{r}} d\mathbf{k}, \tag{2.50}$$

where $\mathbf{k}$ is the wave vector. When the function depends only on the magnitude $r = |\mathbf{r}|$, then we have the following simpler expressions for $d = 1$, 2, and 3:

$$\tilde{f}(k) = 2\int_0^\infty f(r)\cos kr\,dr, \qquad f(r) = \frac{1}{\pi}\int_0^\infty \tilde{f}(k)\cos kr\,dk, \qquad d = 1, \tag{2.51}$$

$$\tilde{f}(k) = 2\pi\int_0^\infty f(r) r J_0(kr)\,dr, \qquad f(r) = \frac{1}{2\pi}\int_0^\infty \tilde{f}(k) k J_0(kr)\,dk, \qquad d = 2, \tag{2.52}$$

$$\tilde{f}(k) = \frac{4\pi}{k}\int_0^\infty f(r) r \sin kr\,dr, \qquad f(r) = \frac{1}{2\pi^2 r}\int_0^\infty \tilde{f}(k) k \sin kr\,dk, \qquad d = 3, \tag{2.53}$$

where $k = |\mathbf{k}|$ and $J_0(x)$ is the zeroth-order Bessel function of the first kind.

The nonnegativity condition (2.45) holds for any wave vector $\mathbf{k}$. In particular, it holds for $\mathbf{k} = \mathbf{0}$, i.e., the real-space volume integral of $\chi(\mathbf{r})$ must be positive semidefinite or

$$\int [S_2(\mathbf{r}) - \phi_1^2] d\mathbf{r} \geq 0. \tag{2.54}$$

The integral condition (2.54) holds for statistically homogeneous but anisotropic media. This nonnegativity condition could also have been obtained immediately from the work of Lu and Torquato (1990a) on the *coarseness*, or standard deviation of the local volume fraction. In particular, it can be obtained from the asymptotic expression (11.20) for large window sizes and the fact that the coarseness is positive semidefinite (see Chapter 11).

If the medium is also statistically isotropic, then the two-point correlation function depends only on the magnitude $r \equiv |\mathbf{r}|$, and (2.54) simplifies as

$$\int_0^\infty [S_2(r) - \phi_1^2] r^{d-1} dr \geq 0. \tag{2.55}$$

Here we have used the fact that $d\mathbf{r} = \Omega(d) r^{d-1} dr$ in a d-dimensional spherical coordinate system, where

$$\Omega(d) = \frac{2\pi^{d/2}}{\Gamma(d/2)} \tag{2.56}$$

is the positive d-dimensional solid angle and $\Gamma(x)$ is the gamma function. If we let

$$M_n = \int_0^\infty [S_2(r) - \phi_1^2] r^n \, dr \tag{2.57}$$

denote the nth moment of the function $S_2(r) - \phi_1^2$, then (2.55) states that the moment M_{d-1} is positive semidefinite for isotropic two-phase random media in d dimensions.

Thus far, we have shown that there is one nonnegativity condition for a d-dimensional homogeneous medium, namely, condition (2.45). However, the symmetry possessed by isotropic media enables one to obtain d different nonnegativity conditions. In particular, Torquato (1999) demonstrated that for an isotropic autocorrelation function $S_2(r)$ in d dimensions, the one-, two-, ..., and d-dimensional Fourier transforms of $\chi(r)$ are all positive semidefinite. Let $\tilde{\chi}(k; m)$ denote the m-dimensional Fourier transform of $\chi(r)$. Then, for all values of the wave number k (i.e., $k \geq 0$), we have that

$$\tilde{\chi}(k; m) \geq 0, \qquad m = 1, 2, \ldots, d. \tag{2.58}$$

This is easily proved by recalling that for d-dimensional isotropic media, $S_2(r)$ can be extracted from a cut of the d-dimensional medium with an m-dimensional subspace ($m = 1, 2, \ldots, d-1$). The m-dimensional subspace represents a lower-dimensional random medium but with the same S_2 as in d dimensions. Thus, the nonnegativity condition (2.45) applies to this lower-dimensional random medium, yielding (2.58).

It follows immediately from (2.55) and (2.58) that

$$M_n \geq 0, \qquad n = 0, 1, \ldots, d-1. \tag{2.59}$$

Thus, for three-dimensional isotropic media, the zeroth, first, and second moments of $S_2(r) - \phi_1^2$ must be positive semidefinite. For two-dimensional isotropic media, the zeroth and first moments must be positive semidefinite, whereas for one-dimensional media, only the zeroth moment need be positive semidefinite. The real-space conditions (2.54) and (2.59) are special cases of the more general and restrictive integral conditions (2.45) and (2.58), respectively.

Algorithms have been developed recently to construct realizations of two-phase random media with specified microstructural correlation functions (see Chapter 12). One can use the integral nonnegativity conditions (2.45) and (2.58) as well as the "binary" conditions (2.46)– (2.48) to test whether hypothetical autocorrelation or autocovariance function meet necessary realizability conditions. The zero-wave number integral conditions (2.54) and (2.59) may first be checked, since they are easier to compute than the full Fourier transform; if they are negative, then there is no need to compute the Fourier transform. We note that nonnegativity conditions on certain integrals involving the three- and two-point probability functions have also been obtained (Torquato 1980, Milton 1981b, Milton and McPhedran 1982, Torquato 1999, Markov 1999).

2.3 Surface Correlation Functions

Surface correlation functions contain information about the random interface $\partial\mathcal{V}$ and are of basic importance in the trapping and flow problems. In this context, we will let phase 1 denote the fluid or "void" phase, and phase 2 the "solid" phase. The simplest surface correlation function is the specific surface $s(\mathbf{x})$ (interface area per unit volume) at point $\mathbf{x}$, which is a one-point correlation function for statistically inhomogeneous media, i.e.,

$$s(\mathbf{x}) = \langle \mathcal{M}(\mathbf{x}) \rangle, \tag{2.60}$$

where $\mathcal{M}(\mathbf{x})$ is the interface indicator function given by (2.3). Note that the nonnegative specific surface cannot be interpreted as a probability, since the chance that a point at $\mathbf{x}$ lands on the interface is zero. For homogeneous media, it is a constant everywhere, which we will denote simply by s.

Two-point surface correlation functions for statistically inhomogeneous media are defined by

$$F_{sv}(\mathbf{x}_1, \mathbf{x}_2) = \langle \mathcal{M}(\mathbf{x}_1) \mathcal{I}(\mathbf{x}_2) \rangle, \tag{2.61}$$

$$F_{ss}(\mathbf{x}_1, \mathbf{x}_2) = \langle \mathcal{M}(\mathbf{x}_1) \mathcal{M}(\mathbf{x}_2) \rangle, \tag{2.62}$$

where $\mathcal{I}(\mathbf{x}) \equiv \mathcal{I}^{(1)}(\mathbf{x})$ is the indicator function for the void phase. These functions are called the surface–void and surface–surface correlation functions, respectively, and they arise in rigorous bounds on the trapping constant (Doi 1976, Rubinstein and Torquato 1988) and fluid permeability (Doi 1976, Rubinstein and Torquato 1989). For homogeneous media they depend only on the displacement $\mathbf{r} = \mathbf{x}_2 - \mathbf{x}_1$, and for isotropic media they depend only on the distance $r = |\mathbf{r}|$. The functions F_{sv} and F_{ss} can be obtained from any plane cut through a medium that is isotropic. Figure 2.6 shows events that contribute to these functions. When the two points are far from one another in systems without long-range order, $F_{sv}(\mathbf{x}_1, \mathbf{x}_2) \to s(\mathbf{x}_1) S_1(\mathbf{x}_2)$ and $F_{ss}(\mathbf{x}_1, \mathbf{x}_2) \to s(\mathbf{x}_1) s(\mathbf{x}_2)$. In the case of homogeneous media (of special interest to us in subsequent chapters), these asymptotic results for $|\mathbf{r}| \to \infty$ reduce to

$$F_{sv}(\mathbf{r}) \to \langle \mathcal{M} \rangle \langle \mathcal{I} \rangle = s\phi_1, \qquad F_{ss}(\mathbf{r}) \to \langle \mathcal{M} \rangle^2 = s^2, \tag{2.63}$$

where $\phi_1 = \langle \mathcal{I} \rangle$ is the *porosity*, or the volume fraction of the void phase.

The generalization to an n-point surface correlation function in which a subset of m of the n points is associated with the interface and the remaining $n - m$ points are associated with the void space is obvious:

$$F_{ss\cdots svv\cdots v}(\mathbf{x}^m; \mathbf{x}^{n-m}) = \left\langle \left[\prod_{i=1}^{m} \mathcal{M}(\mathbf{x}_i) \right] \left[\prod_{j=m+1}^{n} \mathcal{I}(\mathbf{x}_j) \right] \right\rangle, \tag{2.64}$$

where $\mathbf{x}^{n-m} \equiv \mathbf{x}_{m+1}, \mathbf{x}_{m+2}, \ldots, \mathbf{x}_n$. As we have emphasized, surface correlation functions are not probability functions. However, by associating with the two-phase interface a finite thickness, a probabilistic interpretation can be given in the limit that the thickness tends to zero; see Sections 4.1, 4.2.1, and 12.4.3. Observe that since the indicator functions in expression (2.64) are nonnegative, the surface correlation function is also nonnegative, i.e.,

$$F_{ss\ldots svv\ldots v}(\boldsymbol{x}^m; \boldsymbol{x}^{n-m}) \geq 0, \qquad \text{for all } \boldsymbol{x}^n \text{ in } \mathcal{V}. \tag{2.65}$$

Such correlation functions and their generalizations have been studied for particle systems (Torquato 1986a), as discussed in Chapter 4. In Chapters 4–6 and 12 we discuss the determination of F_{sv} and F_{ss} for particle models.

2.4 Lineal-Path Function

Another interesting and useful statistical measure is what we call the *lineal-path function* $L^{(i)}$ (Lu and Torquato 1992a). For statistically isotropic media, it is defined as follows:

$$L^{(i)}(z) = \text{Probability that a line segment of length } z \text{ lies wholly in phase } i \text{ when randomly thrown into the sample.} \tag{2.66}$$

In stochastic geometry, the quantity $\phi_i[1 - L^{(i)}(z)]$ is sometimes referred to as the *linear contact distribution function* (Stoyan et al. 1995). Figure 2.6 shows an event that contributes to the lineal-path function. We see that $L^{(i)}(z)$ contains a coarse level of *connectedness* information about phase i, albeit only along a *lineal path* of length z in phase i. The lineal-path function is a *lower-order* microstructural function, since it is a lower-order case of the canonical n-point correlation function discussed in Section 4.4.

The lineal-path function is a monotonically decreasing function of z, since the space available in phase i to a line segment of length z decreases with increasing z. At the extreme values of $L^{(i)}(z)$, we have that

$$L^{(i)}(0) = \phi_i, \qquad L^{(i)}(\infty) = 0,$$

where ϕ_i is the volume fraction of phase i. The "tail" of $L^{(i)}(z)$ (i.e., large z behavior) provides information about the largest lineal paths in phase i. If we define $L^{(12)}(z)$ to be the probability that a line segment of length z intersects any parts of the two-phase interface when randomly thrown into the sample, then it is clear that

$$L^{(1)}(z) + L^{(2)}(z) + L^{(12)}(z) = 1.$$

For three-dimensional media, we observe that $L^{(i)}(z)$ is equivalent to the area fraction of phase i measured from the projected image of a three-dimensional slice of thickness z onto a plane, as depicted in Figure 2.8. It is a problem of long-standing interest in stereology to find the projected area fraction or, equivalently, the lineal-path function $L^{(i)}(z)$, for three-dimensional particle systems. Its evaluation for nontrivial microstructures remains a challenging theoretical problem because of, in the language of

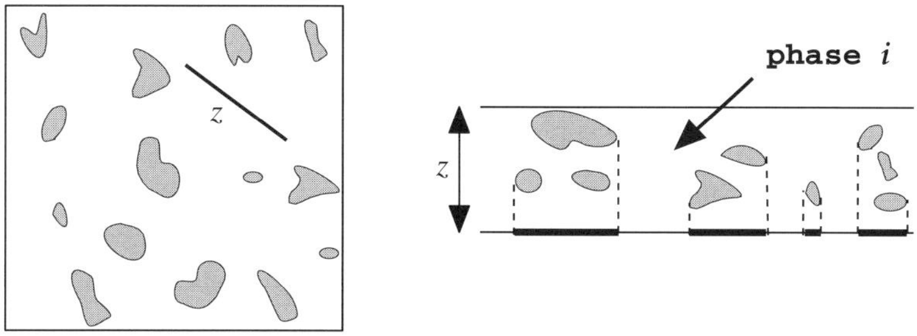

Figure 2.8 In two dimensions, the lineal-path function is the fraction of phase *i* obtained from a projection of a slab of thickness *z* onto a line.

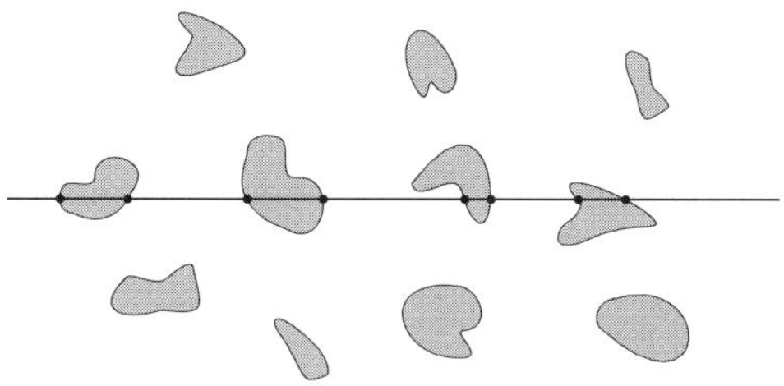

Figure 2.9 Chords are the line segments between the intersections of an infinitely long line with the two-phase interface.

Underwood (1970), "overlap" effects due to projection of the three-dimensional image and "truncation" effects due to slicing the system (see Figure 2.8).

For statistically homogeneous but anisotropic media, $L^{(i)}(\mathbf{z})$ will depend not only on the magnitude of vector $\mathbf{z}$ but on its orientation. For statistically inhomogeneous media, $L^{(i)}(\mathbf{x}_1, \mathbf{x}_2)$ will depend on the absolute positions $\mathbf{x}_1$ and $\mathbf{x}_2$ of the end points of the vector $\mathbf{z} = \mathbf{x}_2 - \mathbf{x}_1$.

2.5 Chord-Length Density Function

A quantity related to the lineal-path function $L^{(i)}(z)$ is the *chord-length probability density function* $p^{(i)}(z)$ (Matheron 1975, Torquato and Lu 1993). (The latter has been also called the chord-length "distribution" function.) Chords are all of the line segments between intersections of an infinitely long line with the two-phase interface (see Figure 2.9). The density function $p^{(i)}(z)$ is defined for statistically isotropic media as follows:

$$p^{(i)}(z)dz = \text{Probability of finding a chord of length between } z \text{ and } z+dz \text{ in phase } i. \tag{2.67}$$

Since it is a probability density function (having dimensions of inverse length), $p^{(i)}(z) \geq 0$ for all z, and it normalizes to unity, i.e.,

$$\int_0^\infty p^{(i)}(z)dz = 1. \tag{2.68}$$

Knowledge of the chord-length density function is of basic importance in transport problems involving "discrete free paths" and thus has application in Knudsen diffusion and radiative transport in porous media (Ho and Strieder 1979, Tokunaga 1985, Tassopoulos and Rosner 1992). The function $p^{(i)}(z)$ has also been measured for sedimentary rocks (Thompson, Katz and Krohn 1987) for the purpose of studying fluid flow through such porous media. The chord-length density function $p^{(i)}(z)$ is also a quantity of great interest in stereology (Underwood 1970). For example, the *mean chord* (or *intercept*) *length* is the first moment of $p^{(i)}(z)$.

We now show that $p^{(i)}(z)$ is related to the lineal-path function $L^{(i)}(z)$ using a simple probability argument (Torquato and Lu 1993). First, we observe that the lineal-path function $L^{(i)}(z)$ can be obtained by counting the relative number of times that a line segment of length z is wholly in phase i when thrown randomly onto an infinite line in the system. Denote by A the midpoint of the line segment. The probability that point A is in phase i is simply ϕ_i, the volume fraction of phase i. Second, given that the point A is in phase i (it is then on a chord), what is the probability that point A is on a chord with length between y and $y + dy$? Since the *length fraction* of a chord with length between y and $y + dy$ is

$$\frac{yp^{(i)}(y)dy}{\int_0^\infty yp^{(i)}(y)dy},$$

then the probability that the point A is on a chord with length between y and $y + dy$ is this length fraction multiplied by ϕ_i, i.e.,

$$\frac{\phi_i yp^{(i)}(y)dy}{\int_0^\infty yp^{(i)}(y)dy}.$$

Third, just because point A of a line segment of length z (*distinct from the length* y) is in phase i does not mean that the whole line segment is in phase i. The probability that a line segment of length z is on a chord of length y under the condition that the point A is on that chord is

$$\frac{(y - z)\Theta(y - z)}{y},$$

where $\Theta(x)$ is the Heaviside step function defined to be

$$\Theta(x) = \begin{cases} 0, & \text{if } x < 0, \\ 1, & \text{if } x \geq 0. \end{cases} \tag{2.69}$$

Now $L^{(i)}(z)$, the probability that the line segment of length z is entirely in phase i, can be obtained by combining the results given immediately above. Integrating the probability that the line segment is on chords with length between y and $y+dy$ over all possible y, we obtain

$$L^{(i)}(z) = \frac{\phi_i \int_0^\infty (y-z)p^{(i)}(y)\Theta(y-z)dy}{\int_0^\infty yp^{(i)}(y)dy}. \tag{2.70}$$

Differentiating (2.70) yields

$$\frac{dL^{(i)}(z)}{dz} = -\frac{\phi_i}{\ell_C}\int_z^\infty p^{(i)}(y)dy, \tag{2.71}$$

where $\ell_C^{(i)}$ is the mean chord length for phase i given by

$$\ell_C^{(i)} = \int_0^\infty zp^{(i)}(z)dz. \tag{2.72}$$

The first derivative of the lineal-path function is related to the *cumulative* distribution function $G(z)$ associated with $p(z)$, i.e., $G(z) = \mathcal{P}\{z \le Z\} = -(\ell_C/\phi_i)dL^{(i)}/dz$ (where Z is the associated continuous random variable). Differentiation of (2.71) and rearrangement of terms gives

$$p^{(i)}(z) = \frac{\ell_C}{\phi_i}\frac{d^2L^{(i)}(z)}{dz^2}. \tag{2.73}$$

Formula (2.73) establishes the connection between the chord-length probability density function $p^{(i)}(z)$ and the lineal-path function $L^{(i)}(z)$. The determination of both of these quantities for particle systems as well as digitized samples of real media is dealt with in Chapters 5, 6, and 12.

It is important to note that the above relations are valid for statistically isotropic systems of arbitrary microstructure. For such media it is simple to show that the mean chord length $\ell_C^{(i)}$ is related to the slope of the two-point probability function $S_2^{(i)}$ at the origin via the expression

$$\ell_C^{(i)} = \frac{\phi_i}{-\left.\dfrac{dS_2^{(i)}}{dr}\right|_{r=0}} = \frac{\omega_d}{\omega_{d-1}}\frac{\phi_i d}{s}, \tag{2.74}$$

where we have used (2.34). For the first three space dimensions, we have

$$\ell_C^{(i)} = \begin{cases} \dfrac{2\phi_i}{s}, & d=1, \\[2mm] \dfrac{\pi\phi_i}{s}, & d=2, \\[2mm] \dfrac{4\phi_i}{s}, & d=3. \end{cases} \tag{2.75}$$

The results (2.75) are well known in stereology (Underwood 1970).

For a three-dimensional isotropic medium we can use (2.75) to relate the specific surface in three dimensions $s(d=3)$ to the interface perimeter per unit area $s(d=2)$ (measured from a planar cut through the medium) and the number of interface points per unit length $s(d=1)$ (measured from a linear cut through the medium). Since $\ell_C^{(i)}$ and ϕ_i remain invariant when determined from $(d-1)$-dimensional cuts through a d-dimensional isotropic medium, then from (2.75) we immediately obtain

$$s(d=3) = \frac{4}{\pi} s(d=2) = 2\, s(d=1). \tag{2.76}$$

These results are also well known in stereology, albeit using the notation $S_V \equiv s(d=3)$, $L_A \equiv s(d=2)$ and $P_L \equiv s(d=1)$ (Underwood 1970).

2.6 Pore-Size Functions

The *pore-size probability density function* $P(\delta)$ (also referred to as pore-size "distribution" function) first arose to characterize the void or "pore" space in porous media (Prager 1963a). Actually, $P(\delta)$ can be used to probe either phase 1 or phase 2 of general random media consisting of two material phases. For simplicity, we will define $P(\delta)$ for phase 1, keeping in mind that it is equally well defined for phase 2. The function $P(\delta)$ for isotropic media is defined as follows:

$P(\delta)d\delta =$ Probability that a randomly chosen point in $\mathcal{V}_1(\omega)$ lies at a distance between δ and $\delta + d\delta$ from the *nearest* point on the pore-solid interface. (2.77)

Since it is a probability density function (having dimensions of inverse length), $P(\delta) \geq 0$ for all δ and it normalizes to unity, i.e.,

$$\int_0^\infty P(\delta) d\delta = 1. \tag{2.78}$$

At the extreme values of $P(\delta)$, we have that

$$P(0) = \frac{s}{\phi_1}, \qquad P(\infty) = 0, \tag{2.79}$$

where s/ϕ_1 is the interfacial area per unit pore volume. The associated *complementary* cumulative distribution function $F(\delta) = \mathcal{P}\{\Delta \geq \delta\}$ (where Δ is the associated continuous random variable)

$$F(\delta) = \int_\delta^\infty P(r) dr \tag{2.80}$$

is a nonincreasing function of δ such that

$$F(0) = 1, \qquad F(\infty) = 0. \tag{2.81}$$

Thus, $F(\delta)$ is the fraction of pore space that has a pore radius larger than δ.

Figure 2.6 shows an event that contributes to the pore-size density function. In stochastic geometry, the quantity $1 - F(\delta)$ is sometimes referred to as the *spherical contact distribution function* (Stoyan et al. 1995).

The moments of $P(\delta)$, defined as

$$\langle \delta^n \rangle = \int_0^\infty \delta^n P(\delta) d\delta, \tag{2.82}$$

provide useful characteristic length scales of the random medium. Integrating by parts and using (2.80) gives the alternative representation of the moments in terms of the cumulative distribution function:

$$\langle \delta^n \rangle = n \int_0^\infty \delta^{n-1} F(\delta) d\delta. \tag{2.83}$$

Lower-order moments of $P(\delta)$ arise in bounds on the mean survival and principal relaxation times (Prager 1963a, Torquato and Avellaneda 1991).

For a three-dimensional system, $P(\delta)$ is related to the probability of inserting a sphere of radius δ into the system. Thus, it contains a coarse level of *three-dimensional connectedness* information about phase 1. The pore-size function, therefore, cannot be extracted from a two-dimensional cross-section of the material; it is an *intrinsically three-dimensional descriptor*. It is noteworthy that the mathematically well-defined function $P(\delta)$ is not the usual pore-size "distribution" function obtained experimentally from mercury porosimetry (Scheidegger 1974, Dullien 1979).

The quantities $P(\delta)$ and $F(\delta)$ are actually trivially related to the "void" nearest-neighbor probability density function $H_V(r)$ and "void" exclusion probability $E_V(r)$, respectively, studied by Torquato, Lu and Rubinstein (1990) for systems of spherical inclusions and defined in Section 2.8. For example, consider any system of interacting identical spheres of radius R. Then using the definitions (2.88) and (2.90) for $H_V(r)$ and $E_V(r)$, it is clear that $\delta = r - R$, and so

$$P(\delta) = \frac{H_V(\delta + R)}{\phi_1}, \qquad \delta \geq 0, \tag{2.84}$$

$$F(\delta) = \frac{E_V(\delta + R)}{\phi_1}, \qquad \delta \geq 0. \tag{2.85}$$

Similarly, for spheres with a polydispersivity in size, P and F are related to the "void" nearest-surface functions h_V and e_V (described in Section 2.8) via the relations

$$P(\delta) = \frac{h_V(\delta)}{\phi_1}, \qquad \delta \geq 0, \tag{2.86}$$

$$F(\delta) = \frac{e_V(\delta)}{\phi_1}, \qquad \delta \geq 0. \tag{2.87}$$

We note that the pore-size functions are *lower-order* microstructural functions, since the void nearest-neighbor and nearest-surface functions are as well (see Section 2.8). In Chapters 4–6 and 12 we discuss the determination of the pore-size functions for particle models, as well as digitized media.

2.7 Percolation and Cluster Functions

The formation of very large "clusters" of a phase in a heterogeneous material (on the order of the system size) can have a dramatic influence on its macroscopic properties. A cluster of phase i is defined as the part of phase i that can be reached from a point in phase i without passing through phase $j \neq i$. A *critical point*, known as the *percolation threshold*, is reached when a sample-spanning cluster first appears. Unfortunately, any of the *lower-order* microstructural functions described thus far do not adequately reflect information about nontrivial cluster formation in the system. Torquato, Beasley and Chiew (1988) have introduced and represented the so-called two-point cluster function $C_2^{(i)}(\boldsymbol{x}_1, \boldsymbol{x}_2)$, defined to be the probability of finding two points at $\boldsymbol{x}_1$ and $\boldsymbol{x}_2$ in the same cluster of phase i . Thus, $C_2^{(i)}$ is the analogue of $S_2^{(i)}$, but unlike its predecessor, it contains nontrivial topological "connectedness" information. The measurement of $C_2^{(i)}$ for a three-dimensional material sample cannot be made from a two-dimensional cross-section of the material, since it is an intrinsically three-dimensional microstructural function.

Further mathematical details about $C_2^{(i)}$ and other existing percolation-sensitive quantities have been deferred until Chapters 9 and 10. Chapter 12 describes, among other considerations, the evaluation of $C_2^{(i)}$ from computer simulations.

2.8 Nearest-Neighbor Functions

All of the aforementioned statistical descriptors are defined for random media of *arbitrary microstructure*. In the special case of random media composed of *particles* (phase 2) distributed randomly throughout another material (phase 1), there is a variety of natural morphological descriptors. We describe some of them below for statistically isotropic media composed of identical spherical particles of diameter D (or radius $R = D/2$) at number density ρ distributed throughout another phase. (The reader is referred to Chapter 3 for a treatment of the statistical mechanics of particle systems.) We begin by defining nearest-neighbor functions.

In considering a many-body system of interacting particles, a key fundamental question to ask is the following: What is the effect of the nearest neighbor on some reference particle in the system? The answer to this query requires knowledge of the probability associated with finding the nearest neighbor at some given distance from a *reference* particle, i.e., the *"particle" nearest-neighbor probability density function* H_P. (This has been also called the nearest-neighbor "distribution" function.) Knowing H_P is of importance in a host of problems in the physical and biological sciences, including transport processes in heterogeneous materials (Keller, Rubenfeld and Molyneux 1967, Rubinstein and Torquato 1988, Rubinstein and Torquato 1989), stellar dynamics (Chandrasekhar 1943), spatial patterns in biological systems (McNally and Cox 1989), and the molecular physics of liquids and amorphous solids (Reiss, Frisch and Lebowitz 1959, Bernal

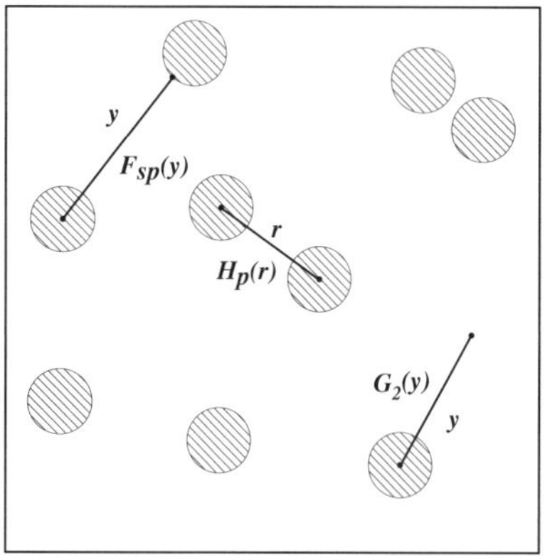

Figure 2.10 A schematic showing events that contribute to lower-order functions for random arrays of spheres (phase 2). Shown is the "particle" nearest-neighbor probability density H_P defined by (2.89), point/particle function $G_2 \equiv G_2^{(1)}$ defined by (2.120), and the surface-particle function F_{sp} defined by (2.122).

1964, Finney 1970, Zallen 1983, Torquato et al. 1990). Hertz (1909) was the first to consider its evaluation for a system of spatially *uncorrelated* "point" particles, i.e., particles whose centers are *Poisson* distributed (see Section 3.1.2). The calculation of H_P for nonoverlapping particles is nontrivial.

A different nearest-neighbor function, H_V, arises in the *scaled-particle* theory of liquids (Reiss et al. 1959, Hefland, Reiss, Frisch and Lebowitz 1960). This quantity (defined more precisely below) essentially characterizes the probability of finding a nearest-neighbor particle center at a given distance from an arbitrary point in the system. Since H_V is nontrivial when the point is located in the space *exterior* to the particles, we refer to it as the *"void" nearest-neighbor probability density function*.

There are other quantities closely related to H_V and H_P that we also consider. These are the so-called exclusion probabilities E_V and E_P and the conditional pair distributions G_V and G_P as defined below.

The nearest-neighbor functions $H_V(r)$ and $H_P(r)$ are defined as follows:

$$H_V(r)dr = \text{Probability that at an arbitrary point in the system the center of the nearest particle lies at a distance between } r \text{ and } r+dr. \tag{2.88}$$

$$H_P(r)dr = \text{Probability that at an arbitrary particle center in the system the center of the nearest particle lies at a distance between } r \text{ and } r+dr. \tag{2.89}$$

Note that since both $H_V(r)$ and $H_P(r)$ are probability density functions, they are nonnegative for all r, normalize to unity, and have dimensions of inverse length. Observe further that for statistically *inhomogeneous* media, $H_V(r)$ and $H_P(r)$ will depend also upon the position of the arbitrary point and the location of the central particle, respectively. Figure 2.10 shows an event that contributes to $H_P(r)$.

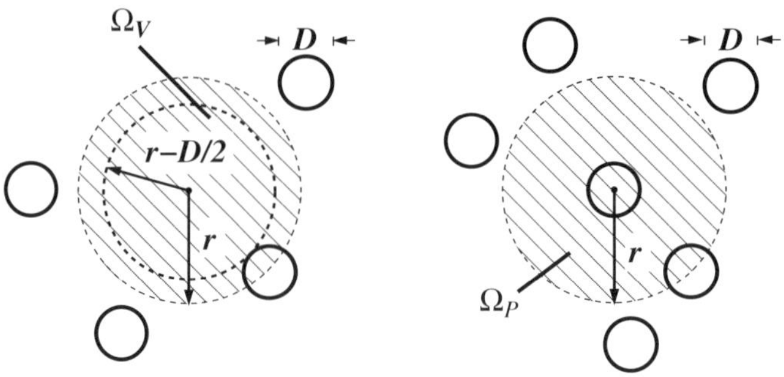

Figure 2.11 Schematic representations of the regions $\Omega_V(r)$ and $\Omega_P(r)$. Left panel: $\Omega_V(r)$ is the spherical region of radius r. The sphere of radius $r - D/2$ can be interpreted as a "test" particle of the same radius. Right panel: $\Omega_P(r)$ is a sphere of radius r surrounding some reference particle.

It is useful to introduce the associated dimensionless "exclusion" probabilities $E_V(r)$ and $E_P(r)$ defined as follows:

$E_V(r) =$ Probability of finding a region $\Omega_V(r)$ (which is a d-dimensional spherical cavity of radius r centered at some arbitrary point) empty of particle centers.
$=$ Expected fraction of space available to a "test" sphere of radius $r - D/2$ inserted into the system. (2.90)

$E_P(r) =$ Probability of finding a region $\Omega_P(r)$ (which is a d-dimensional spherical cavity of radius r centered at some arbitrary particle center) empty of other particle centers. (2.91)

Figure 2.11 gives a schematic representation of the regions $\Omega_V(r)$ and $\Omega_P(r)$. The first and second lines of (2.90) are equivalent, since the region excluded to a particle center of radius D by a "test" particle of radius $r - D/2$ is a sphere of radius r (see also Chapter 4). Thus, the *test* particle serves to probe the space available to it. For this reason, the density function $H_V(r)$ can also be interpreted to be the expected surface area per unit volume of the interface between available and unavailable spaces.

It follows that the exclusion probabilities are *complementary cumulative distribution functions* associated with the density functions and thus are related to the latter via

$$E_V(r) = 1 - \int_0^r H_V(x)dx \tag{2.92}$$

and

$$E_P(r) = 1 - \int_0^r H_P(x)dx. \tag{2.93}$$

Thus, both of these functions are monotonically decreasing functions of r. The integrals of (2.92) and (2.93) respectively represent the probabilities of finding at least one particle center in regions $\Omega_V(r)$ and $\Omega_P(r)$. Differentiating the exclusion-probability relations with respect to r gives

$$H_V(r) = \frac{-\partial E_V}{\partial r} \tag{2.94}$$

and

$$H_P(r) = \frac{-\partial E_P}{\partial r}. \tag{2.95}$$

Note that generalizations of these quantities describing events in which exactly n particle centers are contained within the regions $\Omega_V(r)$ and $\Omega_P(r)$ have been studied by Vezzetti (1975), Ziff (1977), and Truskett, Torquato and Debenedetti (1998) and by Truskett et al. (1998), respectively.

It is helpful to write the probability density functions as a product of two different functions. Specifically, for d-dimensional particles, let

$$H_V(r) = \rho s_1(r) G_V(r) E_V(r) \tag{2.96}$$

and

$$H_P(r) = \rho s_1(r) G_P(r) E_P(r), \tag{2.97}$$

where

$$s_1(r) = \frac{2\pi^{d/2} r^{d-1}}{\Gamma(d/2)} \tag{2.98}$$

is the surface area of a *single* d-dimensional sphere of radius r. For example, for $d = 1$, 2, and 3, $s_1(r)$ equals 2, $2\pi r$, and $4\pi r^2$, respectively.

Given definitions (2.88)–(2.92), the *conditional* pair "distribution" functions G_V and G_P must have the following interpretations:

$\rho s_1(r) G_V(r) dr$ = Given that region $\Omega_V(r)$ (spherical cavity of radius r) is empty of particle centers, the probability of finding particle centers in the spherical shell of volume $s_1(r)dr$ encompassing the cavity.

= Average number of particles at a radial distance between r and $r+dr$ from the center of $\Omega_V(r)$, given that this region is empty of particle centers. (2.99)

$\rho s_1(r) G_P(r) dr$ = Given that region $\Omega_P(r)$ (sphere of radius r encompassing any particle centered at some arbitrary position) is empty of particle centers, the probability of finding other particle centers in the spherical shell of volume $s_1(r)dr$ surrounding the central particle.

= Average number of particles at a radial distance between r and $r+dr$ from the center of $\Omega_P(r)$, given that this region is empty of particle centers. (2.100)

The exclusion probabilities are related to the pair distribution functions via the expressions

$$E_V(r) = \exp\left[-\int_0^r \rho s_1(y) G_V(y)dy\right], \tag{2.101}$$

$$E_P(r) = \exp\left[-\int_0^r \rho s_1(y) G_P(y)dy\right], \tag{2.102}$$

which are obtained by use of (2.94)–(2.97). Combination of (2.94), (2.95), (2.101), and (2.102) yields

$$H_V(r) = \rho s_1(r) G_V(r) \exp\left[-\int_0^r \rho s_1(y) G_V(y)dy\right] \tag{2.103}$$

and

$$H_P(r) = \rho s_1(r) G_P(r) \exp\left[-\int_0^r \rho s_1(y) G_P(y)dy\right]. \tag{2.104}$$

We see that once any one of the triplet H_V, E_V, G_V (H_P, E_P, G_P) is known, any of the other the nearest-neighbor functions can be ascertained via the interrelations (2.92)–(2.97) and (2.101)–(2.104). The nearest-neighbor functions are *lower-order* microstructural functions, since they are lower-order cases of the canonical n-point correlation function discussed in Section 4.4

We note that there are exact conditions that the void quantities must obey when r equals the sphere radius R for any statistically homogeneous and isotropic system of identical spheres. By definitions (2.88) and (2.90), we have that

$$H_V(R) = s, \qquad E_V(R) = \phi_1, \tag{2.105}$$

where s and ϕ_1 are the specific surface and volume fraction of phase 1, respectively. This expression combined with (2.96) yields

$$G_V(R) = \frac{s}{\rho s_1(R)\phi_1}. \tag{2.106}$$

These relations are true even if the spheres overlap to varying degrees. Most of the void quantities at their extreme values are known exactly:

$$E_V(0) = G_V(0) = 1,\; H_V(0) = \rho s_1(0) \qquad E_V(\infty) = H_V(\infty) = 0.$$

Some of the particle quantities at their extreme values are known exactly:

$$E_P(0) = 1, \qquad E_P(\infty) = H_P(\infty) = 0.$$

The behavior of the functions H_P and G_P at $r = 0$ and of G_V and G_P at $r = \infty$ are microstructure-dependent (see Chapters 5 and 6).

Consider the spatial moments of H_V and H_P. The moments of H_V are trivially related to moments of the pore-size function $P(\delta)$ for the special case of spheres (see

Section 2.6). The nth moment of $H_P(r)$ is defined as

$$m_n = \int_0^\infty r^n H_P(r)dr. \tag{2.107}$$

The lower limit of zero in the integral allows for particles that can get arbitrarily close to one another, such as in a Poisson distribution of spheres. A particularly important moment is the first moment $\ell_P \equiv m_1$, which is just the *mean nearest-neighbor distance between particles*. In the special case of ensembles of statistically isotropic *impenetrable* spheres of diameter D, the mean nearest-neighbor distance is given as

$$\ell_P = \int_D^\infty rH_P(r)dr, \tag{2.108}$$

which is equivalent to

$$\begin{aligned} \ell_P &= D + \int_D^\infty E_P(r)dr \\ &= D + \int_D^\infty \exp\left[-\int_0^r \rho s_1(y)G_P(y)dy\right]dr. \end{aligned} \tag{2.109}$$

Since, as we will see in Chapter 5, $E_P \geq 0$ for impenetrable spheres, it follows that $\ell_P \geq D$.

Finally, we would like to describe related nearest-neighbor functions. The nearest-neighbor functions discussed thus far have involved finding nearest *centers* of particles at given locations. One can instead define nearest-neighbor functions in the same way as before but in terms of finding nearest *surfaces* of particles (Lu and Torquato 1992b). Let us denote the surface counterparts by h_V, e_V, and g_V in the case of the void quantities and by h_P, e_P, and g_P in the case of the particle quantities. For spheres that are monodispersed in size (i.e., identical), the "surface" quantities contain the same information as the "center" quantities. Indeed, for identical spheres of radius R, we have that

$$h_V(r) = H_V(r+R), \qquad e_V(r) = E_V(r+R). \tag{2.110}$$

However, for spheres with a *polydispersivity in size*, the surface quantities are more meaningful, since the sphere with the nearest surface may not be the sphere with the nearest center.

As already remarked, the surface quantities are defined similarly to the center quantities except that the former are concerned with nearest surfaces. For example, following Lu and Torquato (1992b), the probability densities for polydisperse sphere systems are defined as follows:

$h_V(r)dr$ = Probability that the nearest particle surface lies at a distance between r and $r+dr$ from an arbitrary point in the system. (2.111)

$$h_P(r)dr = \text{Probability that the nearest particle surface lies at a distance between } r \text{ and } r+dr \text{ from the center of an arbitrary particle of radius } R. \tag{2.112}$$

It is important to emphasize that the radius R of the reference particle in the particle nearest-surface quantity h_P must be specified.

The corresponding exclusion probabilities are, as before, *complementary* cumulative distribution functions associated with h_V and h_P, i.e.,

$$e_V(r) = 1 - \int_{-\infty}^{r} h_V(x)\,dx \tag{2.113}$$

and

$$e_P(r) = 1 - \int_{-\infty}^{r} h_P(x)\,dx. \tag{2.114}$$

In each case the lower integration limit is $-\infty$ to allow for polydispersivity with sizes ranging to the infinitely large. Accordingly, r will generally lie in the interval $(-\infty, \infty)$ because the reference point may sometimes lie in the *particle phase* itself. This rather bizarre notion can be readily understood by appealing to Section 4.2, which describes the space available to "test" particles when added to a system of spheres of radius R. Allowing a test particle to have a negative radius r (down to $-R$) enables it to penetrate into the particle phase. It follows from (2.113) and (2.114) that

$$h_V(r) = -\frac{\partial e_V}{\partial r}, \qquad h_P(r) = -\frac{\partial e_P}{\partial r}. \tag{2.115}$$

The conditional pair functions g_V and g_P are defined through the following relations:

$$h_V(r) = g_V(r)e_V(r), \qquad h_P(r) = g_P(r)e_P(r). \tag{2.116}$$

Notice that surface quantities g_V and g_P are defined differently from G_V and G_P in that the former absorb the surface area terms not contained in the latter. Moreover, for any polydisperse system of spheres, the void quantities evaluated at the origin are, by definition, given as

$$h_V(0) = s, \qquad e_V(0) = \phi_1, \qquad g_V(0) = s/\phi_1. \tag{2.117}$$

The quantity s/ϕ_1 is the interface area per unit volume of phase 1.

One can compute spatial moments of either h_V or h_P. The moments of h_V are trivially related to moments of the pore-size density function for systems of spheres (see Section 2.6). The natural generalization of the first moment of H_P given by (2.108) for monodisperse systems is the following definition for polydisperse systems for a reference particle of radius R:

$$\lambda_P = \int_0^{\infty} r h_P(r)dr \; - R. \tag{2.118}$$

For the special case of impenetrable spheres, this definition, after integration by parts, is equivalent to

$$\lambda_P = \int_R^\infty e_P(r)dr. \tag{2.119}$$

We refer to λ_P as the *mean surface–surface distance.*

The determination of the nearest-neighbor functions for monodisperse and polydisperse sphere systems is taken up in Chapters 4–6.

2.9 Point/q-Particle Correlation Functions

Consider statistically inhomogeneous media composed of N identical spherical particles of radius R (phase 2) distributed throughout another phase (phase 1). Let $\mathbf{r}^q \equiv \{\mathbf{r}_1, \ldots, \mathbf{r}_q\}$ denote the positions of q sphere centers and let $d\mathbf{r}^q \equiv d\mathbf{r}_1 d\mathbf{r}_2 \cdots d\mathbf{r}_q$. The point/$q$-particle correlation (or "distribution") function $G_n^{(i)}(\mathbf{x}; \mathbf{r}^q)$ is defined as follows (Torquato 1986b):

$$G_n^{(i)}(\mathbf{x}; \mathbf{r}^q)d\mathbf{r}^q = \text{Probability of finding a point in phase } i \text{ at } \mathbf{x} \text{ and the center of a sphere in volume element } d\mathbf{r}_1 \text{ about } \mathbf{r}_1\text{, the center of another sphere in volume element } d\mathbf{r}_2 \text{ about } \mathbf{r}_2, \ldots, \text{ and the center of another sphere in volume element } d\mathbf{r}_q \text{ about } \mathbf{r}_q\text{, where } n = 1 + q. \tag{2.120}$$

The correlation function $G_n^{(i)}(\mathbf{x}; \mathbf{r}^q)$ is a hybrid quantity: It is a probability function with respect to the position $\mathbf{x}$ and a joint probability density function (up to a trivial factor) with respect to the positions $\mathbf{r}^q$ of the q particles. In light of this nature, it obeys the normalization condition

$$\int G_n^{(i)}(\mathbf{x}; \mathbf{r}^q)d\mathbf{r}^q = \frac{N!}{(N-q)!} S_1^{(i)}(\mathbf{x}), \tag{2.121}$$

where $S_1^{(i)}(\mathbf{x})$ is the one-point probability function for phase i defined in Section 2.2.1. Note that $G_n^{(i)}(\mathbf{x}; \mathbf{r}^q)$ divided by the right side of (2.121) is indeed a probability density function, since it is nonnegative and normalizes to unity. Originally, $G_n^{(i)}$ was denoted as $G_q^{(i)}$ by Torquato (1986b).

For statistically homogeneous media, $G_n^{(i)}$ depends only on the relative displacements $\mathbf{y}_1, \ldots, \mathbf{y}_q$, where $\mathbf{y}_k = \mathbf{x} - \mathbf{r}_k$. For isotropic media, it depends only on the distances between all of the n points. Figure 2.10 shows an event that contributes to the two-point quantity $G_2(y)$, where $y = |\mathbf{x} - \mathbf{r}_1|$. The point/$q$-particle correlation function arises in bounds on the effective conductivity (Torquato 1986b), effective elastic moduli (Quintanilla and Torquato 1995), trapping constant (Rubinstein and Torquato 1988), and fluid permeability (Rubinstein and Torquato 1989).

Torquato (1986b) showed that the point/q-particle correlation function can be expressed as a special ensemble average of the indicator function $\mathcal{I}^{(i)}(\mathbf{x})$ for phase i given by (2.1) [see also (4.46) and (4.74)]. Using this expression, it is easy to demonstrate that

$$G_n^{(1)}(\boldsymbol{x};\boldsymbol{r}^q)+G_n^{(2)}(\boldsymbol{x};\boldsymbol{r}^q)=\rho_q(\boldsymbol{r}^q),$$

where $\rho_q(\boldsymbol{r}^q)$ is the q-particle probability density function associated with finding q spheres with configuration $\boldsymbol{r}^q$, described more fully in Chapter 3. We see that since $G_1^{(i)}=S_1^{(i)}$, then we define $\rho_0\equiv 1$. It is clear that

$$G_n^{(1)}(\boldsymbol{x};\boldsymbol{r}^q)=0 \quad \text{if} \quad |\boldsymbol{x}-\boldsymbol{r}_k|<R, \quad k=1,\ldots,n,$$

since the point $\boldsymbol{x}$ cannot be in any sphere for the $G_n^{(1)}$. The last two expressions then give

$$G_n^{(2)}(\boldsymbol{x};\boldsymbol{r}^q)=\rho^q(\boldsymbol{r}^q) \quad \text{if} \quad |\boldsymbol{x}-\boldsymbol{r}_k|<R, \quad k=1,\ldots,n.$$

The asymptotic properties of the $G_n^{(i)}(\boldsymbol{x};\boldsymbol{r}^q)$ have been given by Torquato (1986b). Chapters 4–6 discuss the determination of the point/q-particle correlation function for monodisperse and polydisperse sphere systems.

2.10 Surface/Particle Correlation Function

The surface/particle correlation function $F_{sp}(\boldsymbol{x};\boldsymbol{r}_1)$ for statistically inhomogeneous systems of N identical spheres is defined as follows:

$$F_{sp}(\boldsymbol{x};\boldsymbol{r}_1) = \text{Correlation function associated with a point being on the interface at } \boldsymbol{x} \text{ and the probability of finding the center of a sphere in volume element } d\boldsymbol{r}_1 \text{ about } \boldsymbol{r}_1. \tag{2.122}$$

This function obeys the normalization condition

$$\int F_{sp}(\boldsymbol{x};\boldsymbol{r}_1)d\boldsymbol{r}_1=Ns(\boldsymbol{x}), \tag{2.123}$$

where $s(\boldsymbol{x})$ is the local specific surface defined by (2.60). The n-point generalization of this function is discussed in Chapter 4.

For homogeneous media, F_{sp} depends only on the displacement $\boldsymbol{y}=\boldsymbol{x}-\boldsymbol{r}_1$. For isotropic media, it depends only on the distance $y=|\boldsymbol{y}|$. Figure 2.10 shows an event that contributes to $F_{sp}(y)$. The surface/particle function F_{sp} arises in rigorous bounds on the fluid permeability of random beds of spheres (Torquato and Beasley 1987).

CHAPTER 3

Statistical Mechanics of Many-Particle Systems

Statistical mechanics is the branch of theoretical physics that attempts to predict, by starting at the level of atoms, molecules, spins, or other small "particles," the bulk properties of systems in which a large number of these particles interact with one another. In other words, it links the microscopic properties of matter (molecular interactions and structure), as determined from the laws of quantum or classical mechanics, to its macroscopic properties (e.g., pressure of a liquid). The province of statistical physics is more general, extending to any situation in which one is interested in the collective behavior of interacting *entities*, from population dynamics through solids, liquids, and gases to cosmology as well as *random heterogeneous materials*. Systems composed of many interacting particles (albeit much larger than molecular dimensions) are often useful models of random heterogeneous materials, and thus one can exploit the powerful machinery of statistical mechanics to study such materials. Moreover, as we will see in subsequent chapters, the formalism of statistical mechanics can be extended to nonparticulate systems.

The purpose of this chapter is to introduce foundational statistical-mechanical notions and results that we will heavily employ in the remaining chapters of Part I. Readers interested in a more general and comprehensive treatment of statistical mechanics are referred to the books by Tolman (1979), Hansen and McDonald (1986), Hill (1987), and Huang (1987).

The chapter begins with a review of well-established concepts and techniques, such as n-particle probability density functions to characterize structure, interparticle potentials, and the Ornstein–Zernike integral equation formalism. Many of these ideas are applied in subsequent chapters. Since the structure of particle systems at high densities is primarily determined by repulsive interactions between the particles, we devote an appreciable portion of this chapter to the discussion of *hard-particle* systems

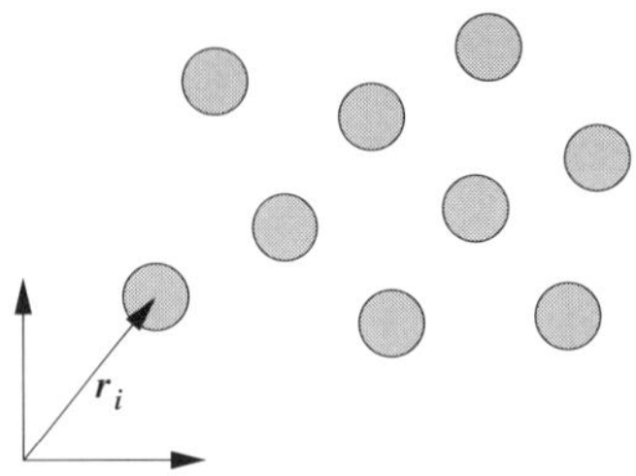

Figure 3.1 Definition of center-of-mass coordinate $\boldsymbol{r}_i$ for the ith particle.

in equilibrium as well as a special nonequilibrium ensemble called the *random sequential addition process*. Here we summarize some basic results but also describe the latest developments.

Finally, we close the chapter by discussing a new notion to describe random packings of spheres, namely, the *maximally random jammed state*. This idea has been introduced to replace the traditional one of a *random close packed state,* which has been recently demonstrated to be ill-defined.

3.1 Many-Particle Statistics

Any *ensemble* of many-particle systems (i.e., a large collection of replicas of the system all having the same macroscopic parameters) is completely spatially characterized classically by the probability density function associated with finding a particular configuration of particles. We define and discuss such many-particle statistics below. Since the many-particle probability density function is strongly affected by the interparticle potential, we also review several commonly used model potentials.

3.1.1 n-Particle Probability Densities

For simplicity, we will begin by considering a classical, *closed* system of N interacting identical spherical particles in the space $\mathcal{V} \in \Re^d$ of volume V, keeping in mind that we can generalize the ensuing discussion to treat spheres with a polydispersivity in size as well as oriented, nonspherical particles. Let $\boldsymbol{r}_i$ designate the position of the center of the ith particle (see Figure 3.1). Let $\Phi_N(\boldsymbol{r}^N)$ be the N-body (or N-particle) potential for particles with configuration $\boldsymbol{r}^N \equiv \{\boldsymbol{r}_1, \boldsymbol{r}_2, \ldots, \boldsymbol{r}_N\}$. The spatial arrangement of the particles is strongly influenced by the potential Φ_N, which can be decomposed into 1-particle terms, 2-particle terms, 3-particles, etc., as follows:

$$\Phi_N(\boldsymbol{r}^N) = \sum_{i=1}^{N} \varphi_1(\boldsymbol{r}_i) + \sum_{i<j}^{N} \varphi_2(\boldsymbol{r}_i, \boldsymbol{r}_j) + \sum_{i<j<k}^{N} \varphi_3(\boldsymbol{r}_i, \boldsymbol{r}_j, \boldsymbol{r}_k) + \cdots . \tag{3.1}$$

Here φ_n is the intrinsic n-particle potential *in excess of the contributions* from $\varphi_1, \varphi_2, \ldots, \varphi_{n-1}$.

In many physical systems (e.g., liquids and colloids), three-particle and higher-order potentials are negligibly small compared to one- and two-particle terms, i.e., to

$$\Phi_N(\mathbf{r}^N) \approx \sum_{i=1}^{N} \varphi_1(\mathbf{r}_i) + \sum_{i<j}^{N} \varphi_2(\mathbf{r}_i, \mathbf{r}_j). \tag{3.2}$$

The quantity φ_1 represents the one-particle potential due to an external field. For example, $\varphi_1(z) = mgz$ for a gravitational field, and the associated force field is $\boldsymbol{F} = -\nabla\varphi_1 = -mg\hat{\boldsymbol{k}}$, where $\hat{\boldsymbol{k}}$ is the unit vector in the positive (vertical) z-direction. In the absence of an external field, we may assume pairwise additivity, i.e.,

$$\Phi_N(\mathbf{r}^N) = \sum_{i<j}^{N} \varphi_2(r_{ij}). \tag{3.3}$$

In general, the microstate of the system (positions and velocities of the particles) at any instant will be determined, in a classical description, by Newton's or Hamilton's equations of motion. The *phase space* is the space of all possible positions and velocities of the particles. However, for our purposes, we are interested in the spatial configuration of the particles, regardless of their velocities. Note that for nonspherical particles, $\mathbf{r}_i$ would represent the center-of-mass coordinate as well as the particle orientation.

The configuration of the particles is statistically characterized by the *specific* N-particle probability density function $P_N(\mathbf{r}^N, t)$ defined as follows:

$$P_N(\mathbf{r}^N, t)\, d\mathbf{r}^N = \begin{array}{l}\text{Probability of finding the center of particle 1 in volume element } d\mathbf{r}_1 \text{ about } \mathbf{r}_1, \text{ the center of particle 2 in}\\ \text{volume element } d\mathbf{r}_2 \text{ about } \mathbf{r}_2 \ldots, \text{ and the center of}\\ \text{particle } N \text{ in volume element } d\mathbf{r}_N \text{ about } \mathbf{r}_N \text{ at time}\\ t,\end{array} \tag{3.4}$$

where $d\mathbf{r}^N \equiv d\mathbf{r}_1 d\mathbf{r}_2 \cdots d\mathbf{r}_N$. Since P_N is a probability density function, it normalizes to unity, i.e.,

$$\int P_N(\mathbf{r}^N, t)\, d\mathbf{r}^N = 1, \tag{3.5}$$

and therefore has dimensions of V^{-N}.

The N-particle function $P_N(\mathbf{r}^N, t)$ for a *nonequilibrium* ensemble generally depends upon the N-particle potential $\Phi_N(\mathbf{r}^N)$ and the particular dynamical process involved to create the system; in shorthand notation,

$$P_N(\Phi_N, \text{ dynamics}).$$

We use the term "dynamics" in a broad sense: It may signify the time evolution of the system (the subject of *kinetic theory*) or the nonequilibrium nature of a "frozen" state of the system, i.e., a time-independent *quenched* nonequilibrium system.

If the system is in *thermodynamic equilibrium* (roughly speaking, we mean that the particles sample the phase space uniformly), then P_N is independent of time and is uniquely determined by Φ_N. In particular, in the *canonical* ensemble (closed systems with fixed N and V in contact with a heat bath at absolute temperature T), one has (Hansen and McDonald 1986)

$$P_N(\boldsymbol{r}^N) = \frac{e^{-\beta\Phi_N(\boldsymbol{r}^N)}}{Q_N} \qquad \text{with } Q_N = \int e^{-\beta\Phi_N(\boldsymbol{r}^N)}\, d\boldsymbol{r}^N, \tag{3.6}$$

where Q_N is the canonical *configurational partition function*, $\beta = 1/(kT)$ is a reciprocal temperature, and k is Boltzmann's constant (trivially related to the ideal gas constant). By contrast, there is an infinite number of nonequilibrium ensembles consistent with an N-particle potential.

We recall that in the context of random heterogeneous materials, our interest is in microstructures that can be taken to be independent of time. In practice, this requirement restricts us to equilibrium systems or, more generally, *quenched* nonequilibrium systems.

For time-independent ensembles consisting of *indistinguishable* particles, it is convenient to introduce the *generic* n-particle probability density function $\rho_n(\boldsymbol{r}^n)$, defined as

$$\rho_n(\boldsymbol{r}^n) = \frac{N!}{(N-n)!}\int P_N(\boldsymbol{r}^N) d\boldsymbol{r}^{N-n}, \tag{3.7}$$

where $d\boldsymbol{r}^{N-n} \equiv d\boldsymbol{r}_{n+1} d\boldsymbol{r}_{n+2} \cdots d\boldsymbol{r}_N$. In words, $\rho_n(\boldsymbol{r}^n)d\boldsymbol{r}^n$ is proportional to the probability of finding *any* subset of n particles with configuration $\boldsymbol{r}^n$ in volume element $d\boldsymbol{r}^n$. Even though it follows from (3.5) and (3.7) that ρ_n is not normalized to unity but rather

$$\int \rho_n(\boldsymbol{r}^n) d\boldsymbol{r}^n = \frac{N!}{(N-n)!}, \tag{3.8}$$

it is still commonly referred to as a "probability density function," since it can be made so trivially by dividing it by the normalization constant $N!/(N-n)!$.

For statistically homogeneous media, $\rho_n(\boldsymbol{r}^n)$ is translationally invariant and hence depends only on the relative displacements of the positions with respect to some chosen origin, say $\boldsymbol{r}_1$:

$$\rho_n(\boldsymbol{r}^n) = \rho_n(\boldsymbol{r}_{12}, \boldsymbol{r}_{13}, \ldots, \boldsymbol{r}_{1n}), \tag{3.9}$$

where $\boldsymbol{r}_{ij} = \boldsymbol{r}_j - \boldsymbol{r}_i$. In particular, the one-particle function ρ_1 is just equal to the constant *number density* of particles ρ, i.e.,

$$\rho_1(\boldsymbol{r}_1) = \rho \equiv \lim_{N,V\to\infty} \frac{N}{V}. \tag{3.10}$$

The limit indicated in (3.10) is referred to as the *thermodynamic limit*.

For statistically homogeneous media, it is convenient to define the so-called *n-particle correlation function*

$$g_n(\boldsymbol{r}^n) = \frac{\rho_n(\boldsymbol{r}^n)}{\rho^n}. \tag{3.11}$$

In systems without long-range order and in which the particles are mutually far from one another (i.e., $r_{ij} = |\boldsymbol{r}_{ij}| \to \infty$, $1 \le i < j \le N$), $\rho_n(\boldsymbol{r}^n) \to \rho^n$ and we have from (3.11)

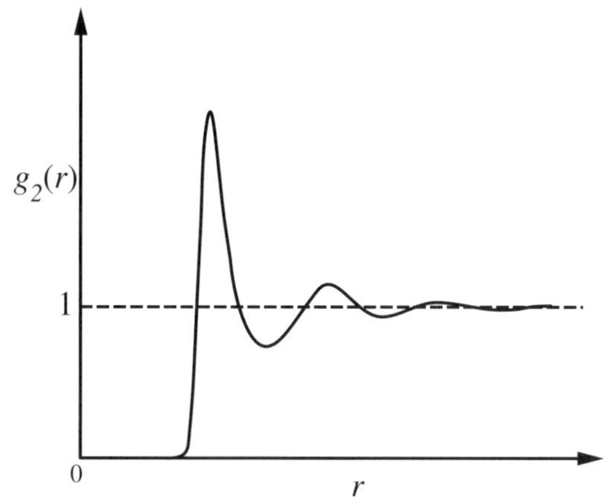

Figure 3.2 Radial distribution function $g_2(r)$ for a typical disordered system of interacting particles.

that $g_n(\mathbf{r}^n) \to 1$. Thus, the deviation of g_n from unity provides a measure of the degree of spatial correlation between the particles, with unity corresponding to no spatial correlation.

The important two-particle quantity

$$g_2(\mathbf{r}_{12}) = \frac{\rho_2(\mathbf{r}_{12})}{\rho^2} \tag{3.12}$$

is usually referred to as the *pair correlation function*. In the statistically isotropic case, it depends on the radial distance r_{12} only, i.e.,

$$g_2(\mathbf{r}_{12}) = g_2(r_{12}), \tag{3.13}$$

and is referred to as the *radial distribution function*. From (3.12), we see that $\rho g_2(r)d\mathbf{r} = \rho s_1(r)g_2(r)dr$ is proportional to the *conditional probability* of finding a particle center in a spherical shell of volume $s_1(r)dr$, given that there is another at the origin. Here $s_1(r)$ is the surface area of a d-dimensional sphere of radius r, as given by (2.98). Hence, for a finite system, integrating $\rho g_2(r)$ over the volume yields $N-1$, i.e., all the particles except the one at the origin. Alternatively, $\rho s_1(r)g_2(r)dr$ is the average number of particles at a radial distance between r and $r + dr$ from a reference particle. For homogeneous and isotropic systems, the general quantity g_n depends only on the relative distances between the n particle centers.

Figure 3.2 shows a plot of a radial distribution function for a typical dense *disordered* system of interacting particles, i.e., a pair potential with both short-range repulsive and attractive interactions (see Section 3.1.2). The probability of two particles having a very small separation is very low because of large repulsive forces; thus, for values of r of order less than that of the diameter D, $g_2(r)$ is zero. Since the density is large, there is a strong likelihood that a first neighbor shell will be at a distance of around $r = D$ from

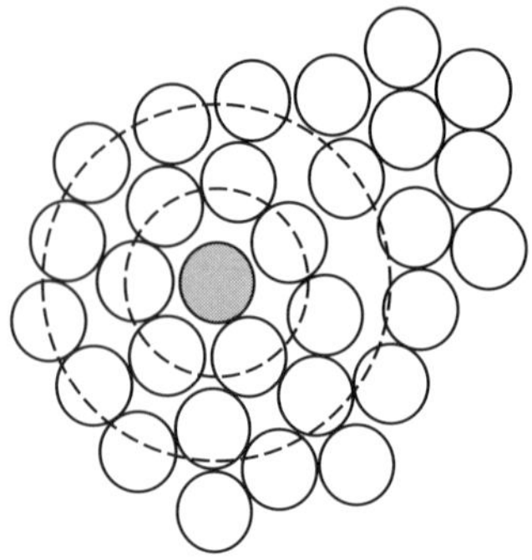

Figure 3.3 Near-neighbor shell contributions to the radial distribution function for a disordered system of identical particles.

a particle located at the origin (see Figure 3.3). This first peak corresponds roughly to the minimum in the pair potential. The nearest neighbors, which make up the first coordination shell, tend to exclude the *next-nearest* neighbors from distances on the order of $r \approx 3D/2$, and thus $g_2(r)$ is less than unity here. The second peak in $g_2(r)$ corresponds to the most probable location for the next-nearest neighbors. Since the system is disordered, the amplitudes of the oscillations dampen as r increases, i.e., the system exhibits short-range order for several diameters. Since there is no long-range order, $g_2(r) \to 1$ as $r \to \infty$.

The pair correlation function g_2 is of central importance in the study of the liquid state for several reasons. First, for particles that interact through pairwise additive central forces (see discussion below), knowledge of g_2 allows one to compute the thermodynamic (equilibrium) properties of the system (Hansen and McDonald 1986). For example, the pressure p of a d-dimensional isotropic system is given by

$$p = \rho kT - \frac{\rho^2}{2d} \int_0^\infty s_1(r) g_2(r) r \frac{d\varphi_2}{dr} dr, \tag{3.14}$$

where φ_2 is the pair potential defined above and $s_1(r)$ is given by (2.98). Second, for any physical many-body potential, the pair correlation function determines the isothermal compressibility $\kappa_T \equiv \rho^{-1}(\partial\rho/\partial p)_T$ of a homogeneous system via the expression

$$\rho kT\kappa_T = 1 + \rho \int [g_2(\mathbf{r}) - 1] d\mathbf{r}. \tag{3.15}$$

We note that this *compressibility* relation is derived for an "open" system (*grand canonical ensemble*) and is directly related to the *density fluctuations* (specifically, the variance in density). Third, g_2 is accessible in the laboratory via wave scattering experiments from the *structure factor*, which is proportional to the Fourier transform of $g_2(r) - 1$. Fourth, it plays an important role in the study of nonequilibrium systems. Lastly, the higher-order functions, g_3, g_4, etc. are more difficult to compute (see Section 3.2).

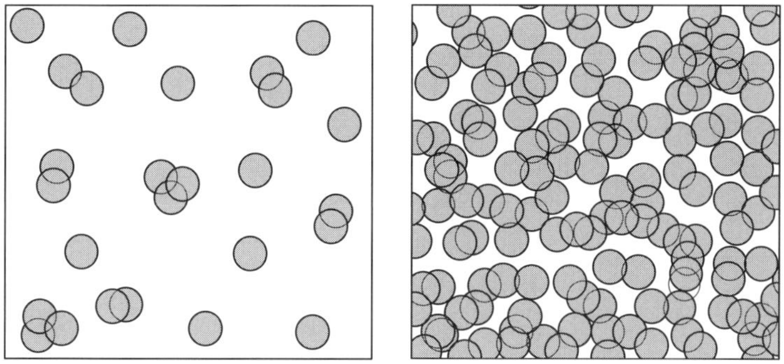

Figure 3.4 Two-dimensional overlapping-particle systems at a low density with at most three particle overlaps (left panel) and at a high density above the percolation threshold (right panel).

3.1.2 Pair Potentials

Several model pair potentials are described below. Important definitions that we will use are the *Boltzmann factor* associated with pair potential φ_2, given by

$$e(r) \equiv \exp[-\beta\varphi_2(r)], \tag{3.16}$$

and the *Mayer function*

$$f(r) \equiv e(r) - 1. \tag{3.17}$$

Later in the chapter we will show that the Boltzmann factor $e(r)$ is the dilute limit of the radial distribution function $g_2(r)$ for equilibrium systems.

Noninteracting or Overlapping Particles

By definition, for a system of noninteracting particles we have that $\Phi_N = 0$. From the point of view of statistical thermodynamics, this is the trivial case of an *ideal gas*. However, this is a *nontrivial* model of a heterogeneous material, since the lack of spatial correlation implies that the particles may overlap to form complex clusters, as shown in Figure 3.4. At low sphere densities, the particle phase is a dispersed, disconnected phase, but above a critical value (called the percolation threshold), the particle phase becomes connected. For $d = 2$ and $d = 3$, this threshold occurs at a sphere volume fraction of about 0.68 and 0.29, respectively. The reader is referred to Chapter 10 for details regarding continuum percolation.

In the study of disordered media this is a very popular model and goes by a variety of names, including "fully penetrable spheres," "randomly overlapping spheres," "spatially uncorrelated spheres," "Poisson distributed spheres," and the "Swiss-cheese model." It is a special case of the "Boolean" model used in stochastic geometry (Stoyan et al. 1995). We will usually refer to this model simply as *overlapping spheres*, but on occasion will call it *fully penetrable spheres* when using it as a special limit of interpenetrable-sphere

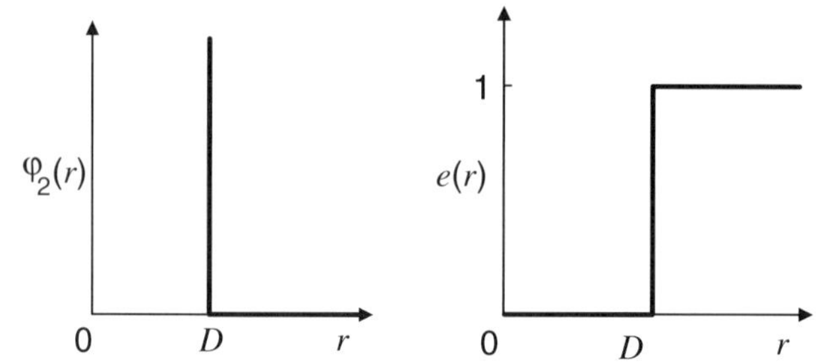

Figure 3.5 The hard-sphere potential (left) and the corresponding Boltzmann factor (right).

models as described below. Interpenetrable-sphere systems in general are useful models of consolidated media, such as sandstones and other rocks, and sintered materials.

For a statistically homogeneous system of overlapping spheres in equilibrium, we see from relation (3.6) that $P_N = 1/V^N$, and hence, from expressions (3.7) and (3.11), the n-particle statistics in the thermodynamic limit are simple constants for any configuration $\mathbf{r}^n$, i.e.,

$$\rho_n(\mathbf{r}^n) = \rho^n, \qquad g_n(\mathbf{r}^n) = 1. \tag{3.18}$$

We note that the probability $P(n)$ that exactly n sphere centers are contained in some region of space of volume W within a homogeneous system of overlapping spheres at number density ρ is known analytically (in contrast to systems of *interacting* spheres) and given by

$$P(n) = \frac{(\rho W)^n}{n!}\exp(-\rho W), \qquad n = 0, 1, 2, \ldots, \tag{3.19}$$

which is the *Poisson* distribution with mean ρW.

Hard-Sphere Potential

In the hard-sphere pair potential (see Figure 3.5), the particles do not interact for interparticle separation distances greater than the sphere diameter D but experience an infinite repulsive force for distances less than or equal to D, i.e.,

$$\varphi_2(r) = \begin{cases} +\infty, & 0 \le r \le D, \\ 0, & r > D, \end{cases} \tag{3.20}$$

and therefore the Boltzmann factor is just the step function

$$\exp[-\beta\varphi_2(r)] = \begin{cases} 0, & 0 \le r \le D, \\ 1, & r > D. \end{cases} \tag{3.21}$$

Note that pairwise additivity [cf. (3.3)] is exact for the hard-sphere potential.

Hard-sphere systems have received considerable attention, since they serve as a useful model for a number of physical systems, such as simple liquids (Reiss et al. 1959, Hansen and McDonald 1986, Rintoul and Torquato 1996a), glasses (Zallen 1983),

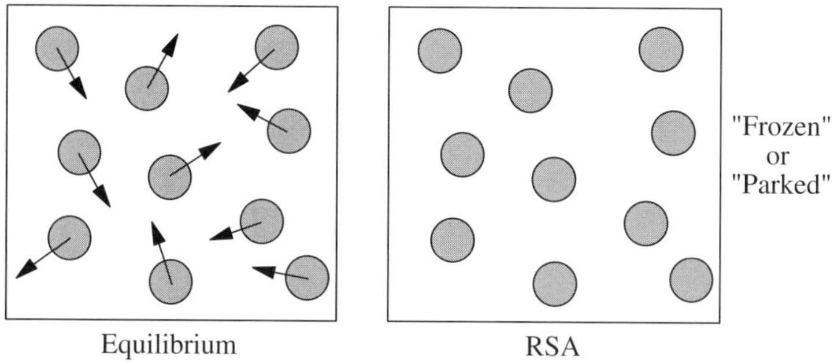

Figure 3.6 "Snapshot" of an equilibrium system of hard particles (left) and a realization of hard particles assembled according to an RSA process (right). In the former the particles are free to sample the configuration space subject to impenetrability of the other particles, but in the latter the particles are frozen at their initial positions.

colloidal dispersions (Russel, Saville and Schowalter 1989), fiber-reinforced composites (Torquato and Lado 1988a), particulate composites (Torquato and Lado 1986), packed beds (Georgiadis and Catton 1988), and granular media (Metcalfe, Shinbrot, McCarthy and Ottino 1995, Jenkins and Hanes 1998, Shinbrot, Alexander and Muzzio 1999, Mehta and Barker 2000). The hard-sphere model approximates well the structure of dense-particle systems with more complicated potentials (e.g., the Lennard–Jones model discussed below) because short-range repulsion between the particles is the primary factor in determining the spatial arrangement of the particles. The hard-sphere system is also appealing, since despite its simplicity, there are many interesting open questions regarding its structure (see Section 3.3).

It is important to note that for hard-sphere systems the impenetrability constraint does not uniquely specify the statistical ensemble. The hard-sphere system can be in thermal equilibrium or in one of the infinitely many nonequilibrium states, such as the random sequential addition (or adsorption) (RSA) process that is produced by randomly, irreversibly, and sequentially placing nonoverlapping objects into a volume (Widom 1966). While particles in equilibrium have thermal motion such that they sample the configuration space uniformly (with an average kinetic energy per particle equal to $3kT/2$), particles in an RSA process do not sample the configuration space uniformly, since their positions are forever "frozen" (i.e., do not diffuse) after they have been placed into the system (see Figure 3.6). For identical d-dimensional RSA spheres, the filling process terminates at the *saturation limit,* which is substantially lower than the maximum density for random hard spheres in equilibrium. Denoting the maximum sphere volume fraction by $\phi_2^{\max}$, it turns out that for identical hard spheres in an RSA process in the thermodynamic limit, $\phi_2^{\max} \approx 0.75, 0.55$, and 0.38 for $d = 1, 2$, and 3, respectively (Reńyi 1963, Feder 1980, Cooper 1988). In contrast, for identical *disordered* hard spheres in equilibrium, $\phi_2^{\max}$ is exactly unity for $d = 1$, and for $d = 2$ and 3, $\phi_2^{\max} \approx 0.83$ and 0.64, respectively (see Section 3.3). It is important to note

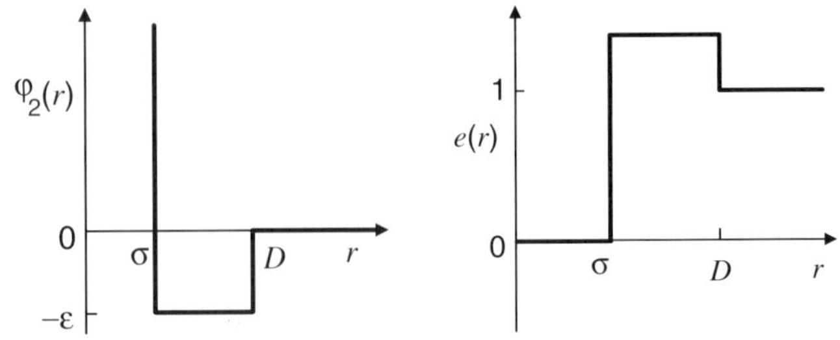

Figure 3.7 The square-well potential (left) and the corresponding Boltzmann factor (right).

that unlike what we see in the case of overlapping spheres, the n-particle probability density function ρ_n for general ensembles of hard spheres is nontrivial. The subsequent sections will treat both equilibrium and RSA hard-sphere systems in greater detail.

Square-Well Potential

Thus far we have considered only repulsive interactions. If attractive forces are included as well, the system will generally exhibit a first-order liquid–vapor phase transition. A useful model that incorporates both repulsive and attractive interactions is the *square-well* potential (see Figure 3.7), defined by

$$\varphi_2(r) = \begin{cases} +\infty, & 0 \le r \le \sigma, \\ -\epsilon, & \sigma < r \le D, \\ 0, & r > D, \end{cases} \tag{3.22}$$

where ϵ is the attractive well depth. The associated Boltzmann factor is

$$\exp[-\beta\varphi_2(r)] = \begin{cases} 0, & 0 \le r \le \sigma, \\ \exp[\beta\epsilon], & \sigma < r \le D, \\ 1, & r > D. \end{cases} \tag{3.23}$$

The *sticky-sphere* model proposed by Baxter (1968) is actually a special case of the square-well potential. The sticky-sphere potential is defined by (3.22) in which $\exp(\beta\epsilon) = 12\tau(1 - \sigma/D)$ and the limit $\sigma \to D$ is taken such that the parameter τ remains a fixed constant. Thus, the Boltzmann factor becomes

$$\exp[-\beta\varphi_2(r)] = \begin{cases} \dfrac{D}{12\tau}\delta(r - D), & 0 \le r \le D, \\ 1, & r > D, \end{cases} \tag{3.24}$$

developing a Dirac delta contribution at contact, i.e., the attractive well now has the nature of a surface adhesion or "stickiness." Here τ^{-1} is a *dimensionless stickiness parameter* that can vary between zero (for perfectly nonsticky hard-spheres) to ∞ (for perfectly sticky hard-spheres). A schematic configuration is shown in Figure 3.8. The sticky-sphere potential is a useful means of modeling agglomeration of particles such as

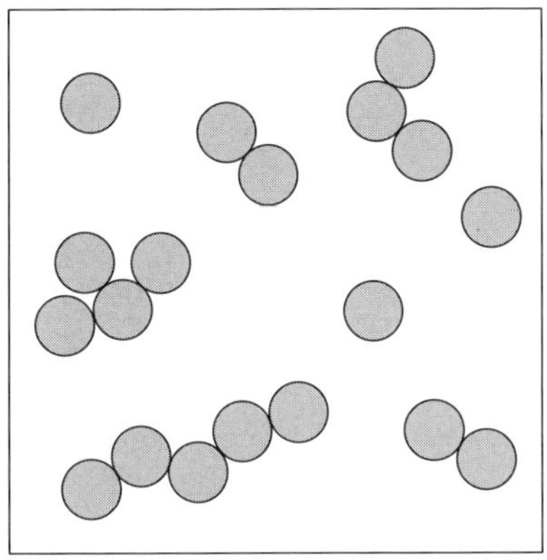

Figure 3.8 A realization of a sticky-particle system.

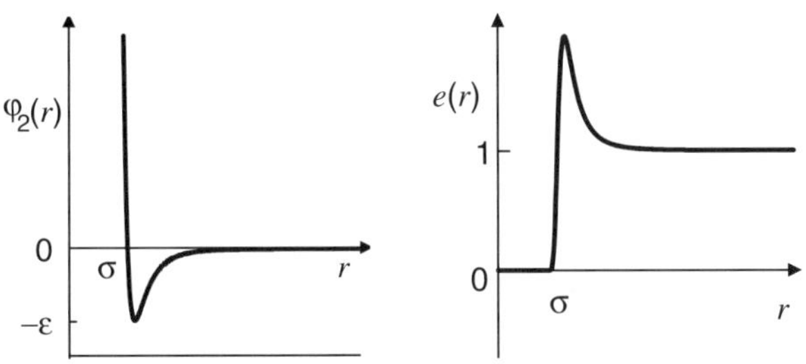

Figure 3.9 The Lennard–Jones potential (left) and the corresponding Boltzmann factor (right).

occurs in colloids. Baxter showed that this system undergoes a first-order liquid–vapor phase transition.

Lennard–Jones Potential

A well-known potential that involves both repulsive and attractive interactions and possesses a continuous first derivative is the Lennard-Jones potential (see Figure 3.9), given by

$$\varphi_2(r) = 4\epsilon\left[\left(\frac{\sigma}{r}\right)^{12} - \left(\frac{\sigma}{r}\right)^{6}\right], \tag{3.25}$$

where σ is the distance at which $\varphi_2(r) = 0$, and ϵ is the attractive well depth. The first term on the right side represents a strongly repulsive core and the second term represents a short-ranged attraction due to "dispersion" forces. The Lennard-Jones model is a prototypical potential for classical simple liquids (Hansen and McDonald 1986), such as argon, and for colloidal dispersions (Russel et al. 1989).

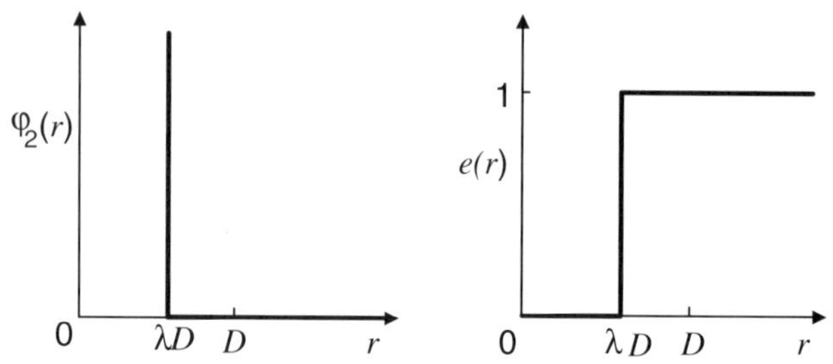

Figure 3.10 The penetrable-concentric-shell potential (cherry-pit model) (left) and the corresponding Boltzmann factor (right).

Concentric-Shell Model

A versatile potential introduced by Torquato (1984) to study random media is the *concentric-shell* model, defined by

$$\varphi_2(r) = \begin{cases} +\infty, & 0 \le r \le \lambda D, \\ \lambda\varphi(r), & r > \lambda D, \end{cases} \tag{3.26}$$

where λ is an *impenetrability* parameter that lies in the interval $[0, \lambda_o]$ and λ_o is a bounded real number such that $\lambda_o \ge 1$. Here $\varphi(r)$ is any sufficiently short-ranged pair potential that incorporates attractive as well as finite repulsive interactions. Note that $\varphi_2(r) \to 0$ when $\lambda \to 0$, and thus one recovers the spatially uncorrelated case of overlapping spheres in this limit.

When the impenetrability parameter satisfies the condition $0 \le \lambda \le 1$ and φ is nonzero, each particle of diameter D may be regarded as possessing an inner impenetrable core of diameter λD, encompassed by a *penetrable* or "soft" concentric shell. If $\lambda > 1$, then the impenetrable core extends beyond the physical extent of the particle.

A popular choice for φ_2 is to "turn off" interactions beyond a separation distance of λD. Hence, the spheres of diameter D are distributed throughout space subject only to the condition of a mutually impenetrable core of diameter λD. In this special instance, the pair potential is given by

$$\varphi_2(r) = \begin{cases} +\infty, & 0 \le r \le \lambda D, \\ 0, & r > \lambda D, \end{cases} \tag{3.27}$$

and hence the Boltzmann factor (see Figure 3.10) is the step function

$$\exp[-\beta\varphi_2(r)] = \begin{cases} 0, & 0 \le r \le \lambda D, \\ 1, & r > \lambda D. \end{cases} \tag{3.28}$$

When $0 \le \lambda \le 1$, each sphere of diameter D may be thought of as being composed of an impenetrable core of diameter λD encompassed by a *perfectly* penetrable concentric shell of thickness $(1-\lambda)D/2$ (see Figure 3.11). By varying the impenetrability parameter λ between 0 and 1, one can continuously pass between fully penetrable (overlapping)

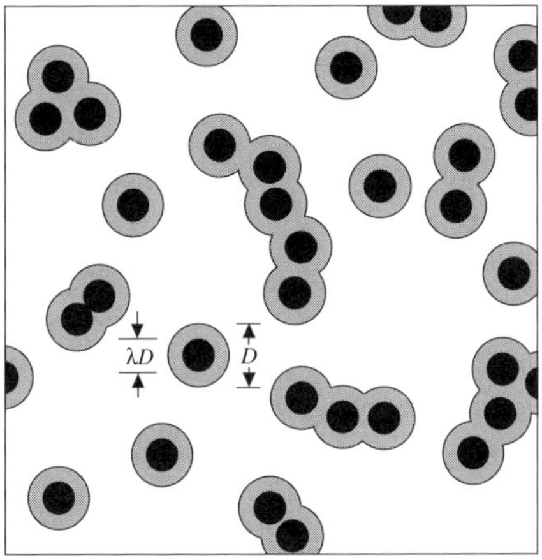

Figure 3.11 Particles in the penetrable-concentric-shell (cherry-pit) model.

spheres and totally impenetrable spheres, respectively. This interpenetrable-sphere model is referred to as the *penetrable-concentric-shell* model or, more colloquially, the *cherry-pit* model.

Interpenetrable-sphere models enable one to study the effect of the degree of connectivity of the particle phase on the effective properties. The degree of connectivity of the individual phases may greatly influence the effective properties of two-phase media particularly when the phase property values differ significantly. Another interpenetrable-sphere system is the *permeable-sphere* model (Salacuse and Stell 1982), which is not defined through a potential. Here spheres of diameter D are assumed to be noninteracting when nonintersecting, with the probability of intersection given by a radial distribution function $g_2(r)$ that is $1-\epsilon$, independent of r, when $r < D$, where $0 \le \epsilon \le 1$.

For statistically homogeneous and isotropic systems in the cherry-pit model, the n-particle distribution function for arbitrary values of the impenetrability parameter λ and density ρ can be related to the corresponding function for the extreme limit of totally impenetrable spheres ($\lambda = 1$) as follows (Torquato 1984):

$$g_n(\mathbf{r}_{12}/D, \ldots, \mathbf{r}_{1n}/D;\, \rho, \lambda) = g_n(\mathbf{r}_{12}/\lambda D, \ldots, \mathbf{r}_{1n}/\lambda D;\, \rho\lambda^d, 1). \tag{3.29}$$

For the special case $n = 2$, this relation states that the radial distribution function $g_2(r_{12}/D;\, \rho, \lambda)$ at density ρ is equal to the radial distribution function $g_2(r_{12}/\lambda D;\, \rho\lambda^d, 1)$ of a totally impenetrable system at the larger reduced distance $\mathbf{r}_{12}/\lambda D$ and smaller density $\rho\lambda^d$. Clearly, as $\lambda \to 0$, the pairs of particles become spatially uncorrelated ($g_2 \to 1$); i.e., we recover g_2 for the fully penetrable limit.

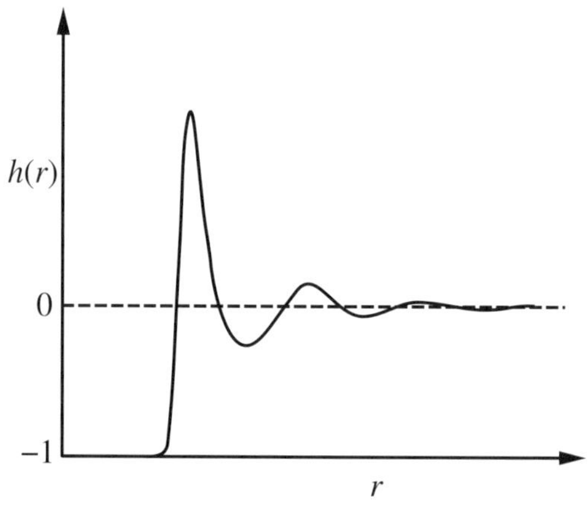

Figure 3.12 The total correlation function $h(r)$ for a typical disordered system of interacting particles.

3.2 Ornstein–Zernike Formalism

In their study of density fluctuations in fluid systems near the critical point, Ornstein and Zernike (1914) introduced the direct correlation function $c(\mathbf{r})$ via an integral equation that linked it to the pair correlation function $g_2(\mathbf{r})$. This integral equation serves as a basis for obtaining estimates of $g_2(\mathbf{r})$.

Consider statistically homogeneous systems of spherical particles. It is convenient to define the *total correlation function* $h(\mathbf{r})$ as

$$h(\mathbf{r}) = g_2(\mathbf{r}) - 1. \tag{3.30}$$

For disordered systems, $h(\mathbf{r})$ tends to zero when $r \to \infty$ (see Figure 3.12). In analogy with the results of Section 2.2.5, the *structure factor* $S(\mathbf{k})$, alluded to in Section 3.1.1, must be nonnegative, i.e., $S(\mathbf{k}) \equiv 1 + \rho\tilde{h}(\mathbf{k}) \geq 0$ for all $\mathbf{k}$, where $\tilde{h}(\mathbf{k})$ is the Fourier transform of $h(\mathbf{r})$. Ornstein and Zernike (1914) proposed a decomposition of h into a "direct" part and "indirect" part:

$$h(\mathbf{r}_{12}) = c(\mathbf{r}_{12}) + \rho \int h(\mathbf{r}_{23})c(\mathbf{r}_{13})d\mathbf{r}_3, \tag{3.31}$$

where $\mathbf{r}_{ij} = \mathbf{r}_j - \mathbf{r}_i$. The Ornstein–Zernike integral equation (3.31) may be considered to be a definition of the *direct correlation function* $c(\mathbf{r})$. We note that this definition is valid even for *nonequilibrium* systems. Note that in contrast to P_N and ρ_n defined by (3.4) and (3.7), the quantities g_2, h, and c are dimensionless.

In words, the Ornstein–Zernike equation (3.31) states that the total correlation between particles at $\mathbf{r}_1$ and $\mathbf{r}_2$ can be separated into two contributions: (i) a *direct* effect of a particle at $\mathbf{r}_1$ on one at $\mathbf{r}_2$, which is generally short-ranged (having roughly the same range as the pair potential φ_2) and is characterized by $c(\mathbf{r})$, and (ii) an *indirect* effect, in which a particle at $\boldsymbol{r}_1$ influences some other particle at $\boldsymbol{r}_3$, which in turn affects a particle at $\boldsymbol{r}_2$, directly or indirectly through other particles. This indirect effect is weighted by the density and averaged over all possible $\boldsymbol{r}_3$. It is clear that the indirect effect, and hence h, will be generally of longer range than c. By successive substitutions, we can reexpress (3.31) as

$$h(\boldsymbol{r}_{12}) = c(\boldsymbol{r}_{12}) + \rho \int c(\boldsymbol{r}_{13})c(\boldsymbol{r}_{23})d\boldsymbol{r}_3 + \rho^2 \int\int c(\boldsymbol{r}_{13})c(\boldsymbol{r}_{24})c(\boldsymbol{r}_{34})d\boldsymbol{r}_3 d\boldsymbol{r}_4 + \cdots .$$

The term $c(\boldsymbol{r}_{12})$ is the direct contribution; the second term is the indirect contribution comprising the direct effect between a particle at $\boldsymbol{r}_1$ and one at $\boldsymbol{r}_3$ and the direct effect between the particle at $\boldsymbol{r}_3$ and one at $\boldsymbol{r}_2$, averaged over the position $\boldsymbol{r}_3$, and so on.

The integral of (3.31) is a convolution of the functions h and c, and hence (3.31) can be written more tersely as

$$h(\boldsymbol{r}_{12}) = c(\boldsymbol{r}_{12}) + \rho h(\boldsymbol{r}_{23}) \otimes c(\boldsymbol{r}_{13}), \tag{3.32}$$

where for two arbitrary functions $f_1(\boldsymbol{r})$ and $f_2(\boldsymbol{r})$, the symbol $\otimes$ denotes a convolution

$$f_1(\boldsymbol{r}_{23}) \otimes f_2(\boldsymbol{r}_{13}) \equiv \int f_1(\boldsymbol{r}_{13} - \boldsymbol{r}_{12}) f_2(\boldsymbol{r}_{13}) d\boldsymbol{r}_3. \tag{3.33}$$

It will be convenient to use diagrammatic notation to represent so-called *cluster integrals* (Stell 1964, Hansen and McDonald 1986). For example, the convolution integral of (3.31) or (3.32) can be written as

$$\rho \int h(\boldsymbol{r}_{23})c(\boldsymbol{r}_{13})d\boldsymbol{r}_3 =$$

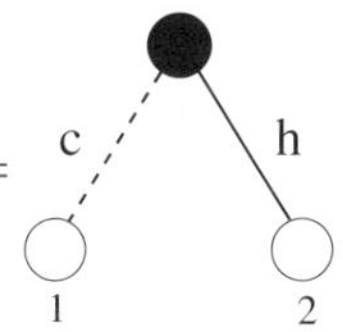

. (3.34)

A diagram consists of a certain number of *circles* and *bonds* connecting the circles. There are two types of circles: white circles that are labeled and associated with the positions not integrated over [e.g., 1 and 2 in (3.34)], and black circles associated with the integrated positions that are unlabeled, since they are dummy variables [e.g., the black circle in (3.34)]. A function $F(i,j)$ of the positions i and j is associated with each bond connecting the circles i and j. For example, in (3.34), the dashed and solid lines denote a c-bond and h-bond, respectively. A function $G(i)$ is also associated with each circle i. In the case of (3.34), constant factors of ρ and unity are associated with the black and white circles, respectively. Finally, the value of a diagram is a function of the coordinates attached to its white circles and includes a combinatorial factor that depends on its topological structure.

It is well known that the Fourier transform of a convolution integral of two functions is simply the product of the Fourier transforms of the individual functions. Thus, taking the Fourier transform of the Ornstein–Zernike equation yields

$$\tilde{h}(\boldsymbol{k}) = \tilde{c}(\boldsymbol{k}) + \rho\tilde{c}(\boldsymbol{k})\tilde{h}(\boldsymbol{k}) = \frac{\tilde{c}(\boldsymbol{k})}{1 - \rho\tilde{c}(\boldsymbol{k})}, \tag{3.35}$$

where $\tilde{F}(\boldsymbol{k})$ is the Fourier transform of a function $F(\boldsymbol{r})$ (see Section 2.2.5). We see that the compressibility equation (3.15) can be rewritten as

$$\rho kT\kappa_T = 1 + \rho\tilde{h}(0) = [1 - \rho\tilde{c}(0)]^{-1}. \tag{3.36}$$

The compressibility diverges to infinity as the *critical point* of a fluid is approached. According to (3.36), this implies that $\tilde{h}(0)$ diverges and thus $h(\boldsymbol{r})$ becomes long-ranged (Stanley 1971, Binney, Dowrick, Fisher and Newman 1992). For an isotropic system at the critical point, this means that for large $r = |\boldsymbol{r}|$,

$$h(r) \sim \frac{1}{r^\alpha}, \qquad 0 \le \alpha \le d.$$

This behavior is a reflection of the slow decay of the correlations in density fluctuations between different points in the fluid. We also see from (3.36) that $\tilde{c}(0)$ is finite and thus $c(r)$ remains short-ranged at this critical point. Interestingly, there are percolation relations involving the *connectedness function* that are analogous to relation (3.36) (see Chapters 9 and 10).

Mayer and Montroll (1941) obtained the general density expansion of the pair correlation function for systems with pairwise additive potentials. Diagrammatically, the cluster expansion of g_2 can be written as (Stell 1964, Hansen and McDonald 1986)

$$\begin{aligned} g_2(\boldsymbol{r}_{12}) &= h(\boldsymbol{r}_{12}) + 1 \\ &= \text{the sum of all distinct (at least doubly connected) diagrams} \\ &\quad\ \text{consisting of two white circles (labeled 1 and 2) that are} \\ &\quad\ \text{linked by an } e\text{-bond, black } \rho\text{-circles, and } f\text{-bonds,} \end{aligned} \tag{3.37}$$

where e and f refer, respectively, to the Boltzmann factor (3.16) and Mayer function (3.17). This can be recast as the density expansion

$$g_2(\boldsymbol{r}_{12}) = e(r_{12})\left[1 + \sum_{n=1}^{\infty} \rho^n g_2^{(n)}(\boldsymbol{r}_{12})\right]. \tag{3.38}$$

For example, the first-order term is given by

$$g_2^{(1)}(\boldsymbol{r}_{12}) =$$

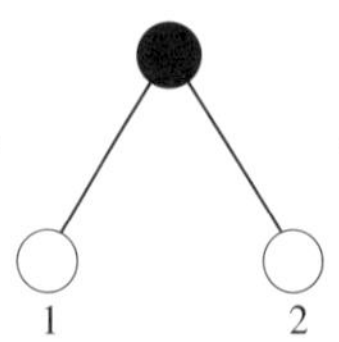

$$= \int d\boldsymbol{r}_3 f(r_{13}) f(r_{23}). \tag{3.39}$$

Here the black circle carries the factor of unity (not ρ as before), and the solid line denotes an f-bond. Rushbrooke and Scoins (1953) found the corresponding density expansion of the direct correlation function $c(\mathbf{r})$. Through first-order density, one has

$c(\mathbf{r}_{12}) =$  $+$ 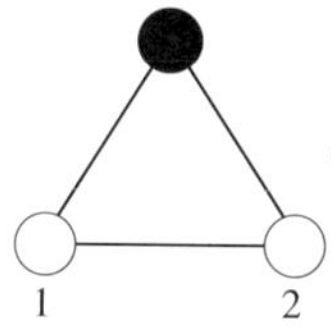 , (3.40)

where the black circle carries a factor of ρ. We see from (3.38) that in the zero-density limit $\rho \to 0$, $g_2(\mathbf{r})$ tends to the Boltzmann factor $e(r) = \exp[-\beta\varphi_2(r)]$, and both $c(\mathbf{r})$ and $h(\mathbf{r})$ tend to the Mayer function $f(r)$.

We note in passing that integral equations for the three-particle and higher-order equilibrium probability densities ($n \geq 3$) have been obtained (Hansen and McDonald 1986, Hill 1987), but they become increasingly difficult to solve as n increases for general interparticle potentials for $d \geq 2$. Often, approximations for the g_n for $n \geq 3$ are given in terms of the two-particle densities. For example, the well-known *superposition approximation* for g_3 is given by

$$g_3(\mathbf{r}_{12}, \mathbf{r}_{13}, \mathbf{r}_{23}) \approx g_2(\mathbf{r}_{12})g_2(\mathbf{r}_{13})g_2(\mathbf{r}_{23}). \tag{3.41}$$

Relation (3.41) is known to be accurate for low densities and for equilateral triangular configurations, especially at high densities. In the low-density limit ($\rho \to 0$), the g_n are exactly given by products of Boltzmann factors, i.e.,

$$g_n(\mathbf{r}_{12}, \ldots, \mathbf{r}_{1n}) = \prod_{i<j}^{n} \exp[-\beta\varphi_2(r_{ij})].$$

In the special case of one-dimensional systems in which the particles interact with a nearest-neighbor potential, the g_n are given exactly in terms of products of g_2's (Salsburg, Zwanzig and Kirkwood 1953).

3.3 Equilibrium Hard-Sphere Systems

Given the importance of the hard-sphere potential in capturing the salient structural features of systems of interacting particles with more complex potentials, we now turn our attention to the most well-studied hard-sphere model, namely, hard spheres in equilibrium. Statistically *homogeneous* ensembles of *equilibrium* hard spheres are also necessarily statistically *isotropic*. Despite the apparent simplicity of such systems, they exhibit surprisingly rich behavior, and there remain many open questions regarding their structure over the full density range. For example, there is overwhelming computer-simulation and experimental evidence that isotropic equilibrium hard-sphere systems for $d \geq 2$ undergo a disorder–order phase transition at sufficiently high densities (Reiss and Hammerich 1986). This phase transition is most easily described by examining the pressure p of identical d-dimensional hard spheres of diameter D at number density ρ. This is obtained from the general pressure equation (3.14) and the fact that $d\phi_2/dr = -kT\exp[\beta\phi_2(r)]\delta(r-D)$ for the hard-sphere potential (3.20):

$$\frac{p}{\rho kT} = 1 + 2^{d-1}\eta g_2(D^+). \tag{3.42}$$

In this formula, $g_2(D^+)$ is the contact value (from the right side) of the radial distribution function, and η is a dimensionless *reduced density* defined by

$$\eta = \rho v_1(D/2), \tag{3.43}$$

where $v_1(a)$ is the d-dimensional volume of a single sphere of radius a given by

$$v_1(a) = \frac{\pi^{d/2}}{\Gamma(1+d/2)}a^d \tag{3.44}$$

and $\Gamma(x)$ is the gamma function. For example, for the first three space dimensions, we explicitly have

$$v_1(a) = \begin{cases} 2a, & d=1, \\ \pi a^2, & d=2, \\ 4\pi a^3/3, & d=3. \end{cases} \tag{3.45}$$

For hard spheres, $\eta = \phi_2$, where ϕ_2 is the sphere volume fraction or *packing fraction*. Via the Ornstein–Zernike equation, we can rewrite (3.42) in terms of the direct correlation function $c(r)$ as

$$\frac{p}{\rho kT} = 1 + 2^{d-1}\eta[c(D^+) - c(D^-)]. \tag{3.46}$$

Observe that the equilibrium hard-sphere system is *athermal* in the sense that temperature enters its phase diagram in a trivial manner, i.e., the pressure p divided by the temperature T is independent of T [cf. (3.42) or (3.46)]. This athermal nature is a consequence of the fact that the N-body potential (3.3) of the hard-sphere system is exactly either zero or infinite.

Although a rigorous proof for the existence of a first-order liquid-to-solid phase transition in three dimensions is not yet available, computer simulations since the early work of Hoover and Ree (1968) strongly support such a transformation. Experiments on colloidal hard-sphere systems also confirm such a phase transition (Pusey and van Megan 1986, Rutgers, Dunsmuir, Xue, Russel and Chaikin 1996). Figure 3.13 schematically shows the phase diagram for three-dimensional hard spheres in the density-pressure plane. There is a thermodynamically stable *liquid* or *disordered* branch that begins at $\eta = 0$ and continues up to the *freezing-point* volume fraction η_f, which occurs at $\eta_f \approx 0.494$ (Hoover and Ree 1968). At η_f, the phase diagram splits into two portions. One part is the *metastable* extension of the liquid branch (Speedy 1994), which remains amorphous (disordered) , and ends at the "critical" or "singular" point $\eta_c \approx 0.644$ (Rintoul and Torquato 1996a) where the pressure diverges to infinity; see also Section 12.2.4. There is now strong evidence that there is no *glass transition* of *thermodynamic origin* along this metastable branch (Rintoul and Torquato 1996a, Torquato 2000b), in contrast to previous findings. Roughly speaking, the glass transition can be

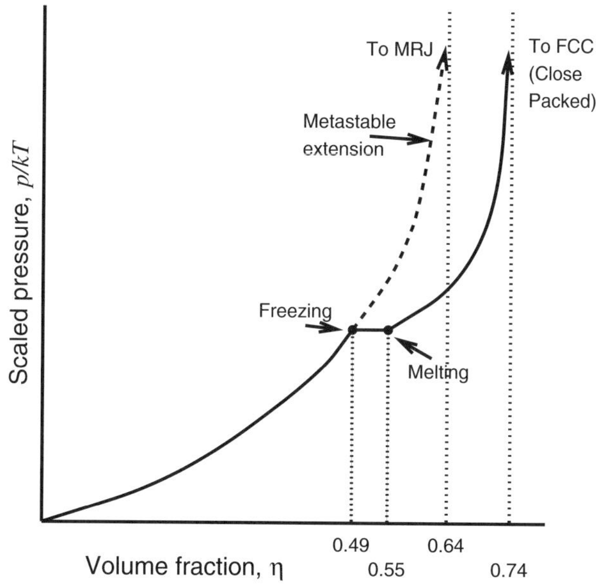

Figure 3.13 A schematic of the phase diagram for the three-dimensional hard-sphere system in equilibrium.

viewed as the point at which a liquid becomes structurally arrested (on macroscopic, experimental time scales) upon rapid cooling (Zallen 1983, Debenedetti 1996). The state of matter below this transition is termed a "glass," which is a disordered material that lacks the periodicity of a crystal but behaves mechanically like a solid.

The other portion of the phase diagram (solid curve) represents the thermodynamically stable part of the phase diagram for $\eta > \eta_f$. Along the horizontal *tieline*, both liquid and solid can coexist until the *melting point* $\eta_m \approx 0.545$ is reached. This is an *entropically* driven first-order phase transition, i.e., at high enough densities, there are many more ordered states than disordered states available to the system, and hence the system begins to freeze (i.e., crystallize). Thus, the spheres increase in entropy by ordering. (This is akin to the entropy gained by passengers on a crowded subway car by arranging themselves approximately equidistantly from one another to maximize the space in their vicinity!) The counterintuitive notion that entropy creates order (rather than disorder) arises because of the aforementioned athermal nature of hard spheres. The portion of the curve that continues above the melting point is the *solid* or *ordered* branch, ending at the close-packed crystal at a volume fraction of $\pi/\sqrt{18} \approx 0.740$. This is the maximum volume fraction for a packing of spheres in three dimensions for which the pressure also diverges to infinity. It is still an open question whether this *equilibrium* state is a face-centered cubic (FCC) lattice or other packings consistent with $\pi/\sqrt{18}$ (e.g., hexagonal close packing), although numerical evidence points to an FCC lattice (Mau and Huse 1999).

The complete phase diagram for two-dimensional hard-disk systems has still not been fully established numerically. There is a liquid or disordered branch that ends at $\eta_f \approx 0.69$ (Hoover and Ree 1968) and an ordered branch that ends at the hexagonal closed-packed volume fraction of $\pi/\sqrt{12} \approx 0.907$. However, there are conflicting opinions as to whether there is a first-order liquid-to-solid phase transition or a Kosterlitz–Thouless–Halperin–Nelson–Young (KTHNY) (Kosterlitz and Thouless 1973, Halperin and Nelson 1978, Young 1979) type transition to the ordered phase. The latter scenario predicts that a two-dimensional crystal undergoes a continuous melting transition via the unbinding of *dislocations* into a *hexatic* phase (short-range translational order and quasi-long-range orientational order). The hexatic phase is predicted to undergo a second continuous transition to the disordered liquid via the unbinding of *disclinations*. The KTHNY scenario does not rule out the possibility of a first-order melting by another mechanism, however. Some numerical studies rule out a KTHNY transition and support a first-order transition (Weber, Marx and Binder 1995), while others are consistent with a KTHNY transition but do not rule out a first-order transition entirely (Jaster 1998).

In contrast to hard spheres, it is difficult to create via numerical simulations long-lasting metastable hard-disk systems above the freezing point, i.e., there is a propensity for the system to crystallize when $\eta > \eta_f$. This difference is due to the fact that for $d = 2$, the densest local packing is consistent with the densest global packing. The maximum number of d-dimensional spheres that can be locally packed such that each sphere contacts the others is $d+1$. The d-dimensional polyhedron that results by taking the sphere centers as vertices is a *simplex*: line segment, equilateral triangle, and regular tetrahedron for $d = 1, 2$, and 3, respectively. Whereas for $d = 1$ and $d = 2$, identical simplices can tile space, it is not possible to tile three-dimensional space with identical nonoverlapping regular tetrahedra (see Section 8.1.1): the system is geometrically "frustrated" in that local optimal packing rules are inconsistent with global packing constraints. In three dimensions, one can create *suboptimal* global packings with finite-sized clusters of spheres (e.g., icosahedral clusters) with local densities higher than the global maximum at the expense of having large voids elsewhere.

In summary, if the metastable extension of the liquid branch exists in two dimensions, the barrier to crystallization is small. The fact that hard-disk systems can be compressed to disordered jammed states, albeit with ordered domains (Lubachevsky and Stillinger 1990), suggests that the two-dimensional metastable extension exists. Under this scenario, it is reasonable to assume that this metastable branch ends at the density $\eta_c \approx 0.83$ (Quickenden and Tan 1974). Table 3.1 summarizes important packing fractions for equilibrium hard spheres for $d = 1, 2$, and 3.

It is noteworthy that the "critical" packing fraction η_c, corresponding to the maximum value along the metastable extension of the liquid branch, is approximately equal to what is commonly regarded to be the *random close packing* (RCP) density of identical hard spheres. However, as discussed in Section 3.5, it has been recently shown that the traditional idea of the RCP state is ill defined and should be replaced by a new notion called the *maximally random jammed* (MRJ) state. The density η_c is actually a

Table 3.1 Some important densities for d-dimensional equilibrium hard-sphere systems.

STATE	$d=1$	$d=2$	$d=3$
Freezing density, η_f	-	0.69	0.494
Maximum metastable density, η_c	-	0.83	0.644
Maximum density	1.0	$\pi/\sqrt{12} \approx 0.907$	$\pi/\sqrt{18} \approx 0.740$

singular point in the sense that it is the only point along the metastable extension with a nonzero average coordination number Z [i.e., $g_2(r)$ develops a delta-function contribution at $r = D$ as in (10.42)]. Hence, it is more accurate to refer to η_c as the MRJ state for equilibrium hard spheres.

We now describe different analytical methods to obtain the radial distribution function $g_2(r)$ for identical statistically isotropic d-dimensional hard spheres of diameter D at reduced density η in thermal equilibrium.

3.3.1 Low-Density Expansions

Before obtaining the low-density expansion of the radial distribution function $g_2(r)$, we first state an important geometrical result that will be of use to us here and in the subsequent discussions. Specifically, we note that the overlap or intersection volume v_n^{int} of n d-dimensional spheres of radii $a_1, a_2, \ldots, a_n$ centered at positions $\boldsymbol{x}_1, \boldsymbol{x}_2, \ldots, \boldsymbol{x}_n$ is given by the convolution integral

$$v_n^{\text{int}}(\boldsymbol{x}_1, \boldsymbol{x}_2, \ldots, \boldsymbol{x}_n; a_1, \ldots, a_n) = \int d\boldsymbol{y} \prod_{i=1}^{n} \Theta(a_i - |\boldsymbol{y} - \boldsymbol{x}_i|), \tag{3.47}$$

where $\Theta(x)$ is the Heaviside step function

$$\Theta(x) = \begin{cases} 0, & x < 0, \\ 1, & x \geq 0, \end{cases} \tag{3.48}$$

and $v_1^{\text{int}}(a)$ is just the volume $v_1(a)$ of a d-dimensional sphere of radius a. We denote by $v_n^{\text{int}}(\boldsymbol{x}_1, \boldsymbol{x}_2, \ldots, \boldsymbol{x}_n; a)$ the intersection volume when the spheres have the same size ($a_i = a$ for all i). For example, for the first three space dimensions, the intersection volumes of two identical spheres (divided by v_1) whose centers are separated by the distance $r = |\boldsymbol{x}_2 - \boldsymbol{x}_1|$ are given by

$$\frac{v_2^{\text{int}}(r; a)}{v_1(a)} = (1 - \frac{r}{2a})\Theta(2a - r), \qquad d = 1, \tag{3.49}$$

$$\frac{v_2^{\text{int}}(r; a)}{v_1(a)} = \frac{2}{\pi}\left[\cos^{-1}\left(\frac{r}{2a}\right) - \frac{r}{2a}(1 - \frac{r^2}{4a^2})^{1/2}\right]\Theta(2a - r), \qquad d = 2, \tag{3.50}$$

$$\frac{v_2^{\text{int}}(r; a)}{v_1(a)} = \left[1 - \frac{3}{4}\frac{r}{a} + \frac{1}{16}\left(\frac{r}{a}\right)^3\right]\Theta(2a - r), \qquad d = 3. \tag{3.51}$$

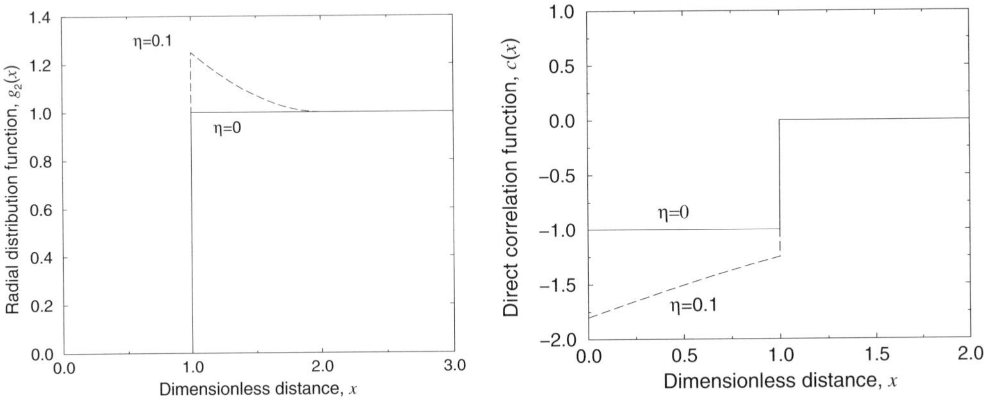

Figure 3.14 Radial distribution function and direct correlation function for a system of hard spheres versus dimensionless distance $x = r/D$ at dilute conditions as predicted by (3.52) and (3.53), respectively, with $d = 3$.

These convolution integrals can be evaluated by Fourier transform techniques; however, because of the discontinuous nature of the integrands, they can be easier to evaluate in real space. For example, for $d = 3$, it is convenient to transform to a *bipolar* coordinate system (Hansen and McDonald 1986).

According to relation (3.38), the radial distribution function $g_2(r)$ through first order in density, for any dimension d, is given by

$$
\begin{aligned}
g_2(r) &= \Theta(r-D)\left[1 + \rho \int d\mathbf{r}_3 \Theta(D - r_{13})\Theta(D - r_{23})\right] \\
&= \Theta(r-D)\left[1 + \eta \frac{v_2^{\text{int}}(r; D)}{v_1(D/2)}\right],
\end{aligned}
\tag{3.52}
$$

where we have used relations (3.17) [i.e., $f(r) = -\ \Theta(D - r)$] and (3.47). Moreover, $r \equiv r_{12}$ and η is given by (3.43). We can see that in the infinitely dilute limit $\eta \to 0$, $g_2(r)$ becomes the simple step function $\Theta(r - D)$ as shown in Figure 3.14 (solid lines). The next contribution in density is simply related to the intersection volume of two spheres of radius D, which results in $g_2(r)$ developing a peak at $r = D$ (dashed curve in Figure 3.14). This means that more pairs of particles are nearly touching one another than are not, a characteristic which persists at higher densities and dimensionalities (see, for example, Figure 3.15). According to relation (3.40), the corresponding density expansion for the direct correlation function is given by

$$
c(r) = -\ \Theta(D - r)\left[1 + \eta \frac{v_2^{\text{int}}(r; D)}{v_1(D/2)}\right]. \tag{3.53}
$$

Figure 3.14 also includes the corresponding plots of $c(r)$.

3.3.2 Arbitrary Fluid Densities

In the case of one-dimensional hard spheres (i.e., hard rods) of diameter D, the radial distribution function was determined analytically by Zernike and Prins (1927). They obtained the exact result

$$\eta g_2(x) = \sum_{k=1}^{\infty} \Theta(x-k) \frac{\eta^k (x-k)^{k-1}}{(1-\eta)^k (k-1)!} \exp\left[-\frac{\eta(x-k)}{1-\eta}\right], \tag{3.54}$$

where $x = r/D$ is a dimensionless distance. The corresponding exact direct correlation function $c(x)$ was found by Percus (1964) and is given by

$$c(x) = -\Theta(1-x)\frac{(1-\eta x)}{(1-\eta)^2}. \tag{3.55}$$

We note that the higher-order functions g_n can be expressed as products of g_2's (Salsburg et al. 1953).

For dimensions $d \geq 2$, exact analytical solutions of the radial distribution function $g_2(r)$ do not exist for arbitrary density. Therefore, the Ornstein–Zernike equation has been solved in various approximations. The most popular closure is the so-called *Percus–Yevick* approximation (Percus and Yevick 1958) in which one applies the boundary conditions

$$h(r) = -1, \qquad 0 \leq r \leq D, \tag{3.56}$$

$$c(r) = 0, \qquad r > D, \tag{3.57}$$

to solve the Ornstein–Zernike equation. The condition (3.56) enforces the exact hard-core constraint, whereas the condition (3.57) assumes that $c(r)$ is exactly equal to the pair potential $\varphi_2(r)$, which vanishes for $r > D$. For odd values of d, the Ornstein–Zernike relation is analytically soluble in the Percus–Yevick approximation. For example, for $d = 1$, it yields the exact result (3.55), and for $d = 3$, the direct correlation function is given by

$$c(x) = \begin{cases} -a_1 - 6\eta a_2 x - \frac{1}{2}\eta a_1 x^3, & x \leq 1, \\ 0, & x > 1, \end{cases} \tag{3.58}$$

where

$$a_1 = \frac{(1+2\eta)^2}{(1-\eta)^4}, \qquad a_2 = -\frac{(1+\eta/2)^2}{(1-\eta)^4}, \tag{3.59}$$

$\eta = \rho\pi D^3/6$, and $x = r/D$. This was first solved independently by Thiele (1963) and by Wertheim (1963) using Laplace transform methods. Subsequently, Baxter (1968) devised a powerful factorization technique to solve the Ornstein–Zernike equation. The real-space total correlation function h (or, equivalently, g_2) is obtained by Fourier transforming c, substituting $\tilde{c}$ into (3.35), and then inverting the latter to yield

$$h(r) = \frac{1}{2\pi^2 r}\int_0^{\infty} dk \frac{\tilde{c}(k)}{1-\rho\tilde{c}(k)} k \sin kr, \tag{3.60}$$

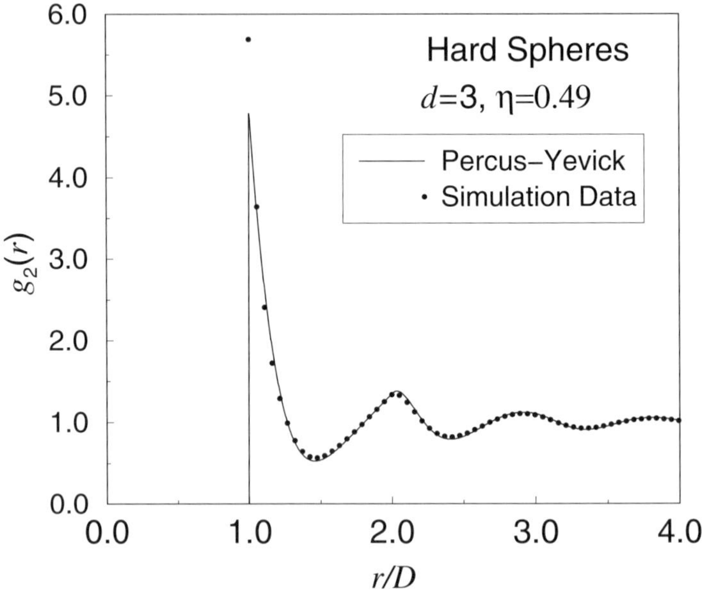

Figure 3.15 Comparison between "exact" molecular dynamics and Percus–Yevick results for the radial distribution function of a system of hard spheres in equilibrium at $\eta = 0.49$.

where

$$\tilde{c}(k) = -\frac{4\pi}{q^3}\left\{ a_1[\sin q - q\cos q] + \frac{6\eta a_2}{q}[2q\sin q + (2-q^2)\cos q - 2] \right.$$
$$\left. + \frac{\eta a_1}{2q^3}[4q(q^2-6)\sin q - (24 - 12q^2 + q^4)\cos q + 24] \right\} \tag{3.61}$$

and $q = kD$ is a dimensionless wave number.

The Percus–Yevick solution approximately sums all of the diagrams in the exact expansion (3.37) (Stell 1964, Hansen and McDonald 1986). Indeed, it is exact at low densities and provides a very good approximation even at moderately high liquid densities. It is least accurate at the highest liquid density, i.e., at the freezing point $\eta_f = 0.494$. Figure 3.15 compares the Percus–Yevick approximation for the radial distribution function near freezing ($\eta = 0.49$) to the corresponding "exact" result as obtained from molecular dynamics simulations that we have performed. The Percus–Yevick result is obtained by numerically evaluating (3.60) with (3.61) using efficient Fourier-transform techniques (Lado 1971). Although the Percus–Yevick solution provides a reasonable approximation at this large density, it has two defects. First, it predicts $g_2(D^+) = (1+\eta/2)/(1-\eta)^2$, which underestimates the contact value and therefore the pressure of the hard-sphere system. [Highly accurate expressions for $g_2(D^+)$ are given in Section 5.2.5.] The second defect is that $g_2(r)$ oscillates slightly out of phase with the simulation results, and the amplitudes of these oscillations decrease too slowly with increasing r, so that the structure factor [which is proportional to $\tilde{h}(k)$] is too high. These deficiencies at high densities led Verlet and Weis (1972) to develop a semiempirical modification of the Percus–Yevick g_2 that is in excellent agreement with numerical data.

For even values of d, the Ornstein–Zernike equation in the Percus–Yevick approximation must be solved numerically. Such a numerical solution was given by Lado (1968) for hard disks ($d = 2$). For $d = 2$, a semiempirical proposal by Leutheusser (1986) yields excellent agreement with the numerically determined direct correlation function. Baus and Colot (1987) devised a simpler proposal for the direct correlation function that is applicable in any dimension. The Baus–Colot approximation for the direct correlation function yields the exact relation (3.55) for $d = 1$ and the Percus–Yevick approximation (3.58) for $d = 3$, depending on the level of approximation used for the dimensionless pressure $z(\eta) \equiv p/(\rho kT)$. For the special case $d = 2$, they obtain

$$c(x) = -\frac{\partial}{\partial\eta}\left[\eta z(\eta)\right]\Theta(1-x)\left[1 - b^2\eta + b^2 v_2^{\text{int}}(x/b; D/2)\eta\right], \tag{3.62}$$

where $b = b(\eta)$ is obtained by solving the equation

$$\frac{2}{\pi}\left[b^2(b^2-4)\sin^{-1}\left(\frac{1}{b}\right) - (b^2+2)(b^2-1)^{1/2}\right] = \frac{1}{\eta^2}\left[1 - 4\eta - \left\{\frac{\partial}{\partial\eta}[\eta z(\eta)]\right\}^{-1}\right]. \tag{3.63}$$

As before, h or g_2 is found by inverting (3.35). A simple but relatively accurate approximation for z is given by

$$z(\eta) = \frac{1 + \left(\dfrac{7\pi - 12\sqrt{3}}{3\pi}\right)\eta^2}{(1-\eta)^2}.$$

More sophisticated approximations are given in Baus and Colot (1987).

3.4 Random Sequential Addition Processes

The RSA process, introduced in Section 3.1.2, is an example of a nonequilibrium hard-particle ensemble whose statistics can be studied analytically in some cases. The RSA process is produced by randomly, irreversibly, and sequentially placing nonoverlapping objects into a volume (Widom 1966, Feder 1980, Cooper 1988, Viot, Tarjus and Talbot 1993). As the process continues, it becomes more difficult to find available regions into which the objects can be added, and eventually (in the saturation limit) no further additions are possible. The adsorption of proteins on solid surfaces (Feder 1980), certain coagulation processes (Cooper 1988), and the settlement of territories by animals (Tanemura and Hasegawa 1980) are modeled well by the RSA process, for example.

We have already pointed out that the RSA process at the saturation limit has a dramatically different structure than the corresponding equilibrium system at the same density. To further illustrate the differences between RSA and equilibrium ensembles of hard spheres, we follow the arguments of Widom (1966), who computed the generic three-particle probability density function $\rho_3(\mathbf{r}_1,\mathbf{r}_2,\mathbf{r}_3)$ for simple three-particle systems. He considered the insertion of three spheres into a system of volume V, neglecting boundary effects. The volume from which the first sphere excludes the center of the second sphere is just the volume $v_1(D)$ of a sphere of radius D. Similarly, $v_2(r_{ij};D)$ is the volume from which the center of a third sphere is excluded by the spheres, one centered at $\mathbf{r}_i$, the other at $\mathbf{r}_j$, where $r_{ij}=|\mathbf{r}_j-\mathbf{r}_i|$. The quantity $v_2(r_{ij};D)$ is just the union volume of two spheres of radius D separated by a distance r_{ij} and is equal to $2v_1(D)-v_2^{\rm int}(r_{ij};D)$, where $v_2^{\rm int}(r;D)$ is the intersection volume of two exclusion spheres defined by (3.47). Therefore, from the definition of the RSA process, we have that

$$\rho_3(\mathbf{r}_1,\mathbf{r}_2,\mathbf{r}_3)=2V^{-1}(V-v_1)^{-1}\{[V-v_2(r_{12})]^{-1}+[V-v_2(r_{13})]^{-1}+[V-v_2(r_{23})]^{-1}\}.$$

Note that this probability density is not uniform in the space of allowable configurations, but instead depends on the configuration through the union volumes. The density function ρ_3 is favored by those triplet configurations in which pairs of spheres are so distant that their associated exclusion spheres do not overlap. By contrast, the equilibrium distribution, which allows all of the particles to move subject to their mutual impenetrability, has a probability density function ρ_3 that is uniform in the space of allowable configurations and is given by

$$\rho_3(\mathbf{r}_1,\mathbf{r}_2,\mathbf{r}_3)=6V^{-1}(V-v_1)^{-1}(V-\Omega)^{-1},$$

where

$$\Omega=(V-v_1)^{-1}\int v_2(r_{12})d\mathbf{r}_2.$$

Observe that the RSA and equilibrium distributions are identical at the two-particle level.

It is of basic importance in RSA processes to understand the rate equation governing the time evolution of the number density $\rho(t)$ of the system at time t. If we use the common designation $\Phi(t)$ for the fraction of the available space for a new particle at time t, then is clear that the coverage rate is proportional to $\Phi(t)$, i.e.,

$$\frac{d\rho(t)}{dt}=k\Phi(t), \tag{3.64}$$

where k is a rate constant ("attempt frequency") per unit volume that may be set equal to unity without loss of generality. Interestingly, the geometrical quantity $\Phi(t)$ is a special case of the *void exclusion probability* $E_V(r,\rho)$, defined to be the fraction of the space available to a test particle of radius $r-D/2$ added to a system at density ρ or, equivalently, time t (see Section 2.8). Specifically, setting $r=D$, we have that

$$\Phi(t)=E_V(D,t). \tag{3.65}$$

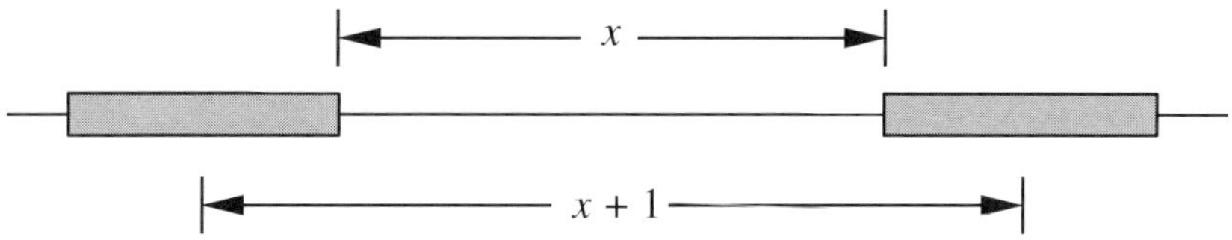

Figure 3.16 A gap of length x for $x \geq 1$. Insertion of a rod into this gap leads to its destruction.

3.4.1 One-Dimensional Identical Hard Rods

It is instructive to study in some detail the case of the RSA process for one-dimensional hard rods, since a variety of useful statistics, including the gap distribution function (Mackenzie 1962, Reńyi 1963, Widom 1966, Gonzalez, Hemmer and Hoye 1974) and the pair correlation function (Bonnier, Boyer and Viot 1994), can be determined exactly. This one-dimensional case is commonly known as the *car-parking problem*. Rods of length D are to be placed on a line of length L at a rate of α per unit length and per unit time. An important problem is to find the expected value for coverage at the saturation limit as $L \to \infty$. A naive guess might be to assume that the size r of the gaps (i.e., space between the rods) is uniformly distributed, leading to the conclusion that the saturation particle fraction is $\eta_c = \eta(t = \infty) = 2/3$, where t is the time.

Let us define the dimensionless gap distance $x = r/D$ and the dimensionless time $\tau = t\alpha D$. Moreover, we define the quantity $n(x, \tau)dx$ to be the dimensionless number density of gaps with length between x and $x + dx$ at time τ. Then the total reduced number density of gaps $\eta(\tau)$ at time τ is given by

$$\eta(\tau) = \int_0^\infty dx n(x, \tau). \tag{3.66}$$

The time evolution of $n(x, \tau)$ is governed by a rate equation that takes into account the destruction and creation of gaps of length x (Gonzalez et al. 1974, Evans 1993, Viot et al. 1993). For $x \geq 1$, gaps can be both created and destroyed. Intervals of length x are destroyed if a rod is added to the available space $x - 1$ between the rods (see Figure 3.16). Similarly, intervals of length x are created if a rod is added to intervals of length $y \geq x + 1$. The associated rate equation can be written as

$$\frac{\partial n(x, \tau)}{\partial \tau} = -(x - 1)n(x, \tau) + 2\int_{x+1}^\infty dy n(y, \tau), \qquad x \geq 1. \tag{3.67}$$

In the case of $x < 1$, the gaps cannot be destroyed, since a particle cannot be added to the gap. Therefore, one need only consider gap intervals of length x that are created when a rod is added to intervals of length $y \geq x + 1$. The rate equation is then given by

$$\frac{\partial n(x, \tau)}{\partial \tau} = 2\int_{x+1}^\infty dy n(y, \tau), \qquad x < 1. \tag{3.68}$$

For $x \geq 1$, we introduce the function $H(\tau)$ such that

$$n(x,\tau) = \exp\Big[- (x-1)\tau \Big] H(\tau). \tag{3.69}$$

Substituting this into (3.67) yields the first-order differential equation

$$\frac{dH}{d\tau} = \frac{2e^{-\tau}}{\tau} H(\tau). \tag{3.70}$$

The initial condition $H(0) = 0$ can be determined from the previous relations by noting that $\eta(0) = 0$. Integrating (3.70) gives

$$H(\tau) = \tau^2 \exp\left[-2\int_0^\tau \frac{1-e^{-u}}{u} du\right] = \exp\left[-2(\gamma + \mathrm{Ei}(\tau))\right], \tag{3.71}$$

where $\mathrm{Ei}(\tau)$ is the exponential integral and $\gamma = 0.57721\ldots$ is Euler's constant.

For $x < 1$, the gap distribution function is given by

$$n(x,\tau) = 2\int_0^\tau du \frac{H(u)}{u} e^{-xu}. \tag{3.72}$$

The density of adsorbed rods is calculated from (3.66) by substituting the expression for $n(x,\tau)$ in (3.72) and integrating with respect to x:

$$\eta(\tau) = \int_0^\tau du \frac{H(u)}{u^2}. \tag{3.73}$$

At the saturation limit (i.e., as $\tau \to \infty$), the maximum density is exactly

$$\eta_c = \eta(\tau = \infty) = \int_0^\infty du \frac{H(u)}{u^2} \approx 0.7476. \tag{3.74}$$

Moreover, we can expand the reduced density of (3.73) about $\tau^{-1} = 0$ to give

$$\eta(\tau) \sim \eta(\infty) - \frac{\exp(-2\gamma)}{\tau},$$

and hence for large τ we have the asymptotic algebraic behavior

$$\eta(\infty) - \eta(\tau) \sim \tau^{-1}. \tag{3.75}$$

Evaluating the gap distribution function (3.72) at the final state $\tau = \infty$ shows that it diverges logarithmically as $x \to 0$, implying the same logarithmic singularity in the pair correlation function $g_2(x+1,\infty)$, i.e.,

$$g_2(x+1,\infty) = -2[\eta(\infty)]^{-2} \exp(-2\gamma) \ln x, \quad x \to 0. \tag{3.76}$$

Indeed, an analytical integral expression for the pair correlation function $g_2(x+1,\tau)$ at any τ or density has been given recently by Bonnier et al. (1994) that at the saturation limit exhibits the expected logarithmic divergence. Calculation of the integral relation for g_2 becomes progressively difficult as x increases from zero. However, owing to the fact that the g_2 decays to its long-range value very rapidly after $x \approx 2$ (even at the saturation limit), one need not evaluate the expression for large distances.

Bonnier et al. (1994) also showed that the pair correlation function has *super-exponential* decay. Specifically, they found that at any finite time τ or density η,

$$g_2(x+1,\eta) \sim \frac{1}{\Gamma(x+1)}\left(\frac{2}{\ln x}\right)^x, \quad x \to \infty. \tag{3.77}$$

Thus, g_2 is a short-ranged function at any density. This turns out to be true in higher dimensions as well.

It can be shown that the aforementioned exclusion probability $E_V(x,\tau)$ for $x \geq 1$ at time τ is given by (Rintoul, Torquato and Tarjus 1996)

$$E_V(x,\tau) = \frac{H(\tau)}{\tau^2}e^{-2(x-1)\tau}, \quad x \geq 1. \tag{3.78}$$

Hence, according to (3.65), the fraction of available space when a test particle of unit length is added to the system at some fixed time is given by

$$E_V(1,\tau) = \frac{H(\tau)}{\tau^2}. \tag{3.79}$$

Since small times imply small densities, we can manipulate the expansions of (3.71) and (3.73) in powers of τ to give the small-density expansion of the fraction of available space as

$$E_V(1,\eta) = 1 - 2\eta + \frac{1}{2}\eta^2 + \frac{2}{9}\eta^3 + \mathcal{O}(\eta^4). \tag{3.80}$$

This last expansion is to be compared with the corresponding expansion of the exact expression $E_V(1,\eta) = (1-\eta)\exp[-2\eta/(1-\eta)]$ for an equilibrium system of hard rods (Section 5.2.5), which expands as

$$E_V(1,\eta) = 1 - 2\eta + \frac{1}{2}\eta^2 + \frac{1}{3}\eta^3 + \mathcal{O}(\eta^4). \tag{3.81}$$

We see that the RSA and equilibrium expansions are the same up through order η^2 but differ at third- and higher-order terms, behavior consistent with the opening remarks of this section.

3.4.2 Identical Hard Spheres in Higher Dimensions

In contrast to the one-dimensional instance, there are no analytical solutions for the coverage at the saturation limit for identical d-dimensional hard spheres for $d \geq 2$. Consequently, RSA processes at high densities for $d \geq 2$ have been primarily investigated using computer simulation techniques. Numerical experiments have yielded that $\eta(\infty) \approx 0.547$ for $d = 2$ (hard disks) (Feder 1980) and $\eta(\infty) \approx 0.38$ for $d = 3$ (hard spheres) (Cooper 1988) in the thermodynamic limit.

In his numerical study of RSA hard disks, Feder (1980) postulated that the asymptotic coverage for d-dimensional hard spheres follows the algebraic behavior

$$\eta(\infty) - \eta(\tau) \sim \tau^{-1/d}. \tag{3.82}$$

Theoretical arguments supporting Feder's law (3.82) have been put forth by Pomeau (1980) and Swendsen (1981). Not surprisingly, the saturation limit is approached more slowly as the space dimension increases. Moreover, similar arguments lead to the conclusion that the pair correlation function $g_2(x+1,\infty)$ at the saturation limit possesses a logarithmic singularity as $x \to 0$, independent of dimension (Pomeau 1980, Swendsen 1981), i.e.,

$$g_2(x+1,\infty) \sim \ln x, \quad x \to 0. \tag{3.83}$$

3.4.3 General Hard-Particle Systems

Both theoretical (Talbot and Schaaf 1989) and numerical (Meakin and Jullien 1992) work on RSA disks of *different sizes* indicate that Feder's law is generally not valid, but the coverage still saturates asymptotically as

$$\eta(\infty) - \eta(\tau) \sim \tau^{-\alpha}, \tag{3.84}$$

where α is some size-dependent positive exponent that can be less than $1/d$. Interestingly, fractal patterns can arise in RSA processes when the particles possess a power-law size distribution, $f(R) \sim R^{\beta-1}$, for $R \le R_{\max}$ (Brilliantov, Andrienko, Krapivsky and Kurths 1996). The patterns become ordered as β increases, and the *Apollonian packing* is obtained in the limit $\beta \to \infty$.

For oriented and unoriented squares in two dimensions, the saturation-limit coverages are, respectively, $\eta(\infty) \approx 0.562$ (Feder 1980) and $\eta(\infty) \approx 0.530$ (Viot, Tarjus, Ricci and Talbot 1992). For nonspherical objects, the asymptotic behavior for coverage is still described by (3.84), but generally the exponent α depends on the shape of the particles. For example, see the work of Viot et al. (1992), who studied RSA of randomly oriented spherocylinders, ellipses, rectangles, and needles in two dimensions.

3.5 Maximally Random Jammed State

Bernal (1965) has remarked that "heaps (random close-packed arrangements of particles) were the first things that were ever measured in the form of basketfuls of grain for the purpose of trading or of collection of taxes." Biologists, materials scientists, engineers, chemists, and physicists have studied random packings of identical spheres to understand the structure of living cells, liquids, granular media, glasses, and amorphous solids, to mention but a few examples. Despite its long history, there are many fundamental conundrums concerning random packings of spheres that remain elusive, including a precise definition of *random close packing* (RCP). Torquato, Truskett and Debenedetti (2000) have recently demonstrated that the venerable concept of the RCP state is ill-defined and should be abandoned. To replace this notion, they have introduced the new concept of a *maximally random jammed* state, which can be made precise. The ensuing discussion follows their treatment of the problem.

3.5.1 Random Close Packing Is Ill-Defined

It is only recently that a putatively air-tight rigorous proof has been devised for *Kepler's conjecture*: the densest possible packing fraction η for identical spheres in three dimensions is $\pi/\sqrt{18} \approx 0.7405$, corresponding to the close-packed face-centered cubic (FCC) lattice or its stacking variants (Hales 1998). If a precise definition of the RCP state could be presented, then one might go about quantifying the problem with the same rigor that has been used to prove Kepler's conjecture. The prevailing notion of the RCP state is that it is the *maximum density that a large, random collection of spheres can attain and that this density is a well-defined and unique quantity*. An anonymous author summarizes this traditional view as follows: "Ball bearings and similar objects have been shaken, settled in oil, stuck with paint, kneaded inside rubber balloons–and all with no better result than (a packing fraction of) ... 0.636" (Anonymous 1972).

Torquato et al. (2000) have observed that there has existed ample evidence in the literature, in the form of actual and computer experiments, to suggest strongly that the RCP state is ill-defined and, unfortunately, dependent on the protocol employed to produce the random packing and on other system characteristics. In a classic experiment, Scott and Kilgour (1969) obtained the RCP value $\eta_c \approx 0.637$ by pouring ball bearings into a large container, vertically vibrating the system for sufficiently long times to achieve maximum densification, and extrapolating the measured volume fractions to eliminate finite-size effects. Presumably each sphere in this state is "locked" in by contacts with, on average, six neighbors such that there is no way to increase density further to achieve the highest crystalline close-packing value. Important dynamical parameters for this experiment include the pouring rate and both the amplitude and frequency of vibration. The key interactions are interparticle forces, including (ideally) repulsive hard-sphere interactions, friction between the particles (which inhibits densification), and gravity. It is clear that the final volume fraction will be sensitive to these system characteristics. Indeed, in a recent experimental study (Pouliquen, Nicolas and Weidman 1997) it was shown that one can achieve denser (partially crystalline) packings when the particles are poured at low rates into horizontally shaken containers.

There has been a temptation to compare experimentally observed statistics of RCP configurations (volume fraction, radial distribution functions, Voronoi statistics) to those generated on a computer. One must be careful in making such comparisons because it is difficult to simulate the features of real systems, such as the method of preparation and system characteristics (e.g., shaking, friction). Computer algorithms are valuable because they can be used to generate and study idealized random packings, but the final states are clearly *protocol-dependent*. For example, a popular rate-dependent densification algorithm (Jodrey and Tory 1985, Jullien, Sadoc and Mosseri 1997) achieves η_c between 0.642 and 0.649, a Monte Carlo scheme (Tobochnik and Chapin 1988) gives $\eta_c \approx 0.68$, a differential-equation densification algorithm produces

$\eta_c \approx 0.64$ (Zinchenko 1994), and a "drop and roll" algorithm (Visscher and Bolsterli 1972) yields $\eta_c \approx 0.60$. In contrast to the last two algorithms, the first two algorithms produce configurations in which either the majority or all of the particles are not in contact with one another. Current algorithms still cannot truly account for friction between the spheres.

3.5.2 Definition of Maximally Random Jammed State

Torquato et al. (2000) have argued that the aforementioned inconsistencies and deficiencies of RCP arise because it is an ill-defined state, explaining why, to this day, there is no rigorous prediction of the RCP density, although such attempts have been made (Gotoh and Finney 1974, Berryman 1983). The term "close packed" implies that the spheres are in contact with one another with the highest possible coordination number on average. This is consistent with the aforementioned traditional view that RCP is the highest possible density that a random packing of close-packed spheres can possess. However, the terms "random" and "close packed" (never clearly defined) are at odds with one another. Increasing the degree of coordination, and thus the bulk system density, comes at the expense of disorder. The precise proportion of each of these competing effects is *arbitrary*, and therein lies the problem.

A precise mathematical definition that supplants the inadequate RCP state should apply to any statistically homogeneous and isotropic system of identical spheres (with specified interactions) in any space dimension d. Although we discard the term "close packed," we must retain the idea that the particles are in contact with one another, while maintaining the greatest generality. We say that a particle is *jammed* if it cannot be translated while holding fixed the positions of all of the other particles in the system. The system itself is jammed if each particle is jammed in the aforementioned local sense and there can be no *collective* motion of any contacting subset of particles that leads to unjamming. This definition eliminates so-called *loose packings* as well as systems with "rattlers" (movable but caged particles) in the infinite-volume limit. Jammed structures created via computer algorithms (Lubachevsky and Stillinger 1990) or actual experiments will contain a very small concentration of such rattler particles, the precise concentration of which is protocol-dependent. Thus, in practice, one may wish to accommodate this type of a jammed structure, although the ideal limit described above is the precise mathematical definition of a jammed state that we have in mind. Nevertheless, it should be emphasized that it is the overwhelming majority of spheres that compose the underlying "jammed" network that confers global rigidity to the particle packing and, in any case, the "rattlers" could be removed without disrupting the jammed remainder.

The definition of the maximally random jammed (MRJ) state introduced by Torquato et al. (2000) is based on the minimization of an order metric described below. The most challenging fundamental problem that arises here is the quantification of randomness or its antithesis: order. We recall that a classical many-particle system is completely characterized statistically by the N-particle probability density function $P_N(\boldsymbol{r}^N)$ associated with finding the particles with configuration $\boldsymbol{r}^N$. Clearly, such complete information is never available, and in practice, one *must settle for reduced information*. For example, from functionals of higher-order correlation functions, one can extract a set of scalar *order* metrics $\psi_1, \psi_2, \ldots, \psi_n$ such that $0 \leq \psi_i \leq 1\ \forall i$, where 0 corresponds to the absence of order (maximum disorder) and 1 corresponds to maximum order (absence of disorder). The set of order metrics that one selects (see Section 3.5.3) is *unavoidably subjective*, given that there is no single and complete scalar measure of order.

Nonetheless, within these *necessary limitations*, there is a systematic way to choose the best order metrics to be used in the objective function (the quantity to be minimized). The most general objective function consists of weighted combinations of order metrics. The set of all jammed states will define a certain region in the n-dimensional space of order metrics. In this region of jammed structures, the order metrics can be divided into two categories: those that share a common minimum and those that do not. The strategy is clear: Retain those order metrics that share a common minimum and discard those that do not, since they are conflicting measures of order. [Note that two order metrics that are positively correlated (i.e., $\partial\psi_i/\partial\psi_j > 0$) will share a common minimum.] Moreover, since all of the metrics sharing a common minimum are essentially equivalent measures of order (there exists a jammed state in which all order metrics are minimized), choose from among these the one that is the most sensitive measure, which we will simply denote by ψ. (The objective function ψ could also represent some judicious weighted combination of order metrics, but we will not consider such ψ below.)

Consider all possible configurations of a d-dimensional system of identical spheres, with specified interactions, at a sphere volume fraction η in the infinite-volume limit. For every η, there will be a minimum and maximum value of the order metric ψ. By varying η between zero and its maximum value (corresponding to the triangular lattice for $d = 2$ and FCC lattice for $d = 3$), the locus of such extrema define upper and lower bounds within which all structures of identical spheres must lie. Figure 3.17 shows a schematic plot of the order metric versus volume fraction. Note that at $\eta = 0$, the most disordered ($\psi = 0$) configurations of sphere centers can be realized. As the packing fraction is increased, the hard-core interaction prevents access to the most random configurations of sphere centers (gray region). Hence, the lower boundary of ψ, representing the most disordered configurations, increases monotonically with η. The upper boundary of ψ corresponds to the most ordered structures at each volume fraction, e.g., perfect (generally) open lattice structures ($\psi = 1$). Of course, the details of the figure will depend on the particular choice of ψ. Nevertheless, the salient features of this plot are as follows: (1) All sphere structures must lie within the allowable (white) region, and (2) the jammed structures are a special subset of the allowable structures. We define the MRJ state to be the *configuration that minimizes ψ among all statistically homogeneous and isotropic jammed structures*. We denote by η_M the MRJ packing fraction.

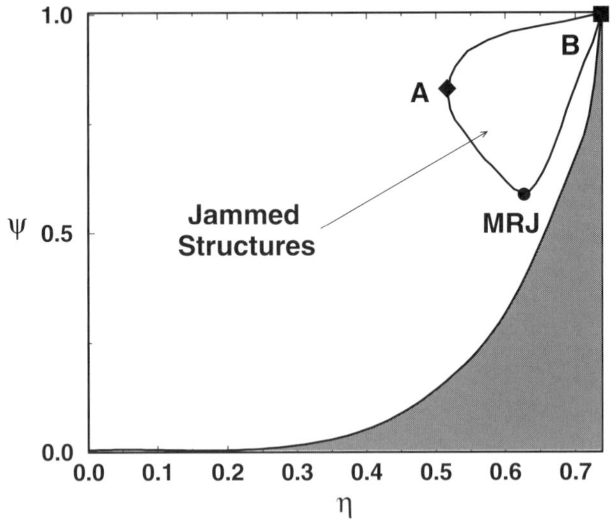

Figure 3.17 A schematic plot of the order metric ψ versus volume fraction η for a system of identical spheres with prescribed interactions. All structures at a given packing fraction η must lie between the upper and lower bounds (white region). The gray region is inaccessible. The boundary containing the subset of jammed structures is shown. The jammed structure that minimizes the order metric ψ is the maximally random jammed (MRJ) state.

The points A and B in Figure 3.17 represent the jammed structures with the lowest and highest densities, respectively. The former structures are of great interest, since they represent the lowest-density packings that are still globally rigid. However, they have yet to be identified for $d > 1$. Examples of low-density jammed structures for $d = 2$ have been noted by Lubachevsky, Stillinger and Pinson (1991). For $d = 3$, the close-packed simple cubic lattice (contained by rigid boundaries) with a packing fraction of $\eta = \pi/6 \approx 0.52$ is an obvious example of a low-density jammed structure, but this is almost surely not the lowest-density jammed structure. Point B for $d = 3$, for example, represents the close-packed FCC or hexagonal lattice, depending on the choice for ψ.

3.5.3 Order Metrics

To quantify the order (or disorder) in jammed sphere systems, Torquato et al. (2000) chose to examine two simple but fundamentally important measures of order: *bond-orientational* order and *translational* order. The first is obtainable in part from the metric Q_6, and the second is obtainable in part from the radial distribution function $g_2(r)$. To each nearest-neighbor bond emanating from a sphere for $d = 3$, one can associate the spherical harmonics $Y_{lm}(\theta, \varphi)$ using the bond angles as arguments. Then Q_6 is defined by (Steinhardt, Nelson and Ronchetti 1983)

$$Q_6 \equiv \left(\frac{4\pi}{13} \sum_{m=-6}^{6} \left| \overline{Y_{6m}} \right|^2 \right)^{1/2}, \tag{3.85}$$

where $\overline{Y_{6m}}$ denotes an average over all bonds. For a completely disordered system in the infinite-volume limit, Q_6 equals zero, whereas Q_6 attains its maximum value for space-filling jammed structures ($Q_6^{\rm FCC} \approx 0.575$) in the perfect FCC crystal. Thus, Q_6 provides a global measure of FCC crystallite formation in the system. For convenience, we normalize the orientational order metric by its value in the perfect FCC crystal, i.e., $Q \equiv Q_6/Q_6^{\rm FCC}$.

Scalar measures of translational order have not been well studied. Torquato et al. (2000) have introduced a simple translational order metric T that measures the degree of spatial ordering, relative to the perfect FCC lattice at the same volume fraction. Specifically,

$$T = \left| \frac{\sum_{i=1}^{N_C} (n_i - n_i^{\rm ideal})}{\sum_{i=1}^{N_C} (n_i^{\rm FCC} - n_i^{\rm ideal})} \right|, \tag{3.86}$$

where n_i, for the system of interest, denotes the average occupation number for the shell of width $a\delta$ centered at a distance from a reference sphere that equals the ith nearest-neighbor separation for the open FCC lattice at that density. Moreover, a is the first nearest-neighbor distance for that FCC lattice, and N_C is the total number of shells. Similarly, $n_i^{\rm ideal}$ and $n_i^{\rm FCC}$ are the corresponding shell occupation numbers for an ideal gas (spatially uncorrelated spheres) and the open FCC lattice. Observe that $T = 0$ for an ideal gas (perfect randomness), and $T = 1$ for perfect FCC spatial ordering.

Both Q and T are crystal-dependent measures in that they measure order with respect to the FCC lattice. Other reasonable choices for order metrics have been tested, including crystal-independent ones, such as an information-theoretic *entropy*, another translational-order metric, and a "local" version of Q. Importantly, the evaluations of these order metrics resulted in the same qualitative behavior as that given by Q and T for the configurations discussed immediately below. All of these results, as well as the utility of other, more sophisticated, order metrics for many-particle systems in general, are described by Truskett, Torquato and Debenedetti (2000) and Kansal, Truskett and Torquato (2000a).

3.5.4 Molecular Dynamics Simulations

To support the aforementioned arguments, Torquato et al. (2000) carried out molecular dynamics simulations (Chapter 12) using systems of 500 identical hard spheres with periodic boundary conditions. Starting from an equilibrium liquid configuration at a volume fraction $\eta = 0.3$, the system was compressed to a jammed state by the well-known method of Lubachevsky and Stillinger (1990), which allows the diameter of the particles to grow linearly in time with a dimensionless rate Γ. The jammed state

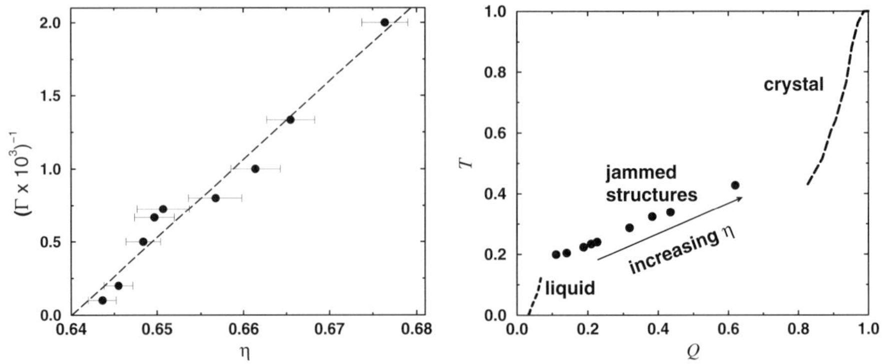

Figure 3.18 Left panel: The reciprocal compression rate Γ^{-1} versus the volume fraction η of the final jammed state of hard spheres obtained using the molecular dynamics compression protocol of Lubachevsky and Stillinger (1990). Right panel: The Q-T plane for the hard-sphere system, where T and Q are translational and orientational order metrics, respectively (where $\delta = 0.196$ and $N_C = 7$). Shown are the average values for the jammed states, as well as states along the equilibrium liquid (dotted) and crystal (dashed) branches.

occurs when the diameters can no longer increase in time, the sphere collision rate diverges, and no further compression can be achieved after relaxing the configuration at the jammed volume fraction. Figure 3.18 shows that the volume fraction of the final jammed states is inversely proportional to the compression rate Γ. A linear extrapolation of the data to the infinite compression rate limit yields $\eta \approx 0.64$, which is close to the supposed RCP value reported by Scott and Kilgour (1969). Figure 3.19 shows a jammed configuration at $\eta \approx 0.64$.

The relationship between translational and bond-orientational ordering was characterized for the first time by Torquato et al. (2000). Figure 3.18 shows their results for the aforementioned jammed structures in the Q-T plane. This order plot reveals several key points. First, T and Q are positively correlated and therefore are essentially equivalent measures of order for the jammed structures. Second, they found that the MRJ packing fraction η_M is approximately equal to 0.64 for 500-sphere systems using the Lubachevsky–Stillinger protocol. It should be noted, however, that a systematic study of other protocols may indeed find jammed states with a lower degree of order as measured by Q or some other order metric. Lastly and most importantly, the degree of order increases monotonically with the jammed packing fraction. These results demonstrate that the notion of RCP as the highest possible density that a random sphere packing can attain is ill-defined, since one can achieve packings with arbitrarily small increases in volume fraction at the expense of small increases in order.

For purposes of comparison, the order plot of Figure 3.18 includes results for the equilibrium hard-sphere system for densities along the stable liquid branch and stable crystal branch (ending at the maximum close-packed FCC state). Interestingly, the equilibrium structures exhibit the same monotonicity properties as the jammed struc-

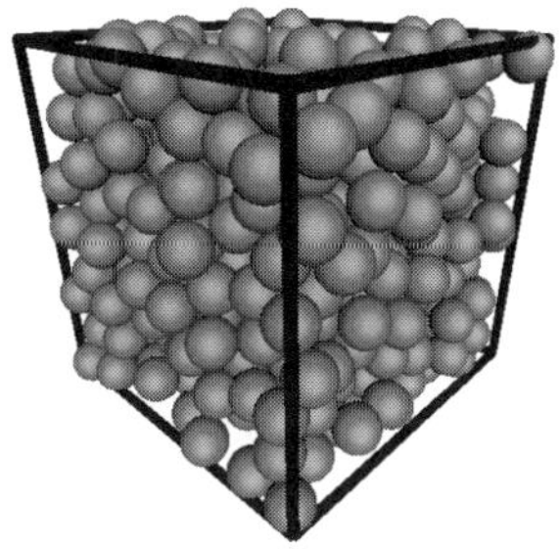

Figure 3.19 A jammed configuration of 500 spheres at $\eta \approx 0.64$ generated from the compression algorithm.

tures; i.e., T increases with increasing Q and the degree of order increases with the packing fraction. Note that neither Q nor T is equal to unity along the equilibrium crystal branch because of thermal motion. Recall that since hard spheres along the metastable branch (not shown in Figure 3.18) do not jam until the "critical" (highest) density η_c, it follows that η_c corresponds to the MRJ state for equilibrium hard spheres (see Section 3.3).

It is of interest to ask whether a close-packed crystal composed of a random sequence of FCC and HCP layer placements should be classified as a "random packing" according to the above criteria. In other words, is the point (Q, T) for such a system close to the minimum for the jammed structures shown in the order plot of Figure 3.18? Simulation results (not shown) indicate that the answer is a definitive no; not only are both Q and T higher, but Q lies in the range $0.84 < Q < 1$, depending on the stacking arrangement (Torquato et al. 2000).

3.5.5 Concluding Remarks

To summarize, the notion of the RCP state is not well-defined mathematically. To replace this idea, a new concept has been introduced: the maximally random jammed state, which can be defined precisely once an order metric ψ is chosen. This lays the mathematical groundwork for studying randomness in packings of particles and initiates the search for the MRJ state in a quantitative way not possible before. Nevertheless, significant challenges remain. First, new and efficient protocols (both experimental and computational) that generate jammed states must be developed. Second, since the characterization of randomness in sphere packings is in its infancy, the systematic investigation of better order metrics is *crucial*.

Finally, we note that the use of scalar order metrics to characterize the degree of randomness of heterogeneous materials of arbitrary microstructure has heretofore not been investigated. The appropriate order metrics will depend on the types of heterogeneous media of interest. The quantification of microstructures in order-metric space may potentially be employed to categorize classes of random heterogeneous materials.

CHAPTER 4

Unified Approach to Characterize Microstructure

A unified and general formalism has been developed by Torquato (1986c) to characterize statistically the microstructure of heterogeneous materials composed of d-dimensional spherical inclusions distributed throughout a matrix phase. He accomplished this by introducing a general n-point correlation function H_n and by deriving two different series representations for it in terms of the probability density functions that characterize the particle configurations (quantities that are known, in principle). This methodology provides a means of calculating any of the various types of microstructural correlation functions that have arisen in rigorous expressions for the transport, electromagnetic, and transport properties (see Chapter 2 and Part II) for such nontrivial models of two-phase disordered media as well as generalizations of these correlation functions. For this reason, H_n is referred to as the *canonical* n-point correlation function.

Since the derivation of these series representations of the H_n is rather involved technically, we will begin by considering the derivation of simple one-point correlation functions: the matrix volume fraction ϕ_1 and specific surface s. Having established the expressions for ϕ_1 and s, the derivation of the general H_n follows along similar lines of reasoning. Asymptotic and bounding properties of the canonical function H_n are described. We also obtain representations of the H_n for dispersions of spheres with a size distribution. These representations are easily extended to identical inclusions whose configuration is completely specified by center-of-mass coordinates (e.g., oriented ellipsoids, cubes, squares). Moreover, one can also generalize the results to inclusions of arbitrary shape and to nonparticulate systems, such as cell models (Chapter 8).

4.1 Volume Fraction and Specific Surface

Consider a statistically *inhomogeneous* ensemble of N identical possibly overlapping d-dimensional spheres of radius R centered at positions $\mathbf{r}^N \equiv \{\mathbf{r}_1, \mathbf{r}_2, \ldots, \mathbf{r}_N\}$ in the space $\mathcal{V} \in \Re^d$ of volume V (see Figure 4.1). The sphere positions are statistically characterized according to the "specific" N-body probability density function $P_N(\mathbf{r}^N)$ for the ensemble [cf. (3.4)]. Phase 1 is taken to be the space $\mathcal{V}_1$ exterior to the space $\mathcal{V}_2$ occupied by the particles (phase 2). We will generically refer to phase 1 as the *matrix* phase, which *may or may not be a connected* phase. Note that in the case of porous media, the *void*, or *pore*, phase may be taken to be either the matrix phase or particle phase.

Before discussing the canonical n-point correlation function H_n, the object of general interest in this chapter, it is useful to begin by considering special limiting cases of it at the one-point level. Thus, our focus in this section will be on simple one-point correlation functions.

According to Chapter 2, the one-point correlation, or *probability*, function $S_1^{(1)}(\mathbf{x})$ for phase 1 for general two-phase media is defined as the ensemble average of $\mathcal{I}^{(1)}(\mathbf{x})$, the indicator function for phase 1 given by (2.1). For sphere systems we will suppress the superscript 1 to designate the matrix phase, and thus denote the one-point function by $S_1(\mathbf{x}) \equiv \langle \mathcal{I}(\mathbf{x}) \rangle$, where

$$\mathcal{I}(\mathbf{x}) = \begin{cases} 1, & \text{if } \mathbf{x} \text{ is in matrix phase,} \\ 0, & \text{if } \mathbf{x} \text{ is in particle phase.} \end{cases} \tag{4.1}$$

Recall that for statistically homogeneous media, $S_1(\mathbf{x})$ is independent of position $\mathbf{x}$ and is just equal to the matrix volume fraction ϕ_1.

The "local" (position-dependent) matrix volume fraction $S_1(\mathbf{x})$ has the following equivalent probabilistic interpretations for systems of identical spheres of radius R (Torquato and Stell 1982, Torquato and Stell 1983a):

$$\begin{aligned} S_1(\mathbf{x}) &\equiv \langle \mathcal{I}(\mathbf{x}) \rangle \\ &= \text{Probability that a "point" test particle at } \mathbf{x} \text{ lies in the} \\ &\quad \text{"available" space in a system of } N \text{ spheres.} \\ &= \text{Probability that no sphere center is within a sphere of radius} \\ &\quad R \text{ of } \mathbf{x}. \\ &= \text{Probability that a spherical region of radius } R \text{ centered at } \mathbf{x} \\ &\quad \text{is empty of sphere centers.} \end{aligned} \tag{4.2}$$

The *available space* to a point particle introduced into a system of N spheres (that are impenetrable to the point particle) is precisely the matrix space $\mathcal{V}_1$ (see Figure 4.3). This notion of the available space to a "test" particle added to a system of inclusions will prove to be valuable in obtaining the general function H_n later.

To obtain an explicit expression for the local volume fraction $S_1(\mathbf{x})$ for sphere systems, we need to express $\mathcal{I}(\mathbf{x})$ in terms of the sphere positions $\mathbf{r}^N$, which are characterized probabilistically. Following Torquato and Stell (1982), the indicator function

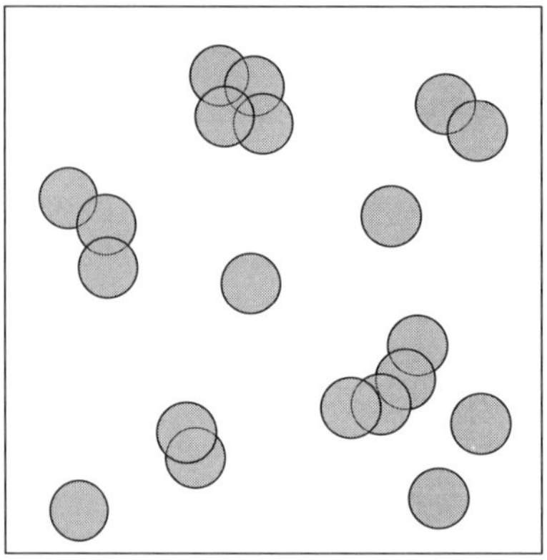

Figure 4.1 Portion of a realization of possibly overlapping two-dimensional spheres (i.e., circular disks). Phase 1 is the space $\mathcal{V}_1$ exterior to the particles (unshaded region). Phase 2 is the space $\mathcal{V}_2$ occupied by the particles (shaded region).

$\mathcal{I}(\boldsymbol{x})$ for the matrix phase is given by

$$\begin{aligned}\mathcal{I}(\boldsymbol{x};\boldsymbol{r}^N) &= \prod_{i=1}^{N}[1-m(|\boldsymbol{x}-\boldsymbol{r}_i|;R)] \\ &= 1-\sum_{i=1}^{N} m(|\boldsymbol{x}-\boldsymbol{r}_i|;R)+\sum_{i<j}^{N} m(|\boldsymbol{x}-\boldsymbol{r}_i|;R)m(|\boldsymbol{x}-\boldsymbol{r}_j|;R) \\ &\quad -\sum_{i<j<k}^{N} m(|\boldsymbol{x}-\boldsymbol{r}_i|;R)m(|\boldsymbol{x}-\boldsymbol{r}_j|;R)m(|\boldsymbol{x}-\boldsymbol{r}_k|;R)+\cdots,\end{aligned} \tag{4.3}$$

where the *exclusion-region* indicator function $m(r;R)$ is defined by

$$m(r;R)=\Theta(R-r)=\begin{cases}1, & r\le R,\\ 0, & r>R,\end{cases} \tag{4.4}$$

and $\Theta(x)$ is the Heaviside step function (3.48). The volume *excluded* to the center of a spherical inclusion of radius R by a point "test" particle is itself a spherical region of radius R. This is why we refer to $m(r;R)$ as an exclusion-region indicator function. Importantly, if the test particle had nonzero size, then the exclusion region would not simply be a spherical region of radius R (see Figure 4.3). Note that the kth sum in (4.3) is over all distinguishable k-tuplets of possibly overlapping particles and thus contains $N!/[(N-k)!k!]$ terms. The first sum accounts for regions of space occupied by N spheres (without any overlap); the second sum accounts for possible overlap between pairs of spheres; the third sum accounts for possible overlap between triplets of spheres, etc.

For statistically *inhomogeneous* media, we can explicitly find an expression for $S_1(\boldsymbol{x})$ by ensemble averaging (4.3), i.e.,

$$
\begin{aligned}
S_1(\mathbf{x}) &= \int \mathcal{I}(\mathbf{x};\mathbf{r}^N)P_N(\mathbf{r}^N)d\mathbf{r}^N \\
&= \int \prod_{i=1}^{N}[1-m(|\mathbf{x}-\mathbf{r}_i|)]P_N(\mathbf{r}^N)d\mathbf{r}^N \\
&= 1-\sum_{i=1}^{N}\int m(|\mathbf{x}-\mathbf{r}_i|)P_N(\mathbf{r}^N)d\mathbf{r}^N \\
&\quad +\sum_{i<j}^{N}\int m(|\mathbf{x}-\mathbf{r}_i|)m(|\mathbf{x}-\mathbf{r}_j|)P_N(\mathbf{r}^N)d\mathbf{r}^N - \cdots \\
&= 1-N\int m(|\mathbf{x}-\mathbf{r}_1|)P_N(\mathbf{r}^N)d\mathbf{r}^N \\
&\quad +\frac{N(N-1)}{2}\int m(|\mathbf{x}-\mathbf{r}_1|)m(|\mathbf{x}-\mathbf{r}_2|)P_N(\mathbf{r}^N)d\mathbf{r}^N - \cdots \\
&= 1+\sum_{k=1}^{\infty}\frac{(-1)^k}{k!}\int \rho_k(\mathbf{r}^k)\prod_{j=1}^{k}m(|\mathbf{x}-\mathbf{r}_j|)d\mathbf{r}_j,
\end{aligned}
\tag{4.5}
$$

where we have suppressed indication of the dependence of m on R. The fourth line of (4.5) follows since P_N (for identical particles) is invariant under interchange of particles. The last line follows from the definition (3.7) of the n-particle probability density function ρ_n. The desired general expression (4.5) for $S_1(\mathbf{x})$ was derived by Torquato and Stell (1982).

Let us now consider obtaining the corresponding general expression for the specific surface. Recall from Section 2.3 that the position-dependent specific surface is the one-point correlation function defined by

$$
s(\mathbf{x}) = \langle \mathcal{M}(\mathbf{x}) \rangle, \tag{4.6}
$$

where $\mathcal{M}(\mathbf{x}) = |\nabla \mathcal{I}(\mathbf{x})|$ is the indicator function for the interface [cf. (2.3)]. If the system is statistically homogeneous, then $s(\mathbf{x})$ is just a constant s independent of $\mathbf{x}$. For identical spheres of radius R, $\mathcal{M}(\mathbf{x})$ can be expressed as a simple derivative of $\mathcal{I}(\mathbf{x};R)$ via

$$
\mathcal{M}(\mathbf{x};R) = -\frac{\partial \mathcal{I}(\mathbf{x};R)}{\partial R}. \tag{4.7}
$$

Relations (4.3) and (4.7) and the identity $\delta(R-r) = \partial m(r;R)/\partial R$ yield

$$
\begin{aligned}
\mathcal{M}(\mathbf{x};R) &= \sum_{i=1}^{N}\delta(R-|\mathbf{x}-\mathbf{r}_i|) - \sum_{i<j}^{N}\delta(R-|\mathbf{x}-\mathbf{r}_i|)m(|\mathbf{x}-\mathbf{r}_j|) \\
&\quad - \sum_{i<j}^{N}\delta(R-|\mathbf{x}-\mathbf{r}_j|)m(|\mathbf{x}-\mathbf{r}_i|) + \cdots,
\end{aligned}
\tag{4.8}
$$

where $\delta(\mathbf{x})$ is the Dirac delta function. Averaging (4.8) yields

$$
\begin{aligned}
s(\boldsymbol{x}) &= \sum_{i=1}^{N} \int \delta(R - |\boldsymbol{x} - \boldsymbol{r}_i|) P_N(\boldsymbol{r}^N) d\boldsymbol{r}^N \\
&\quad - \sum_{i \neq j}^{N} \int \delta(R - |\boldsymbol{x} - \boldsymbol{r}_i|) m(|\boldsymbol{x} - \boldsymbol{r}_j|) P_N(\boldsymbol{r}^N) d\boldsymbol{r}^N \\
&\quad + 3 \sum_{i<j<k}^{N} \int \delta(R - |\boldsymbol{x} - \boldsymbol{r}_i|) m(|\boldsymbol{x} - \boldsymbol{r}_j|) m(|\boldsymbol{x} - \boldsymbol{r}_k|) P_N(\boldsymbol{r}^N) d\boldsymbol{r}^N - \cdots \\
&= \int \delta(R - |\boldsymbol{x} - \boldsymbol{r}_1|) \rho_1(\boldsymbol{r}_1) d\boldsymbol{r}_1 - \int \delta(R - |\boldsymbol{x} - \boldsymbol{r}_1|) m(|\boldsymbol{x} - \boldsymbol{r}_2|) \rho_2(\boldsymbol{r}_1, \boldsymbol{r}_2) d\boldsymbol{r}_1 d\boldsymbol{r}_2 \\
&\quad + \frac{1}{2} \int \delta(R - |\boldsymbol{x} - \boldsymbol{r}_1|) m(|\boldsymbol{x} - \boldsymbol{r}_2|) m(|\boldsymbol{x} - \boldsymbol{r}_3|) \rho_3(\boldsymbol{r}_1, \boldsymbol{r}_2, \boldsymbol{r}_3) d\boldsymbol{r}_1 d\boldsymbol{r}_2 d\boldsymbol{r}_3 - \cdots \\
&= \sum_{k=1}^{\infty} \frac{(-1)^{k-1}}{(k-1)!} \int \rho_k(\boldsymbol{r}^k)\, \delta(R - |\boldsymbol{x} - \boldsymbol{r}_1|)\, d\boldsymbol{r}_1 \prod_{j=2}^{k} m(|\boldsymbol{x} - \boldsymbol{r}_j|) d\boldsymbol{r}_j. \qquad (4.9)
\end{aligned}
$$

This expression was derived by Torquato and Stell (1984) and by Chiew and Glandt (1984).

Note that ensemble averaging (4.7) yields that $s(\boldsymbol{x})$ is simply related to the first derivative of S_1 with respect to R:

$$
s(\boldsymbol{x}) = -\frac{\partial S_1(\boldsymbol{x}; R)}{\partial R} = \frac{\partial S_1^{(2)}(\boldsymbol{x}; R)}{\partial R}. \qquad (4.10)
$$

The second equality follows from the trivial relation $S_1^{(2)} = 1 - S_1$, where $S_1^{(2)}$ is the one-point probability function for phase 2 (particle phase). As we observed in Section 2.3, the specific surface $s(\boldsymbol{x})$ cannot be interpreted as the probability of finding a point at $\boldsymbol{x}$ on the interface, since this is always zero. However, we see from (4.10) that the product $s(\boldsymbol{x})\partial R$ can be interpreted as a probability, namely, the probability of finding a point at $\boldsymbol{x}$ in any of the infinitesimal concentric shells of radius $R + \partial R/2$ and thickness ∂R surrounding the particles (see Figure 4.2). Thus, when the spheres of radius R are uniformly and radially *"dilated"* by a differential amount ∂R, we can say that $s(\boldsymbol{x})$ is proportional to the probability that $\boldsymbol{x}$ lies in the dilated space. This interpretation of s can be used to calculate it from computer simulations (see Section 12.4.3).

4.1.1 Bounding Properties

For general systems of identical spheres, Torquato and Stell (1983a) showed that there are successive bounds on the matrix n-point probability functions S_n. For simplicity, consider homogeneous ensembles of spheres and $n = 1$. The first few successive bounds on the position-independent matrix volume fraction $S_1 = \phi_1$ are given by

$$
\phi_1 \leq 1, \qquad (4.11)
$$

$$
\phi_1 \geq 1 - \int m(r_{12}) \rho_1(\boldsymbol{r}_2) d\boldsymbol{r}_2, \qquad (4.12)
$$

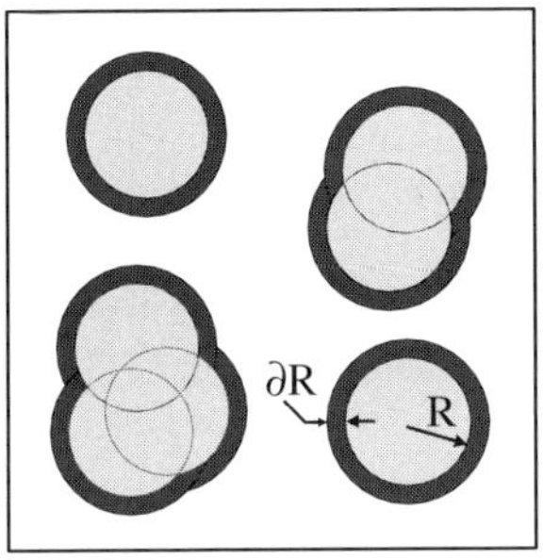

Figure 4.2 Schematic of the infinitesimal dilation process in a portion of a system of possibly overlapping equisized spheres of radius R. Here ∂R is the differential thickness. The specific surface s is proportional to the probability of finding a point in the dilated space (indicated as black regions).

$$\phi_1 \leq 1 - \int m(r_{12})\rho_1(\mathbf{r}_2)d\mathbf{r}_2 + \frac{1}{2!}\int m(r_{12})m(r_{13})\rho_2(\mathbf{r}_{23})d\mathbf{r}_2 d\mathbf{r}_3, \tag{4.13}$$

which are the first three partial sums of the series (4.5). Moreover, we have changed the labels $\mathbf{x}, \mathbf{r}_1, \ldots, \mathbf{r}_n$ to $\mathbf{r}_1, \mathbf{r}_2, \ldots, \mathbf{r}_{n+1}$ and let $r_{ij} \equiv |\mathbf{r}_j - \mathbf{r}_i|$. More generally, if $q^{(\ell)}$ denotes the partial sum of the series (4.5) from $k = 0$ to $k = \ell$ (with $q^{(0)} \equiv 1$), then we have

$$\left.\phi_1\right\} \begin{matrix} \geq \\ \leq \end{matrix} \left\{ q^{(\ell)} \quad \begin{matrix} \text{for } \ell \text{ odd}, \\ \text{for } \ell \text{ even}. \end{matrix} \right. \tag{4.14}$$

The physical interpretation of these bounds is easy to see by using an analysis due to Torquato and Stell (1983a), which was based on a related analysis given by Boltzmann (1898) and later developed by Stell (1966); see also Stell (1985). The fraction of the available space to a point test particle in a system of N identical spheres is at most equal to unity and thus corresponds to the first bound (4.11). Clearly, we have overestimated ϕ_1 and thus must subtract off the expected fraction of space occupied by the exclusion spheres, given by the second integral in (4.12). However, the first two terms in (4.12) underestimate ϕ_1, since they neglect possible overlap between pairs of exclusion spheres. Thus, we must add the expected fraction of space occupied by the overlap volume between all distinguishable pairs of exclusion spheres given by the third integral in (4.13). But now we have overestimated the effect of overlap, since we have overcounted the overlap whenever three or more exclusion spheres overlap. Continuing along this line of reasoning leads to the general expression (4.14).

Similarly, one has successive bounds on the specific surface s for analogous reasons. The first few successive bounds on s are given by

$$s \leq \int \delta(R - r_{12})\rho_1(\mathbf{r}_2)d\mathbf{r}_2,$$

$$s \geq \int \delta(R - r_{12})\rho_1(\mathbf{r}_2)d\mathbf{r}_2 - \int \delta(R - r_{12})m(r_{13})\rho_2(\mathbf{r}_{23})d\mathbf{r}_2 d\mathbf{r}_3,$$

$$s \leq \int \delta(R - r_{12})\rho_1(\mathbf{r}_2)d\mathbf{r}_2 - \int \delta(R - r_{12})m(r_{13})\rho_2(\mathbf{r}_{23})d\mathbf{r}_2 d\mathbf{r}_3$$
$$+ \frac{1}{2}\int \delta(R - r_{12})m(r_{13})m(r_{14})\rho_3(\mathbf{r}_{23}, \mathbf{r}_{24})d\mathbf{r}_2 d\mathbf{r}_3 d\mathbf{r}_4,$$

which are the first three partial sums of (4.9) using the aforementioned change of variables. More generally, if $w^{(\ell)}$ denotes the partial sum of the series (4.9) from $k = 1$ up to $k = \ell$, then we have

$$s \left.\begin{matrix} \geq \\ \leq \end{matrix}\right\} w^{(\ell)} \quad \begin{matrix} \text{for } \ell \text{ even,} \\ \text{for } \ell \text{ odd.} \end{matrix} \tag{4.15}$$

4.1.2 Example Calculations

It is instructive to describe how one evaluates the one-point expressions (4.5) and (4.9) for S_1 and s for some statistically homogeneous model microstructures given the n-particle functions ρ_n. Calculations of higher-order correlation functions are given in the next several chapters.

Consider homogeneous ensembles of d-dimensional identical spheres of radius R at number density ρ. Recall from Chapter 3 that the dimensionless, or *reduced*, density η is defined by

$$\eta = \rho v_1(R), \tag{4.16}$$

where $v_1(R)$ is d-dimensional volume of a *single* sphere is given by

$$v_1(R) = \frac{\pi^{d/2}}{\Gamma(1 + d/2)} R^d, \tag{4.17}$$

and $\Gamma(x)$ is the gamma function. For example, for $d = 1, 2,$ and 3, $v_1(R) = 2R$, πR^2, and $4\pi R^3/3$, respectively. The corresponding d-dimensional surface area of a *single* sphere, denoted by $s_1(R)$, is just

$$s_1(R) = \frac{\partial v_1}{\partial R} = \frac{d\pi^{d/2}}{\Gamma(1 + d/2)} R^{d-1}. \tag{4.18}$$

For $d = 1, 2,$ and 3 this yields $s_1(R) = 2$, $2\pi R$, and $4\pi R^2$, respectively.

First, we evaluate relations (4.5) and (4.9) for the extreme limits of the cherry-pit model [cf. (3.27) and Figure 3.10], i.e., $\lambda = 0$ (fully penetrable spheres) and $\lambda = 1$ (totally impenetrable spheres). Then we discuss calculations for arbitrary values of the impenetrability parameter λ.

Fully Penetrable Spheres, $\lambda = 0$

For homogeneous fully penetrable spheres, we recall that ρ_n is trivially equal to the constant ρ^n [cf. (3.18)]. Substitution of this relation into (4.5) with $S_1 = \phi_1$ yields

$$\phi_1(R) = 1 + \sum_{k=1}^{\infty} \frac{(-\rho)^k}{k!} \int \prod_{j=1}^{k} m(|\mathbf{x} - \mathbf{r}_j|; R)d\mathbf{r}_j = 1 + \sum_{k=1}^{\infty} \frac{(-\rho)^k}{k!} v_1^k(R),$$

and hence the position-independent matrix volume fraction is

$$\phi_1(R) = 1 - \phi_2(R) = \exp(-\eta). \tag{4.19}$$

Thus, we see that because of overlap, the reduced density η is always greater than or equal to the sphere volume fraction ϕ_2 ($\eta \geq \phi_2$), in contrast to totally impenetrable spheres, where $\eta = \phi_2$, as shown below. Observe that (4.19) could also have been obtained by setting $n = 0$ and $W = v_1(R)$ in the Poisson distribution formula (3.19).

Similarly, using (4.9), the specific surface is the constant given by

$$\begin{aligned} s(R) &= \rho\phi_1 \int \delta(R - |\mathbf{x} - \mathbf{r}_1|)\, d\mathbf{r}_1 \\ &= \rho\phi_1 \int \delta(R - y)\, d\mathbf{y} \\ &= \rho\phi_1 s_1(R) \\ &= \frac{d}{R}\eta\phi_1. \end{aligned} \tag{4.20}$$

The first line of (4.20) follows in light of (4.19), and last line follows from (4.18). Note that result (4.20) could have been obtained by simply differentiating (4.19) with respect to R according to (4.10).

Totally Impenetrable Spheres, $\lambda = 1$

At the opposite extreme of totally impenetrable spheres, the matrix volume fraction for homogeneous media is trivial, since it is just the complement of the particle volume fraction given trivially by $\eta = \phi_2 = \rho v_1(R)$, i.e.,

$$\phi_1(R) = 1 - \phi_2(R) = 1 - \eta. \tag{4.21}$$

Similarly, the specific surface for this system is given by

$$s(R) = \rho s_1(R) = \frac{d}{R}\phi_2. \tag{4.22}$$

However, it is still instructive (albeit laborious) to obtain these results via the general expressions (4.5) and (4.9).

To do so, we first introduce the n-body correlation function $g_n = \rho_n/\rho^n$, as defined by (3.11). By totally impenetrable particles, we mean that the hard core imposes the constraint

$$g_n(\mathbf{r}_{12}, \ldots, \mathbf{r}_{1n}) = 0 \quad \text{for} \quad r_{ij} < 2R, \quad 1 \leq i < j \leq n, \tag{4.23}$$

where $\mathbf{r}_{ij} = \mathbf{r}_j - \mathbf{r}_i$. For other interparticle separation distances, the g_n are nontrivial.

How, then, do we get the simple result (4.21)? Using (4.5) and the diagrammatic notation introduced in Chapter 3, the first few terms are

$$\begin{aligned} S_1 = \phi_1 = 1 &- \rho \int m(r_{12}; R) d\mathbf{r}_2 + \frac{\rho^2}{2!} \int m(r_{12}; R) m(r_{13}; R) g_2(\mathbf{r}_{23}) d\mathbf{r}_2 d\mathbf{r}_3 \\ &- \frac{\rho^3}{3!} \int m(r_{12}; R) m(r_{13}; R) m(r_{14}; R) g_3(\mathbf{r}_{23}, \mathbf{r}_{24}) d\mathbf{r}_2 d\mathbf{r}_3 d\mathbf{r}_4 + \cdots \end{aligned}$$

$= 1 -$ 

$+$ 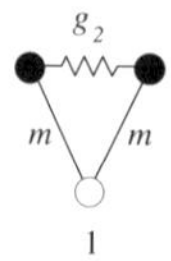

$-$ 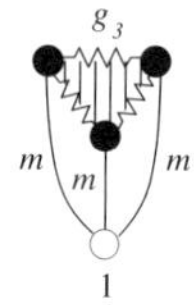

$+ \cdots,$ (4.24)

where we have again changed variables from $\boldsymbol{x}, \boldsymbol{r}_1, \ldots, \boldsymbol{r}_n$ to $\boldsymbol{r}_1, \boldsymbol{r}_2, \ldots, \boldsymbol{r}_{n+1}$. Here a *solid line* is an m-bond, a *wavy line* is a g_2-bond, and the *cross-hatched triangle with wavy edges* is a g_3-bond. However, since

$$m(r_{ij}) = 0 \quad \text{for} \quad r_{ij} > R, \quad 1 \le i < j \le n, \tag{4.25}$$

while the condition (4.23) must be maintained, any term containing the subdiagram

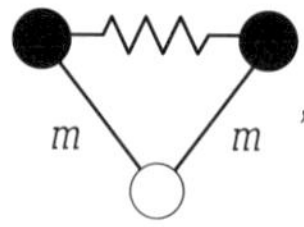

, (4.26)

i.e., terms containing g_n for $n \ge 3$, will be identically zero. Therefore,

$$\begin{aligned} \phi_1 &= 1 - \rho \int m(r_{12}; R) d\boldsymbol{r}_2 \\ &= 1 - \rho v_1, \end{aligned}$$

which, of course, agrees with (4.21).

Similarly, since any term in (4.9) involving the subdiagram

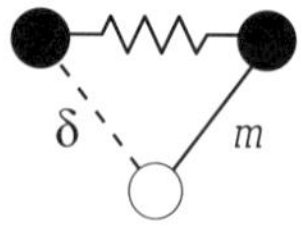

is identically zero, we have

$$s = \rho \int \delta(R - r_{12}) d\boldsymbol{r}_2,$$

which yields (4.22). Here a dashed line is a δ-bond.

Cherry-Pit Model for Arbitrary λ

For arbitrary λ and $d \ge 2$, it is not possible to evaluate S_1 exactly for all ϕ_2, since the g_n are not known exactly. However, excellent approximations can be obtained in these instances, as explained in Chapter 5.

4.2 Canonical Correlation Function H_n

The key idea employed by Torquato (1986c) to define and derive series representations of the canonical n-point correlation function H_n is the *available space* and *available surface* to a "test" particle that is inserted into the system. The available space and surface concepts go back to Boltzmann (1898); see also Stell (1966), Torquato and Stell (1983a), and Stell (1985).

4.2.1 Definitions

Consider inserting a test particle of radius b into an inhomogeneous system of N inclusions of radius R whose positions $\mathbf{r}^N$ are known statistically according to the specific N-body probability density function $P_N(\mathbf{r}^N)$. We observed in Section 4.1 that for a point test particle ($b = 0$) that is impenetrable to the N inclusions, the available space to it when it is inserted into such a system is the matrix space, as shown in Figure 4.3. The available space to a test particle when $b > 0$ (which is impenetrable to the N inclusions) is found by circumscribing *exclusion spheres* of radius $a = b + R$ around each inclusion. In general, these exclusion spheres can overlap, and hence the available space is the complex space exterior to the exclusion spheres as depicted in Figure 4.3. For a point test particle ($b = 0$), the exclusion spheres can actually be made smaller than the inclusions (i.e., $0 \le a < R$) by generally allowing them to penetrate the inclusions.

Now imagine adding p test particles to the system of N identical spherical inclusions such that $p \ll N$. It is natural to associate with the ith test particle a subdivision of the space $\mathcal{V} \in \Re^d$ occupied by the random medium into two disjoint regions: the available space D_i to the ith test particle (i.e., the space outside N spheres of radius $a_i = b_i + R$ centered at $\mathbf{r}^N$) and the complementary unavailable space D_i^*. Thus, $\mathcal{V} = D_i \cup D_i^*$. Furthermore, we let $\mathcal{S}_i$ denote the available surface between D_i and D_i^*. We define the indicator functions for the available space and surface, respectively, to the ith test particle as

$$\mathcal{I}(\mathbf{x}; a_i) = \begin{cases} 1, & \text{if } \mathbf{x} \in D_i, \\ 0, & \text{otherwise}, \end{cases} \tag{4.27}$$

and

$$\mathcal{M}(\mathbf{x}; a_i) = |\nabla \mathcal{I}(\mathbf{x}; a_i)|. \tag{4.28}$$

Since $p \ll N$, the test particles have no influence on one another.

For the system at hand, the indicator function for D_i can be related to the positions of the inclusions $\mathbf{r}^N$ in the following manner:

$$\begin{aligned} \mathcal{I}(\mathbf{x}; a_i) &= \prod_{j=1}^{N} [1 - m(|\mathbf{x} - \mathbf{r}_j|; a_i)] \\ &= 1 - \sum_{j=1}^{N} m(|\mathbf{x} - \mathbf{r}_j|; a_i) + \sum_{j<k}^{N} m(|\mathbf{x} - \mathbf{r}_j|; a_i) m(|\mathbf{x} - \mathbf{r}_k|; a_i) - \cdots, \end{aligned} \tag{4.29}$$

where

$$m(r; a) = \Theta(a - r) = \begin{cases} 1, & r \le a, \\ 0, & r > a, \end{cases} \tag{4.30}$$

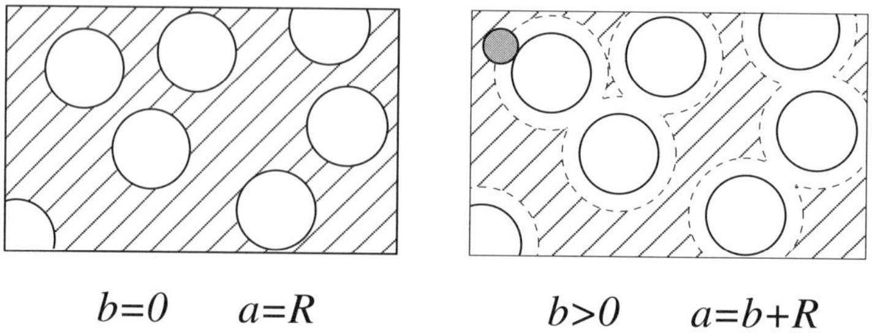

Figure 4.3 The same system of identical spheres of radius R is shown in both panels. However, the available spaces (cross-hatched regions) to a point test particle (left panel) and a nonzero-sized test particle (right panel) are clearly different. Here b and $a = b + R$ are the radii of the test particle and exclusion sphere, respectively.

is the *exclusion sphere indicator function*. Similarly, we have that the indicator function for the available surface $\mathcal{S}_i$ is given by

$$\begin{aligned}\mathcal{M}(\boldsymbol{x}; a_i) &= -\frac{\partial \mathcal{I}(\boldsymbol{x}; a_i)}{\partial a_i} \\ &= \sum_{j=1}^{N} \delta(|\boldsymbol{x} - \boldsymbol{r}_j| - a_i) - \sum_{j<k}^{N} \delta(|\boldsymbol{x} - \boldsymbol{r}_j| - a_i) m(|\boldsymbol{x} - \boldsymbol{r}_k|; a_i) \\ &\quad - \sum_{j<k}^{N} \delta(|\boldsymbol{x} - \boldsymbol{r}_k| - a_i) m(|\boldsymbol{x} - \boldsymbol{r}_j|; a_i) + \cdots, \end{aligned} \tag{4.31}$$

where $\delta(x)$ is the Dirac delta function. Clearly, $\mathcal{M}(\boldsymbol{x}; a_i)$ is a function that is zero everywhere except when $\boldsymbol{x}$ describes a position on $\mathcal{S}_i$. Equations (4.29) and (4.31) generalize the corresponding relations derived by Torquato and Stell (1982), and by Torquato and Stell (1984) and Chiew and Glandt (1984), respectively, for the special case in which all $b_i = 0$ and all $a_i = R$.

The ensemble averages of $\mathcal{I}(\boldsymbol{x}; a_i)$ and $\mathcal{M}(\boldsymbol{x}; a_i)$ are simply the volume fraction $S_1(\boldsymbol{x}; a_i) = \phi_1(\boldsymbol{x}; a_i)$ and specific surface $s(\boldsymbol{x}; a_i)$ associated with the available space $\mathcal{D}_i$ and the available surface $\mathcal{S}_i$, respectively. It is only when all $b_i = 0$ and all $a_i = R$ that these two averages are, respectively, equal to the usual position-dependent matrix volume fraction $S_1(\boldsymbol{x}; R) = \phi_1(\boldsymbol{x}; R)$ and the specific surface of the particle–matrix interface $s(\boldsymbol{x}; R)$.

We now define a more general n-point function G_n in terms of available-space indicator functions for the case of a mixture of n test particles of radii $b_1, \ldots, b_n$ at positions $\boldsymbol{x}^n \equiv \{\boldsymbol{x}_1, \boldsymbol{x}_2, \ldots, \boldsymbol{x}_n\}$, and N identical spherical inclusions of radius R:

$$
\begin{aligned}
G_n(\boldsymbol{x}^n; \varnothing) &= \left\langle \prod_{i=1}^{n} \mathcal{I}(\boldsymbol{x}_i; a_i) \right\rangle \\
&= \int \left[\prod_{i=1}^{n} \mathcal{I}(\boldsymbol{x}_i; a_i) \right] P_N(\boldsymbol{r}^N) d\boldsymbol{r}^N \\
&= \text{Probability that a test particle of radius } b_1 \text{ centered at } \boldsymbol{x}_1 \\
&\quad \text{lies in } D_1, \text{ another test particle of radius } b_2 \text{ centered at} \\
&\quad \boldsymbol{x}_2 \text{ lies in } D_2, \ldots, \text{ and another test particle of radius } b_n \\
&\quad \text{centered at } \boldsymbol{x}_n \text{ lies in } D_n. \\
&= \text{Probability that } n \text{ spheres of radii } a_1, \ldots, a_n \text{ centered at} \\
&\quad \text{positions } \boldsymbol{x}^n, \text{ respectively, are empty of inclusion centers.}
\end{aligned}
\tag{4.32}
$$

It is immediately clear that when all of the test particles shrink to zero such that $a_i \to R$ for all i, $G_n(\boldsymbol{x}^n; \varnothing) \to S_n(\boldsymbol{x}^n)$, where $S_n(\boldsymbol{x}^n)$ is the n-point probability function (defined generally in Section 2.2) for the matrix phase of sphere systems. Thus, the correlation function $G_n(\boldsymbol{x}^n; \varnothing)$ generalizes the quantity $S_n(\boldsymbol{x}^n)$ for sphere systems.

We are interested in defining the still more general n-point correlation function $G_n(\boldsymbol{x}^p; \boldsymbol{r}^q)$ in which p of the n points refer to the positions $\boldsymbol{x}^n$ of the test particles, and the remaining $q = n - p$ points refer to the positions $\boldsymbol{r}^q \equiv \{\boldsymbol{r}_1, \boldsymbol{r}_2, \ldots, \boldsymbol{r}_q\}$ of the inclusions. Given (4.32), it follows that

$$
\begin{aligned}
G_n(\boldsymbol{x}^p; \boldsymbol{r}^q) &= \frac{N!}{(N-q)!} \int \left[\prod_{i=1}^{p} \mathcal{I}(\boldsymbol{x}_i; a_i) \right] P_N(\boldsymbol{r}^N) d\boldsymbol{r}^{N-q} \\
&= \text{Probability that a test particle of radius } b_1 \text{ centered at} \\
&\quad \boldsymbol{x}_1 \text{ lies in } D_1, \text{ another test particle of radius } b_2 \text{ centered} \\
&\quad \text{at } \boldsymbol{x}_2 \text{ lies in } D_2, \ldots, \text{ and another test particle of radius} \\
&\quad b_p \text{ centered at } \boldsymbol{x}_p \text{ lies in } D_p, \text{ and of finding the center of} \\
&\quad \text{a sphere in volume element } d\boldsymbol{r}_1 \text{ about } \boldsymbol{r}_1, \text{ the center of} \\
&\quad \text{another sphere in volume element } d\boldsymbol{r}_2 \text{ about } \boldsymbol{r}_2, \ldots, \text{ and} \\
&\quad \text{the center of another sphere in volume element } d\boldsymbol{r}_q \text{ about} \\
&\quad \boldsymbol{r}_q, \text{ where } n = p + q.
\end{aligned}
\tag{4.33}
$$

In contrast to $G_n(\boldsymbol{x}^n; \varnothing)$ defined by (4.32), $G_n(\boldsymbol{x}^p; \boldsymbol{r}^q)$ is not a pure probability function, but rather is a hybrid quantity: It is a probability function with respect to the positions $\boldsymbol{x}^p$ and, when properly scaled, is a probability density function with respect to the positions $\boldsymbol{r}^q$ of the q inclusions. The dependence on $\boldsymbol{r}^q$ arises because the integrations are over the positions of all of the inclusions that are not included in the subset of the q inclusions. The factor $N!/(N-q)!$ in (4.34) accounts for the indistinguishability of the inclusions. Note that for the case $q = 0$, (4.34) reduces to (4.32). Moreover, in the instance $p = 0$, $G_n(\varnothing; \boldsymbol{r}^n)$ is identical to the n-particle probability density $\rho_n(\boldsymbol{r}^n)$ [i.e., (4.34) reduces to (3.7)].

We are now in the position of being able to define the canonical n-point correlation function H_n. [This quantity was originally called the n-point "distribution" function (Torquato 1986c).] As we stated in the introduction of this chapter, H_n provides a unified means of characterizing the structure of the random medium. Specifically, H_n is defined as the following derivatives of the aforementioned function $G_n(\boldsymbol{x}^p; \boldsymbol{r}^q)$:

$$\begin{aligned} H_n(\boldsymbol{x}^m; \boldsymbol{x}^{p-m}; \boldsymbol{r}^q) &= (-1)^m \frac{\partial}{\partial a_1} \cdots \frac{\partial}{\partial a_m} G_n(\boldsymbol{x}^p; \boldsymbol{r}^q) \\ &= \frac{N!}{(N-q)!} \int \left[\prod_{i=1}^{m} \mathcal{M}(\boldsymbol{x}_i; a_i) \right] \left[\prod_{i=m+1}^{p} \mathcal{I}(\boldsymbol{x}_i; a_i) \right] P_N(\boldsymbol{r}^N) d\boldsymbol{r}^{N-q}, \end{aligned} \quad (4.34)$$

where R is held fixed and $n = p + q$. As a result of the appearance of the indicator functions for the available surfaces $\mathcal{S}_1, \mathcal{S}_2, \ldots, \mathcal{S}_m$ in (4.34), the first set of arguments $\boldsymbol{x}^m$ of H_n refers to the center $\boldsymbol{x}_1$ of a test particle of radius b_1 being on $\mathcal{S}_1$, the center $\boldsymbol{x}_2$ of another test particle of radius b_2 being on $\mathcal{S}_2, \ldots$, and the center $\boldsymbol{x}_m$ of another test particle of radius b_m being on $\mathcal{S}_m$. In other words, the function H_n, unlike the less general quantity G_n, contains interfacial information about the available surfaces associated with the first m test particles. The second set of arguments $\boldsymbol{x}^{p-m} \equiv \{\boldsymbol{x}_{m+1}, \boldsymbol{x}_{m+2}, \ldots, \boldsymbol{x}_p\}$ of H_n refers to the centers of the test particles of radii $b_{m+1}, b_{m+2}, \ldots, b_p$ being in their respective available spaces $D_{m+1}, D_{m+2}, \ldots, D_p$. The third set of arguments $\boldsymbol{r}^q$ refers to a particular configuration of the centers of any q inclusions. It is clear that G_n is a special case of H_n, i.e., $H_n(\boldsymbol{x}^m; \boldsymbol{x}^{p-m}; \boldsymbol{r}^q) = G_n(\boldsymbol{x}^p; \boldsymbol{r}^q)$ when $m = 0$. Note that since all of the quantities in the integrand of (4.34) are nonnegative, H_n is also nonnegative, i.e.,

$$H_n(\boldsymbol{x}^m; \boldsymbol{x}^{p-m}; \boldsymbol{r}^q) \geq 0, \qquad \text{for all } \boldsymbol{x}^p \text{ and } \boldsymbol{r}^q \text{ in } \mathcal{V}. \quad (4.35)$$

As with any surface correlation derived from an indicator function of the type (4.31), the interfacial component of H_n (i.e., when $m \neq 0$) does not have a probabilistic interpretation. However, it can be interpreted probabilistically by considering infinitesimal dilations of the exclusion spheres, as discussed in Section 4.1 (see also Figure 4.2), i.e., $H_n(\boldsymbol{x}^n; \emptyset; \emptyset)$ is proportional to the probability that the points $\boldsymbol{x}^n$ lie in the dilated regions surrounding the interfaces $\mathcal{S}_1, \mathcal{S}_2, \ldots, \mathcal{S}_m$, respectively.

Since H_n has the character of a joint probability density function with respect to the arguments $\boldsymbol{r}^q$, one has the normalization condition

$$\int H_n(\boldsymbol{x}^m; \boldsymbol{x}^{p-m}; \boldsymbol{r}^q) d\boldsymbol{r}^q = \frac{N!}{(N-q)!} H_p(\boldsymbol{x}^m; \boldsymbol{x}^{p-m}; \emptyset), \quad (4.36)$$

where $q = n - p$. This is obtained by integrating the second line of (4.34) over the positions $\boldsymbol{r}^q$. We see that the n-point canonical function $H_n(\boldsymbol{x}^m; \boldsymbol{x}^{p-m}; \boldsymbol{r}^q)$ divided by the right-hand side of (4.36) is indeed a probability density function, since it is nonnegative and normalizes to unity.

As we will see in Section 4.4, from the single function H_n, one can obtain all of the various types of correlation functions that arise in rigorous structure-property relations (by setting $b_i = 0$ and $a_i = R$, $i = 1, \ldots, p$) and their generalizations. For statistically homogeneous media, $P_N(\boldsymbol{r}^N)$ is translationally invariant, implying that the quantities G_n and H_n are functions only of relative positions.

4.2.2 Asymptotic Properties

Here we develop certain asymptotic properties of G_n and H_n. It suffices to focus our attention on H_n alone, since G_n is a special case of H_n.

Let the set $(\boldsymbol{x}^m; \boldsymbol{x}^{p-m}; \boldsymbol{r}^q)$ be partitioned into $L = L(\alpha)$ disjoint subsets, where $\{\alpha\} = \{\alpha_1; \alpha_2; \ldots; \alpha_L\}$ is any partition of the aforementioned set, $\{\alpha_i\}$ is the ith subset, and n_i is the number of elements in $\{\alpha_i\}$. The subset $\{\alpha_i\}$ is in turn partitioned into three subsets $\{\alpha_i^{(1)}; \alpha_i^{(2)}; \alpha_i^{(3)}\}$, where $\{\alpha_i^{(1)}\}$ contains elements of the subset $\boldsymbol{x}^m$, $\{\alpha_i^{(2)}\}$ contains elements of the subset $\boldsymbol{x}^{p-m}$, and $\{\alpha_i^{(3)}\}$ contains elements of the subset $\boldsymbol{r}^q$. Let $n_i^{(j)}$ $(j = 1, 2, 3)$ denote the number of elements in $\{\alpha_i^{(j)}\}$, so that $\sum_{i=1}^{L} n_i^{(1)} = m$, $\sum_{i=1}^{L} n_i^{(2)} = p - m$, and $\sum_{i=1}^{L} n_i^{(3)} = q$. Let all of the relative distances between the n_i elements of the subset $\{\alpha_i\}$ remain bounded, and let F_{α_i} be the polyhedron with n_i vertices located at the positions associated with the subset $\{\alpha_i\}$. We denote the centroid of the F_{α_i} by $\boldsymbol{R}_i$. If the system possesses no long-range order (see Section 2.2.4), then H_n factorizes into L products as follows:

$$\lim_{\text{all } R_{jk}\to\infty} H_n(\boldsymbol{x}^m; \boldsymbol{x}^{p-m}; \boldsymbol{r}^q) = \prod_{i=1}^{L} H_{n_i}\left(\alpha_i^{(1)}; \alpha_i^{(2)}; \alpha_i^{(3)}\right), \tag{4.37}$$

where $R_{jk} = |\boldsymbol{R}_k - \boldsymbol{R}_j|$ is the relative distance between the centroids of F_{α_j} and F_{α_k}, such that $1 \le j < k \le L$.

From relation (4.34), it follows that when $k+1$ of the elements (points) of $\boldsymbol{x}^m$ coincide and $l+1$ of the elements (points) of $\boldsymbol{x}^{p-m}$ coincide, then

$$H_n(\boldsymbol{x}^m; \boldsymbol{x}^{p-m}; \boldsymbol{r}^q) \to H_{n-k-l}(\beta_1; \beta_2; \boldsymbol{r}^q), \tag{4.38}$$

where $\{\beta_1\}$ and $\{\beta_2\}$ denote the subsets associated with the elements of $\boldsymbol{x}^m$ and $\boldsymbol{x}^{p-m}$, respectively, that do not coincide.

Note that in the cherry-pit model [cf. (3.27)], $H_n(\boldsymbol{x}^m; \boldsymbol{x}^{p-m}; \boldsymbol{r}^q)$ is identically zero for certain $\boldsymbol{r}^q$. In particular, for an impenetrability parameter λ, we have

$$H_n(\boldsymbol{x}^m; \boldsymbol{x}^{p-m}; \boldsymbol{r}^q) = 0 \quad \text{if} \quad |\boldsymbol{r}_i - \boldsymbol{r}_j| < 2\lambda R \tag{4.39}$$

for $1 \le i < j \le q$.

4.3 Series Representations of H_n

Following Torquato (1986c), we now derive series representations of both G_n and H_n for the inhomogeneous particle mixture (inclusions and test particles) described in the previous section. Once the series representation of the G_n is derived, we obtain the corresponding series for the H_n from definition (4.34). Recall that the inclusions are generally distributed with an arbitrary degree of penetrability. We proceed by obtaining two equivalent but diagrammatically different representations of the G_n and the H_n. The key idea behind these derivations is the judicious expansion of the product $\Pi_{i=1}^{p} \mathcal{I}(\boldsymbol{x}_i; a_i)$ appearing in the definition (4.34).

4.3.1 Mayer Representation

We begin by recognizing that the product of the indicator functions $\Pi_{i=1}^{p}\mathcal{I}(\mathbf{x}_i; a_i)$ appearing in (4.34) may also be rewritten, with the aid of (4.29), in the following manner:

$$\begin{aligned}\prod_{i=1}^{p}\mathcal{I}(\mathbf{x}_i; a_i) &= \left[\prod_{l=1}^{q}\prod_{k=1}^{p} e(y_{kl}; a_k)\right]\prod_{l=q+1}^{N}\prod_{k=1}^{p}\left[1 - m(y_{kl}; a_k)\right] \\ &= \prod_{l=1}^{q}\prod_{k=1}^{p} e(y_{kl}; a_k)\left[1 - \sum_{i=q+1}^{N} m^{(p)}(\mathbf{x}^p; \mathbf{r}_i)\right. \\ &\qquad \left. + \sum_{\substack{i=q+1 \\ i<j}}^{N} m^{(p)}(\mathbf{x}^p; \mathbf{r}_i) m^{(p)}(\mathbf{x}^p; \mathbf{r}_j) - \cdots\right],\end{aligned} \tag{4.40}$$

where

$$m^{(p)}(\mathbf{x}^p; \mathbf{r}_j) = 1 - \prod_{i=1}^{p}[1 - m(y_{ij}; a_i)], \tag{4.41}$$

$$e(r; a) = 1 - m(r; a) = \Theta(r - a) = \begin{cases} 0, & r < a, \\ 1, & r \geq a, \end{cases} \tag{4.42}$$

and $y_{ij} = |\mathbf{x}_i - \mathbf{r}_j|$.

For statistically *inhomogeneous* media, relations (4.34) and (4.40) lead to

$$\begin{aligned}\frac{G_n(\mathbf{x}^p; \mathbf{r}^q)}{\prod_{l=1}^{q}\prod_{k=1}^{p} e(y_{kl}; a_k)} &= \rho_q(\mathbf{r}^q) - \int m^{(p)}(\mathbf{x}^p; \mathbf{r}_{q+1})\rho_{q+1}(\mathbf{r}^{q+1})d\mathbf{r}_{q+1} \\ &+ \frac{1}{2!}\int m^{(p)}(\mathbf{x}^p; \mathbf{r}_{q+1}) m^{(p)}(\mathbf{x}^p; \mathbf{r}_{q+2})\rho_{q+2}(\mathbf{r}^{q+2})d\mathbf{r}_{q+1}d\mathbf{r}_{q+2} - \cdots.\end{aligned}$$

In general, these relations yield

$$G_n(\mathbf{x}^p; \mathbf{r}^q) = \sum_{s=0}^{\infty}(-1)^s G_n^{(s)}(\mathbf{x}^p; \mathbf{r}^q), \qquad n = p + q, \tag{4.43}$$

where

$$G_n^{(s)}(\mathbf{x}^p; \mathbf{r}^q) = \frac{\prod_{l=1}^{q}\prod_{k=1}^{p} e(y_{kl}; a_k)}{s!}\int \rho_{q+s}(\mathbf{r}^{q+s})\prod_{j=q+1}^{q+s} m^{(p)}(\mathbf{x}^p; \mathbf{r}_j)d\mathbf{r}_j \tag{4.44}$$

for $s \geq 1$, and we define

$$G_n^{(0)}(\mathbf{x}^p; \mathbf{r}^q) \equiv \rho_q(\mathbf{r}^q)\prod_{l=1}^{q}\prod_{k=1}^{p} e(y_{kl}; a_k). \tag{4.45}$$

Expression (4.43) applies to general nonequilibrium ensembles for the multicomponent mixture described earlier. In the context of liquid-state theory, an expression of this type was derived by (Mayer 1947) for a *single-component* fluid in thermal equilibrium. Accordingly, we refer to (4.43) as the Mayer representation of the G_n. For an equilibrium ensemble of particles, (4.43) could have been derived using the hierarchies of Baer and Lebowitz (1964) for mixtures under infinite dilution of the test-particle "species" (Torquato 1986c). The connection between certain statistical-mechanical hierarchies [Kirkwood and Salsburg (1953) and Mayer and Montroll (1941) equations] and the n-point probability functions S_n that arise in random-media theory was first shown by Torquato and Stell (1983a). Note that when $q = 0$ and $a_i = R$ $(i = 1, \ldots, p)$, the Mayer representation of the G_n becomes the Mayer–Montroll equations for the S_n derived by Torquato and Stell. Moreover, when $p = 1$, $a_1 = R$, and $q > 0$, (4.43) reduces to the Kirkwood–Salsburg equations for the point/q-particle correlation functions derived by Torquato (1986b).

Combination of (4.34) and (4.43) yields what we refer to as the Mayer representation of the H_n:

$$H_n(\boldsymbol{x}^m; \boldsymbol{x}^{p-m}; \boldsymbol{r}^q) = \sum_{s=0}^{\infty}(-1)^s H_n^{(s)}(\boldsymbol{x}^m; \boldsymbol{x}^{p-m}; \boldsymbol{r}^q), \qquad n = p + q, \tag{4.46}$$

where

$$H_n^{(s)}(\boldsymbol{x}^m; \boldsymbol{x}^{p-m}; \boldsymbol{r}^q) = (-1)^m \frac{\partial}{\partial a_1} \cdots \frac{\partial}{\partial a_m} G_n^{(s)}(\boldsymbol{x}^p; \boldsymbol{r}^q), \tag{4.47}$$

and $G_n^{(s)}$ is given by (4.44).

4.3.2 Kirkwood–Salsburg Representation

Without derivation, we state here an alternative but *equivalent* expression for the canonical function H_n obtained by Torquato (1986c) that he referred to as the Kirkwood–Salsburg representation:

$$H_n(\boldsymbol{x}^m; \boldsymbol{x}^{p-m}; \boldsymbol{r}^q) = \sum_{s=0}^{\infty}(-1)^s H_n^{(s)}(\boldsymbol{x}^m; \boldsymbol{x}^{p-m}; \boldsymbol{r}^p), \qquad n = p + q, \tag{4.48}$$

where

$$H_n^{(s)}(\boldsymbol{x}^m; \boldsymbol{x}^{p-m}; \boldsymbol{r}^q) = (-1)^m \frac{\partial}{\partial a_1} \cdots \frac{\partial}{\partial a_m} G_n^{(s)}(\boldsymbol{x}^p; \boldsymbol{r}^q), \tag{4.49}$$

$$G_n^{(s)}(\boldsymbol{x}^p; \boldsymbol{r}^q) = \frac{\prod_{k=1}^{q} e(y_{1k}; a_1)}{s!} \int G_{n+s-1}(\boldsymbol{x}^{p-1}; \boldsymbol{r}^{q+s}) \prod_{j=q+1}^{q+s} m(y_{1j}; a_1) d\boldsymbol{r}_j \tag{4.50}$$

for $s \geq 1$, and we define

$$G_n^{(0)}(\boldsymbol{x}^p; \boldsymbol{r}^q) \equiv G_{n-1}(\boldsymbol{x}^{p-1}; \boldsymbol{r}^q) \prod_{k=1}^{q} e(y_{1k}; a_1). \tag{4.51}$$

Moreover, the corresponding expression for G_n is given by

$$G_n(\mathbf{x}^p; \mathbf{r}^q) = \sum_{s=0}^{\infty}(-1)^s G_n^{(s)}(\mathbf{x}^p; \mathbf{r}^q), \qquad n = p+q, \tag{4.52}$$

Note that unlike the sth term of the Mayer representation (4.44), which involves integrands containing the $(q+s)$-particle probability density associated with the inclusions $\rho_{q+s}(\mathbf{r}^{q+s})$, the corresponding term in (4.50) involves integrands containing the quantity G_{n+s-1}. Since the ρ_n are, in principle, known functions, the Mayer representation (4.44) will prove to be superior to the Kirkwood–Salsburg representation for the purposes of evaluating the G_n exactly. However, the latter representation provides a means of bounding the G_n that is more powerful than the former.

4.3.3 Bounding Properties

It is well known that hierarchies of the type given by (4.43) and (4.52) allow one to bound rigorously the corresponding correlation functions (Baer and Lebowitz 1964). This property was exploited by Torquato and Stell (1983a) to yield successive bounds on the S_n.

Torquato (1986c) obtained corresponding bounds on the G_n and H_n, which we now summarize. In particular, we have

$$G_n \left.\begin{matrix} \geq \\ \leq \end{matrix}\right\} Q_n^{(\ell)} \quad \begin{cases} \text{for } \ell \text{ odd}, \\ \text{for } \ell \text{ even}, \end{cases} \tag{4.53}$$

where the $Q_n^{(\ell)}$ is the partial sum

$$Q_n^{(\ell)} = \sum_{s=0}^{\ell}(-1)^s G_n^{(s)}. \tag{4.54}$$

Here the $G_n^{(s)}$ are given by either relation (4.44) or (4.50).

Similarly, from expressions (4.46) and (4.48), we have for even m

$$H_n \left.\begin{matrix} \geq \\ \leq \end{matrix}\right\} W_n^{(\ell)} \quad \begin{cases} \text{for } \ell \text{ odd}, \\ \text{for } \ell \text{ even}, \end{cases} \tag{4.55}$$

and for odd m

$$H_n \left.\begin{matrix} \geq \\ \leq \end{matrix}\right\} W_n^{(\ell)} \quad \begin{cases} \text{for } \ell \text{ even}, \\ \text{for } \ell \text{ odd}, \end{cases} \tag{4.56}$$

where the $W_n^{(\ell)}$ are the partial sums

$$W_n^{(\ell)} = \sum_{s=0}^{\ell}(-1)^s H_n^{(s)}. \tag{4.57}$$

The $H_n^{(s)}$ are given by either relation (4.47) or (4.49).

The fact that we have a succession of upper and lower bounds on G_n and H_n does not necessarily imply that $G_n^{(s)}$ and $H_n^{(s)}$ uniformly decrease in absolute value as s increases, i.e., that the bounds are successively better and better. However, there will be instances in which the series for G_n and H_n will be truncated, so that for certain distributions of the inclusions, there exists some $k > 0$ such that $G_n^{(s)} = H_n^{(s)} = 0$ for all $s > k$.

As an example, consider G_n for a system consisting of p point test particles (where $a_i = R$ for all i) and N *totally impenetrable* spherical inclusions of radius R. For such a medium, the terms of the Kirkwood-Salsburg infinite series (4.52) are identically zero for all $s > 1$, i.e.,

$$G_n(\mathbf{x}^p; \mathbf{r}^q) = G_{n-1}(\mathbf{x}^{p-1}; \mathbf{r}^q) - \int G_n(\mathbf{x}^{p-1}; \mathbf{r}^{q+1}) m(y_{1,q+1}; R) d\mathbf{r}_1 \,. \tag{4.58}$$

This result follows from the fact that the product

$$m(y_{1,q+1}; R) m(y_{1,q+2}; R) G_{n+s-1}(\mathbf{x}^{p-1}; \mathbf{r}^{q+s}) \tag{4.59}$$

appearing in the sth term in (4.52) is identically zero for all $s > 1$, since $m(y; R) = 0$ for $y > R$ and $G_n(\mathbf{x}^p; \mathbf{r}^q) = 0$ for all $|\mathbf{r}_i - \mathbf{r}_j| < 2R$. Similarly, the terms of the Mayer series (4.43) are identically zero for all $s > p$, since the product

$$\rho_{q+s}(\mathbf{r}^{q+s}) \prod_{j=q+1}^{q+s} m^{(p)}(\mathbf{x}^p; \mathbf{r}_j) \tag{4.60}$$

appearing in the sth term in (4.43) is identically zero for all $s > p$, so that

$$G_n(\mathbf{x}^p; \mathbf{r}^q) = \sum_{s=0}^{p} (-1)^s G_n^{(s)}(\mathbf{x}^p; \mathbf{r}^q). \tag{4.61}$$

For such a system, the corresponding Mayer series for the H_n terminates after the same number of terms.

The Kirkwood–Salsburg representation provides a more powerful means of bounding the correlation functions than the Mayer representation. To illustrate this point, we examine the bounds on G_2 and G_3 for a mixture of point test particles (where $a_i = R$ for all i) and N *totally impenetrable* inclusions when $q = 1$. Below we define $e(y) \equiv e(y; R)$ and $m(y) \equiv m(y; R)$. When $n = 2$, the Kirkwood–Salsburg and Mayer representations are identical, and hence application of (4.53) yields

$$G_2(\mathbf{x}_1; \mathbf{r}_1) \le e(y_{11})\, \rho_1(\mathbf{r}_1) \tag{4.62}$$

and

$$G_2(\mathbf{x}_1; \mathbf{r}_1) = e(y_{11}) \Big[\rho_1(\mathbf{r}_1) - \int \rho_2(\mathbf{r}_1, \mathbf{r}_2) m(y_{12}) d\mathbf{r}_2 \Big]. \tag{4.63}$$

For homogeneous media, $\rho_1(\mathbf{r}_1)$ is simply the number density ρ. For $n = 3$, however, the two series representations are not diagrammatically identical. The inequalities (4.53) in the Kirkwood–Salsburg representation give

$$G_3(\mathbf{x}_1, \mathbf{x}_2; \mathbf{r}_1) \le e(y_{11})\, G_2(\mathbf{x}_2; \mathbf{r}_1) \tag{4.64}$$

and

$$G_3(\boldsymbol{x}_1, \boldsymbol{x}_2; \boldsymbol{r}_1) = e(y_{11})\Big[G_2(\boldsymbol{x}_2; \boldsymbol{r}_1) - \int G_3(\boldsymbol{x}_2; \boldsymbol{r}_1, \boldsymbol{r}_2) m(y_{12}) d\boldsymbol{r}_2\Big]. \tag{4.65}$$

On the other hand, the inequalities (4.53) in the Mayer representation yield

$$G_3(\boldsymbol{x}_1, \boldsymbol{x}_2; \boldsymbol{r}_1) \le e(y_{11})\, e(y_{21})\, \rho_1(\boldsymbol{r}_1), \tag{4.66}$$

$$G_3(\boldsymbol{x}_1, \boldsymbol{x}_2; \boldsymbol{r}_1) \ge e(y_{11})\, e(y_{21})\Big[\rho_1(\boldsymbol{r}_1) - \int \rho_2(\boldsymbol{r}_1, \boldsymbol{r}_2) m^{(2)}(\boldsymbol{x}_1, \boldsymbol{x}_2; \boldsymbol{r}_2) d\boldsymbol{r}_2\Big], \tag{4.67}$$

$$G_3(\boldsymbol{x}_1, \boldsymbol{x}_2; \boldsymbol{r}_1) = e(y_{11})\, e(y_{21})\Big[\rho_1(\boldsymbol{r}_1) - \int \rho_2(\boldsymbol{r}_1, \boldsymbol{r}_2) m^{(2)}(\boldsymbol{x}_1, \boldsymbol{x}_2; \boldsymbol{r}_2) d\boldsymbol{r}_2 + \frac{1}{2}\int \rho_3(\boldsymbol{r}_1, \boldsymbol{r}_2, \boldsymbol{r}_3) m^{(2)}(\boldsymbol{x}_1, \boldsymbol{x}_2; \boldsymbol{r}_2) m^{(2)}(\boldsymbol{x}_1, \boldsymbol{x}_2; \boldsymbol{r}_3) d\boldsymbol{r}_2 d\boldsymbol{r}_3\Big]. \tag{4.68}$$

In light of inequality (4.62), it is clear that (4.64) provides a better upper bound on G_3 than does (4.66). When $\boldsymbol{x}_1$ coincides with $\boldsymbol{x}_2$, relation (4.64), unlike (4.66), is exact. When $\boldsymbol{x}_1, \boldsymbol{x}_2$, and $\boldsymbol{r}_1$ are far apart from one another in a homogeneous system, then $G_2 \to \phi_1 \rho$ and $G_3 \to \phi_1^2 \rho$ [cf. (4.37)], and hence, the upper bound (4.64) although not exact in this case, is sharper than (4.66). The exact relation (4.65) obviously provides a better lower bound on G_3 than does inequality (4.67). In general, for any n, the Kirkwood–Salsburg, rather than the Mayer, representation will provide sharper bounds on the G_n or H_n.

4.4 Special Cases of H_n

Given the $\rho_n(\boldsymbol{r}^n)$, one can now compute H_n using the representation (4.46) or (4.48). Using these expressions for the canonical function H_n, we can now obtain any of the microstructural correlation functions that have arisen in the random-media literature (see Chapter 2) and their generalizations.

In many applications we will be interested in H_n in the limit that the radii of the test particles become zero so that $a_i = R$ for all i. For example, in this limit, the n-point matrix probability function is given by

$$S_n(\boldsymbol{x}^n) = \lim_{a_i \to R, \forall i} H_n(\emptyset; \boldsymbol{x}^n; \emptyset). \tag{4.69}$$

Similarly, the lower-order surface correlation functions are given by

$$s = \lim_{a_1 \to R} H_1(\boldsymbol{x}_1; \emptyset; \emptyset), \tag{4.70}$$

$$F_{sv}(\boldsymbol{x}_1, \boldsymbol{x}_2) = \lim_{a_i \to R, \forall i} H_2(\boldsymbol{x}_1; \boldsymbol{x}_2; \emptyset), \tag{4.71}$$

$$F_{ss}(\boldsymbol{x}_1, \boldsymbol{x}_2) = \lim_{a_i \to R, \forall i} H_2(\boldsymbol{x}_1, \boldsymbol{x}_2; \emptyset; \emptyset), \tag{4.72}$$

$$F_{sp}(\boldsymbol{x}_1, \boldsymbol{r}_1) = \lim_{a_1 \to R} H_2(\boldsymbol{x}_1; \emptyset; \boldsymbol{r}_1). \tag{4.73}$$

Three-point surface correlation functions (e.g., $F_{ssv}, F_{ssp}, F_{svp}$) and their n-point generalizations can be obtained from the H_n in a similar fashion. The point/q-particle correlation function is expressible as

$$G_n(\mathbf{x}_1; \mathbf{r}^q) = \lim_{a_1 \to R} H_n(\emptyset; \mathbf{x}_1; \mathbf{r}^q). \tag{4.74}$$

It is noteworthy that using the formula for H_n, one can also establish relationships between these different types of correlation functions (Torquato 1986a).

In some cases, the sizes of the test particles that one wishes to introduce in a random medium are not always negligible compared to the characteristic lengths of the phases, and hence the correlation functions will depend upon the *relative sizes* of the test particles and characteristic length scales of the phases. Such generalized quantities are of practical importance in understanding gel size-exclusion chromatography (Yau, Kirkland and Bly 1979) and transport of nonzero-sized particles in porous media (Torquato 1991c, Kim and Torquato 1992a). The general series representations of the H_n are clearly capable of characterizing the microstructure in such instances, since the b_i are, in general, *nonzero*.

In particular, to obtain the "void" nearest-neighbor functions defined in Section 2.8, one must consider the addition of one test particle of radius $b_1 = r - R$ (or $a_1 = r$). For example, the exclusion probability function and associated probability density functions are respectively given by

$$E_V(r) = H_1(\emptyset; \mathbf{x}_1; \emptyset), \tag{4.75}$$

$$H_V(r) = H_1(\mathbf{x}_1; \emptyset; \emptyset). \tag{4.76}$$

Moreover, the "particle" nearest-neighbor functions (Section 2.8) are also special cases of the canonical function H_n. Specifically, setting $b_1 = r - R$ (or $a_1 = r$), the exclusion probability is given by

$$E_P(r) = \lim_{|\mathbf{x}_1 - \mathbf{r}_1| \to 0} \frac{H_2(\emptyset; \mathbf{x}_1; \mathbf{r}_1)}{\rho_1(\mathbf{r}_1)}. \tag{4.77}$$

Using this expression and relation (2.95) yields the corresponding expression for the nearest-neighbor probability density $H_P(r)$.

For isotropic media, the lineal-path function $L(z) \equiv L^{(1)}(z)$ for the matrix phase (Section 2.4) is a special type of "void" exclusion probability function E_V (Lu and Torquato 1992a). Specifically, it is the probability that a randomly placed "test" *line segment* of length z lies in the matrix, which is equal to the probability of finding an excluded region Ω_E empty of inclusion centers. Thus, Ω_E is the volume excluded to a spherical inclusion of radius R by a line of length z, i.e., a d-dimensional spherocylinder of cylindrical length z and radius R with hemispherical caps of radius R on either end (see Figure 4.4). The exclusion region indicator function for the spherocylinder is

$$m(\mathbf{x}; z, R) = \begin{cases} 1, & \mathbf{x} \in \Omega_E(z, R) \\ 0, & \text{otherwise.} \end{cases} \tag{4.78}$$

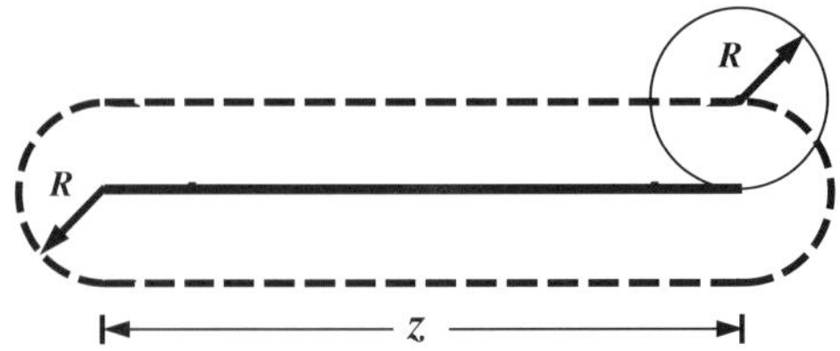

Figure 4.4 A spherocylinder of cylindrical length z and radius R with hemispherical caps of radius R is the volume excluded to a sphere center of radius R by a line of length z.

In summary, the exact series representation of $L(z)$ is obtained from the void exclusion probability function E_V (and thus H_1) by a simple replacement of the exclusion region indicator function by (4.78).

The specific series representations of all of the aforementioned functions for systems of monodisperse and polydisperse spheres will be given in Chapters 5 and 6, respectively. Evaluations of the microstructural functions for various sphere models will also be given there.

4.5 Polydispersivity

The formalism of Torquato (1986c) employed to obtain series representations of the canonical n-point correlation function H_n for monodisperse spheres of variable penetrability has been generalized to the polydisperse case by Lu and Torquato (1991). For purposes of generality, one considers spheres with a continuous distribution in radius R characterized by a probability density function $f(R)$ that normalizes to unity:

$$\int_0^\infty f(R)dR = 1. \tag{4.79}$$

One can obtain corresponding results for spheres with M different sizes from the continuous case by letting

$$f(R) = \sum_{i=1}^{M} \frac{\rho_i}{\rho}\delta(R - R_i), \tag{4.80}$$

where ρ_i and R_i are number density and radius of type-i particles, respectively, and ρ is the *total number density*.

Since the derivation of the series for H_n for polydisperse spheres follows closely the one for monodisperse spheres, we will just state the main result here:

$$H_n(\boldsymbol{x}^m; \boldsymbol{x}^{p-m}; \boldsymbol{r}^q) = \sum_{s=0}^{\infty}(-1)^s H_n^{(s)}(\boldsymbol{x}^m; \boldsymbol{x}^{p-m}; \boldsymbol{r}^q), \quad n = p + q, \tag{4.81}$$

where

$$H_n^{(s)}(\boldsymbol{x}^m; \boldsymbol{x}^{p-m}; \boldsymbol{r}^q) = (-1)^m \frac{\partial}{\partial b_1} \cdots \frac{\partial}{\partial b_m} G_n^{(s)}(\boldsymbol{x}^p; \boldsymbol{r}^q), \tag{4.82}$$

holding the inclusion radii fixed, and, in the Mayer representation,

$$G_n^{(s)}(\boldsymbol{x}^p; \boldsymbol{r}^q) = \frac{\prod_{l=1}^{q} \prod_{k=1}^{p} e(y_{kl}; a_k^{(l)})}{s!} \int \cdots \int dR_1 \cdots dR_{q+s} f(R_1) \cdots f(R_{q+s})$$
$$\times \rho_{q+s}(\boldsymbol{r}^{q+s}; R_1, \ldots, R_{q+s}) \prod_{j=q+1}^{q+s} m^{(p)}(\boldsymbol{x}^p; \boldsymbol{r}_j) d\boldsymbol{r}_j \tag{4.83}$$

and

$$m^{(p)}(\boldsymbol{x}^p; \boldsymbol{r}_j) = 1 - \prod_{i=1}^{p} [1 - m(y_{ij}; a_i^{(j)})]. \tag{4.84}$$

Here $f(R_1) \cdots f(R_{q+s}) \rho_{q+s}(\boldsymbol{r}^{q+s}; R_1, \ldots, R_{q+s})$ is the probability density associated with finding an inclusion with radius R_1 at $\boldsymbol{r}_1$, another inclusion with radius R_2 at $\boldsymbol{r}_2$, etc. Moreover, $a_i^{(j)} = b_i + R_j$, where b_i is the radius of the ith test particle and R_j is the radius of the inclusion at $\boldsymbol{r}_j$. Since the differential operations in (4.82) are performed while holding R_j fixed, $\partial b_i = \partial a_i^{(j)}$, and thus (4.82) indeed yields (4.47) for monodisperse inclusions.

We can again obtain series representations of any of the correlation functions that have arisen in the random-media literature (see Chapter 2) by considering special limits of the H_n. For example, we have

$$S_n(\boldsymbol{x}^n) = \lim_{a_i^{(j)} \to R_j, \forall i} H_n(\emptyset; \boldsymbol{x}^n; \emptyset), \tag{4.85}$$

$$s = \lim_{a_1 \to R} H_1(\boldsymbol{x}_1; \emptyset; \emptyset), \tag{4.86}$$

$$F_{sv}(\boldsymbol{x}_1, \boldsymbol{x}_2) = \lim_{a_i^{(j)} \to R_j, \forall i} H_2(\boldsymbol{x}_1; \boldsymbol{x}_2; \emptyset), \tag{4.87}$$

$$F_{ss}(\boldsymbol{x}_1, \boldsymbol{x}_2) = \lim_{a_i^{(j)} \to R_j, \forall i} H_2(\boldsymbol{x}_1, \boldsymbol{x}_2; \emptyset; \emptyset), \tag{4.88}$$

$$G_n(\boldsymbol{x}_1; \boldsymbol{r}^q) = \lim_{a_1^{(j)} \to R_j} H_n(\emptyset; \boldsymbol{x}_1; \boldsymbol{r}^q). \tag{4.89}$$

Importantly, the limiting processes indicated in all of the definitions above are to be taken before any averaging over the sphere radii. Moreover, the exclusion nearest-surface quantities are given by

$$e_V(r) = H_1(\emptyset; \boldsymbol{x}_1; \emptyset), \tag{4.90}$$

$$e_P(r) = \lim_{|\boldsymbol{x}_1 - \boldsymbol{r}_1| \to 0} \frac{H_2(\emptyset; \boldsymbol{x}_1; \boldsymbol{r}_1)}{\rho_1(\boldsymbol{r}_1)}, \tag{4.91}$$

where $b_1 = r$ (or $a_1^{(j)} = r + R_j$). As in the monodisperse case, the expression for $L(z)$ is obtained from the void exclusion function e_V (and thus H_1) when the exclusion region indicator function is taken to be given by (4.78).

4.6 Other Model Microstructures

The formal results obtained here are also easily extended to inclusions whose configuration is fully specified by center-of-mass coordinates (e.g., oriented ellipsoids, cubes, squares, etc.), by appropriately generalizing the indicator functions $\mathcal{I}(\boldsymbol{x})$, relation (4.27); and $\mathcal{M}(\boldsymbol{x})$, relation (4.28). Calculations of correlation functions for such statistically anisotropic models are presented in Chapter 7. The formalism also extends to nonparticulate systems, such as cell or lattice models, as described in Chapter 8.

CHAPTER 5

Monodisperse Spheres

In this chapter we will evaluate the series representations of the canonical function H_n, derived in the previous chapter, for various assemblies of identical (i.e., monodisperse) spheres of radius R. Such models are not as restrictive as one might initially surmise. For example, one can vary the connectedness of the particle phase (and therefore its percolation threshold) by allowing the spheres to interpenetrate one another in varying degrees. We saw in Chapter 3 that one extreme of this interpenetrable-sphere model is the case of spatially uncorrelated (i.e., Poisson distributed) spheres that we call fully penetrable (or overlapping) spheres. Overlapping spheres, at low sphere densities, are useful models of nonpercolating dispersions (see Figure 3.4). At high densities, overlapping spheres can be used to model consolidated media such as sandstones and sintered materials (Torquato 1986b). Figure 5.1 shows a distribution of identical overlapping disks at a very high density that resembles the sandstone depicted in Figure 1.3.

In the opposite extreme of totally impenetrable spheres, each sphere experiences an infinite repulsive force when the interparticle separation distance r is less than r_0, where $r_0 \geq 2R$, and an attractive force (in general) when $r \geq r_0$. This type of assembly is a useful model of unconsolidated media, such as fiber-reinforced materials in two dimensions (Torquato and Lado 1988a), colloidal dispersions (Russel et al. 1989), packed beds (Scheidegger 1974), and particulate composites (Torquato and Lado 1986). Figure 5.1 also illustrates a realization of a totally impenetrable disk system.

In the ensuing sections we will evaluate many of the different types of microstructural descriptors defined in Chapter 2 from series representation of the canonical n-point function H_n for the following statistically homogeneous systems of monodisperse particles:

- fully penetrable spheres

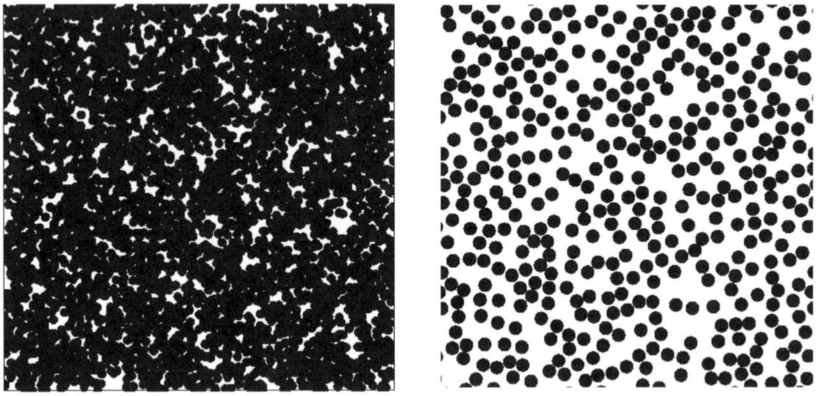

Figure 5.1 Left panel: A system of identical overlapping disks at high density. Right panel: A system of identical nonoverlapping disks.

- totally impenetrable spheres
- interpenetrable spheres

We will specifically focus on calculating the following functions:

- n-point probability functions
- surface correlation functions
- lineal-path function
- chord-length density function
- pore-size functions
- nearest-neighbor functions
- point/q-particle correlation functions

The chapter closes with a brief discussion on the evaluation of the canonical function for statistically inhomogeneous particle systems.

5.1 Fully Penetrable Spheres

The simplicity of the n-body statistics in the case of a statistically homogeneous system of identical d-dimensional fully penetrable spheres enables us to express the canonical n-point correlation function H_n analytically in terms of *purely geometrical information*. We know from Chapter 3 that $g_n(\mathbf{r}^n) = 1$ or, equivalently, $\rho_n(\mathbf{r}^n) = \rho^n$ for this model, where ρ is the sphere number density [cf. (3.18)]. This result in combination with the series representation (4.46) for H_n yields the analytical result

$$H_n(\mathbf{x}^m; \mathbf{x}^{p-m}; \mathbf{r}^q) = (-1)^m \frac{\partial}{\partial a_1} \cdots \frac{\partial}{\partial a_m} \Bigg\{ \rho^q \prod_{\ell=1}^{q} \prod_{k=1}^{p} e(|\mathbf{x}_k - \mathbf{r}_\ell|; a_k) \times \exp[-\rho\, v_p(\mathbf{x}^p; a_1, \ldots, a_p)] \Bigg\}. \tag{5.1}$$

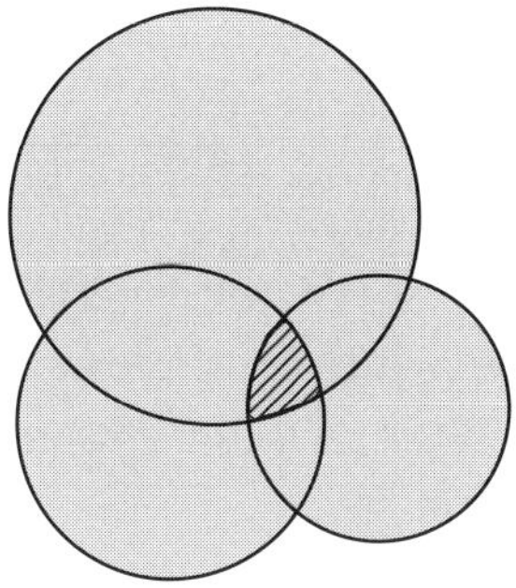

Figure 5.2 Union and intersection areas of three different-sized circles shown as shaded and cross-hatched regions, respectively.

Here $e(y;a) = \Theta(y-a)$ and $v_p(\boldsymbol{x}^p; a_1, \ldots, a_p)$ is the union volume of p d-dimensional spheres of radii $a_1, \ldots, a_p$, centered at $\boldsymbol{x}_1, \ldots, \boldsymbol{x}_p$ defined by

$$\begin{aligned} v_p(\boldsymbol{x}^p; a_1, \ldots, a_p) &= \int m^{(p)}(\boldsymbol{x}^p; \boldsymbol{r}_1) d\boldsymbol{r}_1 \\ &= \int \left\{ 1 - \prod_{i=1}^{p} [1 - m(|\boldsymbol{r}_1 - \boldsymbol{x}_i|; a_i)] \right\} d\boldsymbol{r}_1, \end{aligned} \tag{5.2}$$

where $m^{(p)}(\boldsymbol{x}^p; \boldsymbol{r}_1)$ is defined by (4.41) and $m(y;a) = \Theta(a-y)$.

With the help of the intersection-volume formula (3.47), the union volume v_p can be expressed in terms of the intersection volume $v_p^{\text{int}}(\boldsymbol{x}^p; a_1, \ldots, a_p)$ of p d-dimensional spheres of radii $a_1, \ldots, a_p$, centered at $\boldsymbol{x}_1, \ldots, \boldsymbol{x}_p$, respectively, as well as all lower-order intersection volumes $v_{p-1}^{\text{int}}, \ldots, v_1^{\text{int}}$:

$$\begin{aligned} v_p(\boldsymbol{x}^p; a_1, \ldots, a_p) &= \sum_{i=1}^{p} v_1(a_i) - \sum_{i<j}^{p} v_2^{\text{int}}(r_{ij}; a_i, a_j) + \sum_{i<j<k}^{p} v_3^{\text{int}}(r_{ij}, r_{ik}, r_{jk}; a_1, a_2, a_3) \\ &\quad - \cdots + (-1)^{p+1} v_p^{\text{int}}(\boldsymbol{x}^p; a_1, \ldots, a_p), \end{aligned} \tag{5.3}$$

where $v_1 \equiv v_1^{\text{int}}$ and $r_{ij} = |\boldsymbol{x}_j - \boldsymbol{x}_i|$. For example, for $p = 2$ and $p = 3$,

$$v_2(r_{12}; a_1, a_2) = v_1(a_1) + v_1(a_2) - v_2^{\text{int}}(r_{12}; a_1, a_2), \tag{5.4}$$

$$\begin{aligned} v_3(r_{12}, r_{13}, r_{23}; a_1, a_2, a_3) &= v_1(a_1) + v_1(a_2) + v_1(a_3) - v_2^{\text{int}}(r_{12}; a_1, a_2) \\ &\quad - v_2^{\text{int}}(r_{13}; a_1, a_3) - v_2^{\text{int}}(r_{23}; a_2, a_3) \\ &\quad + v_3^{\text{int}}(r_{12}, r_{13}, r_{23}; a_1, a_2, a_3). \end{aligned} \tag{5.5}$$

The distinction between the union and intersection areas of three different-sized circles is illustrated in Figure 5.2.

For $d = 2$, Kratky (1978) has shown that the area of intersection of four or more equal-sized circles can be reduced to a linear combination of the areas of intersection of two and three circles. An analytical expression for the intersection area of three circles was given by Rowlinson (1964). For $d = 3$, Kratky (1981) also showed that the intersection volumes of five or more spheres of equal radius can be expressed as a linear combination of intersection volumes of two, three, and four spheres. The intersection volumes of three spheres (Powell 1964) and four spheres (Helte 1994) are known analytically. Therefore, in principle, the exact intersection (and hence union) area of n disks or volume of n spheres of equal radius can be exactly computed.

5.1.1 n-Point Probability Functions

The n-point probability function S_n for phase 1 (matrix phase) is obtained from the general definition (4.69) in terms of the canonical function, i.e, $S_n(\mathbf{x}^n) = H_n(\emptyset; \mathbf{x}^n; \emptyset)$ in the limit that all of the radii of the test particles shrink to zero (or $a_i \to R, \forall i$). This definition and expression (5.1) yield the exact relation

$$S_n(\mathbf{x}^n) = \exp\Big[-\rho v_n(\mathbf{x}^n; R)\Big] \tag{5.6}$$

for d-dimensional fully penetrable spheres, where $v_n(\mathbf{x}^n; R)$ denotes the union volume of n identical spheres of radius R. The general expression (5.6) was derived first by Torquato and Stell (1983b), who referred to S_n as the n-point *matrix* probability functions.

When $n = 1$, we see that this expression gives

$$S_1 = \phi_1 = \exp(-\eta),$$

which is in agreement with relation (4.19) for the "matrix" volume fraction ϕ_1 given earlier. Recall that $\eta = \rho v_1(R)$ is a reduced density [cf. (4.16)], where $v_1(R)$ is given by (4.17) in d dimensions.

When $n = 2$, we see that relation (5.6) gives

$$S_2(r) = \exp\Big[-\eta \frac{v_2(r; R)}{v_1(R)}\Big], \tag{5.7}$$

where $r = |\mathbf{x}_{12}|$. The union volume of two spheres for the first three space dimensions is obtained by combining (3.49)–(3.51) with (5.4) to give

$$\frac{v_2(r; R)}{v_1(R)} = 2\Theta(r - 2R) + \Big(1 + \frac{r}{2R}\Big)\Theta(2R - r), \qquad d = 1, \tag{5.8}$$

$$\frac{v_2(r; R)}{v_1(R)} = 2\Theta(r - 2R) + \frac{2}{\pi}\left[\pi + \frac{r}{2R}\Big(1 - \frac{r^2}{4R^2}\Big)^{1/2} - \cos^{-1}\Big(\frac{r}{2R}\Big)\right]\Theta(2R - r), \quad d = 2, \tag{5.9}$$

$$\frac{v_2(r; R)}{v_1(R)} = 2\Theta(r - 2R) + \left[1 + \frac{3}{4}\frac{r}{R} - \frac{1}{16}\Big(\frac{r}{R}\Big)^3\right]\Theta(2R - r), \qquad d = 3. \tag{5.10}$$

Note that because the union volume $v_2(r; R)$ becomes constant after one diameter ($r = 2R$), S_2 attains its asymptotic value of ϕ_1^2 for $r \geq 2R$, *independent of the density*. Thus,

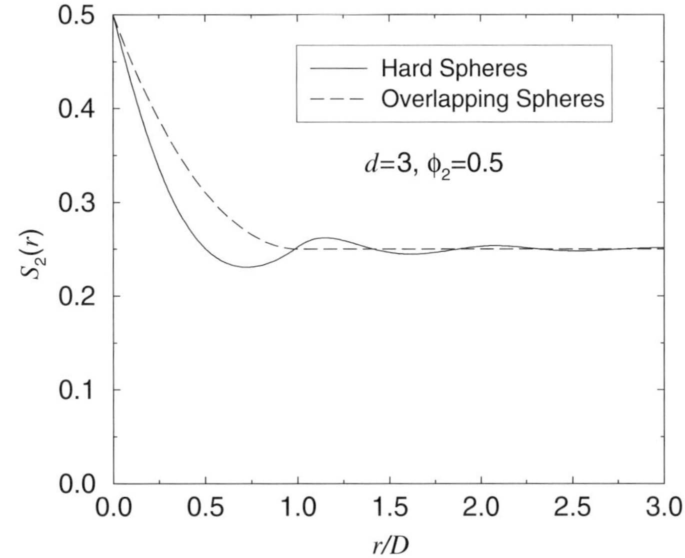

Figure 5.3 The matrix two-point probability function $S_2(r)$ versus the dimensionless distance r/D for two models of isotropic distributions of spheres of diameter $D = 2R$ at a sphere volume fraction $\phi_2 = 0.5$.

even though this system percolates at $\phi_{2c} \approx 0.68$ for $d = 2$ (Quintanilla, Torquato and Ziff 2000) and $\phi_{2c} \approx 0.29$ for $d = 3$ (Rintoul and Torquato 1997a), the function S_2 is insensitive to the long-range phenomenon of clustering near the percolation threshold (see Section 9.2.2).

Figure 5.3 shows S_2 for three-dimensional overlapping spheres at a sphere volume fraction $\phi_2 = 0.5$. Included in the figure is the corresponding plot of S_2 for equilibrium hard (totally impenetrable) spheres, which, unlike overlapping spheres, exhibits short-range order as discussed in Section 5.2.1.

When $n = 3$, relation (5.6) gives the three-point probability function as

$$S_3(r,s,t) = \exp\left[-\eta \frac{v_3(r,s,t;R)}{v_1(R)}\right], \tag{5.11}$$

where r, s, and t are the lengths of the sides of a triangle. For $d = 3$, for example, the intersection volume of three identical spheres of *unit radius* in three dimensions (Powell 1964) is given by

$$\begin{aligned} v_3^{\text{int}}(r,s,t;1) = \frac{Q}{6}rst + \frac{4}{3}\tan^{-1}\left(\frac{Q \cdot rst}{r^2+s^2+t^2-8}\right) \\ + r(1-r^2/12)\tan^{-1}\left(\frac{2Q \cdot st}{-r^2+s^2+t^2}\right) \end{aligned}$$

$$+ s(1 - s^2/12) \tan^{-1}\left(\frac{2Q \cdot rt}{r^2 - s^2 + t^2}\right)$$
$$+ t(1 - t^2/12) \tan^{-1}\left(\frac{2Q \cdot rs}{r^2 + s^2 - t^2}\right), \quad L < 1, \tag{5.12}$$

where L is the circumradius of the aforementioned triangle given by

$$L = \frac{rst}{[(r+s+t)(-r+s+t)(r-s+t)(r+s-t)]^{1/2}}, \tag{5.13}$$

and $Q = \sqrt{1 - L^2}/L$. We choose the branch of $\tan^{-1} x$ that takes values between 0 and π. Expression (5.12) is valid only if a common volume exists, i.e., $L < 1$. If $L > 1$, then either there is no common volume to three spheres or else the common volume is expressible in terms of the intersection volume of *two* spheres. Relation (5.12) when combined with (5.5) yields the union volume of three spheres $v_3(r, s, t; 1)$, which can be substituted into (5.11) to yield S_3 for fully penetrable spheres. Helte (1994) found S_4 for the same model. Indeed, S_n for $n \geq 4$ can be obtained in principle, since as noted earlier in Section 5.1, v_n for $n \geq 4$ can be expressed in terms of v_1, v_2, v_3, and v_4.

We note in passing that since the S_n for fully penetrable spheres are determined by intersection volumes of $n, n-1, \ldots,$ and 1 spheres, bounds on S_n are easily found in terms of lower-order quantities S_m $(m < n)$ (Torquato and Stell 1983b). For example, it was found that

$$\frac{S_2(x)S_2(y)}{\phi_1} \leq S_3(r, s, t) \leq \frac{S_2(r)S_2(s)S_2(t)}{\phi_1^3}, \tag{5.14}$$

where x and y are the smallest and next smallest distances among r, s, and t. We see that the lower bound satisfies all of the limiting conditions on S_3 as given by (2.22) and (2.23). The upper bound also satisfies conditions (2.23), but it does not satisfy conditions (2.22) which specify situations in which the arguments of S_3 vanish.

5.1.2 Surface Correlation Functions

We now consider obtaining the specific surface s for d-dimensional spheres using the above formalism. In general, for any monodisperse system of spheres, we have from (4.70) that

$$s = \lim_{a_1 \to R} H_1(\mathbf{x}_1; \emptyset; \emptyset).$$

Using this relation and expression (5.1) yields

$$s = \rho \phi_1 s_1(R) = \frac{\eta \phi_1 d}{R},$$

which is in agreement with relation (4.20) given earlier. Here $s_1(R) = \partial v_1(R)/\partial R$ is the surface area of a sphere [cf. (4.18)].

Similarly, the surface–void and surface–surface correlation functions follow from (5.1) by taking appropriate derivatives of the quantities within brackets with respect to the radii a_i, as specified by relations (4.71) and (4.72). We obtain

$$F_{sv}(r) = -\lim_{a_1 \to R} \frac{\partial}{\partial a_1} \exp[-\rho v_2(r; a_1, R)], \tag{5.15}$$

$$F_{ss}(r) = \lim_{a_1, a_2 \to R} \frac{\partial}{\partial a_1} \frac{\partial}{\partial a_2} \exp[-\rho v_2(r; a_1, a_2)]. \tag{5.16}$$

Thus, we generally need the union volume of two spheres of radii a_1 and a_2, the nontrivial part of which is the intersection volume.

For $d = 3$, for example, this intersection volume, when $a_1 < a_2$, is

$$v_2^{\text{int}}(r; a_1, a_2) = \begin{cases} \frac{4\pi}{3} a_1^3, & 0 \le r \le a_2 - a_1, \\ \frac{4\pi}{3} f(r; a_1, a_2), & a_2 - a_1 \le r \le a_2 + a_1, \\ 0, & r \ge a_2 + a_1, \end{cases} \tag{5.17}$$

where

$$f(r; a_1, a_2) = \frac{-3(a_2^2 - a_1^2)^2}{16r} + \frac{a_2^3 + a_1^3}{2} - \frac{3r}{8}(a_2^2 + a_1^2) + \frac{r^3}{16}.$$

Relation (5.17) when combined with (5.4) yields the union volume of two spheres $v_2(r; a_1, a_2)$, which can be substituted into (5.15) and (5.16) to give the two-point surface correlation functions for fully penetrable spheres:

$$F_{sv}(r) = \frac{3\eta}{R}\Big[1 - \Big(\frac{1}{2} - \frac{r}{4R}\Big)\Theta(2R - r)\Big] S_2(r),$$

$$F_{ss}(r) = \left\{ \frac{9\eta^2}{R^2}\Big[1 - \Big(\frac{1}{2} - \frac{r}{4R}\Big)\Theta(2R - r)\Big]^2 + \frac{3\eta}{2rR}\Theta(2R - r) \right\} S_2(r),$$

where $S_2(r)$ is the two-point probability function for the "void" phase given by (5.7). These relations were first given by Doi (1976).

5.1.3 Lineal-Path Function

According to Section 4.4, we can obtain the exact series expression for the lineal-path function $L(z) \equiv L^{(1)}(z)$ for the matrix phase (phase 1) using the series expression for E_V [obtained from (4.46) and (4.75)] and taking the "exclusion region" indicator function m to be that given by (4.78). Thus, for any homogeneous and isotropic system of interacting identical spheres

$$L(z) = 1 + \sum_{k=1}^{\infty} \frac{(-1)^k}{k!} \int \rho_k(\mathbf{r}^k) \prod_{j=1}^{k} m(\mathbf{x} - \mathbf{r}_j; z, R) d\mathbf{r}_j, \tag{5.18}$$

which was first given by Lu and Torquato (1992a).

For the case of d-dimensional fully penetrable spheres, one exactly has that $\rho_k(\mathbf{r}^k) = \rho^k$, and thus from (5.18) we obtain the exact result

$$L(z) = \exp[-\rho v_E(z)], \tag{5.19}$$

where $v_E(z)$ is the d-dimensional volume of the spherocylindrical exclusion region Ω_E (see Figure 4.4) given by

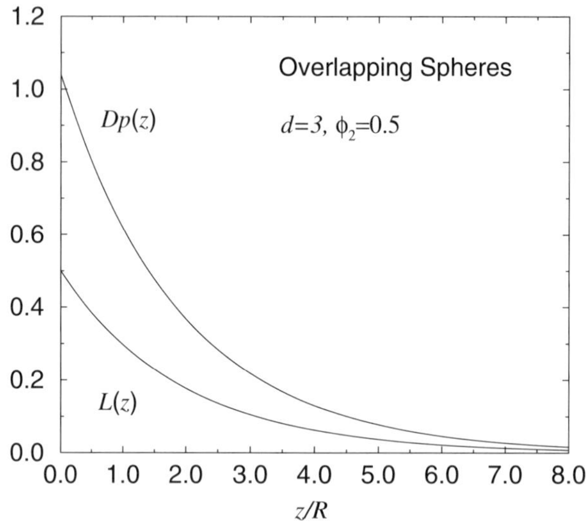

Figure 5.4 Matrix lineal-path function $L(z)$ and chord-length density function $p(z)$ for a three-dimensional system of overlapping spheres of radius R at a sphere volume fraction $\phi_2 = 0.5$, as obtained from (5.23) and (5.26) with $d = 3$.

$$v_E = v_1(R; d) + v_1(R; d-1)z, \tag{5.20}$$

where $v_1(R; d)$ is the d-dimensional volume of a sphere of radius R given by (4.17). [Note that (5.19) could also have been obtained by setting $n = 0$ and $W = v_E$ in the Poisson distribution formula (3.19).] Since the matrix volume fraction for fully penetrable spheres is given by $\phi_1 = \exp(-\eta)$, the lineal-path function becomes

$$L(z) = \phi_1^{\;1 + \frac{\omega_{d-1}}{\omega_d}\frac{z}{R}}, \tag{5.21}$$

where

$$\omega_d \equiv v_1(1; d) = \frac{\pi^{d/2}}{\Gamma(1 + d/2)} \tag{5.22}$$

is the d-dimensional volume of a sphere of unit radius. For the first three space dimensions (Lu and Torquato 1992a), we have

$$L(z) = \begin{cases} \phi_1^{\;1 + \frac{1}{2}\frac{z}{R}}, & d = 1, \\ \phi_1^{\;1 + \frac{2}{\pi}\frac{z}{R}}, & d = 2, \\ \phi_1^{\;1 + \frac{3}{4}\frac{z}{R}}, & d = 3. \end{cases} \tag{5.23}$$

The case $d = 3$ is plotted in Figure 5.4.

The lineal-path function for the particle phase $L^{(2)}(z)$ in an overlapping-sphere system is more difficult to obtain than its matrix-phase counterpart, since it cannot be obtained from an exclusion probability. The quantity $L^{(2)}(z)$ actually contains a coarse level of information about particle clustering. Indeed, the two-point cluster function $C_2(z)$ (Section 9.2.2), which is the probability that two points lie in the same cluster of connected particles, provides the following upper bound (Quintanilla and Torquato 1996a):

$$L^{(2)}(z) \leq C_2(z), \tag{5.24}$$

where the equality applies for one-dimensional systems. Thus, for the case $d = 1$, an exact analytical expression exists for $L^{(2)}(z)$, since $C_2(z)$ is known [cf. (10.11)]. Quintanilla and Torquato (1996a) have provided an exact methodology to compute $L^{(2)}(z)$ in higher dimensions that amounts to inverting a Laplace transform.

5.1.4 Chord-Length Density Function

The matrix chord-length density function $p(z) \equiv p^{(1)}(z)$ is easily derived given the expression (5.21) for the lineal-path function, normalization condition (2.68), and the interrelation (2.73). This gives the exact relation

$$p(z) = \frac{\eta\, \omega_{d-1}}{\omega_d R} \phi_1^{\frac{\omega_{d-1}}{\omega_d} \frac{z}{R}}. \tag{5.25}$$

For the first three space dimensions (Torquato and Lu 1993), we have

$$p(z) = \begin{cases} \dfrac{\eta}{2R} \phi_1^{\frac{z}{2R}}, & d = 1, \\ \dfrac{2\eta}{\pi R} \phi_1^{\frac{2z}{\pi R}}, & d = 2, \\ \dfrac{3\eta}{4R} \phi_1^{\frac{3z}{4R}}, & d = 3. \end{cases} \tag{5.26}$$

The function $p(z)$ for $d = 3$ is shown in Figure 5.4.

The corresponding mean chord length ℓ_C is obtained by applying formula (2.72). We find that

$$\ell_C = \frac{\omega_d}{\eta\, \omega_{d-1}} R = \frac{\omega_d\, \phi_1 d}{\omega_{d-1}} \frac{1}{s}, \tag{5.27}$$

where we have used formula (4.20) for the specific surface s. Expression (5.27) agrees with the exact relation (2.74) connecting ℓ_C to s for arbitrary microstructures. For the first three space dimensions, we have

$$\ell_C = \begin{cases} \dfrac{2}{\eta}R = \dfrac{2\phi_1}{s}, & d = 1, \\ \dfrac{\pi}{2\eta}R = \dfrac{\pi\phi_1}{s}, & d = 2, \\ \dfrac{4}{3\eta}R = \dfrac{4\phi_1}{s}, & d = 3. \end{cases} \tag{5.28}$$

5.1.5 Nearest-Neighbor Functions

Since the centers of fully penetrable spheres are spatially uncorrelated, there is no difference between the "void" and "particle" nearest-neighbor functions. Thus, in this special instance, we drop the subscripts normally used to distinguish these quantities. From relation (4.76), we see that setting $m = 0$, $p = 1$, and $q = 0$ in (5.1) gives the nearest-neighbor exclusion probability in d dimensions to be

$$E(r) = \exp[-\rho v_1(r)], \tag{5.29}$$

and thus from interrelations (2.94) and (2.96) the nearest-neighbor density function H and pair conditional function G are given by

$$H(r) = \rho s_1(r) \exp[-\rho v_1(r)], \tag{5.30}$$

$$G(r) = 1. \tag{5.31}$$

Here $v_1(r)$ and $s_1(r) = dv_1/dr$ are, respectively, the volume and surface area of a d-dimensional sphere of radius r given by (4.17) and (4.18). These results generalize the three-dimensional results first given by Hertz (1909) for a Poisson distribution of "point" particles. Indeed, formula (5.29) could also have been obtained by setting $n = 0$ and $W = v_1(r)$ in the Poisson distribution formula (3.19). Figure 5.5 depicts $H(r)$ for the first three space dimensions at a reduced density $\eta = \rho v_1(D/2) = 0.5$.

The mean nearest-neighbor distance ℓ_P between particles is easily found using the definition (2.107) with $n = 1$ and relation (5.30). We find in any space dimension d that

$$\frac{\ell_P}{D} = \frac{\Gamma(1 + 1/d)}{2\eta^{1/d}}, \tag{5.32}$$

where we recall that η is related to the sphere volume fraction via $\eta = -\ln(1 - \phi_2)$.

5.1.6 Pore-Size Functions

As we noted earlier, the pore-size density function $P(\delta)$ for identical spheres of radius R is trivially obtained from the void nearest-neighbor density function $H_V(r)$ via interrelation (2.84). Using this relation in combination with (5.30) gives

$$P(\delta) = \frac{\rho s_1(\delta + R)}{\phi_1} \exp[-\rho\, v_1(\delta + R)] \tag{5.33}$$

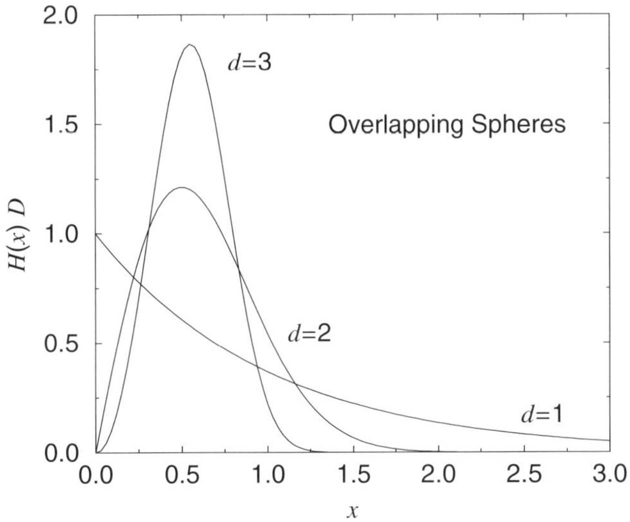

Figure 5.5 Dimensionless nearest-neighbor probability density function $H(x)D$ for systems of fully penetrable spheres of diameter D [cf. (5.30)] versus $x = r/D$ at a reduced density $\eta = -\ln(1 - \phi_2) = 0.5$.

for d-dimensional overlapping spheres. Figure 5.16 on page 152 compares $P(\delta)$ for overlapping spheres ($d = 3$) at $\phi_2 = 0.5$ to two other sphere models that possess hard cores.

5.1.7 Point/q-Particle Correlation Functions

Employing the general definition (4.74) connecting the point/q-particle correlation function G_n to H_n and expression (5.1) gives, for d-dimensional spheres, the exact relation

$$G_n(\boldsymbol{x}_1; \boldsymbol{r}^q) = \rho^q \phi_1 \prod_{\ell=1}^{q} e(y_{1\ell}; R), \tag{5.34}$$

where $y_{1\ell} = |\boldsymbol{x}_1 - \boldsymbol{r}_\ell|$. This expression was first given by Torquato (1986b).

5.2 Totally Impenetrable Spheres

We now consider statistically homogeneous ensembles of totally impenetrable spheres. The canonical function H_n is generally more difficult to evaluate for such spheres than for overlapping spheres, even though in some instances the series exactly terminates after a finite number of terms. The reason is that spatial correlations in the impenetrable-sphere systems result in nontrivial n-particle probability density functions. However, when exact evaluations are not possible, one can approximately and accurately sum the series using certain schemes.

5.2.1 n-Point Probability Functions

From the series representation (4.46) for the canonical function H_n and the definition (4.69), we can recover the series expression for the matrix n-point probability function S_n for impenetrable-sphere systems first given by Torquato and Stell (1982). In doing so, we will change variables from $\mathbf{r}_1, \mathbf{r}_2, \ldots, \mathbf{r}_n$ to $\mathbf{r}_{n+1}, \mathbf{r}_{n+2}, \ldots, \mathbf{r}_{n+k}$ and from $\mathbf{x}_1, \mathbf{x}_2, \ldots, \mathbf{x}_n$ to $\mathbf{r}_1, \mathbf{r}_2, \ldots, \mathbf{r}_n$. We find for statistically homogeneous ensembles of impenetrable spheres of radius R that

$$S_n(\mathbf{r}_{12}, \mathbf{r}_{13}, \ldots, \mathbf{r}_{1n}) = 1 + \sum_{k=1}^{n} (-1)^k \frac{\rho^k}{k!} \int \cdots \int g_k(\mathbf{r}_{n+1}, \mathbf{r}_{n+2}, \ldots, \mathbf{r}_{n+k}) \times \prod_{j=n+1}^{n+k} \left\{ 1 - \prod_{i=1}^{n} [1 - m(r_{ij}; R)] \right\} d\mathbf{r}_j, \tag{5.35}$$

where $\mathbf{r}_{ij} = \mathbf{r}_j - \mathbf{r}_i$, $r_{ij} = |\mathbf{r}_{ij}|$, $g_k = \rho_k/\rho^k$, and

$$m(r; R) = \Theta(R - r) = \begin{cases} 1, & r \leq R, \\ 0, & r > R. \end{cases} \tag{5.36}$$

We see that the representation (5.35) is no longer an infinite series but terminates exactly after terms involving the n-body correlation function g_n, defined by (3.11). This occurs because of the conflicting demands of the impenetrability condition on g_k for $k > n$ [cf. (4.23)] and the indicator function m [cf. (5.36)]. Indeed, we have already seen in obtaining the one-point function S_1 for impenetrable spheres (Section 4.1.2) that any term involving the subdiagram

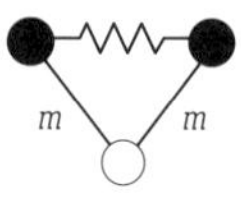

(5.37)

is identically zero. Here the wavy line indicates either a g_2-bond or the edge of a g_n-polyhedron with $n \geq 3$. We see that (5.35) for $n = 1$ yields the expected result

$$S_1 = 1 - \eta = 1 - \phi_2 = \phi_1, \tag{5.38}$$

in agreement with (4.21).

Some of the two-body terms in the series representation (5.35) for $n = 2$ contain the zero subdiagram (5.37) (Torquato and Stell 1982). After simplification, we obtain

$$S_2(\mathbf{r}_{12}) = 1 - \rho \int m(r_{13}; R) d\mathbf{r}_3 - \rho \int m(r_{23}; R) d\mathbf{r}_3 + \rho \int m(r_{13}; R) m(r_{23}; R) d\mathbf{r}_3 + \rho^2 \int \int g_2(\mathbf{r}_{34}) m(r_{13}; R) m(r_{24}; R) d\mathbf{r}_3 d\mathbf{r}_4$$

$$= 1 - 2\phi_2 + \rho m \otimes m + \rho^2 g_2 \otimes m \otimes m$$

$$= 1 - 2\phi_2 +$$ 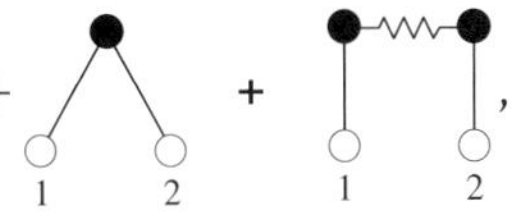 (5.39)

where the operation $\otimes$ denotes a convolution defined by (3.33).

The diagrams in (5.39) have simple probabilistic interpretations. This is most easily seen by examining the two-point probability function for the particle phase, which we will denote by $S_2^{(2)}$. Recall from (2.29) that $S_2^{(2)} = 1 - 2\phi_1 + S_2$. Combining this expression with (5.39) then yields

$$S_2^{(2)}(\boldsymbol{r}_{12}) =$$ 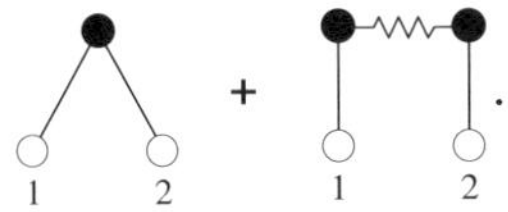 (5.40)

Thus, we see that the probability of finding two points, with a displacement $\boldsymbol{r}_{12}$, in the particle phase can be written as the sum of two different probabilities. The first diagram, equal to $\rho v_2^{\text{int}}(r_{12}; R)$, is the probability that both points fall in a *single* sphere. The two-body (or second) diagram, equal to $\rho^2 g_2 \otimes m \otimes m$, is the probability that one point falls in one particle and the other point falls in a different particle. For particles that can interpenetrate one another, however, such simple interpretations are generally not valid because of the allowability of overlap.

Since $v_2^{\text{int}}(r_{12}; R) = m \otimes m$, we can rewrite (5.39) as

$$S_2(\boldsymbol{r}_{12}) = 1 - \rho v_2(r_{12}; R) +$$ 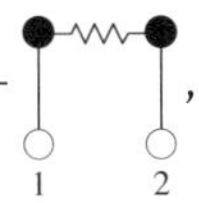 (5.41)

where $v_2(r_{12}; R)$ is the union volume of two spheres given by (5.8), (5.9), and (5.10) in one, two, and three dimensions, respectively. While $v_2(r_{12}; R)$ is a purely geometric quantity, the two-body diagram is model-dependent by virtue of its dependence on the pair correlation function g_2.

Before discussing calculations of S_2 for specific models, it is useful to make some remarks about the three-point probability function S_3 for the matrix phase. Its explicit representation for impenetrable spheres in terms of two- and three-body statistics is obtained from (5.35). However, the particle-phase counterpart $S_3^{(2)}$, which can be expressed in terms of matrix probability functions via relation (2.8), has the more succinct form

$$S_3^{(2)}(\boldsymbol{r}_{12}, \boldsymbol{r}_{13}) =$$ 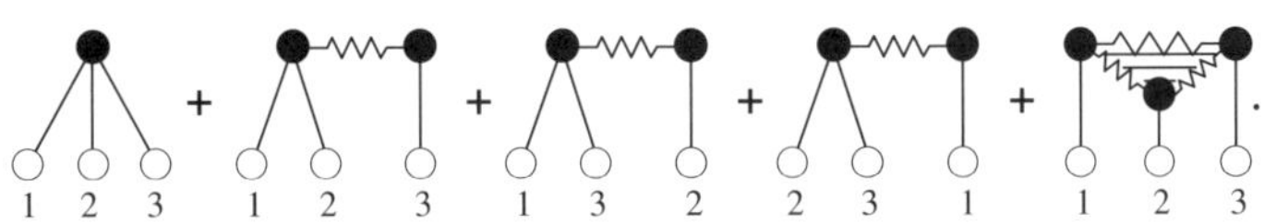 (5.42)

The probability of finding three points in the particle phase can be written as the sum of the probabilities of three different events. The first diagram in (5.42) is the probability that all three points fall in a single sphere. The next three two-body diagrams constitute the probability that one point falls in one particle and the other two points fall in a different particle. These two-body diagrams are not difficult to evaluate using existing accurate analytical approximations for g_2 or via Monte Carlo simulations. Finally, the last diagram is the probability that each point falls in different spheres. It is more problematic to compute, since g_3 is known less precisely than g_2. One can resort to superposition-type approximations for g_3 (Section 3.2) or evaluate the three-body diagram using Monte Carlo methods. The evaluation of S_n for $n \geq 4$ becomes increasingly unwieldy because of the difficulty involved in estimating the g_n (see Section 3.2).

We note in passing that the probability of finding n points at positions $\mathbf{r}^n$ in a single sphere of a homogeneous totally impenetrable sphere system is given by

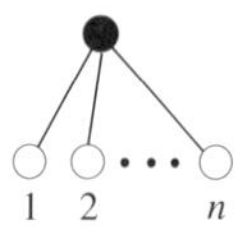

$$= \rho v_n^{\text{int}}(\mathbf{r}^n; R) = \rho \int d\mathbf{x} \prod_{i=1}^{n} m(|\mathbf{r}_i - \mathbf{x}|; R), \tag{5.43}$$

where $v_n^{\text{int}}(\mathbf{r}^n; R)$ is the intersection volume of n spheres of radius R at positions $\mathbf{r}^n$; see also (3.47). The one-body contribution to the n-point probability function $S_n^{(2)}$ is exactly given by diagram (5.43).

Torquato and Lado (1985) found an exact analytical expression for the matrix two-point function $S_2(r_{12})$ for an equilibrium system of hard rods of length D. This was done by direct integration of the two-body diagram in (5.41) and use of the analytical expression for g_2 given by (3.54). After some simplification (Quintanilla and Torquato 1996b), their expression can be written in terms of the dimensionless distance $x = r_{12}/D$ as

$$S_2(x) = (1-\eta) \sum_{k=0}^{M} \frac{1}{k!} \left[\frac{(x-k)\eta}{1-\eta} \right]^k \exp\left[-\frac{(x-k)\eta}{1-\eta} \right], \tag{5.44}$$

where $M \leq x \leq M+1$.

In higher dimensions, one can evaluate (5.41) exactly through the first few terms in a density expansion (Torquato and Stell 1985a). However, since g_2 is not known analytically for arbitrary densities for $d \geq 2$, one must necessarily utilize approximations to evaluate the two-body diagram in (5.41). Since the two-body diagram is a double convolution integral, then we can express it, for isotropic media, as

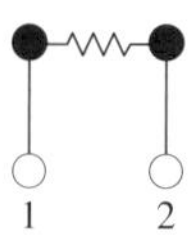

$$= \phi_2^2 + \rho^2 \mathcal{F}^{-1}[\tilde{m}^2 \tilde{h}], \tag{5.45}$$

where $\mathcal{F}^{-1}$ denotes the inverse Fourier transform defined by (2.50), $\tilde{f}(k)$ is the Fourier transform of a function $f(r)$ (see Section 2.2.5), and $h(r) = g_2(r) - 1$ is the total correlation function defined by (3.30). For example, when $d = 3$, we have that

Table 5.1 The matrix two-point probability function $S_2(r)$ versus r/D for three-dimensional isotropic equilibrium systems of hard spheres of diameter D for different values of the sphere volume fraction ϕ_2 as computed from the Verlet–Weis correction to the Percus–Yevick approximation.

	$S_2(r)$				
r/D	$\phi_2 = 0.1$	0.2	0.3	0.4	0.5
0.0	0.900	0.800	0.700	0.600	0.500
0.1	0.886	0.770	0.656	0.541	0.427
0.2	0.871	0.742	0.613	0.487	0.363
0.3	0.857	0.715	0.575	0.440	0.313
0.4	0.844	0.691	0.543	0.403	0.275
0.5	0.833	0.670	0.516	0.374	0.250
0.6	0.823	0.654	0.496	0.355	0.236
0.7	0.816	0.642	0.482	0.344	0.231
0.8	0.810	0.634	0.476	0.341	0.233
0.9	0.807	0.631	0.475	0.345	0.240
1.0	0.807	0.633	0.481	0.355	0.252
1.1	0.808	0.636	0.488	0.365	0.261
1.2	0.809	0.639	0.493	0.368	0.261
1.3	0.810	0.641	0.494	0.368	0.256
1.4	0.810	0.641	0.494	0.365	0.251
1.5	0.810	0.641	0.493	0.361	0.246
1.6	0.810	0.641	0.491	0.358	0.245
1.7	0.810	0.641	0.490	0.357	0.245
1.8	0.810	0.640	0.489	0.357	0.248
1.9	0.810	0.640	0.488	0.358	0.251
2.0	0.810	0.640	0.489	0.360	0.253

$$\tilde{m}(k) = \frac{4\pi R^3}{kR}\left[\frac{\sin kR}{(kR)^2} - \frac{\cos kR}{kR}\right] \tag{5.46}$$

and

$$\mathcal{F}^{-1}[\tilde{m}^2\tilde{h}] = \frac{1}{2\pi^2 r}\int_0^\infty \tilde{m}^2(k)\tilde{h}(k)k\sin(kr)\,dk. \tag{5.47}$$

For the case of equilibrium ensembles of statistically isotropic hard d-dimensional spheres, it is convenient to utilize approximate expressions for the Fourier transform of the *direct* correlation function $\tilde{c}(k)$, which is related to $\tilde{h}(k)$ through the Ornstein–Zernike equation (3.35). For $d = 3$, Torquato and Stell (1985a) used the Percus–Yevick approximation for $\tilde{c}$ [given by (3.61)] and the correction due to Verlet and Weis (1972) to evaluate (5.47) and thus S_2. Table 5.1 summarizes this evaluation of $S_2(r)$ for several packing fractions along the stable, disordered fluid branch (although the highest reported value $\phi_2 = 0.5$ is just slightly above the freezing point). These three-dimensional

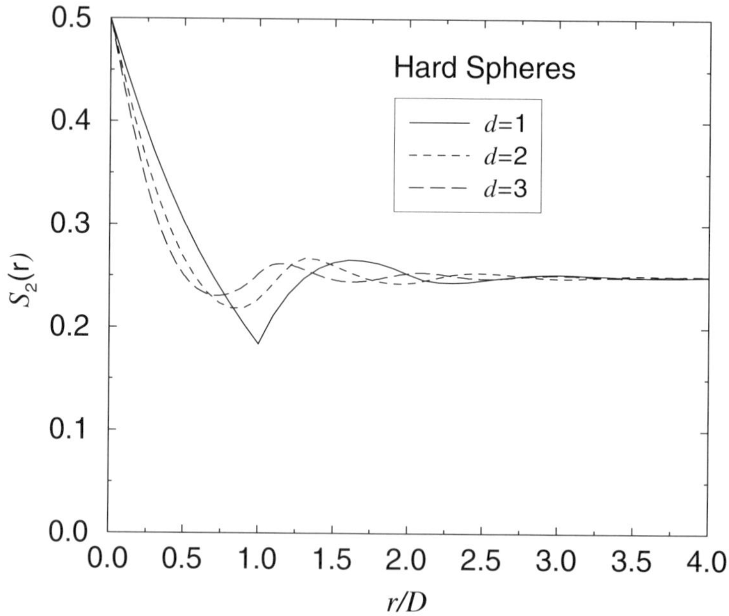

Figure 5.6 The matrix two-point probability function $S_2(r)$ versus the dimensionless distance r/D for isotropic, equilibrium systems of hard spheres of diameter D at a packing fraction $\phi_2 = 0.5$ for the first three space dimensions.

results were shown to be in excellent agreement with computer-simulation data (Haile, Massobrio and Torquato 1985), which can be considered to be exact.

Torquato and Lado (1985) used the Percus–Yevick approximation for $\tilde{c}$ to compute S_2 for disordered equilibrium hard disks ($d = 2$). Figure 5.6 compares the aforementioned results for $d = 2$ and $d = 3$ with the exact one-dimensional result (5.44). We see that S_2 oscillates about its large-distance asymptotic value for several diameters before attaining its asymptotic value, reflecting short-range order due to exclusion-volume (hard-core) effects. The short-range order exhibited in this model at high densities is to be contrasted with the behavior of S_2 for fully penetrable spheres, which decays to its asymptotic value after one sphere diameter (see Figure 5.3). Of course, as the density goes to zero, both hard- and overlapping-sphere models become identical, and hence oscillations in S_2 vanish.

5.2.2 Surface Correlation Functions

One can recover any n-point surface correlation function (in which a subset of the n points refers to positions on the interface) from the canonical function H_n (Torquato 1986c). Again, we find that the series (4.46) terminates exactly after terms involving the n-body correlation function g_n. Indeed, in evaluating in Section 4.1.2 the specific surface function s, a one-point correlation function, we found that any term involving the subdiagram

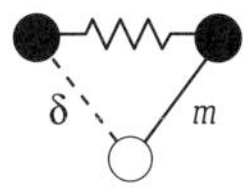 (5.48)

is identically zero for impenetrable spheres. This observation combined with definition (4.70) and series (4.46) yields for homogeneous media that

$$s = \frac{d}{R}\eta,$$

which of course agrees with the earlier result (4.22).

Using the definitions (4.71) and (4.72) and the series representation (4.46) for H_n enabled Torquato (1986c) to find the following expressions for the two-point surface correlation functions F_{sv} and F_{ss}:

$$\begin{aligned} F_{sv}(\boldsymbol{r}) &= \lim_{a_1 \to R} H_2(\boldsymbol{x}_1; \boldsymbol{x}_2; \emptyset) \\ &= s - \rho\delta \otimes m - \rho^2 g_2 \otimes \delta \otimes m \end{aligned}$$

$= s -$ 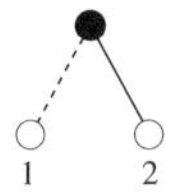 $-$ 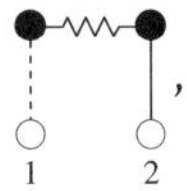 , (5.49)

$$\begin{aligned} F_{ss}(\boldsymbol{r}) &= \lim_{a_1, a_2 \to R} H_2(\boldsymbol{x}_1, \boldsymbol{x}_2; \emptyset; \emptyset) \\ &= \rho\delta \otimes \delta + \rho^2 g_2 \otimes \delta \otimes \delta \end{aligned}$$

$=$ 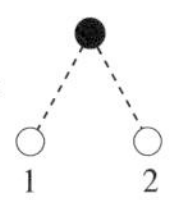 $+$ 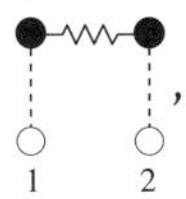 , (5.50)

where $\boldsymbol{r} = \boldsymbol{x}_2 - \boldsymbol{x}_1$.

As in the case of the two-point function $S_2(\boldsymbol{r})$, the diagrammatic representations of the integrals help to elucidate their physical meaning. For example, referring to (5.50), the one-body diagram is the contribution to F_{ss} when both points 1 and 2 describe positions on the surface of the same sphere. The two-body diagram is the contribution to F_{ss} when point 1 and point 2 describe positions on the surfaces of two different spheres.

The one-body diagrams in (5.49) and (5.50) are easily evaluated in real space. For example, for $d = 3$, using bipolar coordinates (Torquato 1980), we obtain

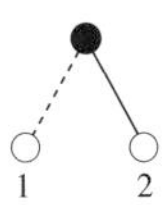 $= \frac{s}{2}\left(1 - \frac{r}{2R}\right)\Theta(2R - r),$ (5.51)

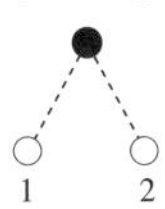 $= \frac{s}{2r}\Theta(2R - r).$ (5.52)

Since two-body diagrams are double convolution integrals, then we have the following Fourier representations:

$F^*_{sv} \equiv$ 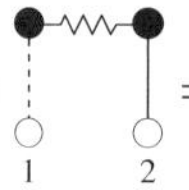 $= s\eta + \rho^2 \mathcal{F}^{-1}[\tilde{m}\tilde{\delta}\tilde{h}], \quad (5.53)$

$F^*_{ss} \equiv$ 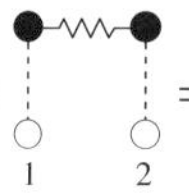 $= s^2 + \rho^2 \mathcal{F}^{-1}[\tilde{\delta}^2\tilde{h}], \quad (5.54)$

where $h(r) = g_2(r) - 1$ is the total correlation function. For example, when $d = 3$, we have, for isotropic media, that

$$\tilde{\delta}(k) = \frac{4\pi R}{k}\sin kR, \tag{5.55}$$

$$\mathcal{F}^{-1}[\tilde{m}\tilde{\delta}\tilde{h}] = \frac{1}{2\pi^2 r}\int_0^\infty \tilde{m}(k)\tilde{\delta}(k)\tilde{h}(k)k\sin(kr)dk, \tag{5.56}$$

$$\mathcal{F}^{-1}[\tilde{\delta}^2\tilde{h}] = \frac{1}{2\pi^2 r}\int_0^\infty \tilde{\delta}^2(k)\tilde{h}(k)k\sin(kr)dk. \tag{5.57}$$

For $d = 3$, Torquato (1986a) used the Percus–Yevick approximation for $\tilde{c}$ [given by (3.61)] and the Verlet–Weis correction to evaluate the two-body terms (5.56) and (5.57), and thus F_{sv} and F_{ss}, for statistically isotropic systems of hard spheres in equilibrium. The resulting two-body diagrams (5.53) and (5.54) (scaled by their large-distance asymptotic values) are plotted in Figure 5.7 for a packing fraction $\phi_2 = 0.5$. Seaton and Glandt (1986) showed that these theoretical results were in excellent agreement with computer-simulation data.

5.2.3 Lineal-Path Function

In the instance of totally impenetrable or hard spheres, the exact series representation (5.18) of $L(z) \equiv L^{(1)}(z)$, and thus $p(z) \equiv p^{(1)}(z)$ for the matrix phase, can only be evaluated exactly for the case $d = 1$ (i.e., hard rods). For $d = 1$, this is possible, since the ρ_n can be expressed in terms of products of the exact pair function ρ_2. It is impossible to evaluate the series for $d \geq 2$ exactly because the n-particle probability densities $\rho_n(\mathbf{r}^n)$ are not known exactly. One must therefore devise approximate schemes to evaluate and sum the series. Lu and Torquato (1992a) obtained accurate approximations for $L(z)$ for isotropic equilibrium hard-sphere systems in the cases $d = 2$ and $d = 3$ using *scaled-particle theory*. Their results are easily generalized to any space dimension d, namely,

$$L(z) = \phi_1 \exp\Big[-\frac{\eta\,\omega_{d-1}}{\phi_1\omega_d}\frac{z}{R}\Big], \tag{5.58}$$

where ω_d is given by (5.22). For the first three space dimensions, we have

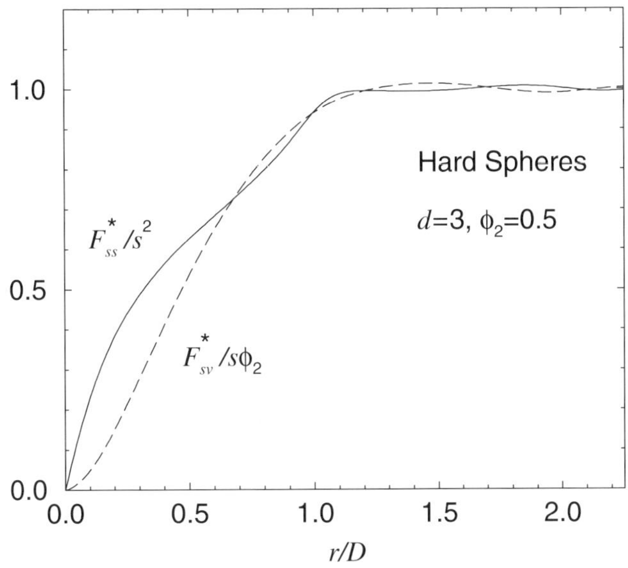

Figure 5.7 The scaled two-body contributions to the surface correlation functions $F_{sv}(r)$ and $F_{ss}(r)$ for an isotropic equilibrium system of hard spheres of diameter D at a packing fraction $\phi_2 = 0.5$. Here $F^*_{sv} = \rho_2 \otimes \delta \otimes m$ and $F^*_{ss} = \rho_2 \otimes \delta \otimes \delta$.

$$L(z) = \begin{cases} \phi_1 \exp\Big[-\dfrac{\eta}{2\phi_1}\dfrac{z}{R}\Big], & d = 1, \\ \phi_1 \exp\Big[-\dfrac{2\eta}{\pi\phi_1}\dfrac{z}{R}\Big], & d = 2, \\ \phi_1 \exp\Big[-\dfrac{3\eta}{4\phi_1}\dfrac{z}{R}\Big], & d = 3. \end{cases} \tag{5.59}$$

The result for $d = 1$ is exact. The case $d = 3$ is depicted in Figure 5.8. Note the difference between (5.59) and the corresponding result (5.23) for overlapping spheres.

5.2.4 Chord-Length Density Function

The chord-length density function $p(z) \equiv p^{(1)}(z)$ for the matrix phase in equilibrium systems of hard spheres of radius R is easily obtained from relation (5.58) for L, normalization condition (2.68), and interrelation (2.73):

$$p(z) = \frac{\eta\,\omega_{d-1}}{\phi_1\,\omega_d R}\exp\Big[-\frac{\eta\,\omega_{d-1}}{\phi_1\,\omega_d}\frac{z}{R}\Big]. \tag{5.60}$$

For the first three space dimensions (Torquato and Lu 1993), we have

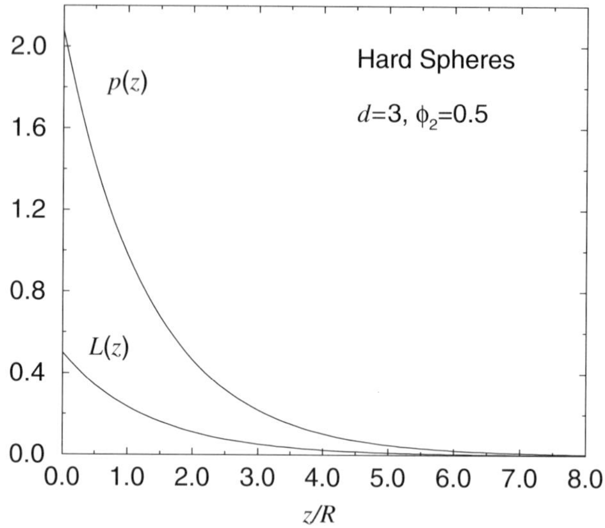

Figure 5.8 Matrix lineal-path function $L(z)$ and chord-length density function $p(z)$ for a three-dimensional equilibrium system of hard spheres of radius R at a sphere volume fraction $\phi_2 = 0.5$, as obtained from (5.59) and (5.61) with $d = 3$.

$$p(z) = \begin{cases} \dfrac{\eta}{2\phi_1 R} \exp\left[-\dfrac{\eta}{2\phi_1}\dfrac{z}{R}\right], & d = 1, \\ \dfrac{2\eta}{\pi\phi_1 R} \exp\left[-\dfrac{2\eta}{\pi\phi_1}\dfrac{z}{R}\right], & d = 2, \\ \dfrac{3\eta}{4\phi_1 R} \exp\left[-\dfrac{3\eta}{4\phi_1}\dfrac{z}{R}\right], & d = 3. \end{cases} \tag{5.61}$$

The result for $d = 1$ is exact. The instance $d = 3$ is plotted in Figure 5.8.

Application of the formula (2.72) gives the corresponding mean chord length ℓ_C:

$$\ell_C = \frac{\phi_1\, \omega_d}{\eta\, \omega_{d-1}} R = \frac{\omega_d\, \phi_1 d}{\omega_{d-1}} \frac{1}{s}. \tag{5.62}$$

The second equality is obtained using formula (4.22) for the specific surface s. Expression (5.62) agrees with the exact relation (2.74) connecting ℓ_C to s for arbitrary microstructures. For the first three space dimensions, we have

$$\ell_C = \begin{cases} \dfrac{2\phi_1}{\eta} R = \dfrac{2\phi_1}{s}, & d = 1, \\ \dfrac{\pi\phi_1}{2\eta} R = \dfrac{\pi\phi_1}{s}, & d = 2, \\ \dfrac{4\phi_1}{3\eta} R = \dfrac{4\phi_1}{s}, & d = 3. \end{cases} \tag{5.63}$$

5.2.5 Nearest-Neighbor Functions

Employing the series representation (4.46) for the canonical n-point correlation function H_n and relation (4.75), we have that the void nearest-neighbor exclusion probability function, for *any statistically homogeneous ensemble of identical, interacting spheres*, is exactly given by

$$E_V(r) = 1 + \sum_{k=1}^{\infty}(-1)^k \frac{1}{k!} \int \rho_k(\mathbf{r}^k) \prod_{j=1}^{k} m(|\mathbf{x} - \mathbf{r}_j|; r) d\mathbf{r}_j, \tag{5.64}$$

where m is the indicator function defined by (4.30), i.e., $m(y; r) = \Theta(r - y)$. This result was first given by Reiss et al. (1959) using a different approach. The other nearest-neighbor quantities, H_V and G_V, can be found from (5.64) and relations (2.94) and (2.96). Similarly, the series expression for the particle exclusion probability function is obtained from (4.77) and (4.46):

$$E_P(r) = 1 + \sum_{k=1}^{\infty}(-1)^k \frac{1}{k!} \int \frac{\rho_{k+1}(\mathbf{r}^{k+1})}{\rho} \prod_{j=2}^{k+1} m(|\mathbf{r}_1 - \mathbf{r}_j|; r) d\mathbf{r}_j. \tag{5.65}$$

This result is independent of the position $\mathbf{r}_1$ for statistically homogeneous media. The other particle nearest-neighbor quantities, H_P and G_P, can be found from (5.65) and interrelations (2.95) and (2.97).

Before presenting calculations of nearest-neighbor functions, we first state exact conditions that these functions must obey for any hard-sphere system. Since one particle excludes another from occupying the same space, it is obvious from the definitions (2.89), (2.92), and (2.100) that

$$E_P(r) = 1, \quad H_P(r) = G_P(r) = 0, \quad \text{for} \quad 0 \le r < D. \tag{5.66}$$

Torquato et al. (1990) found rigorous successive upper and lower bounds on the nearest-neighbor functions. One of the simplest inequalities worth noting here is the following lower bound, which is valid for *any* statistically homogeneous and isotropic ensemble of identical spheres:

$$G_P(r) \ge g_2(r), \tag{5.67}$$

where $g_2(r)$ is the radial distribution function defined in Section 3.1. For hard-sphere systems in particular, $G_P(D^+) = g_2(D^+)$.

The "void" functions must satisfy certain exact conditions for any statistically homogeneous ensemble of hard spheres (Reiss et al. 1959, Hefland et al. 1960). For example, a spherical cavity of radius r and volume $v_1(r)$ can contain at most one particle center if $r \le D/2$. Thus, for statistically homogeneous media, the exclusion probability is then given by

$$E_V(r) = 1 - \rho v_1(r), \quad \text{for} \quad 0 \le r \le D/2. \tag{5.68}$$

Hence, by (2.92) and (2.96), we also have

$$H_V(r) = \rho s_1(r), \quad \text{for} \quad 0 \le r \le D/2, \tag{5.69}$$

and

$$G_V(r) = \frac{1}{1 - \rho v_1(r)}, \quad \text{for} \quad 0 \le r \le D/2. \tag{5.70}$$

Recall from (2.105) that for $r = D/2$, we have $E_V(D/2) = 1 - \eta = \phi_1$ and $H_V(D/2) = s = \eta d/R$. Note that for $r < D/2$, the "test" particle [cf. (2.90)] may be regarded as a "point" particle that is capable of penetrating the mutually impenetrable particles. Hence, for $r < D/2$, decreasing r then increases E_V, according to (5.68), until E_V reaches its maximum value of unity at $r = 0$. For possibly overlapping particles, relations (5.66) and (5.68)–(5.70) will not hold [see, for example, (5.30)–(5.31)]. A trivial lower bound that is valid for any ergodic or homogeneous ensemble of spheres is the following (Torquato et al. 1990):

$$G_V(r) \ge 1. \tag{5.71}$$

Although the void and particle quantities are not the same for $r < D$, they are in fact related to one another for $r \ge D$ for the special case of an isotropic *equilibrium* ensemble of hard spheres. Specifically, it has been shown (Torquato et al. 1990) that

$$E_P(r) = \frac{E_V(r)}{E_V(D)}, \qquad r \ge D. \tag{5.72}$$

Combination of (5.72) with the relations (2.94)–(2.97) gives the following expressions relating the different nearest-neighbor functions:

$$H_P(r) = \frac{H_V(r)}{E_V(D)}, \qquad r \ge D, \tag{5.73}$$

$$G_P(r) = G_V(r), \qquad r \ge D. \tag{5.74}$$

In other words, roughly speaking, the environment around a hard sphere of radius $D/2$ is the same as the environment around a spherical void region of radius D, resulting in equality of the conditional pair functions $G_P = G_V$ for $r \ge D$. However, for the range $0 \le r \le D$, $G_V(r) \ne G_P(r)$.

Equilibrium Ensemble

For an equilibrium distribution of d-dimensional hard spheres, one can relate the void pair distribution function at $r = D^+$ to this function at $r = \infty$ (Reiss et al. 1959, Torquato et al. 1990):

$$a_0 \equiv G_V(\infty) = 1 + 2^{d-1}\eta\, G_V(D^+), \tag{5.75}$$

where

$$a_0 = \frac{p}{\rho k T}, \tag{5.76}$$

where (5.75) is the reduced equation of state [cf. (3.42)], p is the pressure, T is absolute temperature, and k is Boltzmann's constant. Hence, in light of (5.74), one may replace

G_V by G_P in (5.75). For an equilibrium hard-sphere ensemble, it also has been shown (Reiss et al. 1959, Hefland et al. 1960) that $G_V(r)$ is generally a monotonically increasing function of r and, for large distances, is given by the asymptotic form

$$G_V(x) = a_0 + \frac{a_1}{x} + \frac{a_2}{x^2} + \cdots + \frac{a_{d-1}}{x^{d-1}} + \mathcal{O}\left(\frac{1}{x^d}\right), \quad x \gg 1, \tag{5.77}$$

for arbitrary dimension d, where $x = r/D$ is a dimensionless distance.

For an equilibrium distribution of hard rods ($d = 1$), one can evaluate the series (5.64) through all orders in density. As discussed in Section 3.2, this is true because in one dimension the ρ_n can be expressed in terms of products of the exact pair function ρ_2. Using these results, it is found (Torquato et al. 1990) that

$$E_V(x) = \phi_1 \exp\left[\frac{-\eta(2x-1)}{1-\eta}\right], \quad x \geq \frac{1}{2}, \tag{5.78}$$

$$H_V(x) = \frac{2}{D}\eta \exp\left[\frac{-\eta(2x-1)}{1-\eta}\right], \quad x \geq \frac{1}{2}, \tag{5.79}$$

$$G_V(x) = \frac{1}{\phi_1}, \quad x \geq \frac{1}{2}. \tag{5.80}$$

Here $x = r/D$ is a dimensionless distance. These results were first given by Hefland et al. (1960) using physical arguments. Another way of getting these results is to recognize that the lineal path function $L(z)$ for one-dimensional particle systems is trivially related to the exclusion probability $E_V(r)$ according to the expression

$$E_V(r) = L(2r - D). \tag{5.81}$$

This relation combined with hard-rod relation (5.59) for L gives (5.78). Indeed, for *any* ensemble of identical spheres in any dimension, we have the inequality

$$E_V(r) \leq L(2r - D), \tag{5.82}$$

the equality holding when $d = 1$.

The corresponding particle quantities for hard rods are easily obtained (Torquato et al. 1990) by combining the equilibrium interrelations (5.72)–(5.74) with the void relations (5.78)–(5.80) to give

$$E_P(x) = \exp\left[\frac{-2\eta(x-1)}{1-\eta}\right], \quad x \geq 1, \tag{5.83}$$

$$H_P(x) = \frac{2\eta}{D(1-\eta)} \exp\left[\frac{-2\eta(x-1)}{1-\eta}\right], \quad x \geq 1, \tag{5.84}$$

$$G_P(x) = \frac{1}{\phi_1}, \quad x \geq 1. \tag{5.85}$$

From the definition of the mean nearest-neighbor distance ℓ_P, given by (2.108), and the results above, it is easily found that

$$\ell_P = 1 + \frac{1-\eta}{2\eta}. \tag{5.86}$$

We note in passing that the nearest-neighbor functions for nonequilibrium RSA rods have been given analytically by Rintoul et al. (1996). As discussed below, the RSA quantities have substantially different functional forms from the corresponding equilibrium quantities at sufficiently high densities.

For $d > 1$, an exact evaluation of the series (5.64) and (5.65) is out of the question, since the n-particle probability densities are not exactly known. The well-known scaled-particle theory of Reiss et al. (1959) bypasses this problem by providing a means to sum the series approximately for isotropic *equilibrium* systems of hard spheres. The key quantity in scaled-particle theory is the asymptotic form of the pair conditional function G_V given by (5.77). The important observation made by these authors is that the form (5.77) is also a good approximation for small x and thus can provide a good approximation for the entire range between $x = 1/2$ and $x = \infty$. For $d = 3$, Reiss et al. determined the unknown coefficients a_0, a_1, and a_2 by utilizing three of the exact conditions mentioned earlier, namely, the "infinity" condition (5.75) and the continuity of G_V and its first derivative at $x = 1/2$, given relation (5.70). The corresponding two-dimensional results were given by Hefland et al. (1960). Torquato et al. (1990) subsequently employed different conditions to ascertain the unknown coefficients a_i to get somewhat more accurate approximations for the void quantities as well as corresponding approximations for the particle quantities. All of the aforementioned approximations were intended to be used for densities along the disordered fluid branch only, i.e., up to the freezing point η_f.

Torquato (1995a) further refined these approximations for both the void and particle nearest-neighbor functions and also found accurate expressions for them along the *metastable extension of the liquid branch*, i.e., $\eta_f \leq \eta \leq \eta_c$ (see Section 3.3 and Figure 3.13). Specifically, he noted that the *void* quantities are fundamentally different from the *particle* quantities along the metastable extension. The former, unlike the latter, do not vanish or diverge at the "singular" point η_c, which we now know to be the *maximally random jammed* (MRJ) state for equilibrium hard spheres (see Section 3.5). (For example, as $\eta \to \eta_c$, G_P diverges to infinity, whereas G_V remains finite.) Thus, the functional nature of the *particle quantities* below and above the freezing point η_f must be different. Moreover, since the void quantities are *nonsingular* when $\eta = \eta_c$, an ansatz was made that they could be analytically continued above the freezing point using the void expressions for the fluid branch ($0 \leq \eta \leq \eta_f$) to provide useful estimates along the aforementioned metastable branch up to η_c. This ansatz was tested against simulation results (Rintoul and Torquato 1998) and was shown to provide a very good approximation.

We report Torquato's results below for both $d = 2$ and $d = 3$ without derivations. In what follows, we use the phrase "*equilibrium hard spheres* (*disks*)" and its variants to mean systems along the disordered stable liquid branch or the disordered metastable extension of the liquid branch.

In the case of equilibrium hard disks ($d = 2$), he obtained

$$E_V(x) = \phi_1 \exp[-\eta(4a_0x^2 + 8a_1x + a_2)], \qquad x \geq \frac{1}{2}, \tag{5.87}$$

$$H_V(x) = \frac{8\eta}{D}(a_0 x + a_1)\, E_V(x), \qquad x \geq \frac{1}{2}, \tag{5.88}$$

$$G_V(x) = a_0 + \frac{a_1}{x}, \qquad x \geq \frac{1}{2}, \tag{5.89}$$

where the coefficients a_0, a_1 and a_2 are given by

$$a_0 = \frac{1 + 0.128\eta}{(1-\eta)^2}, \qquad a_1 = \frac{-0.564\eta}{(1-\eta)^2}, \qquad a_2 = -(a_0 + 4a_1).$$

In the case of equilibrium hard spheres ($d = 3$), he obtained

$$E_V(x) = \phi_1 \exp[-\eta(8a_0 x^3 + 12a_1 x^2 + 24a_2 x + a_3)], \quad x \geq \frac{1}{2}, \tag{5.90}$$

$$H_V(x) = \frac{24\eta}{D}(a_0 x^2 + a_1 x + a_2)\, E_V(x), \qquad x \geq \frac{1}{2}, \tag{5.91}$$

$$G_V(x) = a_0 + \frac{a_1}{x} + \frac{a_2}{x^2}, \qquad x \geq \frac{1}{2}, \tag{5.92}$$

where the coefficients a_0, a_1, a_2, and a_3 are given by

$$a_0 = \frac{1 + \eta + \eta^2 - \eta^3}{(1-\eta)^3}, \qquad a_1 = \frac{\eta(3\eta^2 - 4\eta - 3)}{2(1-\eta)^3},$$

$$a_2 = \frac{\eta^2(2-\eta)}{2(1-\eta)^3}, \qquad a_3 = -(a_0 + 3a_1 + 12a_2).$$

The void relations (5.87)–(5.89) and (5.90)–(5.92) are valid for all densities along the disordered fluid and metastable branches, and thus for the range $0 \leq \eta \leq \eta_c$. In the case $d = 2$, this assumes the existence of the metastable extension (Section 3.3); otherwise, the expressions are valid only for $\eta \leq \eta_f$.

Figure 5.9 compares the three-dimensional theoretical results (5.90) and (5.91) for E_V and H_V to computer simulation results (Rintoul and Torquato 1998) for the range $x \geq 0.5$ at the high packing fraction of $\phi_2 = 0.6$. It is seen that the theory matches the simulation data very well. Figure 5.10 provides a plot of the nearest-neighbor probability density function H_V as obtained from the exact relation (5.69) for $x \leq 0.5$ and the approximate relation (5.91) for $x \geq 0.5$ for a low and a high density.

For two-dimensional hard disks, the *particle* quantities are given by

$$E_P(x) = \exp\{-\eta[4a_0(x^2 - 1) + 8a_1(x-1)]\}, \quad x \geq 1, \tag{5.93}$$

$$H_P(x) = \frac{8\eta}{D}(a_0 x + a_1)\, E_P(x), \qquad x \geq 1, \tag{5.94}$$

$$G_P(x) = a_0 + \frac{a_1}{x}, \qquad x \geq 1, \tag{5.95}$$

where the coefficients a_0 and a_1 are given by

$$a_0 = \begin{cases} \dfrac{1 + 0.128\eta}{(1-\eta)^2}, & 0 \leq \eta \leq \eta_f, \\[2ex] 1 + 2\eta g_f(1)\dfrac{(\eta_c - \eta_f)}{(\eta_c - \eta)}, & \eta_f \leq \eta \leq \eta_c, \end{cases}$$

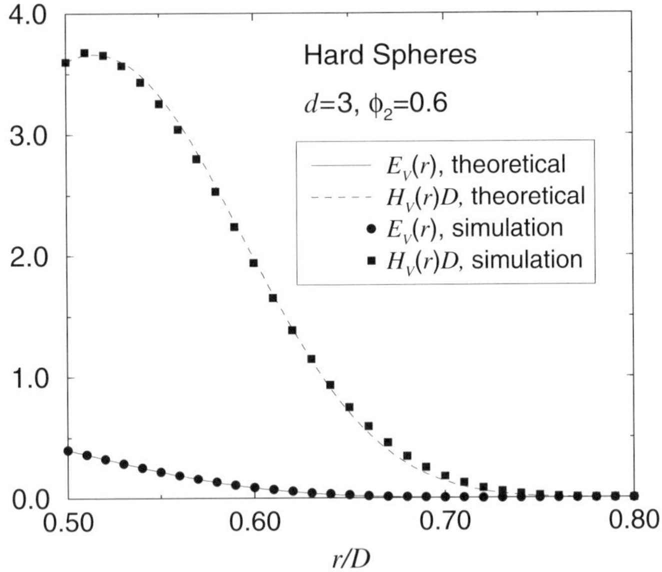

Figure 5.9 Void nearest-neighbor functions $E_V(r)$ and $H_V(r)D$ versus $x = r/D$ for an equilibrium system of hard spheres of diameter D at a packing fraction $\phi_2 = 0.6$. Comparison of simulations and theory as obtained from (5.90) and (5.91).

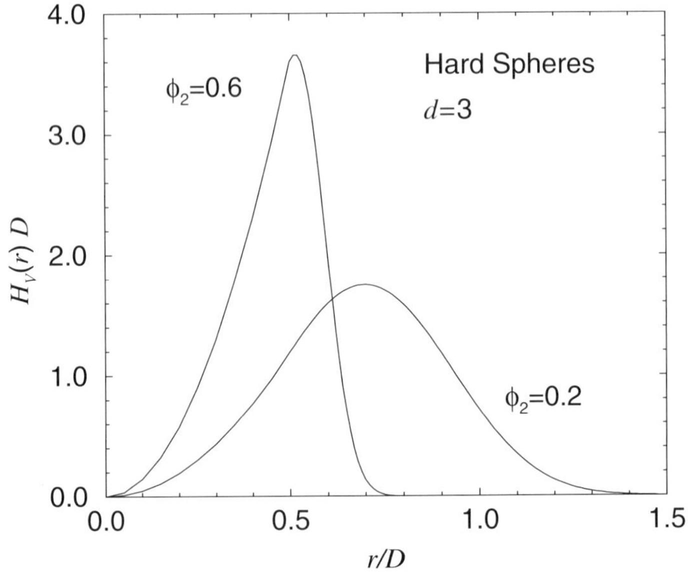

Figure 5.10 Dimensionless void nearest-neighbor probability density function $H_V(r)D$ versus $x = r/D$ for equilibrium systems of hard spheres of diameter D at $\phi_2 = 0.2$ and $\phi_2 = 0.6$ as obtained from (5.69) and (5.91). Note that the range of r/D here is larger than in Figure 5.9.

$$a_1 = \begin{cases} \dfrac{-0.564\eta}{(1-\eta)^2}, & 0 \le \eta \le \eta_f, \\ -g_f(1)\dfrac{(\eta_c - \eta_f)}{(\eta_c - \eta)} + \dfrac{1}{1-\eta}, & \eta_f \le \eta \le \eta_c. \end{cases}$$

Here $g_f(1) \equiv g_f(1^+) = (1-0.436\eta_f)/(1-\eta_f)^2$ is the contact value of the radial distribution function at $\eta = \eta_f$. Table 3.1 provides estimates of η_f and η_c .

For three-dimensional hard spheres, the particle quantities are given by

$$E_P(x) = \exp\{-\eta[8a_0(x^3-1) + 12a_1(x^2-1) + 24a_2(x-1)]\}, \quad x \ge 1, \tag{5.96}$$

$$H_P(x) = 24\eta(a_0x^2 + a_1x + a_2)\, E_P(x), \qquad x \ge 1, \tag{5.97}$$

$$G_P(x) = a_0 + \frac{a_1}{x} + \frac{a_2}{x^2}, \qquad x \ge 1, \tag{5.98}$$

where the coefficients a_0, a_1, and a_2 are given by

$$a_0 = \begin{cases} \dfrac{1+\eta+\eta^2-\eta^3}{(1-\eta)^3}, & 0 \le \eta \le \eta_f, \\ 1 + 4\eta g_f(1)\dfrac{(\eta_c - \eta_f)}{(\eta_c - \eta)}, & \eta_f \le \eta \le \eta_c, \end{cases}$$

$$a_1 = \begin{cases} \dfrac{\eta(3\eta^2 - 4\eta - 3)}{2(1-\eta)^3}, & 0 \le \eta \le \eta_f, \\ \dfrac{3\eta - 4}{2(1-\eta)} + 2(1-3\eta)g_f(1)\dfrac{(\eta_c - \eta_f)}{(\eta_c - \eta)}, & \eta_f \le \eta \le \eta_c, \end{cases}$$

$$a_2 = \begin{cases} \dfrac{\eta^2(2-\eta)}{2(1-\eta)^3}, & 0 \le \eta \le \eta_f, \\ \dfrac{2-\eta}{2(1-\eta)} + (2\eta - 1)g_f(1)\dfrac{(\eta_c - \eta_f)}{(\eta_c - \eta)}, & \eta_f \le \eta \le \eta_c. \end{cases}$$

Here $g_f(1) \equiv g_f(1^+) = (1-\eta_f/2)/(1-\eta_f)^3$ is the contact value of the radial distribution function at $\eta = \eta_f$. Table 3.1 provides estimates of η_f and η_c.

Figure 5.11 compares the three-dimensional theoretical results (5.97) for H_P to computer simulation results (Torquato and Lee 1990) at a low and a high density. Agreement between theory and simulation data is excellent. Figure 5.12 shows the conditional pair function $G_P(x)$ for the first three space dimensions at $\phi_2 = 0.4$ as computed from (5.85), (5.95), and (5.98). Note that G_P for equilibrium systems is a monotonically increasing function of x,except in the case of $d = 1$, where it is the constant ϕ_1^{-1}.

What is the effect of impenetrability of the spheres on the particle nearest-neighbor density function H_P? As illustrated in Figure 5.13 for the case $d = 2$, the particle nearest-neighbor density function H_P for hard disks is substantially different from the corresponding quantity for fully penetrable spheres [cf. (5.30)]. Obviously, inside the hard core ($x < 1$), H_P for hard sphcres must be zero, whereas H_P for penetrable spheres is nonzero, independent of the dimension. For hard spheres, $H_P(x)$ attains its maximum value at $x = 1$ and then monotonically decreases with increasing x, in contrast to the behavior for overlapping spheres. Moreover, since density fluctuations in the fully-

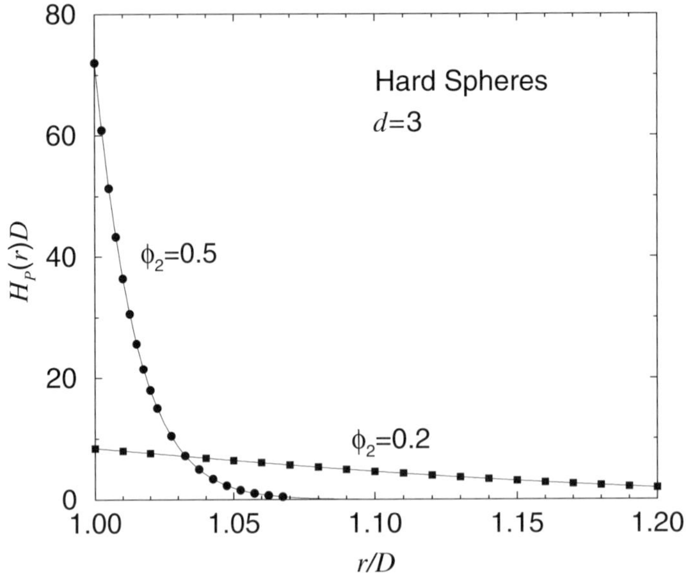

Figure 5.11 Dimensionless particle nearest-neighbor density function $H_P(r)D$ versus r/D for equilibrium systems of hard spheres of diameter D at packing fractions $\phi_2 = 0.2$ and $\phi_2 = 0.5$. Comparison of simulations (black circles and squares) and theory as computed from (5.97). The range of r/D here is relatively small.

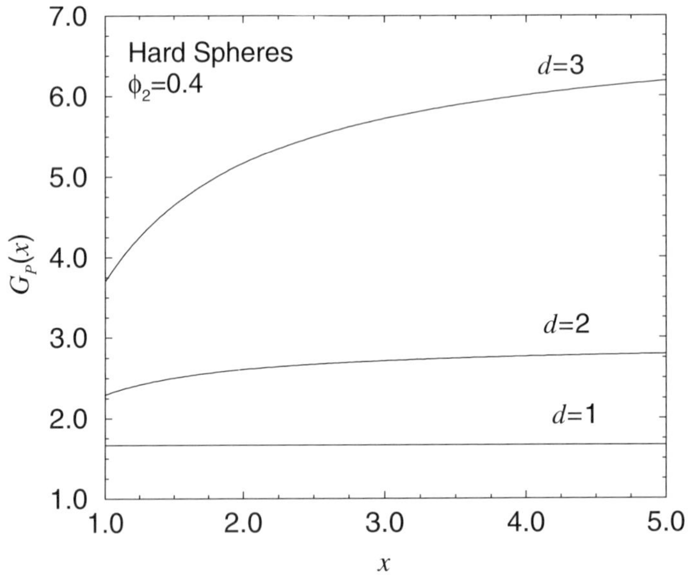

Figure 5.12 Particle nearest-neighbor conditional function $G_P(x)$ versus $x = r/D$ for equilibrium systems of hard spheres of diameter D at packing fraction $\phi_2 = 0.4$ for $d = 1, 2$, and 3 as found from (5.85), (5.95), and (5.98), respectively.

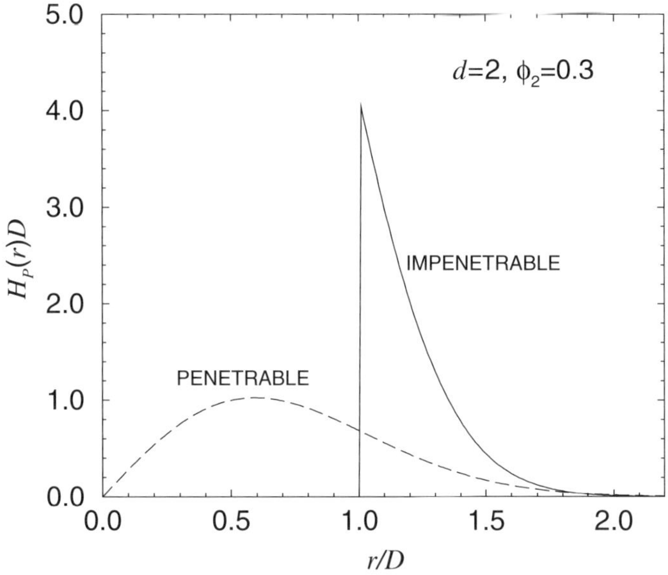

Figure 5.13 Comparison of the dimensionless nearest-neighbor probability density function $H_P(r)D$ for a system of fully penetrable disks [cf. (5.30)] and for an equilibrium system of hard disks [cf. (5.94)] of diameter D at $\phi_2 = 0.3$.

penetrable-sphere system are larger than in the hard-sphere case in any dimension, $H_P(x)$ for large x in the former will be larger than in the latter.

Mean Nearest-Neighbor Distance

In contrast to *ordered sphere packings* in d dimensions (Conway and Sloane 1993), few rigorous results concerning the structure of random packings of hard d-dimensional spheres have been established. Torquato (1995b) has established some rigorous bounds on the mean nearest-neighbor distance ℓ_P in *ergodic*, isotropic d-dimensional hard-sphere ensembles that depend on the packing fraction η and contact value $G_P(1) \equiv G_P(1^+)$. These results are stated in the form of three theorems and corollaries that immediately follow from them.

Theorem 5.1 *For any ergodic ensemble of isotropic packings of identical d-dimensional hard spheres, if* $G_P(1) \leq G_P(x)$ *for* $1 \leq x < \infty$, *then*

$$\ell_P \leq 1 + \frac{1}{d2^d \eta G_P(1)}. \tag{5.99}$$

Proof: Since $G_P(1) \leq G_P(x)$ for $1 \leq x < \infty$, the definition (2.109) for ℓ_P leads to the upper bound

$$\ell_P \le 1 + \int_1^\infty dx \exp[-2^d \eta G_P(1)(x^d - 1)]. \tag{5.100}$$

The integral of (5.100) can be further simplified by transforming to the variable $u = x - 1$, giving

$$\ell_P \le 1 + \int_0^\infty du \exp[-2^d \eta G_P(1)(u^d + du^{d-1} + \cdots + du)]. \tag{5.101}$$

Since each term of the polynomial $u^d + du^{d-1} + \cdots + du$ is nonnegative, the integral of (5.101) is bounded from above by retaining only the linear term du, yielding bound (5.99). ∎

Corollary 5.1 *In the special case of an equilibrium ensemble of isotropic packings of identical d-dimensional hard spheres, the mean nearest-neighbor distance ℓ_P is related to the thermodynamic pressure p, absolute temperature T, and Boltzmann's constant k by the inequality*

$$\ell_P \le 1 + \frac{1}{2d(p/\rho kT - 1)}. \tag{5.102}$$

Proof: This follows from three results: Theorem 5.1, the reduced equation of state $p/\rho kT = 1 + 2^{d-1}\eta G_P(1)$ [cf. (5.75)], and the the fact that $G_P(x)$ is a nondecreasing function of x for equilibrium ensembles. ∎

We now apply Theorem 5.1 to another ergodic ensemble, namely, the nonequilibrium random sequential addition (RSA) process discussed in Chapter 3. For identical d-dimensional spheres at the *saturation-limit* density (i.e., where the filling process terminates), the distance between nearest-neighbor surfaces must be less than unity (a sphere diameter). Clearly, this density will be less than the maximally-random-jammed (MRJ) density for equilibrium hard spheres where ℓ_P is exactly unity. However, since the radial distribution function at contact $g(1)$ [or equivalently $G_P(1)$] diverges as η approaches the RSA saturation limit, Theorem 5.1 leads to the contradictory result that $\ell_P = 1$ at the saturation limit. Thus, since statistically homogeneous systems of RSA spheres are ergodic and isotropic, one must conclude that $G_P(x)$ at or near the saturation limit is not always less than the contact value $G_P(1)$ for $1 \le x < \infty$. On physical grounds, it is clear that for sufficiently large x, $G_P(x)$ must be larger than $G_P(1)$. In summary, $G_P(x)$ is generally a *nonmonotonic* function of r for RSA spheres at sufficiently high densities, in contrast to equilibrium spheres. Indeed, subsequent to the work of Torquato (1995b), it was shown (Rintoul et al. 1996) that the exact solution for $G_P(x)$ for RSA one-dimensional hard rods exhibits this nonmonotonicity at high densities. We recall that as $\eta \to 0$, RSA and equilibrium ensembles become identical.

Theorem 5.2 *For any ergodic ensemble of isotropic packings of identical d-dimensional hard spheres, if $(1-\eta)^{-1} \le G_P(x)$ for $1 \le x < \infty$, then*

$$\ell_P \le 1 + \frac{1-\eta}{d 2^d \eta}. \tag{5.103}$$

Proof: The proof of this theorem proceeds in the same fashion as for Theorem 5.1. ∎

Remark:

1. The condition $(1-\eta)^{-1} \leq G_P(x)$ is true for a large class of ergodic ensembles, including the equilibrium ensemble. We note that for equilibrium hard rods ($d = 1$), the upper bound (5.103) is exact, since $G_P(r) = (1-\eta)^{-1}$ and hence $\ell_P = 1 + (1-\eta)/2\eta$.

To illustrate the utility of Theorem 5.2, we again examine the RSA process. For RSA rods ($d = 1$) at $\eta = 0.5$, Monte Carlo simulations have yielded $\ell_P = 1.53$. Theorem 5.2, however, states that $\ell_P \leq 1.5$ at $\eta = 0.5$. We conclude that $G_P(x)$ for RSA rods at $\eta = 0.5$ is not always larger than $(1-\eta)^{-1} = 2$, *in contrast to equilibrium rods*. This conclusion is true for $\eta > 0.5$ as well.

Theorem 5.3 *For any ergodic ensemble of isotropic packings of identical d-dimensional hard spheres,*

$$\ell_P \leq 1 + \frac{1}{d2^d\eta}. \tag{5.104}$$

Proof: For any ergodic, isotropic hard-sphere ensemble, it is always true that $G_P(x) \geq 1$ for $1 \leq x < \infty$, since $G_P(x) = 1$ applies to "point" particles, i.e., spatially uncorrelated spheres. Using this fact, the proof proceeds in the same fashion as for Theorem 5.1. ∎

Remark:

1. Theorem 5.1 is an ensemble-dependent result in that ℓ_P is given in terms of the contact value $G_P(1)$. By contrast, although the inequalities of Theorems 5.2 and 5.3 are weaker than (5.99), they are also more general, since they depend only on the packing fraction η. Figure 5.14 shows how the upper bound on ℓ_P of Theorem 5.3 dramatically drops off as d is increased.

Theorem 5.3, the most general bound, has some interesting corollaries, which we now state.

Corollary 5.2 *Any packing of identical d-dimensional hard spheres in which the mean nearest-neighbor distance obeys the relation*

$$\ell_P > 1 + \frac{1}{d2^d\eta} \tag{5.105}$$

cannot be ergodic and isotropic.

Remarks:

1. Relation (5.105) defines a region in the η-ℓ_P plane that is prohibited to ergodic, isotropic packings, and thus Corollary 5.2 provides a quantitative and experimentally measurable criterion to ascertain when a hard-sphere system is *definitely not ergodic and isotropic*.

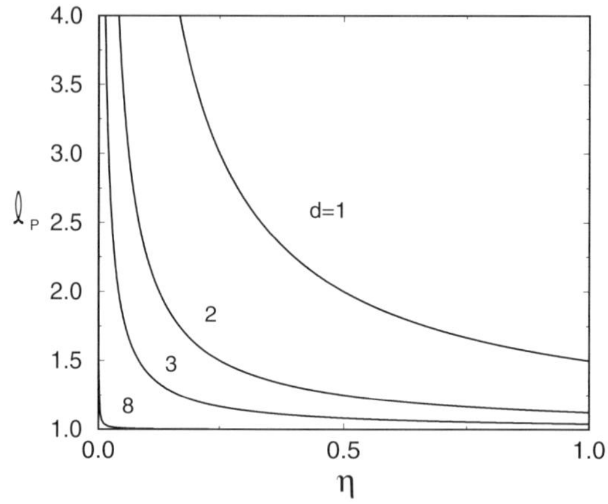

Figure 5.14 Upper bound on ℓ_P of Theorem 5.3 versus packing fraction η for several d.

2. Examples of *nonergodic, anisotropic* ensembles that obey (5.105) are periodic cubic arrays of particles at sufficiently small packing fractions. For example, for periodic identical hard rods ($d = 1$), $\ell_P = 1 + (1 - \eta)/\eta$, and hence this system satisfies (5.105) for all $\eta < 1/2$. Figure 5.15 depicts the region prohibited to ergodic, isotropic systems for $d = 3$.

Corollary 5.3 *As the dimension d of any ergodic ensemble of isotropic packings of identical hard spheres increases, the mean distance drops off at least as fast as $(d2^d)^{-1}$ and approaches unity for nonzero η in the limit $d \to \infty$. The maximum packing fraction η_c in turn approaches zero in the limit $d \to \infty$.*

Remark

1. This provides a rigorous proof that $\eta_c \to 0$ as $d \to \infty$ for ergodic isotropic hard-sphere systems. Thus, all such ensembles (equilibrium or not) lose their distinction in so far as packing efficiency is concerned as d is made large.

In the special case of isotropic equilibrium systems of hard particles (stable liquid branch and metastable extension), bound (5.99) of Theorem 5.1 can be written explicitly for $d = 2$ and $d = 3$ using the aforementioned approximations for $G_P(1)$. For $d = 2$, using (5.95) and (5.99), we obtain

$$\ell_P \leq \begin{cases} 1 + \dfrac{(1-\eta)^2}{8\eta(1-0.436\eta)}, & 0 \leq \eta \leq \eta_f, \\ 1 + \dfrac{(\eta_c - \eta)}{8\eta g_f(1)(\eta_c - \eta_f)}, & \eta_f \leq \eta \leq \eta_c. \end{cases} \tag{5.106}$$

For $d = 3$, using (5.98) and (5.99), we have

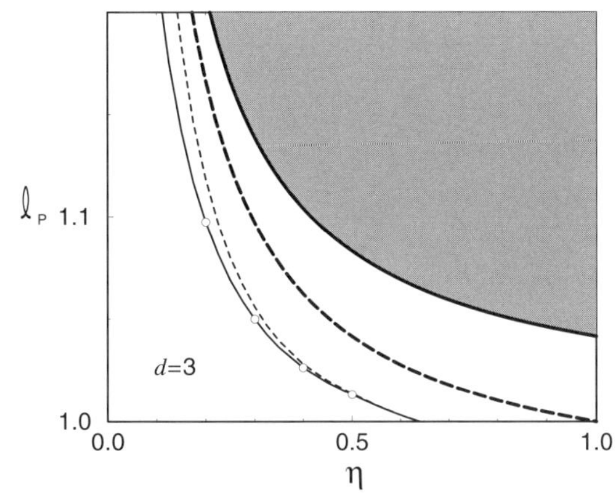

Figure 5.15 Mean nearest-neighbor distance ℓ_P (in units of diameter) versus packing fraction η for a system of hard spheres ($d = 3$). Thin solid line is equilibrium prediction from (2.109) and (5.98); open circles are corresponding simulation data. Thin dashed line is upper bound of Theorem 5.1 for an equilibrium ensemble from (5.107). Thick dashed and solid lines are upper bounds of Theorems 5.2 and 5.3, respectively. Shaded region is prohibited to ergodic, isotropic systems of hard spheres according to Corollary 5.2.

$$\ell_P \le \begin{cases} 1 + \dfrac{(1-\eta)^3}{24\eta(1-\eta/2)}, & 0 \le \eta \le \eta_f, \\ 1 + \dfrac{(\eta_c - \eta)}{24\eta g_f(1)(\eta_c - \eta_f)}, & \eta_f \le \eta \le \eta_c. \end{cases} \tag{5.107}$$

Figure 5.15 also depicts the prediction of the mean nearest-neighbor distance ℓ_P for equilibrium hard spheres ($d = 3$) versus the packing fraction η as computed from definition (2.109) and formula (5.98). This prediction is seen to be in excellent agreement with available simulation data (Torquato and Lee 1990, Rintoul and Torquato 1998). In the limit $\eta \to \eta_c$, this prediction of ℓ_P correctly goes to unity. Included in the figure are the bounds of Theorems 5.1, 5.2, and 5.3. The upper bound of Theorem 5.1 is very sharp (i.e., nearly exact) for packing fractions between freezing and the MRJ state, becoming exact in the limit $\eta \to \eta_c$.

5.2.6 Pore-Size Functions

Torquato and Avellaneda (1991) used interrelation (2.84) and the approximate expression for $H_V(r)$ for hard spheres given by Torquato et al. (1990) to get the pore-size functions $P(\delta)$ and $F(\delta)$. Slightly more accurate expressions for these functions can be obtained using (5.91). The results from this latter approximation for $P(\delta)$ are plotted in Figure 5.16 for $\phi_2 = 0.5$. Included in the figure are corresponding results for two other values of the impenetrability parameter λ in the cherry-pit model (Section 3.1.2): $\lambda = 0$

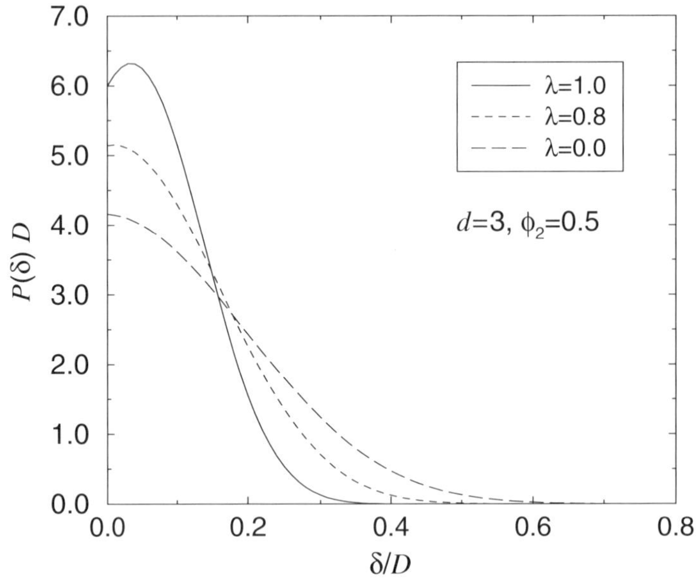

Figure 5.16 The pore-size density function $P(\delta)$ versus the dimensionless distance δ/D for equilibrium ensembles of spheres of radius $R = D/2$ in the *cherry-pit* model for three different values of the impenetrability parameter λ.

(overlapping spheres) and $\lambda = 0.8$. The method used to find $P(\delta)$ for the intermediate case $\lambda = 0.8$ is described in Section 5.3.4.

5.2.7 Point/q-Particle Correlation Functions

From the series representation (4.46) for the canonical function H_n and the definition (4.74), we can recover the series expression for the matrix point/q-particle function $G_n(\boldsymbol{x}; \boldsymbol{r}^q) \equiv G_n^{(1)}(\boldsymbol{x}; \boldsymbol{r}^q)$ for impenetrable-sphere systems, first given by Torquato (1986a):

$$G_n(\boldsymbol{x}; \boldsymbol{r}^q) = \prod_{i=1}^{n} e(|\boldsymbol{x} - \boldsymbol{r}_i|; R)\rho_q(\boldsymbol{r}^q) - \int \rho_{q+1}(\boldsymbol{r}^{q+1}) m(|\boldsymbol{x} - \boldsymbol{r}_{q+1}|; R) d\boldsymbol{r}_{q+1}. \tag{5.108}$$

We see that the series for hard spheres terminates exactly after two terms because all of the remaining terms in the series contain the subdiagram (5.37), which is identically zero.

In the special case $n = 2$ and letting $r = |\boldsymbol{x} - \boldsymbol{r}_1|$, we have

$$G_2(r) = e(r)[\rho - \rho g_2 \otimes m]$$

$$= \rho e(r)\Big[\, 1 - \quad \Big]. \tag{5.109}$$

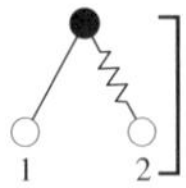

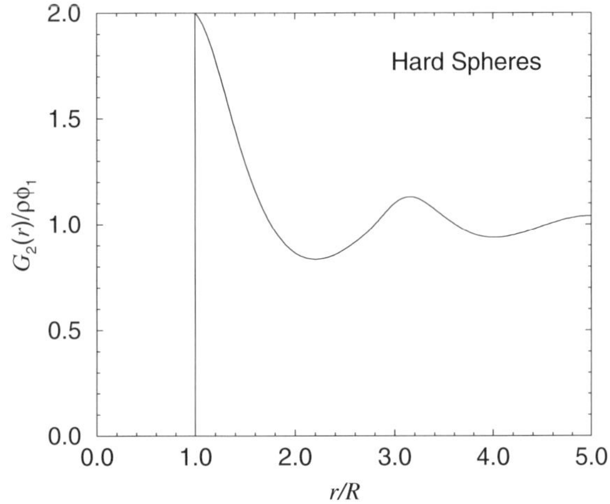

Figure 5.17 The point/1-particle function $G_2(r)$, scaled by its asymptotic value $\rho\phi_1$, for an equilibrium system of hard spheres of radius R at a packing fraction $\phi_2 = 0.5$.

Since the two-body diagram is a double convolution integral, it has the following Fourier representation:

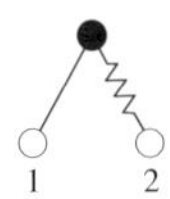

$$= \eta + \rho\mathcal{F}^{-1}[\tilde{m}\tilde{h}], \tag{5.110}$$

where $h(r) = g_2(r) - 1$ is the total correlation function. For example, when $d = 3$, we have that

$$\mathcal{F}^{-1}[\tilde{m}\tilde{h}] = \frac{1}{2\pi^2 r}\int_0^\infty \tilde{m}(k)\tilde{h}(k)k\sin(kr)dk. \tag{5.111}$$

For $d = 3$, Torquato (1986a) used the Percus–Yevick approximation for $\tilde{c}$ [given by (3.61)] and the Verlet–Weis correction to evaluate the two-body diagram (5.110) and thus G_2. This calculation of G_2 (scaled by its asymptotic value) for a packing fraction $\phi_2 = 0.5$ is shown in Figure 5.17.

5.3 Interpenetrable Spheres

The expression (4.46) for the canonical function H_n in the appropriate limits leading to the correlation functions of interest in random media is in general an infinite series for interpenetrable spheres. This implies that higher-order n-particle density functions ρ_n are required, which are not known analytically for $d \geq 2$. Thus, analytical expressions of the density functions for interpenetrable spheres are generally obtained using various approximation schemes.

In the case of the penetrable-concentric-shell (or cherry-pit) model, described in Chapter 3, there is a useful isomorphism relating the void nearest-neighbor functions of systems of totally impenetrable spheres ($\lambda = 1$) to that of systems with an arbitrary impenetrability parameter λ. These results lead trivially to results for the porosity ϕ_1, specific surface s, and pore-size density function $P(\delta)$ for the cherry-pit model. Moreover, one can get the porosity and specific surface for the space and surface available, respectively, to finite-size test particles in the cherry-pit model. All of these results are described below as well as remarks on how to find approximate expressions for the two-point probability function S_2 for the cherry-pit model.

5.3.1 Nearest-Neighbor Functions

Torquato et al. (1990) observed that there is a simple isomorphism between the void nearest-neighbor functions E_V, H_V, and G_V of totally-impenetrable-sphere systems and corresponding quantities for systems with an arbitrary impenetrability parameter λ. Specifically, given such expressions for $\lambda = 1$, one can obtain corresponding results for $\lambda < 1$ by replacing D with λD, since it is only the hard core that has meaning for the nearest-neighbor quantities. This implies the transformations $x \to x/\lambda$ and $\eta \to \eta\lambda^d$, where $x = r/D$. Thus, we have the following exact mapping:

$$E_V(x;\eta,\lambda) = E_V\Big(\frac{x}{\lambda};\eta\lambda^d,1\Big), \tag{5.112}$$

$$H_V(x;\eta,\lambda) = H_V\Big(\frac{x}{\lambda};\eta\lambda^d,1\Big), \tag{5.113}$$

$$G_V(x;\eta,\lambda) = G_V\Big(\frac{x}{\lambda};\eta\lambda^d,1\Big). \tag{5.114}$$

These results are valid for any homogeneous ensemble of spheres in the cherry-pit model. In particular, one can apply the equilibrium expressions (5.78)–(5.80) for $d = 1$, (5.87)–(5.89) for $d = 2$, and (5.90)–(5.92) for $d = 3$.

Recall that $E_V(r)$ and $H_V(r)$ are the volume fraction and specific surface available to a test particle of radius $b = r - D/2$, respectively (see Section 2.8). Thus, if the test particle is a point particle $b = 0$ that cannot penetrate the sphere phase, then we can get the standard matrix-phase volume fraction (porosity) $\phi_1(\eta,\lambda)$ and specific surface $s(\eta,\lambda)$ for the cherry-pit model by evaluating the expressions above at $x = 1/2$, i.e.,

$$\phi_1(\eta,\lambda) = E_V\Big(\frac{1}{2\lambda};\eta\lambda^d,1\Big), \tag{5.115}$$

$$s(\eta,\lambda) = H_V\Big(\frac{1}{2\lambda};\eta\lambda^d,1\Big). \tag{5.116}$$

More generally, the volume fraction $\phi_1(\eta,\lambda,\beta)$ and specific surface $s(\eta,\lambda,\beta)$ available to a test particle of radius $b = \beta D/2$ are given by

$$\phi_1(\eta,\lambda,\beta) = E_V\Big(\frac{1+\beta}{2\lambda};\eta\lambda^d,1\Big), \tag{5.117}$$

$$s(\eta,\lambda,\beta) = H_V\Big(\frac{1+\beta}{2\lambda};\eta\lambda^d,1\Big). \tag{5.118}$$

The results for a finite-size test particle are of relevance in understanding gel size-exclusion chromatography (Yau et al. 1979, Rikvold and Stell 1985) and transport of macromolecules in disordered porous media (Torquato 1991c, Kim and Torquato 1992a). The isomorphism described above that leads to the rescaling of the totally-impenetrable-sphere results for ϕ_1 and s was first utilized by Rikvold and Stell (1985), although they obtained different approximations, as discussed below.

5.3.2 Volume Fraction

Using the mapping (5.115) and expressions (5.78), (5.87), and (5.90), we have the following relations for the porosity $\phi_1(\eta, \lambda)$ for equilibrium spheres in the cherry-pit model in the first three space dimensions:

$$\phi_1(\eta, \lambda) = (1 - \eta\lambda^d) \exp\left[-\frac{(1-\lambda^d)\eta}{(1-\eta\lambda^d)^d} \right] A(\eta, \lambda), \tag{5.119}$$

where

$$A(\eta, \lambda) = 1, \qquad d = 1,$$

$$A(\eta, \lambda) = \exp\left[-\frac{\eta^2\lambda^2(0.128 - 2.256\lambda + 2.128\lambda^2)}{(1-\eta\lambda^2)^2} \right], \qquad d = 2,$$

$$A(\eta, \lambda) = \exp\left\{ -\frac{\eta^2\lambda^3(\lambda - 1)}{2(1-\eta\lambda^3)^3}[(7\lambda^2 + 7\lambda - 2) - 2\eta\lambda^3(7\lambda^2 - 5\lambda + 1) + \eta^2\lambda^6(5\lambda^2 - 7\lambda + 2)] \right\}, \qquad d = 3.$$

In the extreme limits $\lambda = 0$ and $\lambda = 1$, relation (5.119) yields the exact results (4.19) and (4.21), respectively. The result for $d = 1$ is exact for all λ and was first given by Rikvold and Stell (1985). The results for $d = 2$ and $d = 3$ are in excellent agreement with the computer-simulation results of Lee and Torquato (1988a). The present results are summarized in Figures 5.18 and 5.19, where we actually plot the sphere volume fraction ϕ_2 versus the hard-core volume fraction $\eta\lambda^d$ (which attains its maximum value at the MRJ state of about 0.83 and 0.64 for $d = 2$ and $d = 3$) for several values of λ. Rikvold and Stell (1985) obtained scaled-particle approximations for $d = 2$ and $d = 3$ that are only slightly less accurate than (5.119). It should be noted that Chiew and Glandt (1984) and Torquato and Stell (1984) also obtained analytical approximations for the porosity ϕ_1 and specific surface in the "permeable-sphere" model (Section 3.1.2).

5.3.3 Specific Surface

Application of the mapping (5.116) and expressions (5.79), (5.88), and (5.91) gives the following relations for the specific surface $s(\eta, \lambda)$ for equilibrium spheres in the cherry-pit model in the first three space dimensions:

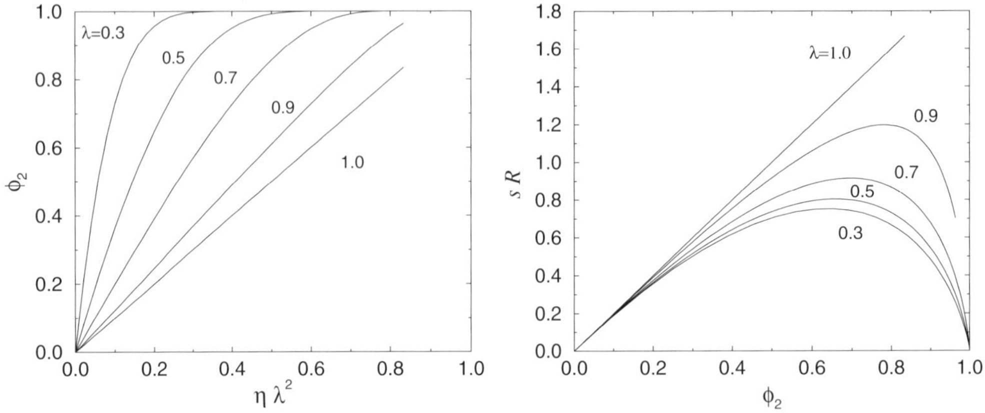

Figure 5.18 Left panel: Area fraction of disks ϕ_2 versus hard-core area fraction $\eta\lambda^2$ for given values of the impenetrability parameter λ as obtained from expression (5.119) for the porosity $\phi_1 = 1 - \phi_2$ with $d = 2$. Right panel: Corresponding specific surface s versus ϕ_2.

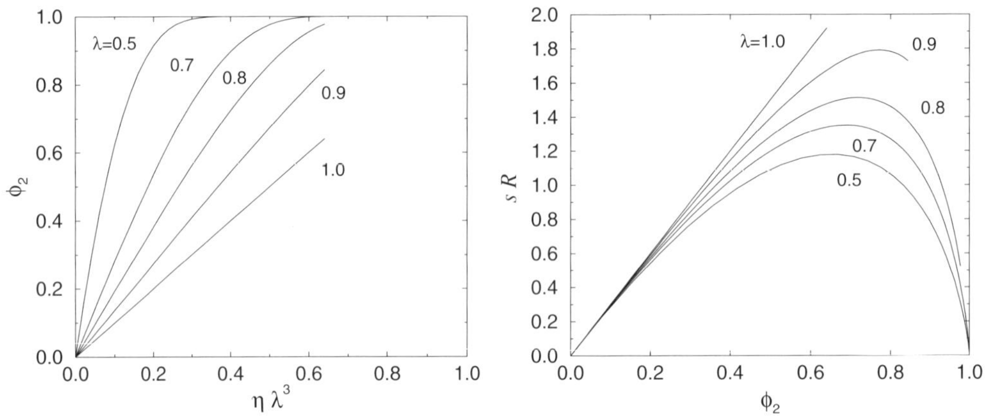

Figure 5.19 Left panel: Volume fraction of spheres ϕ_2 versus hard-core volume fraction $\eta\lambda^3$ for given values of the impenetrability parameter λ as obtained from expression (5.119) for the porosity $\phi_1 = 1 - \phi_2$ with $d = 3$. Right panel: Corresponding specific surface s versus ϕ_2.

$$s(\eta, \lambda) = \frac{d}{R} \frac{\eta}{(1 - \eta\lambda^d)^d} B(\eta, \lambda)\, \phi_1(\eta, \lambda), \tag{5.120}$$

where

$$B(\eta, \lambda) = 1, \qquad d = 1,$$
$$B(\eta, \lambda) = 1 + \eta\lambda^2(0.128 - 1.128\lambda), \qquad d = 2,$$
$$B(\eta, \lambda) = 1 + \eta\lambda^3(1 - 3\lambda) + \eta^2\lambda^6(1 - 2\lambda^2)^2 - \eta^3\lambda^9(1 - 3\lambda + 2\lambda^2), \quad d = 3,$$

$\phi_1(\eta, \lambda)$ is given by (5.119), and $R = D/2$. In the extreme limits, $\lambda = 0$ and $\lambda = 1$, relation (5.120) yields the exact results (4.20) and (4.22). The result for $d = 1$ is exact for all λ and was first found by Rikvold and Stell (1985). Figures 5.18 and 5.19 depict the present results for $d = 2$ and $d = 3$, respectively.

5.3.4 Pore-Size Functions

We recall that for monodisperse spheres, the pore-size probability density function $P(\delta)$ and its associated complementary cumulative distribution function $F(\delta)$ are trivially determined from the functions H_V and E_V via interrelations (2.84) and (2.85). Thus, in light of the discussion of Section 5.3.1 and relations (5.112) and (5.113), we find that the pore-size functions for the cherry-pit model are given by

$$P(\delta; \eta, \lambda) = \frac{H_V\left(\frac{\beta+1}{2\lambda}; \eta\lambda^d, 1\right)}{E_V\left(\frac{1}{2\lambda}; \eta\lambda^d, 1\right)}, \tag{5.121}$$

$$F(\delta; \eta, \lambda) = \frac{E_V\left(\frac{\beta+1}{2\lambda}; \eta\lambda^d, 1\right)}{E_V\left(\frac{1}{2\lambda}; \eta\lambda^d, 1\right)}, \tag{5.122}$$

where $\beta = \delta/R$ and $R = D/2$. Figure 5.16 shows $P(\delta)$ for three-dimensional equilibrium spheres in the intermediate case $\lambda = 0.8$ as obtained from (5.121) and expressions (5.90) and (5.91).

5.3.5 Other Statistical Descriptors

Unfortunately, the simple rescaling of the arguments of the void nearest-neighbor functions is no longer valid when one must consider spatial correlation between two or more test particles or between a single test particle and a subset of the actual inclusions. This means, for example, that the n-point probability function S_n for $n \geq 2$, point/q-particle function $G_n(\mathbf{x}; \mathbf{r}^q)$ (for $q \geq 1$) and two-point surface correlation functions F_{sv} and F_{ss} for the cherry-pit model cannot be obtained by a simple rescaling of the corresponding totally-impenetrable-sphere results.

However, an important observation made by Torquato and Stell (1983a) can be exploited here that enables one to compute these more complicated correlation functions in the cherry-pit model or any interpenetrable-sphere model. They showed that the matrix n-point probability function S_n is equivalent to the n-body correlation function g_n of "point" particles at infinite dilution in a sea of the actual inclusions of radius R at density. In particular, for a statistically homogeneous medium, they proved that

$$g_n(\mathbf{x}^n) = \frac{S_n(\mathbf{x}^n)}{\phi_1^n}, \tag{5.123}$$

where $\mathbf{x}^n$ denotes the positions of the n point particles. Thus, one can take advantage of existing statistical-mechanical results for particle statistics of hard-sphere mixtures.

For example, Lebowitz (1964) has found the solution of the mixture Ornstein–Zernike equation for the radial distribution function g_2 in the Percus–Yevick approximation. Application of this solution for a binary mixture of point particles and spheres in the cherry-pit model in the limit of infinite dilution gives, among other quantities, the radial distribution function g_2 for the point-particle species. The function S_2 is obtained after use of the equivalence result (5.123) with $n = 2$. The accuracy of the approximation increases as the value of the impenetrability parameter λ approaches unity. This mixture picture can be used to get other correlation functions, such as F_{sv}, F_{ss}, and G_2. Indeed, Given and Stell (1991) used this mapping and approximate solutions of Ornstein–Zernike equations to find all of the aforementioned two-point functions for other interpenetrable-sphere models.

We mention that the two-point probability function $S_2(r)$ has been determined from computer simulations by Smith and Torquato (1988) for distributions of disks in the cherry-pit model. It was found that for the range $0 < \lambda < 0.5$, S_2 was negligibly different from S_2 for fully penetrable disks ($\lambda = 0$) at the same ϕ_2. For $0.5 \le \lambda < 1$, the amplitude of the oscillations in S_2, for fixed ϕ_2, increased as λ increased. The specific surface for this two-dimensional model was also evaluated in this study.

5.4 Statistically Inhomogeneous Systems

Considerable progress has been made in recent years in characterizing the microstructure of statistically *homogeneous* two-phase random media. However, significantly less research has been devoted to the study of statistically *inhomogeneous* two-phase media. In such instances, the system has a *preferred origin*, and ergodicity is lost, i.e., one cannot equate ensemble and volume averages. Examples of statistically inhomogeneous two-phase media include composites in which the heterogeneity length scale is not much smaller than the macroscopic size of the sample, porous media with spatially variable fluid permeability (Gelhar 1993), distributions of galaxies (Bond, Kofman and Pogosyan 1996), and functionally graded materials (Cherradi, Kawasaki and Gasik 1994). Figure 2.2 shows realizations of models of statistically inhomogeneous media.

Following the development of the study of the microstructure and properties of homogeneous random media, Quintanilla and Torquato (1997a) introduced a model for particulate, statistically inhomogeneous two-phase random media. This model is a two-phase system consisting of an *inhomogeneous* distribution of fully penetrable spheres in space whose particle density obeys any specified variation in the local volume fraction. This inhomogeneous model is nontrivial in that cluster formation naturally arises and it permits significantly more complicated microstructures than simple layered models for graded materials.

Quintanilla and Torquato obtained analytical expressions for various correlation functions, including the canonical n-point function H_n, the nearest-neighbor functions E and H, and the lineal-path function L. These expressions were derived using both the above machinery and also measures on nonstationary Poisson processes (Stoyan et al. 1995). The form of these microstructural functions closely resembles those given in Section 5.1, but unlike those in the homogeneous case, these functions will depend on the *absolute as opposed to relative positions* of their arguments.

CHAPTER 6

Polydisperse Spheres

Polydispersivity in the size of the particles constitutes a fundamental feature of the microstructure of a wide class of dispersions of technological importance, including composite solid propellant combustion (Kerstein 1987), sintering of powders (Rahaman 1995), colloids (Russel et al. 1989), transport and mechanical properties of particulate composite materials (Christensen 1979), and flow in packed beds (Scheidegger 1974). The effect of particle-size distribution on the microstructure and effective properties of dispersions can be dramatic and therefore is of great interest.

Starting with the exact series representation (4.81) of the canonical n-point function H_n, we will obtain explicit expressions for various microstructural descriptors and compute them for systems of d-dimensional *fully penetrable* and *totally impenetrable* spheres with a *polydispersivity* in size. We will obtain results for the n-point probability functions, surface correlation functions, lineal-path function, chord-length density function, pore-size functions, nearest-surface functions, and point/q-particle correlation functions. We will show that the polydisperse results are relatively simple generalizations of the monodisperse results of Chapter 5.

The spheres generally possess a continuous distribution in radius R characterized by a (normalized) probability density $f(R)$ [cf. (4.79)]. Recall that the discrete case is contained in $f(R)$ via relation (4.80). The average of any function $w(R)$ is defined by

$$\langle w(R)\rangle = \int_0^\infty w(R)f(R)dR. \tag{6.1}$$

The reduced density η of the system is defined as

$$\eta = \rho\langle v_1(R)\rangle, \tag{6.2}$$

where ρ is the total number density and the average sphere volume is

$$\langle v_1(R)\rangle = \frac{\pi^{d/2}}{\Gamma(1+d/2)}\,\langle R^d\rangle. \tag{6.3}$$

The average surface area of the spheres is given by

$$\langle s_1(R)\rangle = \frac{d\pi^{d/2}}{\Gamma(1+d/2)}\,\langle R^{d-1}\rangle. \tag{6.4}$$

There is a variety of choices for the size distribution $f(R)$ that one can employ. Two probability densities that have been widely used to characterize physical phenomena are the Schulz (1939) and log-normal (Cramer 1954) distributions. The Schulz distribution is defined as

$$f(R) = \frac{1}{\Gamma(m+1)}\left(\frac{m+1}{\langle R\rangle}\right)^{m+1} R^m \exp\left[\frac{-(m+1)R}{\langle R\rangle}\right], \tag{6.5}$$

where $\Gamma(x)$ is the gamma function. When the parameter m is restricted to integer values, m lies in the interval $[0,\infty)$, $\Gamma(m+1)=m!$, and the nth moment of this distribution is given by

$$\langle R^n\rangle = \frac{(m+n)!}{m!}\frac{1}{(m+1)^n}\,\langle R\rangle^n. \tag{6.6}$$

By increasing m, the variance decreases, i.e., the distribution becomes sharper. In the monodisperse limit $m\to\infty$, $f(R)\to\delta(R-\langle R\rangle)$. The case $m=0$ gives an exponential distribution in which many particles have extremely small radii.

The log-normal distribution is defined as

$$f(R) = \frac{1}{R\sqrt{2\pi\beta^2}}\exp\left\{-\frac{[\ln(R/\langle R\rangle)]^2}{2\beta^2}\right\}, \tag{6.7}$$

where $\beta^2 = \langle(\ln R)^2\rangle - \langle\ln R\rangle^2$. It is seen that the quantity $\ln R$ is normally or Gaussian distributed. The nth moment is given by

$$\langle R^n\rangle = \exp(n^2\beta^2/2)\,\langle R\rangle^n. \tag{6.8}$$

As $\beta^2\to 0$, $f(R)\to\delta(R-\langle R\rangle)$. Figure 6.1 shows the Schulz and log-normal size distributions.

6.1 Fully Penetrable Spheres

Allowing overlapping spheres to have a polydispersivity in size enables one to model a richer class of microstructures than that of the monodisperse case, as shown in Figure 6.2. This example qualitatively resembles the microstructure of a cermet, also shown in Figure 6.2.

Since the n-particle probability density function is given by $\rho_n(\mathbf{r}^n)=\rho^n$ for homogeneous systems of d-dimensional polydisperse fully penetrable spheres (where ρ is the

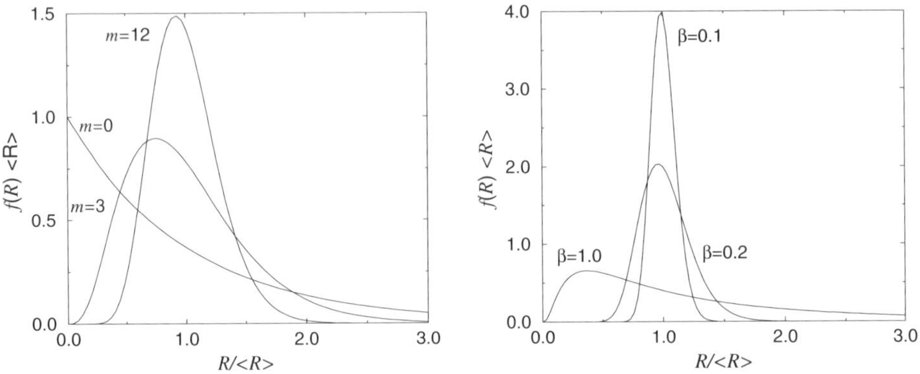

Figure 6.1 Left panel: Schulz size distribution for several values of m. Right panel: Log-normal size distribution for several values of β.

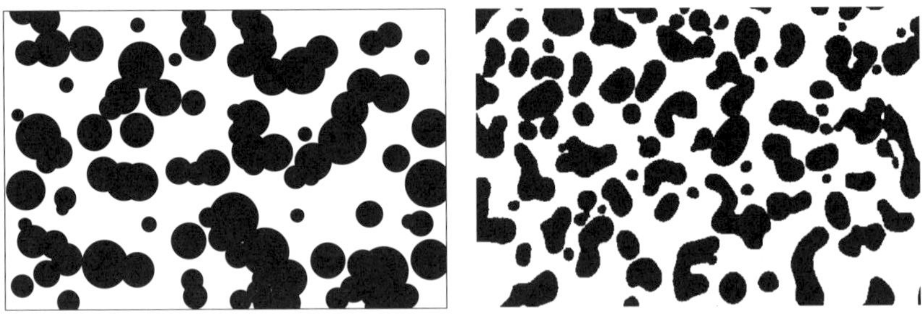

Figure 6.2 Left panel: Polydisperse system of overlapping spheres. Right panel: Micrograph of a silver-magnesium fluoride cermet (Poladian 1990). The inclusions are made of silver and form a variety of shapes from simple globules to long filaments.

total number density), the series representation (4.81) for the canonical function H_n yields the analytical result

$$H_n(\mathbf{x}^m; \mathbf{x}^{p-m}; \mathbf{r}^q) = (-1)^m \frac{\partial}{\partial b_1} \cdots \frac{\partial}{\partial b_m} \left\{ \rho^q \prod_{\ell=1}^{q} \prod_{k=1}^{p} e(|\mathbf{x}_k - \mathbf{r}_\ell|; a_k^{(\ell)}) \right.$$
$$\left. \times \exp[-\rho \, \langle v_p(\mathbf{x}^p; a_1^{(1)}, \ldots, a_p^{(1)}) \rangle] \right\}, \tag{6.9}$$

where

$$\langle v_p(\mathbf{x}^p; a_1^{(1)}, \ldots, a_p^{(1)}) \rangle = \int \int \Big\{ 1 - \prod_{i=1}^{p} [1 - m(|\mathbf{r}_1 - \mathbf{x}_i|; a_i^{(1)})] \Big\} f(R_1) dR_1 d\mathbf{r}_1 \tag{6.10}$$

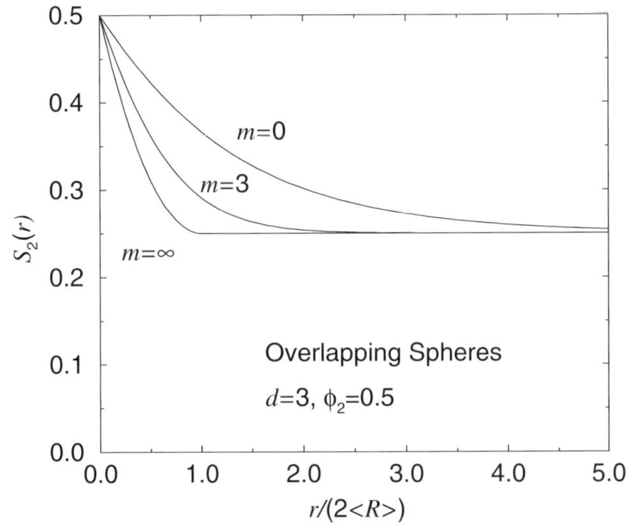

Figure 6.3 The matrix two-point function $S_2(r)$ versus dimensionless distance $r/(\langle 2R\rangle)$ for three-dimensional polydisperse systems of overlapping spheres for a sphere volume fraction $\phi_2 = 0.5$. The Schulz size distribution $f(R)$ is used here. The monodisperse case is $m = \infty$.

and $v_p(\mathbf{x}^p; a_1^{(1)}, \ldots, a_p^{(1)})$ is defined by (5.2). Also, $a_i^{(j)} = b_i + R_j$, b_i is the radius of the ith test particle, and R_j is the radius of the inclusion at $\mathbf{r}_j$. As in the monodisperse case, H_n for this polydisperse model is expressible analytically in terms of purely geometrical information.

6.1.1 n-Point Probability Functions

Utilizing the overlapping-sphere result (6.9) for H_n and definition (4.85) gives the n-point probability function S_n for phase 1 as

$$S_n(\mathbf{x}^n) = \exp[-\rho \, \langle v_n(\mathbf{x}^n; R)\rangle \,], \tag{6.11}$$

where v_n is the union volume of n d-dimensional spheres of radius R defined by (5.3). Recall that v_n and thus S_n may be computed in principle for any n; see Section 5.1. Relation (6.11) was given by Stell and Rikvold (1987) for $d = 3$ and by Joslin and Stell (1986a) for $d = 2$; see also Stoyan et al. (1995). When $n = 1$, we see that (6.11) yields the volume fraction

$$S_1 = \phi_1 = \exp[-\eta] \tag{6.12}$$

for homogeneous media, where η is given by (6.2). This expression was given by Chiew, Stell and Glandt (1985) for the case $d = 3$. A graph of $S_2(r) = \exp[-\rho\langle v_2(r; R)\rangle]$ is shown in Figure 6.3 for the Schulz size distribution, where $v_2(r; R)$ is given by (5.10) and $2\langle R\rangle$ is the average diameter.

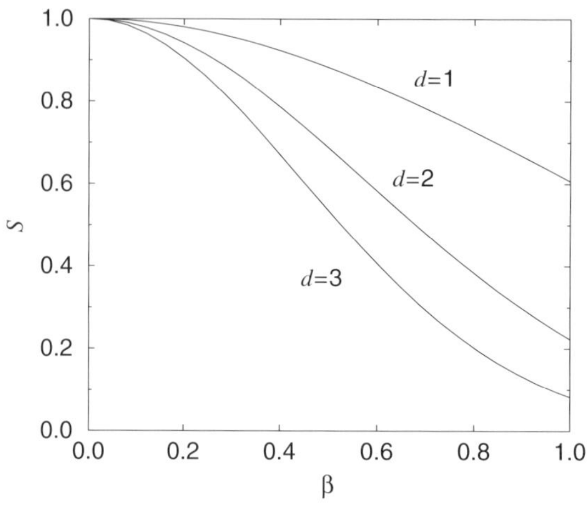

Figure 6.4 The surface area ratio S versus the log-normal polydispersivity parameter β for the first three space dimensions.

6.1.2 Surface Correlation Functions

Combination of the definition (4.86) as well as results (6.4) and (6.9) yields the following simple expression for the specific surface:

$$s = \rho\phi_1 \langle s_1(R)\rangle = \eta\phi_1 d\,\frac{\langle R^{d-1}\rangle}{\langle R^d\rangle}, \tag{6.13}$$

where ϕ_1 and η are given by (6.12) and (6.2), respectively. If we let S represent the ratio of the specific surface of a polydisperse system to that of a monodisperse system at the same volume fraction with radius $\langle R\rangle$, then we find upon use of (6.13) and (4.20) that

$$S = \frac{\langle R^{d-1}\rangle}{\langle R^d\rangle}\,\langle R\rangle. \tag{6.14}$$

The key ratio $\langle R^d\rangle/\langle R^{d-1}\rangle$ is proportional to the ratio of average volume to average surface area of d-dimensional spheres. For the Schulz distribution

$$S = \frac{m+1}{m+d}, \tag{6.15}$$

and for the log-normal distribution

$$S = \exp[(1-2d)\beta^2/2]. \tag{6.16}$$

In both cases, increasing the degree of polydispersivity decreases the surface area ratio S. In contrast to the log-normal case, S for the Schulz distribution is the constant unity for $d = 1$, i.e., the specific surface of the polydisperse one-dimensional system is the same as the monodisperse one. The ratio S is shown in Figure 6.4 for the log-normal distribution for $d = 1, 2,$ and 3.

Analytical expressions for the two-point surface correlation functions F_{sv} and F_{ss} for polydisperse overlapping spheres are obtained by applying (6.9) and definitions (4.87) and (4.88). For $d = 3$, we find that

$$F_{sv}(r) = 4\pi\rho \left\langle \left[R^2 - \left(\frac{R^2}{2} - \frac{rR}{4} \right) \Theta(2R - r) \right] \right\rangle S_2(r) \tag{6.17}$$

and

$$F_{ss}(r) = \left\langle 16\pi^2\rho^2 \left[R^2 - \left(\frac{R^2}{2} - \frac{rR}{4} \right) \Theta(2R - r) \right]^2 + \frac{2\pi\rho}{r} R^2 \Theta(2R - r) \right\rangle S_2(r), \tag{6.18}$$

where S_2 is given by (6.11) with $n = 2$. Miller and Torquato (1989) obtained expressions for the functions F_{sv} and F_{ss} of binary mixtures of overlapping spheres. This discrete case is easily found from the continuous results given immediately above and the size distribution (4.80) with $M = 2$. These discrete results were subsequently generalized to the continuous case by Torquato and Lu (1990) and by Given, Blawzdziewicz and Stell (1990a).

6.1.3 Lineal-Path Function

According to Section 4.5, we can obtain an exact expression for the "matrix" lineal-path function $L(z) \equiv L^{(1)}(z)$ of any homogeneous and isotropic system of interacting polydisperse spheres by using the series expression for e_V [obtained from (4.81) and (4.90)] and taking the "exclusion region" indicator function m to be given by (4.78). This gives

$$L(z) = 1 + \sum_{k=1}^{\infty} \frac{(-1)^k}{k!} \int dR_1 \cdots dR_k f(R_1) \cdots f(R_k) \rho_k(\mathbf{r}^k; R_1, \ldots, R_k) \times \prod_{j=1}^{k} m(\mathbf{x} - \mathbf{r}_j; z, R_j) d\mathbf{r}_j, \tag{6.19}$$

which was originally found by Lu and Torquato (1992c).

For d-dimensional polydisperse overlapping spheres, we find from (6.19), using the result $\rho_k(\mathbf{r}^k) = \rho^k$ and the same arguments as in the monodisperse case (Section 5.1.3), the exact relation

$$L(z) = \phi_1^{\,1 + \frac{\omega_{d-1}\langle R^{d-1}\rangle}{\omega_d \langle R^d \rangle} z}, \tag{6.20}$$

where ω_d is given by (5.22). For the first three space dimensions (Lu and Torquato 1992c), we have

$$L(z) = \begin{cases} \phi_1^{\,1 + \frac{1}{2\langle R \rangle} z}, & d = 1, \\ \phi_1^{\,1 + \frac{2\langle R \rangle}{\pi \langle R^2 \rangle} z}, & d = 2, \\ \phi_1^{\,1 + \frac{3\langle R^2 \rangle}{4\langle R^3 \rangle} z}, & d = 3. \end{cases} \tag{6.21}$$

It is interesting to observe that the polydisperse results are very similar in form to the monodisperse results (5.21). Indeed, if one scales the distance z in the polydisperse results by the length $\langle R^d\rangle/\langle R^{d-1}\rangle$, proportional to the average volume-to-surface ratio of a sphere, results for $L(z)$ for any degree of polydispersivity collapse to the monodisperse result [cf. (5.21)]. This also turns out to be true for relation (6.22) for the chord-length distribution function $p(z)$. Thus, Figure 5.4 for the monodisperse case can yield the corresponding polydisperse results when interpreted in this manner.

6.1.4 Chord-Length Density Function

The chord-length density function $p(z) \equiv p^{(1)}(z)$ for the matrix phase is trivially obtained given expression (6.20) for L, normalization condition (2.68), and interrelation (2.73). We find that in any space dimension d,

$$p(z) = \frac{\eta\, \omega_{d-1}\langle R^{d-1}\rangle}{\omega_d\langle R^d\rangle}\phi_1^{\frac{\omega_{d-1}\langle R^{d-1}\rangle}{\omega_d\langle R^d\rangle}z}. \tag{6.22}$$

For the first three space dimensions (Torquato and Lu 1993), this yields

$$p(z) = \begin{cases} \dfrac{\eta}{2\langle R\rangle}\phi_1^{\frac{1}{2\langle R\rangle}z}, & d=1,\\[2ex] \dfrac{2\eta\langle R\rangle}{\pi\langle R^2\rangle}\phi_1^{\frac{2\langle R\rangle}{\pi\langle R^2\rangle}z}, & d=2,\\[2ex] \dfrac{3\eta\langle R^2\rangle}{4\langle R^3\rangle}\phi_1^{\frac{3\langle R^2\rangle}{4\langle R^3\rangle}z}, & d=3. \end{cases} \tag{6.23}$$

The corresponding mean chord length ℓ_C is obtained by applying formula (2.72). We find that

$$\ell_C = \frac{\omega_d\langle R^d\rangle}{\eta\, \omega_{d-1}\langle R^{d-1}\rangle} = \frac{\omega_d\, \phi_1 d}{\omega_{d-1}}\frac{1}{s}, \tag{6.24}$$

where we have used formula (6.13) for the specific surface s.

6.1.5 Nearest-Surface Functions

Recall from Section 2.8 that for polydisperse systems it is more meaningful to consider distances between nearest surfaces than between nearest particle centers. As in the monodisperse case, there is no distinction between the void and particle quantities for polydisperse overlapping spheres. From relations (2.115), (4.90), and (6.9) with $a_1^{(1)} = r+R$, where r is the radius of the "test" particle, we can obtain explicit expressions for the nearest-surface quantities. Specifically, the nearest-surface exclusion probability and density function in d dimensions for both positive r and allowable negative r (determined by the largest inclusion radius) are respectively

$$e_V(r) = e_P(r) = \exp[-\rho\langle v_1(r+R)\Theta(r+R)\rangle], \qquad \text{for all } r, \tag{6.25}$$

$$h_V(r) = h_P(r) = \rho\langle s_1(r+R)\Theta(r+R)\rangle e_V(r), \qquad \text{for all } r, \tag{6.26}$$

where $s_1(r)$ is the surface area of a single sphere of radius r given by (2.98) and the appearance of the Heaviside step function $\Theta(r+R)$ ensures that the argument $r+R$ remains nonnegative. Recall from Sections 2.8 and 4.5 that a negative value of r corresponds to a trivial instance in which the center of the test particle penetrates into an inclusion or, equivalently, the "reference" point is in the particle phase. These expressions were given by Lu and Torquato (1992b).

6.1.6 Pore-Size Functions

Recall from Section 2.6 that for the special case of homogeneous polydisperse spheres, overlapping or not, the pore-size probability density function $P(\delta)$ and its associated complementary cumulative distribution function $F(\delta)$ are trivially obtained from the nearest-surface functions via the interrelations

$$P(\delta) = \frac{h_V(\delta)}{\phi_1}, \qquad F(\delta) = \frac{e_V(\delta)}{\phi_1}, \qquad \delta \geq 0. \tag{6.27}$$

Thus, the use of (6.25) and (6.26) in these relations gives the pore-size functions for d-dimensional polydisperse overlapping spheres.

6.1.7 Point/q-Particle Correlation Functions

Utilizing the expression (6.9) and the general definition (4.89) connecting the point/q-particle correlation function G_n to H_n yields, for d-dimensional overlapping spheres, the exact relation

$$G_n(\boldsymbol{x}_1; \boldsymbol{r}^q) = \rho^q \phi_1 \prod_{\ell=1}^{q} \langle e(y_{1\ell}; R)\rangle, \tag{6.28}$$

where $y_{1\ell} = |\boldsymbol{x}_1 - \boldsymbol{r}_\ell|$.

6.2 Totally Impenetrable Spheres

Let us now consider totally impenetrable (hard) spheres with a polydispersivity in size (Figure 6.5). Such systems exhibit intriguing microstructural features, some of which are only beginning to be understood. It is known, for example, that a relatively small degree of polydispersivity can completely suppress the disorder-order phase transition seen in monodisperse hard-sphere systems (Henderson, Mortensen, Underwood and van Megen 1996); see also Section 3.3. Interestingly, equilibrium mixtures of small and large hard spheres can "phase separate" (i.e., the small and large spheres demix) at sufficiently high densities but the precise nature of such phase transitions has not yet

Figure 6.5 Hard disks with a polydispersivity in size.

been established and is a subject of intense interest; see Dijkstra, van Roij and Evans (1999) and references therein.

Very little is rigorously known about the packing characteristics of polydisperse hard spheres. For example, the *highest achievable* overall packing fraction $\eta_{\max}^{(2)}$ of even a binary mixture of hard spheres in $\Re^d$ for arbitrary values of the radii R_1 and R_2 is unknown, not to mention the determination of the corresponding structures. However, it is intuitively clear that $\eta_{\max}^{(2)}$ is bounded from below by the maximum packing fraction $\eta_{\max}^{(1)}$ for a monodisperse system of hard spheres. The lower bound $\eta_{\max}^{(2)} \geq \eta_{\max}^{(1)}$ is independent of the radii and corresponds to the case when the two components are completely *phase separated* (demixed), each at the packing fraction $\eta_{\max}^{(1)}$. Moreover, one can bound $\eta_{\max}^{(2)}$ from above in terms of the monodisperse value $\eta_{\max}^{(1)}$ for arbitrary values of R_1 and R_2. Specifically, consider a wide separation of sizes ($R_1 \ll R_2$) and imagine a *sequential process* in which the larger spheres are first packed at the maximum density $\eta_{\max}^{(1)}$ for a monodisperse system. The remaining interstitial space between the larger spheres can now be packed with the smaller spheres at the packing fraction $\eta_{\max}^{(1)}$ provided that $R_1/R_2 \to 0$. The overall packing fraction in this limit is given by $1-(1-\eta_{\max}^{(1)})^2$, which is an upper bound for *any binary system*. Thus, $\eta_{\max}^{(2)} \leq 1-(1-\pi/\sqrt{12})^2 \approx 0.991$ for $d=2$ and $\eta_{\max}^{(2)} \leq 1-(1-\pi/\sqrt{18})^2 \approx 0.933$ for $d=3$, where $\eta_{\max}^{(1)}$ corresponds to the close-packed lattice in two or three dimensions (see Table 3.1). The same arguments extend to systems of M different hard spheres with radii $R_1, R_2, \ldots, R_M$ in $\Re^d$. Therefore, the overall maximum packing fraction $\eta_{\max}^{(M)}$ of such a general mixture in $\Re^d$ [where η is defined by (6.2) with (4.80)] is bounded from above and below by

$$\eta_{\max}^{(1)} \leq \eta_{\max}^{(M)} \leq 1-(1-\eta_{\max}^{(1)})^M. \tag{6.29}$$

The lower bound corresponds to the case when the M components completely demix, each at the packing fraction $\eta_{\max}^{(1)}$. The upper bound corresponds to the generalization of the aforementioned ideal sequential packing process for arbitrary M in which we take the limits $R_1/R_2 \to 0$, $R_2/R_3 \to 0, \cdots, R_{M-1}/R_M \to 0$. Specific *nonsequential* protocols (algorithmic or otherwise) that can generate structures that approach the upper bound (6.29) for arbitrary values of M are currently unknown and thus the development of such protocols is an open area of research. We see that in the limit $M \to \infty$, the upper bound approaches unity, corresponding to space filling polydisperse spheres with an infinitely wide separation in sizes. Of course, one can also imagine constructing space-filling polydisperse spheres with a continuous size distribution with sizes ranging to the infinitesimally small. In Chapter 16, we discuss such a construction called the Hashin-Shtrikman coated-spheres model. Not surprisingly, the determination of the maximally random jammed (MRJ) state (Section 3.3) for a polydisperse hard-sphere system is a wide open question.

The aforementioned conundrums serve to illustrate the richness of polydisperse hard-sphere systems but further discussion about them is beyond the scope of this book. In what follows, we discuss the evaluation of microstructural correlation functions for these systems.

6.2.1 n-Point Probability Functions

The series representation of the matrix n-point probability function S_n in the case of polydisperse hard spheres is obtained from the series (4.81) for the canonical function H_n and the definition (4.85). It is found that

$$S_n(\mathbf{x}^n) = 1 + \sum_{k=1}^{n} (-1)^k \frac{1}{k!} \int \cdots \int dR_1 \cdots dR_k f(R_1) \cdots f(R_k) \rho_k(\mathbf{r}_1, ..., \mathbf{r}_k; R_1, ..., R_k)$$
$$\times \prod_{j=1}^{k} \left\{ 1 - \prod_{i=1}^{n} [1 - m(|\mathbf{x}_i - \mathbf{r}_j|; R_j)] \right\} d\mathbf{r}_j. \tag{6.30}$$

Note that the original infinite series terminates exactly after terms involving ρ_n for precisely the same reasons as in the monodisperse case (see Section 5.2.1). Moreover, the expression (6.30) further simplifies whenever any term contains the analogue of the subdiagram (5.37).

Thus, for a statistically homogeneous system, the volume fraction of phase 1 is given by

$$S_1 = \phi_1 = 1 - \rho \langle v_1(R) \rangle = 1 - \eta = 1 - \phi_2, \tag{6.31}$$

where $\eta = \phi_2$ is the reduced density (equivalent to the packing fraction) defined by (6.2). The corresponding two-point function is given by

$$S_2(\mathbf{r}) = 1 - \rho \, \langle v_2(r; R) \rangle + \int dR_1 f(R_1) \int dR_2 f(R_2) \int d\mathbf{r}_1 \int d\mathbf{r}_2 \rho_2(\mathbf{r}_{12}; R_1, R_2)$$
$$\times \; m(|\mathbf{x}_1 - \mathbf{r}_1|; R_1) \, m(|\mathbf{x}_2 - \mathbf{r}_2|; R_2), \tag{6.32}$$

where $v_2(r; R)$ is the union volume of two spheres of radius R [cf. (5.8), (5.9), and (5.10)] and $\mathbf{r} = \mathbf{x}_1 - \mathbf{x}_2$. In the process of computing conductivity and elasticity bounds for polydisperse systems of hard spheres, Thovert, Kim, Torquato and Acrivos (1990) determined the two- and three-point probabilities S_2 and S_3, but did not state these results explicitly. The result S_2 was explicitly given independently by Given et al. (1990a) and Lu and Torquato (1991).

The analytical solution of the Ornstein–Zernike equation for ρ_2 in the Percus–Yevick approximation was extended by Blum and Stell (1979, 1980) to polydisperse hard spheres. Thus, this solution and the Fourier transform techniques described in Section 5.2.1 enable one to compute S_2 for polydisperse hard spheres in equilibrium. Given et al. (1990a) used S_2 to compute bounds on the trapping constant of such polydisperse systems but did not explicitly give the spatial dependence of S_2.

6.2.2 Surface Correlation Functions

The series expression (4.81) for the canonical function H_n yields the surface correlation functions of interest under special limits. For hard polydisperse spheres, the infinite series terminates for exactly the same reasons as for the monodisperse case. In particular, from the definition (4.86), the specific surface for homogeneous media is found to be

$$s = \rho\langle s_1(R)\rangle = \eta d\,\frac{\langle R^{d-1}\rangle}{\langle R^d\rangle}. \tag{6.33}$$

If we again let S denote the ratio of the specific surface of a polydisperse system to a monodisperse system at the same volume fraction with radius $\langle R\rangle$, then we find that S is identical to the overlapping result (6.14). Recall that S decreases with the degree of polydispersivity.

Similarly, use of definitions (4.87) and (4.88) and the series (4.81) gives

$$\begin{aligned} F_{sv}(\boldsymbol{r}) = s &- \rho\int dR_1 f(R_1)\int d\boldsymbol{r}_1\delta(|\boldsymbol{x}_1-\boldsymbol{r}_1|-R_1)m(|\boldsymbol{x}_2-\boldsymbol{r}_1|;R_1)\\ &- \int dR_1 f(R_1)\int dR_2 f(R_2)\int d\boldsymbol{r}_1\int d\boldsymbol{r}_2\rho_2(\boldsymbol{r}_{12};R_1,R_2)\\ &\times\ \delta(|\boldsymbol{x}_1-\boldsymbol{r}_1|-R_1)m(|\boldsymbol{x}_2-\boldsymbol{r}_2|;R_2) \end{aligned} \tag{6.34}$$

and

$$\begin{aligned} F_{ss}(\boldsymbol{r}) = \rho &\int dR_1 f(R_1)\int d\boldsymbol{r}_1\delta(|\boldsymbol{x}_1-\boldsymbol{r}_1|-R_1)\delta(|\boldsymbol{x}_2-\boldsymbol{r}_1|-R_1)\\ &+ \int dR_1 f(R_1)\int dR_2 f(R_2)\int d\boldsymbol{r}_1\int d\boldsymbol{r}_2\rho_2(\boldsymbol{r}_{12};R_1,R_2)\\ &\times\ \delta(|\boldsymbol{x}_1-\boldsymbol{r}_1|-R_1)\delta(|\boldsymbol{x}_2-\boldsymbol{r}_2|-R_2), \end{aligned} \tag{6.35}$$

where $\boldsymbol{r}=\boldsymbol{x}_1-\boldsymbol{x}_2$. These results were first obtained independently by Given et al. (1990a) and Lu and Torquato (1991).

6.2.3 Lineal-Path Function

For the same reasons as in the monodisperse case (see Section 5.2.3), the series (6.19) cannot be evaluated exactly for $d \geq 2$. Lu and Torquato (1992c) found *scaled-particle* approximations of the lineal-path function $L(z) \equiv L^{(1)}(z)$ for the matrix phase of a system of polydisperse hard spheres in equilibrium. These results in any dimension d are given by

$$L(z) = \phi_1 \exp\left[-\frac{\eta\, \omega_{d-1}\langle R^{d-1}\rangle}{\phi_1 \omega_d \langle R^d\rangle} z\right]. \tag{6.36}$$

For the first three space dimensions (Lu and Torquato 1992c), we have

$$L(z) = \begin{cases} \phi_1 \exp\left[-\dfrac{\eta}{2\phi_1\langle R\rangle} z\right], & d = 1, \\ \phi_1 \exp\left[-\dfrac{2\eta\langle R\rangle}{\pi\phi_1\langle R^2\rangle} z\right], & d = 2, \\ \phi_1 \exp\left[-\dfrac{3\eta\langle R^2\rangle}{4\phi_1\langle R^3\rangle} z\right], & d = 3. \end{cases} \tag{6.37}$$

Whereas the expression for $d = 1$ is exact, the other expressions are accurate approximations. As in the case of overlapping spheres, we see that if one scales the distance z in (6.36) by the length $\langle R^d\rangle/\langle R^{d-1}\rangle$ (proportional to the average volume-to-surface ratio of a sphere), results for $L(z)$ for any polydispersivity collapse onto the monodisperse curve [cf. (5.58)]. This turns out to be true for relation (6.38) for the chord-length distribution function $p(z)$. Therefore, Figure 5.8 for the monodisperse case also depicts the corresponding polydisperse results when z is scaled in this manner.

6.2.4 Chord-Length Density Function

The chord-length density function $p(z) \equiv p^{(1)}(z)$ for the matrix phase of systems of polydisperse equilibrium hard spheres is easily obtained using expressions (2.68), (2.73), and (6.36):

$$p(z) = \frac{\eta\, \omega_{d-1}\langle R^{d-1}\rangle}{\phi_1 \omega_d \langle R^d\rangle} \exp\left[-\frac{\eta\, \omega_{d-1}\langle R^{d-1}\rangle}{\phi_1\, \omega_d \langle R^d\rangle} z\right]. \tag{6.38}$$

Thus, in the first three space dimensions (Torquato and Lu 1993), we have

$$p(z) = \begin{cases} \dfrac{\eta}{2\phi_1\langle R\rangle} \exp\left[-\dfrac{\eta}{2\phi_1\langle R\rangle} z\right], & d = 1, \\ \dfrac{2\eta\langle R\rangle}{\pi\phi_1\langle R^2\rangle} \exp\left[-\dfrac{2\eta\langle R\rangle}{\pi\phi_1\langle R^2\rangle} z\right], & d = 2, \\ \dfrac{3\eta\langle R^2\rangle}{4\phi_1\langle R^3\rangle} \exp\left[-\dfrac{3\eta\langle R^2\rangle}{4\phi_1\langle R^3\rangle} z\right], & d = 3. \end{cases} \tag{6.39}$$

The corresponding explicit expressions for the mean chord lengths are

$$\ell_C = \frac{\phi_1\, \omega_d \langle R^d \rangle}{\eta\, \omega_{d-1} \langle R^{d-1} \rangle} = \frac{\omega_d\, \phi_1 d}{\omega_{d-1}} \frac{1}{s}, \tag{6.40}$$

where we have used formula (6.33) for the specific surface s.

6.2.5 Nearest-Surface Functions

Using the definition (4.90) and series (4.81) for the canonical function H_n gives the following series representation of the void nearest-surface exclusion probability function for any statistically homogeneous ensemble of totally impenetrable polydisperse spheres:

$$e_V(r) = 1 + \sum_{k=1}^{\infty} (-1)^k \frac{1}{k!} \int \cdots \int dR_1 \cdots dR_k f(R_1) \cdots f(R_k)$$
$$\times\, \rho_k(\mathbf{r}^k; R_1, \ldots, R_k) \prod_{j=1}^{k} m(|\mathbf{x} - \mathbf{r}_j|; r) d\mathbf{r}_j, \tag{6.41}$$

where $m(y; r) = \Theta(r + R - y)$. Similarly, the corresponding relation for the particle nearest-surface exclusion probability is obtained from (4.91) and (4.81), yielding

$$e_P(r) = 1 + \sum_{k=1}^{\infty} (-1)^k \frac{1}{\rho k!} \int \cdots \int dR_1 \cdots dR_{k+1} f(R_1) \cdots f(R_{k+1})$$
$$\times\, \rho_{k+1}(\mathbf{r}^{k+1}; R_1, \ldots, R_{k+1}) \prod_{j=2}^{k+1} m(|\mathbf{r}_1 - \mathbf{r}_j|; r) d\mathbf{r}_j. \tag{6.42}$$

The results (6.41) and (6.42) were first given by Lu and Torquato (1992b). The series expressions for the other nearest-surface quantities are found using the above expressions for the exclusion probabilities and interrelations (2.115) and (2.116).

The nearest-surface quantities must obey certain exact conditions for totally impenetrable spheres *in any ensemble*. In the case of the particle quantities, the *reference* particle of radius R excludes the surface of any other particles from a sphere of radius R, and therefore we have

$$e_P(r) = 1, \qquad h_P(r) = g_P(r) = 0, \qquad 0 \le r \le R. \tag{6.43}$$

In the case of the void quantities, the reference point can only be inside one of the particles, and hence the void nearest-surface exclusion probability and probability density function for all allowable negative values of r (see Sections 2.8 and 6.1.5) are given respectively by

$$e_V(r) = 1 - \rho\langle v_1(r+R)\Theta(r+R)\rangle, \qquad r < 0, \tag{6.44}$$
$$h_V(r) = \rho\langle s_1(r+R)\Theta(r+R)\rangle, \qquad r < 0. \tag{6.45}$$

The function g_V is just the ratio h_V/e_V, and henceforth for brevity we will just report results for e_V and h_V.

In the special case of an equilibrium ensemble of polydisperse hard spheres, it can be shown (Lu and Torquato 1992b) that the particle quantities are obtained from the void quantities via the following formulas:

$$e_P(r) = \frac{e_V(r)}{e_V(R)}, \qquad r \geq R, \tag{6.46}$$

$$h_P(r) = \frac{h_V(r)}{e_V(R)}, \qquad r \geq R. \tag{6.47}$$

Using the same methods already discussed in Section 5.2.5 for monodisperse spheres, Lu and Torquato (1992b) found analytical expressions for the nearest-surface functions of statistically isotropic systems of polydisperse hard spheres in *equilibrium* for the first three space dimensions for $r > 0$. These expressions (summarized below) are exact for polydisperse hard rods ($d = 1$) and are accurate approximations for $d = 2$ and $d = 3$.

Consider the void quantities first. For $d = 1$, they obtained

$$e_V(x) = \phi_1 \exp\left[-\frac{2\eta x}{1-\eta}\right], \qquad x \geq 0, \tag{6.48}$$

$$h_V(x) = \frac{2\eta}{\langle D\rangle} \exp\left[-\frac{2\eta x}{1-\eta}\right], \qquad x \geq 0, \tag{6.49}$$

where $x = r/\langle D\rangle$ is a dimensionless distance, and $\langle D\rangle = 2\langle R\rangle$ is the average diameter. For polydisperse hard disks ($d = 2$) in equilibrium, they obtained

$$e_V(x) = \phi_1 \exp\left[-4\eta S(a_0x^2 + a_1x)\right], \qquad x \geq 0, \tag{6.50}$$

$$h_V(x) = \frac{4\eta S}{\langle D\rangle}(2a_0x + a_1)e_V(x), \qquad x \geq 0, \tag{6.51}$$

where S is the surface area ratio given by (6.14) with $d = 2$, and the coefficients a_0 and a_1 are given by

$$a_0 = \frac{1 + \eta(S-1)}{(1-\eta)^2}, \qquad a_1 = \frac{1}{1-\eta}. \tag{6.52}$$

For $d = 3$, they obtained three different approximations. Here we state their Carnahan–Starling approximation:

$$e_V(x) = \phi_1 \exp\left[-2\eta S(a_0x^3 + a_1x^2 + a_2x)\right], \qquad x \geq 0, \tag{6.53}$$

$$h_V(x) = \frac{2\eta S}{\langle D\rangle}(3a_0x^2 + 2a_1x + a_2)e_V(x), \qquad x \geq 0, \tag{6.54}$$

where S is the surface area ratio given by (6.14) with $d = 3$, and the coefficients a_0, a_1, and a_2 are given by

$$a_0 = \frac{4[\langle R\rangle^2/\langle R^2\rangle](1-\eta)(1-\eta+3\eta S) + 8\eta^2S^2}{(1-\eta)^3},$$

$$a_1 = \frac{6[\langle R\rangle^2/\langle R^2\rangle](1-\eta) + 9\eta S}{(1-\eta)^2}, \qquad a_2 = \frac{3}{1-\eta}.$$

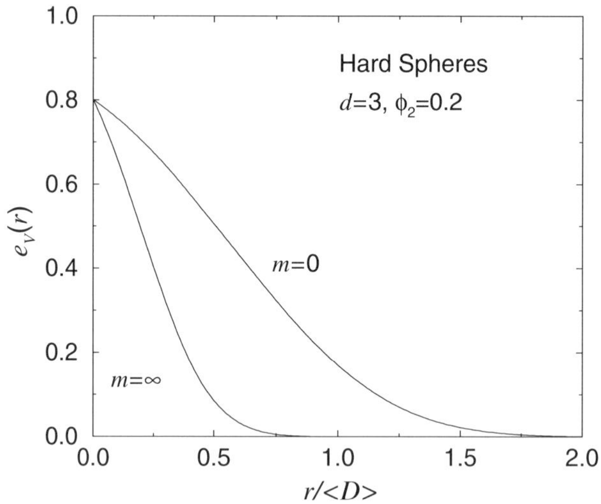

Figure 6.6 Void nearest-surface exclusion probability $e_V(r)$ versus dimensionless distance $r/\langle D\rangle$ for three-dimensional polydisperse systems of hard spheres in equilibrium as computed from (6.53) at a packing fraction $\phi_2 = 0.2$. Schulz distribution (6.5) for $m = 0$ and $m = \infty$ (monodisperse limit).

The relation for the coefficient a_0 corrects a misprint contained in Lu and Torquato (1992b) as first noted by Snyder (1998). Synder used the above three-dimensional expressions for the void nearest-surface functions and the particle nearest-surface functions (given below) to study the distribution of void space in concrete. In the monodisperse limit, the two-dimensional relations (6.50) and (6.51) reduce to the scaled-particle approximations for E_V and H_V given in Torquato et al. (1990) upon use of the formulas in (2.110). In the same limit, the three-dimensional expressions (6.53) and (6.54) reduce to the Carnahan-Starling approximations for E_V and H_V obtained by Torquato et al. (1990) upon use of the formulas in (2.110).

Figure 6.6 compares the void nearest-surface exclusion probability $e_V(r)$ for three-dimensional polydisperse spheres to monodisperse spheres at the same packing fraction ϕ_2. The polydisperse system is characterized by a Schulz size distribution with $m = 0$, and only positive values of the argument r are shown. It is seen that the effect of polydispersivity is to increase $e_V(r)$ relative to the monodisperse result for $r \geq 0$. The reason for such behavior can be seen from (6.2), and noting that, we see that at fixed η or ϕ_2, the number density ρ will be smaller in the polydisperse system than in the monodisperse system, since the average sphere volume in the former is larger than in the latter. Thus, the probability of finding a large void region in the polydisperse system is larger than in the monodisperse system.

The corresponding equilibrium *particle* nearest-surface functions were obtained by Lu and Torquato (1992b) using the formulas above for the void quantities and the interrelations (6.46) and (6.47). For example, for $d = 1$, they obtained the exact results

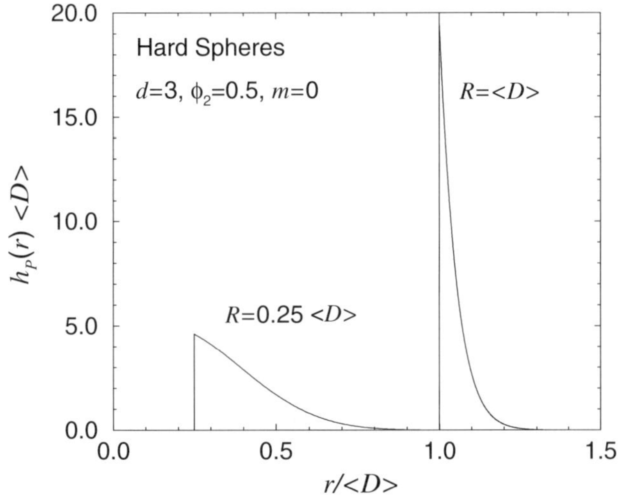

Figure 6.7 Dimensionless particle nearest-surface density function $h_P(r)\langle D\rangle$ versus dimensionless distance $r/\langle D\rangle$ for a three-dimensional polydisperse system of hard spheres in equilibrium as computed from (6.54) and (6.47) at a packing fraction $\phi_2 = 0.5$. Schulz distribution (6.5) for $m = 0$. Reference particle radii are $0.25\langle D\rangle$ and $\langle D\rangle$.

$$e_P(x) = \exp\left[-\frac{2\eta(x - R/\langle D\rangle)}{1-\eta}\right], \qquad x \geq R/\langle D\rangle, \tag{6.55}$$

$$h_P(x) = \frac{2\eta}{\langle D\rangle\phi_1}\exp\left[-\frac{2\eta(x - R/\langle D\rangle)}{1-\eta}\right], \qquad x \geq R/\langle D\rangle. \tag{6.56}$$

Their two- and three-dimensional expressions for the particle quantities are obtained using the void formulas (6.50), (6.51), (6.53), and (6.54), and interrelations (6.46) and (6.47).

Figure 6.7 compares the particle nearest-surface probability $h_P(r)$ for three-dimensional polydisperse spheres for two different reference particles: one with radius $R = 0.25\langle D\rangle$ and the other with $R = \langle D\rangle$. Both results are for the Schulz distribution with $m = 0$ and $\phi_2 = 0.5$. The larger the reference particle, the larger the value of h_P is at the surface of the reference particle, but h_P decreases rapidly with increasing r. As the reference particle increases in size, it is more likely to have other particles in the immediate vicinity of its surface.

The mean nearest surface–surface distance λ_P can be computed using the definition (2.119) and expression (6.55) for e_P (Lu and Torquato 1992b). For the case $d = 1$, one obtains the simple result

$$\frac{\lambda_P}{\langle D\rangle} = \frac{\phi_1}{2\eta},$$

which is independent of the type of reference particle. In higher dimensions, the distance λ_P is inversely related to the radius R of the reference particle. On average, bigger

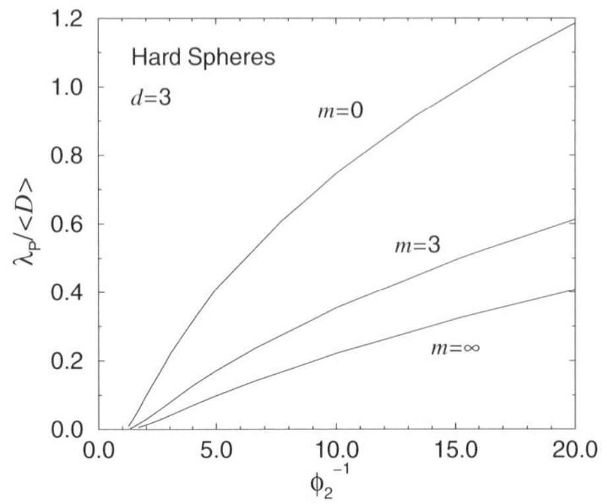

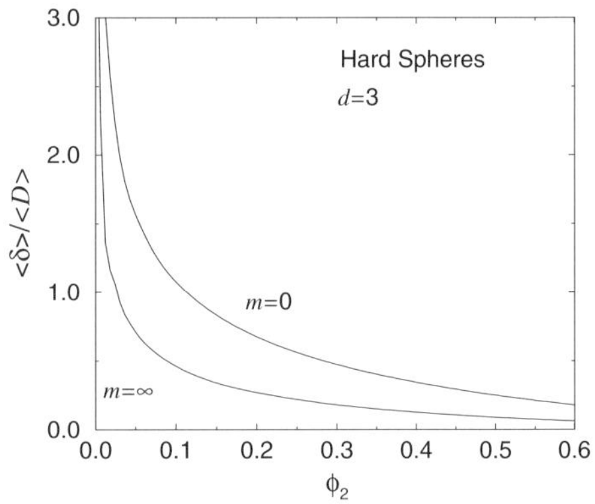

Figure 6.8 Left panel: Mean surface–surface distance λ_P (scaled by average diameter $\langle D\rangle$) versus ϕ_2^{-1} for three-dimensional polydisperse systems of hard spheres in equilibrium. Schulz distribution (6.5) for several values of m. The radius of the reference particle is $0.5\langle D\rangle$. Right panel: Mean pore size $\langle\delta\rangle/\langle D\rangle$ versus ϕ_2 for the same system.

reference particles have more neighbors than smaller ones and thus have a better chance of having a nearest particle a shorter distance away from them. Figure 6.8 depicts the dimensionless distance $\lambda_P/\langle D\rangle$ in three dimensions as a function of the inverse packing fraction ϕ_2^{-1} for several values of the Schulz parameter m. This is computed from relations (2.119), (6.53), and (6.46). Assuming that all other parameters are equal, the effect of polydispersivity is to increase the mean nearest surface–surface distance.

6.2.6 Pore-Size Functions

The results of Section 6.2.5 in combination with the interrelations of (6.27) give the pore-size functions $P(\delta)$ and $F(\delta)$ for d-dimensional polydisperse hard spheres in equilibrium. Included in Figure 6.8 is the mean pore size $\langle\delta\rangle$ as found from (2.83), (2.87), and (6.53). As expected, the average pore size increases with polydispersivity, assuming that all other parameters are equal.

6.2.7 Point/q-Particle Correlation Functions

The series representation (4.81) for the canonical function H_n and the definition (4.89) give the series expression for the matrix point/q-particle function for polydisperse hard spheres to be

$$\begin{aligned} G_n(\mathbf{x};\mathbf{r}^q) &= \prod_{i=1}^{q} e(|\mathbf{x}-\mathbf{r}_i|;R_i)f(R_1)\cdots f(R_q)\rho_q(\mathbf{r}^q) \\ &\quad - \int dR_{q+1}f(R_1)\cdots f(R_{q+1})\rho_{q+1}(\mathbf{r}^{q+1};R_1,\ldots,R_{q+1}) \\ &\qquad \times m(|\mathbf{x}-\mathbf{r}_{q+1}|;R_{q+1})d\mathbf{r}_{q+1}. \end{aligned} \tag{6.57}$$

As in the monodisperse instance, the series terminates exactly after two terms.

CHAPTER 7

Anisotropic Media

Statistically homogeneous but *anisotropic* media will generally be endowed with anisotropic effective properties (see Part II of this book) and thus represent an important class of random heterogeneous materials. Examples include stratified geological media, aligned short and long fiber composites, and laminates (Postma 1955, Dullien 1979, Christensen 1979, Adler 1992). The optimization of the properties of anisotropic media rests on a precise description of the microstructure.

We will show that the formalism described in earlier chapters to characterize the microstructure of sphere systems can be generalized to models of anisotropic media. In particular, we will examine systems of unidirectionally oriented nonspherical inclusions in a matrix and certain hierarchical laminates that possess, in specific limits, optimal effective conductivities and elastic moduli. The microstructural functions studied below will be used to compute rigorous estimates for the conductivity tensor (Willis 1977, Sen and Torquato 1989), stiffness tensor (Willis 1977, Torquato 1997) and mean survival time (Torquato and Lado 1991) for statistically anisotropic media. Note that sphere systems can be statistically anisotropic (e.g., crystal-like arrangements) but we do not study such models here.

7.1 General Considerations

We begin by considering the class of statistically homogeneous but anisotropic media composed of arrays of *unidirectionally* oriented identical particles of *arbitrary shape* in a matrix. Torquato and Sen (1990) obtained series representations of the matrix n-point probability function S_n for such media by a simple reinterpretation of the relations derived by Torquato and Stell (1982) for spheres. More generally, one can map all of

Figure 7.1 Arrays of oriented rectangles or cylinders of arbitrary but constant cross-section (left panel) and oriented ellipses or ellipsoids (right panel). Both examples exhibit the structure of an idealized *nematic liquid crystal.*

the sphere results developed for the *canonical* n-point function H_n (Chapter 4) into corresponding results for the aforementioned class of anisotropic media.

An important first step in carrying out this mapping is to recognize that the Torquato-Stell results are applicable not only to spheres but also [after an appropriate generalization of the *inclusion indicator function* $m(r)$, given by (4.4)] to inclusions in which each configurational coordinate is fully specified by its center-of-mass position. This class of materials includes oriented rectangles, ellipses, etc., for $d = 2$ (see Figure 7.1) and oriented rectangular parallelepipeds, ellipsoids, cylinders, etc., for $d = 3$. In all of these examples, the microstructure is that of an idealized *nematic liquid crystal,* i.e., there is long-range orientational order but only short-range positional order (Chaikin and Lubensky 1995). For this class of microstructures, the inclusion indicator function generalizes as

$$m(\boldsymbol{r}) = \begin{cases} 1, & \boldsymbol{r} \in D_I, \\ 0, & \text{otherwise}, \end{cases} \tag{7.1}$$

where D_I is the inclusion region and $\boldsymbol{r}$ is a position vector measured with respect to the centroid of the inclusion. We will let x_i $(i = 1, \ldots, d)$ denote the Cartesian components of $\boldsymbol{r}$ in the principal axes coordinate frame.

We now give explicit representations for $m(\boldsymbol{r})$ for several two-dimensional inclusion shapes. For a rectangle with sides of lengths $2a$ and $2b$,

$$m(\boldsymbol{r}) = \begin{cases} 1, & |x_1| \le a, \ \ |x_2| \le b, \\ 0, & \text{otherwise}. \end{cases} \tag{7.2}$$

Of course, squares are just the special case $a = b$. For an ellipse with semiaxes of lengths a and b,

$$m(\mathbf{r}) = \begin{cases} 1, & \dfrac{x_1^2}{a^2} + \dfrac{x_2^2}{b^2} \leq 1, \\ 0, & \text{otherwise.} \end{cases} \tag{7.3}$$

Now let us consider three-dimensional examples. The inclusion indicator function for a rectangular parallelepiped with sides of lengths $2a$, $2b$, and $2c$ is given by

$$m(\mathbf{r}) = \begin{cases} 1, & |x_1| \leq a, \ |x_2| \leq b, \ |x_3| \leq c, \\ 0, & \text{otherwise.} \end{cases} \tag{7.4}$$

The cube is just the case $a = b = c$. For an ellipsoidal inclusion with semiaxes of lengths a, b, and c,

$$m(\mathbf{r}) = \begin{cases} 1, & \dfrac{x_1^2}{a^2} + \dfrac{x_2^2}{b^2} + \dfrac{x_3^2}{c^2} \leq 1, \\ 0, & \text{otherwise.} \end{cases} \tag{7.5}$$

Finally, as a last example, the inclusion indicator function for a circular cylinder of diameter $2a$ and length $2b$ is

$$m(\mathbf{r}) = \begin{cases} 1, & x_1^2 + x_2^2 \leq a^2, \ |x_3| \leq b, \\ 0, & \text{otherwise.} \end{cases} \tag{7.6}$$

To summarize, replacing the inclusion indicator function in the series representation of S_n for spheres with (7.1) yields the corresponding series expression for identical inclusions whose configurational coordinate is fully specified by its center-of-mass position (i.e., an idealized nematic-liquid-crystal structure). The expression for spheres is given by the series (4.46) for the canonical function H_n in the limit specified by (4.69). Actually, using this simple replacement, one can obtain not only the S_n for such anisotropic media but the canonical n-point correlation function $H_n(\mathbf{x}^m; \mathbf{x}^{p-m}; \mathbf{r}^q)$ with $m = 0$, i.e.,

$$\lim_{a_i \to R, \, \forall i} H_n(\emptyset; \mathbf{x}^p; \mathbf{r}^q), \quad n = p + q.$$

Thus, one can immediately obtain all of the H_n, except those involving surface information. While this includes $G_n(\mathbf{x}; \mathbf{r}^q)$ and its generalizations, we limit ourselves here to the determination of S_n for simplicity.

7.2 Fully Penetrable Oriented Inclusions

Given the prescription described above and relation (5.6), the n-point matrix probability functions S_n for fully penetrable inclusions at number density ρ with liquid-crystal structure are again easily given:

$$S_n(\mathbf{x}^n) = \exp[-\rho v_n(\mathbf{x}^n)]. \tag{7.7}$$

Here $\upsilon_n(\boldsymbol{x}^n)$ denotes the union volume of n oriented identical objects of arbitrary shape centered at $\boldsymbol{x}^n$.

For example, in the case $n = 2$, we require the union volume $\upsilon_2(\boldsymbol{r})$ of two inclusion regions whose centroids are separated by the displacement $\boldsymbol{r} \equiv \boldsymbol{x}_2 - \boldsymbol{x}_1$. We now state $\upsilon_2(\boldsymbol{r})$ for the two-dimensional examples discussed in Section 7.1. Let x and y denote the Cartesian components of $\boldsymbol{r}$, let θ be the polar angle between $\boldsymbol{r}$ and the x-axis, and define $r = |\boldsymbol{r}|$. For oriented rectangular inclusions, we have that

$$\upsilon_2(\boldsymbol{r}) = 8ab - (2a - |x|)(2b - |y|)\Theta(2a - |x|)\Theta(2b - |y|), \tag{7.8}$$

where $\Theta(x)$ is the Heaviside step function [cf. (3.48)]. For oriented elliptical inclusions whose symmetry axis is aligned with the y-axis, we have

$$\upsilon_2(\boldsymbol{r}) = 2\pi ab - \upsilon_2^{\text{int}}(\boldsymbol{r}), \tag{7.9}$$

where

$$\upsilon_2^{\text{int}}(\boldsymbol{r}) = 2ab\left[\cos^{-1}\left(\frac{r}{2R(\theta)}\right) - \frac{r}{2R(\theta)}\left(1 - \frac{r^2}{4R(\theta)^2}\right)^{1/2}\right]\Theta[2R(\theta) - r] \tag{7.10}$$

and

$$R(\theta) = \frac{a}{[1 - (1 - a^2/b^2)\sin^2\theta]^{1/2}}. \tag{7.11}$$

Comparing (7.10) to the circular-inclusion result (3.50) enables us to see that $R(\theta)$ is effectively an "angle-dependent" radius (see also Section 7.3).

Let us now state $\upsilon_2(\boldsymbol{r})$ for the three-dimensional examples discussed in Section 7.1. Let x, y, and z denote the Cartesian components of $\boldsymbol{r}$, and let θ be the polar angle between $\boldsymbol{r}$ and the z-axis. For oriented rectangular parallelepipeds, we have that

$$\upsilon_2(\boldsymbol{r}) = 16abc - (2a - |x|)(2b - |y|)(2c - |z|)\Theta(2a - |x|)\Theta(2b - |y|)\Theta(2c - |z|). \tag{7.12}$$

For *spheroidal* inclusions in which $a = c$ and whose symmetry axis is aligned with the z-axis, we have that

$$\upsilon_2(\boldsymbol{r}) = \frac{8\pi a^2 b}{3} - \upsilon_2^{\text{int}}(\boldsymbol{r}), \tag{7.13}$$

where

$$\upsilon_2^{\text{int}}(\boldsymbol{r}) = \frac{4\pi a^2 b}{3}\left[1 - \frac{3}{4}\frac{r}{R(\theta)} + \frac{1}{16}\left(\frac{r}{R(\theta)}\right)^3\right]\Theta[2R(\theta) - r] \tag{7.14}$$

and

$$R(\theta) = \frac{a}{[1 - (1 - a^2/b^2)\cos^2\theta]^{1/2}}. \tag{7.15}$$

Comparing (7.14) to the spherical-inclusion result (3.51) enables us to conclude that $R(\theta)$ is effectively an "angle-dependent" radius (see also Section 7.3). For oriented circular cylindrical inclusions, we have that

$$v_2(\boldsymbol{r}) = 4\pi a^2 b - (2b - |x\cos\theta|)A(|x\sin\theta|)\Theta(2a - |x\sin\theta|)\Theta(2b - |x\cos\theta|), \tag{7.16}$$

where

$$A(r) = 2a^2\left[\cos^{-1}\frac{r}{2a} - \frac{r}{2a}\sqrt{1-\frac{r^2}{4a^2}}\right]\Theta(2a-r). \tag{7.17}$$

7.3 Impenetrable Oriented Inclusions

The evaluation of the relation (5.35) for S_n (with the aforementioned interpretation of the inclusion indicator function) for nematic liquid-crystal structures consisting of oriented rectangular parallelepipeds or oriented circular cylinders that are mutually impenetrable is not analytically tractable for arbitrary volume fraction. Lower-order S_n can be computed relatively easily using the simulation techniques described in Chapter 12. However, in the case of oriented impenetrable (hard) spheroids, there are simplifications that enable one to exploit the same analytical techniques available for hard spheres (see Section 5.2.1). Indeed, this simplification extends to interpenetrable spheroids. Systems of oriented spheroids are useful anisotropic models, since they allow one to span the continuum from oriented needlelike inclusions to oriented platelike inclusions (Onsager 1949).

The task of computing the S_n for oriented *hard* spheroids is made reasonably straightforward by exploiting an observation of Lebowitz and Perram (1983) that an affine (linear) scale transformation to coordinates

$$\mathbf{R} \equiv (X, Y, Z) = (x, y, \frac{a}{b}z) \tag{7.18}$$

converts spheroids of shape

$$\frac{x^2+y^2}{a^2} + \frac{z^2}{b^2} = 1 \tag{7.19}$$

and volume fraction ϕ_2 into spheres of radius a at the *same volume fraction* (see Figure 7.2). Thus, the determination of the structure of aligned hard spheroids reduces to an equivalent problem involving hard spheres (Chapter 3).

Lado and Torquato (1990) used this observation to show that the two-point function $S_2(\boldsymbol{r}; b/a)$ for oriented hard spheroids of aspect ratio b/a at inclusion volume fraction ϕ_2 [where $\boldsymbol{r} = (x, y, z)$] scales as

$$S_2(\boldsymbol{r}; b/a) = S_2\left(\frac{a}{R(\theta)}r; 1\right), \tag{7.20}$$

where $R(\theta)$, defined by relation (7.15), is an angle-dependent "sphere radius", θ is the polar angle between the z-axis and $\boldsymbol{r}$, and $r = |\boldsymbol{r}|$. Thus, (7.20) states that S_2 for hard spheroids of aspect ratio b/a and volume fraction ϕ_2 can be obtained from S_2 for hard-sphere systems of volume fraction ϕ_2 in which distances are scaled by $R(\theta)/a$. Since the

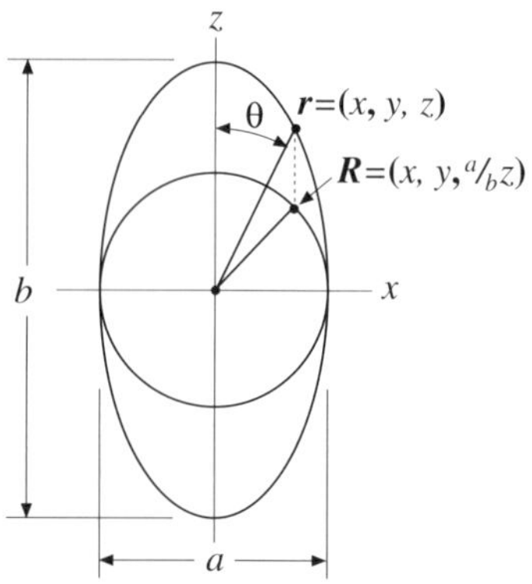

Figure 7.2 Linear transformation taking a spheroid into a sphere (looking along the y-axis). Here θ is the polar angle between the z-axis and $\boldsymbol{r} = (x, y, z)$.

volume fraction ϕ_2 is invariant under the transformation (7.18), the number density of the spheroid system $\rho(b/a)$ and number density of the sphere system $\rho(1)$ are related by the expression

$$\rho(b/a) = \frac{a}{b}\rho(1). \tag{7.21}$$

Of course, the volume fraction of the hard spheroid system is

$$\phi_2 = \rho(b/a)v_1 = \rho(b/a)\frac{4\pi a^2 b}{3}. \tag{7.22}$$

Subsequently, Torquato and Lado (1991) proved that relation (7.20) applies not only to hard spheroids but to spheroids with an *arbitrary degree of impenetrability*, i.e., possibly overlapping spheroids. Moreover, they demonstrated that an analogous scaling relation holds in the case of such spheroidal systems for S_n for any value of n, i.e.,

$$S_n(\boldsymbol{r}_{12}, \ldots, \boldsymbol{r}_{1n}; b/a) = S_n\left(\frac{a}{R(\theta_{12})}r_{12}, \ldots, \frac{a}{R(\theta_{1n})}r_{1n}; 1\right). \tag{7.23}$$

Note that the scaling relation (7.23) holds also in two dimensions for ellipses of shape $x^2/a^2 + y^2/b^2 = 1$ oriented in the y-direction with semiaxes a and b. However, $R(\theta)$ is now given by (7.11), and a is the disk radius.

Lado and Torquato (1990) have computed S_2 for *hard* oriented prolate and oblate spheroids in equilibrium at various spheroid volume fractions ϕ_2 and aspect ratios b/a by interpolation using the scaling relation (7.20) and the hard-sphere equilibrium results for S_2 given in Section 5.2.1. Figure 7.3 shows the effects of anisotropy for the prolate case $b/a = 5$ and $\phi_2 = 0.5$ in the form of cross-sections through $S_2(r, \theta)$ for $\theta = 0^\circ$, 45°, and 90°, reading from right to left on the main peak of the curves. Note that distances are in units of the major axis $2b$, so that the curve $\theta = 0$ (the outermost of the three) is identical to the hard-sphere limit. This figure was generated using the corrected Percus–Yevick results for S_2 given in Table 5.1.

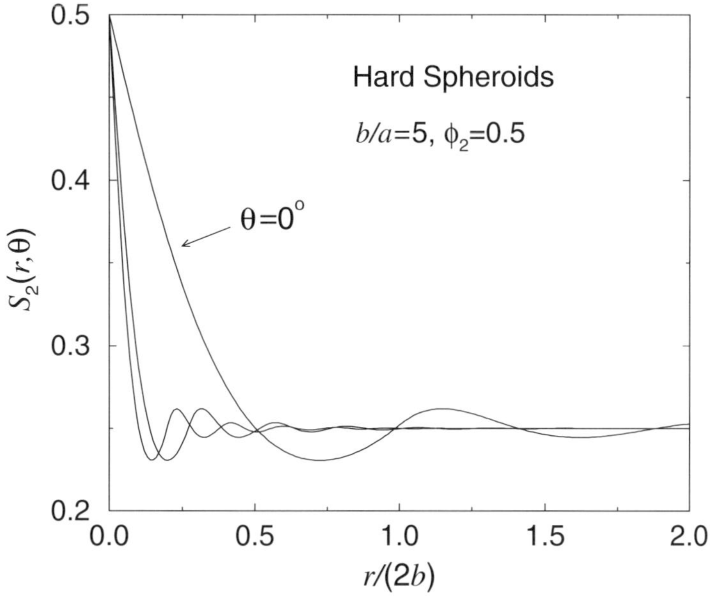

Figure 7.3 Cross sections through the matrix two-point probability function for a system of oriented hard spheroids in equilibrium with an aspect ratio $b/a = 5$ and $\phi_2 = 0.5$. The curves are for $\theta = 0^\circ$, 45°, and 90°, reading from right to left on the main peak of the curves.

We note in passing that finding the S_n for *randomly oriented* impenetrable nonspherical inclusions is a challenging problem. Quintanilla (1999) has found a means to evaluate the series expression (5.35) for S_n of such systems (with an appropriately generalized function m), provided that the inclusions are contained within "security spheres."

7.4 Hierarchical Laminates

Certain hierarchical laminates, with an infinite separation of length scales, are known theoretically to have optimal transport and mechanical properties (see Chapter 16 for details). This construction turns out to be a very useful theoretical tool. The term *nth-rank laminate* refers to a laminate with n different structural levels. Quintanilla and Torquato (1996b) derived analytical expressions for the n-point probability function S_n for more general hierarchical laminates that have a *finite* separation of length scales and arbitrary rank. For simplicity, they focused on two-dimensional laminates. Generalizations of the results to three dimensions is straightforward, albeit at the cost of additional algebraic complexity. In what follows we will limit ourselves to first- and second-rank laminates in two dimensions.

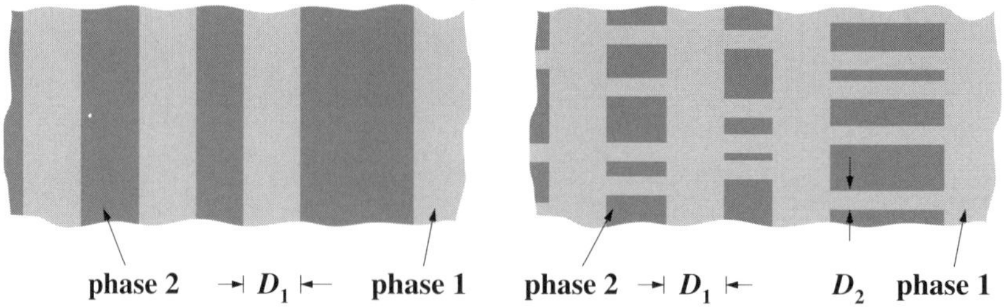

Figure 7.4 Schematics of two-dimensional laminates. Left panel: A portion of a random rank 1 laminate. Right panel: A portion of a random rank 2 laminate.

A *random* first-rank laminate is made by alternating layers of phase 1 and phase 2 materials according to some random process, as shown in Figure 7.4. A random second-rank laminate is constructed in two stages. For simplicity, the first stage is taken to be a series of parallel slabs of *fixed* width D_1 oriented in the y-direction generated by some one-dimensional random process. We define $\phi_1^{(1)}$ and $\phi_2^{(1)}$ to be the volume fractions of the matrix phase (phase 1) and the included phase (phase 2), respectively. The second stage of lamination adds *perpendicular* slabs of width D_2 in the gaps of the first stage. We define $\phi_1^{(2)}$ and $\phi_2^{(2)}$ to be the volume fractions of phases 1 and 2, respectively, for the second-stage process (see Figure 7.4). Clearly,

$$\phi_1^{(1)} + \phi_2^{(1)} = \phi_1^{(2)} + \phi_2^{(2)} = 1. \tag{7.24}$$

Moreover, a point lies in phase 2 of the *entire laminate* exactly when its x-coordinate lies in phase 2 of the first stage and its x-coordinate lies in phase 2 of the second stage of lamination. Since these events are independent, we see that the volume fraction of phase 2 of the entire laminate is given by

$$\phi_2 = 1 - \phi_1 = \phi_2^{(1)}\phi_2^{(2)}. \tag{7.25}$$

Useful one-dimensional models from which laminates can be constructed include fully penetrable rods, totally impenetrable rods, and one-dimensional "random checkerboards." Realizations of these three systems are shown in Figure 7.5; the systems depicted have equal rod lengths and volume fractions of the phases. By a one-dimensional "checkerboard" process we mean systems in which the line is divided into equisized sections of width D. Each section, independent of the other sections, belongs to phase 1 with probability ϕ_1 and phase 2 with probability ϕ_2. (The checkerboard model in higher dimensions is discussed in detail in Section 8.1.4.) By extending these one-dimensional systems into higher dimensions, hierarchical laminates are constructed.

Quintanilla and Torquato (1996b) showed that the two-point probability function for phase 1 for two-dimensional second-rank laminates of arbitrary construction is given by

$$S_2(\mathbf{r}) = L^{(1)}(x)S_2^{(2)}(y) + [S_2^{(1)}(x) - L^{(1)}(x)](\phi_1^{(2)})^2. \tag{7.26}$$

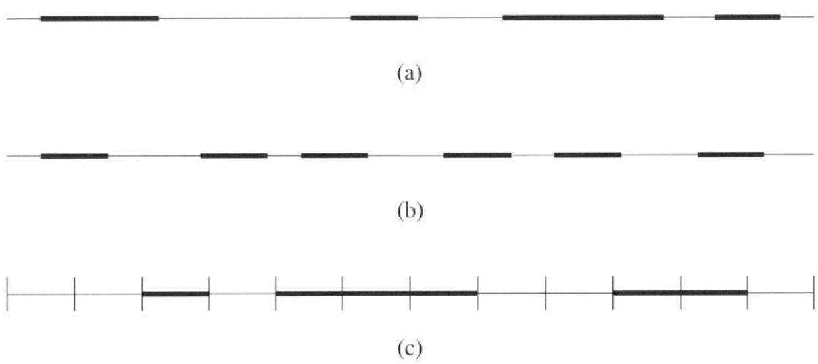

Figure 7.5 Three examples of one-dimensional random systems: (a) fully penetrable rods, (b) totally impenetrable rods, and (c) a one-dimensional "random checkerboard" process. The rods of the three systems have a common width D; however, the particle clusters in systems (a) and (c) can be longer than D. Laminates are generated by extending such systems into higher dimensions.

Here $S_2^{(i)}$ is the two-point probability function for phase 1 for the *one-dimensional process* that determines the ith stage of lamination, and $L^{(i)}$ is the corresponding lineal-path function (Section 2.4). Moreover, $\mathbf{r} = (x,y)$ is the displacement between the two points. Formula (7.26) is derived under the mild assumption that the first and second stages of lamination are generated independently of one another.

We observe that the two-point function of a first-rank laminate is easily obtained from expression (7.26) by letting the volume fraction $\phi_1^{(2)}$ equal 1 [implying $S_2^{(2)}(y) = 1$]. This limit yields

$$S_2(\mathbf{r}) = S_2^{(1)}(x) = S_2^{(1)}(r\cos\theta). \tag{7.27}$$

Finally, we can use the aforementioned analysis to calculate S_n for laminates that do not have orthogonal stages. For example, a second-rank laminate whose second stage is at angle θ from the horizontal can be transformed to a topologically equivalent orthogonal laminate via the linear transformation

$$(x,y) \to (x, y - x\tan\theta). \tag{7.28}$$

To calculate S_n for this "slanted" laminate, we would first project the points to their images under the above transformation and then calculate S_n for this orthogonal laminate.

We now obtain S_2 for *fully penetrable* second-rank laminates in terms of the generating one-dimensional processes. To apply formula (7.26), we need the one-dimensional fully penetrable rod results for S_2, obtained from (5.7) with $d = 1$, and the corresponding lineal-path function, obtained from (5.23) with $d = 1$. Accordingly, for the full laminate, it is found that

$$S_2(x,y) = \begin{cases} \phi_1 e^{-\eta_1|x|/D_1 - \eta_2|y|/D_2}, & |x| \le D_1, |y| \le D_2, \\ \phi_1 e^{-\eta_1|x|/D_1 - \eta_2}, & |x| \le D_1, |y| > D_2, \\ \phi_1[e^{-\eta_1|x|/D_1}(e^{-\eta_2|y|/D_2} - e^{-\eta_2}) + \phi_1], & |x| > D_1, |y| \le D_2, \\ \phi_1^2, & |x| > D_1, |y| > D_2, \end{cases} \tag{7.29}$$

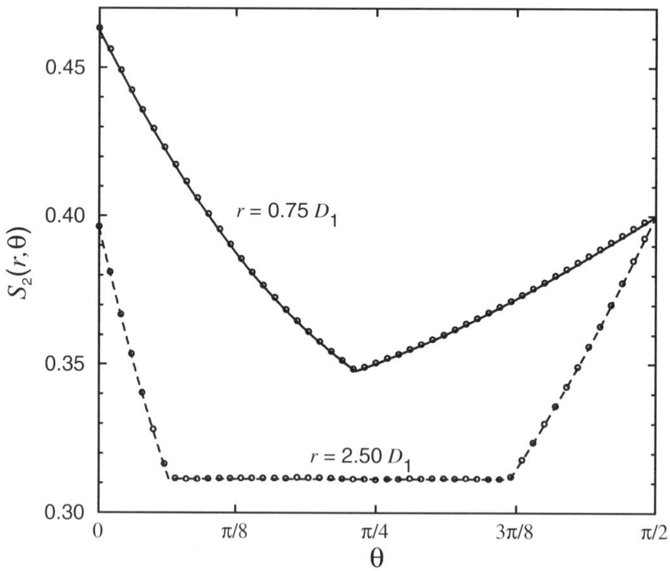

Figure 7.6 The two-point probability function $S_2(r, \theta)$ for a second-rank fully penetrable laminate for two different values of r with $D_2 = D_1/2$, $\eta_1 = 1/4$, and $\eta_2 = 1/3$. Computer simulation data are represented by circles (Quintanilla and Torquato 1996b).

where $\eta_1 = \rho_1 D_1$ and $\eta_2 = \rho_2 D_2$ are the reduced densities for the first and second stages of lamination, respectively, and ρ_i is the number density for the ith stage.

In Figure 7.6 we plot $S_2(r, \theta)$ in polar coordinates ($x = r\cos\theta$ and $y = r\sin\theta$) as computed from (7.29) for two different values of r over $0 \le \theta \le \pi/2$. From geometric considerations, $S_2(r, \theta)$ is symmetric about $\theta = \pi/2$ and is periodic with period π. We also note that S_2 is not symmetric about $\theta = \pi/4$, as expected. It is seen that computer simulation data (also shown) are in excellent agreement with the exact result (7.29).

The two-point probability function for second-rank laminates constructed from systems of identical *hard rods* in thermal equilibrium is obtained by substituting into (7.26) the one-dimensional formulas (5.44) and (5.59) (with $d = 1$) for S_2 and L, respectively. Similarly, the corresponding formula for second-rank checkerboard laminates requires the one-dimensional checkerboard formula (8.37) for S_2 as well as the one-dimensional checkerboard formula for the lineal-path function (Quintanilla and Torquato 1996b), which is given by

$$L(z) = (1 - r)\phi_1^n + r\phi_1^{n+1}, \tag{7.30}$$

where the dimensionless distance is decomposed as $z/D = (n - 1) + r$ with $0 \le r < 1$ for some integer n.

Finally, we note that S_2 can reflect the processes that construct the laminate and thus microstructural information about the composite on its different length scales. In Figure 7.7, $S_2(r, \pi/6)$ is plotted as a function of r and a fixed value of $\theta = \pi/6$ for

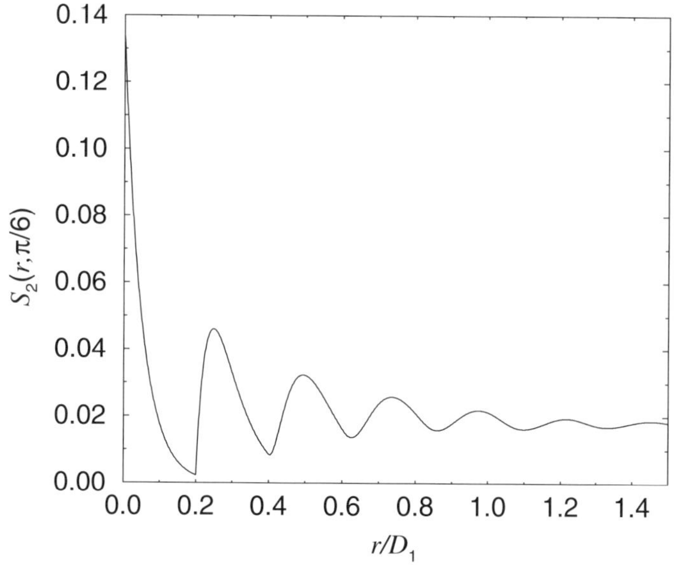

Figure 7.7 $S_2(r, \pi/6)$ in a mixed second-rank laminate. The first stage is generated by *fully penetrable rods* with $\eta_1 = 0.4$. The second stage is generated by *totally impenetrable rods* in thermal equilibrium with $\eta_2 = 0.8$ and $D_2 = 0.1D_1$.

a laminate constructed by *two different* one-dimensional processes. The first stage is generated by a system of fully penetrable rods with reduced density $\eta_1 = 0.4$. The second stage is generated by systems of hard rods in thermal equilibrium with $\eta_2 = 0.8$ and $D_2 = 0.1D_1$. We see that the structure of the laminate on both length scales is reflected in S_2. On the length scale $r = \mathcal{O}(D_1)$, S_2 decays exponentially and then more or less flattens, just as it does for "pure" fully penetrable rods [cf. (5.7) with $d = 1$]. However, on the length scale $r = \mathcal{O}(D_2)$, we see a sharp cusp and dampened oscillations, just as we did with S_2 for "pure" totally impenetrable rods in equilibrium [cf. (5.44) and Figure 5.6].

CHAPTER 8

Cell and Random-Field Models

There is a vast class of random heterogeneous materials whose microstructures cannot be modeled as a distribution of inclusions or cavities of well-defined shape in a matrix. This classification includes animal and plant tissue (which have a cellular structure), bone, foams, froths, polycrystals, block copolymers, and microemulsions, to mention but a few examples. In this chapter we will consider analytical methods to characterize the microstructure of two different nonparticulate models: *cell* and *random-field* models. For simplicity, we focus primarily on the determination of the n-point probability functions for these models.

This chapter begins with a short introduction to Voronoi and Delaunay tessellations of space. Certain cell statistics are subsequently described. This is followed by a discussion of symmetric-cell materials and the evaluation of the n-point probability functions for such materials. A general framework to generate microstructures using the random-field approach is then described. The chapter ends with a discussion on practical methods to compute the n-point probability functions for random-field models.

8.1 Cell Models

Examples of random heterogeneous materials with *cellular* microstructures are plentiful: cell biology (Lewis 1946), plant organs (Jeune and Barabe 1998), wood (Gibson and Ashby 1997), polycrystals (Aboav 1980), foams (Gibson and Ashby 1997, Weaire and Hutzler 2000), and micelles (Noever 1992). These structures may be thought of as being composed of *space-filling cells*. In some of these cases, the material is actually a porous solid with cellular pores characterized by high porosities (Gibson and Ashby

1997). The characterization of the microstructure of random cellular materials is an interesting problem and has been attacked from a variety of viewpoints.

Below we define and discuss the Voronoi and Delaunay tessellations. This is followed by a brief description of certain cell statistics. We then discuss the n-point probability functions for *symmetric-cell* materials in general and the *random checkerboard* model in particular.

8.1.1 Voronoi and Delaunay Tessellations

A *Voronoi tessellation* is a certain fundamental partitioning (tiling) of d-dimensional space into d-dimensional polyhedral cells (polytopes). This construction is also known to mathematicians as a *Dirichlet tessellation*. Condensed-matter physicists refer to the cells as *Wigner–Seitz* cells. The Voronoi tessellation and its dual (the Delaunay tessellation) have been reinvented, given different names, generalized, and applied in numerous different fields, including biology, meteorology, metallurgy, crystallography, forestry, ecology, archaeology, geology, geography, astrophysics, physics, computer science, and engineering (Aurenhammer 1991, Okabe, Boots and Sugihara 1992).

Consider a set of N points or sites with positions $\mathbf{r}_1, \mathbf{r}_2, \ldots, \mathbf{r}_N$ in volume V in d spatial dimensions. This set of positions may be deterministic (e.g., regular lattice) or random. Associated with the ith point at $\mathbf{r}_i$ is the *Voronoi cell*, which is defined to be the region of space nearer to the point at $\mathbf{r}_i$ than to any other point in the set. In two dimensions such a cell is a convex polygon, whereas in three dimensions such a cell is a convex polyhedron. In two dimensions, the boundary of the Voronoi polygon is composed of segments of the perpendicular bisectors of each line (edge) that connects the point at $\mathbf{r}_i$ to its *nearest-neighbor* sites (points that share a Voronoi edge). An example is indicated in Figure 8.1. In three dimensions, the boundary of the Voronoi polyhedron is composed of planes that perpendicularly bisect each edge that connects the point at $\mathbf{r}_i$ to its nearest-neighbor sites (points that share a Voronoi face). We refer to the set of all Voronoi cells associated with the N points as a *Voronoi diagram*. A Voronoi diagram becomes a *Voronoi tessellation* when it extends to all of space. This extension can be accomplished by taking the *thermodynamic limit* (see Section 3.1) or by the use of periodic boundary conditions so that the volume V is surrounded by images of itself *ad infinitum*.

Before discussing the Delaunay tessellation, it is useful to recall a few *topological* definitions and results. Generally speaking, a *graph* is a *topological* object composed of vertices and lines (or edges) connecting some subset of the vertices. A *polyhedral* graph G is one in which each edge belongs to the *face* of a polyhedron. Therefore, the network of edges (or bonds) that constitute the Voronoi tessellation is a special polyhedral graph. In any graph, the number of edges joined to a particular vertex is its *coordination number* Z. The *dual* graph G^* of a polyhedral graph G has vertices each of which corresponds to a face of G and each of whose faces corresponds to a vertex of G. Finally, for any subdivision of a domain into a finite number of polyhedral cells C, one can relate the number of vertices V, the number of edges E, and the number of

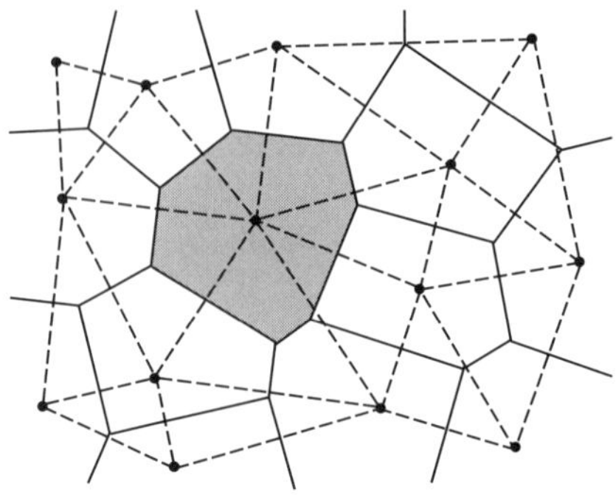

Figure 8.1 Voronoi and Delaunay tessellations for points (black dots) in the plane. The Voronoi polygons are indicated with solid lines, while the Delaunay triangles are indicated with dashed lines.

faces F to each other via Euler's formula:

$$F - E + V = 1, \qquad \text{(two-dimensional Euclidean space)}, \tag{8.1}$$

$$F - E + V - C = 1, \qquad \text{(three-dimensional Euclidean space)}. \tag{8.2}$$

Euler's formula has also been generalized to non-Euclidean spaces (e.g., the surface of a sphere or torus). It can be viewed as a *conservation* law, since it applies even if the structure is evolving with time.

Given a Voronoi graph, the *Delaunay graph* results from joining all pairs of sites that share a Voronoi face (nearest-neighbor sites) and divides all space into polyhedra. This subdivision of space is called the *Delaunay tessellation*, and it is the unique dual of the Voronoi tessellation. If the Delaunay tessellation consists only of simplices, i.e., simplest polyhedra (triangles in two dimensions and a tetrahedra in three dimensions), then we call it a *Delaunay triangulation*. If not, the Delaunay tessellation can still be triangulated, but the resulting triangulation is no longer unique. We note that for random point patterns, the Delaunay tessellation is almost always a triangulation of the space, except for some rare symmetric subset of points.

Let us now consider some specific point patterns that lead to Voronoi tessellations and may be used to model cellular materials. We will first give examples of *regular* point patterns (in two and three dimensions) that result in cells of regular shape. In two dimensions, there are many regular point patterns that one can choose from, but for simplicity we will limit ourselves to some well-known patterns resulting in cells of identical shape and size that tile the plane. As shown in Figure 8.2, points arranged on a regular *triangular lattice* lead to Voronoi cells that are *regular hexagons*. (Note that the corresponding Delaunay tessellation is the regular triangular graph.) The Voronoi cells associated with points arranged on a *regular honeycomb lattice* are *regular triangles*. Thus, the triangular-lattice graph is the dual of the honeycomb-lattice graph. (Observe that the Delaunay tessellation of the honeycomb lattice is not a triangulation.) Since

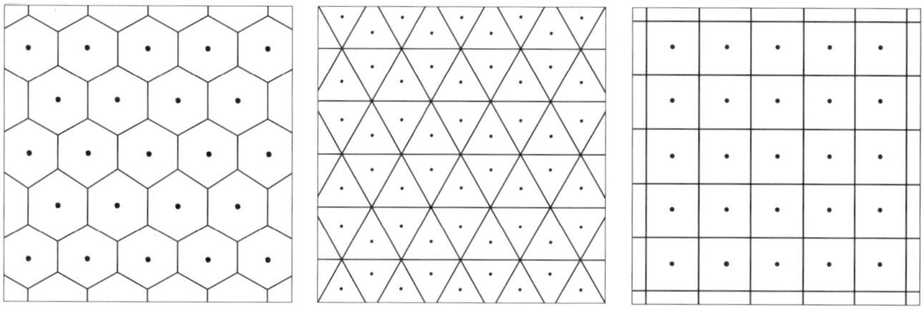

Figure 8.2 Three regular lattices (dots) and their associated Voronoi cells. Left panel: Regular triangular lattice and hexagonal cells. Middle panel: Regular honeycomb lattice and regular triangular cells. Right panel: Square lattice and square cells. The triangular- and honeycomb-lattice graphs are duals of each other. The square-lattice graph is self-dual.

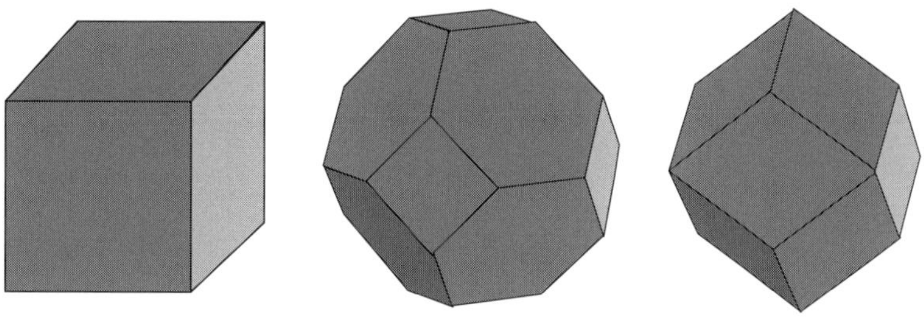

Figure 8.3 Voronoi unit cells for simple cubic, body-centered cubic, and face-centered cubic lattices are the cube (left), truncated octahedron (middle), and rhombic dodecahedron (right). The truncated octahedron is composed of six square and eight regular hexagonal faces. The rhombic dodecahedron is composed of twelve rhombus-shaped faces.

the square lattice leads to square Voronoi cells, the square-lattice graph is self-dual (the dual is another square lattice).

In three dimensions, it is natural to consider points arranged on cubic lattices. The Voronoi cells associated with simple cubic, body-centered cubic, and face-centered cubic lattices are identical *cubes* ($F = 6$, $E = 12$, $V = 8$), identical *truncated octahedra* ($F = 14$, $E = 36$, $V = 24$), and identical *rhombic dodecahedra* ($F = 12$, $E = 24$, $V = 14$), respectively (see Figure 8.3). By definition, these polyhedra individually fill space as specified by the underlying lattice. Although even Aristotle proclaimed that the regular tetrahedron fills space, the cube is the only Platonic solid (tetrahedron, cube, octahedron, dodecahedron, and icosahedron) that possesses this property. However, a certain combination of tetrahedra and octahedra can fill space (Wells 1991). Other identical polyhedral cells that can fill space include regular triangular prisms, rectangular prisms, and hexagonal prisms.

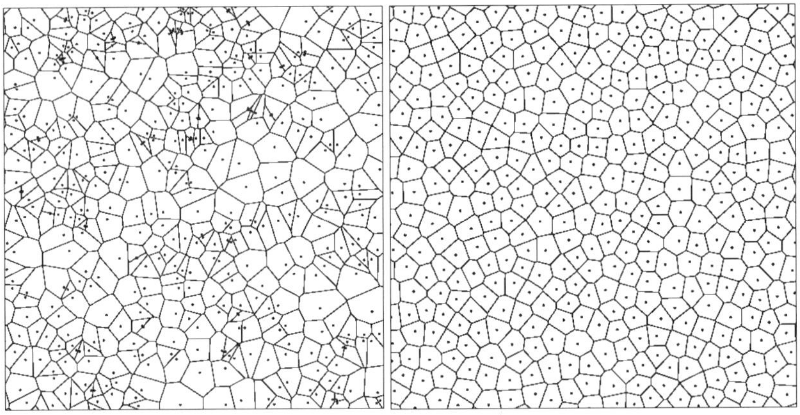

Figure 8.4 Random Voronoi tessellations. Left panel: Sites are fully-penetrable-disk centers (spatially uncorrelated points). Right panel: Sites are RSA hard-disk centers.

To generate random point patterns, we can exploit the statistical-mechanical tools that we developed in this book to produce random distributions of particles (see Chapters 3–7 for theoretical considerations and Chapter 12 for computer-simulation techniques). The sites for the Voronoi tessellations are taken to be the centroids of the particles whose spatial distribution is determined by the interaction potential and the dynamics. The Voronoi/Delaunay representation of particle systems is a convenient means of identifying a particle's *nearest neighbors*, i.e., particles that share faces with a given particle.

The Voronoi tessellation for the centers of fully penetrable disks (i.e., spatially uncorrelated or Poisson distributed points) is illustrated in Figure 8.4. It is seen that there is great variability in the size and shapes of the cells owing to the large spatial fluctuations possessed by fully-penetrable-sphere systems (Chapter 11). On the other hand, spatial correlations between points can be introduced by allowing the spheres to have some repulsive potential so that there is always some minimum distance between the spheres, such as hard spheres in equilibrium or in nonequilibrium (Chapter 3). Figure 8.4 also shows a Voronoi tessellation for a set of points determined by the centers of identical hard disks generated by the random sequential addition (RSA) process. Compared to the spatially uncorrelated case, the RSA tessellation has much less spatial variability, as expected. The corresponding three-dimensional RSA tessellation has been used to study the dynamics of tumor growth in a cellular automaton (Kansal, Torquato, Harsh, Chiocca and Deisboeck 2000b).

8.1.2 Cell Statistics

An immediate consequence of Euler's formula (8.1) for random tessellations of the plane into polygons is that the *average number of edges per cell* $\langle n \rangle$ is six, i.e.,

$$\langle n \rangle = 6, \qquad d = 2, \tag{8.3}$$

provided that all vertices have coordination number $Z = 3$. In many examples in nature, $Z = 3$ and $Z = 4$ for all vertices in two- and three-dimensional tessellations, respectively. Such instances are referred to as *topologically stable structures*, since their topological properties are invariant under small deformations (Weaire and Rivier 1984). [Note that the Voronoi tessellation of a homogeneous pattern of spatially uncorrelated points, as shown in Figure 8.4, is a topologically stable structure, and thus $Z = 3$ exactly (Stoyan et al. 1995).] Relation (8.3) is easily obtained from (8.1) (in the infinite-system limit) using the following relations:

$$3V = 2E = \sum_n nF_n, \qquad d = 2, \tag{8.4}$$

where F_n is the number of faces with n sides. Relation (8.4) expresses the fact that $Z = 3$ and an edge joins two vertices and separates two faces. In practice, the distribution of the number of edges per cell is sharply spiked around $\langle n \rangle = 6$. More generally, for any coordination number Z, the same analysis yields

$$\langle n \rangle = \frac{2Z}{Z-2}, \qquad d = 2. \tag{8.5}$$

The analogue of (8.3) for a three-dimensional tessellation with $Z = 4$ is less restrictive and is given by

$$\langle f \rangle = \frac{12}{6 - \langle n \rangle}, \qquad d = 3, \tag{8.6}$$

where $\langle f \rangle$ is the average number of faces per cell. This relation also applies to each individual cell of the tessellation. For most random tessellations, including those generated from centers of jammed random hard-sphere systems (Finney 1970), $\langle f \rangle \approx 14$. However, this is neither an exact result nor an upper bound.

Quantifying fluctuations in local topological/geometrical properties is a popular way to account for cell variability. For simplicity, we will restrict ourselves here to two dimensions (the extension to higher dimensions is obvious). It is common to consider $p(n)$, the probability of finding a cell with n edges, and $P(A)$, the probability density associated with finding a cell with area A. Both quantities normalize to unity, i.e.,

$$\sum_n p(n) = 1, \qquad \int_0^\infty P(A)dA = 1.$$

It is useful to define the associated moments:

$$\mu_k = \sum_n (n - \langle n \rangle)^k p(n), \qquad k = 1, 2, \ldots, \tag{8.7}$$

$$M_k = \int_0^\infty (A - \langle A \rangle)^k P(A)\, dA, \qquad k = 1, 2, \ldots, \tag{8.8}$$

where $\langle A \rangle$ is the average cell area defined by

$$\langle A \rangle = \int_0^\infty AP(A)dA. \tag{8.9}$$

It is often more convenient to determine the average linear intercept (mean chord length), rather than $\langle A \rangle$, as a measure of the average cell size. The reader is referred to the excellent survey of Weaire and Rivier (1984) for calculations of the functions $p(n)$ and $P(A)$ for specific structures, as well as other aspects of cell networks. In addition to such single-cell statistics, one can also study certain two-cell statistics (Delannay and Le Caer 1994).

It is generally true that a cell with more sides (faces) than the average has neighbors that, on average, have less than the average number of sides (faces). In two dimensions, this empirical correlation has come to be known as the Aboav–Weaire law (Weaire and Rivier 1984), given by

$$m(n) = (6 - a) + \frac{6a + \mu_2}{n}, \tag{8.10}$$

where $m(n)$ is the average number of sides of neighbors of n-sided cells, a is a parameter that is on the order of unity for natural structures, and μ_2 is the second moment obtained from (8.7).

8.1.3 Symmetric-Cell Materials

Two-phase *symmetric-cell materials* are constructed by partitioning space into cells of arbitrary shapes and sizes, with cells being randomly designated as phases 1 and 2 with probability ϕ_1 and ϕ_2, respectively. Miller (1969) introduced this useful class of models as well as *asymmetric-cell materials* that Brown (1974) showed were not physically realizable. Note that symmetric-cell materials, in contrast to distributions of particles in matrix, possess *phase-inversion symmetry* , i.e., the morphology of phase 1 at volume fraction ϕ_1 is statistically identical to that of phase 2 when the volume fraction of phase 1 is $1 - \phi_1$ (see Section 2.2.2). Bruno (1990) has shown that symmetric-cell materials represent a large class of so-called *infinitely interchangeable* materials.

The Voronoi procedure outlined in Section 8.1.1 provides a systematic means of constructing symmetric-cell materials with polyhedral cells. Figure 8.5 shows realizations of two-dimensional symmetric-cell materials using four different Voronoi tessellations: square, triangular, hexagonal, and randomly shaped cells. The square tessellation is also known as the *random checkerboard* model. The cell shapes are not limited to Voronoi polyhedra, however. Figure 8.6 illustrates a tessellation of space into circles, which necessarily must have a continuous range of diameters ranging to the infinitesimally small.

Below we derive general expressions for the n-point probability function $S_n^{(i)}$ of phase i for a symmetric-cell two-phase material of "white" cells (phase 1) with volume fraction ϕ_1 and "black" cells (phase 2) with volume fraction ϕ_2. This analysis generalizes the one presented by Lu and Torquato (1990b) for the random checkerboard model. [Their starting point was the series representation (4.46) in the limit given by (4.69) with appropriately generalized expressions for the indicator function m and the n-particle

Figure 8.5 Four examples of two-dimensional two-phase symmetric-cell materials at the area fraction $\phi_1 = \phi_2 = 0.5$. Top left: Square cells (random checkerboard model). Top right: Triangular cells. Bottom left: Hexagonal cells. Bottom right: Randomly shaped cells generated from RSA hard-disk centers.

probability density function ρ_n.] We will also make some specific remarks about the two- and three-point quantities. The derivation below is considerably simpler than the one given by Miller (1969) and more general (he considered only the three-point quantity).

n-Point Probability Function

Let $S_n(\boldsymbol{x}^n) \equiv S_n^{(1)}(\boldsymbol{x}^n)$ and $\hat{S}_n(\boldsymbol{x}^n) \equiv S_n^{(2)}(\boldsymbol{x}^n)$ denote the probability that n points at positions $\boldsymbol{x}^n \equiv \{\boldsymbol{x}_1, \boldsymbol{x}_2, \ldots, \boldsymbol{x}_n\}$ fall in the white phase and black phase, respectively. Then we can express the n-point quantities as

$$S_n(\boldsymbol{x}^n) = \sum_{k=1}^{n} P_n^{(k)}(\boldsymbol{x}^n), \qquad \text{for all } n, \tag{8.11}$$

$$\hat{S}_n(\boldsymbol{x}^n) = \sum_{k=1}^{n} \hat{P}_n^{(k)}(\boldsymbol{x}^n), \qquad \text{for all } n, \tag{8.12}$$

where, for any $1 \le k \le n$,

$$P_n^{(k)} = \text{Probability that } n \text{ points fall in } k \text{ different white cells,} \tag{8.13}$$

$$\hat{P}_n^{(k)} = \text{Probability that } n \text{ points fall in } k \text{ different black cells.} \tag{8.14}$$

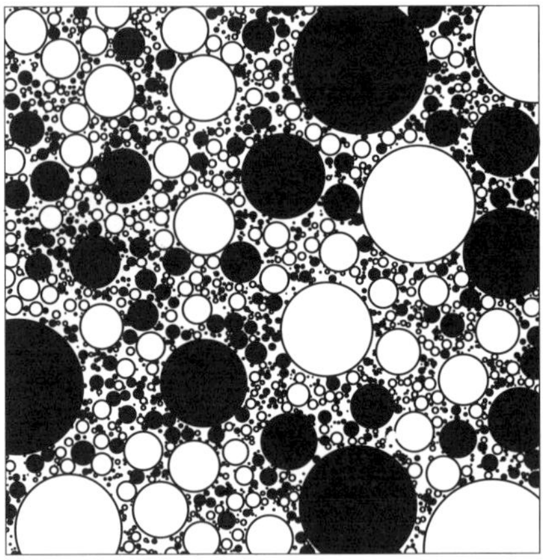

Figure 8.6 A schematic of a symmetric-cell material consisting of space-filling circles at the area fraction $\phi_1 = \phi_2 = 0.5$.

We emphasize that the index k in (8.13) and (8.14) does *not* indicate the phase (as in most other instances).

For a statistically homogeneous symmetric-cell material, the lack of any spatial correlations (i.e., statistical independence) enables us to express $P_n^{(k)}$ and $\hat{P}_n^{(k)}$ in terms of *weights* $W_n^{(k)}$ according to the relations

$$P_n^{(k)}(\boldsymbol{x}^n) = W_n^{(k)}(\boldsymbol{x}^n)\,\phi_1^k, \tag{8.15}$$

$$\hat{P}_n^{(k)}(\boldsymbol{x}^n) = W_n^{(k)}(\boldsymbol{x}^n)\,\phi_2^k, \tag{8.16}$$

where

$$W_n^{(k)}(\boldsymbol{x}^n) = \begin{array}{l}\text{Probability that } n \text{ points at positions } \boldsymbol{x}^n \text{ are in } k \text{ cells,}\\ \text{regardless of their color.}\end{array} \tag{8.17}$$

Note that the weights are *independent of the volume fraction*, bounded between zero and one, and must sum to unity, i.e.,

$$0 \le W_n^{(k)} \le 1, \qquad \sum_{k=1}^{n} W_n^{(k)} = 1. \tag{8.18}$$

As expected, relations (8.11) and (8.12) for the n-point probability functions, in conjunction with (8.15) and (8.16), satisfy the property of phase-inversion symmetry specified by (2.15).

Moreover, the asymptotic properties of the S_n discussed in Section 2.2 impose certain asymptotic properties on the weights. For example, if all of the points coincide, then $W_n^{(1)} = 1$ and $W_n^{(k)} = 0$ for $2 \le k \le n$; if the points are mutually far from one another, then $W_n^{(n)} = 1$ and $W_n^{(k)} = 0$ for $1 \le k \le n-1$.

The weights depend on the cell *shapes* via the indicator function m, the cell *size distribution* via a probability density function $f(\boldsymbol{R})$ (where $\boldsymbol{R}$ represents the vector of characteristic lengths of a cell), and *spatial distribution* of the cells via the generalized probability density function ρ_n. A particularly important weight is the *single-cell* weight $W_n^{(1)}(\boldsymbol{x}^n)$, the probability of finding all n points in any single cell. This quantity is only a function of cell shapes and sizes, depending on a dimensionless size-averaged intersection volume of n cells defined below. If we let $\boldsymbol{r}_i$ denote *both the centroid and orientation* of the ith cell in the system, then we explicitly have for cells of identical but general shape that

$$W_n^{(1)}(\boldsymbol{x}^n) = \frac{\langle v_n^{\text{int}}(\boldsymbol{x}^n;\boldsymbol{R})\rangle}{\langle v_1(\boldsymbol{R})\rangle}, \tag{8.19}$$

where

$$\langle v_n^{\text{int}}(\boldsymbol{x}^n;\boldsymbol{R})\rangle = \int v_n^{\text{int}}(\boldsymbol{x}^n;\boldsymbol{R}) f(\boldsymbol{R})\, d\boldsymbol{R}, \tag{8.20}$$

$$v_n^{\text{int}}(\boldsymbol{x}^n;\boldsymbol{R}) = \int \prod_{i=1}^{n} m(\boldsymbol{x}_i - \boldsymbol{r}_1;\boldsymbol{R})\, d\boldsymbol{r}_1, \tag{8.21}$$

$$m(\boldsymbol{x}-\boldsymbol{r}_i;\boldsymbol{R}) = \begin{cases} 1, & \boldsymbol{x}-\boldsymbol{r}_i \in D_i, \\ 0, & \text{otherwise.} \end{cases} \tag{8.22}$$

Here D_i is the ith cell region and $v_1(\boldsymbol{R}) \equiv v_1^{\text{int}}(\boldsymbol{R})$. Relation (8.19) is just the generalization of the one-body result (5.43) for finding n points in a single impenetrable particle. Intersection volumes of the type given in (8.19) were encountered in our study of the n-point probability functions for particle systems (see Chapters 5, 6, and 7).

Two-Point Probability Function

For $n = 2$, the above analysis enables us to write the two-point quantities (for statistically homogeneous but possibly anisotropic media) as follows:

$$S_2(\boldsymbol{r}) = W_2^{(1)}(\boldsymbol{r})\phi_1 + W_2^{(2)}(\boldsymbol{r})\phi_1^2, \tag{8.23}$$

$$\hat{S}_2(\boldsymbol{r}) = W_2^{(1)}(\boldsymbol{r})\phi_2 + W_2^{(2)}(\boldsymbol{r})\phi_2^2, \tag{8.24}$$

where $\boldsymbol{r} = \boldsymbol{x}_2 - \boldsymbol{x}_1$. The fact that the weights sum to unity enables us to express the two-point quantities in terms of only the single-cell two-point weight $W_2^{(1)}$ and volume fractions, i.e.,

$$S_2(\boldsymbol{r}) = W_2^{(1)}(\boldsymbol{r})\phi_1\phi_2 + \phi_1^2, \tag{8.25}$$

$$\hat{S}_2(\boldsymbol{r}) = W_2^{(1)}(\boldsymbol{r})\phi_1\phi_2 + \phi_2^2. \tag{8.26}$$

The asymptotic properties of the two-point probability functions given by (2.21) impose the following properties on the single-cell weight $W_2^{(1)}(\boldsymbol{r})$: $W_2^{(1)}(0) = 1$ and $W_2^{(1)}(\infty) = 0$.

Three-Point Probability Function

For $n = 3$, the general analysis given above enables us to express the three-point probabilities as

$$S_3(\boldsymbol{r},\boldsymbol{s}) = W_3^{(1)}\phi_1 + W_3^{(2)}\phi_1^2 + W_3^{(3)}\phi_1^3, \tag{8.27}$$

$$\hat{S}_3(\boldsymbol{r},\boldsymbol{s}) = W_3^{(1)}\phi_2 + W_3^{(2)}\phi_2^2 + W_3^{(3)}\phi_2^3, \tag{8.28}$$

where $\boldsymbol{r} = \boldsymbol{x}_2 - \boldsymbol{x}_1$ and $\boldsymbol{s} = \boldsymbol{x}_3 - \boldsymbol{x}_1$. We now show that the three-point functions can be written in terms of only the volume fractions, the single-cell three-point weight $W_3^{(1)}$, and a volume-fraction-independent function of single-cell two-point weights given by

$$\begin{aligned} f_2(\boldsymbol{r},\boldsymbol{s}) &= \frac{S_2(\boldsymbol{r}) + S_2(\boldsymbol{s}) + S_2(\boldsymbol{t})}{\phi_1\phi_2} - 3\frac{\phi_1}{\phi_2}, \\ &= W_2^{(1)}(\boldsymbol{r}) + W_2^{(1)}(\boldsymbol{s}) + W_2^{(1)}(\boldsymbol{t}), \end{aligned} \tag{8.29}$$

where $\boldsymbol{t} = \boldsymbol{r} - \boldsymbol{s}$.

To prove this proposition, we first substitute the sum rule of relation (8.18) into expressions (8.27) and (8.28) to obtain

$$S_3(\boldsymbol{r},\boldsymbol{s}) = W_3^{(1)}\phi_1 + W_3^{(2)}\phi_1^2 + \left[1 - W_3^{(1)} - W_3^{(2)}\right]\phi_1^3, \tag{8.30}$$

$$\hat{S}_3(\boldsymbol{r},\boldsymbol{s}) = W_3^{(1)}\phi_2 + W_3^{(2)}\phi_2^2 + \left[1 - W_3^{(1)} - W_3^{(2)}\right]\phi_2^3. \tag{8.31}$$

We now eliminate the three-point weight $W_3^{(2)}$ in favor of the weight $W_3^{(1)}$ and the function f_2. To do this, we use the general expression (2.8) (with $n = 3$) that relates the three-point probability function for one phase to the corresponding function for the other phase:

$$\hat{S}_3(\boldsymbol{r},\boldsymbol{s}) = 1 - 3\phi_1 + S_2(\boldsymbol{r}) + S_2(\boldsymbol{s}) + S_2(\boldsymbol{t}) - S_3(\boldsymbol{r},\boldsymbol{s}). \tag{8.32}$$

This expression in combination with relations (8.30) and (8.31) yields

$$W_3^{(2)}(\boldsymbol{r},\boldsymbol{s}) = -3W_3^{(1)}(\boldsymbol{r},\boldsymbol{s}) + f_2(\boldsymbol{r},\boldsymbol{s}), \tag{8.33}$$

where f_2 is given by (8.29). Finally, substitution of relation (8.33) into (8.30) and (8.31) enables us to write

$$S_3(\boldsymbol{r},\boldsymbol{s}) = \phi_1\phi_2(\phi_2 - \phi_1)W_3^{(1)}(\boldsymbol{r},\boldsymbol{s}) + \phi_1^2\phi_2 f_2(\boldsymbol{r},\boldsymbol{s}) + \phi_1^3, \tag{8.34}$$

$$\hat{S}_3(\boldsymbol{r},\boldsymbol{s}) = \phi_1\phi_2(\phi_1 - \phi_2)W_3^{(1)}(\boldsymbol{r},\boldsymbol{s}) + \phi_2^2\phi_1 f_2(\boldsymbol{r},\boldsymbol{s}) + \phi_2^3, \tag{8.35}$$

which proves the proposition. Note that the asymptotic properties of the three-point probability functions given by (2.22) and (2.23) impose the following properties on the single-cell weight $W_3^{(1)}(\boldsymbol{r},\boldsymbol{s})$ under all permutations of the distances r, s, and t: $W_3^{(1)}(0,0) = 1$, $W_3^{(1)}(\boldsymbol{r},0) = W_2^{(1)}(\boldsymbol{r})$, and $W_3^{(1)}(\boldsymbol{r},\infty) = W_3^{(1)}(\infty,\infty) = 0$.

Note that when $\phi_1 = \phi_2 = 0.5$, we find from (8.34) and (8.35) that $S_3 = \hat{S}_3$, and therefore S_3 depends only on two-point statistics, i.e.,

$$S_3(\boldsymbol{r},\boldsymbol{s}) = \hat{S}_3(\boldsymbol{r},\boldsymbol{s}) = \frac{S_2(\boldsymbol{r}) + S_2(\boldsymbol{s}) + S_2(\boldsymbol{t})}{2} - \frac{1}{4}. \tag{8.36}$$

As expected, a symmetric-cell material at this volume fraction obeys the symmetry property (2.17).

8.1.4 Random Checkerboard

As noted earlier, Lu and Torquato (1990b) studied the n-point probability functions for the random checkerboard model for arbitrary volume fractions (see the top left example of Figure 8.5). In particular, here we report simplified versions of their results for the two-point probability function $S_2(\mathbf{r})$ of translationally invariant systems. This *special symmetric-cell material* is statistically anisotropic and, for general dimensionality, will depend on the magnitude and orientation of the vector $\mathbf{r} = \mathbf{x}_2 - \mathbf{x}_1$. By averaging over solid angles, we can also obtain the corresponding rotationally invariant two-point function. The results below are given for the first three space dimensions and the distances are expressed in *units of the length of the side of a cubical cell.* As discussed in Section 8.1.3, the key quantity is the single-cell weight $W_2^{(1)}(\mathbf{x}_1, \mathbf{x}_2)$, which is the size-averaged intersection volume of two identical cells at positions $\mathbf{x}_1$ and $\mathbf{x}_2$ divided by the size-averaged cell volume. For the random checkerboard, however, there is no size distribution, and so $W_2^{(1)}(\mathbf{x}_1, \mathbf{x}_2)$ is particularly simple.

For $d = 1$, the two-point function depends only on the distance $r = |\mathbf{r}| = |\mathbf{x}_1 - \mathbf{x}_2|$. From (8.25), we have that

$$S_2(r) = W_2^{(1)}(r)\phi_1\phi_2 + \phi_1^2. \tag{8.37}$$

It is easy to verify that the probability of finding two points in the same one-dimensional cell is given by

$$W_2^{(1)}(r) = \begin{cases} 1 - r, & 0 \le r \le 1, \\ 0, & r \ge 1. \end{cases} \tag{8.38}$$

For $d = 2$, we have that

$$S_2(r, \theta) = W_2^{(1)}(r, \theta)\phi_1\phi_2 + \phi_1^2, \tag{8.39}$$

where

$$W_2^{(1)}(r, \theta) = [1 - r\cos\theta][1 - r\sin\theta]\Theta(1 - r\cos\theta)\Theta(1 - r\sin\theta), \tag{8.40}$$

$\Theta(x)$ is the Heaviside step function (3.48), and θ is the angle that the vector $\mathbf{r}$ makes with the horizontal. Note that $S_2(r, \theta)$ possesses reflection symmetry about $\theta = \pi/4$.

The rotationally invariant two-point probability function $S_2(r)$ can be obtained from the result immediately above by an appropriate angular average, i.e.,

$$S_2(r) = W_2^{(1)}(r)\phi_1\phi_2 + \phi_1^2, \tag{8.41}$$

where

$$W_2^{(1)}(r) = \frac{4}{\pi}\int_0^{\pi/4} W_2^{(1)}(r, \theta)d\theta. \tag{8.42}$$

The step function conditions imply that $\theta \geq \cos^{-1}(1/r)$. For $r \leq 1$, the step functions are always nonzero, and hence $0 \leq \theta \leq \pi/4$. However, for $1 \leq r \leq \sqrt{2}$, $\cos^{-1}(1/r) \leq \theta \leq \pi/4$. Hence, we find explicitly that

$$W_2^{(1)}(r) = \begin{cases} 1 + \dfrac{r^2 - 4r}{\pi}, & 0 \leq r \leq 1, \\ 1 - \dfrac{2 + r^2}{\pi} + \frac{4}{\pi}\left[\sqrt{r^2 - 1} - \cos^{-1}(1/r)\right], & 1 \leq r \leq \sqrt{2}, \\ 0, & r \geq \sqrt{2}. \end{cases} \tag{8.43}$$

For $d = 3$, the two-point probability function is given by

$$S_2(r, \theta, \varphi) = W_2^{(1)}(r, \theta, \varphi)\phi_1\phi_2 + \phi_1^2, \tag{8.44}$$

where

$$\begin{aligned} W_2^{(1)}(r, \theta, \varphi) = {} & [1 - r\cos\theta][1 - r\sin\theta\sin\varphi][1 - r\sin\theta\cos\varphi]\Theta(1 - r\cos\theta) \\ & \times \Theta(1 - r\sin\theta\sin\varphi)\Theta(1 - r\sin\theta\cos\varphi), \end{aligned} \tag{8.45}$$

and θ and φ are the spherical polar and azimuthal angles, respectively, associated with the vector $\boldsymbol{r}$. Because of cubic symmetry, we need only consider the angles in the range $0 \leq \theta \leq \pi/2$ and $0 \leq \varphi \leq \pi/2$.

From the angle-dependent expression for $S_2(r, \theta, \varphi)$, we can get the associated rotationally invariant quantity $S_2(r)$ by the angular average

$$S_2(r) = W_2^{(1)}(r)\phi_1\phi_2 + \phi_1^2, \tag{8.46}$$

where

$$W_2^{(1)}(r) = \frac{2}{\pi}\int_0^{\pi/2}\int_0^{\pi/2} W_2^{(1)}(r, \theta)\sin\theta d\theta d\varphi. \tag{8.47}$$

Therefore, we have exactly that

$$W_2^{(1)}(r) = \begin{cases} 1 - \frac{3}{2}r + \frac{2}{\pi}r^2 - \frac{1}{4\pi}r^3, & 0 \leq r \leq 1, \\ a(r) + f(r), & 1 \leq r \leq \sqrt{2}, \\ b(r) + g(r), & \sqrt{2} \leq r \leq \sqrt{3}, \\ 0, & r \geq \sqrt{3}, \end{cases} \tag{8.48}$$

where

$$\begin{aligned} a(r) = \frac{1}{12\pi r}\Big[& -3 + 18(1 + \pi)r^2 + 6r^4 - 4(8 + 3\pi + 4r^2)(r^2 - 1)^{1/2} \\ & - 16(r^2 - 1)^{3/2} - 48r^2\sin^{-1}(1/r)\Big], \end{aligned}$$

$$b(r) = \frac{1}{12\pi r}\left[25 - 18r^2 - 3r^4 + (32 + 8r^2)(r^2 - 2)^{1/2}\right],$$

Table 8.1 The rotationally averaged weight $W_2^{(1)}(r)$ as a function of r for the random checkerboard for the first three space dimensions.

r	$W_2^{(1)}(r)$		
	$d=1$	$d=2$	$d=3$
0.0	1.0	1.0	1.0
0.1	0.9	0.87586	0.85629
0.2	0.8	0.75808	0.72483
0.3	0.7	0.64668	0.60515
0.4	0.6	0.54163	0.49677
0.5	0.5	0.44296	0.39921
0.6	0.4	0.35065	0.31199
0.7	0.3	0.26470	0.23465
0.8	0.2	0.18513	0.16669
0.9	0.1	0.11192	0.10765
1.0	0.0	0.04507	0.05704
1.1		0.01458	0.02404
1.2		0.00387	0.00857
1.3		0.00051	0.00229
$\sqrt{2}$		0.0	0.00032
1.5			0.00005
1.6			0.00000
$\sqrt{3}$			0.0

$$f(r) = -\frac{1}{\pi}\int_{\sin^{-1}(\frac{1}{r})}^{\pi/2} \left\{4\left[\sin^{-1}\left(\frac{1}{y}\right) + \sqrt{y^2-1}\right] - \pi\right\}(1 - r\cos\theta)\sin\theta d\theta,$$

$$g(r) = -\frac{1}{\pi}\int_{\cos^{-1}(\frac{1}{r})}^{\sin^{-1}(\frac{\sqrt{2}}{r})} \left\{4\left[\sin^{-1}\left(\frac{1}{y}\right) + \sqrt{y^2-1}\right] - \pi\right\}(1 - r\cos\theta)\sin\theta d\theta,$$

and $y = r\sin\theta$. The functions $f(r)$ and $g(r)$ are easily evaluated using standard numerical quadratures. Consistent with our discussion in Section 2.2.5, the presence of edges and corners at the interface in this model means that $S_2(r)$ possesses a quadratic term r^2 for small r.

The two-point weight for a single cell is given as a function of r in Table 8.1 for the first three space dimensions. Figure 8.7 plots the two-point function for the random checkerboard model in the one-dimensional case (8.37) and the three-dimensional rotationally invariant case (8.46). Except for the cusp at $r = 1$ for $d = 1$, the functions are not drastically different from one another.

8.1.5 Ising Model

The Ising model is a simple interacting spin model (Stanley 1971, Binney et al. 1992) that can be conveniently employed to study cooperative (correlated) phenomena and thus represents an alternative to the symmetric-cell material in which spatial correla-

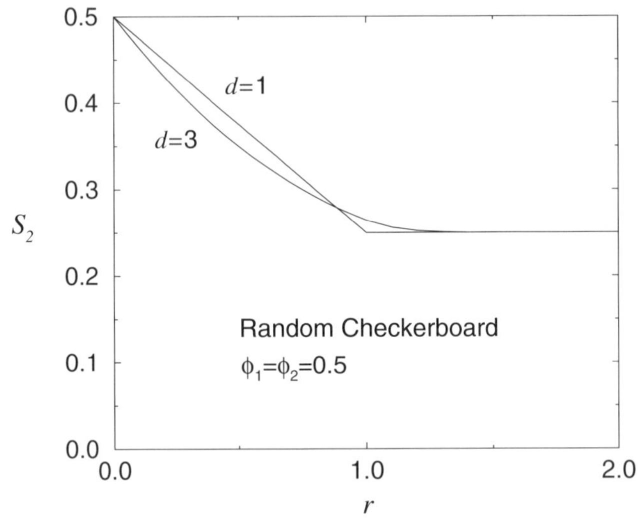

Figure 8.7 The isotropic two-point probability function S_2 versus scaled distance r for the random checkerboard model for $d = 1$ and $d = 3$. The distance is measured in units of a cell length.

tions are absent. The Ising model was originally introduced to model the spontaneous magnetization of a ferromagnet in the absence of an external field. In its simplest form, one considers two different types of spins ("up" or "down" spins) placed at the sites of a lattice (typically a square lattice in two dimensions) that interact via nearest neighbors (Onsager 1944). Let s_i denote the spin value on site i ($s_i = +1$ for an up spin and $s_i = -1$ for a down spin). The Hamiltonian or energy for a system of N Ising spins with configuration specified by $\{s\} = \{s_1, s_2, \ldots\}$ and an external field H is given by

$$E\{s\} = -J\sum_{\langle i,j\rangle} s_i s_j - H\sum_{i=1}^{N} s_i, \tag{8.49}$$

where J is the coupling constant and $\langle i,j\rangle$ denotes summation over nearest neighbor pairs of spins. The coupling constant is a measure of the strength of the interactions between nearest neighbor spins. Notice that when $J > 0$, it is energetically favorable for neighboring spins to be aligned. For sufficiently low temperatures, therefore, this leads to a cooperative phenomenon referred to as *spontaneous magnetization*. In other words, even when $H = 0$, long-range correlations between spins are associated with long-range order in which the lattice has a net magnetization M given by

$$M = \left\langle \sum_{i=1}^{N} s_i \right\rangle,$$

where angular brackets indicate a thermal average. The *critical temperature* T_c is the highest temperature for which there can be nonzero magnetization. The equilibrium thermodynamic properties of the Ising model depend on the temperature T, external field H, number of spins N, type of lattice, and dimension.

The Ising model is trivially mapped into a random-medium cell model by associating up and down spins with black and white phases, respectively. For statistically homogeneous media, if the black phase is designated as phase 2, then the volume fraction of this phase is related to the spontaneous magnetization (in the absence of an external magnetic field) via the relation

$$M = 1 - 2\phi_2.$$

In general, the magnetization (or volume fraction) depends on the temperature T and field H. Note that in the high-temperature limit ($T \to \infty$), this model is identical to the random checkerboard at a volume fraction of 50 percent. In Figure 12.7 we compare two realizations of the Ising model generated from Monte Carlo simulations (see Section 12.2.6): one at a finite temperature just above the critical temperature T_c and the other in the high-temperature limit. Note that the long-range correlations in the system near T_c result in clusters that are significantly larger than those in the high-temperature system.

8.2 Random-Field Models

In the random-field approach to modeling the microstructure of random media, the interface between the phases is defined by level cuts of random fields (Berk 1987, Berk 1991, Teubner 1991, Crossley, Schwartz and Banavar 1991, Blumenfeld and Torquato 1993, Roberts and Teubner 1995). As will be described below, there is great flexibility in the choice of the random field and hence in the class of microstructures that can be produced. This approach is particularly useful in modeling *bicontinuous* media (Section 9.1) such as microemulsions (Berk 1987), carbonate rocks (Crossley et al. 1991), Vycor glass (Crossley et al. 1991), amorphous alloys (Roberts and Teubner 1995), and aerogels (Roberts 1997).

8.2.1 General Considerations

To describe random-field models, we will follow the general formalism of Blumenfeld and Torquato (1993). According to this description, the generation of a random-field model consists of three steps: (i) generation of the source image, (ii) coarse-graining the source image, and (iii) performing level cuts on the coarse-grained image. We begin by assuming that there are N *source points* located at positions $\boldsymbol{x}_1, \ldots, \boldsymbol{x}_N$ in a d-dimensional volume. The global intensity or source function $f(\boldsymbol{x})$ (some measurable quantity of interest) at an arbitrary position $\boldsymbol{x}$ is defined as

$$f(\boldsymbol{x}) = \sum_{n=1}^{N} f_n \delta(\boldsymbol{x} - \boldsymbol{x}_n), \tag{8.50}$$

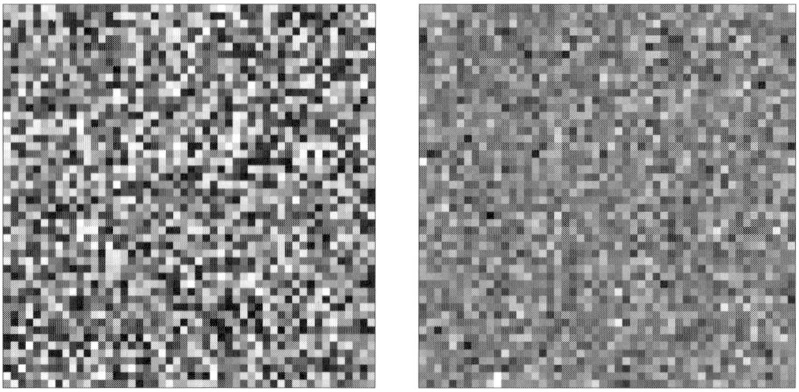

Figure 8.8 Two discrete (digitized) gray-scale source images [discrete examples of (8.50)]: uniform distribution of integers in the interval [0, 255] (left panel) and a Gaussian distribution (right panel). Each image is 50 × 50 pixels.

where f_n is the scalar intensity of the nth point and $\delta(\mathbf{x})$ is the Dirac delta function. In the most general case, the intensities f_n can be chosen from N different probability density functions $p_n(f_n)$, where $p_n df_n$ is the probability of finding the intensity of the nth point between f_n and $f_n + df_n$. For example, the intensities can be chosen from a uniform distribution or Gaussian distribution, as illustrated in Figure 8.8 for discrete (digitized) system analogues.

In the second step of the process, the intensity of the N source points are "smoothed out" by convolving $f(\mathbf{x})$ with a kernel or filter $K(\mathbf{x})$, producing

$$F(\mathbf{x}; \{C\}) = \int f(\mathbf{x}')K(\mathbf{x} - \mathbf{x}'; \{C\})d\mathbf{x}', \tag{8.51}$$

where $F(\mathbf{x})$ is the distribution of the intensity of the smoothed image and $\{C\}$ represents the parameters of the kernel. This convolution operation is assumed to be linear [i.e., K does not depend on $f(\mathbf{x})$] and is therefore sometimes called a *linear filter*. For simplicity, $K(\mathbf{x}; \{C\})$ is assumed to be an even function of $\mathbf{x}$. Examples of kernels include the *Gaussian* filter,

$$K(\mathbf{x}; a) = \exp\Big(-\frac{|\mathbf{x}|^2}{a^2}\Big), \qquad d \geq 1, \tag{8.52}$$

and the d-dimensional *Laplacian–Gaussian* filter,

$$K(\mathbf{x}; a) = \Delta \exp\Big(-\frac{|\mathbf{x}|^2}{a^2}\Big) = \Big(-\frac{2d}{a^2} + \frac{4|\mathbf{x}|^2}{a^4}\Big) \exp\Big(-\frac{|\mathbf{x}|^2}{a^2}\Big), \qquad d \geq 1, \tag{8.53}$$

where Δ is the Laplacian operator, d is the Euclidean dimension, and a is a length parameter that controls the size of the "influence" region of the filter. A smoothed digitized image is shown in the left panel of Figure 8.9. Note that when $K = \delta(\mathbf{x} - \mathbf{x}')$, we recover the original source function $f(\mathbf{x})$.

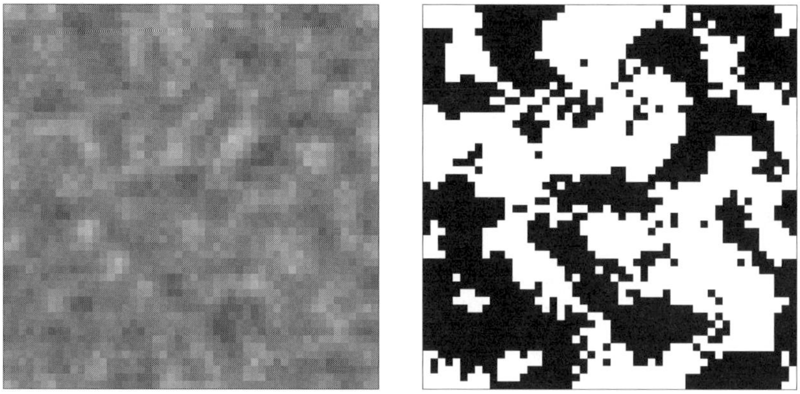

Figure 8.9 Left panel: Smoothed digitized image using formula (8.51) with a Gaussian kernel (8.52) (influence region equal to 7×7 pixels) generated from the uniform distribution shown in the left panel of Figure 8.8. Right panel: The corresponding level cut through the smoothed image. Here $F_o = 130$ and $\phi_2 = 0.5$.

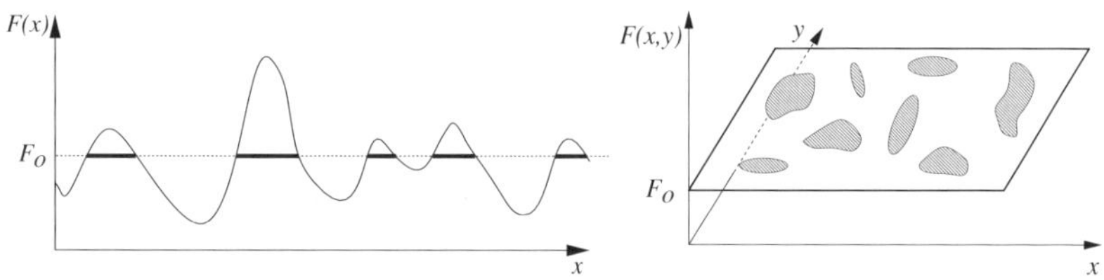

Figure 8.10 Schematics of one-dimensional (left panel) and two-dimensional (right panel) media produced by level cuts through F.

The final image is constructed by performing a level *cut* through $F(\boldsymbol{x})$ at a threshold value F_o. Such a cut defines an interface, i.e., $(d-1)$-dimensional contours, where F intersects the cutting hyperplane. Phase 1 is the region defined by $F < F_o$, and phase 2 is the region defined by $F > F_o$ (see Figure 8.10). This is the well-known "islands within lakes" picture (Zallen 1983). Therefore, the indicator function for phase 2 is given by

$$\mathcal{I}(\boldsymbol{x}) = \Theta[F(\boldsymbol{x}) - F_o], \tag{8.54}$$

where $\Theta(x)$ is the Heaviside step function (3.48). A level cut through the aforementioned smoothed image is also shown in Figure 8.9, which is to be contrasted with the original system depicted in the left panel of Figure 8.8.

Consequently, the n-point probability function S_n for phase 2 is given by the ensemble average

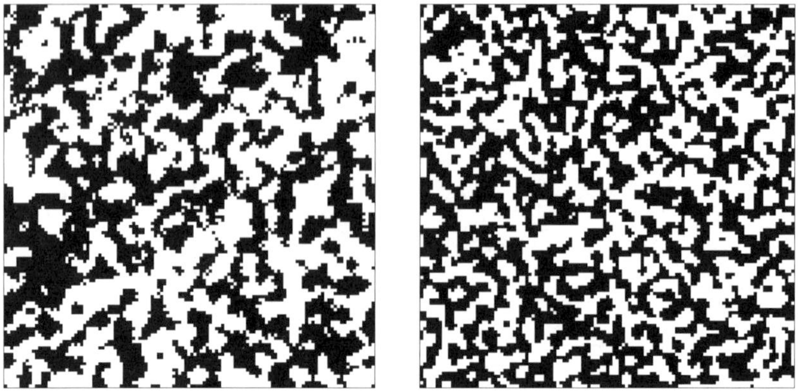

Figure 8.11 Random-field model microstructures at $\phi_2 = 0.5$ that we generated using a Gaussian kernel (left panel) and a Laplacian–Gaussian kernel (right panel) as described in text. Both systems are 120×120 pixels and $a = 5$ pixels. The Gaussian and Laplacian–Gaussian constructions mimic well the microstructures of Dolomite and Vycor glass, respectively.

$$S_n(\boldsymbol{x}_1, \boldsymbol{x}_2, \ldots, \boldsymbol{x}_n) = \left\langle \prod_{i=1}^{n} \Theta[F(\boldsymbol{x}_i) - F_o] \right\rangle$$
$$= \int_{-\infty}^{\infty} \cdots \int_{-\infty}^{\infty} \left\{ \prod_{i=1}^{n} \Theta[F(\boldsymbol{x}_i) - F_o] \right\} P_n[F(\boldsymbol{x}_1), \ldots, F(\boldsymbol{x}_n)] dF_1 \cdots dF_n, \quad (8.55)$$

where $P_n[F(\boldsymbol{x}_1), \ldots, F(\boldsymbol{x}_n)]$ is the joint probability density of $F(\boldsymbol{x})$. For statistically homogeneous media, the constant volume fraction $\phi_2 = S_1$ is specified by a particular value of F_o.

In practice, one works with digitized images, and so we grid-discretize the source by associating each source point with a d-dimensional vector whose components are integral multiples of the pixel or voxel size a. Thus, the index n becomes a vectorial quantity $\boldsymbol{n}$, and the positions of the source points are $\boldsymbol{x}_n = (n_1, n_2, \ldots, n_d)a$, where n_j runs over all integer values up to the size of the system. For an $N \times N$ digitized system, the intensity distribution becomes the following sum:

$$F(i, j, \{C\}) = \sum_k \sum_l f(i+k, j+l) K(k, l, \{C\}), \quad (8.56)$$

where K is an even function taken to be zero outside of some *influence region*.

Using intensities f_n that are uniformly distributed over a specific interval, Crossley et al. (1991) generated two-phase random media that mimic well the structures of Dolomite rock and of Vycor glass as illustrated in Figure 8.11. The former is generated using the Gaussian filter (8.52), whereas the latter is produced using the Laplacian–Gaussian filter (8.53). The corresponding two-point probability function $S_2(r)$ for each of these examples is shown in Figure 8.12. This quantity is obtained via Monte Carlo

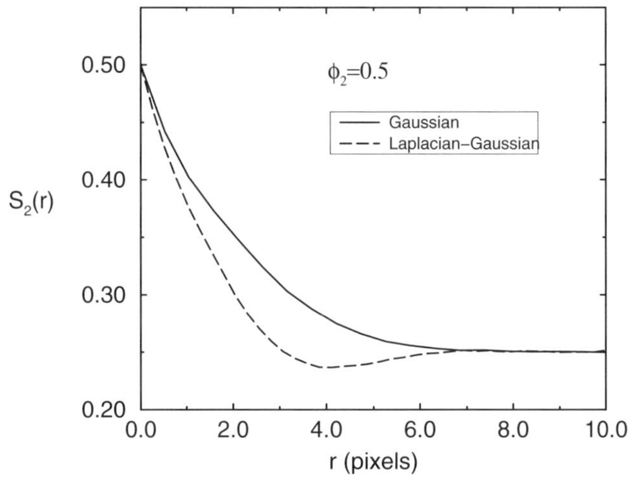

Figure 8.12 Two-point probability function versus r (in pixels) obtained via Monte Carlo simulations from ten realizations for the Gaussian and Laplacian–Gaussian models corresponding to the cases depicted in Figure 8.11.

simulations (as discussed in Chapter 12) using ten realizations for each case. The images of Figure 8.11 represent one such set of realizations.

8.2.2 Gaussian Convolved Intensities

A substantial amount of work has been devoted to instances in which the smoothed intensity $F(\boldsymbol{x})$ is a Gaussian random field (Quiblier 1984, Berk 1987, Berk 1991, Teubner 1991, Adler 1992, Bentz and Martys 1994, Roberts and Teubner 1995, Roberts 1997, Levitz 1998). A complete discussion of this topic is beyond the scope of this book. Here we summarize briefly salient points in so far as they pertain to the determination of the n-point probability functions. When $F(\boldsymbol{x})$ is a Gaussian random field, its joint probability density is exactly given by

$$P_n[F(\boldsymbol{x}_1), \ldots, F(\boldsymbol{x}_n)] = \frac{1}{\sqrt{(2\pi)^n \det \boldsymbol{G}}} \exp\Big(-\frac{1}{2}\boldsymbol{F}^T \boldsymbol{G}^{-1} \boldsymbol{F}\Big), \tag{8.57}$$

where the elements of the matrix $\boldsymbol{G}$ are given by the field–field correlation function

$$G_{ij} \equiv G(x_{ij}) = \langle F(\boldsymbol{x}_i) F(\boldsymbol{x}_j) \rangle, \tag{8.58}$$

$x_{ij} = |\boldsymbol{x}_j - \boldsymbol{x}_i|$, and $\boldsymbol{F}$ is the column vector with elements $F_i \equiv F(\boldsymbol{x}_i)$. For example, the singlet and pair functions are

$$P_1(F_i) = \frac{1}{\sqrt{2\pi}} \exp\Big(\frac{-F_i^2}{2}\Big), \tag{8.59}$$

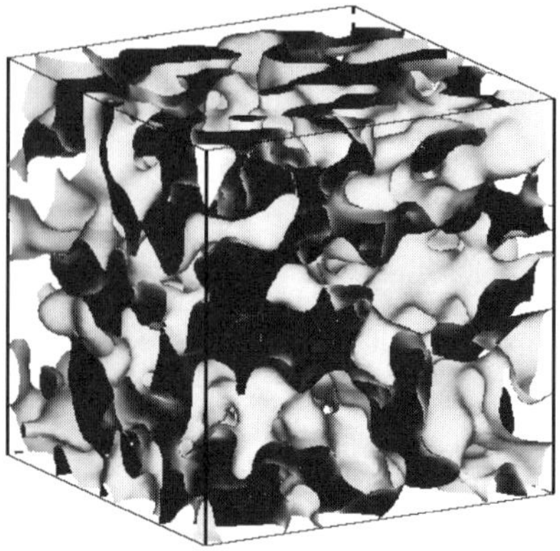

Figure 8.13 A depiction of the interface for a thresholded Gaussian random field (model II) as obtained by Roberts and Teubner (1995).

$$P_2(F_i, F_j) = \frac{1}{2\pi\sqrt{1-G_{ij}^2}} \exp\left[-\frac{F_i^2 + F_j^2 - 2G_{ij}}{2(1-G_{ij}^2)}\right], \quad i \neq j. \tag{8.60}$$

We have assumed without loss of generality that F has zero mean and unit standard deviation. The field–field correlation function is given explicitly in terms of the spectral density of the field $\rho(k)$ by the relation

$$G(x_{ij}) = \int_0^\infty 4\pi k^2 \rho(k) \frac{\sin kx_{ij}}{kx_{ij}} dk. \tag{8.61}$$

Roberts and Teubner (1995) studied various models of thresholded Gaussian random fields. Figure 8.13 shows the three-dimensional interface for the case they refer to as "model II."

Lower-order n-point probability functions S_n of (8.55) may be expressed in terms of quadratures when F is a Gaussian random field. Berk (1987, 1991) and Teubner (1991) have derived the one-point function (volume fraction)

$$S_1 = \phi_2 = \frac{1}{\sqrt{2\pi}} \int_{F_o}^\infty \exp\left(-\frac{1}{2}t^2\right) dt \tag{8.62}$$

and the two-point function

$$S_2(x_{ij}) = \frac{1}{2\pi} \int_0^{G(x_{ij})} \exp\left(-\frac{F_o{}^2}{1+t}\right) \frac{dt}{\sqrt{1-t^2}} + \phi_2^2. \tag{8.63}$$

One can see that relations (8.62) and (8.63) are obtained by combining the general expression (8.55) for S_n with (8.59) and (8.60). The three-point function S_3 was calculated by Roberts and Teubner (1995).

We note in passing that a semianalytic approximation for the chord-length density function $p^{(i)}(z)$ (Section 2.5) of three-dimensional random media has been obtained under the assumption that successive chord lengths are uncorrelated (Roberts and Torquato 1999). This result is exact for Poisson distributed convex inclusions, since the assumption of independent intervals is true. It was shown to be a reasonable approximation to the single level-cut Gaussian random field models of random materials.

CHAPTER 9

Percolation and Clustering

Percolation theory deals with the effects of varying the connectivity of elements (e.g., particles, sites, or bonds) in a random system. A cluster is simply a *connected* group of elements. Roughly speaking, the *percolation transition*, or *threshold*, of the system is the point at which a cluster first spans the system, i.e., the first appearance of *long-range* connectivity. In the thermodynamic limit, the percolation threshold is the point at which a cluster becomes infinite in size. The percolation transition is a wonderful example of a second-order phase transition and critical phenomenon.

Percolation phenomena arise in a variety of applications, including transport and mechanical properties of composites and porous media, spread of diseases and fires, gelation, conductor–insulator transition in metals with disorder, fracture processes in heterogeneous rock formations, circuitry in microchips, the glass transition, sea ice, and even star formation in galaxies. The study of percolation is a vast field, and no attempt will be made to discuss it here comprehensively. Instead, we will first review briefly basic definitions, concepts, and results in the context of discrete (lattice) percolation. This will be followed by a brief introduction to *continuum* percolation. The next chapter deals with specific developments in continuum percolation theory. The reader is referred to the monographs by Stauffer and Aharony (1992), Sahimi (1994), Grimmett (1989), and Meester and Roy (1996) for detailed treatments of various aspects of percolation.

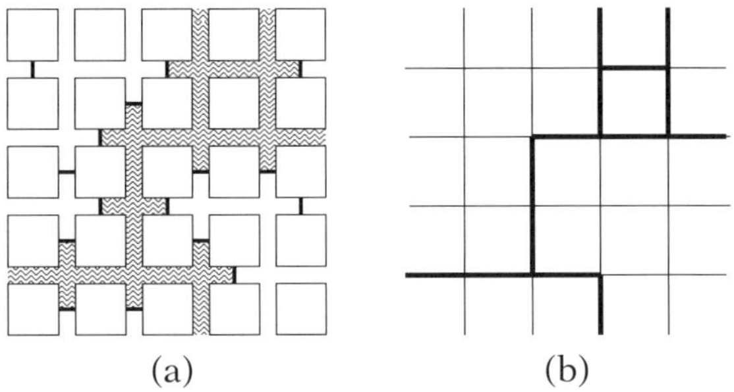

Figure 9.1 (a) A schematic of the passage of a fluid through a model of a porous medium: random network of open channels and blocked channels indicated by "closed valves" (bold lines). (b) Illustration of the corresponding connectivity map. [Adapted after Zallen (1983).]

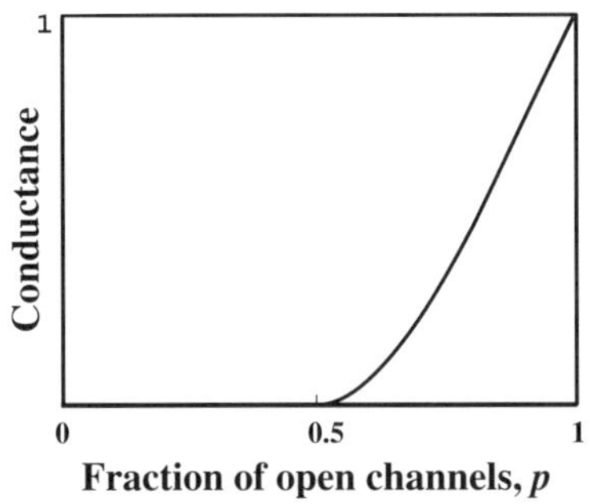

Figure 9.2 A schematic graph of conductance versus p for a square lattice.

9.1 Lattice Percolation

9.1.1 Bond and Site Percolation

The term "percolation" was coined by Broadbent and Hammersley (1957), who had in mind the passage of a fluid through a network of channels (bonds) in a porous medium with some of the channels being blocked at random. A schematic of such a two-dimensional porous medium idealized as a square-lattice channel network with square-shaped grains is shown in Figure 9.1. The corresponding connectivity map in the figure identifies the clusters in the system, i.e., the connected or unblocked pathways. Broadbent and Hammersley showed rigorously that no fluid will flow if the fraction of open (unblocked) channels p is smaller than some nonzero threshold p_c. Above p_c, the flow rate is nonzero. Thus, for $0 \le p \le p_c$, the flow rate is exactly zero, and for $p > p_c$, it increases monotonically with p until it achieves its maximum value at $p = 1$. This behavior is schematically illustrated in Figure 9.2, where the overall flux through the system, generally referred to here as the *conductance*, is plotted as a function of p.

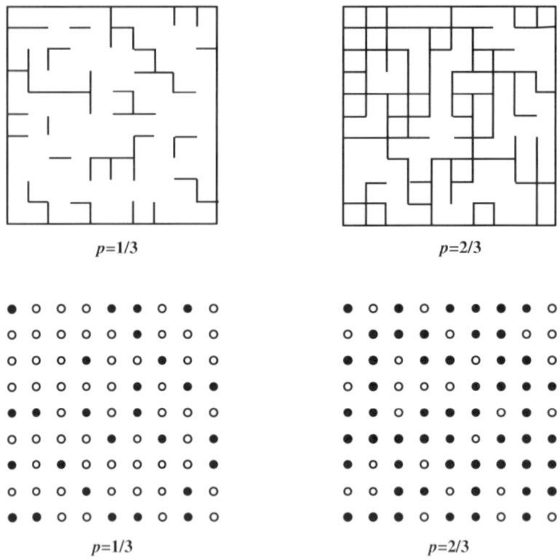

Figure 9.3 Top: Bond networks on a square lattice below and above the percolation threshold: $p_c = 1/2$ for an infinite system. Bottom: Site networks on a square lattice below and above the percolation threshold: $p_c \approx 0.5927$ for an infinite system.

The flow of some generalized fluid through a network of interconnected channels (some of which are blocked) is referred to as a *bond-percolation* process. We could view the blockages as the result of turning off valves placed in the channels, or "pipes" (see Figure 9.1). Carrying this "plumbing" analogy further, instead of placing the valves in the pipes, we could place them at the pipe junctions. This process is referred to as *site percolation*. More precisely, for site percolation on a lattice, each site is occupied with probability p and unoccupied with probability $1 - p$ according to some random process. Similarly, for bond percolation on a lattice, each bond is occupied (open) with probability p and unoccupied (closed) with probability $1 - p$. Unless otherwise stated, we will focus on the simplest random process ("Bernoulli" percolation), in which the bonds (sites) are occupied with probability p, *independently of the other bonds* (*sites*). A *cluster* is a connected group of bonds or sites, and thus what constitutes a cluster depends on the definition of connectedness. The simplest and most commonly used criterion is based on nearest-neighbor connections. For example, for site percolation, one occupied site is connected to another occupied site if it is a nearest neighbor, i.e., if the occupied sites share a common edge. Figure 9.3 illustrates both the bond and site problems. (One can combine features of both types of percolation in what is referred to as *site-bond* percolation.) As noted earlier, there is a critical probability p_c, called the *percolation threshold* (defined more precisely below), at which a sample-spanning cluster first appears. Figure 9.3 depicts systems below p_c (clusters are finite) and above p_c (sample-spanning clusters appear), respectively.

The percolation threshold p_c is a *nonuniversal* quantity in that it depends on the space dimension d, lattice structure, and other microscopic details (e.g., interactions).

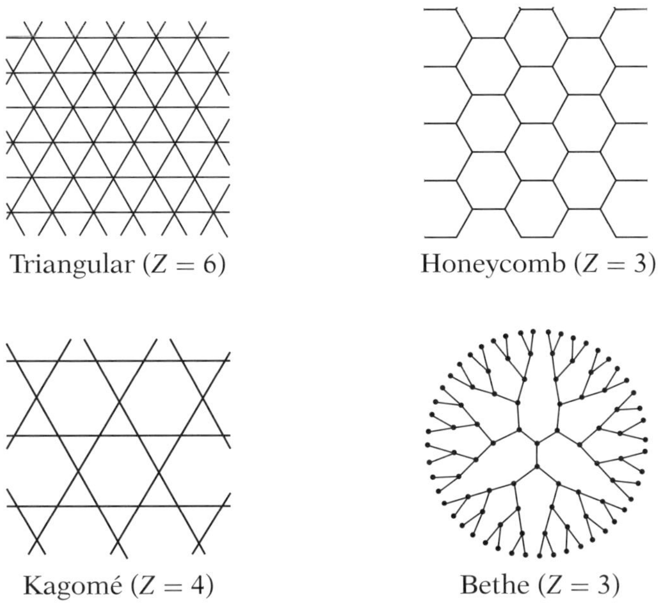

Figure 9.4 Other simple lattices besides the square lattice. The small portion of the Bethe lattice shown in this figure is threefold coordinated.

The threshold p_c can take values between zero and one for $d \geq 2$; for $d = 1, p_c = 1$. The exact determination of p_c is possible for only a few two-dimensional lattices and certain lattices related to the Bethe lattice. The Bethe lattice or tree is an infinite nonperiodic lattice without any closed loops (circuits) in which each site (except for the many surface sites) has a *coordination number* Z, i.e., the number of bonds connected to the same site. The dimensionality of the Bethe lattice is effectively infinite (Grimmett 1989). Fisher and Essam (1961) showed that the percolation threshold for either site or bond percolation on the Bethe lattice is

$$p_c = \frac{1}{Z - 1}. \tag{9.1}$$

Figure 9.4 shows a portion of a threefold coordinated Bethe lattice along with some regular two-dimensional lattices other than the square lattice, including the Kagomé lattice. Note that the sites of the Kagomé lattice correspond to the bonds of the "honeycomb," or hexagonal, lattice.

Table 9.1 gives selected percolation thresholds for two- and three-dimensional networks. For the given two-dimensional examples, the percolation threshold is known exactly for site percolation on a triangular lattice and for bond percolation on the square, triangular, and honeycomb lattices. For two-dimensional bond percolation, one may show that

$$p_c + \bar{p}_c = 1, \tag{9.2}$$

Table 9.1 Selected values of the percolation thresholds for two- and three-dimensional networks. The table includes exact results (see text) as well as numerical results with an error that at worst affects the last decimal place.

Lattice	Z	Site	Bond
Honeycomb	3	0.6970	$1 - 2\sin(\pi/18) \approx 0.6527$
Square	4	0.592746	1/2
Kagomé	4	$1 - 2\sin(\pi/18) \approx 0.6527$	0.52440
Triangular	6	1/2	$2\sin(\pi/18) \approx 0.3473$
Diamond	4	0.4299	0.3886
Simple cubic	6	0.3116	0.24881
BCC	8	0.2464	0.1803
FCC	12	0.198	0.1202

where $\bar{p}_c$ is the threshold of the dual lattice, subject to certain conditions of symmetries of the lattice (Essam 1972). Since the square lattice is self-dual, the exact value $p_c = 1/2$ follows immediately from (9.2). This relation also applies to the triangular and honeycomb lattices (see Table 9.1), since they are duals of one another. Relation (9.2) applies to site percolation when the dual graph is replaced by the *matching* graph (Essam 1972); since the triangular lattice is self-matching, the exact value $p_c = 1/2$ follows. In three dimensions, the percolation thresholds of periodic lattices are known only numerically. Many of the numerical results given in Table 9.1 are taken from the studies of Ziff and coworkers, who developed efficient methods to compute p_c. For two-dimensional lattices, the reader is referred to Ziff and Sapoval (1986) and Suding and Ziff (1999) for a description of the *hull-gradient* method. For three-dimensional lattices, the reader is referred to Lorenz and Ziff (1998). Generally, because lattices in higher dimensions are more highly connected than those in lower dimensions, p_c decreases with increasing d for a given lattice. Since a bond generally has more nearest-neighbor bonds than a site has nearest-neighbor sites for a given lattice, $p_c^{\mathrm{b}} \le p_c^{\mathrm{s}}$, where b and s denote bond and site, respectively.

It has been noted that the *average number of occupied bonds per site* $B_c \equiv Zp_c$ for bond percolation is, to a good approximation, an invariant of percolation networks for a given dimensionality d. We see from Table 9.1 that $B_c \approx 2$ for $d = 2$ and $B_c \approx 1.5$ for $d = 3$. More generally, for any dimension, $B_c \approx d/(d-1)$. Thus, a d-dimensional network will percolate if there are, on average, $d/(d-1)$ occupied bonds per site.

An intriguing but little-studied question is whether both phases of a two-phase material can simultaneously percolate; such a medium is called *bicontinuous*. The answer depends on the dimension and connectivity criteria. For $d = 1$, the answer is trivially no. We can think of the occupied and unoccupied sites (bonds) as constituting the "black" and "white" phases of a two-phase material. Let the probability of finding black and white sites (bonds) be p and $1 - p$, respectively. Clearly, a d-dimensional system possessing phase-inversion symmetry (Section 2.2) will be bicontinuous for $p_c < p < 1 - p_c$ for $d \geq 2$, provided that $p_c < 1/2$. We note that standard Bernoulli lattice percolation models possess phase-inversion symmetry. Thus, all of the three-dimensional lattices listed in Table 9.1 are bicontinuous for $p_c < p < 1 - p_c$ in either site or bond percolation. On the other hand, the only two-dimensional bicontinuous example listed there is bond percolation on a triangular lattice, which is bicontinuous for $2\sin(\pi/18) < p < 1 - 2\sin(\pi/18)$. In general, bicontinuity in two dimensions is considerably more difficult to achieve than in higher dimensions. If $p_c < 1/2$, a system with phase-inversion *asymmetry* may still be bicontinuous. For $d \geq 3$, neither the restriction $p_c < 1/2$ nor phase-inversion symmetry is a necessary condition for bicontinuity; see Section 9.2.

It is important to recognize that systems for $d \geq 2$ can be made bicontinuous by introducing longer-range connectivity criteria (e.g., nearest and next-nearest neighbors). This includes all of the two-dimensional lattice percolation models that are not bicontinuous with nearest-neighbor connectivity criteria given in Table 9.1 (all except bond percolation on a triangular lattice).

9.1.2 Percolation Properties

In the ensuing discussion we will speak primarily in the language of site percolation. There are several key percolation properties that we will define, including the cluster-size distribution n_k, the mean cluster size S, the percolation probability $P(p)$, and the pair-connectedness function $P_2(\mathbf{r})$.

Consider an *infinite* lattice in which each site is occupied with probability p. A k-mer is a cluster that contains k sites. The *cluster-size distribution* n_k is the average number of k-mers per lattice site. Thus, the probability that an arbitrary site is part of a k-mer is kn_k, and hence

$$\sum_{k=1}^{\infty} kn_k = p, \qquad p < p_c. \tag{9.3}$$

Relation (9.3) simply states that the probability that an arbitrary site belongs to *any* cluster is equal to the probability p that it is occupied.

The *mean cluster size* S is the average number of sites in the cluster containing a randomly chosen occupied site. Since the quantity $kn_k/\Sigma_k kn_k$ is the probability that the cluster to which an arbitrary occupied site belongs contains exactly k sites, we have that

$$S = \frac{\sum_{k=1}^{\infty} k^2 n_k}{\sum_{k=1}^{\infty} kn_k}, \qquad p < p_c. \tag{9.4}$$

For $p \ll 1$, S starts at unity, reflecting the complete dominance of monomers at low occupation densities. As p increases, S increases and eventually diverges to infinity at

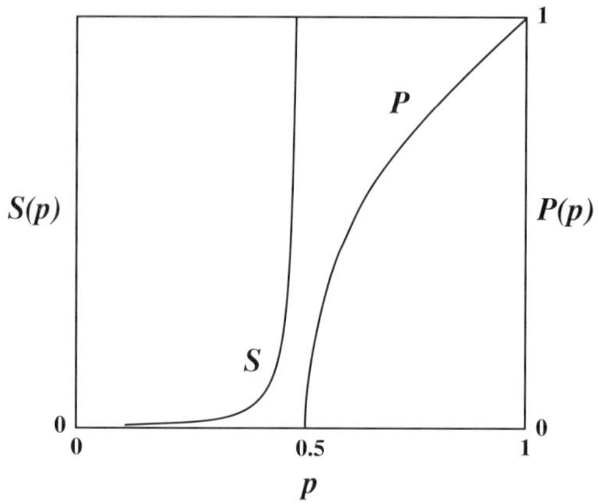

Figure 9.5 A schematic of the mean cluster size S and percolation probability P as a function of occupation probability p for site percolation on a *triangular* lattice.

the percolation threshold p_c. The behavior of S versus p is qualitatively depicted in Figure 9.5.

The *percolation probability* $P(p)$ is the probability that an arbitrarily selected site belongs to an infinite cluster. Clearly, P is always less than p except in the trivial instance when $P = p = 1$. The percolation probability is identically zero for p below p_c in an infinite system. For a very large system consisting of $\mathcal{N}$ sites, the largest cluster is also very large above the threshold, and hence the ratio of the largest cluster size to $\mathcal{N}$ is nonzero in the limit $\mathcal{N} \to \infty$, i.e., $P(p > p_c) > 0$. Thus, for an infinite system,

$$P(p) \begin{cases} = 0 \ , & \text{if} \quad p < p_c, \\ > 0 \ , & \text{if} \quad p > p_c. \end{cases} \tag{9.5}$$

The *critical* or *percolation-threshold value* p_c is defined formally (for an infinite system) as $p_c = \sup\{p : P(p) = 0\}$, where sup indicates the supremum (least upper bound); see also Aizenman and Barsky (1987). The percolation probability rises steeply from its value of zero at $p = p_c$ with increasing p, reflecting the rapidity with which finite clusters link up with the infinite one as p exceeds p_c. Eventually, $P(p)$ approaches p in the limit $p \to 1$, since the infinite cluster consumes the finite ones. The behavior of $P(p)$ is schematically shown in Figure 9.5. It is seen that percolation is a geometrical phase transition with $P(p)$ playing the role of an *order parameter* in a thermodynamic phase transition. For example, for the Ising model of a magnet (Section 8.1.5), the *magnetization* and *susceptibility* are the analogues of the percolation probability $P(p)$ and mean cluster size S, respectively.

The *pair-connectedness function* $P_2(\boldsymbol{r})$ is defined such that $p^2 P_2(\boldsymbol{r})$ gives the probability that a site at the origin and a site j located at position $\boldsymbol{r}$ are both occupied and belong to the same cluster. Essam (1972) showed that the mean cluster size is related to a sum over the pair-connectedness function:

$$S = 1 + p \sum_{\boldsymbol{r}} P_2(\boldsymbol{r}). \tag{9.6}$$

An important concept of scaling theory is the idea of a *correlation length* $\xi(p)$. It is believed that for an isotropic system, the asymptotic behavior of the pair-connectedness function for $r \to \infty$ (Grimmett 1989) is given by

$$P_2(r) \sim \begin{cases} r^{2-d-\eta}, & p = p_c, \\ e^{-r/\xi(p)}, & p \neq 0,\ p_c,\ 1, \end{cases} \tag{9.7}$$

for some η, where $r = |\boldsymbol{r}|$. As the threshold p_c is approached from below, the mean cluster size S and the correlation length $\xi(p)$ must grow, and at the threshold, both of these quantities must diverge to infinity, implying that $\eta < 2$.

We note that in the one-dimensional case, the cluster statistics can be obtained exactly. For example, one has that

$$n_k = p^k(1-p)^2, \tag{9.8}$$

and hence relation (9.4) yields that the mean cluster size is

$$S = \frac{1+p}{1-p}, \qquad p < p_c, \tag{9.9}$$

diverging at the trivial value $p_c = 1$. Clearly, the percolation probability P is equal to 0 for $p < p_c$. Moreover, the pair-connectedness function is given by

$$P_2(r) = p^{r-1}, \qquad r = 1, 2, 3, \ldots . \tag{9.10}$$

For $p < 1$ and $r \to \infty$, we see that

$$pP_2(r) = \exp(-r/\xi), \tag{9.11}$$

where the correlation length is $\xi = -1/\ln p$. Thus, as $p \to p_c = 1$, the correlation length ξ diverges to infinity according to the relation

$$\xi = (p_c - p)^{-1}.$$

Finally, note that substitution of (9.10) into (9.6) yields

$$S = 1 + 2p \sum_{r=1}^{\infty} p^{r-1}, \qquad p < p_c,$$

which agrees with expression (9.9). The factor of 2 arises to account for both positive and negative values of $\boldsymbol{r}$.

9.1.3 Scaling and Critical Exponents

Geometric properties

In the immediate vicinity of the percolation threshold, many percolation quantities have been observed to exhibit power-law scaling behavior. For example, for $|p_c - p| \ll 1$, we have

Table 9.2 Critical exponent and fractal dimension values for selected values of the Euclidean space dimension d. Rational or integer values in the case of $d = 2$ are conjectured to be exact values.

Asymptotic behavior or description	Quantity	$d = 2$	$d = 3$	$d \geq 6$
$P \sim (p - p_c)^{\beta}$	β	5/36	0.41	1
$S \sim \|p_c - p\|^{-\gamma}$	γ	43/18	1.82	1
$\xi \sim \|p_c - p\|^{-\nu}$	ν	4/3	0.88	1/2
$P_2(r) \sim r^{2-d-\eta}(p = p_c)$	η	5/24	−0.068	0
Infinite cluster	d_F	91/48	2.52	4
Backbone	d_F	1.64	1.8	2

$$P \sim (p - p_c)^{\beta}, \qquad p \to p_c^+, \tag{9.12}$$

$$S \sim |p_c - p|^{-\gamma}, \qquad p \to p_c, \tag{9.13}$$

$$\xi \sim |p_c - p|^{-\nu}, \qquad p \to p_c, \tag{9.14}$$

where both S and ξ are defined above p_c if one excludes the infinite cluster. The *critical exponents* β, γ, and ν are *universal*; i.e., they are independent of the lattice structure and its microscopic details, depending only on the dimensionality d. This implies that there is no distinction between site and bond percolation in so far as the critical exponents are concerned, and thus both types of percolation are in the same *universality class*. Indeed, to date, calculations of critical exponents for *continuum* percolation have been found to be in the same universality class as that for lattice percolation. However, the amplitudes (i.e., the implied prefactors in the scaling laws) do depend on the details of the system and hence are *nonuniversal*.

Table 9.2 gives critical exponent values (as well as fractal dimensions as discussed below) for selected values of the dimensionality. Included in the table is the scaling law for the pair-connectedness P_2 for $p = p_c$. Only two of the aforementioned critical exponents can be independently determined, since one has the interrelation

$$d\nu = 2\beta + \gamma. \tag{9.15}$$

Relations of the type (9.15) are referred to as *hyperscaling relations*, since they involve the dimensionality d. It is believed that hyperscaling relations are satisfied when d is below or equal to some *critical dimension* d_c. When $d \geq d_c$, it is believed that critical exponents take on their dimension-independent *mean-field* values obtainable exactly from percolation on an infinite tree such as a Bethe lattice (Fisher and Essam 1961), where it is known that $\beta = \gamma = 1$ and $\nu = 1/2$. Using the hyperscaling relation (9.15), we see that $d_c = 6$, thus explaining the last column in Table 9.2. Hyperscaling first was examined in the context of thermal phase transitions where, in the case of a fluid, mean-field theory corresponds to the well-known Van der Waals equation of state.

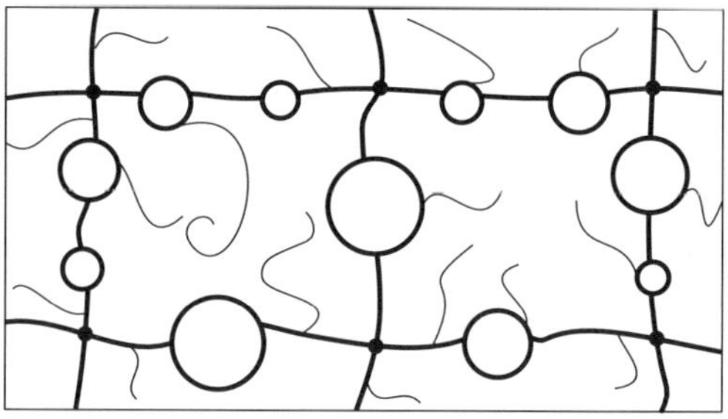

Figure 9.6 A schematic of the nodes-links-blobs picture of the incipient infinite cluster.

There is an interesting issue concerning the existence of the so-called *incipient infinite cluster*, i.e., an infinite cluster at the threshold $p = p_c$. We know that there is no such cluster for $p < p_c$ and that there is one for $p > p_c$. From a strict mathematical viewpoint (Grimmett 1989), the incipient infinite cluster does not exist for $d = 2$, and it is widely believed that the same is true for $d > 2$. However, the event of finding an infinite cluster with nonzero probability at $p = p_c$ can be shown to occur by suitable conditioning (Grimmett 1989); this is consistent with the physical idea that there should be an incipient infinite cluster.

The incipient infinite cluster has an interesting structure: It fills the space that contains it, but does so sparsely. It is a highly ramified structure that schematically may be thought of as being composed of "nodes," "links," and "blobs" (Stanley 1977), as depicted in Figure 9.6. One-dimensional channels, or links, intersect at nodes at distances on the order of the correlation length ξ. There are offshoots from the links that lead to "dead ends" (i.e., dangling ends) in the sense that they cannot carry flow or current. Along the links one may occasionally find denser regions called blobs that have many connections between two points. The blobs have a characteristic size of order ξ. The nodes, links, and blobs together form the *backbone* of the incipient infinite cluster, i.e., the current-carrying portion of the infinite cluster. On the other hand, the dead ends, which comprise most of the mass of the infinite cluster, cannot carry current. Most of the backbone mass resides in the blobs.

Power-law scaling behavior of percolation quantities very near p_c implies that the quantities are *generalized homogeneous* functions of their arguments. [See Stanley (1971) for a definition of such functions.] This in turn implies that the system becomes *scale invariant* as $p \to p_c$, i.e., the system is invariant under scale tranformations, as indicated by the scaling relation for $P_2(r)$ in Table 9.2. Indeed, as we will see in Section 9.1.4, the incipient infinite cluster is a fractal. Therefore, the powerful methods of *renormalzation group theory* (Huang 1987, Binney et al. 1992) can be brought to bear to study such critical phenomena.

Table 9.3 Transport and elasticity critical exponents for selected values of the dimensionality d.

Asymptotic behavior	Exponent	$d=2$	$d=3$	$d\geq 6$
$\sigma_e \sim (p-p_c)^t$	t	1.3	2	3
$k \sim (p-p_c)^e$	e	1.3	2	3
$E_e \sim (p-p_c)^f$	f	3.96	3.75	4

Effective properties

The behavior of effective transport and elastic properties near percolation is different from the purely geometric quantities discussed thus far. Bulk properties of interest include the effective conductivity σ_e, fluid permeability k, and the effective Young's modulus E_e, topics that are comprehensively treated in Part II of the book. Let us assume that we are dealing with a class of percolation networks in which these quantities either vanish or diverge to infinity at the connectivity threshold p_c.

In the conduction problem, imagine a network in which a fraction p of the bonds are conducting and $1-p$ of the bonds are perfect insulators (zero conductance). The conductivity behavior becomes critical at a transition point that is identical to the standard Bernoulli connectivity threshold p_c (Chayes and Chayes 1986). The effective conductivity $\sigma_e(p)$ is zero for $p \leq p_c$, and it is believed that in the vicinity of p_c, $\sigma_e(p)$ obeys the scaling law

$$\sigma_e(p) \sim (p-p_c)^t, \qquad p \to p_c^+, \tag{9.16}$$

where t is the conductivity critical exponent (see Figure 9.2). Although existence of the scaling relation (9.16) has not been rigorously proven, there is an overwhelming number of numerical simulations that support such scaling behavior. In two and three dimensions, numerical simulations yield that $t \approx 1.3$ and $t \approx 2$, respectively (see Table 9.3). Under the assumption of a nodes–links–blobs model of the conducting backbone, Golden (1990) has shown that $\sigma_e(p)$ is convex in p near p_c and that its critical exponent t obeys the inequalities $1 \leq t \leq 2$ for $d = 2, 3$ and $2 \leq t \leq 3$ for $d \geq 4$. Note that the upper bound $t = 2$ for $d = 3$ coincides with the numerical estimate. Investigators have attempted to relate t to geometric exponents such as β, γ, and ν. However, no such rigorous relationship has been established to date, and hence t must presently be regarded as an independent exponent.

If the conducting bonds of the aforementioned network are interpreted as pipes through which a viscous fluid can flow, then one can define the network fluid permeability k as the ratio between the macroscopic fluid flow rate and the applied pressure gradient. The permeability $k(p)$ is zero for $p \leq p_c$, and it is believed that in the vicinity of p_c, $k(p)$ obeys the scaling law

$$k(p) \sim (p-p_c)^e, \qquad p \to p_c^+, \tag{9.17}$$

where e is the permeability critical exponent. In this discrete lattice case in which all occupied bonds are identical, one can map this flow problem exactly onto the conduction problem (current replacing flow rate and electric field replacing pressure gradient). Therefore, one has $e = t$, which (as will be shown) is generally not valid for continuum percolation.

In the elastic problem, imagine a network in which a fraction p of the bonds represent elastic springs that can be stretched (central forces) and bent (bond-bending forces) and $1 - p$ of the bonds are springless (zero forces). For this rotationally invariant model, the elastic behavior is isotropic and becomes critical at the connectivity threshold p_c. [Without bond-bending forces, it turns out that elastic behavior becomes critical at a different threshold than the connectivity threshold p_c; see Sahimi (1994) and references therein.] The effective Young's modulus $E_e(p)$ is defined as the ratio of the macroscopic axial stress divided by the macroscopic axial strain (see Chapter 13 for a more precise definition). For $p \leq p_c$, $E_e(p) = 0$, and it is believed that in the vicinity of p_c, $E_e(p)$ obeys the scaling law

$$E_e(p) \sim (p - p_c)^f, \qquad p \to p_c^+, \tag{9.18}$$

where f is the Young's modulus critical exponent. For a number of years it was thought that the value of f was the same as t, based on certain similarities of the elastic and conduction problems. However, now it is known, both theoretically and experimentally, that f can generally be quite different from t due to the higher tensorial order of the elastic problem. Physically, the difference arises in the above elastic model because of torsional effects that are absent in the conduction problem. Indeed, Torquato (1992) has shown that $f \geq t$ for the more general continuum percolation models (see Chapter 23). Since the lattice model can be regarded as a special case of continuum percolation (see Section 9.2.3), the inequality $f \geq t$ is also valid for lattice percolation. In two and three dimensions, numerical simulations yield $f \approx 3.96$ and $f \approx 3.75$, respectively (see Table 9.3).

Note that for elastically isotropic materials, there are two independent moduli. However, in most situations, other moduli, such as the bulk and shear moduli, will be characterized by the same critical exponent f (see Chapter 23).

Two other critical exponents, s and q, can be defined as well. Consider a network in which a fraction p of the bonds are perfect conductors (zero resistance) and $1 - p$ of the bonds are normal conductors (finite resistance) with *equal strengths*. For p below p_c, the effective conductivity is finite and diverges to infinity at the threshold. Thus, $\sigma_e(p) < \infty$ for $p \leq p_c$, and it is believed that in the vicinity of p_c, $\sigma_e(p)$ obeys the scaling law

$$\sigma_e(p) \sim (p_c - p)^{-s}, \qquad p \to p_c^-, \tag{9.19}$$

where s is the superconductivity critical exponent. One has the dual relation $s = t$ for $d = 2$ (see Corollary 15.1 for a proof), but for $d = 3$, simulations yield $s \approx 0.73$. In the analogous elastic problem, a fraction p of the bonds are perfectly rigid springs (zero compliance) and $1 - p$ of the bonds are normal springs (finite compliance) with *equal strengths*. As before, the springs incorporate central and bond-bending forces. Here it is expected that the effective Young's modulus diverges as p_c is approached from below. Thus, $E_e(p) < \infty$ for $p \le p_c$, and it is believed that in the vicinity of p_c, the effective Young's modulus obeys the scaling law

$$E_e(p) \sim (p_c - p)^{-a}, \qquad p \to p_c^-, \tag{9.20}$$

where a is the superelasticity critical exponent. Simulations yield $a \approx 1.24$ for $d = 2$ and $a \approx 0.65$ for $d = 3$ (Sahimi and Arbabi 1993).

Experimental measurements of critical exponents agree with numerical simulations within the experimental errors. Using a sheet of metal, Benguigui (1984) determined t and f by punching holes in the sheet at random on the sites of a square lattice. He found that $t = 1.2 \pm 0.1$ and $f = 3.5 \pm 0.4$, which are in agreement with numerical results for $d = 2$ (see Table 9.3). Benguigui and Ron (1993) measured the superelastic exponent for a three-dimensional gel/ceramic composite and found that $a = 0.67 \pm 0.05$.

9.1.4 Infinite Cluster and Fractality

The fact that the correlation length ξ is infinite at the percolation threshold implies that the incipient infinite cluster is a fractal, i.e., it is statistically self-similar on all length scales and thus is *scale invariant*. The fractal dimension d_F $(\le d)$ can be found by determining how the mass of a portion of the sample-spanning cluster (its total number of sites or bonds) contained in a sphere of radius R_c scales as a function of R_c, i.e.,

$$M \sim R_c^{d_F}.$$

The backbone of the infinite cluster is also a fractal but, as we will see, has a fractal dimension that is less than that of the entire infinite cluster. Above p_c, the mass $M(R_c)$ of the infinite cluster increases as R_c^d, meaning that it is no longer a fractal but $d_F = d$, just as M is for finite clusters.

Another way of measuring the fractal dimension is by considering the calculation of the mass of the sample-spanning cluster $M(L)$ as a function of the linear size of the system L (for sufficiently large L) at $p = p_c$. Apart from small fluctuations, a plot of log M versus log L at $p = p_c$ (obtained from computer simulations) would yield a straight line with a slope given by the fractal dimension d_F (see Figure 9.7), i.e., $M \sim L^{d_F}$. Note that even for p slightly larger than p_c, the mass of the largest cluster $M(L)$ still scales as L^{d_F}, provided that L is much smaller than the correlation length ξ. However, if $L \gg \xi$, then $M(L) = PL^d$, where $P(p)$ is the percolation probability, which we know scales as $(p - p_c)^\beta$. When L is comparable to ξ, then the two mass expressions for $L \ll \xi$ and $L \gg \xi$ should also be comparable to each other, i.e.,

$$PL^d \propto L^{d_F} \quad \text{with} \quad L = \xi \sim (p - p)^{1/\nu},$$

and hence we have the interrelation

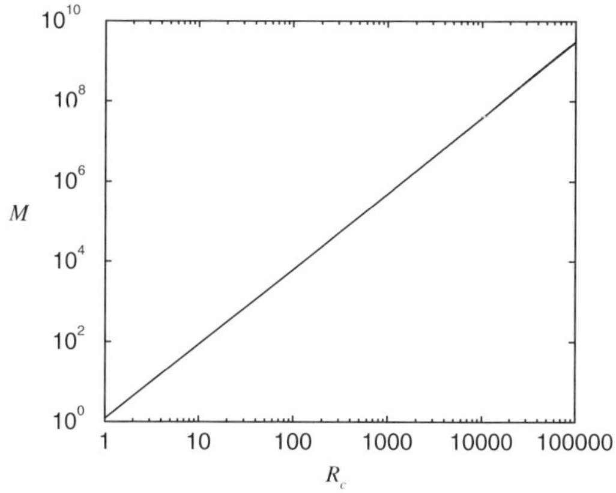

Figure 9.7 Mass of the sample-spanning cluster at $p = p_c$ contained in a circular disk of radius R_c versus R_c for any two-dimensional lattice. The slope of the line determines the fractal dimension as $d_F = 91/48$.

$$d - d_F = \frac{\beta}{\nu}. \tag{9.21}$$

As expected, Table 9.2 shows that this hyperscaling relation is satisfied for $d = 2, 3$, or 6 but is not satisfied for $d > 6$. The fractal dimension d_F of the incipient infinite cluster is equal to 91/48 for $d = 2$ and to 2.52 for $d = 3$. As noted earlier, the backbone has a smaller fractal dimension, as expected; for $d = 2$, $d_F = 1.64$; for $d = 3$, $d_F = 1.8$; and for $d \geq 6$, $d_F = 2$. In passing, we note that a hyperscaling relation involving the exponent η (see Table 9.2) is $d - 2 + \eta = 2\beta/\nu$.

9.1.5 Finite-Size Scaling

Thus far we have focused on infinitely large systems. However, it is important to understand how percolation quantities behave near the threshold in large but finite systems. In practice, one often deals with such systems (e.g., experimental data, computer simulations), and such finite-size scaling analysis also provides an accurate means of determining critical exponents. Finite-size scaling was first investigated by Fisher (1961) in the context of a thermal system near its critical point and can be adapted to study percolation behavior. Following Fisher, a general property $X(L)$ of a system of linear size L near its critical point p_c obeys the relation

$$X(L) = L^{-\delta/\nu} f(z), \tag{9.22}$$

where ν is the exponent defined in Table 9.2, $f(z)$ is an analytic function of

$$z = (p - p_c) L^{-1/\nu}, \tag{9.23}$$

and δ is the appropriate critical exponent for an infinite system, i.e.,

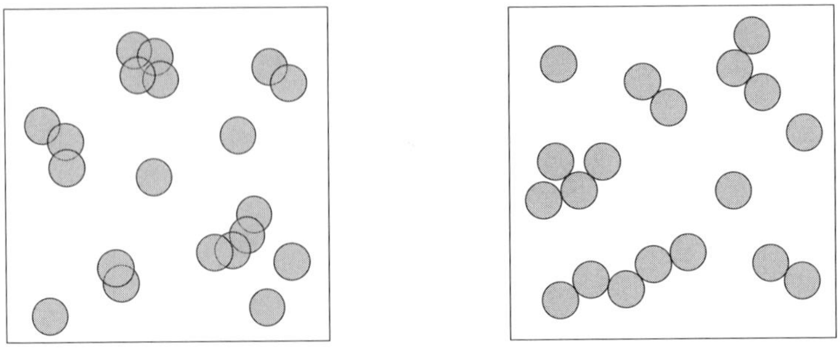

Figure 9.8 Two examples of connectedness criteria: two particles are connected if they overlap (left) or two nonoverlapping particles are connected if they are in contact (right).

$$X(\infty) \sim |p - p_c|^{\delta}. \tag{9.24}$$

To determine the critical exponents, one can in principle calculate X for several values of L and then fit the results to relation (9.22) under the assumption that $f(z)$ can be treated as a constant when $|z| \ll 1$. This step yields the ratio δ/ν. To compute both δ and ν, we use the fact that the percolation threshold of a finite system is shifted according to the relation

$$p_c - p_c(L) \sim L^{-1/\nu}, \tag{9.25}$$

where $p_c(L)$ is the *effective* percolation threshold of a system of size L. Note that $p_c(L)$ is a random variable that is not Gaussian (normally) distributed (Berlyand and Wehr 1997).

9.2 Continuum Percolation

Continuum (off-lattice) percolation is a more realistic model of actual percolation processes that occur in real heterogeneous materials; examples include transport and elastic behavior of a broad class of composites, microemulsions, polymer blends, sintered materials, colloids, the sol-gel transition, and sea ice (Mohanty, Ottino and Davis 1982, Zallen 1983, Brinker and Scherer 1990, McLachlan, Blaszkiewicz and Newnham 1990, Sahimi 1994, Golden, Ackley and Lytle 1998). The volume fraction of one of the phases, say ϕ_2, in continuum percolation plays the same role as the occupation probability p plays in lattice percolation. The *percolation threshold* ϕ_{2c} (defined more precisely below) is the critical volume fraction at which a sample-spanning cluster first appears.

There are a number of different ways of generating continuum percolation models. One popular method is to consider random distributions of inclusions (e.g., spheres, ellipsoids, needles) in a matrix or void. One must then decide on a connectedness criterion, which will depend on the physical or mathematical context. For example, referring to Figure 9.8, two particles are said to be connected if they overlap or if they make point contacts with one another in the cases of nonoverlapping particles.

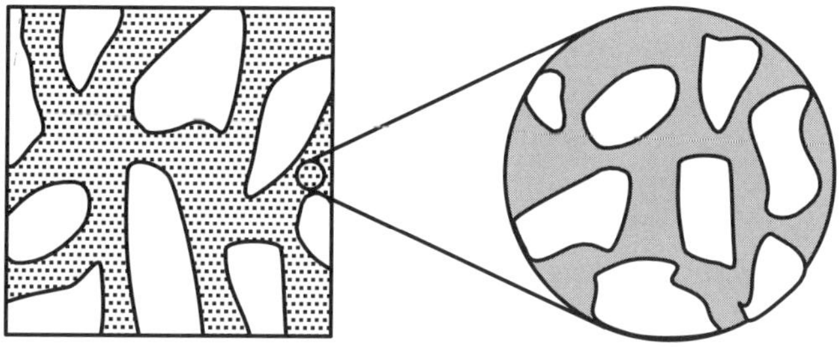

Figure 9.9 First two levels of the self-similar percolating hierarchical construction. The first level is made by mixing phases 1 and 2 such that it is at the percolation threshold of phase 2. The second level is made by mixing, in the same way, phase 1 material with a new phase 2 material composed of the composite used at the first level such that the system is at the percolation threshold of the new phase 2.

Overlapping particles, a prototypical continuum-percolation model, are shown below and above the percolation threshold in Figure 3.4. Another way to construct continuum percolation models is to use cell models, i.e., tessellate space into cells of various shapes and sizes and then assign, according to some random process, material phases (see Figure 8.5). One then defines connectedness according to some near-neighbor rule. A prototypical example of this last class of percolation models is the d-dimensional random checkerboard or, more generally, symmetric-cell materials (Chapter 8).

Values of the percolation thresholds of various continuum models are given in Chapter 10. The percolation threshold can take values between zero and one for $d \geq 2$, depending on the system microstructure and dimension. One can imagine creating a system with a threshold approaching zero by the following *self-similar hierarchical* construction. At the first level of the hierarchy, we mix two phases (phases 1 and 2), according to some specific process, to create a two-phase system that is at $\phi_{2c} < 1$, the percolation threshold of phase 2. At the second level of the hierarchy, we take phase 1 material and mix it (using the same process) with a new phase 2 material that is composed of the composite used in the first level so that the system is at the percolation threshold of the new phase 2 material. The heterogeneity length scale in the second level is much larger than in the first level (see Figure 9.9). The volume fraction of the original phase 2 material at the second level of the hierarchy is $(\phi_{2c})^2$. Continuing in this self-similar hierarchical process gives a system at the nth level of the hierarchy in which phase 2 percolates at $(\phi_{2c})^n$. Thus, in the limit $n \to \infty$, we can create a two-phase system in which the percolation threshold of phase 2 approaches zero. Similarly, we note that even though the volume fraction of phase 1 approaches unity, it does not percolate. Thus, if phase 1 is a conductor and phase 2 is a perfect insulator, one can create a system in the aforementioned limit that is almost entirely a conducting phase but on the whole behaves as an insulator.

Actual systems can be made that have percolation thresholds that are very close to zero. An *aerogel* is a beautiful example of a heterogeneous system that has a very small percolation threshold. *Gelation* is the process by which sufficiently many monomers present in a solvent polymerize at the *gel* point to form a sample-spanning cluster. Thus, the gel point corresponds to the percolation threshold. An aerogel is a special type of gel that is ultra-light as a result of the growth process leading to gelation: the formation of wispy fractal aggregates that link up to form a sample-spanning cluster (Brinker and Scherer 1990). Aerogels can be fabricated that have volume fractions of solid as low as 0.5%, which means that their actual percolation thresholds are even lower. A suspension of very long, randomly oriented fibers in a matrix is another example of a system with a very low threshold (Garboczi, Snyder, Douglas and Thorpe 1995). One can also achieve very low thresholds by compacting binary mixtures of particles of widely different sizes in such a way that the smaller particles form a connected network that resides effectively on the surfaces of the larger particles (Malliaris and Turner 1971, Kusy 1977).

On the other hand, particle suspensions can be made to have percolation thresholds that approach unity by allowing the particles to have a very wide size distribution and endowing the particles with a "soft" repulsive interparticle potential that extends beyond the physical size of the particles. Polydispersivity in size down to the infinitesimally small potentially enables the particles to fill space (in contrast to monodisperse particles; see Section 3.5), and the repulsive interactions forestall the formation of clusters of touching particles until the particle volume fraction approaches unity. Such a manipulation of interparticle forces is possible with colloidal dispersions. Interestingly, a mathematical construction of this type with a threshold approaching unity is the Hashin–Shtrikman coated-spheres assemblage (in the limit $\phi_2 \to 1$), discussed in Chapter 16 and shown in Figure 16.1.

For an ordered or disordered lattice of touching impenetrable spheres at volume fraction ϕ_2, Scher and Zallen (1970) observed that the so-called *critical occupied volume fraction* f_c is an approximate invariant of a d-dimensional system at the site percolation threshold p_c. It is defined according to the relation

$$f_c = \phi_2 \, p_c. \tag{9.26}$$

The quantity f_c is approximately 0.45 and 0.16 for $d = 2$ and $d = 3$, respectively. This correlation is similar to the one noted earlier for bond percolation, where it was observed that $B_c = Zp_c$ is an approximate invariant. As a simple application of relation (9.26), consider, for example, a random packing of contacting impenetrable spheres at the particle volume fraction $\phi_2 = 0.64$. For such a system, relation (9.26) predicts that $p_c = 0.25$, which is slightly above the BCC value given in Table 9.1. We will discuss similar approximate invariants for penetrable particles in the next chapter.

The remarks made about bicontinuity (simultaneous percolation of both phases of a two-phase material) in Section 9.1 apply to continuum percolation with the obvious changes in language. Thus, a d-dimensional two-phase continuum system that possesses phase-inversion symmetry (Section 2.2) will be bicontinuous for $\phi_{2c} < \phi_2 < 1-\phi_{2c}$ for $d \geq 2$, provided that the threshold satisfies $\phi_{2c} < 1/2$. Symmetric-cell materials (Section 8.1.3), which include the random checkerboard, are a class of materials that possess phase-inversion symmetry . As noted earlier, bicontinuity for $d = 2$ is more difficult to achieve than for $d \geq 3$. However, by incorporating longer-range connectivity criteria, one can decrease ϕ_{2c} below 1/2 and induce bicontinuity. For example, percolating clusters of white and black squares cannot coexist in the two-dimensional random checkerboard with nearest-neighbor connections (along the square edges), since the black phase (phase 2) percolates at the standard site value, i.e., $\phi_{2c} = p_c^{\text{site}} \approx 0.5927$. However, if one also allows for next-nearest-neighbor connections (at the corner points), then $\phi_{2c} = 1 - p_c^{\text{site}} \approx 0.4073$ and therefore, according to our general statement, the system is bicontinuous for $1 - p_c^{\text{site}} < \phi_2 < p_c^{\text{site}}$. This specific result was first observed by Sheng and Kohn (1982), who noted the importance of conduction through corner points.

An example of a bicontinuous system without phase-inversion symmetry in which $\phi_{2c} < 1/2$ is identical overlapping spheres (phase 2) in three dimensions, where for $0.29 < \phi_2 < 0.97$ both the spaces occupied and unoccupied by the spheres percolate; see Chapter 10. Examples of bicontinuous systems without phase-inversion symmetry and in which $\phi_{2c} > 1/2$ are close packings of identical spheres in three dimensions. At the sphere percolation threshold $\phi_{2c} = \pi/\sqrt{18} \approx 0.7405$ of a close-packed face-centered-cubic array, for example, the void space exterior to the spheres also percolates.

9.2.1 Percolation Properties

For definiteness, we will consider here continuum percolation of statistically isotropic distributions of spherical inclusions at number density ρ and radius R, keeping in mind that this model can be generalized to inclusions of arbitrary size, shape, and orientation. Recall that the dimensionless reduced density η of the particles in d dimensions is given by (4.16), which is equal to the sphere volume fraction ϕ_2 only if the spheres are impenetrable.

A cluster is defined to be a complete set of particles that are connected, i.e., for any two particles in the set, a path can be drawn between their centers that lies entirely within the particles, and no such path can be drawn between any particle center in the set and any particle center outside the set. A *k-mer* is simply a cluster containing k particles. For example, in the leftmost schematic of Figure 9.8, there are five monomers (1-mers), two dimers (2-mers), one trimer (3-mer), and two 4-mers. We will see that cluster statistics defined earlier for lattice percolation can easily be extended to continuum percolation. For continuum percolation there is a variety of other quantities that we will introduce as well.

The probability s_k that a given particle is part of a k-mer satisfies the relation

$$\sum_{k=1}^{\infty} s_k = 1, \qquad \phi_2 < \phi_{2c}. \tag{9.27}$$

The closely related cluster size distribution n_k is the average number of k-mers per unit number of particles and is related to s_k via the expression

$$s_k = kn_k. \tag{9.28}$$

The average number of k-mers per unit volume is given by

$$\rho_k = \rho n_k, \tag{9.29}$$

and thus the average number of clusters per unit volume is given by

$$n_c = \sum_{k=1}^{\infty} \rho_k. \tag{9.30}$$

Moreover, it is clear that

$$\rho = \sum_{k=1}^{\infty} k\rho_k, \qquad \phi_2 < \phi_{2c}. \tag{9.31}$$

The probability q_k that a randomly chosen *cluster* is a k-mer is given by

$$q_k = \frac{n_k}{\sum_{i=1}^{\infty} n_i}. \tag{9.32}$$

The *average cluster number* Q is the average number of particles in a randomly chosen cluster. Using the above definitions, it is given by

$$Q = \sum_{k=1}^{\infty} kq_k = \frac{\sum_{k=1}^{\infty} kn_k}{\sum_{k=1}^{\infty} n_k} = \left(\sum_{k=1}^{\infty} n_k\right)^{-1}, \qquad \phi_2 < \phi_{2c}. \tag{9.33}$$

This quantity also can be related to the number of particles N and the number of clusters M for a given realization under the thermodynamic limit, i.e., $Q = \lim_{M,N\to\infty} N/M$. We also consider the *particle-averaged cluster number* S (otherwise known as the mean cluster size), which is the average number of particles in the cluster containing a randomly chosen *particle* given by

$$S = \sum_{k=1}^{\infty} ks_k = \frac{\sum_{k=1}^{\infty} k^2 n_k}{\sum_{k=1}^{\infty} kn_k} = \sum_{k=1}^{\infty} k^2 n_k, \qquad \phi_2 < \phi_{2c}. \tag{9.34}$$

We can also define certain expected volumes of clusters. The expected volume of a k-mer is denoted by V_k, and so the particle volume fraction is

$$\phi_2 = \sum_{k=1}^{\infty} \rho_k V_k, \qquad \phi_2 < \phi_{2c}. \tag{9.35}$$

The analogues of Q and S for cluster volume are defined by

$$V_Q = \frac{\sum_{k=1}^{\infty} n_k V_k}{\sum_{k=1}^{\infty} n_k} \tag{9.36}$$

and

$$V_S = \frac{\sum_{k=1}^{\infty} kn_k V_k}{\sum_{k=1}^{\infty} kn_k} = \sum_{k-1}^{\infty} kn_k V_k. \tag{9.37}$$

Here V_Q is the "average cluster volume," or the average volume of a randomly chosen cluster. The quantity V_S is the "particle-averaged cluster volume," or the average volume of the cluster containing a randomly chosen particle.

The percolation probability $P(\phi_2)$ is the probability that an arbitrarily selected particle belongs to the infinite cluster. Therefore, for an infinite system,

$$P(\phi_2) \begin{cases} = 0, & \text{if} \quad \phi_2 < \phi_{2c}, \\ > 0, & \text{if} \quad \phi_2 > \phi_{2c}. \end{cases} \tag{9.38}$$

The *percolation-threshold value* ϕ_{2c} is defined formally as $\phi_{2c} = \sup\{\phi_2 : P(\phi_2) = 0\}$ and, for the same reasons given in Section 9.1.1, is a *nonuniversal* quantity. Note that $P(\phi_2)$ is equivalent to the fraction of the particle-phase volume fraction ϕ_2 that is connected.

The pair-connectedness function $P_2(r)$ can be obtained for a general isotropic system of spheres by decomposing the radial distribution function $g_2(r)$ defined in Chapter 3 into a "connected" part and a "disconnected" part (Coniglio, De Angelis and Forlani 1977), i.e.,

$$g_2(r) = P_2(r) + B_2(r), \tag{9.39}$$

where

$\rho^2 P_2(r)$ = Probability density function associated with finding two particles (whose centers are separated by a distance r) in the same cluster, (9.40)

$\rho^2 B_2(r)$ = Probability density function associated with finding two particles (whose centers are separated by a distance r) *not* in the same cluster. (9.41)

The quantity $B_2(r)$ is referred to as the *pair-blocking* function. The decomposition (9.39) is elaborated on in Section 10.2.

The mean cluster size S is related to the pair-connectedness function P_2 via the expression (Coniglio et al. 1977)

$$S = 1 + \rho \int P_2(r) d\mathbf{r}. \tag{9.42}$$

This relation is completely analogous to the *compressibility equation* (3.15) that arises in the study of the liquid state. Note that the asymptotic form of $P_2(r)$ for large r at the percolation threshold is still given by (9.7) and thus ensures that the mean cluster size (9.42) will diverge when $\phi_2 = \phi_{2c}$.

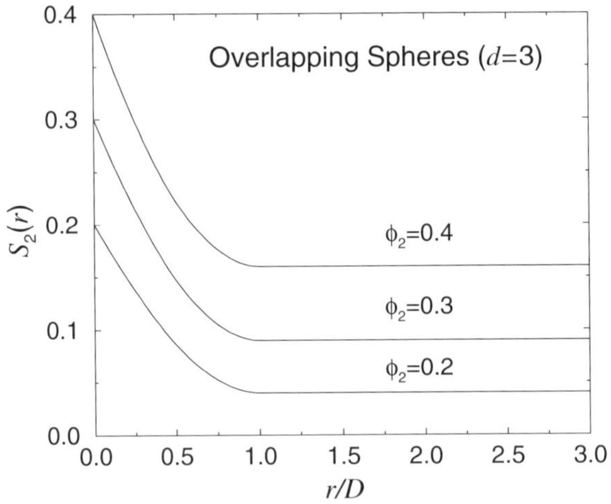

Figure 9.10 Two-point probability $S_2(r)$ versus r/D for particle phase of systems of identical overlapping spheres of diameter D at three different sphere volume fractions. The percolation threshold $\phi_{2c} \approx 0.29$ (Rintoul and Torquato 1997a).

9.2.2 Two-Point Cluster Function

Lower-order n-point correlation functions, such as the two-point probability function S_2, generally do not reflect information about percolating clusters (Section 2.2.5). This point is illustrated in Figure 9.10, where $S_2(r)$ for the *particle* phase of suspensions of identical overlapping spheres in three dimensions (see Chapter 5) is plotted for several values of the sphere volume fraction ϕ_2 below and above the percolation threshold $\phi_{2c} \approx 0.29$ (Rintoul and Torquato 1997a). It is seen that $S_2(r)$ is always a short-ranged function (reaching its asymptotic value at a sphere diameter) and therefore is insensitive to clustering. By contrast, the so-called *two-point cluster function* $C_2(r)$ (Torquato et al. 1988) reflects clustering information, becoming long-ranged as the threshold is approached from below, and can be applied to random media of arbitrary microstructure.

To define C_2 for general statistically *inhomogeneous* media, we decompose the standard two-point probability function for the phase of interest, say phase 2, into a "connected" part and a "disconnected" part:

$$S_2(\mathbf{x}_1,\mathbf{x}_2) = C_2(\mathbf{x}_1,\mathbf{x}_2) + E_2(\mathbf{x}_1,\mathbf{x}_2), \tag{9.43}$$

where

$$C_2(\mathbf{x}_1,\mathbf{x}_2) = \text{Probability of finding two points at positions } \mathbf{x}_1 \text{ and } \mathbf{x}_2 \text{ in the same cluster of phase 2,} \tag{9.44}$$

$$E_2(\mathbf{x}_1,\mathbf{x}_2) = \text{Probability of finding two points at positions } \mathbf{x}_1 \text{ and } \mathbf{x}_2 \textit{ not } \text{in the same cluster of phase 2.} \tag{9.45}$$

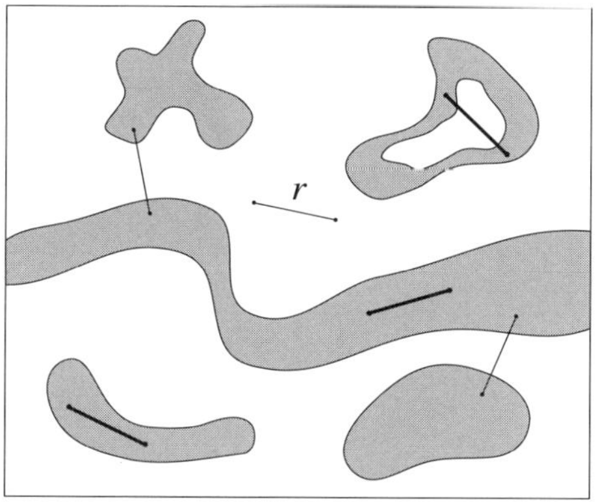

Figure 9.11 Two possible ways in which the end points of a line segment of length r can land in phase 2 (gray region) when randomly tossed into an isotropic medium: (1) end points fall in the same cluster (bold lines) or (2) end points fall in different clusters (lighter lines).

The quantity E_2 is called the *two-point blocking function*.

Clearly, C_2 generally contains nontrivial connectedness information and therefore is a better signature of the microstructure than S_2. To emphasize this point, Figure 9.11 schematically depicts an arbitrary isotropic continuum model in which lines of length $r = |\mathbf{x}_2 - \mathbf{x}_1|$ are shown. The bold lines represent events that contribute only to $C_2(r)$, whereas all of the lines (bold and lighter ones), except the one falling entirely in the matrix, represent events that contribute to $S_2(r)$. This clearly shows that S_2 generally does not distinguish between clustering and nonclustering events. One exception to this would be systems containing a very few but fractal clusters. For a single fractal cluster (e.g., the diffusion-limited aggregation cluster discussed in Chapter 12), $S_2(r) = C_2(r)$ falls off as $r^{d_F - d}$ (Family 1993), and therefore the corresponding Fourier transform or scattering intensity $i(k)$ (Section 2.2.5) scales as k^{-d_F} (Thompson et al. 1987), where d_F is the fractal dimension. The quantity C_2 is discussed further in Chapter 10.

9.2.3 Critical Exponents

Near criticality, the quantities P, S, and ξ have the same type of scaling behavior as in lattice percolation, i.e.,

$$P \sim (\phi_2 - \phi_{2c})^{\beta}, \qquad \phi_2 \to \phi_{2c}^{+}, \tag{9.46}$$

$$S \sim |\phi_2 - \phi_{2c}|^{-\gamma}, \qquad \phi_2 \to \phi_{2c}, \tag{9.47}$$

$$\xi \sim |\phi_2 - \phi_{2c}|^{-\nu}, \qquad \phi_2 \to \phi_{2c}. \tag{9.48}$$

For the model of overlapping particles of various shapes, many previous studies (Haan and Zwanzig 1977, Gawlinski and Stanley 1981, Kertesz and Vicsek 1982, Elam, Kerstein and Rehr 1984, Lorenz, Orgzall and Heuer 1993) have indicated that the geometrical critical exponents, β, γ, and ν, are in the same universality class as lat-

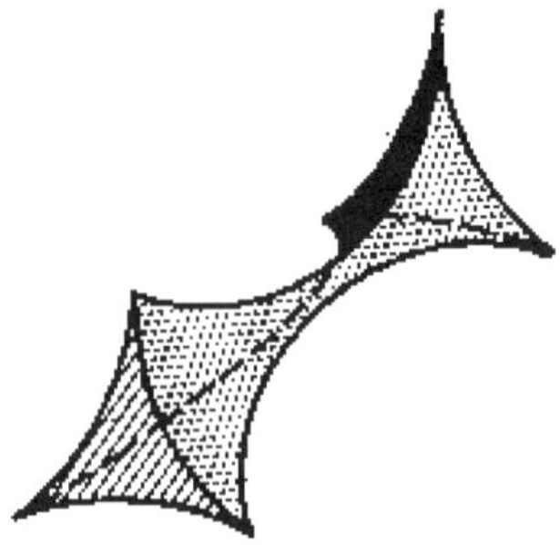

Figure 9.12 Narrow portion of the intervening neck (bond) between three overlapping spherical holes in the three-dimensional Swiss-cheese model. Adapted after Feng et al. (1987).

tice percolation (see Table 9.2). At one time, it was thought that continuum models with short-range correlations, such as exclusion-volume effects in the cherry-pit model (Section 3.1.2), would be in a different universality class. This turns out not to be the case, as a number of investigations have confirmed (Gawlinski and Redner 1983, Lee and Torquato 1990, Lee 1990). Thus, it is believed that continuum and lattice percolation are in the same universality class, provided that correlations are sufficiently short-ranged.

While geometrical critical exponents for continuum percolation remain the same as for lattice percolation, transport and elastic critical exponents generally differ from their lattice values due to a singular distribution of "neck" widths near the threshold (Feng et al. 1987). For simplicity, let phase 1 be the transport or elastic medium and phase 2 be a void phase with vanishing transport and elastic properties. There exists a percolation threshold $\phi_{2c} = 1 - \phi_{1c}$ such that when $\phi_2 > \phi_{2c}$, the system ceases to support transport or mechanical loads, i.e., phase 1 becomes disconnected. The critical exponents for the effective conductivity σ_e, fluid permeability k, and the effective Young's modulus E_e can be defined as follows:

$$\sigma_e \sim (\phi_1 - \phi_{1c})^t, \qquad \phi_1 \to \phi_{1c}^+, \tag{9.49}$$

$$k \sim (\phi_1 - \phi_{1c})^e, \qquad \phi_1 \to \phi_{1c}^+, \tag{9.50}$$

$$E_e \sim (\phi_1 - \phi_{1c})^f, \qquad \phi_1 \to \phi_{1c}^+. \tag{9.51}$$

Feng et al. (1987) showed that the exponents $t, e,$ and f for certain continuum models, including overlapping spheres, are generally different from their lattice values. The continuum models are mapped to percolation networks with randomly occupied bonds (Kerstein 1983) whose strengths of transport or elasticity have a wide distribution. In the case of overlapping spheres, they studied two different scenarios: the "Swiss-cheese" model and the "inverted Swiss-cheese" model. In the Swiss-cheese model, the spheres (phase 2) represent a void or pore phase, and the space exterior to the spheres (phase

Table 9.4 Estimates of the differences Δ_t, Δ_e, and Δ_f between the exponents t, e, and f, respectively, in the Swiss-cheese and standard lattice models as found by Feng et al. (1987). Nonzero entries in the table correspond to sharp upper-bound estimates.

	Conductivity Δ_t	Permeability Δ_e	Elasticity Δ_f
$d = 2$	0	$\frac{3}{2}$	$\frac{3}{2}$
$d = 3$	$\frac{1}{2}$	$\frac{5}{2}$	$\frac{5}{2}$

1) is the transport or elastic medium. For $d = 2$, $\phi_{2c} = 1 - \phi_{1c} \approx 0.68$, and for $d = 3$, $\phi_{2c} = 1 - \phi_{1c} \approx 0.97$ (see Chapter 10).

Near the threshold, the bulk properties of the Swiss-cheese model are limited by many narrow "necks," each of which is bounded by three overlapping spherical holes (see Figure 9.12). Unlike standard percolation models, where all of the bonds are identical, the necks of the Swiss-cheese model have a wide distribution of widths. Indeed, the probability density function of the neck width δ is finite in the limit $\delta \to 0$. Since the transport or elastic strength of a given neck has a power law dependence on δ, such a distribution of neck widths results in a singular distribution of the strengths of the necks. This in turn results in continuum critical exponents t, e, and f that are generally greater than their counterparts in standard lattice percolation. Table 9.4 summarizes the findings of Feng et al. (1987) for the Swiss-cheese model. Note that for $d = 2$, the predictions in Table 9.4 for the conductivity exponent t and the elastic modulus exponent f have been verified experimentally (Benguigui 1986). However, in the *inverted* Swiss-cheese model, where the space exterior to the spheres is the void phase, Feng et al. (1987) found that the discrepancy with lattice exponent values is not as great as it is in the Swiss-cheese model.

As a final remark, we note that, unlike lattice percolation, the conduction threshold and the standard Bernoulli connectivity threshold do not necessarily coincide in continuum percolation. An example is the aforementioned two-dimensional, two-phase checkerboard consisting of a superconducting black phase and a normal conducting white phase. Since conduction can occur through the common corner points of pairs of black squares that are only diagonally connected, the conduction threshold occurs at $\phi_{2c} = 1 - p_c^{\text{site}} \approx 0.4073$ (according to Section 9.2), whereas the Bernoulli connectivity threshold of the black phase occurs at $\phi_{2c} = p_c^{\text{site}} \approx 0.5927$.

CHAPTER 10

Some Continuum Percolation Results

The intent of the present chapter is to derive and discuss some basic results and specific developments in continuum percolation theory. We will begin with a discussion of exact results for cluster statistics and other percolation descriptors for a prototypical model of continuum percolation, namely, identical overlapping spheres in d dimensions. Subsequently, we will describe an Ornstein–Zernike formalism to find the pair-connectedness function $P_2(r)$ for general isotropic models of continuum percolation. The reader should note the beautiful correspondence of this theory to the Ornstein–Zernike formalism for the total correlation function $h(r)$ of equilibrium (or thermal) systems discussed in Chapter 3. This will be followed by a discussion of various approximation schemes to close the resulting integral equation, including the Percus–Yevick approximation. The next topic will be the two-point cluster function $C_2(r)$. First we will present an exact series representation of $C_2(r)$ for dispersions and then discuss its analytical evaluation for certain models. The chapter will conclude with a presentation of percolation thresholds for overlapping sphere systems, overlapping particles of nonspherical shape, and interacting particle systems. The reader is referred to Meester and Roy (1996) for a more mathematical treatment of continuum percolation.

10.1 Exact Results for Overlapping Spheres

Consider statistically homogeneous systems of identical overlapping spheres of diameter D at number density ρ in d dimensions. It is worth repeating here that the volume fraction of the matrix and particle phases are respectively given by

$$\phi_1 = e^{-\eta} \qquad \text{and} \qquad \phi_2 = 1 - e^{-\eta}, \tag{10.1}$$

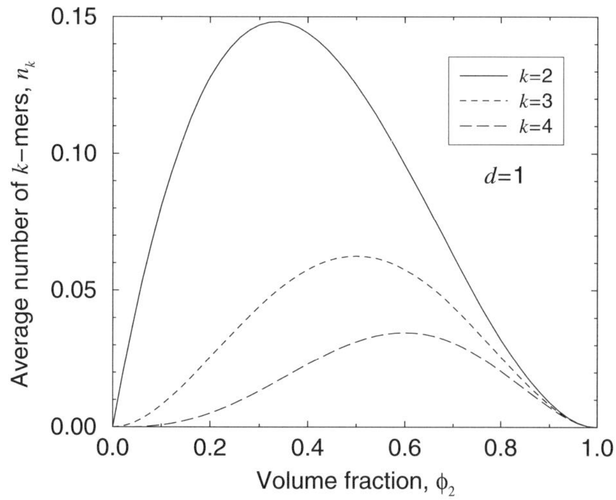

Figure 10.1 Average number n_k of k-mers per unit particle versus particle volume fraction ϕ_2 for a system of overlapping rods.

where $\eta = \rho v_1$ is the reduced density and v_1 is the d-dimensional volume of a sphere [cf. (4.17)]. We begin by considering exact results for one-dimensional systems (overlapping rods) and subsequently progress to higher-dimensional results.

10.1.1 One Dimension

Although one-dimensional models have the trivial percolation threshold $\phi_{2c} = 1$, exact results for cluster and percolation descriptors of such model systems can be used to guide the development of approximations for corresponding quantities in higher dimensions. We specifically obtain results for one-dimensional systems of identical overlapping rods of length D.

k-mer statistics and related quantities

We first consider obtaining the cluster size distribution n_k. Then n_k, also equal to the probability that a given particle is the leftmost particle in a k-mer, is given by (Roach 1968, Quintanilla and Torquato 1996c)

$$n_k = (1 - \phi_2)^2 \phi_2^{k-1}. \tag{10.2}$$

This result, plotted for several values of k in Figure 10.1, was given by Roach (1968). In contrast to Roach, we will prove it using the *arrival process* argument given by Quintanilla and Torquato (1996c), since this paradigm facilitates numerical calculations in higher dimensions. Suppose the center of a given particle is located at r_1. Then this particle is the leftmost particle of a k-mer exactly when the region $[r_1 - D, r_1)$ is empty of centers, the next center after r_1, denoted by r_2, occurs somewhere in $(r_1, r_1 + D]$, the next center after r_2, denoted by r_3, occurs somewhere in $(r_2, r_2 + D]$, ..., the next center after r_{k-1}, denoted by r_k, occurs somewhere in $(r_{k-1}, r_{k-1} + D]$, and, finally, the region $(r_k, r_k + D]$ is empty of centers. These events are independent, since the positions of the centers are generated by a Poisson process. Since this process has density ρ, we conclude that

$$\begin{aligned} n_k &= e^{-2\eta} \int_{r_1}^{r_1+D} dr_2\, \rho e^{-\rho(r_2-r_1)} \int_{r_2}^{r_2+D} dr_3\, \rho e^{-\rho(r_3-r_2)} \cdots \int_{r_{k-1}}^{r_{k-1}+D} dr_k\, \rho e^{-\rho(r_k-r_{k-1})} \\ &= e^{-2\eta} \int_{r_1}^{r_1+D} dr_2 \int_{r_2}^{r_2+D} dr_3 \cdots \int_{r_{k-1}}^{r_{k-1}+D} dr_k\, \rho^{k-1} e^{-\rho(r_k-r_1)} \\ &= (1-\phi_2)^2 \phi_2^{k-1} \end{aligned} \tag{10.3}$$

by induction.

Using (9.28)–(9.30) and (9.32)–(9.34) of the previous chapter, we obtain the following cluster statistics for overlapping rods on the line:

$$s_k = k(1-\phi_2)^2\phi_2^{k-1}, \quad \rho_k = \rho(1-\phi_2)^2\phi_2^{k-1}, \quad q_k = (1-\phi_2)\phi_2^{k-1}, \tag{10.4}$$

$$n_c = \rho(1-\phi_2) = \rho e^{-\eta}, \qquad Q = \frac{1}{1-\phi_2}, \qquad S = \frac{1+\phi_2}{1-\phi_2}. \tag{10.5}$$

Comparing this last equation with (9.47), we see that $\phi_{2c} = \gamma = 1$, where γ is the critical exponent for the mean cluster size. We also see that Q diverges at the percolation threshold; this is consistent with (9.33), since $n_k = 0$ at $\phi_2 = 1$.

The formulas (10.4) and (10.5) were first obtained by Roach (1968), who determined n_k by considering a sequence of Bernoulli trials. Such an argument, however, cannot be rigorously applied to the formation of k-mers in higher dimensions, which we will consider later in this section.

We now consider the average volume V_k of a given k-mer. In any dimension, V_k can be expressed in terms of a conditional expectation on the positions of its k constituent spheres (Quintanilla and Torquato 1996c). Using the aforementioned arrival process paradigm, Quintanilla and Torquato showed that for $d = 1$

$$\begin{aligned} V_k &= \frac{\displaystyle\int_{r_1}^{r_1+D} dr_2 \int_{r_2}^{r_2+D} dr_3 \cdots \int_{r_{k-1}}^{r_{k-1}+D} dr_k (D + r_k - r_1) e^{-\rho(2D+r_k-r_1)}}{\displaystyle\int_{r_1}^{r_1+D} dr_2 \int_{r_2}^{r_2+D} dr_3 \cdots \int_{r_{k-1}}^{r_{k-1}+D} dr_k e^{-\rho(2D+r_k-r_1)}} \\ &= D\left(\frac{k\phi_2 + 1 - k}{\phi_2} + \frac{k-1}{\eta}\right). \end{aligned} \tag{10.6}$$

As expected, this expression is consistent with the known cumulative distribution function of the length of a k-mer in systems of overlapping rods (Roach 1968). The expected volume V_k is plotted in Figure 10.2 for several values of k.

As the density ρ tends to zero, the expected length of a one-dimensional k-mer tends to $D(k+1)/2$. This may be surprising at first glance; after all, at very low densities, k-mers will not exist for $k \geq 2$. However, we are not now calculating the probability

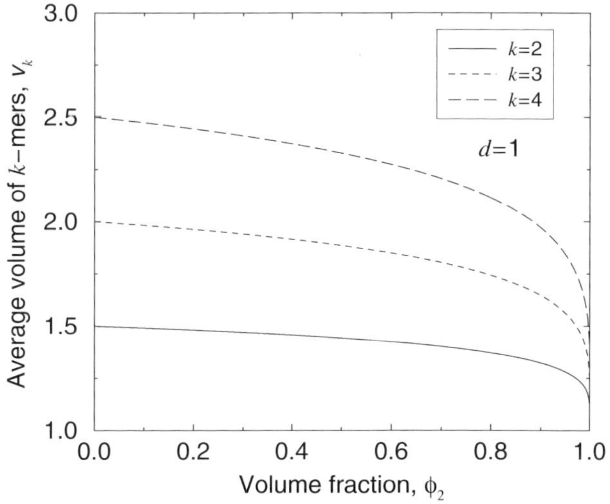

Figure 10.2 Expected volume V_k (in units of $v_1 = D$) of a k-mer versus particle volume fraction ϕ_2 for a system of equisized overlapping rods of length D.

that k-mers exist; instead, this is the expectation of the volume of a k-mer *given that the k-mer exists*.

In one dimension, Q and S were given by (10.5). Using (10.3) and (10.6), we also obtain

$$V_Q = \frac{\phi_2 D}{\phi_1 \eta} = -\frac{\phi_2 D}{\phi_1 \ln(1-\phi_2)} \tag{10.7}$$

and

$$V_S = \frac{2\phi_2 D}{\phi_1 \eta} - D = -\frac{2\phi_2 D}{\phi_1 \ln(1-\phi_2)} - D. \tag{10.8}$$

Notice that V_Q diverges at the percolation threshold for this one-dimensional system. Also, there is a logarithmic contribution in the denominator of V_S. Therefore, in one dimension, V_S does not obey a power law as ϕ_2 approaches the percolation threshold. The presence of η in the denominator, however, would not preclude power-law behavior in higher dimensions since $\eta_c < \infty$ for $d \geq 2$.

Connectedness functions

Here we obtain the two-point cluster function $C_2(r)$ and pair-connectedness function $P_2(r)$, as well as asymptotic formulae for the tail probabilities. We also find the corresponding "blocking" functions.

Before we find these quantities, we first obtain the cumulative distribution of cluster lengths. Each cluster in the system has some random length L_i. The "gaps" (empty intervals between the clusters) have random lengths M_i, so that the gaps and clusters alternate. It is easy to see that the M_i are independent and identically distributed exponential variables with mean $1/\rho$, i.e., $\mathcal{E}(M) = 1/\rho$. The L_i are independent of the M_i and are again independent and identically distributed. The common distribution of the L_i is given by, writing L for L_i,

$$\mathcal{P}\{L > \ell\} = e^{-\rho\ell} \sum_{n=1}^{\infty} \frac{(\rho\ell)^{n-1}}{(n-1)!} \sum_{k=0}^{n} (-1)^k \binom{n}{k} \left(1 - k\frac{D}{\ell}\right)_+^{n-1}, \tag{10.9}$$

where for all integers $n \geq 0$,

$$x_+^n = \begin{cases} 0 & \text{if } x < 0, \\ x^n & \text{if } x \geq 0. \end{cases}$$

This result is obtained by piecing together the results in Feller (1966) under the heading of *waiting for large gaps in Poisson traffic*. In particular, for $\ell < D$, we have $\mathcal{P}\{L > \ell\} = 1$ (since every cluster has at least the length D), and the mean and variance of L are

$$\mathcal{E}(L) = \frac{e^\eta - 1}{\rho}, \qquad \text{Var}\,(L) = \frac{e^{2\eta} - 1 - 2\eta e^\eta}{\rho^2}. \tag{10.10}$$

We note that the mean density of clusters n_c is given by the simple expression

$$n_c = \frac{1}{\mathcal{E}(L) + \mathcal{E}(M)} = \rho e^{-\eta},$$

which agrees with the corresponding expression in (10.5). In one dimension, this clearly is equal to the *mean density of gaps*.

Using the cumulative distribution function (10.9) and renewal theory, Cinlar and Torquato (1995) found that the two-point cluster function is given by

$$C_2(x) = 1 + \sum_{k=1}^{m} (-1)^k \phi_1^k \left\{ \frac{[\eta(x-k+1)]^{k-1}}{(k-1)!} + \frac{[\eta(x-k+1)]^k}{k!} \right\}, \tag{10.11}$$

where m is the positive integer that satisfies $m - 1 \leq x < m$, with $x = r/D$. We recall that $\phi_1 = e^{-\eta}$ for overlapping penetrable rods.

For overlapping rods the pair-connectedness function P_2 is trivially related to the two-point cluster function (Quintanilla and Torquato 1996c):

$$P_2(x) = \begin{cases} 1, & x < 1, \\ C_2(x-1), & x \geq 1. \end{cases} \tag{10.12}$$

This equality is a consequence of the fact that the rods are spatially uncorrelated. For spatially correlated rods, such an equality will generally not hold. Direct integration of P_2 shows that

$$\int_0^\infty P_2(x)dx = \frac{\phi_2}{\phi_1 \eta}, \tag{10.13}$$

and so we obtain, using the "compressibility" relation (9.42) with $d = 1$, that the mean cluster size

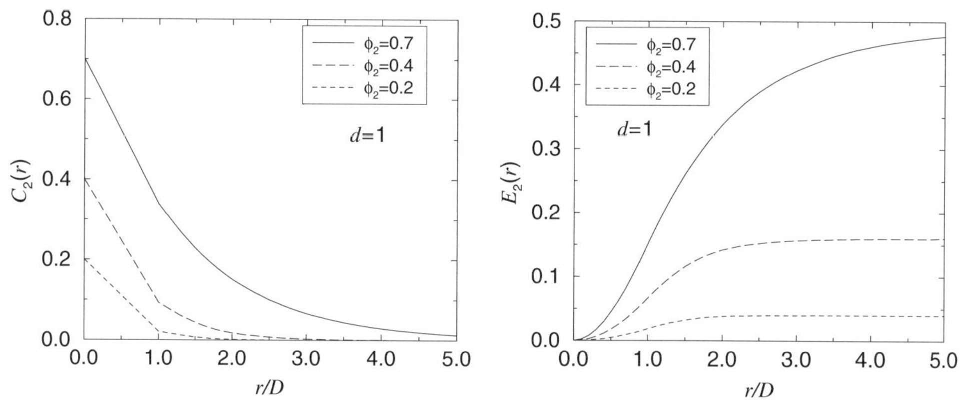

Figure 10.3 Left panel: Two-point cluster function $C_2(r)$ for a system of overlapping rods versus dimensionless distance r/D, as obtained from (10.11), for three values of ϕ_2. Right panel: Corresponding two-point blocking function $E_2(r)$ for a system of overlapping rods, as obtained from (5.7), (9.43), and (10.11).

$$S = 1 + 2\eta \int_0^\infty P_2(x)dx = \frac{1+\phi_2}{1-\phi_2} \tag{10.14}$$

is consistent with that given by (10.5).

In Figure 10.3 we plot the two-point cluster function $C_2(r)$, as obtained from (10.11), for three values of ϕ_2. As expected, $C_2(r)$ is a monotonically decreasing function of r, tending to zero for large r. Its range increases as ϕ_2 increases, reflecting the presence of larger clusters. Figure 10.3 also depicts corresponding results for the two-point blocking function $E_2(r)$. Recall that $E_2(r)$ gives the probability that two points, separated by a distance r, are in *different* clusters. This quantity is easily obtained from (9.43), using relation (5.7) for the conventional two-point probability function $S_2(r)$ (the probability that the two points are in the rod phase), and formula (10.11) for $C_2(r)$. The function $E_2(r)$ must be zero at $r = 0$ and must tend to ϕ_2^2 for large r.

Now we state without proof [which can be found in Cinlar and Torquato (1995)] the asymptotic properties of the tail probabilities. If $\eta = 1$, then

$$C_2(x) \sim 2e^{-1}\, e^{-\eta x}, \qquad x \to \infty. \tag{10.15}$$

If $\eta \neq 1$, then

$$C_2(x) \sim \frac{(c-\eta)e^{-\eta}}{c(c-1)}\, e^{-cx}, \qquad x \to \infty, \tag{10.16}$$

where c is the unique solution of $e^y = e^\eta y/\eta$, $y \neq \eta$.

It is useful here to give the exact expression for the pair-connectedness function $P_2(x)$ for equilibrium rods in the PCS or cherry-pit model (see Section 3.1.2), which was obtained using standard statistical-mechanical techniques for one-dimensional liquids (Vericat, Gianotti and Rodriguez 1987, Pugnaloni, Gianotti and Vericat 1997):

$$P_2(x) = \frac{1}{\eta}\sum_{k=1}^{\infty}\sum_{j=0}^{k}(-1)^j \frac{k!}{k!(k-j)!}\left(\frac{\eta}{1-\eta}\right)^k \frac{[x-k-j(\lambda-1)]^{k-1}}{(k-1)!} \times \Theta[x-k-j(\lambda-1)]\exp\left[-\frac{\eta(x-k)}{1-\eta}\right], \tag{10.17}$$

where $\Theta(x)$ is the Heaviside step function defined by (3.48). Recall that λ is the impenetrability parameter, which lies in the interval $[0, 1]$. Subsequently, the same expression was found by Drory (1997) using a mapping between continuum percolation and a Potts fluid. The corresponding expression for the two-point cluster function C_2 in the cherry-pit model has been given through second order in density (Lee and Torquato 1988b), but the full density-dependent result has yet to be obtained. Note that when $\lambda = 0$, relation (10.17) can be simplified (the double sum reduces to a single sum) and, not surprisingly, turns out to be equal to expression (10.12) for the fully overlapping case.

10.1.2 Higher Dimensions

The aforementioned paradigm can be extended to calculate n_k in higher dimensions. In d dimensions, a given sphere of radius $D/2$ is a monomer exactly when a spherical region of *radius* D centered at the given sphere contains no other sphere centers, and so

$$n_1 = \exp[-2^d \rho v_1] = \phi_1^{2^d}. \tag{10.18}$$

For $k \geq 2$, Given, Kim, Torquato and Stell (1990b) claimed and Penrose (1991) later rigorously proved, in any dimension, that

$$n_k = \frac{\rho^{k-1}}{k!}\int d\mathbf{r}_{12}\int d\mathbf{r}_{13}\cdots\int d\mathbf{r}_{1k}\exp[-\rho v_k(\mathbf{r}^k; D)]I(\mathbf{r}^k; D/2), \tag{10.19}$$

where $v_k(\mathbf{r}^k; D)$ is the *union* volume of k spheres with radius D centered at $\mathbf{r}^k \equiv \mathbf{r}_1, \ldots, \mathbf{r}_k$ and $I(\mathbf{r}^k; D/2)$ is the indicator function for k spheres with centers at $\mathbf{r}_k$ and radius $D/2$ forming a single cluster. Note that $v_1(D) = 2^d v_1(D/2)$, where $v_1(D/2) \equiv v_1$ is the volume of a single sphere.

The integrals determining n_k require knowledge of $v_k(\mathbf{r}^k; D)$, the union volume of k d-dimensional spheres of equal radius. To calculate this quantity, it is sufficient to obtain expressions for the *intersection* volume $v_j^{\text{int}}(\mathbf{r}^j; D)$ of j spheres for $1 \leq j \leq k$ (Section 5.1). Quintanilla and Torquato (1996c) evaluated the integrands of the above integrals exactly for both $d = 2$ and $d = 3$, and then used numerical integration to obtain the n_k for $k = 2, 3$, and 4. Figure 10.4 shows theoretical predictions of n_2, n_3, and n_4 and direct Monte Carlo simulation of these quantities for the case $d = 3$. As we see, simulation and theory are in excellent agreement.

The average volume V_k of a k-mer is given explicitly as the conditional expectation of the volume of k spheres, given that the k spheres indeed form a cluster (Quintanilla and Torquato 1996c). From (10.19), given that k particles form a k-mer, the conditional

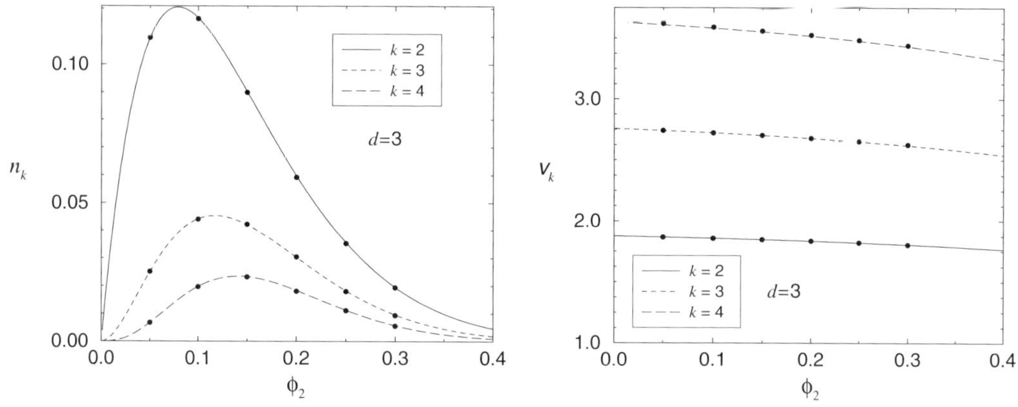

Figure 10.4 Left panel: Average number n_k of k-mers per unit particle versus particle volume fraction ϕ_2 for systems of three-dimensional systems of overlapping spheres. The black circles are simulation data. Right panel: Corresponding expected volume V_k of k-mers in units of υ_1.

probability density function F of the locations of the k particles is given by

$$F(\mathbf{r}_{12}, \ldots, \mathbf{r}_{1k}) = \frac{\exp[-\rho \upsilon_k(\mathbf{r}^k; D)] I(\mathbf{r}^k; D/2)}{\int d\mathbf{r}_{12} \int d\mathbf{r}_{13} \cdots \int d\mathbf{r}_{1k} \exp[-\rho \upsilon_k(\mathbf{r}^k; D)] I(\mathbf{r}^k; D/2)}. \tag{10.20}$$

Since

$$V_k = \int d\mathbf{r}_{12} \cdots \int d\mathbf{r}_{1k} \upsilon_k(\mathbf{r}^k; D/2) F(\mathbf{r}_{12}, \ldots, \mathbf{r}_{1k}), \tag{10.21}$$

from the definition of conditional expectation, we conclude that

$$V_k = \frac{\displaystyle\int d\mathbf{r}_{12} \cdots \int d\mathbf{r}_{1k} \upsilon_k(\mathbf{r}^k; D/2) \exp[-\rho \upsilon_k(\mathbf{r}^k; D)] I(\mathbf{r}^k; D/2)}{\displaystyle\int d\mathbf{r}_{12} \cdots \int d\mathbf{r}_{1k} \exp[-\rho \upsilon_k(\mathbf{r}^k; D)] I(\mathbf{r}^k; D/2)}. \tag{10.22}$$

For $d \geq 2$, we again must settle for numerical evaluation of the V_k. Figure 10.4 depicts lower-order V_k in three dimensions. As we see, the V_k decrease to the volume of a single sphere υ_1 as the sphere volume fraction ϕ_2 increases to 1. This is not surprising: As ϕ_2 increases, we would expect the separation distances of a k-mer to decrease. From the graphs of the V_k for $k \leq 4$ and computer simulations for larger k, a decent empirical approximation for V_k at small volume fractions and arbitrary d is

$$V_k(0) \approx \frac{1 + k(2^d - 1)}{2^d} \upsilon_1. \tag{10.23}$$

However, this does not provide a rigorous upper bound on V_k. A precise determination of V_k and its derivative at $\phi_2 = 0$ should provide an excellent upper bound and approximation of V_k for even intermediate values of ϕ_2.

Table 10.1 The coefficients in the expansions of n_k, V_k, Q, S, V_Q, and V for a system of fully penetrable disks. Recall that $n_1 = e^{-4\eta}$ from (10.18). The series expansions for V_k, V_Q, and V_S are expressed in units of υ_1, the area of a single disk. The expansions for Q, S, V_Q, and V_S are derived from (9.33), (9.34), (9.36), and (9.37), respectively.

	η^0	η^1	η^2	η^3	η^4
n_2		2	−11.3079	32.2915	−62.0415
n_3			4.8720	−35.3346	129.6895
n_4				13.022	−114.823
n_5					36.728
V_2	1.75	−0.1295	−0.0273	0.0028	
V_3	2.5071	−0.257	−0.055		
V_4	3.268	−0.4			
Q	1	2	2.436	2.432	2.2
S	1	4	6.616	8.834	11
V_Q	1	1.5	1.603	1.5	
V_S	1	3	4.548	5.9	

10.1.3 Low-Density Expansions of Cluster Statistics

We will now use this procedure to obtain low-density expansions of the cluster statistics Q, S, V_Q, and V_S, which were defined in Chapter 9. Such low-density expansions are important, since they are exact to accuracy of the above numerical integrations and hence can be used as a benchmark for approximate approaches.

Haan and Zwanzig (1977) used the coefficients in the low-density expansion of S to estimate the percolation threshold and critical exponent for S. They did this by using (9.34) and by obtaining the density expansions of the n_k using the diagrammatic expansion approach. This could also be done from the expression

$$n_k = \sum_{j=0}^{\infty}(-1)^j \frac{\rho^{k-1+j}}{k!j!} \int d\mathbf{r}_{12} \int d\mathbf{r}_{13} \cdots \int d\mathbf{r}_{1k} [\upsilon_k(\mathbf{r}^k; D)]^j I(\mathbf{r}^k; D/2), \tag{10.24}$$

obtained by expanding (10.19) in a Taylor series. Quintanilla and Torquato (1996c) numerically evaluated (10.24) and found the corresponding expansions of the product $n_k V_k$ [obtainable from the numerator of (10.22)]. They then used the coefficients to find the low-density expansions of the cluster statistics Q, S, V_Q, and V_S. We present these expansions in Table 10.1 for $d = 2$ and in Table 10.2 for $d = 3$. As expected, the expansions for the n_k are in agreement with those obtained by Haan and Zwanzig (1977).

Fanti, Glandt and Chiew (1988) obtained a series expansion for V_S for three-dimensional overlapping spheres in terms of the one-point/one-particle connectedness function. They then evaluated V_S to account for binary overlaps under the superposition approximation. They were not, however, concerned with the individual n_k and V_k. Their expression is

Table 10.2 As in Table 10.1, except for a system of fully penetrable spheres. For this system, $n_1 = e^{-8\eta}$ from (10.18).

	η^0	η^1	η^2	η^3	η^4
n_2		4	−49	302.2238	−1250.5030
n_3			22	−359.4203	2959.1209
n_4				139.7867	−2842.60
n_5					964.68
V_2	1.875	−0.1574	−0.1025	−0.0270	
V_3	2.7578	−0.3080	−0.2073		
V_4	3.64	−0.5			
Q	1	4	11	26.743	61.6
S	1	8	34	125.3660	436
V_Q	1	3.5	9.1667	21	140
V_S	1	7	29.0048	102	800

$$V_S = \frac{v_1 S \phi_2}{\eta}. \tag{10.25}$$

If we generalize this expression to d dimensions, we obtain

$$V_S = \frac{S\phi_2}{\eta} \frac{\pi^{d/2}}{\Gamma(1+d/2)} \left(\frac{D}{2}\right)^d \tag{10.26}$$

after using (9.37). We see from (10.5) and (10.8) that this expression, generalized to one dimension, is off by a factor of two.

The asymptotic behavior of V_S near the percolation threshold has not been studied quantitatively with high precision. Such a study would enable one to determine whether V_S shares the same critical exponent as S (i.e., whether $\omega = \gamma$).

10.2 Ornstein–Zernike Formalism

Hill (1955) recognized that the study of the distribution of physical clusters in an isotropic equilibrium system of interacting particles with pair potential $\varphi(r)$ requires an arbitrary separation of the Boltzmann factor $\exp[-\beta\varphi(r)]$ into two parts:

$$\exp[-\beta\varphi(r)] = \exp[-\beta\varphi^+(r)] + \exp[-\beta\varphi^*(r)], \tag{10.27}$$

where $\varphi^+(r)$ and $\varphi^*(r)$ are the pair potentials for bound and unbound particles, respectively, and $\beta = 1/(kT)$. The corresponding Mayer function $f(r) \equiv \exp[-\beta\varphi(r)] - 1$ (Section 3.1.2) can also be separated into contributions for bound and unbound pairs, i.e.,

$$f(r) = f^+(r) + f^*(r), \tag{10.28}$$

where

$$f^+(r) = \exp[-\beta\varphi^+(r)], \qquad f^*(r) = \exp[-\beta\varphi^*(r)] - 1. \tag{10.29}$$

Our definition of connectedness is embodied in how we choose to define bound and unbound pairs. We illustrate this decomposition for two sphere models discussed in Section 3.1.2: the cherry-pit model and Baxter's *sticky* hard-sphere model. The spheres again have diameter D.

For the cherry-pit model, the Boltzmann factor is given by (3.28) and thus

$$f(r) = \begin{cases} -1, & 0 \le r \le \lambda D, \\ 0, & r > \lambda D. \end{cases} \tag{10.30}$$

Therefore, it is natural to define

$$f^+(r) = \begin{cases} 0, & 0 \le r \le \lambda D, \\ 1, & \lambda D \le r \le D, \\ 0, & r > D, \end{cases} \tag{10.31}$$

$$f^*(r) = \begin{cases} -1, & 0 \le r \le D, \\ 0, & r > D. \end{cases} \tag{10.32}$$

For the sticky hard-sphere model, the Boltzmann factor is given by (3.24), and thus

$$f(r) = \begin{cases} \dfrac{D}{12\tau}\delta(r - D) - 1, & 0 \le r \le D, \\ 0, & r > D, \end{cases} \tag{10.33}$$

where the dimensionless stickiness parameter τ^{-1} can vary between zero (for perfectly unsticky hard spheres) to ∞ (for perfectly sticky hard spheres). The parameter τ is effectively a dimensionless temperature. Hence, the bound and unbound Mayer functions are defined as

$$f^+(r) = \frac{D}{12\tau}\delta(r - D), \qquad \text{for all } r, \tag{10.34}$$

$$f^*(r) = \begin{cases} -1, & 0 \le r \le D, \\ 0, & r > D. \end{cases} \tag{10.35}$$

Coniglio et al. (1977) derived the cluster expansion for the isotropic pair-connectedness function $P_2(r)$, defined by (9.40), by replacing every f-bond in the corresponding expansion (3.37) of the radial distribution function $g_2(r)$ with the sum $f(r) = f^+(r) + f^*(r)$. Then $P_2(r)$ is identified as the collection of diagrams having at least one unbroken path of f^+-bonds connecting root points 1 and 2. The graphs that constitute $P_2(r)$ can be divided into the *nodal* or *bridge* graphs and the *direct* or *nonnodal* graphs denoted by $D_2(r)$. The resulting expression is an Ornstein–Zernike relation that is actually a definition of the *direct connectedness function* $D_2(r)$, i.e.,

$$P_2(r) = D_2(r) + \rho D_2(r) \otimes P_2(r), \tag{10.36}$$

where $\otimes$ denotes a convolution integral [cf. (3.33)]. Taking the Fourier transform of (10.36) gives

$$\tilde{P}_2(k) = \frac{\tilde{D}_2(k)}{1 - \rho\tilde{D}_2(k)}, \tag{10.37}$$

or, equivalently,

$$\tilde{D}_2(k) = \frac{\tilde{P}_2(k)}{1 + \rho\tilde{P}_2(k)}. \tag{10.38}$$

The *direct* connectedness function can immediately be used to determine the critical percolation density ρ_c. Using the equivalent Fourier representation of the "compressibility" relation (9.42) for the mean cluster size S and (10.37), we arrive at

$$S = 1 + \rho\tilde{P}_2(0) = [1 - \rho\tilde{D}_2(0)]^{-1}. \tag{10.39}$$

Since S diverges at the threshold,

$$\rho_c = [\tilde{D}_2(0)]^{-1} = \left[\int D_2(r)d\mathbf{r}\right]^{-1}. \tag{10.40}$$

The average coordination number Z for general systems of (possibly overlapping) identical spheres of diameter D is given in terms of the pair-connectedness function via the relation

$$Z = \rho\Omega(d)\int_0^D r^{d-1}P_2(r)dr, \tag{10.41}$$

where $\Omega(d)$ is the total solid angle contained in a d-dimensional sphere and is given by (2.56). Note that hard-sphere systems can possess a nonzero coordination number only if there is a delta function contribution to $P_2(r)$ when the spheres are in contact ($r = D$), i.e., P_2 must be of the form

$$P_2(r) = \frac{Z}{\rho\Omega D^{d-1}}\delta(r - D) + P_2^c(r), \tag{10.42}$$

where $P_2^c(r)$ is the continuous part of $P_2(r)$ for $r > D$.

10.3 Percus–Yevick Approximations

As in the conventional Ornstein–Zernike relation for thermal systems, one can solve the connectedness integral relation (10.36) in various approximations. We discuss below the closure of (10.36) using the Percus–Yevick (PY) approximation.

10.3.1 Permeable-Sphere Model

Chiew and Glandt (1983) solved the Ornstein–Zernike equation (10.36) for the *permeable sphere model* in the PY approximation, i.e., under the conditions that

$$P_2(r) = g_2(r), \qquad 0 \le r \le D, \tag{10.43}$$

$$D_2(r) = 0, \qquad r > D. \tag{10.44}$$

The permeable sphere model (Blum and Stell 1979) is characterized by an impenetrability parameter ϵ, with $\epsilon = 0$ and $\epsilon = 1$ corresponding to overlapping spheres and totally impenetrable spheres, respectively.

For simplicity and to make contact with Monte Carlo simulations, we sketch the solution for the special case of overlapping spheres ($\varepsilon = 0$), which is identical to the cherry-pit model when $\lambda = 0$. Using Baxter's factorization technique (Baxter 1968), one transforms the original OZ equation into two integral equations involving the key function $q(r)$. It turns out that $q(r)$ is the quadratic function given by

$$q(r) = \frac{1}{2}\alpha_1(r^2 - D^2) + \alpha_2(r - D), \tag{10.45}$$

where

$$\alpha_1 = \frac{2\eta - 1}{(1+\eta)^2}, \qquad \alpha_2 = \frac{-3\eta}{2(1+\eta)^2}.$$

It follows from the method of Baxter (1968) that the connectedness quantities are related to $q(r)$ via the relation

$$1 + \rho\tilde{P}_2(k) = \left[1 - \rho\tilde{D}_2(k)\right]^{-1} = \left[\tilde{Q}(k)\tilde{Q}(-k)\right]^{-1}, \tag{10.46}$$

where

$$\tilde{Q}(k) = 1 - 2\pi\rho\int_0^D e^{ikr} q(r)dr. \tag{10.47}$$

Thus, the mean cluster size S, according to relation (10.39), is given by

$$S = [\tilde{Q}(0)]^{-2} = \frac{(1+\eta)^4}{(2\eta - 1)^2}. \tag{10.48}$$

Expression (10.39) yields the percolation threshold $\eta_c = 1/2$, which corresponds to a sphere volume fraction

$$\phi_{2c} = 1 - e^{-\eta_c} \approx 0.393\ . \tag{10.49}$$

Moreover, from expression (10.41), we get the PY prediction of the average coordination number to be

$$Z = 4\ . \tag{10.50}$$

The PY approximation predicts a percolation threshold $\eta_c = 1/2$ ($\phi_{2c} \approx 0.393$) which is well above the accepted value of $\eta_c \approx 0.3418$ ($\phi_{2c} \approx 0.2895$) (Rintoul and Torquato

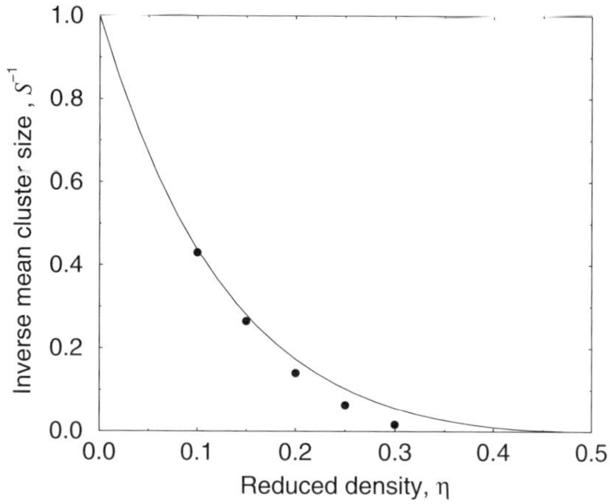

Figure 10.5 Inverse mean cluster size S^{-1} versus reduced density η for a system of overlapping spheres. Comparison of PY prediction (solid curve) with simulation data (Sevick et al. 1988).

1997a) (see Table 10.4). Indeed, it is known from simulations (Sevick, Monson and Ottino 1988, Lee and Torquato 1988b) that the PY prediction consistently underestimates clustering. Figure 10.5 compares the PY result for S^{-1} to corresponding simulation data (Sevick et al. 1988, Lee and Torquato 1988b), revealing that the PY relation (10.48) overestimates S^{-1} (or underestimates S), especially at the higher densities. It also predicts that S should diverge as $\eta \to \eta_c$ with an exponent $\gamma = 2$, in contrast to the accepted value of $\gamma = 1.82$ (see Table 9.2). However, the PY solution enables one to estimate the pair-connectedness function $P_2(r)$ easily via (10.46). This prediction generally lies below corresponding simulation data for $P_2(r)$ (see Figure 10.6) and becomes progressively worse as the density is increased, consistent with an underestimation of S.

The quantitative deficiencies in the PY solution of $P_2(r)$ can be understood diagramatically, as discussed later in the chapter. However, the PY approximation does capture the salient features of clustering and percolation, and hence is a useful starting point in studying more complicated model systems. For example, Chiew et al. (1985) obtained the pair-connectedness function for binary mixtures of permeable spheres to examine, among other things, the effect of size distribution on clustering.

Interestingly, Stell (1984) has shown that there is an isomorphism between the PY solution of the connectedness functions for overlapping spheres and the PY solution of the thermal structure of hard spheres at a *negative* density $-\rho$; specifically,

$$P_2(\rho) \to -h(-\rho), \qquad D_2(\rho) \to -c(-\rho), \tag{10.51}$$

where $h = g_2 - 1$ and c are the total and direct correlation functions for the hard-sphere systems, respectively, discussed in Chapter 3.

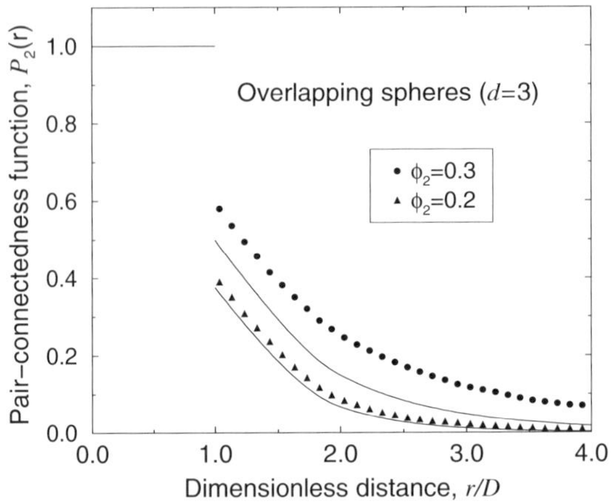

Figure 10.6 Pair-connectedness function P_2 versus dimensionless distance r/D for statistically isotropic systems of identical overlapping spheres for several values of the reduced density η. Comparison of PY prediction (solid curves) with simulation data (filled circles and squares) (Lee and Torquato 1988b).

10.3.2 Cherry-Pit Model

DeSimone, Stratt and Demoulini (1986) solved the Ornstein–Zernike equation in the Percus–Yevick approximation for the penetrable concentric-shell, or cherry-pit, model, i.e., subject to the conditions (10.43) and (10.44) but where $g_2(r)$ in (10.43) is the radial distribution function of hard spheres of diameter λD. Thus, in the limit $\lambda \rightarrow 0$, we have that $P_2(r) = 1$ for $0 \leq r \leq D$ (as it is in the permeable-sphere model when $\varepsilon = 0$) and in the limit $\lambda \rightarrow 1$, $P_2(r) = 0$ for $0 \leq r \leq D$ (as it is in the permeable-sphere model when $\varepsilon = 1$ and the sticky hard-sphere model when $\tau \rightarrow \infty$). For the range of impenetrability parameters $0 < \lambda < 0.5$, one must settle for a numerical solution of the integral equation, but for $0.5 \leq \lambda \leq 1$, it can be solved analytically using Baxter's technique (Baxter 1968).

It turns out that the PY solution predicts that η_c is a nonmonotonic function of λ; i.e., η_c has a minimum at some intermediate value of λ and is largest at $\lambda = 1$ (where it is assumed that the threshold corresponds to the maximally random-jammed density; see Section 3.5). Monte Carlo simulation data (Bug, Safran, Grest and Webman 1985) as well as simple Padé approximants of the mean cluster size (Sen and Torquato 1988) also exhibit this nonmonotonicity property. At first glance, this may appear to be counterintuitive, since one might expect that the critical density η_c should be inversely proportional to the average coordination number and inversely proportional to the excluded volume. However, it is important to observe that this nonmonotonicity property does not extend to the critical volume fraction ϕ_{2c}, as can be seen in Figure 10.8. The reason that ϕ_{2c} is a monotonic function of λ is that the minimum, away from $\lambda = 0$, is not deep enough in the λ-η_c plane to result in a minimum, away from $\lambda = 0$, in the

λ-ϕ_{2c} plane. Recall from Section 5.3.2 that the density of particles η required to achieve a particular volume fraction ϕ_2 at small to moderate values of λ is much larger than it is at values of λ close to unity.

10.3.3 Sticky Hard-Sphere Model

For the sticky hard-sphere system discussed in Section 3.1.2 , Chiew and Glandt (1983) solved the Ornstein–Zernike equation (10.36) in the Percus–Yevick approximation for which the conditions (10.43) and (10.44) also apply. From the condition (10.43), one has the obvious result

$$P_2(r) = \frac{1}{12}\chi D\delta(r - D), \qquad 0 \le r \le D, \tag{10.52}$$

where the dimensionless parameter χ is related to the "stickiness" parameter τ^{-1} via

$$\frac{\phi_2}{12}\chi^2 - \left(\frac{\phi_2}{1-\phi_2} + \tau\right)\chi + \frac{1+\phi_2/2}{(1-\phi_2)^2} = 0. \tag{10.53}$$

The average coordination number obtained from (10.41) is

$$Z = 2\chi\phi_2 \tag{10.54}$$

for sticky spheres in the PY approximation. When there is no stickiness ($\chi = 0$ or $\tau = \infty$), i.e., for equilibrium hard spheres, then $Z = 0$, as expected.

Using Baxter's factorization technique again, Chiew and Glandt obtained the following analytical expression for the Fourier transform of the PY pair-connectedness function in three dimensions:

$$1 + \rho\tilde{P}_2(k) = \left\{1 - 2\chi\phi_2\frac{\sin(kD)}{kD} + 2\left(\frac{\chi\phi_2}{kD}\right)^2[1 - \cos(kD)]\right\}^{-1}. \tag{10.55}$$

According to relation (10.39), the mean cluster size is thus given by

$$S = \frac{1}{(1-\chi\phi_2)^2}, \tag{10.56}$$

and so the percolation threshold corresponds to

$$\phi_{2c} = \eta_c = 1/\chi. \tag{10.57}$$

However, since ϕ_{2c} must be less than the maximum packing fraction $\pi/\sqrt{18}$ in three dimensions (Section 3.3), then (10.57) must be restricted to values of χ such that $\chi \ge \sqrt{18}/\pi$. Equation (10.57) combined with (10.53) yields the locus of the percolation line in the ϕ_2-τ plane as

$$\tau = \frac{19\phi_2^2 - 2\phi_2 + 1}{12(1-\phi_2)^2}. \tag{10.58}$$

10.4 Beyond Percus–Yevick Approximations

The fact that the PY approximations of the connectedness functions are not as quantitatively accurate as the PY solution of the radial distribution function of thermal systems can be understood from an analysis of the cluster diagrams appearing in the exact expression for $P_2(r)$ but omitted in these approximations (Stell 1984). The cluster integrals left out in the PY approximations tend to cancel in the thermal problems successfully but not in the connectedness problems (DeSimone et al. 1986). Xu and Stell (1988) went beyond the PY approximation by developing a so-called generalized mean-spherical approximation scheme by assuming the direct connectedness function $D_2(r)$ to have a Yukawa form ($c_1 \exp[c_2(r - D)]/r$) for $r > D$. This scheme enabled them to obtain results for the overlapping-sphere model that were in excellent agreement with simulation results for $P_2(r)$.

10.5 Two-Point Cluster Function

The two-point cluster function C_2 may be used to characterize continuum percolation systems of arbitrary microstructure, as is evident from the definition (9.44). In the special case of *particle systems*, Torquato et al. (1988) obtained a general series representation of C_2 by decomposing S_2 into "connected" and "disconnected" contributions as prescribed by (9.43). In particular, for statistically isotropic distributions of possibly overlapping identical spheres at number density ρ in d dimensions, $C_2(r)$ can be expressed as the following diagrammatic expansion (Section 3):

$C_2(r_{12}) =$

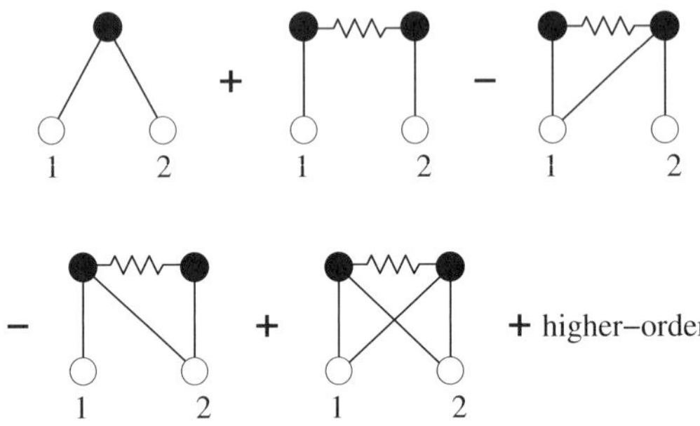

+ higher–order diagrams. (10.59)

Here a solid line indicates an m-bond, where $m(r)$ is the step function (4.4), and a wavy line indicates a P_2-bond. The higher-order diagrams involve three or more inclusions (accounting for particle overlaps) and are given explicitly by Torquato et al. (1988). For the model of overlapping spheres when $d \geq 2$, an exact evaluation of C_2 is not possible, since connectedness functions involving two or more particles are not known exactly.

For the special case of totally impenetrable spheres that can form interparticle contacts (e.g., sticky spheres or jammed hard spheres), only the first two diagrams of relation (10.59) do not vanish identically, and hence it reduces to the exact expression

$$C_2(r_{12}) = \rho v_2^{\rm int}(r_{12}; R) + \rho^2 \int m(r_{13}) m(r_{24}) P_2(r_{34}) d\mathbf{r}_3 d\mathbf{r}_4, \tag{10.60}$$

where $v_2^{\rm int}(r; R) = m \otimes m$ is the intersection volume of two spheres of radius R separated by a distance r (see Section 3.3.1). The first and second terms on the right side of (10.60) give the probabilities that the two points land in a monomer and in two different particles of the same cluster, respectively. Since the Fourier transform of (10.60) is

$$\tilde{C}_2(k) = \rho \tilde{m}^2(k) + \rho^2 \tilde{m}^2(k) \tilde{P}_2(k), \tag{10.61}$$

it immediately follows from definition (10.39) that the mean cluster size S is related to C_2 (Torquato et al. 1988) via the volume integral

$$S = \frac{\rho}{\phi_2^2} \int C_2(r) d\mathbf{r}. \tag{10.62}$$

Since S becomes infinite at the percolation threshold, $C_2(r)$ must be long-ranged at this critical point, as expected. Torquato et al. (1988) showed that the asymptotic behavior of $C_2(r)$ for small r is given by

$$C_2(r) = \phi_2 - \frac{3}{2}\phi_2 \left(\frac{r}{D}\right) + \frac{Z\phi_2}{4}\left(\frac{r}{D}\right)^2 + \mathcal{O}(r^3), \tag{10.63}$$

where Z is given by (10.41).

For the model of sticky hard spheres, Torquato et al. (1988) evaluated C_2 in the Percus–Yevick approximation by Fourier inversion, i.e., by calculating

$$C_2(r) = \frac{1}{2\pi^2 r} \int_0^\infty k \sin(kr) \left\{ 1 - 2\chi\phi_2 \frac{\sin(kD)}{kD} + 2\left(\frac{\chi\phi_2}{kD}\right)^2 [1 - \cos(kD)] \right\}^{-1} dk. \tag{10.64}$$

This expression is obtained by combining (10.55) and (10.61). Figure 10.7 shows that C_2 becomes longer ranged as the percolation threshold is approached from below, as expected.

The function C_2 is also readily ascertained from computer simulations for continuum percolation models. Chapter 12 describes such computations.

10.6 Percolation Threshold Estimates

A variety of techniques exist to estimate the percolation thresholds of continuum models. These include series-expansion analyses, Monte Carlo methods, and real-space renormalization group techniques. For continuum percolation for $d \geq 2$, the percolation threshold is almost never known exactly, and hence numerical simulations are the usual means of accurately estimating percolation thresholds.

Exceptions are certain two-dimensional symmetric-cell materials (Chapter 8). The general claim is that $\phi_{1c} = \phi_{2c} = 1/2$ for any two-dimensional symmetric-cell material

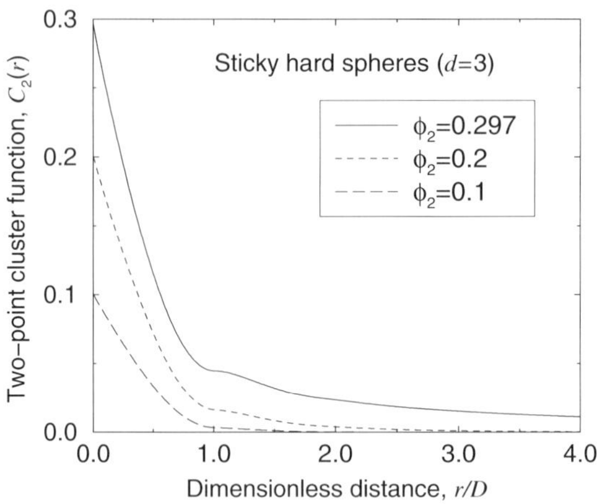

Figure 10.7 Two-point cluster function $C_2(r)$ versus the dimensionless distance r/D for a system of sticky hard spheres of diameter D at several values of the sphere volume fraction ϕ_2. Here $\tau = 0.35$ and $\phi_{2c} = 0.297$.

in which its cell "centers" constitute the sites of a fully triangulated lattice (dual Delaunay triangulation); see Section 8.1.1. The continuum and site percolation thresholds are then identical. Examples of two-dimensional systems in which $\phi_{1c} = \phi_{2c} = 1/2$ include a regular hexagonal tessellation (whose dual is a regular triangular lattice) and Voronoi tessellations of homogeneous Poisson point patterns [numerically verified by Winterfeld, Scriven and Davis (1981)]; see Figure 8.4. There is no rigorous proof of the aforementioned general claim. However, Sykes and Essam (1964) have argued that fully triangulated lattices, under a certain assumption, are self matching and hence, according to Section 9.1, have a percolation threshold of 1/2. Van Den Berg (1981) found exceptions to their assumption and devised a proof under a different assumption.

In what follows, we report values of the percolation thresholds for (i) overlapping disks and spheres, (ii) overlapping particles of nonspherical shape, and (iii) interacting particle systems, including the cherry-pit model.

10.6.1 Overlapping Disks and Spheres

By carrying out a standard ratio analysis (Essam 1972) of the aforementioned series expansions for the mean cluster size (Section 10.1.3), Haan and Zwanzig (1977) obtained estimates of the percolation threshold ϕ_{2c} of overlapping disks and spheres. Many numerical studies have been carried out to determine ϕ_{2c} for these models. Tables 10.3 and 10.4 compare the series-analysis estimates of Haan and Zwanzig (1977) to the currently most accurate numerical evaluations of the percolation thresholds (Rintoul and Torquato 1997a, Quintanilla et al. 2000) given by $\phi_{2c} = 0.67637 \pm 0.00005$ for $d = 2$ and $\phi_{2c} = 0.2895 \pm 0.0005$ for $d = 3$. It is seen that the Haan-Zwanzig estimates are better

Table 10.3 Estimates of the particle-phase percolation threshold ϕ_{2c} for a system of identical overlapping disks ($d = 2$). The value given by Quintanilla et al. (2000) is the best currently available numerical estimate. Recall that $\phi_2 = 1 - e^{-\eta}$.

Haan and Zwanzig (1977)	0.68 ± 0.03
Quintanilla et al. (2000)	0.67637 ± 0.00005

Table 10.4 Estimates of the particle-phase percolation threshold ϕ_{2c} for a system of identical overlapping spheres ($d = 3$). The value given by Rintoul and Torquato (1997a) is the best currently available numerical estimate. Recall that $\phi_2 = 1 - e^{-\eta}$.

Haan and Zwanzig (1977)	0.30 ± 0.02
Rintoul and Torquato (1997a)	0.2895 ± 0.0005

for $d = 2$ than for $d = 3$. We note that for $d = 2$, the numerical estimate of ϕ_{2c} lies well within the rigorous upper and lower bounds obtained by Hall (1985).

What is the effect of particle size distribution on the threshold? It has been rigorously shown that ϕ_{2c} generally depends continuously upon the size distribution and that several alternative definitions of the threshold are equal to one another, provided the radius random variable is bounded (Meester, Roy and Sarkar 1994, Meester and Roy 1996). Some numerical investigations have found that ϕ_{2c} depends weakly on size distribution for both $d = 2$ and $d = 3$ (Pike and Seager 1974, Kertesz and Vicsek 1982, Lorenz et al. 1993) but Phani and Dhar (1984) studied systems of disks of two appreciably different sizes and found a threshold as high as $\phi_{2c} \approx 0.703$. Meester et al. (1994) have conjectured that the monodisperse distribution minimizes the threshold. Piecing together these results, it is natural to ask the following question: Is it possible to achieve a value of ϕ_{2c} arbitrarily close to unity by having a very high concentration of very small spheres and a very low concentration of very large spheres? This is an open question.

Thus far we have been concerned with percolation of the particle phase ϕ_{2c}. However, for $d \geq 3$, the matrix phase (the space exterior to the particles) remains connected for $\phi_2 > \phi_{2c}$ (unlike the two-dimensional case) until it becomes disconnected at a critical value, denoted by $\bar{\phi}_{2c}$, below $\phi_2 = 1$. It turns out that $\bar{\phi}_{2c} \approx 0.97$ for three-dimensional systems of identical overlapping spheres (Kertesz 1981, Elam et al. 1984). Thus, for $d = 3$, overlapping spheres are bicontinuous in the interval $\phi_{2c} \leq \phi_2 \leq \bar{\phi}_{2c}$, where $\phi_{2c} \approx 0.29$ and $\bar{\phi}_{2c} \approx 0.97$. A simulation study for a certain binary mixture of overlapping

Table 10.5 Estimates of the particle-phase percolation threshold ϕ_{2c} for systems of identical overlapping oriented squares and cubes. Recall that $\phi_2 = 1 - e^{-\eta}$.

	Squares	Cubes
Haan and Zwanzig (1977)	0.67 ± 0.01	0.28 ± 0.01
Pike and Seager (1974)	0.67 ± 0.02	

spheres (Rintoul 2000) reveals that $\bar{\phi}_{2c}$ is nearly equal to but statistically different from the monodisperse value.

10.6.2 Nonspherical Overlapping Particles

Using the series-expansion method, Haan and Zwanzig (1977) also estimated percolation thresholds of two-dimensional systems of overlapping oriented equal-sized squares and of three-dimensional systems of overlapping oriented equal-sized cubes. Table 10.5 reports these values along with the numerical estimate for squares by Pike and Seager (1974).

Haan and Zwanzig expressed all of their series expressions for spherical and nonspherical shapes in terms of a dimensionless density variable ρb, where b is just the excluded volume per particle (trivially related to the second virial coefficient of hard-particle systems), in order to make the leading terms identical in each case. This suggested that at the critical number density ρ_c, the quantity $\rho_c\langle v_{ex}\rangle$ is an approximate invariant for overlapping particles of arbitrary shape, where $\langle v_{ex}\rangle$ is some appropriately orientationally averaged exclusion volume (Onsager 1949, Balberg, Anderson, Alexander and Wagner 1984). Note that $\langle v_{ex}\rangle = 4v_1$ for equal-sized circles and oriented squares, and $\langle v_{ex}\rangle = 8v_1$ for equal-sized spheres and oriented cubes, where v_1 is the d-dimensional volume of a single particle.

If $\rho_c\langle v_{ex}\rangle$ were an invariant, then one could estimate the critical reduced density $\eta_c = \rho_c v_1$ (or critical volume fraction $\phi_{2c} = 1 - e^{-\eta_c}$) from a knowledge of the threshold of the prototypical model of d-dimensional overlapping spheres (see Tables 10.3 and 10.4). Comparing Tables 10.3, 10.4, and 10.5, we see that this is approximately the case for d-dimensional spheres and cubes. Other studies indicate that $\rho_c\langle v_{ex}\rangle$ is an approximate invariant for oriented cubes and oriented ellipsoids (Balberg et al. 1984). However, for identical nonspherical objects that are not oriented (e.g., random orientation), there can be significant variation in $\rho_c\langle v_{ex}\rangle$ (Balberg et al. 1984, Xia and Thorpe 1988, Garboczi et al. 1995).

Xia and Thorpe (1988) found that for two-dimensional overlapping systems of identical randomly oriented ellipses with aspect ratio b/a,

$$3.4 \le \rho_c\langle v_{ex}\rangle \le 4.5, \tag{10.65}$$

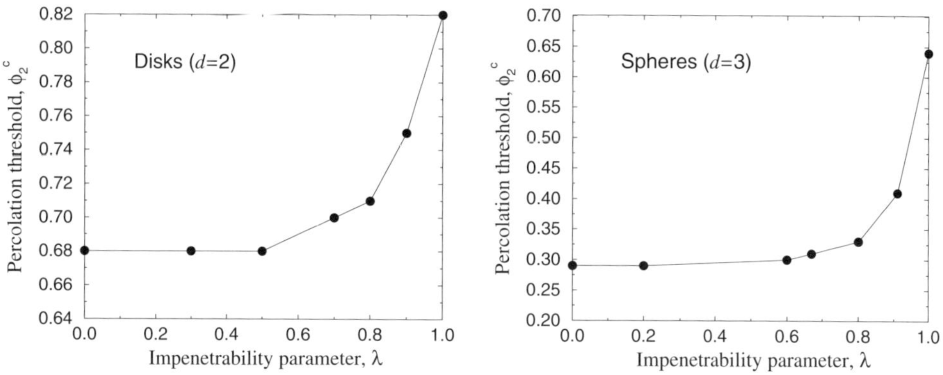

Figure 10.8 Percolation threshold ϕ_2^c versus the impenetrability parameter λ for systems of identical disks (left panel) and identical spheres (right panel) in equilibrium in the cherry-pit model (Lee and Torquato 1988b).

whereas for identical ellipses that can lie only in two orthogonal directions,

$$2.8 \le \rho_c \langle v_{ex} \rangle \le 4.4. \tag{10.66}$$

They also found that their numerical data for the percolation thresholds of these two models were reasonably fit by the empirical formula

$$\phi_{2c} = \frac{18}{19 + 4y}, \tag{10.67}$$

for all values of b/a, where $y = b/a + a/b$. Garboczi et al. (1995) carried out an analogous study for overlapping ellipsoids but concluded that a universal description of ϕ_{2c} for a wide range of aspect ratios was untenable.

10.6.3 Interacting Particle Systems

Let us now consider interacting particle systems, i.e., particles that are spatially correlated due to an interparticle potential. Less is known about the percolation thresholds of such systems.

The simplest examples are d-dimensional spheres in equilibrium in the cherry-pit model (Section 3.1.2). Figure 10.8 plots the percolation threshold ϕ_{2c} versus the impenetrability parameter λ for this model for $d = 2$ and $d = 3$ as obtained from simulations (Lee and Torquato 1988b). We see that the threshold is smallest for fully penetrable spheres ($\lambda = 0$), remains flat for a large range of $\lambda > 0$, and thereafter increases until it achieves its largest value at $\lambda = 1$, corresponding to the maximally random-jammed volume fraction (Section 3.3).

Percolation thresholds have been computed for systems of particles that interact with repulsive as well as attractive forces by Seaton and Glandt (1987) in the case of sticky spheres and by Bug et al. (1985) in the case of permeable objects. The latter study concluded that attractive forces could actually raise the percolation threshold density. However, if the objects are permeable, the relevant parameter is the actual *volume fraction* occupied by the objects, which may not follow the *density* trends (as was discussed in Section 10.3.2). Generally speaking, the effect of attractive forces on *impermeable objects* is to *lower* the percolation-threshold volume fraction.

CHAPTER 11

Local Volume Fraction Fluctuations

One of the most important morphological descriptors of heterogeneous materials is the volume fraction of the phases or, in the case of porous media, the porosity (i.e., the volume fraction of the fluid phase). Although the volume fraction is constant for statistically homogeneous media, on a spatially local level it fluctuates. An interesting question is the following: How does the "local" volume fraction fluctuate about its average value? The answer to this query has relevance to a number of problems, including scattering by heterogeneous media (Debye et al. 1957), the study of noise and granularity of photographic images (O'Neill 1963, Bayer 1964, Lu and Torquato 1990c), transport through porous media (Hilfer 1991, Hilfer 1996), mechanical properties of composites (Ostoja-Starzewski 1993), the properties of organic coatings (Fishman, Kurtze and Bierwagen 1992), and the fracture of composites (Botsis, Beldica and Zhao 1994, Torquato 2000a). It is actually in the context of photographic science that this question of local volume fraction fluctuations was first probed, and here primarily for simple two-dimensional models of photographic emulsions that do not account for impenetrability of the grains (O'Neill 1963, Bayer 1964).

Lu and Torquato (1990a) represented and computed the standard deviation of the *local volume fraction* $\xi(\boldsymbol{x})$ at position $\boldsymbol{x}$ for arbitrary statistically homogeneous two-phase random media in any spatial dimension d. The local volume fraction $\xi(\boldsymbol{x})$ is defined to be the volume fraction of one of the phases, say phase 1, contained in some generally finite-sized "observation window" with position $\boldsymbol{x}$. As illustrated in Figure 11.1, the volume fraction ξ of phase 1 *within a given observation window* is a random variable ranging between 0 and 1, although the macroscopic volume fraction of phase 1 is the constant ϕ_1. The quantity that was specifically studied by Lu and Torquato was the *coarseness* C, defined to be the standard deviation of ξ divided by ϕ_1. The coarseness,

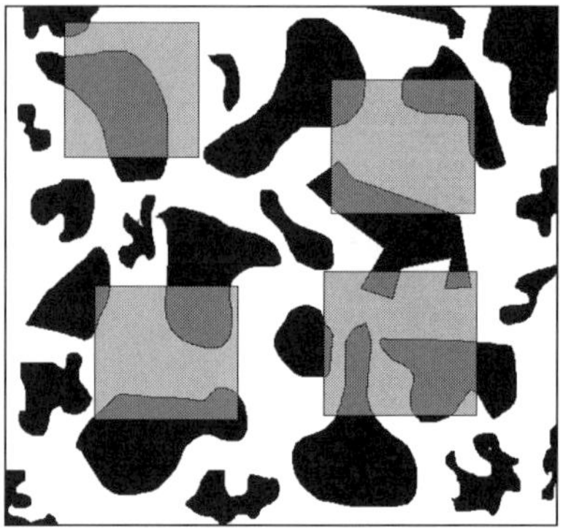

Figure 11.1 A schematic depicting the local volume fraction ξ, defined to be the volume fraction of one of the phases, say phase 1, contained in an observation window. Here square observation windows are shown, but any convenient shape can be used.

for arbitrary d-dimensional two-phase anisotropic media and observation region, can be shown to be related to an integral involving the two-point probability function.

Quintanilla and Torquato (1997b) studied all of the moments of ξ as well as the full distribution of ξ. They developed an expression for the nth moment of ξ for general statistically homogeneous and isotropic media in terms of the n-point probability functions. Alternatively, they also examined the *full distribution* of ξ for several models of random media using analytical and simulation methods.

11.1 Definitions

Consider general two-phase statistically homogeneous media. The volume fraction of phase 1 (or phase 2) fluctuates on a local level. To be specific, consider the volume fraction of phase 1 contained in an arbitrarily shaped *observation region* or *window* Ω_o of d-dimensional volume v_o whose centroid is located at $\mathbf{x}$. As the observation window is moved from point to point in the sample, it is clear that the volume fraction of phase 1 contained in it will fluctuate for finite v_o. Thus, this local volume fraction $\xi(\mathbf{x})$ is a random variable defined to be

$$\xi(\mathbf{x}) \equiv \frac{1}{v_o} \int \mathcal{I}(\mathbf{z})\theta(\mathbf{z} - \mathbf{x})d\mathbf{z}, \tag{11.1}$$

where

$$\theta(\mathbf{x}) = \begin{cases} 1, & \mathbf{x} \in \Omega_o, \\ 0, & \text{otherwise}, \end{cases} \tag{11.2}$$

is the observation window indicator function. Here $\mathcal{I} \equiv \mathcal{I}^{(1)}$ is defined to be the indicator function for phase 1, where we recall that

$$\mathcal{I}(\boldsymbol{x}) \equiv \mathcal{I}^{(1)}(\boldsymbol{x}) = \begin{cases} 1, & \boldsymbol{x} \in \mathcal{V}_1, \\ 0, & \text{otherwise.} \end{cases}$$

It is important to notice that the local volume fraction ξ depends on the *shape* and *size* of the observation window.

In the extreme situations of infinitesimally small and of infinitely large windows, the random variable ξ takes on specific limiting values. In the limit $v_o \to 0$, $\xi(\boldsymbol{x})$ simply becomes the indicator function for phase 1, i.e.,

$$\xi(\boldsymbol{x}) \to \mathcal{I}(\boldsymbol{x}). \tag{11.3}$$

For ergodic media, as $v_o \to \infty$, we have that

$$\xi(\boldsymbol{x}) \to \langle \mathcal{I} \rangle = \phi_1. \tag{11.4}$$

Moreover, the expected value of ξ is clearly given by

$$\langle \xi \rangle = \phi_1. \tag{11.5}$$

A measure of local fluctuations in the volume fraction is given by the *coarseness* C (Lu and Torquato 1990a), defined by

$$C = \frac{\sigma_\xi}{\langle \xi \rangle} = \frac{\sigma_\xi}{\phi_1}, \tag{11.6}$$

where

$$\sigma_\xi^2 = \langle \xi^2 \rangle - \langle \xi \rangle^2 = \langle \xi^2 \rangle - \phi_1^2 \tag{11.7}$$

is the variance associated with $\xi(\boldsymbol{x})$. The coarseness C, a *scaled* standard deviation, provides a quantitative measure of the *nonuniformity of coverage* of the phases for a given observation window size. The dependence of C on v_o is generally nontrivial, but for infinitesimally small and infinitely large windows we respectively have that

$$C = \frac{\sqrt{\phi_1 \phi_2}}{\phi_1} \tag{11.8}$$

and

$$C = 0. \tag{11.9}$$

These limiting relations are easily proved using the definition (11.6) and expressions (11.3) and (11.4).

One can study all of the moments $\langle \xi^n \rangle$ of the local volume fraction or the full distribution of ξ (Quintanilla and Torquato 1997b). The cumulative distribution function of ξ is given by

$$F(x) = \mathcal{P}(\xi \le x). \tag{11.10}$$

We will denote the associated probability density function by $f(x)$, i.e.,

$$f(x) = \frac{dF}{dx}. \tag{11.11}$$

Notice that for systems of particles of fixed size, F will contain a discrete component as well as a continuous component: There is a nonzero probability that the window will be completely devoid of phase 1, completely covered by phase 1, and contain some fixed fraction of phase 1 between 0 and 1. Any discrete contribution to F shows up as a delta-function contribution to the density function f.

11.2 Coarseness

11.2.1 General Formula

Following Lu and Torquato (1990a), we will show that the coarseness C for arbitrary statistically anisotropic d-dimensional media is related to a simple integral over the two-point probability function of phase 1, which we denote without the superscript by $S_2 \equiv S_2^{(1)}$. For the special case of isotropic two-dimensional media ($d = 2$), this formula is equivalent to one derived by O'Neill (1963), who used a different procedure.

From the definitions (11.1) and (11.7), we express the variance for an arbitrarily shaped observation window with some fixed orientation as

$$\sigma_\xi^2 = \frac{1}{\upsilon_o^2}\left\langle \int \mathcal{I}(z)\theta(z-x)dz \int \mathcal{I}(y)\theta(y-x)dy \right\rangle - \phi_1^2. \tag{11.12}$$

We assume that the random medium is statistically homogeneous, and hence the variance is a constant, independent of position. The medium is also assumed to be ergodic. Thus, ensemble and volume averaging commute, enabling us to rewrite the above expression as

$$\sigma_\xi^2 = \frac{1}{\upsilon_o^2}\left[\int dy \int dz \theta(z-x)\theta(y-x)S_2(r)\right] - \phi_1^2, \tag{11.13}$$

where $S_2(r) = \langle \mathcal{I}(y)\mathcal{I}(z)\rangle$ is the two-point probability function for phase 1. The fact that $S_2(r)$ depends on the relative displacement $r = z - y$ follows from homogeneity and enables us to change variables to get

$$\sigma_\xi^2 = \frac{1}{\upsilon_o^2}\left[\int S_2(r)\upsilon_2^{\text{int}}(r; b_o)dr\right] - \phi_1^2, \tag{11.14}$$

where

$$\upsilon_2^{\text{int}}(r; b_o) = \int \theta(y-x)\theta(z-x)dx \tag{11.15}$$

is the intersection volume of two observation regions whose centroids are separated by the displacement r, and b_o denotes all of the geometric parameters associated with the observation window. Dividing (11.15) by υ_o^2 and integrating over r gives

$$\frac{1}{\upsilon_o^2}\int \upsilon_2^{\text{int}}(r; b_o)dr = 1. \tag{11.16}$$

Combination of this relation with (11.6) and (11.14) gives

$$C = \frac{1}{\phi_1 \upsilon_o} \left\{ \int [S_2(\mathbf{r}) - \phi_1^2] \upsilon_2^{\text{int}}(\mathbf{r}; b_o) d\mathbf{r} \right\}^{1/2}. \tag{11.17}$$

This is the desired general expression for the coarseness, valid for d-dimensional statistically anisotropic media of arbitrary topology and given in terms of $S_2(\mathbf{r})$ and $\upsilon_2^{\text{int}}(\mathbf{r}; b_o)$.

Nontrivial information about the geometry of the observation region enters through the intersection volume $\upsilon_2^{\text{int}}(\mathbf{r}; b_o)$. For a d-dimensional *spherical observation window* of radius R_o, we can use the results of Chapter 5 to get $\upsilon_2^{\text{int}}(\mathbf{r}; R_o)$. Using the results from Chapter 7, we ascertain that the intersection volume for d-dimensional rectangular parallelepipeds with sides of lengths $a_1, a_2, \ldots, a_d$ is given by

$$\upsilon_2^{\text{int}}(x_1, \ldots, x_d; a_1, \ldots, a_d) = \prod_{i=1}^{d} (a_i - x_i) \Theta(a_i - x_i), \tag{11.18}$$

where $\Theta(x)$ is the Heaviside step function and $x_1, x_2, \ldots, x_d$ are the Cartesian components of $\mathbf{r}$.

11.2.2 Asymptotic Formula

The quantity $S_2(\mathbf{r}) - \phi_1^2$, which appears in the integrand of (11.17), decays to zero for large r for media without long-range order. We refer to the range over which $S_2(\mathbf{r}) - \phi_1^2$ is nonnegligibly small as the *correlation length* ℓ. In the special case where the characteristic size of the observation region is much larger than ℓ, $\upsilon_2^{\text{int}}(\mathbf{r}; b_o)$ is approximately equal to $\upsilon_2^{\text{int}}(0; b_o) = \upsilon_o$. Thus, (11.17) in such instances yields

$$C = \frac{1}{\phi_1 \upsilon_o^{1/2}} \left\{ \int_{r<\ell} [S_2(\mathbf{r}) - \phi_1^2] d\mathbf{r} \right\}^{1/2}. \tag{11.19}$$

Hence, for large observation windows, we have the asymptotic formula

$$C = K \upsilon_o^{-1/2}, \tag{11.20}$$

where

$$K = \frac{1}{\phi_1} \left\{ \int_{r<\ell} [S_2(\mathbf{r}) - \phi_1^2] d\mathbf{r} \right\}^{1/2} \tag{11.21}$$

is a positive constant that depends upon the volume fraction. The fact that K is positive leads to the integral positivity condition (2.54) on $S_2(\mathbf{r})$. In the limit $\upsilon_o \to \infty$, (11.20) agrees with (11.9).

Equation (11.20) is reminiscent of the *compressibility equation* (3.15) of liquid-state theory, which relates the d-dimensional volume integral involving the *pair correlation function* to density fluctuations in the system. A general treatment of density fluctuations in particle systems has been given by Truskett et al. (1998).

11.2.3 Calculations

The coarseness has been computed for various isotropic distributions of identical d-dimensional spheres of radius R (Lu and Torquato 1990c, Lu and Torquato 1990a). For such models, it is convenient to consider a d-dimensional spherical observation window of radius R_o. Under such conditions, the coarseness expression (11.17) can be simplified as follows:

$$C = \frac{1}{\phi_1}\left\{\frac{d}{\beta^d}\int_0^\infty [S_2(x) - \phi_1^2]\frac{v_2^{\text{int}}(x; R_o)}{v_o} x^{d-1}\,dx\right\}^{1/2}, \tag{11.22}$$

where $\beta = R_o/R$, $x = r/R$ is a dimensionless distance, and

$$v_o = \frac{\pi^{d/2} R_o^d}{\Gamma(1 + d/2)}$$

is the volume of an observation window.

Formula (11.22) was evaluated for the extreme limits of the *cherry-pit* model: fully penetrable spheres and totally impenetrable spheres in equilibrium (Lu and Torquato 1990a). The former is an example of a spatially uncorrelated system and the latter an example of a system with short-range correlations due to exclusion-volume effects. Thus, one can examine the extent to which spatial correlations suppress volume-fraction fluctuations.

The calculation of (11.22) requires the two-point probability functions for the models, which were given in Chapter 5. For the case of d-dimensional fully penetrable spheres of radius R, S_2 is given by (5.6). For impenetrable spheres, the function S_2 is computed from (5.39) using the Percus–Yevick approximation for the radial distribution function [cf. (3.60)].

In Figure 11.2 we display the coarseness for penetrable and impenetrable spheres ($d = 3$) as a function of the sphere volume fraction ϕ_2 for three values of the scaled observation volume v_o/v_1. Here $v_1(R)$ is the volume of a sphere, given by (4.17). For fixed volume ratio v_o/v_1, the coarseness is generally smaller for impenetrable particles than for penetrable particles at the same value of ϕ_2. This effect becomes more pronounced at higher values of the particle volume fraction. Roughly speaking, this is true because exclusion-volume effects associated with impenetrable particles result in a distribution that is *more uniform* than randomly centered particles. For fixed v_o/v_1, the coarseness for fully penetrable particles is a monotonically increasing function of ϕ_2. This functional dependence is in contrast to that of impenetrable spheres, in which C first increases with increasing ϕ_2 for small to moderate ϕ_2, reaches some maximum value, and then decreases with increasing ϕ_2. Moreover, C decreases with increasing volume ratio v_o/v_1 for either penetrable or impenetrable particles.

Lu and Torquato (1990a) showed that increasing the dimensionality for either particle model has the effect of decreasing the coarseness, implying that particle coverage is more uniform in higher dimensions. This point is clearly illustrated in Figure 11.3,

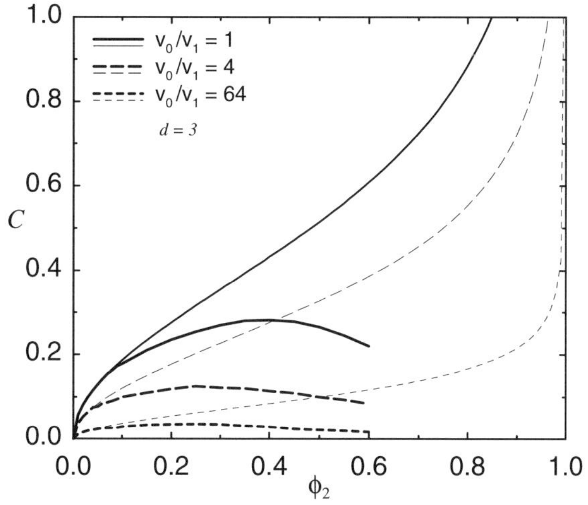

Figure 11.2 The coarseness C versus the particle volume fraction for three different values of the ratio v_o/v_1 for three-dimensional systems of fully penetrable spheres (lighter curves) and of impenetrable spheres (heavier curves) of radius R. Here $v_o = 4\pi R_o^3/3$ and $v_1 = 4\pi R^3/3$.

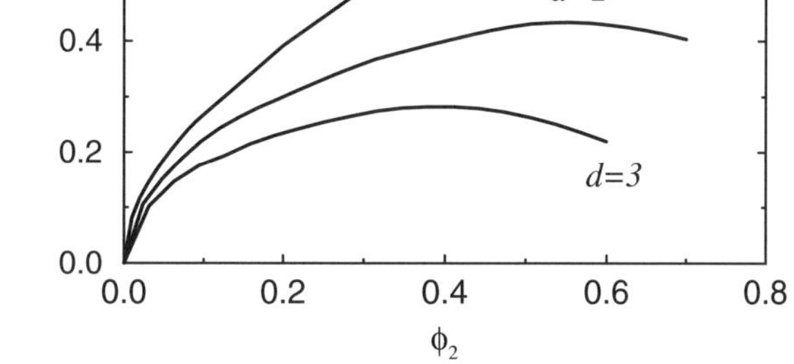

Figure 11.3 Effect of dimensionality d on the coarseness C versus the particle volume fraction ϕ_2 for d-dimensional systems of impenetrable spheres of radius R in equilibrium. The ratio $v_o/v_1 = 1$ is fixed.

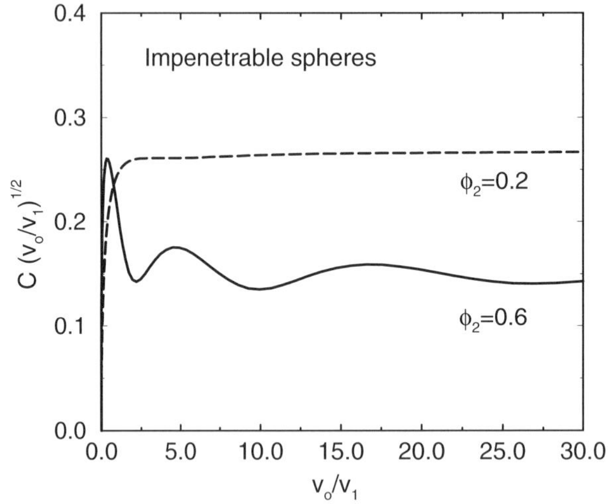

Figure 11.4 The coarseness C, multiplied $(\upsilon_o/\upsilon_1)^{1/2}$, as a function of the scaled observation volume υ_o/υ_1 for three-dimensional systems of impenetrable spheres at $\phi_2 = 0.2$ and 0.6. Here $\upsilon_o = 4\pi R_o^3/3$ and $\upsilon_1 = 4\pi R^3/3$.

where the coarseness for impenetrable particles is plotted as a function of ϕ_2 for $d = 1, 2,$ and 3.

In Figure 11.4 we depict the quantity $C(\upsilon_o/\upsilon_1)^{1/2}$ as a function of the volume ratio υ_o/υ_1 for impenetrable spheres ($d = 3$) in equilibrium for $\phi_2 = 0.2$ and $\phi_2 = 0.6$. Recall that for large observation regions, $C \sim K\upsilon_o^{-1/2}$ [cf. (11.20)]. Thus, $C(\upsilon_o/\upsilon_1)^{1/2}$ approaches the constant $K\upsilon_1^{-1/2}$ for large υ_o. At $\phi_2 = 0.2$, the long-range value of $C(\upsilon_o/\upsilon_1)^{1/2}$ is achieved for relatively small values of the volume ratio ($\upsilon_o/\upsilon_1 \geq 2$). Due to pronounced exclusion-volume effects at $\phi_2 = 0.6$, $C(\upsilon_o/\upsilon_1)^{1/2}$ oscillates about its long-range value of approximately 0.15. Oscillations begin to become noticeable for impenetrable spheres at $\phi_2 = 0.4$ (not shown). The same general trends are observed for impenetrable particles for arbitrary $d > 1$.

11.3 Moments of Local Volume Fraction

Following the arguments used by Lu and Torquato (1990a) to obtain the coarseness formula (11.17), Quintanilla and Torquato (1997b) found the following general formula for the nth moment of $\xi(\mathbf{x})$ of general statistically homogeneous and isotropic two-phase media:

$$\langle \xi^n(\mathbf{x}) \rangle = \frac{1}{\upsilon_o^n} \int S_n(\mathbf{r}_{12}, \ldots, \mathbf{r}_{1n}) \upsilon_n^{\rm int}(\mathbf{r}_{12}, \ldots, \mathbf{r}_{1n}; b_o) d\mathbf{r}_{12} \cdots d\mathbf{r}_{1n}, \tag{11.23}$$

where S_n is the n-point probability function of phase 1 defined by (2.6). Moreover, $v_n^{\text{int}}(\boldsymbol{r}_{12},\ldots,\boldsymbol{r}_{1n};b_o)$ is the *intersection* volume of n observation windows with centroids located at $\boldsymbol{r}_1,\ldots,\boldsymbol{r}_n$ defined by

$$v_n^{\text{int}}(\boldsymbol{r}_{12},\ldots,\boldsymbol{r}_{1n};b_o) = \int d\boldsymbol{r}_{n+1} \prod_{i=1}^{n} \theta(\boldsymbol{r}_{n+1} - \boldsymbol{r}_i). \tag{11.24}$$

The third moment of ξ for fully penetrable rods has been computed by Quintanilla and Torquato (1997b) using (11.23). However, this expression becomes increasingly unwieldy to compute as n increases, even for $d = 1$ or for periodic media (Quintanilla and Torquato 1999).

11.4 Evaluations of Full Distribution

In light of the difficulty in evaluating higher moments of ξ from (11.23) (either analytically or numerically), we now turn to evaluation of the full distribution function F of ξ, given by (11.10), which, at the very least, is easily obtained via numerical simulations. Quintanilla and Torquato (1997b) evaluated F for a wide range of window sizes and volume fractions for the following five models consisting of particles of uniform size:

1. one-dimensional, fully penetrable rods
2. one-dimensional, RSA totally impenetrable rods
3. two-dimensional, fully penetrable aligned squares
4. two-dimensional, RSA totally impenetrable disks
5. three-dimensional, fully penetrable aligned cubes

In all cases, they considered a d-dimensional *cubical* observation window of side of length L. Recall from Chapter 3 that RSA refers to the nonequilibrium random sequential addition process.

A natural question to ask is, under what conditions does a *central limit theorem* exist; i.e., when does ξ follow a normal (Gaussian) distribution? Their primary result is that for *sufficiently large windows* (quantified below), the distribution of ξ for all of the models can be reasonably approximated by a normal distribution whose density function is

$$g(x) = \frac{1}{\sqrt{2\pi\sigma_\xi^2}} \exp\left[-\frac{(x-\phi_1)^2}{2\sigma_\xi^2}\right], \tag{11.25}$$

with variance σ_ξ^2 given by (11.7). Therefore, for sufficiently large windows, knowledge of the first two moments is sufficient to reasonably estimate the distribution of ξ, i.e., information up to the level of the coarseness C is sufficient. Although this result had been rigorously proven for general systems of fully penetrable particles (Hall 1988), it was not determined that it applies to systems with spatial correlations, such as impenetrable particles, until the aforementioned more recent work.

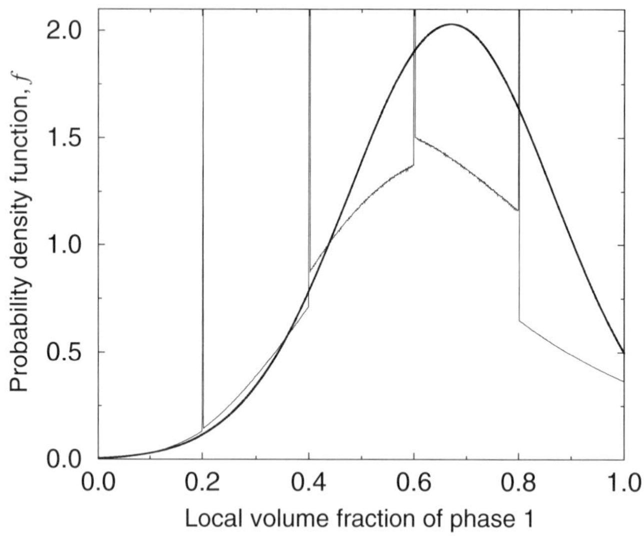

Figure 11.5 The probability density function f of ξ with $L = 5$ for a system of fully penetrable rods of unit length at $\eta = 0.4$ (lighter curve), and a normal distribution (heavier curve) with σ_ξ^2 given by parameters given by (11.7). Note the "spikes" discussed in the text.

For fully penetrable rods, Quintanilla and Torquato (1997b) simplified and numerically inverted an analytical result for the Laplace transform of F due to Domb (1947). For all of the other four models, they evaluated F via Monte Carlo simulations. Some of their results are highlighted below.

In Figure 11.5 the probability density function $f(\xi)$ for fully penetrable rods and a normal (Gaussian) distribution are depicted for $\eta = 0.4$ and $L = 5$ (where $\eta = \rho v_1$ is the usual reduced density). As expected, there are "spikes" in the probability density function at $x = 0, 1/5, \ldots, 1$, since there are analytically known positive probabilities that a given observation window will lie entirely in phase 1, phase 2, or contain some number of nonoverlapping rods (Quintanilla and Torquato 1997b). While ξ has domain [0,1], the normal distribution is defined on the entire real line. However, these distinctions diminish as the window size increases. The discontinuities in f are seen in higher dimensions, albeit less pronounced.

Figure 11.6 shows the associated cumulative distribution function F of ξ obtained from simulations and by numerical inversion at three different window lengths and $\eta = 0.4$. The discontinuities in F have magnitudes that are known analytically (Domb 1947, Quintanilla and Torquato 1997b). The graphs of F approach the normal distribution as L increases: While the graph of F for $L = 5$ is clearly not normal, the graph of F for $L = 50$ cannot be distinguished from a normal distribution on the scale of Figure 11.6.

To assess quantitatively how close F is to a normal distribution, one can measure the *maximum separation* between the sample distribution and the normal distribution, that is,

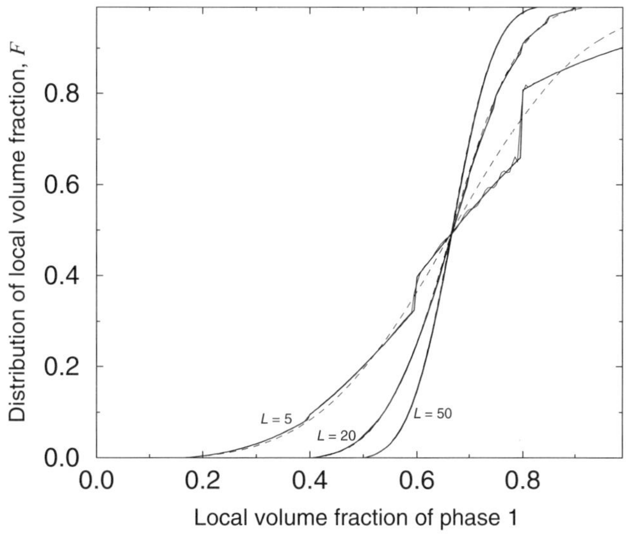

Figure 11.6 The cumulative distribution function F of ξ for a system of fully penetrable rods of unit length with reduced density $\eta = 0.4$. The windows are intervals with lengths $L = 5$, $L = 20$, and $L = 50$. The thick solid lines are simulation data, the thin solid lines are values obtained from numerically inverting the Laplace transform of ξ, and the dashed lines are normal approximations.

$$\delta = \max_{x \in [0,1]} \left[F(x) - \Phi(x) \right], \tag{11.26}$$

where

$$\Phi(x) = \int_{-\infty}^{x} g(t)dt \tag{11.27}$$

is the cumulative distribution function associated with the normal density function (11.25). Clearly, the maximum separation δ is dependent on the window length L and the particle volume fraction ϕ_2.

In Figure 11.7 we plot level curves of δ, defined by (11.26), on the ϕ_2–$(L^2/v_1)^{1/2}$ plane for the two-dimensional systems of fully penetrable squares and RSA totally impenetrable disks. Convergence to the normal distribution is faster in two dimensions than in one. Moreover, the convergence of F for RSA disks is quicker than for fully penetrable squares. This is not surprising, since the variance of ξ for totally impenetrable particles is less than that for fully penetrable particles (Lu and Torquato 1990a). A reasonable empirical condition for closeness to a normal distribution appears to be $\delta \leq 0.02$. Figure 11.7 shows that this occurs for RSA totally impenetrable disks for $(L^2/v_1)^{1/2} \geq 7$ (i.e., $L \geq 12.4$, for disks of unit radius) when $\phi_2 = 0.1$, but for $L \geq 3$ when $\phi_2 = 0.4$.

For all of the one-, two-, and three-dimensional models considered by Quintanilla and Torquato (1997b), the distribution numerically tends to a normal distribution as the size of the window increases, and hence the standard deviation or coarseness C

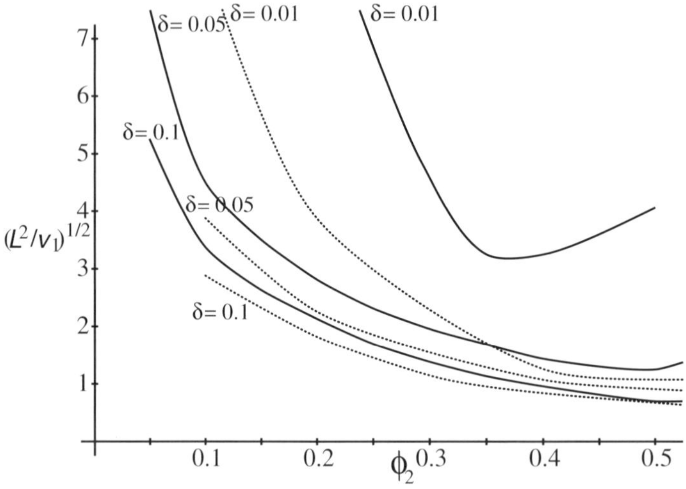

Figure 11.7 Level curves of the maximum separation δ, given by (11.26), for the two-dimensional systems of fully penetrable squares (solid lines) and RSA totally impenetrable disks (dotted lines).

provides a good estimate of the fluctuations. The convergence to a normal distribution increases with the dimension d.

We note that cumulative distribution functions can be computed analytically for periodic heterogeneous media (Quintanilla and Torquato 1999). Although fluctuations are suppressed in periodic media relative to random media, they generally do not vanish.

CHAPTER 12

Computer Simulations, Image Analyses, and Reconstructions

The evaluation of microstructural correlation functions via computer simulations is a two-step process. First, one must generate realizations of the random medium using a particular simulation method. Second, one must sample each realization for the correlation functions of interest and then average over all realizations to get the ensemble-averaged correlation functions. Similar sampling techniques can be used to extract correlation functions from images of real heterogeneous materials.

There is a variety of available simulation techniques for generating realizations of random media. These include *Monte Carlo* methods, *molecular dynamics*, and the *Langevin* method, to mention but a few. Each of the aforementioned techniques can be used to generate Gibbs (equilibrium) as well as either time-independent (quenched) or time-dependent nonequilibrium ensembles. We will focus on time-independent ensembles.

For equilibrium ensembles, the Monte Carlo method [known as the Metropolis scheme in this case (Metropolis, Rosenbluth, Rosenbluth, Teller and Teller 1953)] provides a means of performing a random walk through configuration space. (Note that this walk does not correspond to following the dynamics of the system.) Sequences of configurations of a system of N interacting spherical particles, for example, are generated by successive random displacements and random reorientations (in the case of nonspherical particles). The acceptance or rejection of a configuration is such that it ensures, asymptotically, that the configurational space is sampled according to the probability law appropriate to a particular ensemble.

In contrast to Monte Carlo techniques, the method of molecular dynamics is a deterministic means of generating trajectories in phase space by numerical integration of Newton's equations of motion for a classical dynamical system. The time step interval depends on a number of factors, including the temperature and density of the system as well as the particle masses and potential energies. In the simulation of liquid argon, for example, the time step is typically of order 10^{-14} seconds, and the trajectories are followed for 10^4 to 10^6 steps, depending on the properties of interest. We note that accessing long-time dynamics near the *glass transition* (Zallen 1983, Debenedetti 1996), where structural relaxation times are on the order of 10^4 seconds, is obviously beyond current computer capabilities.

A potential problem with the molecular-dynamics method is that the deterministic evolution of the equations of motion of the system may not always sample the states available to it. A method that has been used to circumvent this problem introduces randomness at every step in the evolution by, for example, integrating the so-called first-order *Langevin* equation of motion. This amounts to introducing a "noise" term in the equations of motion. The Langevin method is more likely to break away from a local region of the configuration space in which the molecular-dynamics method would spend a large amount of time. The descriptions of the molecular-dynamics and Langevin techniques are beyond the scope of this book. The interested reader is referred to the books of Hansen and McDonald (1986), Allen and Tildesley (1987), Binney et al. (1992), and Frenkel and Smit (1996) for a comprehensive discussion of all of the aforementioned methods.

We describe the generation of equilibrium as well as nonequilibrium realizations of random media via the Monte Carlo technique. This is followed by a discussion on how to sample for statistical measures of interest for given realizations. Sampling methods to extract correlation functions from digitized images of heterogeneous materials are then described. Finally, we discuss techniques to reconstruct or construct realizations of random media from a knowledge of limited microstructural information.

12.1 Monte Carlo Simulations

12.1.1 Introduction

We wish to find the area A of an irregularly-shaped region, such as the one shown in Figure 12.1. A Monte Carlo (MC) estimate of A involves randomly placing a large number of points anywhere in the rectangle of area A_{rect} and recording the fraction of points that lie in A, i.e.,

$$A \approx \frac{N_{hits}}{N} A_{rect}.$$

For small N, the estimate is poor, but the estimate becomes exact as $N \to \infty$. Indeed, the error in the estimate can be shown to behave like $N^{-1/2}$.

The same idea can be used to estimate integrals of functions. In one dimension, the integral

$$I = \int_a^b f(x)dx \tag{12.1}$$

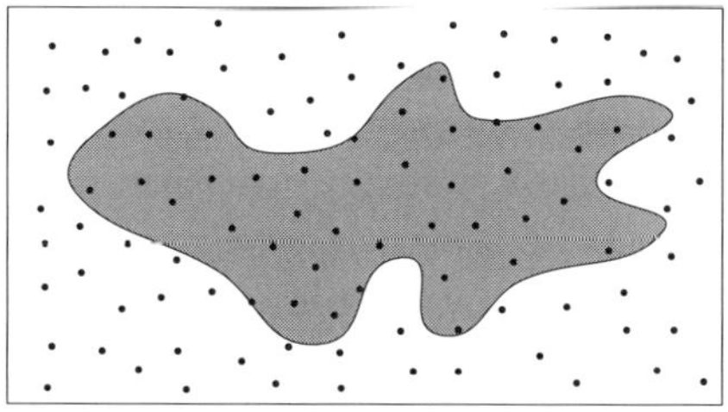

Figure 12.1 A Monte Carlo (MC) estimate of the area A of an irregularly shaped (shaded) region. A_{rect} is the area of the rectangle.

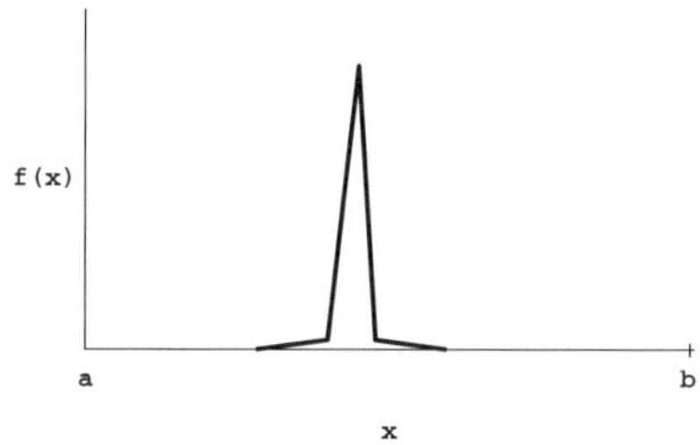

Figure 12.2 Example of a function for which a uniform distribution of random numbers in the interval $[a, b]$ does not work well, unless N is huge.

of the function f can be written approximately as

$$I \approx (b - a)\langle f \rangle, \tag{12.2}$$

where $\langle f \rangle$ is the *unweighted* average of f over the interval $[a, b]$,

$$\langle f \rangle = \frac{1}{N} \sum_{i=1}^{N} f(x_i),$$

and the x values are uniformly distributed random numbers from the interval $[a, b]$. However, this method will not work well for functions such as the one depicted in Figure 12.2. Here N must be very large to achieve reasonable accuracy.

12.1.2 Importance Sampling

Importance sampling methods improve quadrature efficiency by sampling more often in regions of space that make important contributions to the integral [e.g., where $|f(x)|$ is large or rapidly varying]. How then should we distribute our sampling through (configuration) space? Consider rewriting the previous integral (with $a = 0$ and $b = 1$ for convenience) as

$$I = \int_0^1 P(x)\frac{f(x)}{P(x)}\,dx,$$

where $P(x)$ is an arbitrary probability density function. We can rewrite this integral in terms of the nondecreasing cumulative distribution function

$$Q(x) = \int_0^x P(y)dy,$$

for which $Q(0) = 0$ and $Q(1) = 1$, as follows:

$$I = \int_0^1 \frac{f[x(Q)]}{P[x(Q)]}\,dQ = \left\langle \frac{f}{P} \right\rangle.$$

The notation $x(Q)$ indicates that if we consider Q as the integration variable, then x must be expressed in terms of Q. The next step is to generate N random values of Q uniformly distributed in the interval $[0, 1]$, yielding

$$I \approx \frac{1}{N}\sum_{i=1}^{N} \frac{f[x(Q_i)]}{P[x(Q_i)]}.$$

In summary, to integrate a function f with a density function P requires sampling from the nonuniformly distributed random numbers $x(Q)$, where Q is obtained from a uniform distribution in the interval $[0, 1]$. Note that if we choose $P(x)$ to be uniform, i.e., $P(x) = 1/(b - a)$, we recover the aforementioned "brute force" MC sampling. However, one may choose $P(x)$ by minimizing the variance given by

$$\sigma_I^2 = \frac{1}{N}\left[\left\langle \left(\frac{f}{P}\right)^2 \right\rangle - \left\langle \frac{f}{P} \right\rangle^2\right].$$

Thus, the standard deviation (a measure of the error) still behaves like $N^{-1/2}$, but we would ideally want to choose f/P to be a constant.

Consider, for example, the integral

$$I = \int_0^1 \frac{4}{1 + x^2}dx = \pi. \tag{12.3}$$

Table 12.1 summarizes the results of the MC method applied to this integral (along with the associated standard deviations) for two different probability densities P_1 and P_2 as well as various values of N. An evaluation of the integral using an N-step trapezoidal rule is also shown for comparison. The error σ_{I_3} for the trapezoidal rule is just the absolute value of the difference between the exact solution and the quadrature result.

Clearly, the density function P_2 is a better weight than P_1, but they are both inferior to the trapezoidal rule for an equal number of steps. However, the real advantage of the MC method comes in the evaluation of higher-dimensional integrals, such as those integrals over the phase space of an N-particle system in statistical mechanics. Even a crude mesh of 10 points per dimension for a 100-particle system requires 10 function evaluations for each of the 300 coordinates, i.e., 10^{300} function evaluations. This rules

Table 12.1 Monte Carlo evaluation of integral (12.3) using two different probability density functions, for various values of N. The associated standard deviations are σ_{I_1} and σ_{I_2}. Included are the results of the trapezoidal-rule evaluation and the absolute error σ_{I_3} relative to the exact result.

	$P_1 = 1$		$P_2 = \frac{1}{3}(4-2x)$		Trapezoidal rule	
N	I_1	σ_{I_1}	I_2	σ_{I_2}	I_3	σ_{I_3}
10	3.1390	0.2045	3.1414	0.0254	3.1399	0.0017
20	3.1398	0.1451	3.1414	0.0180	3.1412	0.0004
50	3.1418	0.0912	3.1416	0.0114	3.1415	6.7×10^{-5}
100	3.1407	0.0643	3.1416	0.0081	3.1416	1.7×10^{-5}
200	3.1408	0.0456	3.1415	0.0057	3.1416	4.2×10^{-6}
500	3.1411	0.0286	3.1416	0.0036	3.1416	6.7×10^{-7}
1000	3.1413	0.0204	3.1416	0.0026	3.1416	1.7×10^{-7}
2000	3.1414	0.0145	3.1416	0.0018	3.1416	4.2×10^{-8}
5000	3.1415	0.0092	3.1416	0.0011	3.1416	6.7×10^{-9}

out the use of a grid-type quadrature technique for statistical-mechanical applications, since it requires the age of the universe to complete!

The error still decreases like $N^{-1/2}$ for the MC calculation for any dimension. However, if we use the trapezoidal rule in d dimensions and wish to make N function evaluations the error is of order $N^{-2/d}$.

12.2 Metropolis Method for Gibbs Ensembles

12.2.1 Markov Chain

The Metropolis scheme is a means of generating *Gibbs* or *equilibrium* ensembles of systems of interacting particles, spins, etc. For specificity, consider a *canonical ensemble* with a fixed number N of interacting particles, fixed system volume V, and fixed temperature T. The instantaneous particle positions are denoted by $\mathbf{r}^N \equiv \{\mathbf{r}_1, \mathbf{r}_2, \ldots, \mathbf{r}_N\}$. Recall that the probability density function associated with finding a particular configuration of N particles interacting with a total potential energy $\Phi_N(\mathbf{r}^N)$ is given by

$$P_N(\mathbf{r}^N) = \frac{\exp[-\beta\Phi_N(\mathbf{r}^N)]}{Q_N}, \tag{12.4}$$

where $Q_N = \int e^{-\beta\Phi_N(\mathbf{r}^N)} d\mathbf{r}^N$ is the configurational partition function, $\beta = 1/(kT)$, and $d\mathbf{r}^N \equiv d\mathbf{r}_1 \cdots d\mathbf{r}_N$ (see Section 3.1). Often, we are interested in computing the average of some many-body function $f(\mathbf{r}^N)$:

$$\langle f(\mathbf{r}^N)\rangle = \int f(\mathbf{r}^N) P_N(\mathbf{r}^N) d\mathbf{r}^N$$

$$= \int f(\mathbf{r}^N) \frac{e^{-\beta \Phi_N(\mathbf{r}^N)}}{Q_N} d\mathbf{r}^N.$$

Using importance sampling, we can write

$$\langle f \rangle = \left\langle \frac{f P_N}{P} \right\rangle.$$

For most f, the integrand will be significant where P_N is significant. Hence, we make the Metropolis choice $P = P_N$, yielding

$$\langle f \rangle = \langle f \rangle_{trials}.$$

Allen and Tildesley (1987) discuss other choices.

The stumbling block now is that P_N involves the partition function Q_N, which cannot be evaluated with current computers. Metropolis et al. (1953) proposed a procedure that bypasses the need to evaluate Q_N by using Markov chains to generate a sequence of random states such that (by the end of the simulation) each state has occurred with the appropriate probability specified by P_N.

A *Markov chain* is a rule for randomly generating a new configuration of a system from the present one, independently of past configurations. In considering a Markov chain of states (drawn from a finite set of outcomes), it is useful to speak of the states of the chain occurring at "times" $t, t+1, t+2, \ldots$. We define the *transition probability* $p(\alpha \to \alpha')$ to be the probability per unit time that if the system is in state α, it will make a transition to state α' in the next time step. The fact that the system must go somewhere is expressed by

$$\sum_{\alpha'} p(\alpha \to \alpha') = 1.$$

Hence, the probability $P(\alpha')$ of finding the system in state α' is

$$P(\alpha') = \sum_{\alpha} P(\alpha) p(\alpha \to \alpha').$$

If we follow the first-order kinetics associated with $p(\alpha \to \alpha')$, we find that the time rate of change of P, denoted by $\dot{P} \equiv dP/dt$, is given by

$$\dot{P}(\alpha) = \sum_{\alpha'} \left[-P(\alpha) p(\alpha \to \alpha') + P(\alpha') p(\alpha' \to \alpha) \right]. \tag{12.5}$$

Equations of this type are referred to as *master equations*.

At equilibrium in the canonical ensemble, we have that

$$\frac{P(\alpha')}{P(\alpha)} = e^{-\beta \Delta \Phi_N(\alpha \to \alpha')}$$

and $\dot{P}(\alpha) = 0$. This last result combined with the rate equation (12.5) yields the *microreversibility* or *detailed balance* condition

$$\frac{p(\alpha \to \alpha')}{p(\alpha' \to \alpha)} = \frac{P(\alpha')}{P(\alpha)} = e^{-\beta \Delta \Phi_N(\alpha \to \alpha')}. \tag{12.6}$$

Provided that a trajectory algorithm obeys this condition, the statistics acquired from the trajectory will coincide with those of the equilibrium canonical ensemble.

Of course, one can perform MC simulations in other Gibbsian ensembles. Common among these are the *grand canonical* ensemble and the *isothermal–isobaric* ensemble (Allen and Tildesley 1987).

12.2.2 Algorithm

The above results lead to the following Metropolis algorithm for the canonical ensemble:

1. Generate an initial configuration $\alpha \equiv \boldsymbol{r}^N$ of N particles in a fixed volume V at temperature T and calculate the energy $\Phi_N(\alpha)$.
2. Move a particle by displacing it along each axis by amounts randomly and uniformly distributed in the interval $[-\delta, \delta]$, where δ is the maximum step size. Then compute the new energy $\Phi_N(\alpha')$ and change in energy $\Delta\Phi_N = \Phi_N(\alpha') - \Phi_N(\alpha)$.
3. Accept the move from $\alpha \to \alpha'$ with probability

   $$p(\alpha \to \alpha') = \min\left[1, e^{-\beta\Delta\Phi_N}\right]. \tag{12.7}$$

   In words, if the energy change $\Delta\Phi_N$ is negative, the move is always accepted. If $\Delta\Phi_N \geq 0$, then the move is accepted with probability $e^{-\beta\Delta\Phi_N}$ (comparing this to a random number taken from $[0, 1]$).
4. Move each of the remaining particles sequentially according to this prescription.
5. Repeat this process until equilibrium is achieved (see discussion below). Then one can compute configurational averages.

12.2.3 Practical Implementation

The magnitude of the displacement $|\Delta \boldsymbol{r}|$ must be chosen carefully. If it is too large, most steps will be rejected and the system will evolve very slowly. Conversely, if the displacement is too small, the system will again change very slowly. Little is known about the optimum value of the maximum step size δ for general situations. Ultimately, δ should be adjusted so that phase space is explored as efficiently as possible, i.e., particles move the furthest on average per trial move. A rough rule of thumb is to choose δ so that about half of all the trial moves are rejected. A similar criterion may be used for trial reorientations for nonspherical particles.

In practice, many MC cycles (i.e., total number of trial moves per particle) are required in order to drive the system away from the biased initial configuration to equilibrium. Equilibrium is achieved when average measures, such as the pair correlation function, *do not change* (in a statistical sense) *over time*. Figure 12.3 shows three possible starting configurations in two dimensions: a square array, a hexagonal array,

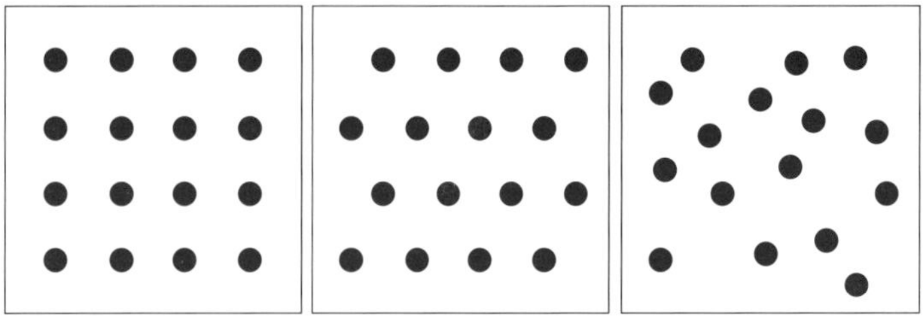

Figure 12.3 Three possible starting configurations in two dimensions.

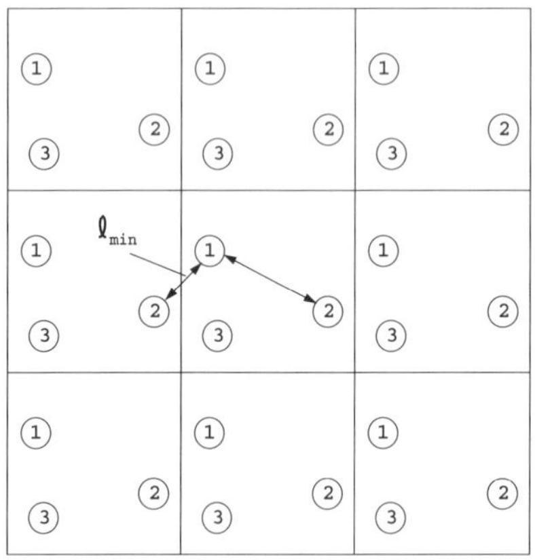

Figure 12.4 Periodic boundary conditions. Here ℓ_{min} represents the minimum image distance between particles 1 and 2.

and a random array. Depending on the system parameters, one initial configuration may be more advantageous to use than the others. For example, at low densities, any of the three initial configurations can be used. In this case, the square array would be used, since it is the easiest to generate. At intermediate densities, the random system will enable one to get to equilibrium fastest. To attain very high densities, the hexagonal array must be used, since it is the most efficient way of packing identical disks in two-dimensional space.

Usually, the number of particles used in simulations is small, e.g., $10 \leq N \leq 100{,}000$. In order to minimize *boundary* effects due to the finite size of the system, *periodic boundary conditions* are typically employed (see Figure 12.4). This means that the unit cell (e.g., cubic box) containing the particles is replicated throughout space to form an infinite lattice. In the course of the simulation, as a particle moves in the central cell or box, its periodic image in each of the neighboring cells moves in exactly the same way. Therefore, as a particle leaves the central cell, one of its images will enter through the opposite face. In two dimensions, movement in the central cell is topologically equivalent to movement on the surface of a torus.

12.2.4 Hard Spheres

To illustrate the Metropolis algorithm, we will outline the key steps to generate canonical equilibrium ensembles of N hard disks of diameter D in a square cell of side of length L at temperature T. The generalization to other space dimensions is obvious. If the particle volume fraction ϕ_2 is specified, then the box length L is determined from the relation

$$\phi_2 = \rho \frac{\pi D^2}{4} = \frac{N}{L^2}\frac{\pi D^2}{4}$$

in units of the diameter. Similarly, one can use this formula to find the diameter D in units of the box length L. Thus, one can choose either L or D as the unit of length in the simulation.

First, an initial configuration $\boldsymbol{r}^N \equiv \{(x_1,y_1),\ldots,(x_i,y_i),\ldots,(x_N,y_N)\}$ of the system is chosen, where the pair (x_i,y_i) gives the Cartesian coordinates for the center of particle i. One attempts to move each of the N disks. A new position for the particle under consideration is chosen by two random numbers Δx and Δy in the range $[-\delta,\delta]$, where the maximum step size δ is adjusted to give about 50% acceptance of the new position. The possible new position is then $(x+\Delta x, y+\Delta y)$.

For the hard-sphere system, the energy change of a move will be either 0 (always accepted) or ∞ (always rejected) according to relation (12.7), and thus the acceptance probability is independent of temperature T. Therefore, the acceptance step of the calculation reduces to checking that a displacement does not cause overlap between spheres. Two particles i and j do not overlap if their separation distance r_{ij} obeys the condition

$$r_{ij}^2 = (x_i - x_j)^2 + (y_i - y_j)^2 > D^2.$$

If the attempted move causes an overlap, the particle is not moved. Each of the remaining particles is in turn moved according to this prescription. The number of MC cycles required to achieve equilibrium is determined as stated in Section 12.2.3.

Periodic boundary conditions are employed throughout the simulation. For example, if the center of particle i changes from (x_i,y_i) to $(x_i+\Delta x_i, y_i)$ and if $x_i+\Delta x_i$ is a position outside the cell, then the new position is taken to be $(x_i+\Delta x_i \pm L, y_i)$. A realization of an equilibrium system of hard disks is shown in Figure 12.5.

Finally, a word of caution is in order regarding simulating hard-disk or hard-sphere systems at densities above the *freezing point* η_f. One must be aware that above η_f, the system undergoes a phase transition to a stable ordered state (see Section 3.3). But for a range of densities above η_f, it is possible for the system to get trapped in any number of metastable states (with varying degrees of disorder), especially in three dimensions. Indeed, one may desire the system to lie along the metastable extension

Figure 12.5 A realization of an equilibrium system of hard disks with an area fraction $\phi_2 = 0.4$. Note that periodic boundary conditions were employed.

of the stable disordered branch. Thus, above the freezing point, one should monitor certain measures of *order* [e.g., the parameter Q_6 defined by (3.85)] to ensure that the system is either in a stable ordered state or in a disordered state along the metastable extension (Rintoul and Torquato 1996b).

12.2.5 Other Particle Systems

Of course, one can generate equilibrium particle systems with other types of interaction potentials using the Metropolis scheme. For example, the cherry-pit model with impenetrability parameter λ (see Chapter 3) is generated in the same way as the aforementioned hard-sphere system but with a nonoverlap condition based on the interparticle separation distance of λD. The Metropolis scheme, however, is not an efficient way to generate the fully-penetrable-sphere limit ($\lambda = 0$). For the special value $\lambda = 0$, it is easier to randomly and sequentially place particles in the unit cell without regard to overlap (see Figure 12.6), the number density ρ being connected to the sphere volume fraction ϕ_2 via relation (4.19).

Attractive interactions can also be simulated with appropriate modifications of the rules given above for hard spheres. An example is an equilibrium ensemble of *sticky* spheres (Chapter 3), which is depicted in Figure 12.6. The extension of the Metropolis method to particles with a polydispersivity in size is straightforward. In the case of particles with nonspherical shape, MC particle moves involve both translations and reorientations.

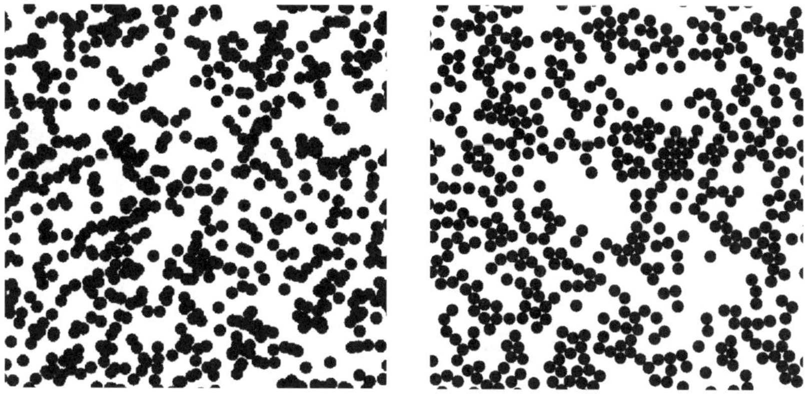

Figure 12.6 Left panel shows a realization of a system of overlapping disks ($\lambda = 0$) with an area fraction $\phi_2 = 0.4$. Right panel shows a realization of a sticky-disk system with a disk area fraction $\phi_2 = 0.4$ and a stickiness parameter $\tau = 0.05$. Both systems are generated using periodic boundary conditions.

12.2.6 Cell Models

Equilibrium realizations of cell models (Chapter 8) can be generated using the Metropolis scheme. The Ising model of a magnetic system is without a doubt the most-studied model of this type (see Section 8.1.5). The temperature T is generally a nontrivial parameter of the ensemble in light of the fact that the Hamiltonian (energy) is given by (8.49). Figure 12.7 compares two different two-dimensional realizations of the Ising model on a square lattice: one in the high-temperature limit ($T \to \infty$) and the other at a finite temperature above but close to the critical temperature T_c. The system near T_c has larger clusters, reflecting the longer-range correlations that are present. Recall that the high-temperature limit of the Ising model is exactly the same as the random checkerboard model at a volume fraction of 50%. The Metropolis scheme is not an efficient way to generate the special limit $T \to \infty$. Rather, the random checkerboard, at *any* volume fraction of the black phase ϕ_2, is easily constructed by first tessellating the unit square into smaller square cells and then coloring each cell black (white) with probability ϕ_2 (ϕ_1), i.e., according to whether a random number in the interval $[0, 1]$ is smaller (larger) than ϕ_2.

12.3 Methods for Generating Nonequilibrium Ensembles

Since there is an infinite number of different nonequilibrium ensembles that one can generate, we focus our discussion here on two simple systems: disks distributed according to the random sequential addition (RSA) process and those distributed according

Figure 12.7 Two snapshots of simulations for the two-dimensional Ising model in the absence of an external field. Here black and white phases are associated with up and down spins, respectively. The left and right panels show systems in which $T = \infty$ and $T = 1.1T_c$, respectively. The system close to T_c has larger clusters.

to the diffusion-limited aggregation (DLA) process. Both systems are generated using MC methods.

The RSA process, discussed in Chapter 3, is easily generated on a computer. One begins with an empty cubic box. Then, using a random number generator (see the subroutine in Appendix A), one places particles randomly and sequentially in the box. If a particle can be placed into the box without overlapping existing particles, then it remains fixed at this location throughout the simulation. If a particle overlaps another existing particle in the box, then another attempt is made until a nonoverlapping location can be found. Periodic boundary conditions are typically employed. The saturation limit is achieved in principle only after an infinite amount of time. The left panel of Figure 12.8 shows a realization of RSA hard disks very near the saturation limit of $\phi_2 \approx 0.55$.

Many clusters in nature grow by aggregation processes, where one particle after another comes into contact with a cluster that grows into a larger cluster (Family 1993). In the diffusion-limited aggregation (DLA) process of Witten and Sander (1981), particles undergo Brownian motion toward the cluster, and stick irreversibly with the cluster when they come into contact with it. Such *fractal* clusters are easily "grown" on a computer. First, a seed particle is placed at an origin. Then, one at a time, random walkers are released from some fixed radius from the origin. When one of these particles makes contact with the aggregate, the next particle is released. The diffusive motion of the particles creates a self-similar cluster, such as the one pictured in Figure 12.8. The fractal dimensions of two- and three-dimensional DLA clusters are approximately 1.71 and 2.5, respectively (Meakin 1988).

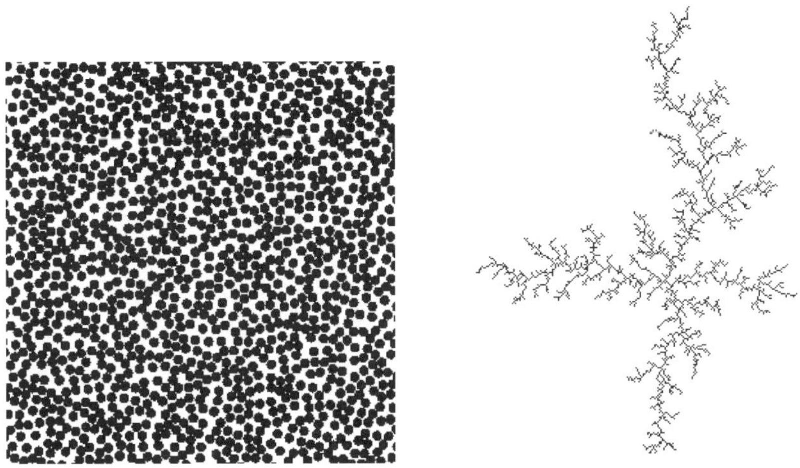

Figure 12.8 Two examples of nonequilibrium systems of disks. The left panel shows a realization of an RSA array of hard disks (with periodic boundary conditions) very near the saturation limit of $\phi_2 \approx 0.55$. The right panel shows a DLA aggregate consisting of many disks.

Note that the random-field models discussed in Chapter 8 generally represent nonequilibrium realizations of heterogeneous media. There we describe computational methodologies to generate realizations of such models.

12.4 Sampling in Particle Systems

Having generated realizations of the random medium, one can now sample the realizations for microstructural descriptors of interest. Here we describe how to compute such statistical quantities.

12.4.1 Radial Distribution Function

For particle systems, the most basic statistical descriptor is the two-particle probability density function or equivalently, for isotropic systems, the radial distribution function $g_2(r)$ (see Section 3.1). For simplicity, consider realizations of statistically isotropic systems of identical interacting d-dimensional spheres at number density ρ. We can obtain $g_2(r)$ from simulations by generating a histogram for the ensemble-average number of particles $n(r)$ contained in a concentric shell of finite thickness ("bin" width) Δr at radial distance r from a reference particle (see Figure 12.9). The radial distance r is assumed to be midway between the inner radius $r - \Delta r/2$ and outer radius $r + \Delta r/2$ of the shell. From the definition of the radial distribution function for a shell of infinitesimal thickness dr given in Section 3.1, we can write for a shell of finite thickness Δr that

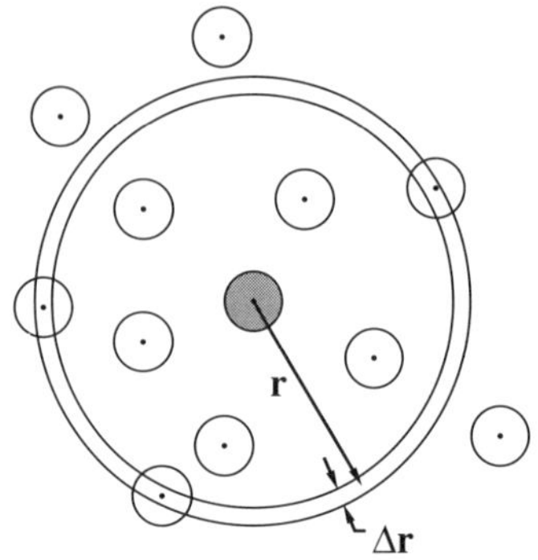

Figure 12.9 Sampling for the radial distribution function.

$$g_2(r) = \frac{n(r)}{\rho v_{shell}(r)}, \tag{12.8}$$

where v_{shell} is the volume of the d-dimensional shell, given by

$$v_{shell}(r) = v_1(r)\left[\frac{(r + \Delta r/2)^d - (r - \Delta r/2)^d}{r^d}\right],$$

and $v_1(r)$ is the volume of a d-dimensional sphere of radius r given by (4.17). The calculation thus reduces to finding the ensemble average $n(r)$.

Finding $n(r)$ is easily done in a simulation by computing the minimum image separations r_{ij} of all pairs of particles i and j from all of the generated configurations (realizations). [The speed of these calculations can be improved appreciably by use of a "neighbor list," i.e., by maintaining a list of neighbors of a particular particle, which is updated at specified intervals (Verlet 1967).] These pair separations are then sorted into a histogram in which each "bin" has width Δr. (The meaningful separations are, of course, limited to distances less than half the box length.) Let bin k, corresponding to distance r, contain $n_k(r)$ pairs of particles. Then the average number of particles whose distance from a reference particle lies in this interval is

$$n(r) = \frac{n_k(r)}{M \cdot N}, \tag{12.9}$$

where M is the total number of configurations and N is the total number of particles. Substitution of (12.9) into (12.8) yields the radial distribution function for a particular value of r. Clearly, the interval Δr must be small compared to the sphere diameter but large enough to yield good statistics. Moreover, to ensure that there are no finite-system-size effects, one should carry out the simulations for successively larger system sizes until convergence is achieved. Figure 12.10 shows g_2 for hard disks as generated from the Metropolis algorithm given in Appendix A.

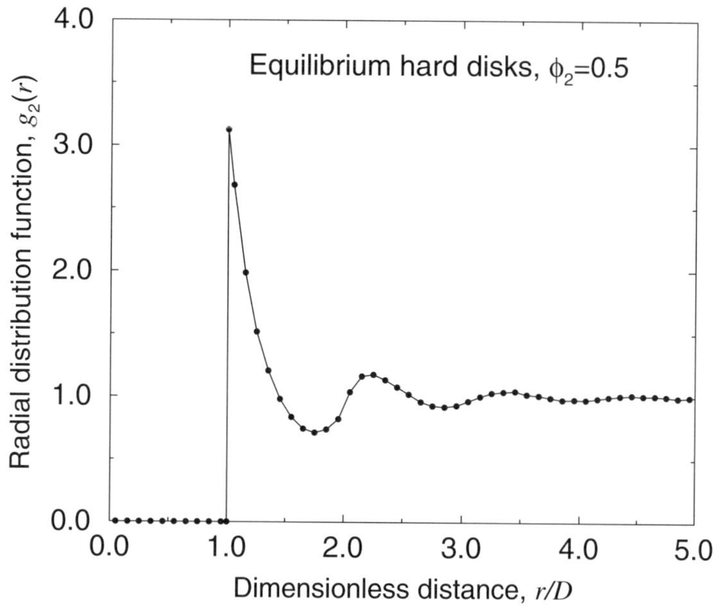

Figure 12.10 The radial distribution function g_2 versus r/D for an equilibrium system of hard disks of diameter D at an area fraction of $\phi_2 = 0.5$ as generated from the program in Appendix A. The filled circles are the simulation data (with negligibly small error bars), and the curves joining the points are linear interpolations. Data were obtained using 576 particles and 500 realizations, with a bin width of $0.1D$.

12.4.2 *n*-point Probability Functions

To calculate the n-point probability function $S_n(\boldsymbol{x}^n) \equiv S_n^{(1)}(\boldsymbol{x}^n)$ for phase 1 (defined in Section 2.2) for statistically inhomogeneous media (whether composed of particles or not), one fixes the points $\boldsymbol{x}^n \equiv \{\boldsymbol{x}_1, \boldsymbol{x}_2, \ldots, \boldsymbol{x}_n\}$ in each realization, records the fraction of times that all points lie in phase 1, and averages over a large number of realizations (Quintanilla and Torquato 1997a). In the limit of a very large number of realizations, this probability will converge to some well-defined limit. However, one must carry out the same procedure for all desired positions $\boldsymbol{x}^n$.

If the system is statistically *homogeneous*, then there are more efficient ways to compute S_n. For example, since there is no longer any preferred origin in the system, the one-point function $S_1(\boldsymbol{x}_1)$ is just the constant volume fraction ϕ_1 of phase 1, which can be evaluated by randomly "tossing" many points into a realization and recording the fraction that fall in phase 1. For the case of N identical spheres of radius R with a configuration $\boldsymbol{r}^N$ in a matrix phase, any point $\boldsymbol{x}$ falls outside the spheres (phase 1) if

$$|\boldsymbol{x} - \boldsymbol{r}_i| > R, \qquad i = 1, 2, \ldots, N. \tag{12.10}$$

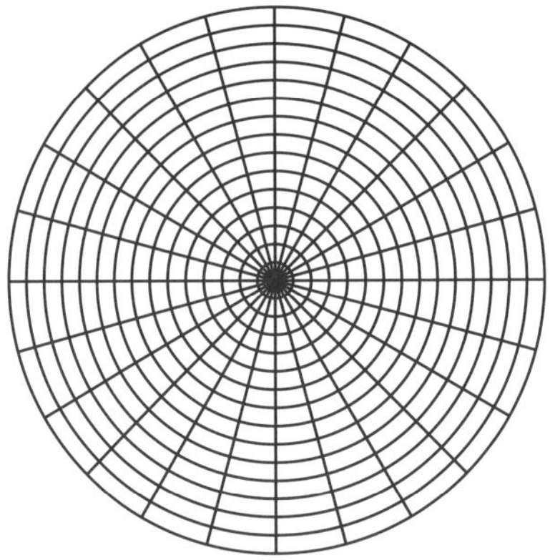

Figure 12.11 A template to sample for S_2 or S_3 in two-dimensional homogeneous media. The grid spacing need not be regular as shown here.

This process is repeated for each realization, and the final result for S_1 is obtained by averaging over all realizations. Execution speed to check the overlap condition (12.10) can be increased significantly using a *cell list* (Haile et al. 1985), which identifies the sphere centers that are in some small subvolume of the system. When the randomly placed point lands in a particular subvolume, then condition (12.10) is checked only for those centers located in the subvolume and its immediate neighboring subvolumes. The one-point function S_1 has been evaluated from simulations for spheres in the cherry-pit model as a function of the impenetrability parameter (Lee and Torquato 1988a).

For statistically homogeneous but anisotropic particle systems (e.g., oriented ellipsoids or cylinders), it is better to compute the two-point function $S_2(\boldsymbol{r})$ (where $\boldsymbol{r} = \boldsymbol{x}_2 - \boldsymbol{x}_1$) by randomly tossing many line segments of length $r = |\boldsymbol{r}|$ with *fixed orientation* (specified by the vector $\boldsymbol{r}$) into each realization, measuring the fraction of times that both ends of the line segment lie outside the particles, and averaging over all realizations. For statistically isotropic media, $S_2(r)$ is found in the same way except that the line segments are randomly oriented.

For both isotropic and anisotropic (but homogeneous) d-dimensional media, efficiency can be improved by randomly tossing a "sampling template" consisting of a grid of many points (Smith and Torquato 1988), as depicted in Figure 12.11 for $d = 2$. Many line segments (in which one end is always at the template origin) of variable lengths and orientations are contained in a template. Sampling templates at many locations (say 10,000 or more) are used to sample each realization. This will yield the two-point function $S_2(\boldsymbol{r})$ that will depend on the vector displacement $\boldsymbol{r}$ for anisotropic media. If the medium is isotropic, then $S_2(r)$ will depend only on the distance $r = |\boldsymbol{r}|$ and hence is obtained by averaging over all orientations of $\boldsymbol{r}$. The function S_2 has been computed for equilibrium hard spheres (Haile et al. 1985) and RSA disks in the cherry-pit model (Smith and Torquato 1988).

The sampling template can also be used to compute the three-point function $S_3(r, s, t)$ for homogeneous and isotropic particle systems (Smith and Torquato 1989), where its arguments are the side lengths of the triangle such that $r = |\mathbf{x}_2 - \mathbf{x}_1|$, $s = |\mathbf{x}_3 - \mathbf{x}_1|$, and $t = |\mathbf{x}_3 - \mathbf{x}_2|$. This is true, since $S_3(r, s, t)$ can be found by measuring the fraction of times that the vertices of such a triangle fall in phase 1 when randomly thrown into the system and averaging over all realizations. One vertex of all the triangles is located at the template origin, and the two other vertices are located at the other grid points. For general isotropic media, one need only consider triangles such that $r \le s \le t$ to get S_3 for all possible r, s and t, since $S_3(r, s, t)$ remains invariant under all permutations of its arguments.

12.4.3 Surface Correlation Functions

For simplicity we discuss the evaluation of the one- and two-point surface correlation functions defined in Section 2.3 for isotropic systems of possibly overlapping equisized spheres. As noted in Section 4.1, the specific surface s, a one-point correlation function, may be computed by exploiting the interpretation that it is proportional to the probability of finding a point in the dilated region around the spheres (see Figure 4.2). Recall that the dilation region is created by uniformly and radially dilating the spheres by a differential amount ∂R. In a simulation, this thickness must be finite, which we denote by ΔR. Once the probability of finding a point in the dilated region is determined using the techniques of Section 12.4.2, an estimate of s is this probability divided by the thickness ΔR. However, according to formula (4.10), s is defined as the aforementioned ratio in the limit that the thickness ΔR tends to zero. Thus, one must consider several values of the thickness and then extrapolate the results for s to the limit that $\Delta R \to 0$.

The surface correlation functions $F_{vs}(r)$ and $F_{ss}(r)$ can be computed using the same idea, as was done by Seaton and Glandt (1986) for both fully penetrable and totally impenetrable spheres. The void-surface correlation function $F_{vs}(r)$ is proportional to the probability of finding one of the end points of a randomly placed line segment of length r in the void region and the other end in the dilated region. The surface-surface correlation function $F_{ss}(r)$ is proportional to the probability of finding both end points of a randomly placed line segment of length r in the dilated region.

12.4.4 Cluster-Type Functions

The pair-connectedness function $P_2(r)$ (Section 9.2.1) is the "connectedness" analogue of the radial distribution function $g_2(r)$ and is computed in the same way as $g_2(r)$, except that one must keep track of the clusters that are found. For simplicity, consider isotropic systems of spheres. Events in which particles are in a concentric shell of thickness Δr at a radial distance r from a particle at the origin and also are members of the same cluster contribute to $P_2(r)$. For the problem of interest, one must first establish a criterion that defines when two particles are "directly" connected. Pairs of particles may be "indirectly" connected by being part of the same cluster. Existing *cluster-counting algorithms* that identify the various clusters in the system include the "cluster-labeling" method (Hoshen and Kopelman 1976) and the "connectivity-matrix" method (Sevick et al. 1988). With such information, one could also compute other cluster statistics such as the cluster size distribution n_k and mean cluster size S (Section 9.1.2).

As usual, periodic boundary conditions are used to minimize finite-system size effects. Such boundary conditions have been employed to find the pair-connectedness function $P_2(r)$ for the sticky-sphere model (Seaton and Glandt 1987) and the concentric-shell model (Sevick et al. 1988). It is important to note that the type of boundary conditions that one employs affects cluster identification. The determination of which particles are members of the same cluster in turn will affect measurements of $P_2(r)$.

Lee and Torquato (1988b) have shown that the application of "free boundary conditions" to the *central cell and replicating cells* is superior to the use of standard periodic boundary conditions in simulating cluster statistics. In particular, the former method leads to a pair-connectedness function that converges more rapidly to the infinite-system limit than when standard periodic boundary conditions are used. Thus, application of free boundary conditions appreciably reduces finite-size errors. The function $P_2(r)$ has been computed using such boundary conditions for identical spheres in the concentric-shell model (Lee and Torquato 1988b) and overlapping spheres of two different sizes (Kim and Torquato 1990a).

In any case, when simulating cluster-type statistics, one should always compute them for various system sizes (i.e., number of particles) to ensure that the results are independent of system size. Near the percolation threshold, this is done using *finite-size scaling* analysis (Section 9.1.5).

The two-point cluster function $C_2(r)$ is the connectedness analogue of the *particle phase* two-point probability function $S_2^{(2)}(r) = 1 - 2\phi_2 + S_2(r)$ (Section 9.2.1). Thus, the simulation of $C_2(r)$ follows the procedure outlined above for $S_2(r)$, except that one must only count those events in which the two points are part of the same cluster. The aforementioned cluster identification algorithms can again be used here. Simulations of the two-point cluster function $C_2(r)$ have been carried out for the concentric-shell, or cherry-pit, model (Lee and Torquato 1989) and sticky-sphere model (Lee 1993). Figure 12.12 shows simulation results for C_2 in the fully penetrable limit ($\lambda = 0$) of the cherry-pit model (Section 10.2) for densities below the percolation threshold $\phi_2^c \approx 0.2895$ (Rintoul and Torquato 1997a). As the density increases, the range over which C_2 is nonnegligibly small increases, indicating the presence of increasingly larger clusters.

12.4.5 Other Correlation Functions

Other statistical functions for particle systems have been evaluated from computer simulations. These numerical studies include:

- nearest-neighbor functions, E_P and H_P (Torquato and Lee 1990)

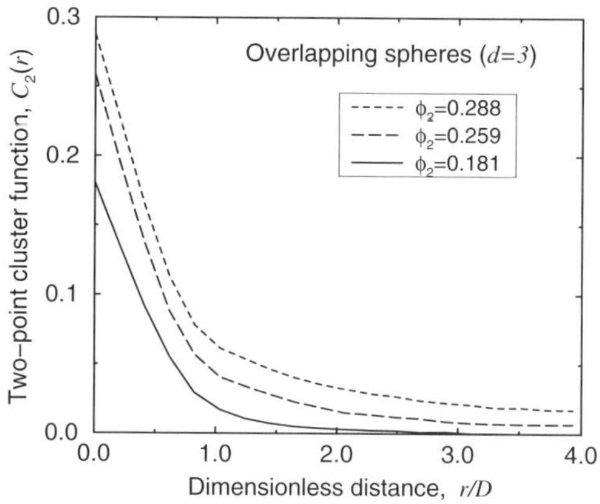

Figure 12.12 Two-point cluster function $C_2(r)$ versus the dimensionless distance r/D for a system of fully penetrable spheres of diameter D at several values of the sphere volume fraction ϕ_2 (Lee and Torquato 1989). Curves are spline fits of the simulation data.

- lineal-path function, $L(z)$ (Lu and Torquato 1992a, Coker and Torquato 1995)
- chord-length density function, $p(z)$ (Coker and Torquato 1995)
- pore-size functions, $P(\delta)$ and $F(\delta)$ (Coker and Torquato 1995)

12.5 Sampling Images and Digitized Media

The measurement of correlation functions from images of real materials has come a long way since the work of Corson (1974a, 1974b) who used an unautomated painstaking procedure to compute three-point functions from photographs of cross-sections of heterogenous materials. There now exist a variety of techniques to obtain two- and three-dimensional images of heterogeneous materials, including transmission electron microscopy (Flegler, Heckman and Klomparens 1993), scanning tunneling electron microscopy (Stroscio and Kaiser 1993), synchrotron-based tomography (Kinney and Nichols 1992), magnetic resonance imaging (Howle, Behringer and Georgiadis 1993), and confocal microscopy (Fredrich, Menendez and Wong 1995). While electron-microscopic techniques only provide two-dimensional images, X-ray-tomographic, magnetic-resonance-imaging, and confocal-microscopic techniques provide full three-dimensional images of materials. All of these imaging methods are nonintrusive, leaving the sample intact and unaltered.

In order to analyze images on the computer, one utilizes digitized representations of the samples, i.e., two- or three-dimensional arrays of gray values spanning the finite system, typically subjected to *periodic boundary conditions*. For a two-phase material, the gray-scale image can be reduced to a binary image by operations such as *thresh-*

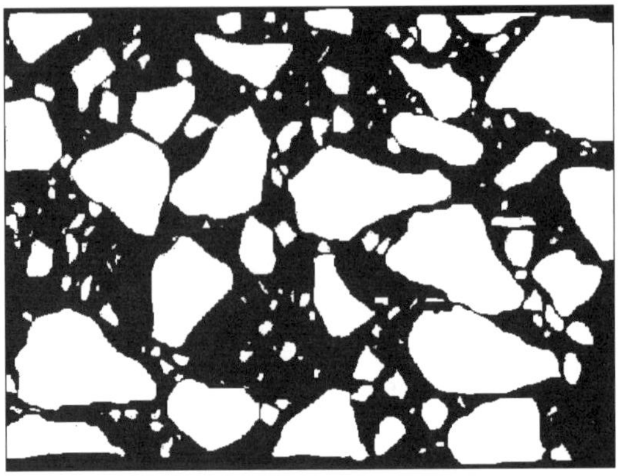

Figure 12.13 Digitized image of concrete. The white phase is the coarse aggregate. The fine aggregate (sand) is contained in the black mortar matrix. The longest side of the rectangular image corresponds to approximately two meters. The image was obtained by Paul Stutzman of the National Institute for Standards and Technology.

olding, in which gray values lighter than a chosen threshold are set to white and the others set to black. The image is thus reduced to an array of *bits* or *pixels* (*voxels*). The same procedure can be applied to multiphase media. The term "pixel" is usually reserved for a two-dimensional element of a regular square array, while a "voxel" is just the three-dimensional counterpart. For simplicity, we will use the term "pixel" to apply to any dimension.

Figure 12.13 shows a digitized image of concrete that was optically obtained and then thresholded. The white phase is the coarse aggregate (gravel), and the black phase is the mortar matrix that contains fine aggregate (sand). Figure 12.14 shows two different representations of the same three-dimensional digitized data set ($128 \times 128 \times 128$ pixels) of a portion of a Fontainebleau sandstone (Yeong and Torquato 1998b). Each pixel is a cubic region of size $7.5 \times 7.5 \times 7.5\ \mu\text{m}^3$. The full data set ($512 \times 512 \times 512$ pixels) was obtained via X-ray microtomography (Coker et al. 1996). The figure depicts the complex three-dimensional pore space as well as a three-dimensional perspective of corresponding surface cuts through the sample.

Digitized representations of heterogeneous materials call for numerical techniques that can directly extract microstructural descriptors of interest. In practice, one often has only a single digitized image to analyze (not an ensemble of them). In such instances, the system size must be sufficiently larger than the correlation length of interest in order for meaningful statistics to be extracted. If one is studying lattice (digitized) models of random media, then of course, one can perform ensemble averages of the relevant microstructural functions over all realizations. In any event, the description below applies to digitized media whether obtained through imaging or simulation techniques. To illustrate the applicability of the sampling techniques for real materi-

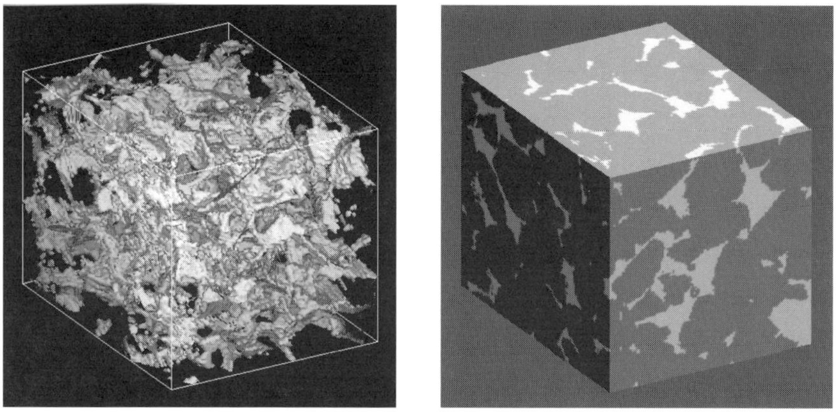

Figure 12.14 Left panel: The pore space of a $128 \times 128 \times 128$ pixel subregion of a Fontainebleau sandstone obtained via X-ray microtomography (each pixel has volume $7.5 \times 7.5 \times 7.5$ μm^3) (Yeong and Torquato 1998b). The pore space is white and opaque, and the grain phase is black and transparent. Right panel: Three-dimensional perspective of surface cuts of the same subregion.

als, results will be given for the aforementioned digitized sandstone image subject to periodic boundary conditions.

We note that the interesting subject of *mathematical morphology* (Serra 1982) will not be covered here. This approach provides a means of extracting certain statistical characteristics of digitized images by subjecting them to repeated transformations (such as *dilation* and *erosion*) and performing measurements on the transformed images.

12.5.1 Two-Point Probability Function

The two-point probability function $S_2(\mathbf{r})$ for either phase 1 or 2 can be computed using a sampling template, such as the one shown in Figure 12.11 (Coker and Torquato 1995). Even though the medium is digitized on a lattice, the template center is chosen on a continuum background. The phase in which the template center lies is determined by choosing the phase of the nearest lattice point. (An interpolation procedure can be used, but this is unnecessary.) The same rule applies for determining the phase of the other template points. As in continuum models, one utilizes many sampling templates and measures the fraction of times that both ends of the line segment of length r, with some specified orientation, lie in phase 1 (phase 2). This will yield the two-point function $S_2(\mathbf{r})$ for generally anisotropic media. Figure 12.15 shows the two-point function $S_2(r)$ for the void phase of the Fontainebleau sandstone (Coker et al. 1996) described earlier.

For an isotropic digitized system, there is a superior way to compute $S_2(r)$ (Yeong and Torquato 1998a). It is convenient to measure the distance r in units of pixels (integral values) such that the end points of r are located at the pixel centers. One need

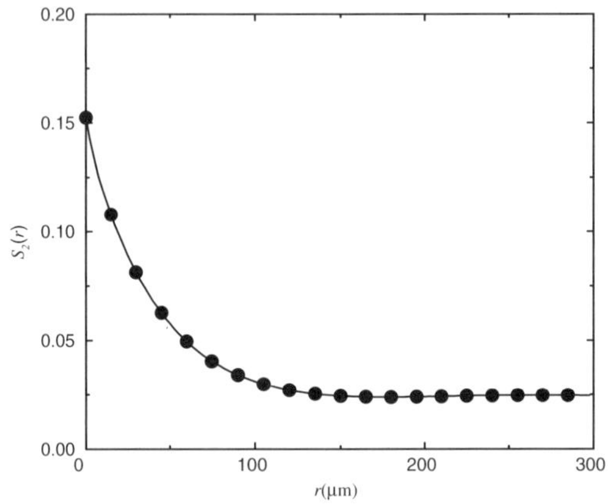

Figure 12.15 Two-point probability function for the void phase of the Fontainebleau sandstone (Coker et al. 1996).

sample for S_2 only along the principal directions of a hypercubic lattice (e.g., rows or columns in two dimensions). Then $S_2(r)$ is a linear function of the distance between adjacent pixels:

$$S_2(r) = (1 - f)S_2(i) + f S_2(i + 1), \quad \text{for} \quad i \leq r < i + 1, \tag{12.11}$$

where i is an integer and f is the fractional part of r. Because of this linear property, the evaluation of $S_2(r)$ at integral values of r is sufficient to determine it for noninteger values of r. Thus, $S_2(r)$ can be evaluated simply by successively translating a line of r $(= i)$ pixels in length (in each of the d orthogonal principal directions) at a distance of one pixel at a time and spanning the whole image. Counting the number of times that the two end points fall in phase 1 (phase 2) and dividing by the total number of "trials" (equal to the system size of the periodically replicated medium) gives $S_2(r)$. This sampling procedure is more accurate and less "noisy" than random sampling, since it uses information contained in every pixel of the entire system. Of course, at additional computational cost, one could sample S_2 in more than d directions (Sheehan and Torquato 2001).

Recall that the slope of $S_2(r)$ at $r = 0$ is related to the specific surface s of the system [cf. (2.34)]. However, this formula is valid when r takes continuous, not discrete, values. For a *digitized* medium on a d-dimensional cubic lattice in which distances are measured in units of pixels, s is related to the discrete number of interfacial faces, and the slope of $S_2(r)$ is easily determined from its values at $r = 0$ and $r = 1$:

$$\left.\frac{d}{dr} S_2(r)\right|_{r=0} = -\frac{s}{2d}. \tag{12.12}$$

It is only for $d = 1$ that (12.12) is the same as the continuum result (2.34).

Another way to compute $S_2(\mathbf{r})$ is to utilize its spectral representation (2.45) and efficient fast Fourier transform techniques (Cule and Torquato 1999). However, the infinite-volume implied in (2.45) is replaced by a finite-volume system under periodic boundary conditions.

Berryman has used image processing techniques to obtain S_2 of synthetic and real porous materials (Berryman 1985a, Berryman and Blair 1986). Specifically, he measured S_2 and estimated s of glass-bead samples and of sandstones. Moreover, he has devised an efficient means of obtaining and visualizing the three-point probability function S_3 (Berryman 1989).

12.5.2 Lineal-Path Function

Computation of the lineal-path function $L(z)$ (Section 2.4) for a digitized sample is straightforward. The following algorithm describes its evaluation for phase 1 (phase 2) (Coker and Torquato 1995):

- Introduce an oriented line into the system.
- Pick a random location on this line; if this point is in phase 1 (phase 2), mark this location as point A; otherwise pick a new random location.
- Move along the line from point A until phase 2 (phase 1) is encountered, which is then labeled point B.
- Increment the counter associated with the distance from point A to point B and all counters associated with distances less than this length; leave counters associated with distances greater than this length unchanged.

This procedure is repeated along the initial oriented line and then repeated over many lines in a particular sample or realization. After many such lines have been investigated, the counters are divided by the total number of random locations chosen. This gives the probability of inserting a line segment of a specified length wholly into a single phase of the system. Figure 12.16 shows the lineal-path function L for the void phase of the Fontainebleau sandstone (Coker et al. 1996).

For an isotropic system, it is more efficient to evaluate $L(z)$ by letting z take integer values and again sampling in d orthogonal directions (Yeong and Torquato 1998a). Let $L(i)$ be the probability of finding a line segment of integer length i entirely in phase 1 (phase 2). To illustrate how to evaluate $L(i)$, first consider a simple one-dimensional system in which only a single phase 1 (phase 2) chord of length n is present. Clearly,

$$L(i) = \begin{cases} (n-i)/N, & \text{when } 0 \le i \le n, \\ 0, & \text{otherwise}, \end{cases} \tag{12.13}$$

where N is the system size in pixels. For the general case of more than one chord or in higher dimensions, the same principle applies to each of the chords, so that L for the entire system is the sum of those functions due to the individual chords. Thus, sampling for L reduces to *identifying the lengths of the chords* of the phase of interest

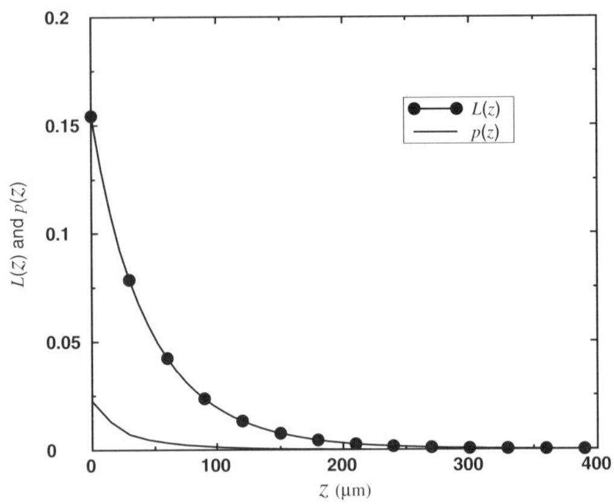

Figure 12.16 Pore-phase lineal-path function $L(z)$ and chord-length density function $p(z)$ for Fontainebleau sandstone (Coker et al. 1996).

in the system. For isotropic media, this method is considerably more efficient than random sampling.

12.5.3 Chord-Length Density Function

Since the chord-length density function $p(z)$ is a probability density function (Section 2.5), it is found for digitized media by binning the chord lengths that are found to compute the lineal-path function $L(z)$. The bins are normalized by the total number of chords measured. This limits the minimum bin size and, thus, chord length to a single lattice spacing. Therefore, to obtain information at zero distance, one can extrapolate to the origin using, for example, a three-point Lagrange extrapolation. Included in Figure 12.16 is the chord-length density function $p(z)$ for the void phase of the Fontainebleau sandstone (Coker et al. 1996).

12.5.4 Pore-Size Functions

The pore-size density function $P(\delta)$ gives the probability of finding phase 2 (phase 1) at a distance between δ and $\delta + d\delta$ from a randomly chosen point in phase 1 (phase 2) (Section 2.6). Unlike many of the previous quantities, $P(\delta)$ is an intrinsically three-dimensional microstructural function and hence cannot be obtained from a two-dimensional slice (Chapter 2). Therefore, it must be determined from a full three-dimensional digitized sample. The algorithm (Coker and Torquato 1995) works as follows:

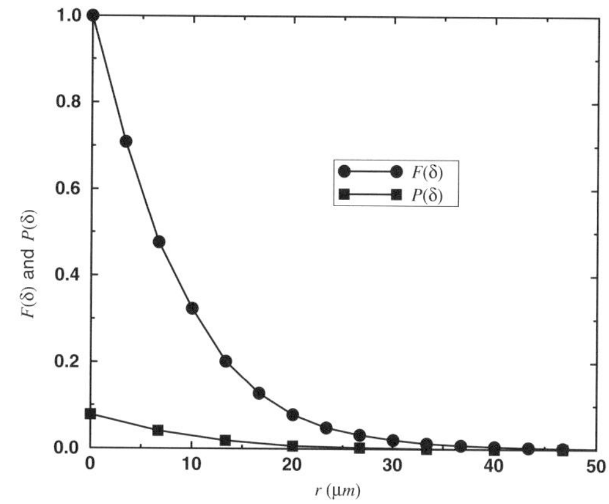

Figure 12.17 Pore-size density function $P(\delta)$ and complementary cumulative distribution function $F(\delta)$ for Fontainebleau sandstone (Coker et al. 1996).

- Choose a random *continuum* location in phase 1 (phase 2).
- Find the largest sphere that just touches phase 2 (phase 1) centered at the above location in phase 1 (phase 2) and record this radius.
- Repeat for many locations and create a list of radii at each location.

The pore-size density function is then obtained by binning the sphere radii found in the second step and dividing by the total number of radii.

The complementary cumulative pore-size distribution function $F(\delta)$ measures the probability of inserting a sphere of radius δ into a specified phase. This function is obtained by taking the list of radii at each location and incrementing all counters associated with radii less than or equal to a given radii. In the end, all counters are divided by the total number of radii. This method of using a list of radii closely parallels the method used to determine the lineal-path function and provides a comparable speedup as the probability of inserting spheres of all possible radii is determined with each measurement. Figure 12.17 shows both $P(\delta)$ and $F(\delta)$ for the void phase of the Fontainebleau sandstone (Coker et al. 1996). The pore-size functions were also computed from a three-dimensional digitized image of a porous gel obtained via X-ray microtomography (Rintoul et al. 1996).

12.5.5 Two-Point Cluster Function

As discussed in Section 12.4.4, the two-point cluster function $C_2(r)$ (Section 9.2.1) is computed in the same manner as $S_2(r)$, except that one must determine when the end points of the line segment of length r fall in the same cluster of phase 1 (phase 2). First one must establish a criterion that defines when two pixels are connected (e.g., a nearest-neighbor connectivity rule) and then the aforementioned cluster identification algorithms can be used. Armed with such information, one could also compute other cluster statistics (e.g., n_k and S); see Section 9.2.1.

12.6 Reconstructing Heterogeneous Materials

The *reconstruction* of realizations of random heterogeneous materials from a knowledge of limited microstructural information (lower-order correlation functions) is an intriguing inverse problem. Clearly, one can never reconstruct the original material perfectly, i.e., such reconstructions are nonunique. Thus, the objective here is *not the same* as that of data decompression algorithms that efficiently restore complete information, such as the gray scale of every pixel in an image.

The generation of realizations of random heterogeneous materials with specified lower-order correlation functions opens up many interesting fundamental and practical questions and issues. Such studies can:

1. Shed light on the nature of the information contained in the various correlation functions that are employed;
2. Ascertain whether the standard two-point probability function S_2, accessible experimentally via scattering, can accurately reproduce the material and, if not, what additional information is required to do so;
3. Identify the class of microstructures that have exactly the same lower-order correlation functions but widely different effective properties;
4. Attempt to reconstruct the full three-dimensional structure of the heterogeneous material from lower-order information extracted from two-dimensional plane cuts through the material: a problem of great practical value, since one often has only two-dimensional information, such as a micrograph or planar image;
5. Probe the interesting issue of nonuniqueness of the generated realizations;
6. *Construct* structures that correspond to specified correlation functions and categorize classes of random media;
7. Provide guidance in ascertaining the mathematical properties that physically realizable correlation functions must possess (Torquato 1999).

An effective reconstruction procedure enables one to generate accurate structures at will, and subsequent analysis can be performed to obtain desired macroscopic properties of the media. This becomes especially useful in generating three-dimensional structures when three-dimensional imaging techniques are not available: a "poor man's" tomography experiment (see item 4 above). Finally, we note that a successful procedure to reconstruct or construct random media can be employed to investigate any physical phenomena where correlation functions play an important role in characterizing spatiotemporal fields, e.g., turbulence (Batchelor 1959).

The first reconstruction procedures were based on thresholding Gaussian random fields. The theory of the statistical topography of Gaussian random fields was originally established in the work of Rice (1945). One approach to reconstructing random media originated with Joshi (1974) and was extended by Quiblier (1984) and Adler (1992). Another related approach, discussed in some detail in Chapter 8, originated with Cahn (1965) and was extended by Berk (1987), Teubner (1991), and Roberts and Teubner (1995). Both approaches are currently limited to the two-point probability function S_2. In addition, such methods are not suitable for extension to non-Gaussian statistics, and hence are model-dependent, i.e., depend on the underlying Gaussian statistics.

Recently, a stochastic optimization technique has been introduced (Rintoul and Torquato 1997b, Yeong and Torquato 1998a) to reconstruct random media. This method is applicable to multidimensional and multiphase media, and is highly flexible in that one can include any set of different types of correlation functions as microstructural information, within computational limits. It is both a generalization and simplification of the aforementioned Gaussian field reconstruction technique. Moreover, it does not depend on any particular statistical model. In what follows, we briefly describe the method and discuss some illustrative problems.

12.6.1 Reconstruction Procedure

A set of *target* correlation functions are prescribed based upon experiments, theoretical models, or some ansatz. Starting from some initial realization of the random medium, the method proceeds to find a realization in which the calculated correlation functions best match the target functions. This is achieved by minimizing the sum of squared differences between the calculated and target functions via stochastic optimization techniques, such as the simulated annealing method (Kirkpatrick, Gelatt and Vecchi 1983). The medium can be a dispersion of particles (Rintoul and Torquato 1997b) or, more generally, a digitized image (Yeong and Torquato 1998a).

For simplicity, we will focus on digitized media here and begin by outlining the reconstruction procedure by considering only a single two-point correlation function for statistically isotropic two-phase media. This is followed by a description of a general procedure that incorporates a set of different n-point correlation functions for anisotropic multiphase systems.

It is desired to generate realizations of two-phase isotropic media that have a target two-point correlation function $f_2(r)$ associated with phase i, where r is the distance between the two points. Let $\hat{f}_2(r)$ be the corresponding function of the reconstructed digitized system (with periodic boundary conditions) at some time step. It is this system that we will attempt to evolve towards $f_2(r)$ from an initial guess of the system configuration.

At any particular time step, we define a fictitious "energy" E such that

$$E = \sum_r \left[\hat{f}_2(r) - f_2(r)\right]^2, \tag{12.14}$$

where the sum is over all *discrete* values of r. To evolve the digitized system towards $f_2(r)$ (i.e., minimizing E), we interchange the states of two arbitrarily selected pixels of different phases, automatically preserving the volume fraction of both phases. After the interchange is performed, we can calculate the new energy E' and the energy difference $\Delta E = E' - E$ between the two successive states. This phase interchange is then accepted with some probability $p(\Delta E)$ that depends on ΔE. One reasonable choice is the Metropolis acceptance rule [cf. (12.7)], i.e.,

$$p(\Delta E) = \begin{cases} 1, & \Delta E \leq 0, \\ \exp(-\Delta E/T), & \Delta E > 0, \end{cases} \tag{12.15}$$

where T is a fictitious "temperature." The concept of finding the lowest-energy state by simulated annealing is based on a well-known physical fact: If a system is heated to a high temperature T and then slowly cooled down to absolute zero, the system equilibrates to its ground state. The *cooling* or *annealing schedule*, which governs the value and the rate of change of T, is chosen to be sufficiently slow to allow the system to evolve to the desired state as quickly as possible without getting trapped in any local energy minima (metastable states). At each annealing step k, the system is allowed to evolve long enough to thermalize at $T(k)$. The temperature is then lowered according to a prescribed annealing schedule $T(k)$ until the energy of the system approaches its ground state value within an acceptable tolerance. This method enables $\hat{f}_2(r)$ to converge gradually to $f_2(r)$. At the ground state, the energy E can viewed as a *least-squares* error.

The generalization to treat a set of different n-point correlation functions for anisotropic multiphase systems is straightforward (Rintoul and Torquato 1997b, Yeong and Torquato 1998a). The energy generalizes as

$$E = \sum_{\boldsymbol{r}_1,\boldsymbol{r}_2,\ldots,\boldsymbol{r}_n} \sum_{\beta} \alpha_\beta \left[\hat{f}_n^{\beta}(\boldsymbol{r}_1,\boldsymbol{r}_2,\ldots,\boldsymbol{r}_n) - f_n^{\beta}(\boldsymbol{r}_1,\boldsymbol{r}_2,\ldots,\boldsymbol{r}_n) \right]^2, \tag{12.16}$$

where the index β denotes the type of correlation function and α_β is the associated weight. There is a variety of functions that can be used in the reconstruction procedure, including the two-point probability function, lineal-path function, two-point cluster function, chord-length density function, and pore-size functions, to name just a few (see Chapter 2).

We note that there are various ways of appreciably reducing computational time. For example, computational cost can be significantly lowered by interchanging pixels only at the *two-phase interface*. Additional cost savings have been achieved by using other stochastic optimization schemes such as the "Great Deluge" algorithm, which can be adjusted to accept only "downhill" energy changes, and the "threshold acceptance" algorithm (Cule and Torquato 1999). Further savings can be attained by developing strategies that exploit the fact that pixel interchanges are *local* and thus one can reuse the correlation functions measured in the previous time step instead of recomputing them fully at any step (Yeong and Torquato 1998a). The reader is referred to the aforementioned references for additional details.

12.6.2 Illustrative Examples

Lower-order correlation functions generally do not contain complete information and thus cannot be expected to yield perfect reconstructions. Of course, the judicious use of combinations of lower-order correlation functions can yield more accurate reconstructions than any single function alone. Yeong and Torquato (1998a,1998b) clearly showed that the two-point function S_2 is not sufficient to reconstruct accurately random media. By also incorporating the lineal-path function L, they were able to obtain better reconstructions. They studied one-, two-, and three-dimensional digitized isotropic media. Each simulation began with an initial configuration of pixels (white for phase 1 and black for phase 2) in the random checkerboard arrangement at a prescribed volume fraction. Here we summarize results for some of the examples that they examined and an example considered by Cule and Torquato (1999).

In the first example, the target system is a one-dimensional equilibrium arrangement of identical hard rods of length D. The target function S_2 for the particle phase is easily extracted from (5.40) and (5.44). The target lineal-path function for the particle phase is trivially given by

$$L(r) = \begin{cases} \phi_2(1 - r/D), & \text{when } 0 \le r < D, \\ 0, & \text{otherwise}. \end{cases} \tag{12.17}$$

Realizations of the target system and S_2-reconstruction are shown in Figure 12.18. While the S_2 profile of the reconstructed system is in excellent agreement with that of the target system, the rods in the former incorrectly have a wide distribution of lengths. This nonuniqueness can be quantified by comparing other correlation functions of the reconstructed and target systems. Indeed, the lineal-path functions of the target and reconstructed systems (included in Figure 12.18) are seen to be significantly different from one another for intermediate to large values of r. A realization of a *hybrid* reconstruction that matches both S_2 and L extremely well is also shown in Figure 12.18. Since L contains connectedness information about the particle phase, the hybrid reconstruction is better able to capture the uniformity in the rod size.

A two-dimensional example illustrating the insufficiency of S_2 in reconstructions is a target system of overlapping disks at a disk volume fraction of $\phi_2 = 0.5$ (see Figure 12.19a). Reconstructions that match S_2 alone, L alone, and both S_2 and L are shown in Figure 12.19. Not surprisingly, the S_2-reconstruction is not very accurate; the cluster sizes are too large, and the system actually percolates (recall from Chapter 10 that overlapping disks percolate at $\phi_2 \approx 0.68$). The L-reconstruction does a better job than the S_2-reconstruction in capturing the clustering behavior. However, the hybrid $(S_2 + L)$-reconstruction is the best.

Debye and coworkers (Debye and Bueche 1949, Debye et al. 1957) claimed (without rigorous proof) that the exponentially decaying two-point probability function for

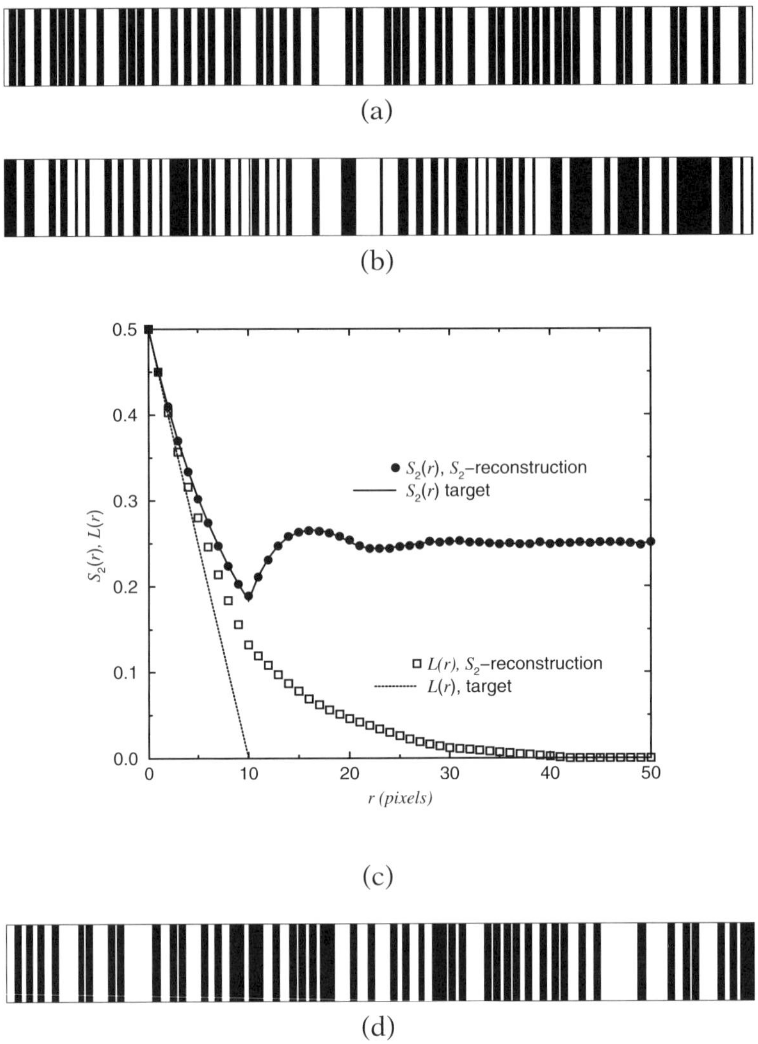

Figure 12.18 (a) Target system: a realization of an equilibrium system of hard rods. System size = 1000 pixels, rod length = 10 pixels, and volume fraction $\phi_2 = 0.5$. Vertical "heights" are given to the "rods" as a visual aid. (b) S_2-reconstruction of equilibrium hard-rod system. (c) S_2 for target and reconstructed systems. Also shown is the lineal-path function L for both systems. (d) Hybrid ($S_2 + L$)-reconstruction of equilibrium hard-rod system.

phase 1 given by

$$\frac{S_2(r) - \phi_1^2}{\phi_1\phi_2} = \exp(-r/a) \tag{12.18}$$

should apply to structures in which one phase consists of "random shapes and sizes." Here the positive parameter a is a correlation length. The optimization method can now

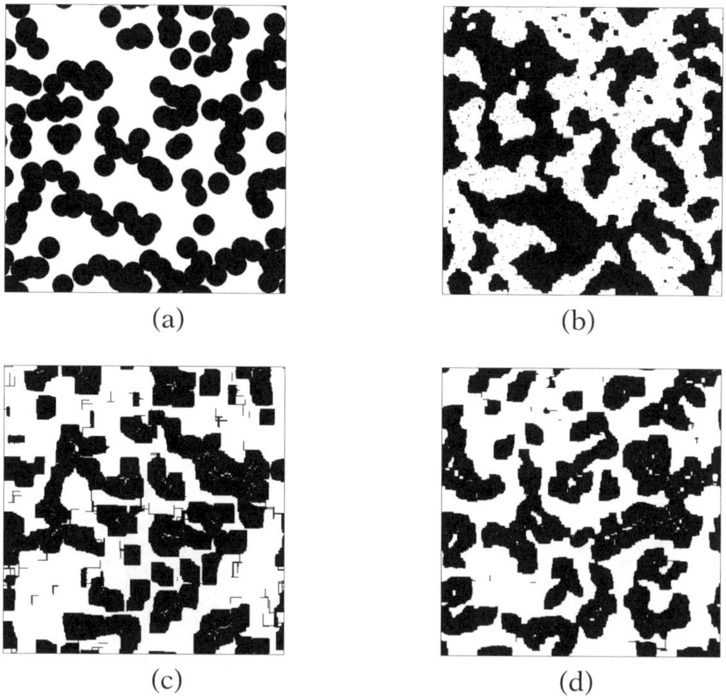

Figure 12.19 (a) Target system: a realization of a system of random overlapping disks. System size = 400 × 400 pixels, disk diameter = 31 pixels, and volume fraction $\phi_2 = 0.5$. (b) S_2-reconstruction. (c) Corresponding L-reconstruction. (d) Corresponding hybrid ($S_2 + L$)-reconstruction.

be used in the *construction* mode to find the specific structures that realize the function given by (12.18). A realization corresponding to (12.18) is shown in Figure 12.20 for $\phi_2 = 0.5$ and $a = 2$ pixels and is seen to be consistent with Debye's intuitive description. Relation (12.18) obeys the necessary nonnegativity condition (2.45) on the spectral function $\tilde{\chi}$ for any d and the "binary" conditions (2.46)–(2.48) (Torquato 1999), leading one to believe that (12.18) is exactly realizable. Indeed, there are specific two-phase microstructures that achieve the "Debye" random-medium function (12.18) in the plane (Stoyan et al. 1995).

A generalization of the Debye random-medium function that possesses a considerable degree of short-range order and *phase-inversion* symmetry (Section 2.2.2) is the following expression (Cule and Torquato 1999):

$$\frac{S_2(r) - \phi_1^2}{\phi_1\phi_2} = e^{-r/a}\,\frac{\sin(qr)}{qr}, \tag{12.19}$$

where $q = 2\pi/b$ and the positive parameter b is a characteristic length that controls oscillations in the term $\sin(qr)/(qr)$, which also decays with increasing r. The one-, two-, and three-dimensional spectral functions satisfy the nonnegativity condition (2.45) and

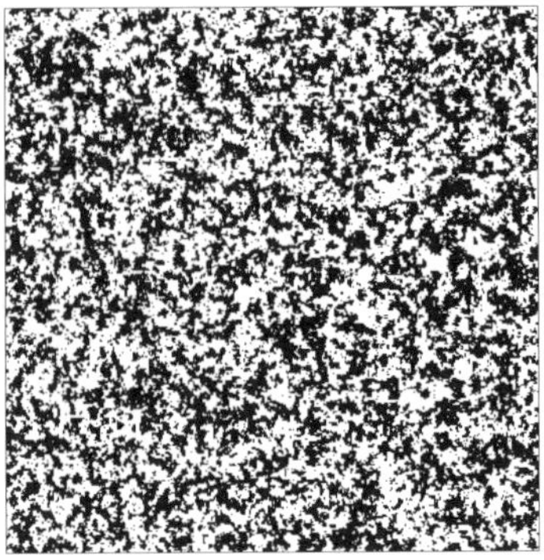

Figure 12.20 S_2-construction of a "Debye" random medium (400×400 pixels). Volume fraction $\phi_2 = 0.5$ and correlation length $a = 2$ pixels.

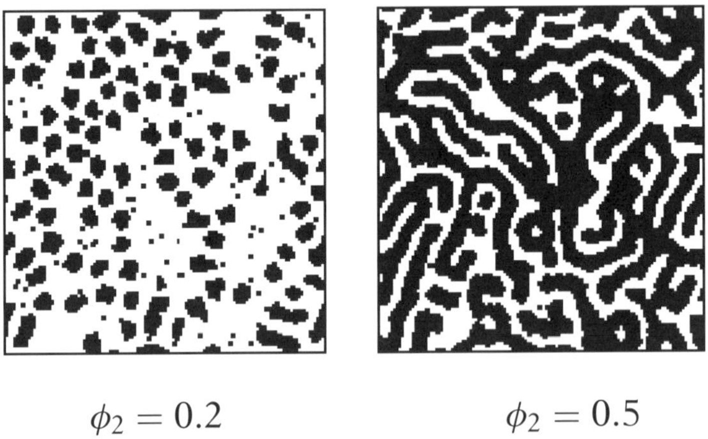

Figure 12.21 Structures corresponding to the target correlation function given by (12.19) for $\phi_2 = 0.2$ and 0.5. Here $a = 32$ pixels and $b = 8$ pixels.

the binary conditions (2.46) and (2.48). Interestingly, although it satisfies the upper bound of binary condition (2.47), it does not necessarily satisfy the lower bound of (2.47) for all ϕ_1, depending on the values of a and b (Torquato 1999). This example serves to illustrate an important point, namely, just because a correlation function meets the nonnegativity requirement of Theorem 2.1, it does not mean that it comes from class B binary processes (see Section 2.2.5).

Two structures possessing the correlation function (12.19) are shown in Figure 12.21 for $\phi_2 = 0.2$ and 0.5, in which $a = 32$ pixels and $b = 8$ pixels. For these sets of parameters, all of the aforementioned necessary conditions on the function are met. At $\phi_2 = 0.2$, the system resembles a *dilute particle suspension* with "particle" diameters of order b. At $\phi_2 = 0.5$, the resulting pattern is *labyrinthine* such that the characteristic sizes of the

Figure 12.22 Hybrid reconstruction of the sandstone shown in Figure 12.14 using both S_2 and L obtained from a single "slice." System size is $128 \times 128 \times 128$ pixels. Left panel: Pore space is white and opaque, and the grain phase is black and transparent. Right panel: Three-dimensional perspective of the surface cuts.

"patches" and "walls" are of order a and b, respectively. Note that $S_2(r)$ was sampled in all directions during the annealing process. In all of the previous two-dimensional examples, however, both S_2 and L were sampled along the two principal directions in order to save computational time. This time-saving step should be implemented only for isotropic media, provided that there is no appreciable short-range order; otherwise, it leads to unwanted anisotropy (Cule and Torquato 1999, Manwart and Hilfer 1999). However, this artificial anisotropy can be significantly reduced by optimizing along four selected directions on a square lattice (Sheehan and Torquato 2001).

To what extent can information extracted from two-dimensional cuts through a three-dimensional isotropic medium, such as S_2 and L, be employed to reproduce intrinsic three-dimensional information, such as connectedness? This question was studied for the aforementioned Fontainebleau sandstone for which we know the full three-dimensional structure via X-ray microtomography (Yeong and Torquato 1998b). The three-dimensional reconstruction that results by using a single slice of the sample and matching both S_2 and L is shown in Figure 12.22. The reconstructions reproduce accurately certain three-dimensional properties of the pore space, such as the pore-size functions, the mean survival time of a Brownian particle, and the fluid permeability. The degree of connectedness of the pore space also compares remarkably well with the actual sandstone, although this is not always the case (Manwart, Torquato and Hilfer 2000).

In summary, the aforementioned stochastic optimization technique is a promising new way to reconstruct or construct random media. However, its full potential and limitations have yet to be fully explored. Many of the fascinating questions and issues identified at the beginning of this section remain open, regardless of the approach taken to address them.

This concludes our treatment of the microstructure of random heterogeneous materials (Part I). The next chapter, on homogenization theory, begins Part II of this book: effective properties of random heterogeneous materials and their link to the microstructure.

PART II

MICROSTRUCTURE/PROPERTY CONNECTION

CHAPTER 13

Local and Homogenized Equations

Homogenization theory is concerned with finding the appropriate *homogenized* (or averaged, or macroscopic) governing partial differential equations describing physical processes occurring in heterogeneous materials when the length scale of the heterogeneities tends to zero. In such instances it is desired that the effects of the microstructure reside wholly in the *macroscopic* or *effective* properties via certain weighted averages of the microstructure. In its simplest form, the method is based on the consideration of two length scales: the macroscopic scale L, characterizing the extent of the system, and the *microscopic* scale ℓ, associated with the heterogeneities. Moreover, it is supposed that some external field is applied that varies on a *characteristic* length scale Λ. If ℓ is comparable in magnitude to Λ or L, then one must employ a microscopic description, i.e., one cannot homogenize the equations.

The limit of interest for purposes of homogenization is

$$L \geq \Lambda \gg \ell.$$

Therefore, there is a small parameter

$$\epsilon = \frac{\ell}{L}$$

associated with rapid fluctuations in the microstructure or local property. Accordingly, the field quantities (e.g., temperature field, electric field, stress field, concentration field, velocity field) depend on two variables: a *global* or *slow* variable $\boldsymbol{x}$ and a *local* or *fast* variable

$$\boldsymbol{y} = \boldsymbol{x}/\epsilon.$$

The slowly varying parts of the fields are imposed by the source, the boundary conditions, or the initial conditions, while the rapidly varying parts are imposed by the local property or microstructure. These variations are schematically shown in Figure 13.1.

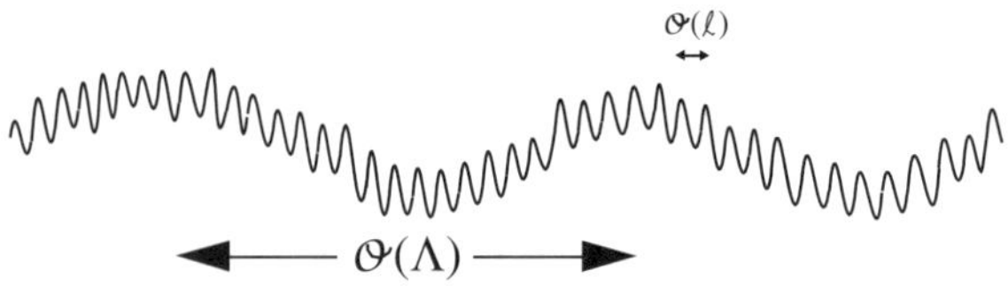

Figure 13.1 A schematic depiction of the slow and rapid parts of the field.

Under these conditions, a complete analysis of the problem involves three steps:

1. One first sets out to find the form of the homogenized or averaged equations, valid on length scales $\mathcal{O}(\Lambda)$, by, for example, performing an asymptotic expansion of the field quantities in terms of the global and local variables. The averaged relations are typically continuum differential equations. This asymptotic analysis is often guided by phenomenology. For example, in the case of electrical (heat) conduction in a heterogeneous material, it is expected that an averaged governing equation applies with an effective conductivity obtainable from an averaged Ohm's (Fourier's) law. Similarly, one seeks an averaged Hooke's law to describe the effective elastic moduli. For flow in porous media, one would like to understand the conditions under which Darcy's law can be derived.
2. Next, one must determine the effective properties that arise in the averaged equations as a function of the microstructure. The quantitative characterization of the microstructure of random media is an enormous subject of research and is dealt with in Part I of this book. The effective properties should mathematically exist as the system volume tends to infinity, independent of the macroscopic boundary conditions.
3. Finally, one must solve the homogenized equations under appropriate boundary or initial conditions.

The remaining chapters in Part II of this book are concerned with methodologies to link the effective properties of a heterogeneous material to its microstructure and the resulting predictions of such structure/property relations.

13.1 Preliminaries

As described in Section 2.1, each realization ω of the two-phase random medium is a domain of space $\mathcal{V} \in \Re^d$. This domain has volume V and is partitioned into two disjoint random sets or phases: phase 1, a domain $\mathcal{V}_1(\omega)$ with volume fraction ϕ_1, and phase 2, a domain $\mathcal{V}_2(\omega)$ with volume fraction ϕ_2. The interface between the two phases is denoted by $\partial\mathcal{V}(\omega)$. Figure 13.2 depicts a schematic of a realization of a two-phase random medium. Ultimately, we are interested in ergodic two-phase random media (Section 2.2.2) and therefore will take the limit $\mathcal{V} \to \Re^d$. Phase i can be a solid, fluid, or void phase characterized by some general constant tensor property $\boldsymbol{K}_i$. Thus, the

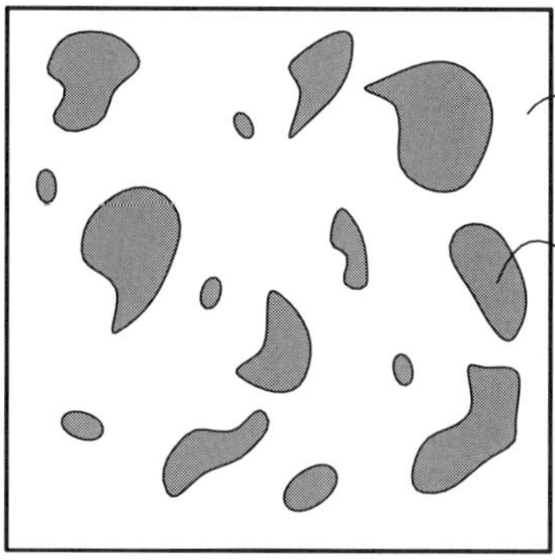

Figure 13.2 A schematic of a realization ω of a two-phase random medium.

associated local property $\boldsymbol{K}(\boldsymbol{x};\omega)$ at position $\boldsymbol{x}$ in the heterogeneous material can be written in terms of the phase indicator function $\mathcal{I}^{(i)}(\boldsymbol{x};\omega)$ [cf. (2.1)] as

$$\boldsymbol{K}(\boldsymbol{x};\omega) = \boldsymbol{K}_1\mathcal{I}^{(1)}(\boldsymbol{x};\omega) + \boldsymbol{K}_2\mathcal{I}^{(2)}(\boldsymbol{x};\omega). \tag{13.1}$$

This book focuses primarily on four different physical processes: conduction, elasticity, diffusion and reaction among traps, and flow in porous media, together with their associated effective properties. The corresponding *steady-state* effective properties of interest are:

- Effective conductivity tensor, $\boldsymbol{\sigma}_e$
- Effective stiffness tensor, $\boldsymbol{C}_e$
- Mean survival time, τ
- Fluid permeability tensor, $\boldsymbol{k}$

It will be shown in this chapter that these effective properties of the random heterogeneous material (under certain assumptions about the microstructure) are determined by *ensemble* averages of local fields that satisfy the appropriate conservation equations, i.e., governing partial differential equations.

The claim from Chapter 1 is that any of these effective properties, which we denote generally by $\boldsymbol{K}_e$, is defined by a linear constitutive relation between an average of a generalized local *flux* $\boldsymbol{F}$ and an average of a generalized local (or applied) *intensity* $\boldsymbol{G}$, i.e.,

$$\boldsymbol{F} \propto \boldsymbol{K}_e \cdot \boldsymbol{G}. \tag{13.2}$$

Referring to Table 1.1, the effective conductivity tensor $\boldsymbol{\sigma}_e$, effective stiffness tensor $\boldsymbol{C}_e$, mean survival time τ, and fluid permeability tensor $\boldsymbol{k}$ fall within the problem classes A, B, C, and D, respectively. The other problems within the classes are either mathematically analogous or related problems, as will be explained below. Although class A and B problems are of different tensorial order, they share many common features and hence can be attacked using similar techniques. This point is exploited in Chapter 23, where we obtain cross-property relations involving the effective conductivity on the one hand and effective elastic moduli on the other.

Similarly, despite the fact that the class C trapping problem is a scalar one and the class D flow problem can be either a vector or second-order tensor one, they also share common features, which will be discussed later in this chapter and in Chapter 23 on cross-property relations. It is important to observe that the class A and B problems are considerably different from class C and D problems. Indeed, whereas both the effective conductivity and effective stiffness tensor are scale-invariant properties, both the survival time and fluid permeability are scale-dependent properties. Moreover, unlike class A and B problems, class C and D problems are not characterized by *local properties*; in other words, there is no such thing as a local survival time or a local fluid permeability. This classification scheme is made more mathematically precise in Section 13.6.

In the last two sections we will examine the equations governing certain *time-dependent* trapping and flow problems. In particular, we will define *macroscopic relaxation times* and discuss their relationship to the steady-state mean survival time τ and steady-state fluid permeability $\boldsymbol{k}$.

13.2 Conduction Problem

In this section we begin by stating the local relations for the fields for the conduction problem (Jackson 1990). To motivate the discussion on homogenization, we then describe a model one-dimensional *periodic medium*. The derivation of the homogenized relations in any space dimension d turns out to be somewhat simpler for periodic media than for random media. Accordingly, since the forms of the resulting averaged relations turn out to be identical in both cases, we derive them first in the periodic setting. This is followed by a discussion of the homogenized relations in the random setting.

13.2.1 Local Relations

Consider the steady-state transport or displacement of a conservable quantity associated with any of the class A problems that are summarized in Table 1.1. To fix ideas, we will speak in the language of electrical or thermal conduction, keeping in mind that the results of this section apply as well to the determination of the effective dielectric constant, magnetic permeability, and diffusion coefficient (see Chapter 1 and Section 13.2.5). Each realization ω of the random heterogeneous material that occupies the space $\mathcal{V}$ is composed of two phases (phases 1 and 2) having constant conductivity tensors $\boldsymbol{\sigma}_1$ and $\boldsymbol{\sigma}_2$. In the ensuing discussion, we will temporarily drop ω from the notation.

Local Differential Equations

Let $\boldsymbol{J}(\boldsymbol{x})$ denote the local electric (thermal) current or flux at position $\boldsymbol{x}$, and let $\boldsymbol{E}(\boldsymbol{x})$ denote the local field intensity. Under steady-state conditions with no source terms, conservation of energy requires that $\boldsymbol{J}$ be solenoidal:

$$\nabla \cdot \boldsymbol{J}(\boldsymbol{x}) = 0 \qquad \text{in } \mathcal{V}, \tag{13.3}$$

for each realization of the ensemble. The intensity field $\boldsymbol{E}$ is taken to be irrotational, i.e.,

$$\nabla \times \boldsymbol{E}(\boldsymbol{x}) = 0 \qquad \text{in } \mathcal{V}, \tag{13.4}$$

which implies the existence of a potential field T, i.e.,

$$\boldsymbol{E} = -\nabla T. \tag{13.5}$$

Thus, $\boldsymbol{E}$ and T represent the electric field (negative of the temperature gradient) and electric potential (temperature) in the electrical (thermal) problem, respectively. We also specify the potential T on the boundary of $\mathcal{V}$. However, we are always ultimately interested in ergodic media and thus in the infinite-volume limit, i.e., $\mathcal{V} \to \Re^d$. Not surprisingly, we will see that the effective conductivity tensor for this large class of heterogeneous materials exists in this limit and is independent of the boundary conditions on the surface of $\mathcal{V}$ (Papanicolaou and Varadan 1979, Golden and Papanicolaou 1983). This result turns out to be true for all of the other effective properties.

Local Constitutive Relation

In order to close the system of (13.3) and (13.4), we will connect $\boldsymbol{J}$ to $\boldsymbol{E}$ by assuming a linear constitutive relation, i.e.,

$$\boldsymbol{J}(\boldsymbol{x}) = \boldsymbol{\sigma}(\boldsymbol{x}) \cdot \boldsymbol{E}(\boldsymbol{x}) \qquad \text{in } \mathcal{V}, \tag{13.6}$$

where, according to (13.1), the local conductivity tensor can be expressed as

$$\boldsymbol{\sigma}(\boldsymbol{x}) = \boldsymbol{\sigma}_1 \mathcal{I}^{(1)}(\boldsymbol{x}) + \boldsymbol{\sigma}_2 \mathcal{I}^{(2)}(\boldsymbol{x}), \tag{13.7}$$

and $\mathcal{I}^{(i)}(\boldsymbol{x})$ is the indicator function for phase i given by (2.1). The flux-intensity relation can be expressed in the inverted form

$$\boldsymbol{E}(\boldsymbol{x}) = \boldsymbol{\rho}(\boldsymbol{x}) \cdot \boldsymbol{J}(\boldsymbol{x}) \qquad \text{in } \mathcal{V}, \tag{13.8}$$

where $\boldsymbol{\rho}$ is the second-order *resistivity* tensor. The conductivity and resistivity tensors are related via

$$\boldsymbol{\rho} \cdot \boldsymbol{\sigma} = \boldsymbol{I}, \tag{13.9}$$

where $\boldsymbol{I}$ is the second-order identity tensor, with components, in a rectangular Cartesian coordinate system, given by the *Kronecker delta*

$$\delta_{ij} = \begin{cases} 1, & i = j, \\ 0, & \text{otherwise.} \end{cases} \tag{13.10}$$

It is important to observe that (13.3), which applies anywhere in the heterogeneous material, implies that the normal component of the flux J_n is continuous across the interface $\partial\mathcal{V}$. Similarly, the curl-free condition (13.4) implies that the potential T or, equivalently, the tangential component of the field E_t is continuous across the interface $\partial\mathcal{V}$. To summarize, the governing equations imply the *ideal* (or *perfect*) interface conditions

$$J_n \text{ and } T \text{ continuous across } \partial\mathcal{V}. \tag{13.11}$$

From a computational viewpoint, the problem can be reformulated as solving (13.3)–(13.6) in *each phase* subject to the interface conditions (13.11) and boundary conditions on the macroscopic sample surface. In general, the tangential component of the flux J_t and the normal component of the field E_n will jump across the interface. Imperfect interfaces (where the potential T jumps across the interface, due to some interfacial resistance, and/or J_n jumps across the interface) were mentioned in Chapter 1.

It is useful to remark on the tensorial nature of the conductivity $\boldsymbol{\sigma}$, which in standard indicial notation is denoted by σ_{ij}. That σ_{ij} is a second-order tensor can be demonstrated by showing that its d^2 components in d dimensions, referred to a coordinate system, transform to d^2 components in another "primed" coordinate system according to the transformation rule for second-order tensors, i.e.,

$$\sigma'_{ij} = l_{ik}l_{jl}\,\sigma_{kl}. \tag{13.12}$$

Here l_{ij} are the direction cosines, i.e., the cosine of the angle between the x_i-axis of the original coordinate system and the x'_j-axis of the new coordinate system. Repeated indices in (13.12) imply summation over all possible values of such indices, i.e.,

$$\sigma'_{ij} = \sum_{k=1}^{d}\sum_{l=1}^{d} l_{ik}l_{jl}\,\sigma_{kl}.$$

If the material admits an *energy density function* w such that the flux (in component form) can be expressed as

$$J_i = \frac{\partial w}{\partial E_i}, \tag{13.13}$$

then it immediately follows from the linear law (13.6) that w, up to an additive constant, is given by

$$w = \frac{1}{2}E_iJ_i = \frac{1}{2}E_i\sigma_{ij}E_j \geq 0, \tag{13.14}$$

and thus the conductivity tensor must be symmetric, i.e.,

$$\sigma_{ij} = \sigma_{ji}. \tag{13.15}$$

From relations (13.9) and (13.15), we see that the resistivity tensor $\boldsymbol{\rho}$ must also be symmetric. The symmetry condition (13.15) reduces the number of independent components from d^2 to $d(d+1)/2$. The argument above leading to (13.15) holds for all of the effective properties in class A (Table 1.1). However, in the case of conduction (electrical or thermal) or diffusion processes, the symmetry of the associated effective tensor (transport property) arises directly from Onsager's reciprocity theorem for irreversible (i.e., dissipative) processes, in which fluxes are linearly related to intensities (Onsager 1931a, Onsager 1931b). This symmetry condition then leads to the energy density function w as given, which physically is the *power dissipation* per unit volume of material. Both the effective dielectric constant and magnetic permeability tensors characterize reversible processes (polarization or magnetization), and therefore the reciprocity theorem does not apply. For such equilibrium properties, w represents the *energy stored* per unit volume of material.

Observe that the nonnegativity of the energy density as expressed by (13.14) places restrictions on the components of $\boldsymbol{\sigma}$ and implies that the conductivity tensor $\boldsymbol{\sigma}$ (or resistivity tensor $\boldsymbol{\rho}$) is *positive definite*. Recall that a symmetric second-order tensor $\boldsymbol{B}$ (B_{ij}) is *positive semidefinite*, if for any vector $\boldsymbol{a}$ (a_i) in $\Re^d$,

$$a_i B_{ij} a_j \geq 0. \tag{13.16}$$

If only the inequality of (13.16) applies for any nonzero vector $\boldsymbol{a} \neq \mathbf{0}$, then $\boldsymbol{B}$ is said to be positive definite. Since the equality of (13.14) holds only for $\boldsymbol{E} = \mathbf{0}$, $\boldsymbol{\sigma}$ is positive definite, and therefore is also positive semidefinite. (Any positive definite tensor $\boldsymbol{B}$ is also positive semidefinite, since either $\boldsymbol{a} \neq \mathbf{0}$ or $\boldsymbol{a} = \mathbf{0}$.) Note that for $\boldsymbol{\sigma}$ and $\boldsymbol{\rho}$ to be positive definite, it is necessary and sufficient for all of their d real eigenvalues to be positive.

13.2.2 Conduction Symmetry

Conduction symmetry is expressed by the property that the components σ_{ij} (σ_{ij}^{-1}) remain invariant under certain transformations of the coordinates. Basic coordinate changes are (a) reflection in a plane, (b) rotation about an axis, and (c) rotation about an axis combined with reflection in a plane that is normal to the axis. If σ_{ij} (σ_{ij}^{-1}) is invariant to reflection in a plane, then that plane is called a *plane of symmetry*. If σ_{ij} (σ_{ij}^{-1}) is invariant to rotation about an axis, then the axis is one of *rotational symmetry*.

We restrict ourselves initially to three dimensions. In light of the symmetry relation (13.15), the conductivity tensor has only six independent constants:

$$\begin{bmatrix} \sigma_{11} & \sigma_{12} & \sigma_{13} \\ \sigma_{12} & \sigma_{22} & \sigma_{23} \\ \sigma_{13} & \sigma_{23} & \sigma_{33} \end{bmatrix}. \tag{13.17}$$

We now consider a number of different cases where the tensor σ_{ij} remains invariant to certain transformations according to the rule (13.12). These symmetries result in a reduction in the number of independent components.

(i) Monoclinic Symmetry

For symmetry with respect to one plane, say the x_1-x_2 plane, it can be shown [using relation (13.12)] that σ_{ij} has 4 independent components:

$$\begin{bmatrix} \sigma_{11} & \sigma_{12} & 0 \\ \sigma_{12} & \sigma_{22} & 0 \\ 0 & 0 & \sigma_{33} \end{bmatrix}. \tag{13.18}$$

This is referred to as *monoclinic* symmetry.

(ii) Orthotropic Symmetry

For symmetry with respect to three orthogonal planes, it can be shown [using relation (13.12)] that σ_{ij} has 3 independent components:

$$\begin{bmatrix} \sigma_{11} & 0 & 0 \\ 0 & \sigma_{22} & 0 \\ 0 & 0 & \sigma_{33} \end{bmatrix}. \tag{13.19}$$

This is referred to as *orthotropic* symmetry.

(iii) Transversely Isotropic Symmetry

For symmetry with respect to a 90° rotation about one axis, say the x_3-axis, it can be shown [using relation (13.12)] that σ_{ij} has 2 independent components:

$$\begin{bmatrix} \sigma_{11} & 0 & 0 \\ 0 & \sigma_{11} & 0 \\ 0 & 0 & \sigma_{33} \end{bmatrix}. \tag{13.20}$$

This is referred to as *transversely isotropic* symmetry. Note that *crystals* with a 6-fold rotational symmetry axis (hexagonal), a 4-fold rotational symmetry axis (tetragonal), or a 3-fold rotational symmetry axis (trigonal) are all transversely isotropic with respect to the conductivity (Nye 1957). (However, neither a tetragonal nor a trigonal crystal has elastic transverse isotropy; see Section 13.3.2.) A crystal has n-fold rotational symmetry if the crystal appears unchanged when it is rotated about an axis through $2\pi/n$ radians, where n is a positive integer.

(iv) Isotropic Symmetry

When the conductivity is independent of the orientation of the coordinate system, it can be shown [using relation (13.12)] that σ_{ij} has only one independent component:

$$\begin{bmatrix} \sigma & 0 & 0 \\ 0 & \sigma & 0 \\ 0 & 0 & \sigma \end{bmatrix}. \tag{13.21}$$

This is referred to as *isotropic* symmetry. Note that the conductivity tensor of a cubic crystal (i.e., one that has four 3-fold rotation axes) is isotropic.

If the material is isotropic in d dimensions, then the conductivity tensor is specified by the scalar σ, i.e,

$$\boldsymbol{\sigma} = \sigma \boldsymbol{I}, \tag{13.22}$$

where

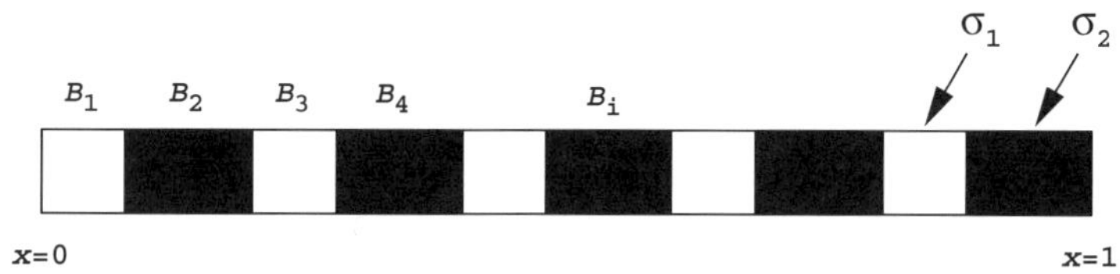

Figure 13.3 One-dimensional periodic composite. The white and shaded regions have conductivities σ_1 and σ_2, respectively.

$$\sigma = \frac{1}{d}\operatorname{Tr}\boldsymbol{\sigma} = \frac{1}{d}\boldsymbol{\sigma} : \boldsymbol{I} \tag{13.23}$$

and $\operatorname{Tr}\boldsymbol{a} \equiv \boldsymbol{a} : \boldsymbol{I} = a_{ii}$ indicates the trace of a second-order tensor $\boldsymbol{a}$.

13.2.3 Model One-Dimensional Problem

It is instructive to consider a simple one-dimensional periodic composite model (depicted in Figure 13.3) as a prelude to the homogenization of arbitrary periodic media in any dimension. The following example is an adaptation of one given by Persson, Persson, Svanstedt and Wyller (1989). Let the composite consist of a periodic arrangement of two phases with volume fractions ϕ_1 and ϕ_2 in the interval $[0, 1]$. Phase 1 lies in regions $B_1, B_3, \ldots, B_{n-1}$, and phase 2 lies in regions $B_2, B_4, \ldots, B_n$, where n is even. The fineness of the microstructure can be adjusted by varying the number of periodic unit (i.e., repeating) cells $m = n/2$. Let us subject the end $x = 0$ to a unit flux and the opposite end $x = 1$ to a potential (temperature) $T = 0$. Therefore, the potential (temperature) profile in the bar is governed by

$$-\frac{d}{dx}\left[\sigma(x)\frac{dT}{dx}\right] = 0$$

subject to the boundary conditions

$$-\sigma_1 \left.\frac{dT}{dx}\right|_{x=0} = 1, \qquad T(1) = 0.$$

According to the discussion of Section 13.2.1, this problem can be reformulated as

$$\frac{d^2T}{dx^2} = 0, \qquad \text{in } B_i, \ \ i = 1, \ldots, n,$$

$$\left.\begin{array}{c} \sigma^+(x)\dfrac{dT^+(x)}{dx} = \sigma^-(x)\dfrac{dT^-(x)}{dx} \\ T^+(x) = T^-(x) \end{array}\right\} \text{ on all interfaces,}$$

with the aforementioned boundary conditions, and where the superscripts + and – on a function denote its limits from the right and left, respectively. The potential $T(x)$ can be written as

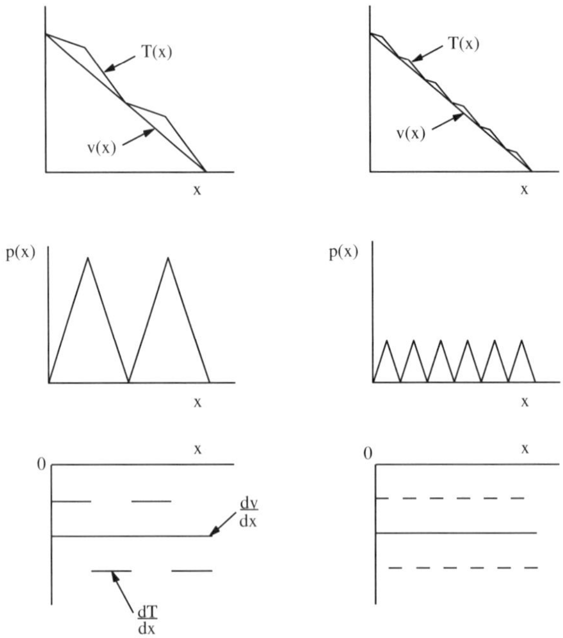

Figure 13.4 The functions $T(x)$, $v(x)$, and $p(x)$ and derivatives dT/dx and dv/dx for $n = 4$ (left panel) and $n = 10$ (right panel) when $\sigma_1 \geq \sigma_2$ and $\phi_1 = \phi_2 = 0.5$.

$$T(x) = v(x) + p(x),$$

where $p(x)$ is a *periodic function* and $v(x)$ is a linear function obeying the boundary value problem

$$\frac{d^2 v}{dx^2} = 0 \qquad \text{for all } x,$$

$$-\sigma_e \left.\frac{dv}{dx}\right|_{x=0} = 1, \qquad v(1) = 0,$$

and the effective conductivity σ_e is given by the harmonic mean of the phase conductivities, i.e.,

$$\sigma_e = \langle \sigma^{-1} \rangle^{-1} = \left(\frac{\phi_1}{\sigma_1} + \frac{\phi_2}{\sigma_2} \right)^{-1}. \tag{13.24}$$

Remarks:

1. It will be shown in Chapter 16 that the harmonic average expression (13.24) is also the correct result for arbitrary random one-dimensional media.
2. It is expected that $p(x)$ converges to zero and, hence, that $T(x)$ converges uniformly to $v(x)$ as $n \to \infty$. Note that while the derivative dT/dx does not converge to dv/dx, the derivatives converge in the *weak* sense; i.e., the average of dT/dx over the domain converges to dv/dx as $n \to \infty$. Figure 13.4 shows the afore-

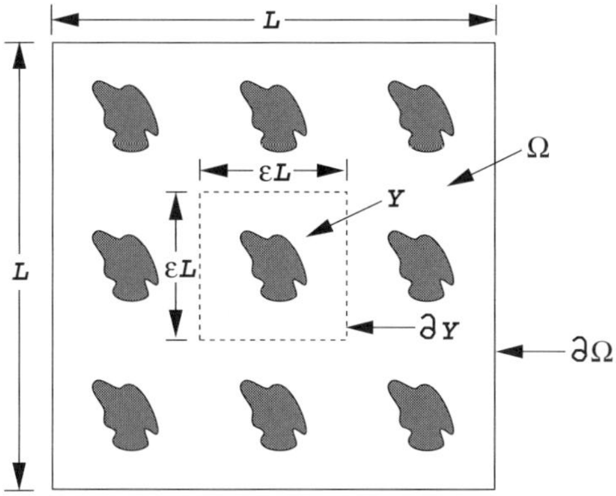

Figure 13.5 Two-dimensional periodic medium with square cells in a domain Ω with boundary $\partial\Omega$. The volume of the domain is L^2. The unit cell Y has a side of length ϵL, where $\epsilon = 1/3$, with a boundary ∂Y.

mentioned quantities plotted for the cases $n = 4$ and $n = 10$ with $\sigma_1 \geq \sigma_2$ and $\phi_1 = \phi_2 = 0.5$.

13.2.4 Homogenization of Periodic Problem in $\Re^d$

Consider now the homogenization problem for a periodic medium contained in some region Ω in $\Re^d$ with a piecewise smooth boundary $\partial\Omega$ (see Figure 13.5). The treatment given here is based on the works of Bensoussan et al. (1978) and Sanchez-Palencia (1980).

Consider an *arbitrary* periodic conductivity tensor $\boldsymbol{\sigma}(\boldsymbol{x})$ and introduce $\epsilon \geq 0$ as a parameter for varying the period by defining $\boldsymbol{\sigma}^\epsilon(\boldsymbol{x}) = \boldsymbol{\sigma}(\boldsymbol{y})$, where $\boldsymbol{y}$ is the fast variable

$$\boldsymbol{y} = \frac{\boldsymbol{x}}{\epsilon}. \tag{13.25}$$

We define a Y*-periodic* function $\boldsymbol{F}(\boldsymbol{y})$ to be one that is periodic in $\boldsymbol{y}$ over some *unit cell* Y. Thus, $\boldsymbol{\sigma}(\boldsymbol{y})$ is a Y-periodic function. To indicate that the local temperature depends on ϵ, we write $T^\epsilon(\boldsymbol{x}) = T(\boldsymbol{x}, \boldsymbol{y})$, where $\boldsymbol{x}$ on the *right side* signifies the slow variable. Also, we allow for a source term $f(\boldsymbol{x})$, which just depends on the slow variable. Thus, the governing equations are

$$-\frac{\partial}{\partial x_i}\left[\sigma_{ij}^\epsilon(\boldsymbol{x})\frac{\partial T^\epsilon}{\partial x_j}\right] = f(\boldsymbol{x}) \qquad \text{in } \Omega, \tag{13.26}$$

$$T^\epsilon = c_1 \quad \text{on } \Gamma_1, \qquad -\sigma_{ij}^\epsilon \frac{\partial T^\epsilon}{\partial x_j} n_i = c_2 \quad \text{on } \Gamma_2, \tag{13.27}$$

where Γ_1 is a portion of the boundary $\partial\Omega$, Γ_2 is the complementary surface, n_i is the unit outward normal, and c_1 and c_2 are constants. Thus, relations (13.27) represent the general situation of mixed boundary conditions, i.e., a Dirichlet condition on Γ_1 and a Neumann condition on Γ_2.

It is natural to perform a two-scale asymptotic expansion of the solution $T^\epsilon(\boldsymbol{x})$ in terms of the global and local variables, i.e.,

$$T^\epsilon(\boldsymbol{x}) = T_0(\boldsymbol{x}) + \epsilon T_1(\boldsymbol{x},\boldsymbol{y}) + \epsilon^2\, T_2(\boldsymbol{x},\boldsymbol{y}) + \cdots\ . \tag{13.28}$$

We remark that for any function $\Psi(\boldsymbol{x}) = \Phi(\boldsymbol{x},\boldsymbol{y})$, the chain rule yields

$$\nabla\Psi = \nabla_x\Phi + \frac{1}{\epsilon}\nabla_y\Phi. \tag{13.29}$$

Substitution of the asymptotic expansion into the differential equation (13.26) yields

$$\begin{aligned}
-\left[\frac{\partial}{\partial x_i} + \frac{1}{\epsilon}\frac{\partial}{\partial y_i}\right]&\left[\sigma_{ij}(\boldsymbol{y})\Big(\frac{\partial}{\partial x_j} + \frac{1}{\epsilon}\frac{\partial}{\partial y_j}\Big)(T_0 + \epsilon T_1 + \epsilon^2 T_2 + \cdots)\right] \\
&= [\epsilon^{-2}A_0 + \epsilon^{-1}A_1 + A_2][T_0 + \epsilon T_1 + \epsilon^2 T_2 + \cdots] = f,
\end{aligned} \tag{13.30}$$

where the differential operators A_0, A_1, and A_2 are given by

$$A_0 = -\frac{\partial}{\partial y_i}\left[\sigma_{ij}(\boldsymbol{y})\frac{\partial}{\partial y_j}\right], \qquad A_1 = -\frac{\partial}{\partial y_i}\left[\sigma_{ij}(\boldsymbol{y})\frac{\partial}{\partial x_j}\right] - \sigma_{ij}(\boldsymbol{y})\frac{\partial^2}{\partial x_i \partial y_j},$$

$$A_2 = -\sigma_{ij}(\boldsymbol{y})\,\frac{\partial^2}{\partial x_i x_j}.$$

Equating powers of ϵ leads to the following three lowest-order equations:

$$A_0 T_0 = 0, \tag{13.31}$$

$$A_0 T_1 + A_1 T_0 = 0, \tag{13.32}$$

$$A_0 T_2 + A_1 T_1 + A_2 T_0 = f. \tag{13.33}$$

Before stating a very useful theorem, we must first introduce a definition. The spatial average of a Y-periodic function $\boldsymbol{F}(\boldsymbol{y})$ over the unit cell Y, denoted by $\langle \boldsymbol{F}(\boldsymbol{y})\rangle$, is defined by

$$\langle \boldsymbol{F}\rangle \equiv \frac{1}{|Y|}\int_Y \boldsymbol{F}(\boldsymbol{y})d\boldsymbol{y}, \tag{13.34}$$

where $|Y|$ is the d-dimensional volume (measure) of Y. The quantity $\boldsymbol{F}(\boldsymbol{y})$ may represent a tensor of arbitrary order.

Theorem 13.1

Let $\boldsymbol{F}(\boldsymbol{y})$ be a Y-periodic function that is square integrable. For the boundary value problem

$$A_0\Phi(\boldsymbol{y}) = \boldsymbol{F}(\boldsymbol{y}) \quad in\ Y, \tag{13.35}$$

where $\Phi(\boldsymbol{y})$ is Y-periodic, the following hold:

(i) There exists a solution Φ if and only if $\langle \boldsymbol{F}(\boldsymbol{y})\rangle = 0$.
(ii) If a solution exists, it is unique up to an additive constant.

The proof of this theorem may be found in Bensoussan et al. (1978).

The function $T_0(\boldsymbol{x})$ automatically satisfies $A_0 T_0 = 0$. Equation (13.32) can be expressed as

$$A_0 T_1 = \frac{\partial \sigma_{ij}}{\partial y_i} \frac{\partial T_0}{\partial x_j},$$

where $\boldsymbol{x}$ is just a parameter here. Now consider the "cell problem"

$$A_0 \chi_j = \frac{\partial \sigma_{ij}}{\partial y_i} \quad \text{in } Y, \tag{13.36}$$

where the function $\chi_j(\boldsymbol{y})$ is Y-periodic. Gauss's divergence theorem gives

$$\int_Y \frac{\partial \sigma_{ij}}{\partial y_i} d\boldsymbol{y} = \int_{\partial Y} \sigma_{ij} n_i dS = 0, \tag{13.37}$$

where ∂Y is the boundary of the unit cell. Periodicity of $\boldsymbol{\sigma}$ renders the surface integral to be zero. Thus, by invoking Theorem 13.1, we get the solution of equation (13.32) as

$$T_1(\boldsymbol{x}, \boldsymbol{y}) = \chi_j(\boldsymbol{y}) \frac{\partial T_0}{\partial x_j} + u(\boldsymbol{x}), \tag{13.38}$$

where $\chi_j(\boldsymbol{y})$ is a solution of (13.36), unique up to a function $u(\boldsymbol{x})$ that depends on $\boldsymbol{x}$ only. By using the same theorem once more, we find that (13.33) has a Y-periodic solution T_2 if and only if

$$\langle f - A_1 T_1 - A_2 T_0 \rangle = 0,$$

or, equivalently,

$$\frac{1}{Y} \int_Y \left[f - \frac{\partial(\sigma_{ik}\chi_j)}{\partial y_k} \frac{\partial^2 T_0}{\partial x_i x_j} + \frac{\partial \sigma_{ij}}{\partial y_i} \frac{\partial u}{\partial x_j} - \sigma_{ik} \frac{\partial \chi_j}{\partial y_k} \frac{\partial^2 T_0}{\partial x_i \partial x_j} + \sigma_{ij} \frac{\partial^2 T_0}{\partial x_i \partial x_j} \right] d\boldsymbol{y} = 0.$$

The functions σ_{ij} and χ_j are Y-periodic, and the functions T_0 and u are independent of $\boldsymbol{y}$. Therefore, the second and third terms in the integral immediately above vanish by Gauss's divergence theorem, and the integral simplifies to

$$-(\sigma_e)_{ij} \frac{\partial^2 T_0}{\partial x_i \partial x_j} = f, \tag{13.39}$$

where

$$(\sigma_e)_{ij} = \langle \sigma_{ij} \rangle + \left\langle \sigma_{ik} \frac{\partial \chi_j}{\partial y_k} \right\rangle. \tag{13.40}$$

The relations (13.39) and (13.40) are the key results: They show that the heterogeneous periodic medium will behave like a homogeneous medium, governed by the steady-state conduction equation (13.39), with a constant conductivity tensor $\boldsymbol{\sigma}_e$ in the limit $\epsilon \to 0$. This is tantamount to the specification of an averaged Ohm's (Fourier's) law, i.e.,

$$\langle \boldsymbol{J}(\boldsymbol{x})\rangle = \boldsymbol{\sigma}_e \cdot \langle \boldsymbol{E}(\boldsymbol{x})\rangle, \tag{13.41}$$

where $\boldsymbol{\sigma}_e$ is given precisely by (13.40). This relation is easily obtained by taking the gradient of expansion (13.28), computing the local fields $\boldsymbol{E}$ and $\boldsymbol{J}$ using definitions (13.5) and (13.6), respectively, averaging the local fields, and utilizing the fact that the periodicity of χ renders $\langle \nabla \chi \rangle = 0$. In the limit that $\epsilon \to 0$, one finally gets the averaged relation (13.41). [The precise manner in which the local fields converge to the homogenized fields in the limit $\epsilon \to 0$ has been investigated by Bensoussan et al. (1978) and Sanchez-Palencia (1980).] The above analysis could also be applied to the situation in which the local conductivity $\boldsymbol{\sigma}(\boldsymbol{x},\boldsymbol{y})$ depends on the slow and fast variables $\boldsymbol{x}, \boldsymbol{y}$, leading to an effective conductivity $\boldsymbol{\sigma}_e$ that varies on macroscopic length scales.

13.2.5 Homogenization of Random Problem in $\Re^d$

The analogue of periodicity in the random setting is statistical homogeneity and ergodicity. Each realization ω of the random medium of space $\mathcal{V} \in \Re^d$ is partitioned into two random sets, or phases, as described in Section 13.1 (see Figure 13.2). Let $\boldsymbol{\sigma}^\epsilon(\boldsymbol{x}) = \boldsymbol{\sigma}(\boldsymbol{y};\omega)$ be the symmetric conductivity tensor, where $\boldsymbol{y} = \boldsymbol{x}/\epsilon$ is the fast variable associated with the microstructure. The conductivity $\boldsymbol{\sigma}(\boldsymbol{y};\omega)$ is a *statistically homogeneous function* (in the strict sense defined in Section 2.2.2) that is *positive definite* and *bounded*. The temperature distribution $T^\epsilon(\boldsymbol{x}) = T(\boldsymbol{x},\boldsymbol{y};\omega)$ satisfies

$$-\frac{\partial}{\partial x_i}\left[\sigma^\epsilon_{ij}(\boldsymbol{x})\frac{\partial T^\epsilon}{\partial x_j}\right] = f(\boldsymbol{x}) \qquad \text{in } \mathcal{V} \tag{13.42}$$

for each realization of the ensemble, and we assume that $T^\epsilon = 0$ is specified on the boundary of $\mathcal{V}$. Ultimately, we consider the limit that the volume $V \to \infty$ such that the volume fraction remains fixed, and then we invoke the ergodic hypothesis, i.e., we study *ergodic media*.

The problem now is to analyze the behavior of $T^\epsilon(\boldsymbol{x})$ as $\epsilon \to 0$. Papanicolaou and Varadan (1979) have shown that when σ^ϵ_{ij} is statistically homogeneous and ergodic, there exists an effective conductivity tensor $(\sigma_e)_{ij}$, independent of the boundary conditions in the infinite-volume limit (Papanicolaou and Varadan 1979, Golden and Papanicolaou 1983), such that if $T_0(\boldsymbol{x})$ is the solution of the *deterministic* heat equation

$$-(\sigma_e)_{ij}\frac{\partial^2 T_0}{\partial x_i \partial x_j} = f(\boldsymbol{x}), \tag{13.43}$$

then

$$\int_{\mathcal{V}} \left\langle |T^\epsilon(\boldsymbol{x}) - T_0(\boldsymbol{x})|^2 \right\rangle d\boldsymbol{x} \to 0, \quad \text{as } \epsilon \to 0. \tag{13.44}$$

Thus, an ergodic heterogeneous medium will behave like a homogeneous deterministic medium with conductivity tensor $(\sigma_e)_{ij}$ when ϵ becomes very small. The result (13.43) is identical in form to result (13.39) for the periodic medium.

To prove the above statement, one again assumes a two-scale asymptotic expansion of $T^\epsilon(\boldsymbol{x};\omega)$ of the type

$$T^\epsilon(\boldsymbol{x};\omega) = T_0(\boldsymbol{x}) + \epsilon T_1(\boldsymbol{x},\boldsymbol{y};\omega) + \epsilon^2 T_2(\boldsymbol{x},\boldsymbol{y};\omega) + \cdots, \tag{13.45}$$

which is substituted into the governing equation (13.42). Since the procedure is similar to the one used for periodic media, the details are not given here, except to say that in the end we find that the effective conductivity is defined by an averaged Ohm's (Fourier's) law

$$\langle \boldsymbol{J}(\boldsymbol{x})\rangle = \boldsymbol{\sigma}_e \cdot \langle \boldsymbol{E}(\boldsymbol{x})\rangle, \tag{13.46}$$

where

$$(\sigma_e)_{ij} = \langle \sigma_{ij}\rangle + \left\langle \sigma_{ik}\frac{\partial \chi_j}{\partial y_k}\right\rangle, \tag{13.47}$$

the function $\chi_j(\boldsymbol{y};\omega)$ solves

$$-\frac{\partial}{\partial y_i}\left[\sigma_{ik}(\boldsymbol{y};\omega)\frac{\partial}{\partial y_k}\chi_j(\boldsymbol{y};\omega)\right] = \frac{\partial}{\partial y_i}\sigma_{ij}(\boldsymbol{y};\omega) \quad \text{in } \Re^d, \tag{13.48}$$

and angular brackets denote an *ensemble* average. Note that the definition (13.46) is consistent with the one given in Table 1.1.

Remarks:

1. There are some important differences between the random and periodic settings. In the random setting, (13.48) always has a solution χ that is not statistically homogeneous but its gradient $\nabla\chi$ is statistically homogeneous and $\langle\chi\rangle = 0$. This is to be contrasted with the periodic setting, where χ itself is a periodic solution. Furthermore, the ensemble average of the right side of (13.48) is zero by ergodicity (as detailed in many of the proofs in Chapter 14), whereas the volume average of the right side of (13.36) in the periodic case is zero by periodicity.
2. The homogenized result (13.46) for the ergodic random setting becomes identical in form to the corresponding periodic result (13.41) in the limit $\epsilon \to 0$ when angular brackets are interpreted as ensemble averages.
3. Macroscopic variation of the effective conductivity tensor could be allowed for, i.e., $\boldsymbol{\sigma}_e(\boldsymbol{x})$ may be assumed to depend on the position $\boldsymbol{x}$, provided that the length scales associated with such variations of $\mathcal{O}(\Lambda)$ are large enough to assume statistical homogeneity at $\boldsymbol{x}$ (local statistical homogeneity).
4. It is shown in Section 14.1.2 that $\boldsymbol{\sigma}_e$ is symmetric and positive definite if the local conductivity tensor $\boldsymbol{\sigma}$ is symmetric and positive definite.
5. In the case of conduction, *macroscopically isotropic* composites are those whose effective conductivity tensor $\boldsymbol{\sigma}_e$ is specified by the scalar σ_e, i.e.,

$$\boldsymbol{\sigma}_e = \sigma_e \boldsymbol{I}.$$

6. For composites consisting of *isotropic phases*, the following general statements can be made:

   - Statistically isotropic composites are always macroscopically isotropic composites [e.g., statistically isotropic arrays of inclusions in a matrix (see Figure 12.6)].

   - However, statistical anisotropy, as measured by correlation functions (Section 2.2.2), does not necessarily imply a *macroscopically anisotropic* composite with an effective tensor $\boldsymbol{\sigma}_e$. For instance, composites with cubic symmetry are statistically anisotropic but are macroscopically isotropic (e.g., cubic lattices of spheres in a matrix).

   - Macroscopically anisotropic composites are necessarily *statistically anisotropic* [e.g., statistically anisotropic arrays of oriented cylinders or ellipsoids in a matrix (see Figure 7.1) or stratified media (see Figure 16.3)].

   Note that macroscopically anisotropic composites are not necessarily *statistically anisotropic* if some of the phases are anisotropic. For instance, systems of statistically isotropic arrays of spherical but anisotropic grains oriented in the same direction in an isotropic matrix are macroscopically anisotropic.
7. The effective relation (13.46) also applies to composites with an arbitrary number of phases that meet the standard two-scale assumption.
8. The effective conductivity tensor $\boldsymbol{\sigma}_e$ of a macroscopically anisotropic composite composed of M isotropic phases is a *homogeneous function of degree one* in its M scalar phase conductivities $\sigma_1, \ldots, \sigma_M$. For two isotropic phases, this means that $\boldsymbol{\sigma}_e(\sigma_1, \sigma_2)$ obeys the relation

$$\boldsymbol{\sigma}_e(\alpha\sigma_1, \alpha\sigma_2) = \alpha\, \boldsymbol{\sigma}_e(\sigma_1, \sigma_2) \qquad \text{for all } \alpha. \tag{13.49}$$

   If we set the constant α equal to $1/\sigma_1$, then we get

$$\frac{\boldsymbol{\sigma}_e(\sigma_1, \sigma_2)}{\sigma_1} = \boldsymbol{\sigma}_e(1, \sigma_2/\sigma_1).$$

   Thus, the homogeneity property (13.49) enables us to reduce the independent variables from two to one (σ_2/σ_1) without any loss of generality. The homogeneity property is trivially proven using the effective relation (13.46) and the observation that $\nabla \cdot \boldsymbol{J}' = 0$ when $\boldsymbol{J}' = \alpha \boldsymbol{J}$ and $\nabla \cdot \boldsymbol{J} = 0$.
9. A given formula for the effective conductivity as a function of the phase conductivities, $\boldsymbol{\sigma}_e(\boldsymbol{\sigma}_1, \boldsymbol{\sigma}_2)$, immediately gives equivalent formulas for the other class A problems by a simple replacement of the conductivities with the constants of interest. For example, letting $\boldsymbol{\epsilon}$ denote the dielectric constant tensor, we can obtain an expression for the effective dielectric constant $\boldsymbol{\epsilon}_e$ by the replacement $\boldsymbol{\sigma}_e \to \boldsymbol{\epsilon}_e$, $\boldsymbol{\sigma}_1 \to \boldsymbol{\epsilon}_1$, and $\boldsymbol{\sigma}_2 \to \boldsymbol{\epsilon}_2$. Similarly, for diffusion past fixed impermeable obstacles (phase 2), the

effective diffusion coefficient $\mathcal{D}_e$ is found by the replacement $\boldsymbol{\sigma}_e \rightarrow \mathcal{D}_e$, $\boldsymbol{\sigma}_1 \rightarrow \mathcal{D}_1$, and $\boldsymbol{\sigma}_2 \rightarrow \mathbf{0}$.

13.2.6 Frequency-Dependent Conductivity

For time-varying electric fields, such as in electromagnetic radiation, the phase conductivities and the effective conductivity $\boldsymbol{\sigma}_e$ are frequency-dependent and complex (real and imaginary parts). The optical properties of a material are often described in terms of the dielectric constant, which is also frequency-dependent and complex. The analysis is simplified if the wavelength is much larger than the heterogeneity length scale, since then Maxwell's equations of electromagnetic wave propagation reduce to the aforementioned steady-state conduction equations but with complex fields and complex properties. The reader is referred to Bergman (1978,1982) and Milton (1980, 1981a, 1981b) and references therein for pertinent literature on the subject of composites with complex properties.

13.3 Elastic Problem

In this section we discuss the local relations for the elasticity problem and elastic symmetry (Sokolnikoff 1956). The homogenized relations follow in the same way as in the previous section. We also discuss other problems that fall within class B.

13.3.1 Local Relations

For each realization ω of the random medium occupying the space $\mathcal{V} \in \Re^d$, we assume that phases 1 and 2 have constant stiffness tensors $\boldsymbol{C}_1$ and $\boldsymbol{C}_2$, respectively. It is desired to obtain the relevant equations governing the local elastostatic fields. In what follows, we temporarily drop ω from the notation.

Local Differential Equations

Let $\boldsymbol{\tau}(\boldsymbol{x})$ and $\varepsilon(\boldsymbol{x})$ denote respectively the symmetric local stress and strain tensors at position $\boldsymbol{x}$. Under steady state without sources, conservation of momentum requires the stress tensor $\boldsymbol{\tau}(\boldsymbol{x})$ to satisfy the equilibrium equations

$$\nabla \cdot \boldsymbol{\tau} = 0 \qquad \text{in } \mathcal{V} \tag{13.50}$$

for each realization of the ensemble. The strain field $\varepsilon(\boldsymbol{x})$ satisfies the compatibility relations

$$\nabla \times [\nabla \times \varepsilon]^{\mathrm{T}} = 0 \qquad \text{in } \mathcal{V}, \tag{13.51}$$

which implies the existence of a displacement field $\boldsymbol{u}$. (Here the superscript T denotes the transpose operation.) In other words, the strain can be written as a symmetrized gradient of displacements

$$\varepsilon(\boldsymbol{x}) = \frac{1}{2}\left[\nabla \boldsymbol{u}(\boldsymbol{x}) + \nabla \boldsymbol{u}(\boldsymbol{x})^{\mathrm{T}}\right]. \tag{13.52}$$

We also specify the displacement $\boldsymbol{u}$ on the boundary of $\mathcal{V}$.

Local Constitutive Relation

In order to obtain the strain field ε or displacement field $\boldsymbol{u}$ from the relations (13.50) and (13.51), we will connect $\boldsymbol{\tau}$ to ε by assuming a linear constitutive relation, i.e.,

$$\boldsymbol{\tau}(\boldsymbol{x}) = \boldsymbol{C}(\boldsymbol{x}) : \varepsilon(\boldsymbol{x}) \qquad \text{in } \mathcal{V}, \tag{13.53}$$

where

$$\boldsymbol{C}(\boldsymbol{x}) = \boldsymbol{C}_1 \mathcal{I}^{(1)}(\boldsymbol{x}) + \boldsymbol{C}_2 \mathcal{I}^{(2)}(\boldsymbol{x}) \tag{13.54}$$

is the local stiffness tensor and $\mathcal{I}^{(i)}(\boldsymbol{x})$ is the indicator function for phase i, given by (2.1). Relation (13.53) is the *generalization* of Hooke's law. Here the symbol : denotes the contraction with respect to two indices. For example,

$$\boldsymbol{a} : \boldsymbol{b} = a_{ij} b_{ji}, \qquad \boldsymbol{B} : \boldsymbol{a} = B_{ijkl} a_{lk}, \tag{13.55}$$

where $\boldsymbol{a}$ and $\boldsymbol{b}$ are second-order tensors and $\boldsymbol{B}$ is a fourth-order tensor. The stress–strain relations can be expressed in inverted form

$$\varepsilon(\boldsymbol{x}) = \boldsymbol{S}(\boldsymbol{x}) : \boldsymbol{\tau}(\boldsymbol{x}), \tag{13.56}$$

where S_{ijkl} is the *compliance* tensor. The stiffness and compliance tensors are related by

$$\boldsymbol{S} : \boldsymbol{C} = \boldsymbol{I}, \tag{13.57}$$

where in component form

$$I_{ijkl} = \frac{1}{2}[\delta_{ik}\,\delta_{jl} \;+\; \delta_{il}\,\delta_{jk}] \tag{13.58}$$

is the *fourth-order identity tensor*. Note that the notation used to designate the full fourth-order identity tensor $\boldsymbol{I}$ is the same as that for the second-order identity tensor with components given by the Kronecker delta δ_{ij} (13.10). Unless otherwise noted, it will be apparent by the context whether the second- or fourth-order identity tensor is being used.

Note that (13.50), which applies anywhere in the heterogeneous material, implies that the traction vector $\boldsymbol{t} = \boldsymbol{\tau} \cdot \boldsymbol{n}$ is continuous across the interface $\partial\mathcal{V}$. Similarly, the compatibility condition (13.51) implies that the displacement $\boldsymbol{u}$ is continuous across the interface $\partial\mathcal{V}$. To summarize, the governing equations imply the *ideal* (or *perfect*) interface conditions

$$\boldsymbol{t} \text{ and } \boldsymbol{u} \text{ continuous across } \partial\mathcal{V}. \tag{13.59}$$

Computationally, the problem can be reformulated as solving (13.50)–(13.53) in *each phase* subject to the interface conditions (13.59) and boundary conditions on the macroscopic sample surface. Imperfect interfaces (where the displacement $\boldsymbol{u}$ jumps across the interface, due to debonding, and/or $\boldsymbol{t}$ jumps across the interface) were noted in Chapter 1.

That C_{ijkl} (S_{ijkl}) is a fourth-order tensor can be demonstrated by showing that its d^4 components in d dimensions, referred to a coordinate system, transform to d^4 components in another "primed" coordinate system according to the transformation rule for fourth-order tensors, i.e.,

$$C'_{ijkl} = l_{im}l_{jn}l_{kp}l_{lq}\, C_{mnpq}. \tag{13.60}$$

Here l_{ij} are the aforementioned direction cosines.

Now, since the stress tensor is symmetric (i.e., $\tau_{ij} = \tau_{ji}$), we must have

$$C_{ijkl} = C_{jikl}. \tag{13.61}$$

Also, since the strain tensor ε_{ij} is symmetric, then

$$C_{ijkl} = C_{ijlk}. \tag{13.62}$$

These two conditions enable one to reduce the 81 independent components of C_{ijkl} to 36 for three-dimensional elasticity. In d dimensions, the d^4 independent components are reduced to $[d(d+1)/2]^2$ components.

When the stress can be expressed as a derivative of the *strain energy density function* u with respect to strain such that

$$\tau_{ij} = \frac{\partial u}{\partial \varepsilon_{ij}}, \tag{13.63}$$

then it immediately follows from the linear Hooke's law (13.53) that u, up to an additive constant, is given by

$$u = \frac{1}{2}\tau_{ij}\varepsilon_{ij} = \frac{1}{2}\varepsilon_{kl}C_{ijkl}\varepsilon_{ij} \geq 0, \tag{13.64}$$

and hence the stiffness tensor must additionally have the symmetry

$$C_{ijkl} = C_{klij}. \tag{13.65}$$

(The scalar quantity u should not be confused with the vector displacement $\boldsymbol{u}$.) In d dimensions, condition (13.65) further reduces the number of independent elastic constants to $d(d+1)(d^2+d+2)/8$. For $d = 3$, this means that the number of independent elastic constants is reduced to 21. From relations (13.57) and (13.61)–(13.65), we see that the compliance tensor S_{ijkl} must possess the same symmetries as the stiffness tensor C_{ijkl}.

Note that the nonnegativity of the strain energy density function u [cf. (13.64)] places restrictions on the components of $\boldsymbol{C}$ and implies that the stiffness tensor $\boldsymbol{C}$ (or compliance tensor $\boldsymbol{S}$) is *positive definite*. Recall that a symmetric fourth-order tensor $\boldsymbol{A}$ (A_{ijkl}) is *positive semidefinite*, if for any second-order tensor $\boldsymbol{b}$ (b_{ij}) in $\Re^d$,

$$b_{ij}A_{ijkl}b_{kl} \geq 0. \tag{13.66}$$

If only the inequality of (13.66) applies for any nonzero tensor $\boldsymbol{b} \neq \mathbf{0}$, then $\boldsymbol{A}$ is said to be positive definite. Since the equality of (13.64) holds only for $\varepsilon = \mathbf{0}$, $\boldsymbol{C}$ is positive definite, and therefore is also positive semidefinite. (Any positive definite tensor $\boldsymbol{A}$ is also positive semidefinite, since either $\boldsymbol{b} \neq \mathbf{0}$ or $\boldsymbol{b} = \mathbf{0}$.) Note that for $\boldsymbol{C}$ and $\boldsymbol{S}$ to be positive definite, it is necessary and sufficient that their $d(d+1)/2$ real eigenvalues be positive.

In summary, the number of independent elastic constants for the general anisotropic linearly elastic material is reduced to 21 for three-dimensional elasticity. Moreover, if certain elastic symmetries exist in the material, this number will be reduced even further, as discussed below.

13.3.2 Elastic Symmetry

Elastic symmetry is expressed by the property that the coefficients $C_{ijkl}(S_{ijkl})$ remain invariant under certain transformations of the coordinates. As in the previous section, basic coordinate changes are (a) reflection in a plane, (b) rotation about an axis, and (c) rotation about an axis combined with reflection in a plane that is normal to the axis.

Consider a linearly elastic homogeneous material. We restrict ourselves initially to three dimensions. In light of the symmetry relations (13.61), (13.62), and (13.65), the generalized Hooke's law (13.53) really represents only six independent equations with 21 elastic constants. This is conveniently represented as a matrix equation expressing a six-element column vector of stresses in terms of a six-element column vector of strains. We write

$$\begin{bmatrix} \tau_1 \\ \tau_2 \\ \tau_3 \\ \tau_4 \\ \tau_5 \\ \tau_6 \end{bmatrix} = \begin{bmatrix} C_{11} & C_{12} & C_{13} & C_{14} & C_{15} & C_{16} \\ C_{12} & C_{22} & C_{23} & C_{24} & C_{25} & C_{26} \\ C_{13} & C_{23} & C_{33} & C_{34} & C_{35} & C_{36} \\ C_{14} & C_{24} & C_{34} & C_{44} & C_{45} & C_{46} \\ C_{15} & C_{25} & C_{35} & C_{45} & C_{55} & C_{56} \\ C_{16} & C_{26} & C_{36} & C_{46} & C_{56} & C_{66} \end{bmatrix} \begin{bmatrix} \varepsilon_1 \\ \varepsilon_2 \\ \varepsilon_3 \\ \varepsilon_4 \\ \varepsilon_5 \\ \varepsilon_6 \end{bmatrix}, \tag{13.67}$$

where the elements of the two column vectors τ_i and ε_i are

$$\tau_1 = \tau_{11}\,,\ \tau_2 = \tau_{22}\,,\ \tau_3 = \tau_{33}\,,\ \tau_4 = \tau_{23}\,,\ \tau_5 = \tau_{13}\,,\ \tau_6 = \tau_{12},$$
$$\varepsilon_1 = \varepsilon_{11}\,,\ \varepsilon_2 = \varepsilon_{22}\,,\ \varepsilon_3 = \varepsilon_{33}\,,\ \varepsilon_4 = 2\varepsilon_{23}\,,\ \varepsilon_5 = 2\varepsilon_{13}\,,\ \varepsilon_6 = 2\varepsilon_{12}.$$

The matrix equation (13.67) can be written more compactly as

$$\tau_i = C_{ij}\varepsilon_j, \tag{13.68}$$

where C_{ij} is the 6×6 symmetric matrix whose elements are given in (13.67). The mapping between the tensor C_{ijkl} and the matrix C_{pq} is accomplished by replacing the subscripts ij (or kl) by p (or q) using the following rules:

$$11 \leftrightarrow 1\,,\ 22 \leftrightarrow 2\,,\ 33 \leftrightarrow 3\,,\ 23 \text{ or } 32 \leftrightarrow 4\,,\ 13 \text{ or } 31 \leftrightarrow 5\,,\ 12 \text{ or } 21 \leftrightarrow 6.$$

The relation (13.68) may be inverted to give

$$\varepsilon_i = S_{ij}\tau_j, \tag{13.69}$$

where the 6×6 symmetric matrix S_{ij} is just the inverse of the matrix C_{ij} given in (13.68). The transformation between the tensor S_{ijkl} and the matrix S_{pq} is similar to that between C_{ijkl} and C_{pq} except for the following conditions: $S_{pq} = S_{ijkl}$ if $1 \leq p, q \leq 3$; $S_{pq} = 2S_{ijkl}$ if $1 \leq p \leq 3$ and $4 \leq q \leq 6$ or if $4 \leq p \leq 6$ and $1 \leq q \leq 3$; $S_{pq} = 4S_{ijkl}$ if $4 \leq p, q \leq 6$. It is noteworthy that the matrices C_{ij} and S_{ij} do not transform as tensors.

In what follows we consider a number of different cases where the tensor C_{ijkl} (S_{ijkl}) remains invariant to certain transformations according to the rule (13.60). We then represent the elastic symmetry class as a 6×6 stiffness or compliance matrix.

(i) Monoclinic Symmetry

For symmetry with respect to one plane, say the x_1-x_2 plane, it can be shown that C_{ij} has 13 independent components:

$$\boldsymbol{C} = \begin{bmatrix} C_{11} & C_{12} & C_{13} & 0 & 0 & C_{16} \\ C_{12} & C_{22} & C_{23} & 0 & 0 & C_{26} \\ C_{13} & C_{23} & C_{33} & 0 & 0 & C_{36} \\ 0 & 0 & 0 & C_{44} & C_{45} & 0 \\ 0 & 0 & 0 & C_{45} & C_{55} & 0 \\ C_{16} & C_{26} & C_{36} & 0 & 0 & C_{66} \end{bmatrix}. \tag{13.70}$$

This is referred to as *monoclinic* symmetry. As in the general anisotropic case, here a pure shear strain can give rise to a normal stress.

(ii) Orthotropic Symmetry

For symmetry with respect to three orthogonal planes, it can be shown that C_{ij} has 9 independent components:

$$\boldsymbol{C} = \begin{bmatrix} C_{11} & C_{12} & C_{13} & 0 & 0 & 0 \\ C_{12} & C_{22} & C_{23} & 0 & 0 & 0 \\ C_{13} & C_{23} & C_{33} & 0 & 0 & 0 \\ 0 & 0 & 0 & C_{44} & 0 & 0 \\ 0 & 0 & 0 & 0 & C_{55} & 0 \\ 0 & 0 & 0 & 0 & 0 & C_{66} \end{bmatrix}. \tag{13.71}$$

This is referred to as *orthotropic* symmetry.

(iii) Transverse Square Symmetry

For symmetry with respect to a 90° rotation about one axis, say the x_1-axis, it can be shown that C_{ij} has 6 independent components:

$$
\boldsymbol{C} = \begin{bmatrix} C_{11} & C_{12} & C_{12} & 0 & 0 & 0 \\ C_{12} & C_{22} & C_{23} & 0 & 0 & 0 \\ C_{12} & C_{23} & C_{22} & 0 & 0 & 0 \\ 0 & 0 & 0 & C_{44} & 0 & 0 \\ 0 & 0 & 0 & 0 & C_{66} & 0 \\ 0 & 0 & 0 & 0 & 0 & C_{66} \end{bmatrix}. \tag{13.72}
$$

This is referred to as *transverse square* symmetry. A tetragonal crystal has such symmetry.

(iv) Transversely Isotropic Symmetry

For symmetry of rotation with respect to one axis, say the x_1-axis, it is easily shown that C_{ij} has 5 independent components:

$$
\boldsymbol{C} = \begin{bmatrix} C_{11} & C_{12} & C_{12} & 0 & 0 & 0 \\ C_{12} & C_{22} & C_{23} & 0 & 0 & 0 \\ C_{12} & C_{23} & C_{22} & 0 & 0 & 0 \\ 0 & 0 & 0 & \frac{1}{2}(C_{22} - C_{23}) & 0 & 0 \\ 0 & 0 & 0 & 0 & C_{66} & 0 \\ 0 & 0 & 0 & 0 & 0 & C_{66} \end{bmatrix}. \tag{13.73}
$$

This is referred to as *transversely isotropic* symmetry. A crystal with a 6-fold rotational symmetry axis (hexagonal) is transversely isotropic with respect to the stiffness (Nye 1957). Although a crystal with a 3-fold rotational symmetry axis (trigonal) is not transversely isotropic (Nye 1957), there are only two independent elastic moduli in the transverse plane. Hence, *two-dimensional* crystals with a 3-fold rotational symmetry axis are elastically isotropic.

It is useful to explicitly write the stress–strain relations here, since the case of transverse isotropy is treated in some detail in subsequent portions of the book. Given C_{ij} above, it follows that

$$
\begin{aligned}
\tau_{11} &= C_{11}\varepsilon_{11} + C_{12}\varepsilon_{22} + C_{12}\varepsilon_{33}, \\
\tau_{22} &= C_{12}\varepsilon_{11} + C_{22}\varepsilon_{22} + C_{23}\varepsilon_{33}, \\
\tau_{33} &= C_{12}\varepsilon_{11} + C_{23}\varepsilon_{22} + C_{22}\varepsilon_{33}, \\
\tau_{23} &= (C_{22} - C_{23})\varepsilon_{23}, \\
\tau_{13} &= 2C_{66}\varepsilon_{13}, \\
\tau_{12} &= 2C_{66}\varepsilon_{12}.
\end{aligned} \tag{13.74}
$$

It is convenient to relate the five constants C_{11}, C_{12}, C_{22}, C_{23}, and C_{66} to more easily measurable *engineering* elastic moduli, such as Young's modulus, Poisson's ratios, and the shear modulus. First consider applying a uniform strain ε in both the x_2-direction and x_3-direction such that axial extension in the x_1-direction is prohibited, i.e.,

$$\varepsilon_{11} = 0, \quad \varepsilon_{22} = \varepsilon_{33} = \varepsilon.$$

Letting

$$\tau_{22} = \tau_{33} = \tau,$$

we find from (13.74) that

$$\tau = 2k_{23}\varepsilon,$$

where

$$k_{23} = \frac{1}{2}(C_{22} + C_{23}) \tag{13.75}$$

is the *plane strain* or *transverse* bulk modulus.

Now consider a simple state of uniaxial stress, i.e.,

$$\tau_{11} \neq 0, \quad \tau_{22} = \tau_{33} = \tau_{12} = \tau_{23} = \tau_{13} = 0.$$

For such a state, (13.74) reveals that

$$\tau_{11} = E_{11}\varepsilon_{11},$$

where

$$E_{11} = C_{11} - \frac{2C_{12}^2}{C_{22} + C_{23}} \tag{13.76}$$

is the *longitudinal* Young's modulus. The Poisson ratios that characterize the typical lateral contraction (expansion) that accompanies uniaxial tension (compression) in the x_1-direction are defined by the relations

$$\nu_{12} = -\frac{\varepsilon_{22}}{\varepsilon_{11}}, \qquad \nu_{13} = -\frac{\varepsilon_{33}}{\varepsilon_{11}}.$$

Generally, ν_{ij} is Poisson's ratio, where the first index i indicates the direction of the imposed stress or strain and the second index j indicates the response direction. For the aforementioned uniaxial stress state, we have from (13.74) that

$$\nu_{12} = \nu_{13} = \frac{C_{12}}{C_{22} + C_{23}}. \tag{13.77}$$

The directly measurable shear moduli are defined in the usual way, i.e.,

$$G_{12} = G_{13} = C_{66}, \tag{13.78}$$

$$G_{23} = \frac{1}{2}(C_{22} - C_{23}). \tag{13.79}$$

Using the relations above, we can express the five coefficients of (13.73) in terms of the directly measurable moduli:

$$C_{11} = E_{11} + 4\nu_{12}^2 k_{23}, \qquad C_{12} = 2k_{23}\nu_{12},$$

$$C_{22} = k_{23} + G_{23}, \qquad C_{23} = k_{23} - G_{23}, \qquad C_{66} = G_{12}.$$

Other engineering moduli could have been measured. For example, for a state of uniaxial stress in the x_2-direction, one would measure E_{22}, ν_{21} ($\neq \nu_{12}$), and ν_{23}, and for uniaxial stress in the x_3-direction, one would measure E_{33}, ν_{31} ($\neq \nu_{13}$), and ν_{32}. Now, since some of these moduli are equal by symmetry, we will henceforth use the following notation:

$$E^L \equiv E_{11}, \quad \nu^L \equiv \nu_{12} = \nu_{13}, \quad G^L \equiv G_{12} = G_{13}, \tag{13.80}$$

$$E^T \equiv E_{22} = E_{33}, \quad \nu^T \equiv \nu_{23} = \nu_{32}, \quad G^T \equiv G_{23}, \quad k^T \equiv k_{23}. \tag{13.81}$$

Here E^L, ν^L, and G^L are the longitudinal Young modulus, Poisson ratio, and shear modulus, respectively, whereas E^T, ν^T, G^T, and k^T are the transverse Young modulus, Poisson ratio, shear modulus, and bulk modulus, respectively.

Since there are only five independent moduli, there are interrelations among the properties. For example, two of them are

$$G^T = \frac{E^T}{2(1+\nu^T)} \tag{13.82}$$

and

$$\frac{4}{E^T} = \frac{1}{k^T} + \frac{1}{G^T} + \frac{4(\nu^L)^2}{E^L}. \tag{13.83}$$

The previous two relations can be combined to give the following relation for the transverse Poisson ratio ν^T:

$$\nu^T = \frac{k^T - G^T - 4k^T G^T (\nu^L)^2 / E^L}{k^T + G^T + 4k^T G^T (\nu^L)^2 / E^L}. \tag{13.84}$$

Other property interrelations have been given by Christensen (1979).

The nonnegativity of the strain energy density function u [cf. (13.64)], implies that the moduli k^T, G^T, and E^L are all positive. Using these nonnegativity conditions and (13.84), it is easily shown that

$$-1 \leq \nu^T \leq 1. \tag{13.85}$$

Note that the transverse Poisson ratio can be negative, with -1 being the lower limit. A negative Poisson's ratio implies that a uniaxial tensile (compressive) load applied in any direction in the transverse plane will lead to an expansion (contraction) of the material in the direction orthogonal to the applied load. The lower limit of -1 is found by taking the limit $G^T/k^T \to \infty$ and $E^L/G^T \to \infty$. The upper limit of 1 is obtained by taking the limit that $k^T/G^T \to \infty$ and $E^L/k^T \to \infty$. By contrast, for a three-dimensional isotropic material, Poisson's ratio must lie in the interval $[-1, 0.5]$, as shown below. It is interesting to note that in the limit $E^L/(\nu^L)^2 \to \infty$, (13.83) and (13.84) respectively reduce to

$$4/E^T = 1/k^T + 1/G^T, \quad \nu^T = (k^T - G^T)/(k^T + G^T),$$

which are the same as the *two-dimensional isotropic elasticity* results, given by (13.103) and (13.104) with $d = 2$.

We can express the aforementioned results for transverse isotropy in terms of the 6×6 compliance matrix S_{ij} as follows:

$$\boldsymbol{S} = \begin{bmatrix} \dfrac{1}{E^L} & -\dfrac{\nu^L}{E^L} & -\dfrac{\nu^L}{E^L} & 0 & 0 & 0 \\ -\dfrac{\nu^L}{E^L} & \dfrac{1}{E^T} & -\dfrac{\nu^T}{E^T} & 0 & 0 & 0 \\ -\dfrac{\nu^L}{E^L} & -\dfrac{\nu^T}{E^T} & \dfrac{1}{E^T} & 0 & 0 & 0 \\ 0 & 0 & 0 & \dfrac{1}{G^T} & 0 & 0 \\ 0 & 0 & 0 & 0 & \dfrac{1}{G^L} & 0 \\ 0 & 0 & 0 & 0 & 0 & \dfrac{1}{G^L} \end{bmatrix}. \tag{13.86}$$

(v) Cubic Symmetry

For symmetry with respect to 90° rotations about two perpendicular axes, say the x_1-axis and x_2-axis, it can be shown that C_{ij} has 3 independent components:

$$\boldsymbol{C} = \begin{bmatrix} C_{11} & C_{12} & C_{12} & 0 & 0 & 0 \\ C_{12} & C_{11} & C_{12} & 0 & 0 & 0 \\ C_{12} & C_{12} & C_{11} & 0 & 0 & 0 \\ 0 & 0 & 0 & C_{44} & 0 & 0 \\ 0 & 0 & 0 & 0 & C_{44} & 0 \\ 0 & 0 & 0 & 0 & 0 & C_{44} \end{bmatrix}. \tag{13.87}$$

This is referred to as *cubic* symmetry.

(vi) Isotropic Symmetry

When the elastic moduli are independent of the orientation of the coordinate system, there are two independent elastic moduli:

$$\boldsymbol{C} = \begin{bmatrix} C_{11} & C_{12} & C_{12} & 0 & 0 & 0 \\ C_{12} & C_{11} & C_{12} & 0 & 0 & 0 \\ C_{12} & C_{12} & C_{11} & 0 & 0 & 0 \\ 0 & 0 & 0 & \frac{1}{2}(C_{11}-C_{12}) & 0 & 0 \\ 0 & 0 & 0 & 0 & \frac{1}{2}(C_{11}-C_{12}) & 0 \\ 0 & 0 & 0 & 0 & 0 & \frac{1}{2}(C_{11}-C_{12}) \end{bmatrix}. \tag{13.88}$$

This is referred to as *isotropic* symmetry. The elastic constants C_{12} and $(C_{11} - C_{12})$ are identified as the Lamé constant λ and shear modulus G, i.e.,

$$C_{12} = \lambda, \qquad \frac{1}{2}(C_{11} - C_{12}) = G, \qquad C_{11} = \lambda + 2G.$$

In indicial notation, the stress–strain relations for a d-dimensional material are written as

$$\tau_{ij} = \lambda \varepsilon_{kk}\, \delta_{ij} + 2G\varepsilon_{ij}, \qquad i,j = 1,\ldots,d. \tag{13.89}$$

The strain–stress relation can be written as

$$\varepsilon_{ij} = -\frac{\nu}{E}\,\tau_{kk}\,\delta_{ij} + \frac{1+\nu}{E}\tau_{ij}, \qquad i,j = 1,\ldots,d, \tag{13.90}$$

where E is Young's modulus (uniaxial stress in any direction divided by the strain in the same direction) and ν is Poisson's ratio (negative of the strain in the direction of a uniaxial stress divided by the associated transverse strain).

Alternatively, the stress–strain relations can be represented in terms of the *deviatoric* and *dilatational* (hydrostatic) components of the stress and strain. Let s_{ij} and e_{ij} be the deviatoric components of stress and strain, defined in any space dimension d as

$$s_{ij} = \tau_{ij} - \frac{1}{d}\,\delta_{ij}\,\tau_{kk}, \tag{13.91}$$

$$e_{ij} = \varepsilon_{ij} - \frac{1}{d}\,\delta_{ij}\varepsilon_{kk}. \tag{13.92}$$

Thus, we have from (13.89) that

$$\begin{aligned} s_{ij} &= 2Ge_{ij}, \qquad \text{(deviatoric relation)},\\ \tau_{kk} &= dK\varepsilon_{kk}, \qquad \text{(dilatational relation)}, \end{aligned} \tag{13.93}$$

where K is the bulk modulus governing volumetric or dilatational changes, which can be expressed in terms of λ and G as

$$K = \lambda + \frac{2}{d}G. \tag{13.94}$$

The nonnegativity of the strain energy density function u [cf. (13.64)] implies that the moduli K, G, and E are all positive.

For subsequent discussion it will be convenient to obtain the appropriate form of the fourth-order tensors C_{ijkl} and S_{ijkl} for the isotropic case. It is seen from the stress–strain relation (13.89) that the stiffness tensor can be written in terms of the projection tensors $\mathbf{\Lambda}_h$ and $\mathbf{\Lambda}_s$ as follows:

$$\mathbf{C} = dK\mathbf{\Lambda}_h + 2G\mathbf{\Lambda}_s, \tag{13.95}$$

where, in component form,

$$(\Lambda_h)_{ijkl} = \frac{1}{d}\delta_{ij}\delta_{kl}, \tag{13.96}$$

$$(\Lambda_s)_{ijkl} = \frac{1}{2}[\delta_{ik}\delta_{jl} + \delta_{il}\delta_{jk}] - \frac{1}{d}\delta_{ij}\delta_{kl}. \tag{13.97}$$

The tensor $\mathbf{\Lambda}_h$ projects onto fields that are everywhere isotropic, i.e., hydrostatic fields, whereas the tensor $\mathbf{\Lambda}_s$ projects onto fields that are everywhere trace-free, i.e., shear fields. We note the following useful identities:

$$(\Lambda_h)_{ijkl}I_{ijkl} = 1, \qquad (\Lambda_s)_{ijkl}I_{ijkl} = \frac{(d-1)(d+2)}{2},$$
$$(\Lambda_h)_{ijkl} + (\Lambda_s)_{ijkl} = I_{ijkl}, \qquad (\Lambda_h)_{ijmn}(\Lambda_h)_{mnkl} = (\Lambda_h)_{ijkl}, \tag{13.98}$$
$$(\Lambda_s)_{ijmn}(\Lambda_s)_{mnkl} = (\Lambda_s)_{ijkl}, \qquad (\Lambda_h)_{ijmn}(\Lambda_s)_{mnkl} = 0.$$

Similarly, the strain–stress relations (13.90) give the compliance tensor $\boldsymbol{S}$ as

$$\boldsymbol{S} = \frac{1+\nu(1-d)}{E}\mathbf{\Lambda}_h + \frac{1+\nu}{E}\mathbf{\Lambda}_s. \tag{13.99}$$

Taking the inverse of relation (13.95) and using the identities (13.98) yields

$$\boldsymbol{S} = \frac{1}{dK}\mathbf{\Lambda}_h + \frac{1}{2G}\mathbf{\Lambda}_s. \tag{13.100}$$

Comparison of expressions (13.99) and (13.100) gives the interrelations

$$G = \frac{E}{2(1+\nu)}, \tag{13.101}$$

$$K = \frac{E}{d[1+\nu(1-d)]}. \tag{13.102}$$

Interrelation (13.101), unlike (13.102), is independent of the space dimension. Combination of these two equations yields the useful interrelations

$$\frac{d^2}{E} = \frac{1}{K} + \frac{d(d-1)}{2G}, \tag{13.103}$$

$$\nu = \frac{dK - 2G}{d(d-1)K + 2G}. \tag{13.104}$$

Using expression (13.104) and the positivity of K and G, it can be shown that Poisson's ratio ν is bounded according to

$$-1 \le \nu \le \frac{1}{(d-1)}. \tag{13.105}$$

The upper limit of $\nu = 1/(d-1)$ is obtained from (13.104) by taking the limit that $K/G \to \infty$, i.e., the *incompressible limit*. In three dimensions, when $\nu \approx 1/2$, as in rubbery solids and liquids, the bulk modulus greatly exceeds the shear modulus. The lower limit of $\nu = -1$ is found by taking the limit $G/K \to \infty$. For materials with $\nu \approx -1$, the shear modulus far exceeds the bulk modulus. Elastically isotropic materials having a negative Poisson's ratio, called *auxetic* materials, are extremely rare in nature. However, as discussed in Section 13.3.4, composite materials with negative Poisson's ratio have been devised and fabricated.

Note that one must specify either *plane-strain* or *plane-stress* elasticity when it is desired to connect the moduli for isotropic two-dimensional (planar) elasticity to the moduli for isotropic three-dimensional elasticity. Plane-strain elasticity is physically relevant in considering a fiber-reinforced material. On the other hand, plane-stress elasticity is physically relevant in considering two-phase composites in the form of thin sheets. The planar shear modulus G (either in plane-strain or plane-stress elasticity) is equal to the three-dimensional shear modulus G. However, the bulk-moduli relations are not as simple. We will follow the convention of denoting the *planar bulk modulus* by k and the three-dimensional bulk modulus by K. The plane-strain bulk modulus k is related to K by the expression

$$k = K + G/3. \tag{13.106}$$

By contrast, the plane-stress bulk modulus k obeys the relation

$$k = \frac{9KG}{3K + 4G}. \tag{13.107}$$

These and other interrelations among the planar and three-dimensional moduli are derived in Appendix B.

13.3.3 Homogenization of Random Problem in $\Re^d$

The homogenization of the elastic problem for periodic media has been given by Sanchez-Palencia (1980). Since the derivation of the homogenized results for random elastic media follows closely the one for periodic elastic media, which in turn is completely analogous to the derivation that we gave previously for the conductivity problem in Section 13.2, we will only sketch the procedure here.

We recall that each realization ω of the random medium occupies the space $\mathcal{V} \in \Re^d$, which is partitioned into two random sets or phases, as described in the beginning of this chapter (see Figure 13.2). The stiffness tensor is a rapidly oscillating random function of position, which we write as $\boldsymbol{C}^\epsilon(\boldsymbol{x}) = \boldsymbol{C}(\boldsymbol{y}; \omega)$, where $\boldsymbol{y} = \boldsymbol{x}/\epsilon$ is the fast variable associated with the fineness of the microstructure. As before, it is a *statistically homogeneous function* that is *positive definite* and *bounded*. The displacement $\boldsymbol{u}^\epsilon(\boldsymbol{x}) = \boldsymbol{u}(\boldsymbol{x}, \boldsymbol{y}; \omega)$ satisfies the equilibrium equation for each realization of the ensemble, i.e.,

$$-\frac{\partial}{\partial x_j}\left[C^\epsilon_{ijkl}(\boldsymbol{x})\varepsilon^\epsilon_{kl}(\boldsymbol{x})\right] = f_i(\boldsymbol{x}) \qquad \text{in } \mathcal{V}, \tag{13.108}$$

where $\varepsilon^\epsilon(\boldsymbol{x}) = \varepsilon(\boldsymbol{x}, \boldsymbol{y})$ is the symmetrized gradient of $\boldsymbol{u}^\epsilon$, which we assume is specified on the boundary of $\mathcal{V}$, and $\boldsymbol{f}(\boldsymbol{x})$ is a body force per unit volume.

When the medium is statistically homogeneous and ergodic, there exists an effective stiffness tensor $(C_e)_{ijkl}$, independent of the boundary conditions in the infinite-volume limit such that if $u_i^{(0)}(\boldsymbol{x})$ is the solution of the *deterministic* equilibrium elasticity equation

$$-(C_e)_{ijkl}\frac{\partial}{\partial x_j}\varepsilon^{(0)}_{kl}(\boldsymbol{x}) = f_i(\boldsymbol{x}), \tag{13.109}$$

where

$$\varepsilon_{ij}^{(0)} = \frac{1}{2}\left[\frac{\partial u_j^{(0)}}{\partial x_i} + \frac{\partial u_i^{(0)}}{\partial x_j}\right], \tag{13.110}$$

then

$$\int_{\mathcal{V}} \langle |\boldsymbol{u}^{\epsilon}(\boldsymbol{x}) - \boldsymbol{u}^{(0)}(\boldsymbol{x})|^2 \rangle d\boldsymbol{x} \to 0, \quad \text{as } \epsilon \to 0. \tag{13.111}$$

To prove this statement, one begins by assuming the two-scale expansion

$$u_i^{\epsilon}(\boldsymbol{x}) = u_i^{(0)}(\boldsymbol{x}) + \epsilon u_i^{(1)}(\boldsymbol{x},\boldsymbol{y};\omega) + \epsilon^2 u_i^{(2)}(\boldsymbol{x},\boldsymbol{y};\omega) + \cdots \tag{13.112}$$

of $u_i^{\epsilon}(\boldsymbol{x})$ and substitute it into (13.108). Proceeding as in the previous section on the effective conductivity, one can show that the effective stiffness tensor is defined by an averaged Hooke's law

$$\langle \boldsymbol{\tau}(\boldsymbol{x}) \rangle = \boldsymbol{C}_e : \langle \boldsymbol{\varepsilon}(\boldsymbol{x}) \rangle, \tag{13.113}$$

where

$$(C_e)_{ijkl} = \langle C_{ijkl} \rangle + \left\langle C_{ijmn} \frac{1}{2}\left(\frac{\partial \chi_{kmn}}{\partial y_l} + \frac{\partial \chi_{lmn}}{\partial y_k}\right)\right\rangle \tag{13.114}$$

and χ_{kmn} is the third-order tensor that solves

$$-\frac{\partial}{\partial y_i}\left[C_{ijkl}(\boldsymbol{y};\omega)\frac{1}{2}\left(\frac{\partial \chi_{kmn}}{\partial y_l} + \frac{\partial \chi_{lmn}}{\partial y_k}\right)\right] = \frac{\partial}{\partial y_i} C_{ijmn}(\boldsymbol{y};\omega) \quad \text{in } \Re^{\mathrm{d}}. \tag{13.115}$$

The definition (13.113) is consistent with the one given in Table 1.1.

Remarks:

1. As in the conductivity problem, macroscopic variation of the effective stiffness tensor could be allowed for; i.e., $\boldsymbol{C}_e(\boldsymbol{x})$ may be assumed to depend on the position $\boldsymbol{x}$, provided that the length scales associated with such variations of $\mathcal{O}(\Lambda)$ are large enough to assume statistical homogeneity at $\boldsymbol{x}$ (local statistical homogeneity).
2. It is shown in Section 14.2.2 that $\boldsymbol{C}_e$ is symmetric and positive definite if the local conductivity tensor $\boldsymbol{C}$ is symmetric and positive definite.
3. In the case of elasticity, *macroscopically isotropic* composites are those whose effective stiffness tensor is specified by

$$\boldsymbol{C}_e = dK_e \boldsymbol{\Lambda}_h + 2G_e \boldsymbol{\Lambda}_s, \tag{13.116}$$

   where $\boldsymbol{\Lambda}_h$ and $\boldsymbol{\Lambda}_s$ are the hydrostatic and shear projection tensors given by (13.96) and (13.97), respectively. Similarly, the effective compliance tensor is given by

$$\boldsymbol{S}_e = \frac{1}{dK_e}\boldsymbol{\Lambda}_h + \frac{1}{2G_e}\boldsymbol{\Lambda}_s. \tag{13.117}$$

4. For composites consisting of *isotropic phases*, the following general statements can be made:

- Statistically isotropic composites are always macroscopically isotropic composites [e.g., statistically isotropic arrays of inclusions in a matrix (see Figure 12.6)].

- However, *statistical anisotropy* (as depicted in Figures 7.1 and 16.3, for example) necessarily implies a *macroscopically anisotropic* composite with an effective tensor $\boldsymbol{C}_e$. This is in contrast to the conduction problem, where statistical anisotropy does not ensure macroscopic anisotropy.

Macroscopically anisotropic elastic composites are not necessarily *statistically anisotropic* if some of the phases are anisotropic. For instance, systems of statistically isotropic arrays of spherical but anisotropic grains oriented in the same direction in an isotropic matrix are macroscopically anisotropic.

5. The effective relation (13.113) applies also to composites with an arbitrary number of phases that meet the standard two-scale assumption.
6. The effective stiffness tensor $\boldsymbol{C}_e$ of a macroscopically anisotropic composite composed of M isotropic phases is a *homogeneous function of degree one* in its $2M$ scalar phase moduli $K_1, G_1, \ldots, K_M, G_M$. For two isotropic phases, this means that $\boldsymbol{C}_e(K_1, G_1, K_2, G_2)$ obeys the relation

$$\boldsymbol{C}_e(\alpha K_1, \alpha G_1, \alpha K_2, \alpha G_2) = \alpha\, \boldsymbol{C}_e(K_1, G_1, K_2, G_2), \qquad \text{for all } \alpha. \tag{13.118}$$

If we set the constant α equal to $1/K_1$, then we get

$$\frac{\boldsymbol{C}_e(K_1, G_1, K_2, G_2)}{K_1} = \boldsymbol{C}_e(1, G_1/K_1, K_2/K_1, G_2/K_1).$$

Thus, the homogeneity property (13.118) enables us to reduce the independent variables from four to three without any loss of generality. The homogeneity property is trivially proven using the effective relation (13.113) and the observation that $\nabla \cdot \boldsymbol{\tau}' = 0$ when $\boldsymbol{\tau}' = \alpha \boldsymbol{\tau}$ and $\nabla \cdot \boldsymbol{\tau} = 0$.

13.3.4 Heterogeneous Materials

It is shown in Section 14.2.2 that if the local stiffness $\boldsymbol{C}$ is both symmetric and positive definite, then the effective stiffness tensor $\boldsymbol{C}_e$ is also symmetric and positive definite. Thus, all of the statements made above concerning the positivity properties and symmetries of the stiffness tensor of homogeneous materials apply as well to $\boldsymbol{C}_e$. Therefore, there is no need to repeat this entire description for the effective stiffness tensor $\boldsymbol{C}_e$. However, for future discussion, we describe briefly the effective moduli for transversely isotropic and for isotropic two-phase materials. In particular, we note that in the transversely isotropic case only three moduli (rather than five as in the homogeneous case) are independent.

Transversely Isotropic Composite

Consider a transversely isotropic composite as schematically indicated in Figure 13.6. Generally, such a fiber-reinforced material has phase boundaries that are cylindrical surfaces of arbitrary shape with generators parallel to one axis. Using the procedure

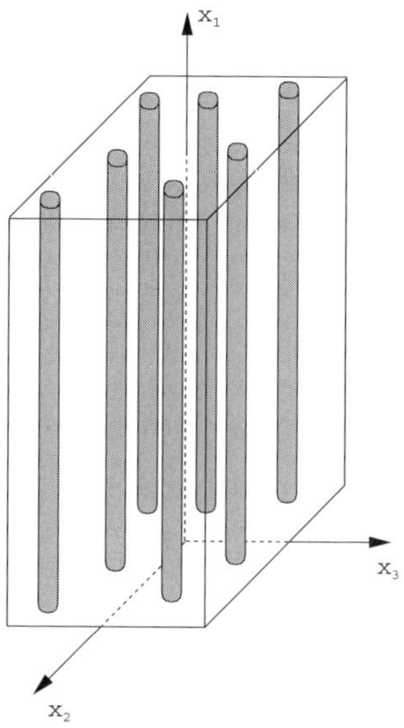

Figure 13.6 A schematic of a fiber-reinforced composite.

given above for homogeneous transversely isotropic materials, it is easily shown that the *average strain–stress* relations are given by

$$\langle\epsilon_{11}\rangle = \frac{1}{E_e^L}\langle\tau_{11}\rangle - \frac{\nu_e^L}{E_e^L}\langle\tau_{22}\rangle - \frac{\nu_e^L}{E_e^L}\langle\tau_{33}\rangle, \tag{13.119}$$

$$\langle\epsilon_{22}\rangle = -\frac{\nu_e^L}{E_e^L}\langle\tau_{11}\rangle + \frac{1}{E_e^T}\langle\tau_{22}\rangle - \frac{\nu_e^T}{E_e^T}\langle\tau_{33}\rangle, \tag{13.120}$$

$$\langle\epsilon_{33}\rangle = -\frac{\nu_e^L}{E_e^L}\langle\tau_{11}\rangle - \frac{\nu_e^T}{E_e^T}\langle\tau_{22}\rangle + \frac{1}{E_e^T}\langle\tau_{33}\rangle, \tag{13.121}$$

$$\langle\epsilon_{12}\rangle = \frac{1}{2G_e^L}\langle\tau_{12}\rangle\ ,\ \langle\epsilon_{13}\rangle = \frac{1}{2G_e^L}\langle\tau_{13}\rangle\ ,\ \langle\epsilon_{23}\rangle = \frac{1}{2G_e^T}\langle\tau_{23}\rangle, \tag{13.122}$$

where E_e^T, G_e^T, and ν_e^T are the effective transverse Young modulus, shear modulus, and Poisson ratio, respectively, and E_e^L, G_e^L, and ν_e^L are the effective longitudinal Young modulus, shear modulus, and Poisson ratio, respectively.

There are only five independent effective constants, and thus there are interrelations among the effective moduli. For example, two of them are

$$G_e^T = \frac{E_e^T}{2(1+\nu_e^T)}, \tag{13.123}$$

$$\frac{4}{E_e^T} = \frac{1}{k_e^T} + \frac{1}{G_e^T} + \frac{4(\nu_e^L)^2}{E_e^L}, \tag{13.124}$$

where k_e^T is the transverse bulk modulus *without axial extension* (i.e., $\langle\epsilon_{11}\rangle = 0$). Combining these two relations and taking appropriate limits [as in (13.84)] shows that the effective transverse Poisson ratio lies in the range $-1 \le \nu_e^T \le 1$. A transversely isotropic porous material with negative Poisson's ratios in certain directions was designed using a topology optimization technique and subsequently fabricated using stereolithography (Sigmund, Torquato and Aksay 1998).

Hill (1964) has shown that for the special case of transversely isotropic composites possessing two phases, there are actually only three independent effective properties. For example, if we are given the transverse properties G_e^T, k_e^T and axial property G_e^L, then we can obtain E_e^L and ν_e^L through the following relations, which involve only the effective property k_e^T:

$$E_e^L = \langle E\rangle + \frac{4(\nu_2-\nu_1)^2}{\left(\dfrac{1}{k_2^T}-\dfrac{1}{k_1^T}\right)^2}\left(\left\langle\frac{1}{k^T}\right\rangle - \frac{1}{k_e^T}\right), \tag{13.125}$$

$$\nu_e^L = \langle\nu\rangle - \frac{(\nu_2-\nu_1)}{\left(\dfrac{1}{k_2^T}-\dfrac{1}{k_1^T}\right)}\left(\left\langle\frac{1}{k^T}\right\rangle - \frac{1}{k_e^T}\right), \tag{13.126}$$

where $k_i^T = K_i + G_i/3$ is the transverse bulk modulus for phase i.

Isotropic Composite

From the discussion above for homogeneous isotropic materials, it is clear that the averaged stress–strain and strain–stress relations in any space dimension d are given respectively by

$$\langle\tau_{ij}\rangle = K_e\langle\epsilon_{kk}\rangle\delta_{ij} + 2G_e\left[\langle\epsilon_{ij}\rangle - \langle\epsilon_{kk}\rangle\frac{\delta_{ij}}{d}\right], \qquad i,j=1,\ldots,d, \tag{13.127}$$

$$\langle\varepsilon_{ij}\rangle = -\frac{\nu_e}{E_e}\langle\tau_{kk}\rangle\delta_{ij} + \frac{1+\nu_e}{E_e}\langle\tau_{ij}\rangle, \qquad i,j=1,\ldots,d, \tag{13.128}$$

where

$$K_e = \lambda_e + \frac{2}{d}G_e \tag{13.129}$$

is the effective bulk modulus, λ_e is the effective Lamé constant, G_e is the effective shear modulus, E_e is the effective Young modulus, and ν_e is the effective Poisson ratio. Of course, these effective Hooke's laws could also have been obtained using (13.113), (13.116), and (13.117). Interrelations between the effective moduli can be obtained in exactly the same way as for a homogeneous isotropic material. For example, we have

$$G_e = \frac{E_e}{2(1+\nu_e)}, \quad K_e = \frac{E_e}{d[1+\nu_e(1-d)]}, \quad \frac{d^2}{E_e} = \frac{1}{K_e} + \frac{d(d-1)}{2G_e}. \tag{13.130}$$

The analogue of (13.104) for an isotropic composite is the expression

$$\nu_e = \frac{dK_e - 2G_e}{d(d-1)K_e + 2G_e} \tag{13.131}$$

for the effective Poisson ratio. Using similar arguments as before, we see that ν_e is bounded by

$$-1 \le \nu_e \le \frac{1}{(d-1)}. \tag{13.132}$$

Elastically isotropic auxetic composites (i.e., $\nu_e < 0$) have been devised and/or fabricated. Hierarchical laminates (Chapter 16) can exhibit extremal Poisson's ratio approaching -1 (Milton 1992). Auxetic foams with *reentrant* (i.e., nonconvex) cell structures have been fabricated (Lakes 1987). In two dimensions, analogous auxetic cellular materials have been designed and fabricated (Sigmund 1994, Larsen, Sigmund and Bouwstra 1997, Xu, Arias, Brittain, Zhao, Gryzbowski, Torquato and Whitesides 1999). One may also achieve a Poisson ratio of -1 by a two-dimensional chiral honeycomb structure (Prall and Lakes 1997).

13.3.5 Relationship Between Elasticity and Viscous Fluid Theory

There is a well-known connection between linear elasticity and slow viscous flow of a fluid that is worth noting here. Consider time-dependent phenomena and let us write the *equations of motion* for a homogeneous material in terms of the displacement $\boldsymbol{u}$, i.e.,

$$(\lambda + G)\,\nabla(\nabla \cdot \boldsymbol{u}) + G\,\Delta \boldsymbol{u} + \boldsymbol{f} = \rho \frac{\partial^2 \boldsymbol{u}}{\partial t^2},$$

where ρ is the mass density and t is time. Now, in the case of an *incompressible* material, we have the solenoidal condition

$$\nabla \cdot \boldsymbol{u} = 0,$$

but incompressibility also means that $K \to \infty$ or $\lambda \to \infty$. Thus, the first term in the momentum equation is indeterminate and can be written in terms of the reactive hydrostatic pressure p, giving

$$-\nabla p + G\,\Delta \boldsymbol{u} + \boldsymbol{f} = \rho\,\frac{\partial^2 \boldsymbol{u}}{\partial t^2}. \tag{13.133}$$

Let us now compare the expression above to the Navier–Stokes equations of motion for an *incompressible Newtonian viscous fluid*, i.e.,

$$-\nabla p + \mu\,\Delta \boldsymbol{v} + \boldsymbol{f} = \rho\left(\frac{\partial \boldsymbol{v}}{\partial t} + \boldsymbol{v} \cdot \nabla \boldsymbol{v}\right),$$

where $\boldsymbol{v}$ is the velocity vector, $\boldsymbol{f}$ is a body force per unit volume, p is the pressure, and μ is the dynamic viscosity. Conservation of mass for an incompressible fluid is expressed by the condition

$$\nabla \cdot \boldsymbol{v} = 0.$$

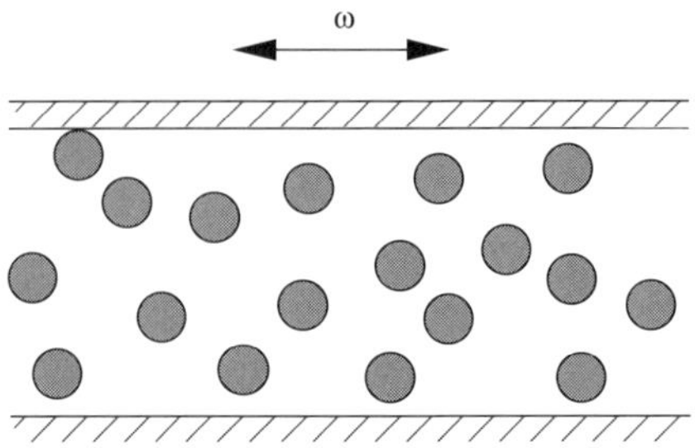

Figure 13.7 Suspension of rigid particles in an incompressible fluid between a plate that harmonically oscillates with frequency ω and a fixed plate. In the limit $\omega \to \infty$, the determination of the effective viscosity μ_e becomes equivalent to finding the steady-state effective shear modulus G_e of rigid particles (with the same configuration) in an incompressible matrix.

For slow viscous flow (Reynolds number going to zero), the nonlinear inertial term $\boldsymbol{v} \cdot \nabla \boldsymbol{v}$ can be shown to be very small compared to the rest of the terms, and thus we obtain the *creeping flow* equations

$$-\nabla p + \mu \, \Delta \boldsymbol{v} + \boldsymbol{f} = \rho \frac{\partial \boldsymbol{v}}{\partial t}. \tag{13.134}$$

Thus, under steady-state conditions, the creeping flow equations (13.134) are identical to the governing elasticity equations (13.133) for incompressible materials when $\boldsymbol{u}$ and G are identified with $\boldsymbol{v}$ and μ, respectively. The time-dependent equations (13.134) will be employed in Section 13.8.

13.3.6 Viscosity of a Suspension

Under certain conditions, obtaining the effective viscosity μ_e of a suspension of perfectly rigid particles ($\mu_2 \to \infty$) in an incompressible fluid with viscosity μ_1 (under creeping flow conditions) is equivalent to finding the *steady-state* effective shear modulus G_e of a composite composed of the same perfectly rigid particles in an incompressible matrix with shear modulus G_1, i.e.,

$$\frac{\mu_e}{\mu_1} = \frac{G_e}{G_1}. \tag{13.135}$$

Relation (13.135) is exact when hydrodynamic interactions between the particles can be neglected (i.e., at infinitely dilute conditions $\phi_2 \to 0$), regardless of the configuration of the particles. At nondilute conditions, (13.135) is also exact, provided that the configurations of the particles in the flow and elasticity problems are identical. This situation is realized for arbitrary particle concentrations when the fluid is subjected to an applied oscillating shear rate with frequency ω in the infinite-frequency limit ($\omega \to \infty$), as depicted in Figure 13.7. However, for finite frequencies in the flow problem at nondilute conditions, the configuration of the particles changes with time and is not known in advance (Russel et al. 1989), and thus relation (13.135) generally does not hold.

13.3.7 Viscoelasticity

Many composite materials, particularly those possessing a polymeric phase, will exhibit a time and rate dependence that is not captured by elasticity theory. In these materials deformations grow (creep), stresses relax, and mechanical vibrations are damped. These effects are magnified at elevated temperatures. Such materials are said to be *viscoelastic*, since they display aspects of both viscous and elastic types of behavior. Under quasi-static conditions (when inertial effects are negligibly small), there is a well-known exact connection between the elastic and viscoelastic problems called the *elastic–viscoelastic correspondence principle* (Hashin 1965a, Christensen 1979). If the time-domain viscoelastic equations are Laplace transformed, then this principle states that the static elastic problem can be converted to the transformed solutions of the viscoelastic problem by simply replacing the static stiffness tensor $\boldsymbol{C}_e$ with $s\hat{\boldsymbol{C}}_e(s)$, where s is the transform variable and $\hat{\boldsymbol{C}}_e(s)$ is the Laplace transform of the time-dependent stiffness $\boldsymbol{C}(t)$. The determination of the quasi-static tensor $\boldsymbol{C}(t)$ is found via a transform inversion.

Under steady-state harmonic conditions at sufficiently low values of the frequency ω, the correspondence principle is even simpler. The complex viscoelastic stiffness $\boldsymbol{C}^*(\omega)$ can be found from formulas for the static effective stiffness $\boldsymbol{C}_e$ by replacing the real elastic moduli with their complex counterparts. For composite materials, the quasi-static condition is satisfied if the heterogeneity length scale is much smaller than the wavelength. Viscoelasticity will not be covered in any detail in this book. The reader is referred to Hashin (1965a) and Christensen (1979) and references therein for pertinent literature on this subject.

13.4 Steady-State Trapping Problem

Consider the problem of diffusion and reaction among *partially absorbing* "traps" in each realization ω of the random medium defined in Section 13.1. Let $\mathcal{V}_1(\omega)$ be the region in which diffusion occurs (i.e., trap-free, or pore, region) and let $\mathcal{V}_2(\omega)$ be the trap region. The concentration field of the reactants $c(\boldsymbol{x}, \boldsymbol{y}; \omega)$ at position $\boldsymbol{x}$ exterior to the traps at time t is generally governed by the mass conservation equation

$$\frac{\partial c}{\partial t} = \mathcal{D}\Delta c - \kappa_B c + G \qquad \text{in } \mathcal{V}_1(\omega) \tag{13.136}$$

with the boundary condition at the pore–trap interface given by

$$\mathcal{D}\frac{\partial c}{\partial n} + \kappa c = 0 \qquad \text{on } \partial\mathcal{V}(\omega) \tag{13.137}$$

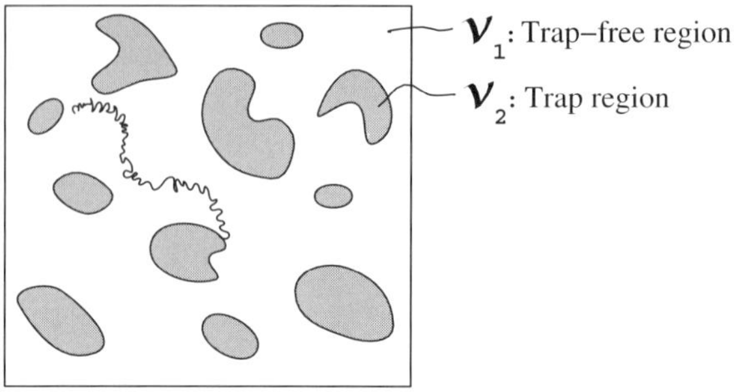

Figure 13.8 A schematic of the "trapping" problem with a static trap phase. Diffusion occurs in the trap-free (or pore) region. A diffusing particle (indicated by the erratic trajectory) is absorbed at the pore–trap interface with a probability related to the surface rate constant κ.

and some specified initial condition and boundary conditions. Here $\mathcal{D}$ is the diffusion coefficient of the reactant, κ_B is a positive bulk rate constant, κ is a positive surface rate constant, G is a generation rate per unit trap-free volume, and $\boldsymbol{n}$ is the unit outward normal from the pore space. Without loss of generality, one can set the bulk rate constant κ_B equal to zero: The solution $c(\boldsymbol{x}, t)$ of (13.136) with $\kappa_B \neq 0$ multiplied by $\exp(\kappa_B t)$ gives the corresponding solution with $\kappa_B = 0$. However, for the steady-state problems mentioned below, the solution $c(\boldsymbol{x}; \omega)$ for $\kappa_B = 0$ is not simply related to the solution $c(\boldsymbol{x}; \omega)$ for $\kappa_B \neq 0$ (Talbot and Willis 1984).

It is useful to introduce the dimensionless surface rate constant

$$\kappa^* = \frac{\kappa \ell}{\mathcal{D}}, \tag{13.138}$$

where ℓ is a characteristic pore length scale, and distinguish between two extreme regimes:

$$\begin{aligned} &\kappa^* \gg 1 \quad \text{(diffusion-controlled)}, \\ &\kappa^* \ll 1 \quad \text{(reaction-controlled)}. \end{aligned} \tag{13.139}$$

In the *diffusion-controlled* regime, the diffusing species takes a long time to diffuse to the pore–trap interface relative to the characteristic time associated with the surface reaction; i.e., the process is governed by diffusion. For infinite surface reaction ($\kappa = \infty$), the traps are perfect absorbers, and thus the interface condition is of the *Dirichlet* kind, with $c = 0$. On the other hand, in the *reaction-controlled* regime, the characteristic time associated with surface reaction is large compared with the diffusion time to the pore–trap interface. For vanishing surface reaction ($\kappa = 0$), the traps are perfect reflectors, and hence the interface condition is of the *Neumann* kind, with $\partial c/\partial n = 0$.

We will study relation (13.136) with condition (13.137) for two different situations: (i) the *steady-state* solution with $\kappa_B = 0$ and (ii) the *time-dependent* solution with $\kappa_B = G = 0$. The quantities of central interest for these problems are respectively (i) the *mean survival time* τ of a Brownian particle and (ii) the *relaxation times* T_n, which are inversely proportional to the eigenvalues. The times τ and T_1 are intimately linked to characteristic length scales of the pore region. Whereas the mean survival time τ is determined by the "average pore size," the *principal* (largest) relaxation time T_1 is governed by diffusion occurring in the largest cavities (pores) in the system. The steady-state trapping will be discussed in the present section. The time-dependent problem will be considered in Section 13.7.

13.4.1 Local Relations

Consider the steady-state problem of diffusion and reaction among perfectly absorbing traps ($\kappa = \infty$). The rate of production of the reactants per unit volume G is exactly compensated by the rate of removal by the traps.

Local Differential Equation

The conservation of mass equation (13.136) under steady-state conditions and without bulk reaction reduces to the following Poisson equation for the concentration field $c(\boldsymbol{x};\omega)$:

$$\mathcal{D}\Delta c = -G \qquad \text{in } \mathcal{V}_1(\omega), \tag{13.140}$$

$$c = 0 \qquad \text{on } \partial\mathcal{V}(\omega). \tag{13.141}$$

Henceforth, we will assume a zero-flux condition at the boundary of $\mathcal{V}$ (macroscopic sample surface). However, as we have emphasized throughout this chapter, the effective property (trapping constant in this case) is independent of the macroscopic boundary conditions for ergodic media.

13.4.2 Homogenization of Random Problem in $\Re^d$

What is the appropriate *macroscopic* constitutive relation for the steady-state trapping problem? In the chemical physics literature (Prager 1963a, Calef and Deutch 1983), it is assumed that the trapping constant γ (a quantity that will be shown to be inversely proportional to τ) obeys the first-order rate equation

$$G = \gamma\mathcal{D}C, \tag{13.142}$$

where C represents an average concentration field. To derive this constitutive relation rigorously from homogenization theory, we follow the treatment of Rubinstein and Torquato (1988) for the case of perfectly absorbing traps ($\kappa = \infty$).

In general, it is assumed that there exists a small parameter $\epsilon = \ell/L$ associated with rapid fluctuations in the structure of $\mathcal{V}_1(\omega)$ and that the concentration field c depends on two scales: a slow scale $\boldsymbol{x}$ and a fast scale $\boldsymbol{y} = \boldsymbol{x}/\epsilon$. Thus, $c^\epsilon(\boldsymbol{x}) = c(\boldsymbol{x},\boldsymbol{y};\omega)$ satisfies

$$\mathcal{D}\Delta c^\epsilon(\boldsymbol{x}) = -G(\boldsymbol{x}) \qquad \text{in } \mathcal{V}_1(\omega), \tag{13.143}$$

$$c^\epsilon(\boldsymbol{x}) = 0 \qquad \text{on } \partial\mathcal{V}(\omega). \tag{13.144}$$

In order to derive the macroscopic behavior, we assume that c^ϵ admits a two-scale expansion of the form

$$c^\epsilon(\boldsymbol{x}) = \epsilon^2 c_0(\boldsymbol{x},\boldsymbol{y};\omega) + \epsilon^3 c_1(\boldsymbol{x},\boldsymbol{y};\omega) + \cdots, \tag{13.145}$$

in contrast to the previous two problems [cf. (13.45) and (13.112)]. Note that the first nontrivial term in this expansion for c^ϵ is $\mathcal{O}(\epsilon^2)$, in contrast to the expansion (13.45) for T^ϵ. Physically, this expansion arises because an $\mathcal{O}(1)$ production rate gives a concentration of $\mathcal{O}(\epsilon^2)$. Substitution of this expansion into relation (13.143) and use of the identities

$$\nabla = \nabla_x + \frac{1}{\epsilon}\nabla_y, \qquad \Delta \equiv \nabla^2 = \Delta_x + \frac{2}{\epsilon}\nabla_x \cdot \nabla_y + \frac{1}{\epsilon^2}\Delta_y \tag{13.146}$$

yields

$$\mathcal{D}\left(\frac{1}{\epsilon^2}\Delta_y + \frac{2}{\epsilon}\nabla_x \cdot \nabla_y + \Delta_x\right)\left(\epsilon^2 c_0 + \epsilon^3 c_1 + \cdots\right) = -G(\boldsymbol{x}).$$

Therefore, the leading-order equation is given by

$$\begin{aligned} \mathcal{D}\Delta_y c_0(\boldsymbol{x},\boldsymbol{y};\omega) &= -G(\boldsymbol{x}) \qquad \text{in } \mathcal{V}_1(\omega), \\ c_0(\boldsymbol{x},\boldsymbol{y};\omega) &= 0 \qquad \text{on } \partial\mathcal{V}(\omega). \end{aligned}$$

Hence, we can write

$$c_0 = \mathcal{D}^{-1} G(\boldsymbol{x}) u(\boldsymbol{y};\omega), \tag{13.147}$$

where the scaled auxiliary concentration field u solves

$$\Delta_y\, u(\boldsymbol{y};\omega) = -1 \qquad \text{in } \mathcal{V}_1(\omega), \tag{13.148}$$

$$u(\boldsymbol{y};\omega) = 0 \qquad \text{on } \partial\mathcal{V}(\omega), \tag{13.149}$$

and we extend u in the trap region $\mathcal{V}_2(\omega)$ to be zero. Assuming that the medium is locally (i.e., on the fast scale) statistically homogeneous and ergodic, the ensemble average of any function $g(\boldsymbol{x},\boldsymbol{y};\omega)$ is simply a function of $\boldsymbol{x}$ only. Averaging (13.147) gives the constitutive relation

$$G(\boldsymbol{x}) = \gamma\mathcal{D}C(\boldsymbol{x}), \tag{13.150}$$

where

$$C(\boldsymbol{x}) = \langle c_0(\boldsymbol{x},\boldsymbol{y};\omega)\rangle$$

and the constant γ, called the *trapping constant*, is given by

$$\gamma^{-1} = \langle u(\boldsymbol{y};\omega)\rangle = \langle u\mathcal{I}^{(1)}\rangle. \tag{13.151}$$

We see that relation (13.150) agrees with (13.142) but additionally allows for slow spatial variations in C and G.

Remarks:

1. The trapping problem is fundamentally different than the previous two problems. Here there is no local constitutive relation. Moreover, the trapping constant γ has

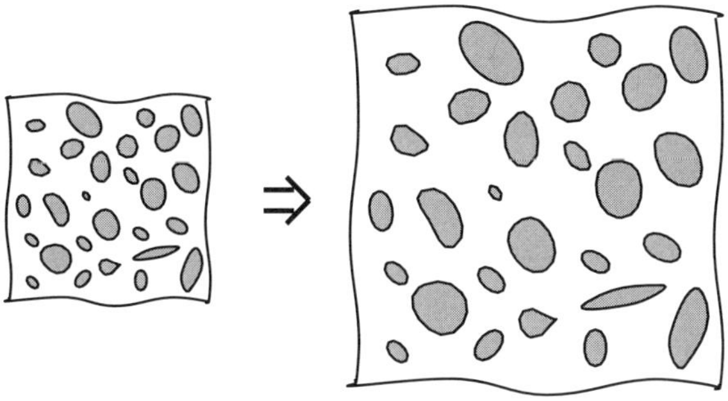

Figure 13.9 A schematic illustrating the scale-dependence of both the trapping constant γ (for perfectly absorbing traps) and fluid permeability tensor $\boldsymbol{k}$. If the linear dimensions of the porous medium are doubled, then the new trapping constant is four times smaller than the original one, whereas the components of the new fluid permeability tensor are four times larger than the original components.

dimensions (length)$^{-2}$ and therefore, unlike the effective conductivity or elastic moduli, is a scale-dependent property. Thus, if we apply an affine transformation to the system in each of the space coordinates (i.e., rescale each coordinate by some constant factor α), then the trapping constant γ' of the new system is related to the trapping constant γ of the original system via $\gamma' = \gamma/\alpha^2$ (see Figure 13.9). As discussed in the next section, the fluid permeability tensor $\boldsymbol{k}$ is also a scale-dependent property.

2. What is the physical interpretation of the trapping constant γ? Dimensional analysis of (13.150) reveals that γ has dimensions of (length)$^{-2}$, since $\mathcal{D}$ has dimensions of (length)^2 per unit time, C has dimensions of inverse volume, and G has dimensions of inverse volume per unit time. Consider a constant production rate G. Now let the total number of diffusing (Brownian) particles created outside of traps per unit time be N and the total number of particles exterior to traps at a given time be N_0. Then the *average trapping rate* k_R (per particle) is given by

$$k_R = \frac{N}{N_0}.$$

The inverse of this quantity, called the *mean survival time* τ of a Brownian particle, is thus given by

$$\tau = k_R^{-1} = \frac{N_0}{N}.$$

It is clear that

$$N = GV_1, \qquad N_0 = CV,$$

where V_1 is the volume of phase 1 and V is the total volume. Use of the relations immediately above and definition (13.151) allows us to write

$$\tau = \frac{1}{\gamma \phi_1 \mathcal{D}} = \frac{\langle u \rangle}{\phi_1 \mathcal{D}}. \tag{13.152}$$

Thus, the trapping constant γ is proportional to the trapping rate k_R or inversely proportional to the mean survival time τ. Roughly speaking, the quantity γ^{-1} provides a measure of the *average pore size*. In subsequent discussions we will refer to both the trapping constant γ and the mean survival time τ. We note that substitution of (13.152) into (13.150) yields the corresponding constitutive relation defining τ, i.e.,

$$C(\boldsymbol{x}) = \tau \phi_1 G(\boldsymbol{x}). \tag{13.153}$$

We see that this definition is consistent with the one given in Table 1.1.

3. Since γ does not vanish when the trap-free region becomes disconnected, the trapping constant does not exhibit *percolation* or *critical* behavior. This is in contrast to the effective conductivity, effective stiffness, and fluid permeability, which are affected by phase connectedness and hence do exhibit percolation behavior.
4. Unlike the effective conductivity, effective stiffness, or fluid permeability, the trapping constant γ remains a scalar quantity even if the microstructure is statistically anisotropic [e.g., arrays of oriented cylinders or ellipsoids (see Figure 7.1)].
5. It is important to note that in the case of finite surface reaction ($\kappa < \infty$), i.e., when u solves

$$\Delta_y\, u(\mathbf{y}; \omega) = -1 \qquad \text{in } \mathcal{V}_1(\omega), \tag{13.154}$$

$$\mathcal{D}\frac{\partial u}{\partial n} + \kappa u = 0 \qquad \text{on } \partial\mathcal{V}(\omega), \tag{13.155}$$

the averaged relation (13.151) still defines the trapping constant.

13.5 Steady-State Fluid Permeability Problem

It is well established that the slow flow of an incompressible viscous fluid through porous media is often described by Darcy's law (Scheidegger 1974):

$$\boldsymbol{U} = -\frac{k}{\mu}\nabla p_0, \tag{13.156}$$

where $\boldsymbol{U}$ is the average fluid velocity, ∇p_0 is the applied pressure gradient, μ is the *dynamic viscosity*, and k is the *fluid permeability*. Figure 13.10 depicts a porous medium and the relevant flow parameters. The permeability k has dimensions of (length)^2 and, roughly speaking, may be regarded as an *effective pore channel area of the dynamically connected part of the pore space*.

We refer to the portion of the *connected* pore space that carries an appreciable part of the flow (say, above some designated flow rate) as the "dynamically connected" part of the pore space. However, there will be regions of connected pore space that carry no appreciable flow, and such regions contribute negligibly to the fluid permeability.

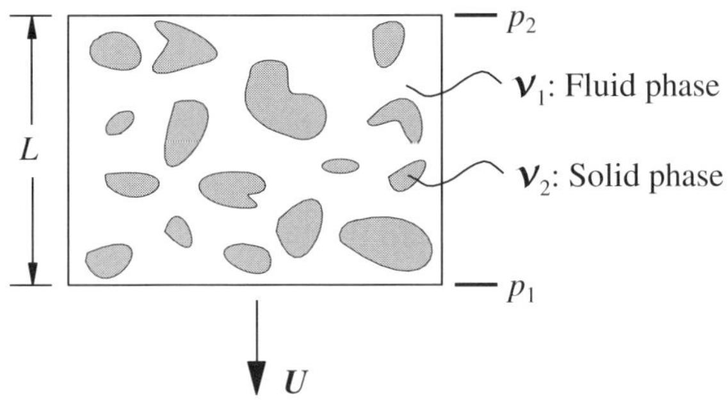

Figure 13.10 A schematic of a porous medium indicating the applied pressure drop $(p_2 - p_1)/L$ and average velocity $\boldsymbol{U}$ that determine the permeability k in Darcy's law.

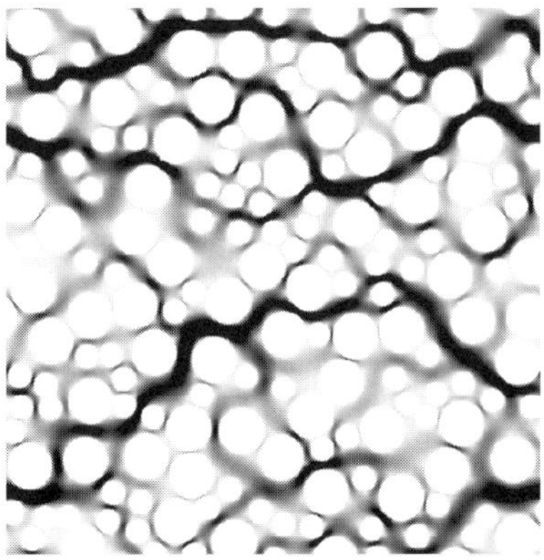

Figure 13.11 Gray-scale image of fluid speeds in a Stokes-flow simulation of a two-dimensional flow past a bed of two different-sized disks as computed by N. Martys at the National Institute of Standards and Technology. Black and white pore regions indicate highest fluid speeds and no flow, respectively. Similar images can be found in Martys and Garboczi (1992).

Therefore, permeability estimates for general porous media based on purely simple pore statistics, such as porosity and specific surface, are inherently fundamentally unsound. Figure 13.11 beautifully illustrates the idea that only a subset of the pore space contributes to the fluid permeability.

13.5.1 Local Relations

For each realization ω of the random porous medium, let $\mathcal{V}_1(\omega)$ be the region through which the fluid flows (i.e., pore, or void, region) and let $\mathcal{V}_2(\omega)$ be the solid region.

Local Differential Equations

The fluid motion satisfies the Stokes equations

$$\mu \Delta \boldsymbol{v} = \nabla p \qquad \text{in } \mathcal{V}_1(\omega), \tag{13.157}$$

$$\nabla \cdot \boldsymbol{v} = 0 \qquad \text{in } \mathcal{V}_1(\omega), \tag{13.158}$$

$$\boldsymbol{v} = 0 \qquad \text{on } \partial\mathcal{V}(\omega), \tag{13.159}$$

where $\boldsymbol{v}$ and p are the local velocity and pressure fields, respectively. The first relation is the steady-state momentum equation in the limit of vanishing Reynolds number, the second relation states that the fluid is incompressible, and the last relation is the no-slip condition on the pore–solid interface. Henceforth, we will assume a zero-traction condition on the boundary of $\mathcal{V}$ (macroscopic sample boundary), but since we are interested ultimately in ergodic media, the permeability will not depend on the details of the boundary conditions.

13.5.2 Homogenization of Random Problem in $\Re^d$

Darcy's law has been derived using the method of homogenization by Sanchez-Palencia (1980) for periodic media and by Rubinstein and Torquato (1989) for random media. Let us consider the *random* setting in which the medium is taken generally to be *macroscopically anisotropic*. The porous medium, which occuipes space $\mathcal{V}(\omega) \in \Re^d$, is assumed to have a microscopic length scale ℓ (e.g., the scale over which $\mathcal{I}^{(1)}$ varies) that is small compared to a typical macroscopic length scale L. Again, there is a small parameter $\epsilon = \ell/L$, and we assume that the velocity $\boldsymbol{v}$ and pressure p depend on two scales: a slow scale $\boldsymbol{x}$ and a fast scale $\boldsymbol{y} = \boldsymbol{x}/\epsilon$. Therefore, $\boldsymbol{v}^\epsilon(\boldsymbol{x}) = \boldsymbol{v}(\boldsymbol{x},\boldsymbol{y};\, \omega)$ and $p^\epsilon(\boldsymbol{x}) = p(\boldsymbol{x},\boldsymbol{y};\, \omega)$ satisfy

$$\mu \Delta \boldsymbol{v}^\epsilon(\boldsymbol{x}) = \nabla p^\epsilon(\boldsymbol{x}) \qquad \text{in } \mathcal{V}_1(\omega), \tag{13.160}$$

$$\nabla \cdot \boldsymbol{v}^\epsilon(\boldsymbol{x}) = 0 \qquad \text{in } \mathcal{V}_1(\omega), \tag{13.161}$$

$$\boldsymbol{v}^\epsilon(\boldsymbol{x}) = 0 \qquad \text{on } \partial\mathcal{V}(\omega). \tag{13.162}$$

To derive the macroscopic equations, we assume a two-scale expansion

$$\boldsymbol{v}^\epsilon(\boldsymbol{x}) = \epsilon^2 \boldsymbol{v}_0(\boldsymbol{x},\boldsymbol{y};\, \omega) \; + \epsilon^3 \boldsymbol{v}_1(\boldsymbol{x},\boldsymbol{y};\, \omega) \; + \cdots ,$$

$$p^\epsilon(\boldsymbol{x}) = p_0(\boldsymbol{x}) + \epsilon\, p_1(\boldsymbol{x},\boldsymbol{y};\, \omega) + \cdots .$$

Note that the first nontrivial term in the expansion for the velocity is $\mathcal{O}(\epsilon^2)$, in contrast to the expansion for the pressure. Physically, this expansion arises because an $\mathcal{O}(1)$ pressure gives a velocity of $\mathcal{O}(\epsilon^2)$. Substitution of these relations into the Stokes equations (13.160)–(13.162) gives the leading-order equations as

$$\mu \Delta_y \boldsymbol{v}_0(\boldsymbol{x},\boldsymbol{y};\, \omega) = \nabla_y\, p_1(\boldsymbol{x},\boldsymbol{y};\, \omega) + \nabla_x\, p_0(\boldsymbol{x}) \quad \text{in } \mathcal{V}_1(\omega), \tag{13.163}$$

$$\nabla_y \cdot \boldsymbol{v}_0(\boldsymbol{x},\boldsymbol{y};\, \omega) = 0 \qquad \text{in } \mathcal{V}_1(\omega), \tag{13.164}$$

$$\nabla_x \cdot \boldsymbol{v}_0(\boldsymbol{x},\boldsymbol{y};\, \omega) + \nabla_y \cdot \boldsymbol{v}_1(\boldsymbol{x},\boldsymbol{y};\, \omega) = 0 \qquad \text{in } \mathcal{V}_1(\omega), \tag{13.165}$$

$$\boldsymbol{v}_0(\boldsymbol{x},\boldsymbol{y};\, \omega) = 0 \qquad \text{on } \partial\mathcal{V}(\omega). \tag{13.166}$$

Here we have used the identities of (13.146). We assume that the medium is locally (i.e., on the fast scale) statistically homogeneous and ergodic, and hence ensemble averaging (13.165) yields

$$\nabla_x \cdot \boldsymbol{U}(\boldsymbol{x}) + \langle \nabla_y \cdot \boldsymbol{v}_1(\boldsymbol{x}, \boldsymbol{y}; \omega) \rangle = 0, \tag{13.167}$$

where

$$\boldsymbol{U}(\boldsymbol{x}) = \langle \boldsymbol{v}_0(\boldsymbol{x}, \boldsymbol{y}; \omega) \rangle. \tag{13.168}$$

By integrating by parts and using Gauss's divergence theorem (see Chapter 14 for details), it can be shown that

$$\langle \nabla_y \cdot \boldsymbol{v}_1(\boldsymbol{x}, \boldsymbol{y}; \omega) \rangle = 0, \tag{13.169}$$

and hence

$$\nabla_x \cdot \boldsymbol{U}(\boldsymbol{x}) = 0. \tag{13.170}$$

Relation (13.170) is just the *macroscopic* incompressibility condition.

In order to analyze (13.163) and (13.164), we introduce the following statistically homogeneous random functions: a second-order tensor "velocity" field $\boldsymbol{w}(\boldsymbol{y}; \omega)$ and a vector "pressure" field $\boldsymbol{\pi}(\boldsymbol{y}; \omega)$. These quantities are solutions of

$$\Delta_y \boldsymbol{w} = \nabla_y \boldsymbol{\pi} - \boldsymbol{I} \qquad \text{in } \mathcal{V}_1(\omega), \tag{13.171}$$

$$\nabla_y \cdot \boldsymbol{w} = 0 \qquad \text{in } \mathcal{V}_1(\omega), \tag{13.172}$$

$$\boldsymbol{w} = 0 \qquad \text{on } \partial\mathcal{V}(\omega), \tag{13.173}$$

where $\boldsymbol{I}$ is the second-order unit tensor. In these equations the scaled tensor velocity field w_{ij} is the jth component of the velocity due to a unit pressure gradient in the ith direction, and π_j is the jth component of the associated scaled pressure. Note that w_{ij} is generally not symmetric. We extend $\boldsymbol{w}$ and $\boldsymbol{\pi}$ in the solid region $\mathcal{V}_2$ to be zero. It is easily seen that $\boldsymbol{v}_0$ and p_1 can be written as

$$\boldsymbol{v}_0(\boldsymbol{x}, \boldsymbol{y}; \omega) = -\frac{1}{\mu} \nabla\, p_0(\boldsymbol{x}) \cdot \boldsymbol{w}(\boldsymbol{y}; \omega), \tag{13.174}$$

$$p_1(\boldsymbol{x}, \boldsymbol{y}; \omega) = \nabla p_0(\boldsymbol{x}) \cdot \boldsymbol{\pi}(\boldsymbol{y}; \omega). \tag{13.175}$$

Averaging (13.174) gives

$$\boldsymbol{U}(\boldsymbol{x}) = -\frac{1}{\mu} \langle \boldsymbol{w}(\boldsymbol{y}; \omega) \rangle \cdot \nabla p_0(\boldsymbol{x}).$$

The second-order permeability tensor $\boldsymbol{k}$ is then defined by

$$\boldsymbol{k} = \langle \boldsymbol{w}(\boldsymbol{y}; \omega) \rangle. \tag{13.176}$$

In summary, the macroscopic equations that govern the slow viscous flow through an anisotropic porous medium are given by

$$\boldsymbol{U}(\boldsymbol{x}) = -\frac{\boldsymbol{k}}{\mu} \cdot \nabla p_0(\boldsymbol{x}), \tag{13.177}$$

$$\nabla \cdot \boldsymbol{U}(\boldsymbol{x}) = 0, \tag{13.178}$$

where $\boldsymbol{k}$ is given by (13.176). Observe that this definition is consistent with the one given in Table 1.1.

Remarks:

1. The fluid permeability shares some similarities with the trapping constant (or mean survival time) but is fundamentally different from either the effective conductivity or effective elastic moduli. As in the trapping problem, there is no local constitutive relation in the flow problem. Furthermore, both the fluid permeability and trapping constant are scale-dependent properties, in contrast to both the effective conductivity and elastic moduli, which are scale-invariant properties. Thus, if we apply an affine transformation to the system in each of the space coordinates, i.e., rescale each coordinate by some constant factor α, then the permeability of the new system, $\boldsymbol{k}'$, is related to that of the original system, $\boldsymbol{k}$, via $\boldsymbol{k}' = \alpha^2 \boldsymbol{k}$ (see Figure 13.9).
2. The permeability tensor $\boldsymbol{k}$ is given in terms of the random boundary value problem (13.171)–(13.173) and is shown to be symmetric and positive definite in Section 14.4.1.
3. If the medium is macroscopically isotropic, then $\boldsymbol{k} = k\boldsymbol{I}$, where $k = \langle \boldsymbol{w} : \boldsymbol{I} \rangle / d$ is a scalar. In such instances, the auxiliary tensor equations (13.171)–(13.173) become vector equations; i.e., the scalar permeability is redefined as

$$k = \langle \boldsymbol{w} \cdot \boldsymbol{e} \rangle, \tag{13.179}$$

where the scaled *vector* velocity $\boldsymbol{w}$ solves

$$\Delta \boldsymbol{w} = \nabla \pi - \boldsymbol{e} \qquad \text{in } \mathcal{V}_1(\omega), \tag{13.180}$$

$$\nabla \cdot \boldsymbol{w} = 0 \qquad \text{in } \mathcal{V}_1(\omega), \tag{13.181}$$

$$\boldsymbol{w} = 0, \qquad \text{on } \partial \mathcal{V}(\omega). \tag{13.182}$$

In these equations, π is the scaled *scalar* pressure field, $\boldsymbol{e}$ is a unit vector, and the subscript y has been dropped.
4. Whereas statistically isotropic porous media are always macroscopically isotropic, statistical anisotropy does not necessarily imply a *macroscopically anisotropic* porous medium with an effective fluid permeability tensor $\boldsymbol{k}$. For instance, porous media with cubic symmetry are statistically anisotropic but are macroscopically isotropic (e.g., cubic lattices of spheres). Macroscopically anisotropic porous media are necessarily *statistically anisotropic* [e.g., statistically anisotropic beds of oriented cylinders or ellipsoids (see Figure 7.1) or stratified media (see Figure 16.3)].

13.5.3 Relationship to Sedimentation Rate

The problem of the sedimentation of macroscopic particles in a viscous liquid is related to that of flow in a porous medium composed of a fixed bed of particles. In the former, one is interested in finding the *mobility*, which is the constant of proportionality in the relation between the average sedimentation velocity (relative to zero-flux axes) and the force acting on a particle (assuming identical particles); see Table 1.1. One can see that this sedimentation constitutive law bears a strong resemblance to Darcy's law, since the pressure gradient is related to the average force acting on a particle (see Sections

18.4 and 19.4). However, the mobility and permeability are generally not related in a simple manner; the physics of the particle interactions are quite different in the two problems. In the sedimentation problem the particles are free to move, and there is backflow, whereas in the porous-medium problem the particles are fixed, resulting in *screened* interactions. We note that if the relative particle positions are kept fixed during sedimentation, the sedimentation velocity is trivially related to the permeability for the same particle configuration in the infinite-volume limit. This idealized situation was studied for the case of spheres fixed on the sites of cubic lattices (Sangani and Acrivos 1982, Zick and Homsy 1982). The reader is referred to Brady and Durlofsky (1988) and references therein for further discussion on the sedimentation problem.

13.6 Classification of Steady-State Problems

The aforementioned steady-state problems have been designated as falling within classes A, B, C, or D in Section 13.1 and Table 1.1. This classification scheme is somewhat vague, and therefore in this section we make it more mathematically precise. First, we will reserve the symbol α to designate processes that are characterized by a local constitutive relation (e.g., A and B problems) and the symbol β to designate processes that are *not* characterized by a local constitutive relation (e.g., C and D problems).

Our classification scheme is based on the fact that all problems are described by an averaged constitutive relation of a particular tensorial order involving a local generalized flux field that satisfies a conservation equation of a certain tensorial order. In the case of conduction, we see that the averaged constitutive relation (13.46) transforms the vector $\langle \boldsymbol{E} \rangle$ (a first-order tensor) into the vector $\langle \boldsymbol{J} \rangle$ (another first-order tensor). The associated conservation equation for the local flux (13.3) is a scalar equation (a zeroth-order tensor). Thus, the conduction problem and all mathematically equivalent problems are said to fall within class $\alpha(1,1;0)$, where the first two arguments indicate a transformation from a first-order tensor to a first-order tensor, and the last argument indicates the order of the conservation equation. In general, class $\alpha(m,n;p)$ describes transformations from an mth-order tensor to an nth-order tensor with a local flux field that satisfies a pth-order tensor conservation equation. The effective property must therefore be a tensor of order $(m+n)$. Thus, classes A and B are more precisely defined according to the following prescription:

$$A \equiv \alpha(1,1;0), \tag{13.183}$$

$$B \equiv \alpha(2,2;1). \tag{13.184}$$

The second statement follows immediately from the averaged Hooke's law (13.113) and the conservation (equilibrium) equation (13.50). We could make the classification scheme even more elaborate (e.g., one can account for equations governing the generalized intensity field, interface conditions, etc.) but do not do so for the sake of simplicity.

Following the same prescription, we see that classes C and D are defined as follows:

$$C \equiv \beta(0,0;0), \tag{13.185}$$

$$D \equiv \beta(1,1;1). \tag{13.186}$$

The first statement follows from the constitutive relation (13.153) and the diffusion equation (13.140). The second statement follows from Darcy's law (13.156) and the momentum relation in the Stokes equations (13.157).

The above classification scheme can be extended to include coupled phenomena, such as thermoelectricity, thermoelasticity, piezoelectricity, and magnetoelasticity.

13.7 Time-Dependent Trapping Problem

The relaxation times associated with the decay of physical quantities such as the concentration field or nuclear magnetization density are related closely to the characteristic length scales of the pore of the fluid region. In what follows we describe the basic equations and note certain connections to the steady-state trapping problem.

13.7.1 Basic Equations

Let $c(\boldsymbol{x},t)$ denote the physical quantity of interest (e.g., concentration, magnetization density) at local position $\boldsymbol{x}$ and time t, obeying the time-dependent diffusion equation

$$\frac{\partial c}{\partial t} = \mathcal{D}\Delta c + c_o\delta(t) \qquad \text{in } \mathcal{V}_1(\omega), \tag{13.187}$$

$$\mathcal{D}\frac{\partial c}{\partial n} + \kappa c = 0 \qquad \text{on } \partial\mathcal{V}(\omega), \tag{13.188}$$

in a *finite* but *large* pore region $\mathcal{V}_1$. Here c_o is the initial constant concentration field and $\delta(t)$ is the Dirac delta function. In all of the ensuing discussion we will assume a zero-flux condition on the boundary of $\mathcal{V}$.

The solution of (13.187) and (13.188) can be expressed as an expansion in orthonormal functions $\{\psi_n\}$:

$$\frac{c(\boldsymbol{x},t)}{c_o} = \sum_{n=1}^{\infty} a_n e^{-t/T_n}\psi_n(\boldsymbol{x}), \tag{13.189}$$

where

$$\Delta\psi_n = -\lambda_n\psi_n \qquad \text{in } \mathcal{V}_1(\omega), \tag{13.190}$$

$$\mathcal{D}\frac{\partial\psi_n}{\partial n} + \kappa\psi_n = 0 \qquad \text{on } \partial\mathcal{V}(\omega). \tag{13.191}$$

The diffusion relaxation times T_n are related to the eigenvalues λ_n by

$$T_n = \frac{1}{\mathcal{D}\lambda_n}. \tag{13.192}$$

At long times, the smallest eigenvalue λ_1, or principal (largest) relaxation time T_1, dominates. The initial condition and the normal mode expansion (13.189) give

$$\sum_{n=1}^{\infty} a_n \psi_n(\boldsymbol{x}) = 1. \tag{13.193}$$

The eigenfunctions $\{\psi_n\}$ are orthonormal, so that

$$\frac{1}{V_1} \int_{\mathcal{V}_1} \psi_m(\boldsymbol{x}) \psi_n(\boldsymbol{x}) d\boldsymbol{x} = \delta_{mn}, \tag{13.194}$$

and therefore the eigenfunction coefficients are given by

$$a_n = \frac{1}{V_1} \int_{\mathcal{V}_1} \psi_n(\boldsymbol{x}) d\boldsymbol{x}. \tag{13.195}$$

Recall that

$$V_1 = \phi_1 V \tag{13.196}$$

is the *total pore volume*. Because the set $\{\psi_n\}$ is complete, we also have

$$\sum_{n=1}^{\infty} a_n^2 = 1. \tag{13.197}$$

The survival probability $S(t)$ in terms of $c(\boldsymbol{x}, t)$ is defined by the relation

$$S(t) = \frac{1}{V_1} \int_{\mathcal{V}_1} \frac{c(\boldsymbol{x}, t)}{c_o} d\boldsymbol{x}. \tag{13.198}$$

This quantity gives the fraction of Brownian particles that survive until time t; clearly, $S(0) = 1$. Substitution of (13.189) into (13.198) gives

$$S(t) = \sum_{n=1}^{\infty} a_n^2 e^{-t/T_n}. \tag{13.199}$$

The survival probability is depicted in Figure 13.12 for partially absorbing traps ($\kappa > 0$). It is a monotonically decreasing function of time; physically, a Brownian particle is more likely to get trapped as time progresses.

Remarks:

1. Interestingly, the problems described above have a direct connection to nuclear magnetic resonance (NMR) measurements in fluid-saturated porous media (Brownstein and Tarr 1979, Banavar and Schwartz 1987, Wilkinson, Johnson and Schwartz 1991). The characteristic times involved in the decay of nuclear magnetization are related to the pore size because of enhanced relaxation at the pore–solid interface. Therefore, NMR is a powerful noninvasive technique to study the microstructure and physical properties of fluid-saturated porous media. The equations governing the decay of the magnetization density $m(\boldsymbol{x}, t)$ (along a particular direction) are precisely (13.187) and (13.188) with $c(\boldsymbol{x}, t)$ and c_o replaced by $m(\boldsymbol{x}, t)$ and m_o, respectively. The dimensionless volume-integrated magnetization $M(t)/M_o$ ($M_o = m_o V_1$) is simply what we have referred to as the survival probability $S(t)$ given by (13.198), i.e.,

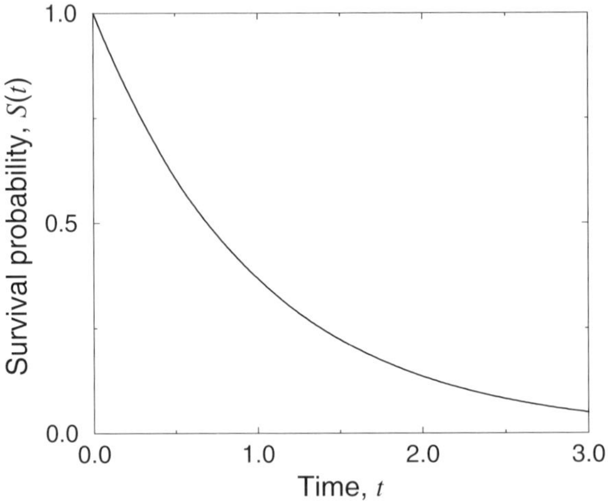

Figure 13.12 A schematic of the survival probability $S(t)$ versus time t for $\kappa > 0$.

$$S(t) = \frac{M(t)}{M_o}. \tag{13.200}$$

The net magnetization $M(t)$ is usually the quantity of principal interest in an NMR experiment. The survival probability has been computed for transport exterior to periodic arrays of spherical traps (Torquato and Kim 1992).

2. Ultimately, we will pass to the limit $V_1, V \to \infty$. In this limit, ergodicity enables us to equate ensemble and volume averages of some stochastic function $f(\boldsymbol{x})$ so that

$$\langle f \rangle = \lim_{V\to\infty} \frac{1}{V} \int f(\boldsymbol{x}) d\boldsymbol{x}. \tag{13.201}$$

The volume integrals (13.194), (13.195), and (13.198) become, respectively,

$$\frac{1}{\phi_1} \langle \psi_m \psi_n \rangle = \delta_{mn}, \tag{13.202}$$

$$a_n = \frac{1}{\phi_1} \langle \psi_n \rangle, \tag{13.203}$$

and

$$S(t) = \frac{1}{\phi_1} \left\langle \frac{c(\boldsymbol{x}, t)}{c_o} \right\rangle. \tag{13.204}$$

Moreover, the spectrum $n(T)$ of the Laplace operator is no longer discrete but continuous, and so sums are replaced by integrals. For example, the series representation (13.199) of the survival probability $S(t)$ is replaced by

$$S(t) = \int_0^\infty n(T) e^{-t/T} dT.$$

3. The long-time behavior of the survival probability $S(t)$ is intimately connected to fluctuations in the pore size. The largest, or principal, relaxation time T_1 (inversely proportional to the smallest eigenvalue λ_1) is determined by the largest pores in the system. Therefore, when fluctuations in the pore size exist that are on the order of the system size, T_1 will diverge to infinity in the infinite-volume limit, and the associated spectrum will be continuous. The corresponding density of states near $\lambda_1 = 0$ is known as the "Lifshitz spectrum" in the theory of disordered systems (Lifshitz, Gredeskul and Pastur 1988). The associated survival probability becomes a stretched exponential, i.e., has the form $\exp[-a\, t^{d/(d+2)}]$ in d dimensions as $t \to \infty$ (Donsker and Varadhan 1975, Lifshitz et al. 1988).

13.7.2 Relationship Between Survival and Relaxation Times

Torquato and Avellaneda (1991) have shown that the mean survival time τ is bounded from above and below in terms of the principal relaxation T_1. Indeed, τ is related to the entire spectrum of the relaxation times (i.e., eigenvalues) or, equivalently, to the area under the survival probability curve $S(t)$. Thus, τ is a *simple but robust parameter* that describes the decay process. These statements are given in the form of two theorems and a corollary.

Theorem 13.2 *For random porous media of arbitrary microstructure at porosity ϕ_1, the following relation holds:*

$$\tau = \sum_{n=1}^{\infty} a_n^2 T_n, \tag{13.205}$$

where the a_n are the averages of the eigenfunctions ψ_n given by (13.203).

This theorem was proved by Torquato and Avellaneda (1991) by taking the Laplace transforms of (13.187) and (13.188) in time and recognizing that the transform of c evaluated at $s = 0$ (where s is the Laplace-transform variable) is trivially related to the steady-state concentration field u that solves (13.154) and (13.155). This proof is very similar to the one used to prove the analogous Theorem 23.6 involving the fluid permeability.

Corollary 13.1 *The mean survival time is also expressible as an integral over the survival probability as follows:*

$$\tau = \int_0^{\infty} S(t)dt. \tag{13.206}$$

This is easily proved by integrating relation (13.199) over all times and using Theorem 13.2.

Theorem 13.3 *For random porous media of arbitrary microstructure at porosity ϕ_1, the mean survival time τ is bounded from above and below in terms of the principal relaxation time T_1 as follows:*

$$a_1^2 T_1 \le \tau \le T_1. \tag{13.207}$$

This theorem follows from elementary properties of the eigenvalues (Torquato and Avellaneda 1991) and Theorem 13.2.

Remarks:

1. The aforementioned corollary and upper bound are valid whether the spectrum is discrete or continuous, and hence they hold for ergodic media.
2. The relaxation times are easy to evaluate for diffusion occurring inside certain simple domains. For example, in the case of transport interior to three-dimensional spherical pores of radius a (Torquato and Avellaneda 1991), the principal relaxation time T_1 obeys the following exact asymptotic expressions:

$$T_1 \sim \frac{a}{3\kappa} + \frac{a^2}{15\mathcal{D}} + \frac{17a^3\kappa}{525\mathcal{D}^2}, \qquad \frac{\kappa a}{\mathcal{D}} \ll 1, \tag{13.208}$$

$$T_1 \sim \frac{a^2}{\pi^2\mathcal{D}} + \frac{2a}{\pi^2\kappa}, \qquad \frac{\kappa a}{\mathcal{D}} \gg 1. \tag{13.209}$$

   Comparing these results to the exact result $\tau = a^2/(15\mathcal{D}) + a/(3\kappa)$ for the survival time, as obtained from (16.83) with $d = 3$, reveals that the upper bound of Theorem 13.3 is very sharp for this simple pore geometry in both the diffusion-controlled and reaction-controlled regimes. This result suggests that the bound may be sharp for more general porous media, provided that they possess a narrow range of pore sizes.
3. Interestingly, Sapoval, Russ, Korb and Petit (1996) have computed the diffusion relaxation times for transport inside certain fractal pores.

13.8 Time-Dependent Flow Problem

In time-dependent flow, the key macroscopic properties are the so-called viscous relaxation times, which reflect information about the pore topology. Below we describe the basic equations and show an interesting connection to the steady-state *conduction* problem (Avellaneda and Torquato 1991).

13.8.1 Basic Equations

Consider the unsteady Stokes equations for the fluid velocity vector field $\boldsymbol{v}(\boldsymbol{x},t)$ at position $\boldsymbol{x}$ and time t in $\mathcal{V}_1$:

$$\frac{\partial \boldsymbol{v}}{\partial t} = -\nabla\left(\frac{p}{\rho}\right) + \nu\Delta\boldsymbol{v} + v_0\boldsymbol{e}\delta(t) \qquad \text{in } \mathcal{V}_1(\omega), \tag{13.210}$$

$$\nabla\cdot\boldsymbol{v} = 0 \qquad \text{in } \mathcal{V}_1(\omega), \tag{13.211}$$

$$\boldsymbol{v} = 0 \qquad \text{on } \partial\mathcal{V}(\omega). \tag{13.212}$$

Here $p(\boldsymbol{x},t)$ is the pressure, ρ is the constant fluid density, ν is the kinematic viscosity, v_0 is a constant speed, $\boldsymbol{e}$ is an arbitrary unit vector in the direction of the applied field, and $\delta(t)$ is the Dirac delta function. It will be implicit in all of the ensuing discussion that we assume a zero-traction condition at the boundary of $\mathcal{V}$.

The solution of (13.210)–(13.212) can be expressed as a sum of normal modes as follows:

$$\frac{\boldsymbol{v}(\boldsymbol{x},t)}{v_0} = \sum_{n=1}^{\infty} b_n e^{-t/\Theta_n} \boldsymbol{\Psi}_n(\boldsymbol{x}), \tag{13.213}$$

where the vector eigenfunctions $\boldsymbol{\Psi}_n$ satisfy

$$\Delta\boldsymbol{\Psi}_n + \nabla Q_n = -\epsilon_n \boldsymbol{\Psi}_n \qquad \text{in } \mathcal{V}_1(\omega), \tag{13.214}$$

$$\nabla\cdot \boldsymbol{\Psi}_n = 0 \qquad \text{in } \mathcal{V}_1(\omega), \tag{13.215}$$

$$\boldsymbol{\Psi}_n = 0 \qquad \text{on } \partial\mathcal{V}(\omega). \tag{13.216}$$

Here the $\Theta_n = 1/(\nu\epsilon_n)$ are viscous relaxation times, and so the nth eigenvalue ϵ_n has dimensions of $(\text{length})^{-2}$. The functions Q_n in (13.214) are the corresponding pressures. The eigenfunctions $\boldsymbol{\Psi}_n$ are orthonormal, so that

$$\frac{1}{V_1}\int_{\mathcal{V}_1} \boldsymbol{\Psi}_m(\boldsymbol{x})\cdot\boldsymbol{\Psi}_n(\boldsymbol{x})d\boldsymbol{x} = \delta_{mn}, \tag{13.217}$$

and the eigenfunction expansion coefficients are given by

$$b_n = \frac{1}{V_1}\int_{\mathcal{V}_1} \boldsymbol{e}\cdot\boldsymbol{\Psi}_n(\boldsymbol{x})d\boldsymbol{x}. \tag{13.218}$$

Here $V_1 = \phi_1 V$ denotes the total pore volume.

Note that the set of orthonormal eigenfunctions $\boldsymbol{\Psi}_n$ is complete in the closed subspace of square integrable divergence-free fields having zero normal component on $\partial\mathcal{V}$ (Temam 1979). According to the classical Hodge decomposition (Temam 1979), we can express the constant unit vector $\boldsymbol{e}$ as the sum of a solenoidal field, with vanishing normal component on the pore–solid interface, and the gradient of a potential, as follows:

$$\boldsymbol{e} = \boldsymbol{E} + \nabla\varphi. \tag{13.219}$$

Here $\boldsymbol{E}$ is a dimensionless field satisfying

$$\nabla\cdot\boldsymbol{E} = 0 \qquad \text{in } \mathcal{V}_1(\omega), \tag{13.220}$$

$$\boldsymbol{E}\cdot\boldsymbol{n} = 0 \qquad \text{on } \partial\mathcal{V}(\omega), \tag{13.221}$$

where $\boldsymbol{n}$ is the unit outward normal from the pore region. Relation (13.219) implies that

$$\nabla\times\boldsymbol{E} = 0 \qquad \text{in } \mathcal{V}_1(\omega). \tag{13.222}$$

We observe that the field $\boldsymbol{E}$ then solves the corresponding electric conduction problem for a porous medium filled with a conducting fluid of conductivity σ_1 and having an insulating solid phase. Hence, $\boldsymbol{E}$ can be physically interpreted as a scaled electric field, i.e., the actual electric field divided by the magnitude of the ensemble-averaged electric field. The field $\boldsymbol{E}$ is related to the scaled effective conductivity of the porous medium σ_e/σ_1 by the energy representation formula (see Theorem 14.1)

$$\sigma_e/\sigma_1 = F^{-1} = \langle \boldsymbol{E} \cdot \boldsymbol{E} \rangle. \tag{13.223}$$

Here $F \equiv \sigma_1/\sigma_e$ is the dimensionless inverse effective conductivity, referred to as the *formation factor*, and angular brackets denote an ensemble average. For statistically homogeneous media, ergodicity enables us to equate ensemble averages with volume averages, and therefore the average of an arbitrary stochastic function $f(\boldsymbol{x})$ is defined by (13.201). Substitution of (13.219) into (13.218) yields, after integration by parts,

$$\begin{aligned} b_n &= \frac{1}{V_1} \int_{\mathcal{V}_1} \boldsymbol{E}(\boldsymbol{x}) \cdot \boldsymbol{\Psi}_n(\boldsymbol{x}) d\boldsymbol{x} \\ &= \frac{1}{\phi_1} \langle \boldsymbol{E} \cdot \boldsymbol{\Psi}_n \rangle. \end{aligned} \tag{13.224}$$

Therefore, the coefficients b_n coincide with the coefficients of the normal mode expansion of the dimensionless field $\boldsymbol{E}$ in the orthonormal set of solenoidal eigenfunctions $\{\boldsymbol{\Psi}_n\}$. Since the $\boldsymbol{\Psi}_n$ are complete in the aforementioned subspace, we have

$$\sum_{n=1}^{\infty} b_n \boldsymbol{\Psi}_n = \boldsymbol{E} \tag{13.225}$$

and

$$\sum_{n=1}^{\infty} b_n^2 = \frac{1}{\phi_1} \langle \boldsymbol{E} \cdot \boldsymbol{E} \rangle = \frac{1}{\phi_1} \frac{\sigma_e}{\sigma_1} = \frac{1}{F\phi_1}. \tag{13.226}$$

The product $F\phi_1$ is referred to as the *tortuosity*. The completion formula (13.226) will prove very useful in deriving cross-property relations between fluid permeability and effective diffusion parameters.

To summarize, we have shown the remarkable result that *the response of the Stokes fluid to a unit applied pressure gradient* $\boldsymbol{e}$ *is identical to the response obtained if* $\boldsymbol{e}$ *is replaced by* $\boldsymbol{E}$, *the dimensionless electric field.* The reason for this is that in steady state, the gradient of the potential, $\nabla\varphi$, in the Hodge decomposition of $\boldsymbol{e}$ corresponds to a pressure fluctuation that does not affect the velocity field.

13.8.2 Relationship Between Permeability and Relaxation Times

Just as the steady-state mean survival time can be related to the diffusion relaxation times T_n, the steady-state permeability k can be related to the viscous relaxation times Θ_n. However, unlike the former case, the latter connection also involves a different effective property of the porous medium, namely, the effective conductivity. Accordingly, this *cross-property* relation will be derived in Chapter 23.

CHAPTER 14

Variational Principles

For random media of arbitrary microstructure, exact analytical solutions of the effective properties are unattainable, and so any rigorous statement about the effective properties must be in the form of rigorous bounds. To get variational bounds on effective properties, one must first express the effective parameter in terms of some functional and then formulate an appropriate variational (extremum) principle for the functional. We shall primarily deal with "energy" functionals. Once the variational principle is established, then specific bounds on the property of interest are obtained by constructing trial, or admissible, fields that conform with the variational principle. Specific bounds derived from trial fields are the subject of Chapter 21. In this chapter we will derive variational principles for the effective conductivity, effective elastic moduli, trapping constant, and fluid permeability.

The use of variational principles to bound the effective properties of heterogeneous materials dates back to the work of Hill (1952), Paul (1960), and Hashin and Shtrikman (1962, 1963). In all of these studies, homogeneous boundary conditions were specified on the surface of the macroscopic samples. However, for random statistically homogeneous media, the variational principles must be modified so that the boundary conditions are replaced with average conditions. Pioneering work on the use of variational principles to bound effective properties of random heterogeneous materials was carried out by Prager and coworkers, who considered the fluid permeability (Prager 1961) and trapping constant (Prager 1963a) of random porous media as well as the diffusion coefficients and viscosity of random suspensions (Prager 1963b). Beran and coworkers made major contributions by focusing on variational bounds on the conductivity (Beran 1965, Beran 1968, Beran and Silnutzer 1971) and the elastic moduli (Beran and Molyneux 1966, Beran 1968) of random heterogeneous materials. Strieder and Aris (1973) developed variational bounds on effective properties associated with diffusion and reaction in heterogeneous media.

In the ensuing discussion we will make heavy use of the *generalized divergence theorem* and *generalized formulas for integration by parts*. Consider a region $\mathcal{V}$ bounded by the surface $\mathcal{S}$ and let $\boldsymbol{n}$ denote the outward unit normal to the surface $\mathcal{S}$. The restrictions on the surface regularity are not severe; it will be sufficient if the surface is piecewise smooth [see Kellogg (1953)]. This mild restriction on the domain, along with continuity requirements of the field variables and their gradients as necessary, enables us to write the generalized divergence theorem as

$$\int_{\mathcal{V}} \nabla * \boldsymbol{A}\, dV = \int_{\mathcal{S}} \boldsymbol{A} * \boldsymbol{n}\, dS, \tag{14.1}$$

where the field variable $\boldsymbol{A}$ may denote a scalar, vector, or tensor. The "star product" $\boldsymbol{A} * \boldsymbol{B}$ includes the usual dot product $\boldsymbol{A} \cdot \boldsymbol{B}$ (single contraction) as well as the dyadic $\boldsymbol{AB}$ (sometimes denoted by $\boldsymbol{A} \otimes \boldsymbol{B}$). Some special cases of (14.1) are written below in indicial notation for rectangular Cartesian components when $\boldsymbol{A}$ is a scalar f, vector $\boldsymbol{v}$, or second-order tensor $\boldsymbol{T}$:

$$\boldsymbol{A} = \text{scalar } f: \qquad \int_{\mathcal{V}} \frac{\partial f}{\partial x_i} dV = \int_{\mathcal{S}} f n_i dS; \tag{14.2}$$

$$\boldsymbol{A} = \text{vector } \boldsymbol{v}: \qquad \int_{\mathcal{V}} \frac{\partial v_i}{\partial x_i} dV = \int_{\mathcal{S}} v_i n_i dS, \tag{14.3}$$

$$\int_{\mathcal{V}} \frac{\partial v_j}{\partial x_i} dV = \int_{\mathcal{S}} v_j n_i dS; \tag{14.4}$$

$$\boldsymbol{A} = \text{tensor } \boldsymbol{T}: \qquad \int_{\mathcal{V}} \frac{\partial T_{jk}}{\partial x_k} dV = \int_{\mathcal{S}} T_{jk} n_k dS, \tag{14.5}$$

$$\int_{\mathcal{V}} \frac{\partial T_{ij}}{\partial x_k} dV = \int_{\mathcal{S}} T_{ij} n_k dS. \tag{14.6}$$

We will employ two generalized formulas for integration by parts. Let $\boldsymbol{B}$ represent another field variable that is a tensor of arbitrary order. The first formula makes use of the identity

$$\nabla \cdot (\boldsymbol{AB}) = \nabla \boldsymbol{A} \cdot \boldsymbol{B} + \boldsymbol{A}(\nabla \cdot \boldsymbol{B}), \tag{14.7}$$

where no contraction is implied in the product $\boldsymbol{AB}$. Taking the volume average of this identity and using the generalized divergence theorem (14.1) yields the first generalized formula for integration by parts:

$$\frac{1}{V} \int_{\mathcal{S}} (\boldsymbol{AB}) \cdot \boldsymbol{n}\, dS = \frac{1}{V} \int_{\mathcal{V}} \nabla \boldsymbol{A} \cdot \boldsymbol{B}\, dV + \frac{1}{V} \int_{\mathcal{V}} \boldsymbol{A}(\nabla \cdot \boldsymbol{B})\, dV. \tag{14.8}$$

The second formula makes use of the identity

$$\nabla \cdot (\boldsymbol{A} \cdot \boldsymbol{B}) = \nabla \boldsymbol{A} : \boldsymbol{B} + \boldsymbol{A} \cdot (\nabla \cdot \boldsymbol{B}), \tag{14.9}$$

where a single contraction is implied in the product $\boldsymbol{A} \cdot \boldsymbol{B}$. Taking the volume average of this identity and using (14.1) gives the second generalized formula for integration by parts:

$$\frac{1}{V}\int_{S}(\boldsymbol{A}\cdot\boldsymbol{B})\cdot\boldsymbol{n}\,dS=\frac{1}{V}\int_{\mathcal{V}}\nabla\boldsymbol{A}:\boldsymbol{B}\,dV+\frac{1}{V}\int_{\mathcal{V}}\boldsymbol{A}\cdot(\nabla\cdot\boldsymbol{B})\,dV. \tag{14.10}$$

14.1 Conductivity

We will derive variational principles that will enable us to bound the effective conductivity from above and below. These variational bounds are based on minimizing the "energy" or, more precisely, minimizing the power dissipation in the system. First, we will make some useful observations concerning fluctuations of the local fields from their average values for ergodic media. Then it will be shown that the effective conductivity has an energy representation. Finally, we will prove the minimum energy principles that lead to rigorous upper and lower bounds on the effective conductivity.

We will consider *ergodic* media, i.e., statistically homogeneous composites that obey the ergodic hypothesis (Section 2.2.2). For such materials, we can equate ensemble and volume averages in the infinite-volume limit. We will assume that the composite is composed of M anisotropic phases characterized by a local conductivity tensor

$$\boldsymbol{\sigma}(\boldsymbol{x})=\sum_{i=1}^{M}\boldsymbol{\sigma}_i\,\mathcal{I}^{(i)}(\boldsymbol{x}), \tag{14.11}$$

where $\boldsymbol{\sigma}_i$ and $\mathcal{I}^{(i)}(\boldsymbol{x})$ are the conductivity and indicator functions for phase i, respectively. Referring to relations (13.3)–(13.6), we see that the local equations that are valid in each phase are

$$\nabla\cdot\boldsymbol{J}(\boldsymbol{x})=0, \tag{14.12}$$

$$\boldsymbol{J}(\boldsymbol{x})=\boldsymbol{\sigma}(\boldsymbol{x})\cdot\boldsymbol{E}(\boldsymbol{x}), \tag{14.13}$$

$$\boldsymbol{E}(\boldsymbol{x})=-\nabla T(\boldsymbol{x}). \tag{14.14}$$

From the last chapter on homogenization theory, we found that the effective conductivity tensor $\boldsymbol{\sigma}_e$ of the composite is defined via the average constitutive relation given by

$$\langle\boldsymbol{J}\rangle=\boldsymbol{\sigma}_e\cdot\langle\boldsymbol{E}\rangle. \tag{14.15}$$

14.1.1 Field Fluctuations

For ergodic media, the average fields in (14.15) are uniform (or constant) vector fields. In such cases, it is useful to decompose the local fields $\boldsymbol{E}(\boldsymbol{x})$ and $\boldsymbol{J}(\boldsymbol{x})$ into their *average* and *fluctuating* components, i.e.,

$$\boldsymbol{E} = \langle \boldsymbol{E} \rangle + \boldsymbol{E}', \tag{14.16}$$

$$\boldsymbol{J} = \langle \boldsymbol{J} \rangle + \boldsymbol{J}', \tag{14.17}$$

where $\boldsymbol{E}'$ and $\boldsymbol{J}'$ are the fluctuating parts of the intensity field and flux, respectively, and angular brackets denote an ensemble average. Thus,

$$\langle \boldsymbol{E}' \rangle = \langle \boldsymbol{J}' \rangle = 0. \tag{14.18}$$

Let the potential T be decomposed as follows:

$$T = -\langle \boldsymbol{E} \rangle \cdot \boldsymbol{x} + T'. \tag{14.19}$$

Taking the gradient of (14.19) (after multiplication by a minus sign) yields $\boldsymbol{E}$, which, when compared to (14.16), gives

$$\boldsymbol{E}' = -\nabla T'. \tag{14.20}$$

Now consider a very large region having volume V and macroscopic surface S within a realization of a composite taken from an ergodic ensemble. Except for infinitesimally small boundary effects, the composite within this region is still statistically homogeneous and hence is virtually ergodic. Thus, ensemble and volume averages are almost identical. In particular, we can write

$$\begin{aligned} \langle \boldsymbol{E}' \rangle &\approx \frac{1}{V} \int_V -\nabla T' dV \\ &= \frac{1}{V} \int_S -T' \boldsymbol{n} dS, \end{aligned} \tag{14.21}$$

where we have used the divergence theorem (14.2). Strictly speaking, one must not consider only the bounding surface of the composite when applying the divergence theorem but the multiphase interface as well. However, since we have continuity of the potential across the interfaces, such surface integrals associated with each interface will sum to exactly zero. Accordingly, we will not bother to write such terms explicitly. Importantly, we see from (14.18) and (14.21) that in the infinite-volume limit,

$$\lim_{V \to \infty} \frac{1}{V} \int_S -T' \boldsymbol{n} dS \to 0; \tag{14.22}$$

i.e., the ratio of the surface integral of (14.22) to the volume V vanishes in the limit $V \to \infty$.

Similarly, for the large region of volume V and surface S, we can express the average of the fluctuating part of the flux as

$$\begin{aligned} \langle \boldsymbol{J}' \rangle &\approx \frac{1}{V} \int_V \boldsymbol{J}' dV \\ &= \frac{1}{V} \int_V \nabla \cdot (\boldsymbol{x} \boldsymbol{J}') dV \\ &= \frac{1}{V} \int_S \boldsymbol{x} (\boldsymbol{J}' \cdot \boldsymbol{n}) dS. \end{aligned} \tag{14.23}$$

The second line of (14.23) follows from the identity (14.7) with $\boldsymbol{A} = \boldsymbol{x}$ and $\boldsymbol{B} = \boldsymbol{J}$ and the fact that $\nabla \cdot \boldsymbol{J} = 0$. The third line is obtained using the divergence theorem (14.5). Since the normal flux is continuous across the multiphase interface, the surface integrals over the interface make no contribution when the divergence theorem is applied, and therefore we do not bother to write such terms explicitly. Hence, we see from (14.18) and (14.23) that in the infinite-volume limit,

$$\lim_{V\to\infty} \frac{1}{V} \int_S \boldsymbol{x}(\boldsymbol{J}' \cdot \boldsymbol{n}) dS \to 0. \tag{14.24}$$

To summarize, ergodicity renders the boundary terms involving the fluctuating quantities T' and $\boldsymbol{J}'$ in (14.22) and (14.24), respectively, to be zero. In subsequent discussion, boundary terms of these types will arise and will be taken to be zero by ergodicity.

14.1.2 Energy Representation

As we saw in Section 13.2.1, the power dissipated per unit volume in a homogeneous linear conductor is a nonnegative quantity that is proportional to the dot product of the intensity field and the flux field (Jackson 1990). In the case of a heterogeneous linear material, we assume that the power dissipated per unit volume $w(\boldsymbol{x})$ at the point $\boldsymbol{x}$ is given by the same but local form, i.e.,

$$w(\boldsymbol{x}) = \frac{1}{2} \boldsymbol{E}(\boldsymbol{x}) \cdot \boldsymbol{J}(\boldsymbol{x}) \geq 0. \tag{14.25}$$

In the dielectric and magnetic context, w is the *energy stored* in the composite per unit volume. It is common to refer to w simply as the "microscopic energy." Use of the constitutive relation (14.13) shows that the microscopic energy is a positive definite quadratic form in either the intensity or the flux; i.e., one has the equivalent but alternative forms

$$w_E(\boldsymbol{x}) = \frac{1}{2} \boldsymbol{E}(\boldsymbol{x}) \cdot \boldsymbol{\sigma}(\boldsymbol{x}) \cdot \boldsymbol{E}(\boldsymbol{x}) \tag{14.26}$$

and

$$w_J(\boldsymbol{x}) = \frac{1}{2} \boldsymbol{J}(\boldsymbol{x}) \cdot \boldsymbol{\sigma}^{-1}(\boldsymbol{x}) \cdot \boldsymbol{J}(\boldsymbol{x}), \tag{14.27}$$

where $\boldsymbol{\sigma}^{-1}$ is the resistivity tensor.

Ensemble averaging the expressions above for the microscopic energies yields the equivalent energy functionals

$$W[\boldsymbol{E}] \equiv \langle w_E(\boldsymbol{x}) \rangle = \frac{1}{2} \langle \boldsymbol{E}(\boldsymbol{x}) \cdot \boldsymbol{\sigma}(\boldsymbol{x}) \cdot \boldsymbol{E}(\boldsymbol{x}) \rangle, \tag{14.28}$$

$$W[\boldsymbol{J}] \equiv \langle w_J(\boldsymbol{x}) \rangle = \frac{1}{2} \langle \boldsymbol{J}(\boldsymbol{x}) \cdot \boldsymbol{\sigma}^{-1}(\boldsymbol{x}) \cdot \boldsymbol{J}(\boldsymbol{x}) \rangle. \tag{14.29}$$

We can also define the "macroscopic energy" $\tilde{W}$ by treating the heterogeneous material as a homogeneous material with uniform fields $\langle \boldsymbol{E}(\boldsymbol{x})\rangle$ and $\langle \boldsymbol{J}(\boldsymbol{x})\rangle$, i.e.,

$$\begin{aligned}\tilde{W} &= \frac{1}{2}\langle \boldsymbol{E}(\boldsymbol{x})\rangle \cdot \langle \boldsymbol{J}(\boldsymbol{x})\rangle \\ &= \frac{1}{2}\langle \boldsymbol{E}(\boldsymbol{x})\rangle \cdot \boldsymbol{\sigma}_e \cdot \langle \boldsymbol{E}(\boldsymbol{x})\rangle \\ &= \frac{1}{2}\langle \boldsymbol{J}(\boldsymbol{x})\rangle \cdot \boldsymbol{\sigma}_e^{-1} \cdot \langle \boldsymbol{J}(\boldsymbol{x})\rangle. \end{aligned} \tag{14.30}$$

It will now be shown that under rather general conditions, the macroscopic energy $\tilde{W}$ is equal to W, the average of the microscopic energy, thus providing an energy representation of the effective conductivity tensor $\boldsymbol{\sigma}_e$ or the effective resistivity tensor $\boldsymbol{\sigma}_e^{-1}$.

Theorem 14.1

For ergodic macroscopically anisotropic multiphase composites with anisotropic constituents, the macroscopic energy $\tilde{W}$ is equal to the average of the microscopic energy W, i.e., the effective conductivity tensor $\boldsymbol{\sigma}_e$ has the energy representation

$$\frac{1}{2}\langle \boldsymbol{E}\rangle \cdot \boldsymbol{\sigma}_e \cdot \langle \boldsymbol{E}\rangle = \frac{1}{2}\langle \boldsymbol{E} \cdot \boldsymbol{\sigma} \cdot \boldsymbol{E}\rangle, \tag{14.31}$$

or, equivalently,

$$\frac{1}{2}\langle \boldsymbol{J}\rangle \cdot \boldsymbol{\sigma}_e^{-1} \cdot \langle \boldsymbol{J}\rangle = \frac{1}{2}\langle \boldsymbol{J} \cdot \boldsymbol{\sigma}^{-1} \cdot \boldsymbol{J}\rangle. \tag{14.32}$$

Proof: Using (14.14) and the integration by parts formula (14.8), we can write the average appearing on the right side of (14.31) as

$$\begin{aligned}\langle \boldsymbol{E}\cdot \boldsymbol{J}\rangle &= -\langle \nabla \cdot (T\boldsymbol{J})\rangle + \langle T(\nabla \cdot \boldsymbol{J})\rangle \\ &= \lim_{V\to\infty} -\frac{1}{V}\int_S T(\boldsymbol{J}\cdot \boldsymbol{n})dS \\ &= \langle \boldsymbol{E}\rangle \cdot \langle \boldsymbol{J}\rangle - \lim_{V\to\infty}\frac{1}{V}\int_S T'(\boldsymbol{J}'\cdot \boldsymbol{n})dS \\ &= \langle \boldsymbol{E}\rangle \cdot \langle \boldsymbol{J}\rangle, \end{aligned} \tag{14.33}$$

which proves the theorem. The second line of (14.33) follows from (14.8) (with $\boldsymbol{A} = T$ and $\boldsymbol{B} = \boldsymbol{J}$) and the use of the divergence-free condition (14.12). The third line follows after the decompositions (14.17) and (14.19) are substituted into the second line and the divergence theorem (14.4) is applied. The fourth line of (14.33) follows from ergodicity. ∎

Remarks:

1. Notice that the result (14.33), valid for ergodic media, implies that the average of the product $\boldsymbol{E}' \cdot \boldsymbol{J}'$ vanishes, i.e.,

$$\langle \boldsymbol{E}' \cdot \boldsymbol{J}'\rangle = 0. \tag{14.34}$$

This means that the fluctuating quantities $\boldsymbol{E}'$ and $\boldsymbol{J}'$ are orthogonal to one another on average. Therefore, it follows that for ergodic media

$$\langle \boldsymbol{E}' \cdot \boldsymbol{J} \rangle = \langle \boldsymbol{E} \cdot \boldsymbol{J}' \rangle = 0,$$

or, equivalently,

$$\langle \boldsymbol{E}' \cdot \boldsymbol{\sigma} \cdot \boldsymbol{E} \rangle = \langle \boldsymbol{J}' \cdot \boldsymbol{\sigma}^{-1} \cdot \boldsymbol{J} \rangle = 0. \tag{14.35}$$

This last identity will prove useful to us in subsequent derivations.

2. It follows from Theorem 14.1 that the effective conductivity tensor $\boldsymbol{\sigma}_e$ is a monotonic function of the local conductivity tensor $\boldsymbol{\sigma}(\boldsymbol{x})$, i.e., $\boldsymbol{\sigma}_e^* > \boldsymbol{\sigma}_e$ when $\boldsymbol{\sigma}^*(\boldsymbol{x}) > \boldsymbol{\sigma}(\boldsymbol{x})$ for all $\boldsymbol{x}$. The proof is simple. In light of the conditions on $\boldsymbol{\sigma}_e^*$ and $\boldsymbol{\sigma}(\boldsymbol{x})$, it follows that $\langle \boldsymbol{E} \cdot \boldsymbol{\sigma}^* \cdot \boldsymbol{E} \rangle > \langle \boldsymbol{E} \cdot \boldsymbol{\sigma} \cdot \boldsymbol{E} \rangle$ for all $\boldsymbol{E}$, which in turn implies that $\langle \boldsymbol{E} \rangle \cdot \boldsymbol{\sigma}_e^* \cdot \langle \boldsymbol{E} \rangle > \langle \boldsymbol{E} \rangle \cdot \boldsymbol{\sigma}_e \cdot \langle \boldsymbol{E} \rangle$ for all $\langle \boldsymbol{E} \rangle$. Therefore, $\boldsymbol{\sigma}_e^* > \boldsymbol{\sigma}_e$.

Theorem 14.2 *The effective conductivity tensor $\boldsymbol{\sigma}_e$ is symmetric and positive definite.*

Proof: To prove the symmetry of $\boldsymbol{\sigma}_e$, we decompose the field $\boldsymbol{E}(\boldsymbol{x})$ into its average and fluctuating parts according to (14.16). The fluctuating part can be written as the scalar product of a second-order tensor $\boldsymbol{e}$ and the average field, i.e., $\boldsymbol{E}' = \boldsymbol{e} \cdot \langle \boldsymbol{E} \rangle$ (see Chapter 20). Using this decomposition and (14.31) yields that the effective conductivity is given by

$$(\sigma_e)_{ij} = \langle \sigma_{ij} \rangle + \langle e_{ik} \sigma_{kl} e_{jl} \rangle. \tag{14.36}$$

It immediately follows that the effective tensor is symmetric, i.e., $(\sigma_e)_{ij} = (\sigma_e)_{ji}$, since $\sigma_{ij} = \sigma_{ji}$.

We observe that since the local conductivity tensor $\boldsymbol{\sigma}$ is positive definite (see Section 13.2.1), the right side of the energy relation (14.31) is positive. Therefore, the effective conductivity tensor $\boldsymbol{\sigma}_e$ is also positive definite. ∎

14.1.3 Minimum Energy Principles

For homogeneous bodies there are two classical minimum energy principles that apply to both the conduction and elasticity problems: the *minimum potential energy principle* and the *minimum complementary energy principle*. The extension of these principles to heterogeneous materials with an *arbitrary number of phases* is relatively straightforward and will be proved here for the case of conduction. The first principle is concerned with the construction of *trial intensity fields* $\hat{\boldsymbol{E}}$ that are irrotational (and satisfy a mild average condition) but whose associated flux fields are not necessarily solenoidal. By contrast, the second principle is concerned with the construction of *trial flux fields* $\hat{\boldsymbol{J}}$ that are solenoidal (and satisfy a mild average condition) but whose associated intensity fields are not necessarily irrotational.

Prager (1963b) was the first to apply variational principles to bound diffusion properties of random media, but he did not prove the principles that he employed. Beran

(1968) demonstrated how to extend variational principles for the conductivity of homogeneous bodies with prescribed homogeneous boundary conditions to variational principles for the effective conductivity of random media. The minimum principles given below follow Beran's approach, but we give more details and treat boundary terms differently.

Theorem 14.3 *Minimum Potential Energy:*
Let A_U be the class of trial intensity fields $\hat{\mathbf{E}}$ defined by the set

$$A_U = \{ergodic\ \hat{\boldsymbol{E}};\ \nabla \times \hat{\boldsymbol{E}} = 0,\ \langle \hat{\boldsymbol{E}} \rangle = \langle \boldsymbol{E} \rangle\}, \tag{14.37}$$

and let

$$W[\hat{\boldsymbol{E}}] = \frac{1}{2} \langle \hat{\boldsymbol{E}}(\boldsymbol{x}) \cdot \boldsymbol{\sigma}(\boldsymbol{x}) \cdot \hat{\boldsymbol{E}}(\boldsymbol{x}) \rangle \tag{14.38}$$

be the trial energy functional, where $\boldsymbol{\sigma}$ is the local conductivity tensor. Then, among all trial fields $\hat{\mathbf{E}}$, the field that makes the associated flux solenoidal is the one that uniquely minimizes the trial energy functional $W[\hat{\boldsymbol{E}}]$. In other words,

$$W[\boldsymbol{E}] \le W[\hat{\boldsymbol{E}}] \qquad \forall \hat{\boldsymbol{E}} \in A_U, \tag{14.39}$$

or, equivalently,

$$\frac{1}{2} \langle \boldsymbol{E} \rangle \cdot \boldsymbol{\sigma}_e \cdot \langle \boldsymbol{E} \rangle \le \frac{1}{2} \langle \hat{\boldsymbol{E}} \cdot \boldsymbol{\sigma} \cdot \hat{\boldsymbol{E}} \rangle \qquad \forall \hat{\boldsymbol{E}} \in A_U, \tag{14.40}$$

where $\mathbf{E}$ *satisfies* (14.14).

Proof: Let the "difference" field $\boldsymbol{G}$ be defined by

$$\boldsymbol{G} = \hat{\boldsymbol{E}} - \boldsymbol{E}$$

such that the trial field $\hat{\boldsymbol{E}}$ satisfies

$$\nabla \times \hat{\boldsymbol{E}} = 0, \quad \langle \hat{\boldsymbol{E}} \rangle = \langle \boldsymbol{E} \rangle.$$

Thus, $\langle \boldsymbol{G} \rangle = 0$, and since both $\boldsymbol{E}$ and $\hat{\boldsymbol{E}}$ are irrotational, $\boldsymbol{G}$ is also irrotational, implying that it can be written as the gradient of a scalar ϕ, i.e.,

$$\boldsymbol{G} = -\nabla \phi.$$

Consider the identity

$$\langle \hat{\boldsymbol{E}} \cdot \boldsymbol{\sigma} \cdot \hat{\boldsymbol{E}} \rangle = \langle \boldsymbol{E} \cdot \boldsymbol{\sigma} \cdot \boldsymbol{E} \rangle + 2 \langle \boldsymbol{G} \cdot \boldsymbol{\sigma} \cdot \boldsymbol{E} \rangle + \langle \boldsymbol{G} \cdot \boldsymbol{\sigma} \cdot \boldsymbol{G} \rangle.$$

We now prove that the middle term of this identity is identically zero. Formula (14.8) for integration by parts with $\boldsymbol{A} = \phi$ and $\boldsymbol{B} = \boldsymbol{J} = \boldsymbol{\sigma} \cdot \boldsymbol{E}$ gives

$$\begin{aligned} \langle \boldsymbol{G} \cdot \boldsymbol{\sigma} \cdot \boldsymbol{E} \rangle &= -\langle \nabla \phi \cdot \boldsymbol{J} \rangle \\ &= -\lim_{V \to \infty} \frac{1}{V} \int_S \phi (\boldsymbol{J} \cdot \boldsymbol{n}) dS + \lim_{V \to \infty} \frac{1}{V} \int_V \phi (\nabla \cdot \boldsymbol{J}) dV \\ &= -\lim_{V \to \infty} \frac{1}{V} \int_S \phi' (\boldsymbol{J}' \cdot \boldsymbol{n}) dS = 0. \end{aligned} \tag{14.41}$$

The third line of (14.41) follows after decomposing ϕ and J into their mean and fluctuating parts (as in the proof of Theorem 14.1), using the condition $\langle \boldsymbol{G} \rangle = 0$, and employing the fact that $\boldsymbol{J}$ is solenoidal. The integral in the third line of (14.41) involving the fluctuating fields is zero by ergodicity. Using this result in combination with energy representation (14.31) gives

$$\langle \hat{\boldsymbol{E}} \cdot \boldsymbol{\sigma} \cdot \hat{\boldsymbol{E}} \rangle = \langle \boldsymbol{E} \rangle \cdot \boldsymbol{\sigma}_e \cdot \langle \boldsymbol{E} \rangle + \langle \boldsymbol{G} \cdot \boldsymbol{\sigma} \cdot \boldsymbol{G} \rangle,$$

and since the second term on the right side is greater than or equal to zero, the statement (14.40) is proved. The equality sign holds only when the difference field $\boldsymbol{G}$ is zero throughout $\mathcal{V}$, i.e., when $\hat{\boldsymbol{E}} = \boldsymbol{E}$.

If there were another field $\boldsymbol{E}^*$ from the set A_U, different from $\boldsymbol{E}$ but with a solenoidal flux, then we could apply the above theorem first to $\boldsymbol{E}^*$ and then to $\boldsymbol{E}$, yielding the contradictory results that $W[\boldsymbol{E}^*] > W[\boldsymbol{E}]$ and $W[\boldsymbol{E}] > W[\boldsymbol{E}^*]$; therefore, $\boldsymbol{E}$ is unique. ∎

Remarks:

1. In words, the minimum potential principle states that the actual macroscopic energy $W[\boldsymbol{E}]$ is bounded from above by the trial macroscopic energy $W[\hat{\boldsymbol{E}}]$ and thus leads to an upper bound $\boldsymbol{\sigma}_e^U$ on the effective conductivity tensor $\boldsymbol{\sigma}_e$ in the positive semidefinite sense [cf. (13.16)].
2. Theorem 14.3 applies as well to periodic media (Bensoussan et al. 1978). Boundary terms are now over the unit cell and vanish because of periodicity.
3. In the parlance of variational calculus, the difference field $\boldsymbol{G}$ (defined in the proof above) is the *first variation* of the trial field, denoted by $\delta\hat{\boldsymbol{E}}$. One seeks the minimizer of the first variation of the trial energy functional, which we have done. Throughout this chapter, however, we will avoid this language and notation, since it is less accessible to the nonspecialist.

Theorem 14.4 *Minimum Complementary Energy:*
Let A_L be the class of trial flux fields $\hat{\boldsymbol{J}}$ defined by the set

$$A_L = \{\text{ergodic } \hat{\boldsymbol{J}};\ \nabla \cdot\ \hat{\boldsymbol{J}} = 0,\ \langle \hat{\boldsymbol{J}} \rangle = \langle \boldsymbol{J} \rangle\}, \tag{14.42}$$

and let

$$W[\hat{\boldsymbol{J}}] = \frac{1}{2} \langle \hat{\boldsymbol{J}}(\boldsymbol{x}) \cdot \boldsymbol{\sigma}^{-1}(\boldsymbol{x}) \cdot \hat{\boldsymbol{J}}(\boldsymbol{x}) \rangle \tag{14.43}$$

be the trial energy functional, where $\boldsymbol{\sigma}$ is the local conductivity tensor. Then among all trial fields $\hat{\boldsymbol{J}}$, the field that makes the associated intensity field irrotational is the one that uniquely minimizes the trial energy functional $W[\hat{\boldsymbol{J}}]$, i.e.,

$$W[\boldsymbol{J}] \le W[\hat{\boldsymbol{J}}] \qquad \forall \hat{\boldsymbol{J}} \in A_L, \tag{14.44}$$

or, equivalently,

$$\frac{1}{2} \langle \boldsymbol{J} \rangle \cdot \boldsymbol{\sigma}_e^{-1} \cdot \langle \boldsymbol{J} \rangle \le \frac{1}{2} \langle \hat{\boldsymbol{J}} \cdot \boldsymbol{\sigma}^{-1} \cdot \hat{\boldsymbol{J}} \rangle \qquad \forall \hat{\boldsymbol{J}} \in A_L, \tag{14.45}$$

where $\boldsymbol{J}$ satisfies (14.12).

Proof: Let the "difference" field $\boldsymbol{Q}$ be defined by

$$\boldsymbol{Q} = \hat{\boldsymbol{J}} - \boldsymbol{J} \tag{14.46}$$

such that

$$\nabla \cdot \hat{\boldsymbol{J}} = 0, \quad \langle \hat{\boldsymbol{J}} \rangle = \langle \boldsymbol{J} \rangle.$$

Therefore, $\langle \boldsymbol{Q} \rangle = 0$, and since both $\boldsymbol{J}$ and $\hat{\boldsymbol{J}}$ are solenoidal, $\boldsymbol{Q}$ is also solenoidal, i.e.,

$$\nabla \cdot \boldsymbol{Q} = 0.$$

Consider the identity

$$\langle \hat{\boldsymbol{J}} \cdot \boldsymbol{\sigma}^{-1} \cdot \hat{\boldsymbol{J}} \rangle = \langle \boldsymbol{J} \cdot \boldsymbol{\sigma}^{-1} \cdot \boldsymbol{J} \rangle + 2\langle \boldsymbol{Q} \cdot \boldsymbol{\sigma}^{-1} \cdot \boldsymbol{J} \rangle + \langle \boldsymbol{Q} \cdot \boldsymbol{\sigma}^{-1} \cdot \boldsymbol{Q} \rangle.$$

Since $\boldsymbol{E} = \boldsymbol{\sigma}^{-1} \cdot \boldsymbol{J} = -\nabla T$, the middle term may be rewritten as

$$\begin{aligned}
\langle \boldsymbol{Q} \cdot \boldsymbol{\sigma}^{-1} \cdot \boldsymbol{J} \rangle &= -\langle \boldsymbol{Q} \cdot \nabla T \rangle \\
&= -\lim_{V\to\infty} \frac{1}{V} \int_S T(\boldsymbol{Q} \cdot \boldsymbol{n}) dS + \lim_{V\to\infty} \frac{1}{V} \int_V T(\nabla \cdot \boldsymbol{Q}) dV \\
&= -\lim_{V\to\infty} \frac{1}{V} \int_S T'(\boldsymbol{Q}' \cdot \boldsymbol{n}) dS = 0.
\end{aligned} \tag{14.47}$$

The second line of (14.47) follows from formula (14.8) with $\boldsymbol{A} = T$ and $\boldsymbol{B} = \boldsymbol{Q}$. The third line results after decomposing T and $\boldsymbol{Q}$ into their mean and fluctuating parts (as in the proof of Theorem 14.1), using the condition $\langle \boldsymbol{Q} \rangle = 0$, and utilizing the fact that $\nabla \cdot \boldsymbol{Q} = 0$. The integral in the third line of (14.47) involving the fluctuating fields is zero by ergodicity. Using this result in combination with energy representation (14.32) gives

$$\langle \hat{\boldsymbol{J}} \cdot \boldsymbol{\sigma}^{-1} \cdot \hat{\boldsymbol{J}} \rangle = \langle \boldsymbol{J} \rangle \cdot \boldsymbol{\sigma}_e^{-1} \cdot \langle \boldsymbol{J} \rangle + \langle \boldsymbol{Q} \cdot \boldsymbol{\sigma}^{-1} \cdot \boldsymbol{Q} \rangle,$$

which proves the statement (14.45). The equality sign holds only when the difference field $\boldsymbol{Q}$ is zero throughout $\mathcal{V}$, i.e., when $\hat{\boldsymbol{J}} = \boldsymbol{J}$. The uniqueness of $\boldsymbol{J}$ follows using the same arguments as for Theorem 14.3. ∎

Remarks:

1. In words, the minimum complementary energy principle states that the actual macroscopic energy $W[\boldsymbol{J}]$ is bounded from above by the trial macroscopic energy $W[\hat{\boldsymbol{J}}]$. Therefore, one can obtain an upper bound on the effective resistivity tensor $\boldsymbol{\sigma}_e^{-1}$ or a lower bound $\boldsymbol{\sigma}_e^L$ on the effective conductivity tensor $\boldsymbol{\sigma}_e$ in the positive semidefinite sense.
2. Theorem 14.4 also applies to periodic media (Bensoussan et al. 1978). Again, boundary terms are over the unit cell and vanish because of periodicity.

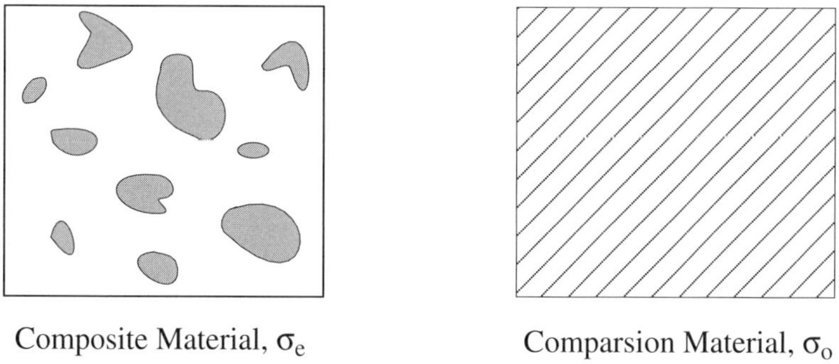

Figure 14.1 Portions of two materials: a composite material with effective conductivity σ_e and a homogeneous comparison material with conductivity σ_0. The composite average field $\langle \boldsymbol{E} \rangle$ is equal to the uniform field $\boldsymbol{E}_0$ in the comparison material.

14.1.4 Hashin–Shtrikman Principle

Consider an ergodic composite body with effective conductivity tensor $\boldsymbol{\sigma}_e$ and a homogeneous *comparison* material (of the same shape and size) with a constant effective conductivity tensor $\boldsymbol{\sigma}_0$ (see Figure 14.1). The intensity field in the comparison material is the uniform field $\boldsymbol{E}_0$, and we require that the average intensity field in the composite obey $\langle \boldsymbol{E} \rangle = \boldsymbol{E}_0$. The composite body is composed of two isotropic phases and described by an effective conductivity tensor $\boldsymbol{\sigma}_e$. For simplicity, we will consider here macroscopically isotropic media, and hence we need only consider the scalar conductivities σ_0 and σ_e, defined respectively by $\boldsymbol{\sigma}_0 = \sigma_0 \boldsymbol{I}$ and $\boldsymbol{\sigma}_e = \sigma_e \boldsymbol{I}$.

Now let the local flux "polarization" field $\boldsymbol{P}$ be defined by

$$\boldsymbol{P} = \boldsymbol{J} - \sigma_0 \boldsymbol{E} = (\sigma - \sigma_0)\boldsymbol{E}, \tag{14.48}$$

where σ represents the local conductivity. Let us also define the difference between the local and comparison fields as

$$\boldsymbol{G} = \boldsymbol{E} - \boldsymbol{E}_0, \tag{14.49}$$

implying that

$$\nabla \times \boldsymbol{G} = 0, \tag{14.50}$$

$$\langle \boldsymbol{G} \rangle = 0. \tag{14.51}$$

Substitution of (14.48) into (14.12) yields

$$\sigma_0 \nabla \cdot \boldsymbol{G} + \nabla \cdot \boldsymbol{P} = 0, \tag{14.52}$$

where we have used (14.49) and the fact that $\nabla \cdot \boldsymbol{E}_0 = 0$.

We now state the Hashin–Shtrikman variational principle (Hashin and Shtrikman 1962). The proof for ergodic media is easily obtained using the same techniques employed to derive the aforementioned classical minimum energy principles, and hence we do not give it here.

Theorem 14.5 *Hashin–Shtrikman Variational Principle:*
Consider any ergodic trial polarization field $\hat{\boldsymbol{P}}$ *subject to the conditions*

$$\sigma_0 \nabla \cdot \hat{\boldsymbol{G}} + \nabla \cdot \hat{\boldsymbol{P}} = 0, \tag{14.53}$$

$$\nabla \times \hat{\boldsymbol{G}} = 0, \tag{14.54}$$

$$\langle \hat{\boldsymbol{G}} \rangle = 0, \tag{14.55}$$

where $\hat{\boldsymbol{G}}$ *is the associated trial difference field and* σ_0 *is a constant. Then*

$$\sigma_* = \sigma_0 + \left\langle \hat{\boldsymbol{P}} \cdot \hat{\boldsymbol{G}} + 2\hat{\boldsymbol{P}} \cdot \boldsymbol{e} - \frac{\hat{\boldsymbol{P}} \cdot \hat{\boldsymbol{P}}}{\sigma - \sigma_0} \right\rangle \tag{14.56}$$

is an upper bound on σ_e *when* $\sigma_0 \geq \sigma_1$ *and a lower bound on* σ_e *when* $\sigma_0 \leq \sigma_2$, *where* $\boldsymbol{E}_0 = \boldsymbol{e}$ *and, without loss of generality,* $\boldsymbol{e}$ *is a unit vector.*

Remarks:

1. The original variational principle of Hashin and Shtrikman was derived for a very large but finite composite body subjected to homogeneous potential boundary conditions on the sample surface (rather than requiring $\langle \hat{\boldsymbol{G}} \rangle = 0$). However, their formulation in the infinite-volume limit is equivalent to Theorem 14.5, since boundary effects then vanish.
2. Theorem 14.5 also can be applied to periodic media (Kohn and Milton 1986).
3. Theorem 14.5 has been generalized to treat anisotropic multiphase random composites with anisotropic constituents (Willis 1977).

14.2 Elastic Moduli

Minimum energy principles will be derived that will enable one to find rigorous upper and lower bounds on the effective elastic moduli. The reader should note that the derivations that follow are completely analogous to those for the conductivity. In particular, we will show that ergodicity renders certain surface integrals, involving fluctuating local fields, to be zero. Next, it will be demonstrated that the effective stiffness tensor has an energy representation. Finally, we will prove the minimum energy principles.

As in the conduction problem, we will consider *ergodic* media (Section 2.2.2). The composite is assumed to be composed of M anisotropic phases characterized by a local stiffness tensor

$$\boldsymbol{C}(\boldsymbol{x}) = \sum_{i=1}^{M} \boldsymbol{C}_i \, \mathcal{I}^{(i)}(\boldsymbol{x}), \tag{14.57}$$

where $\boldsymbol{C}_i$ and $\mathcal{I}^{(i)}(\boldsymbol{x})$ are the stiffness and indicator functions for phase i, respectively. Referring to expressions (13.50)–(13.53), we see that the local equations that are valid in both phases are

$$\nabla \cdot \boldsymbol{\tau}(\boldsymbol{x}) = 0, \tag{14.58}$$

$$\boldsymbol{\tau}(\boldsymbol{x}) = \boldsymbol{C}(\boldsymbol{x}) : \boldsymbol{\varepsilon}(\boldsymbol{x}), \tag{14.59}$$

$$\boldsymbol{\varepsilon}(\boldsymbol{x}) = \frac{1}{2}\Big[\nabla \boldsymbol{u}(\boldsymbol{x}) + \nabla \boldsymbol{u}(\boldsymbol{x})^{\mathrm{T}}\Big]. \tag{14.60}$$

The macroscopically anisotropic composite is characterized by an effective stiffness tensor $\boldsymbol{C}_e$ and is obtainable from the average constitutive relation

$$\langle \boldsymbol{\tau} \rangle = \boldsymbol{C}_e : \langle \boldsymbol{\varepsilon} \rangle. \tag{14.61}$$

14.2.1 Field Fluctuations

Since averages of local fields are uniform for ergodic media, we can decompose the local fields $\boldsymbol{\varepsilon}(\boldsymbol{x})$ and $\boldsymbol{\tau}(\boldsymbol{x})$ into their average and fluctuating components, i.e.,

$$\boldsymbol{\varepsilon} = \langle \boldsymbol{\varepsilon} \rangle + \boldsymbol{\varepsilon}', \tag{14.62}$$

$$\boldsymbol{\tau} = \langle \boldsymbol{\tau} \rangle + \boldsymbol{\tau}', \tag{14.63}$$

where $\boldsymbol{\varepsilon}'$ and $\boldsymbol{\tau}'$ are the fluctuating parts of the intensity field and flux, respectively, and angular brackets denote an ensemble average. Hence, by definition we have

$$\langle \boldsymbol{\varepsilon}' \rangle = \langle \boldsymbol{\tau}' \rangle = 0. \tag{14.64}$$

Let the displacement $\boldsymbol{u}$ be decomposed as follows:

$$\boldsymbol{u} = \langle \boldsymbol{\varepsilon} \rangle \cdot \boldsymbol{x} + \boldsymbol{u}'. \tag{14.65}$$

Taking the symmetrized gradient of (14.65) yields $\boldsymbol{\varepsilon}$, which when compared to (14.62) gives

$$\boldsymbol{\varepsilon}' = \frac{1}{2}\Big[\nabla \boldsymbol{u}' + (\nabla \boldsymbol{u}')^{\mathrm{T}}\Big]. \tag{14.66}$$

We have already noted in Section 14.1 that ensemble and volume averages are essentially equal when one considers a finite but very large region of volume V with surface S within a realization of an ergodic ensemble. In particular, referring to (14.64) and (14.66), we can write

$$\begin{aligned}\langle \boldsymbol{\varepsilon}' \rangle &\approx \frac{1}{V}\int_V \frac{1}{2}\left[\nabla \boldsymbol{u}' + (\nabla \boldsymbol{u}')^{\mathrm{T}}\right] dV \\ &= \frac{1}{V}\int_S \boldsymbol{u}'\boldsymbol{n}\, dS,\end{aligned} \tag{14.67}$$

where we have used the divergence theorem (14.4). Surface integrals over the interface are not written out explicitly, since they are identically zero as a result of continuity of the displacement across the interface. In the infinite-volume limit, we see from (14.64) and (14.67) that

$$\lim_{V\to\infty}\frac{1}{V}\int_S \boldsymbol{u}'\boldsymbol{n}dS \to 0; \tag{14.68}$$

i.e., the ratio of the surface integral of (14.68) to the volume V vanishes in the limit $V \to \infty$.

Similarly, for the large region of volume V and surface S, we can express the average of the fluctuating part of the stress as

$$\begin{aligned}\langle\boldsymbol{\tau}'\rangle &\approx \frac{1}{V}\int_V \boldsymbol{\tau}'dV \\ &= \frac{1}{V}\int_V \nabla\cdot(\boldsymbol{x}\boldsymbol{\tau}')dV \\ &= \frac{1}{V}\int_S \boldsymbol{x}(\boldsymbol{\tau}'\cdot\boldsymbol{n})dS.\end{aligned} \tag{14.69}$$

The second line of (14.69) follows from the identity (14.7) with $\boldsymbol{A} = \boldsymbol{x}$ and $\boldsymbol{B} = \boldsymbol{\tau}$ and the fact that $\nabla\cdot\boldsymbol{\tau} = 0$. The third line is obtained from the divergence theorem (14.1) with $\boldsymbol{A}$ equal to the third-order tensor $\boldsymbol{x}\boldsymbol{\tau}'$. Since the traction is continuous across the multiphase interface, the surface integrals over the interface make no contribution in the application of the divergence theorem, and thus we do not bother to write such terms explicitly. Hence, we see from (14.64) and (14.69) that in the infinite-volume limit,

$$\lim_{V\to\infty}\frac{1}{V}\int_S \boldsymbol{x}(\boldsymbol{\tau}'\cdot\boldsymbol{n})dS \to 0. \tag{14.70}$$

To summarize, ergodicity renders the boundary terms involving the fluctuating quantities $\boldsymbol{u}'$ and $\boldsymbol{\tau}'$ in (14.68) and (14.70), respectively, to be zero. In the discussion that follows, boundary terms of these types will arise and will be taken to be zero by ergodicity.

14.2.2 Energy Representation

We recall from Section 13.3.1 that the energy stored per unit volume in a homogeneous linearly elastic material (without body forces) is a positive quantity that is proportional to the double dot product of the strain tensor and the stress tensor (Sokolnikoff 1956). In the case of a heterogeneous linear material, we assume that the scalar energy stored per unit volume $u(\boldsymbol{x})$ at the point $\boldsymbol{x}$ is given by the same but local form, i.e.,

$$u(\boldsymbol{x}) = \frac{1}{2}\boldsymbol{\varepsilon}(\boldsymbol{x}) : \boldsymbol{\tau}(\boldsymbol{x}) \geq 0. \tag{14.71}$$

We will simply refer to u as the "microscopic energy," which should not be confused with the displacement vector $\boldsymbol{u}$. Use of Hooke's law (14.59) shows that the microscopic energy is a nonnegative quadratic form in either the strain or the stress; i.e., one has the equivalent but alternative forms

$$u_{\boldsymbol{\varepsilon}}(\boldsymbol{x}) = \frac{1}{2}\boldsymbol{\varepsilon}(\boldsymbol{x}) : \boldsymbol{C}(\boldsymbol{x}) : \boldsymbol{\varepsilon}(\boldsymbol{x}) \tag{14.72}$$

and

$$u_{\mathcal{T}}(\boldsymbol{x}) = \frac{1}{2}\boldsymbol{\tau}(\boldsymbol{x}) : \boldsymbol{C}^{-1}(\boldsymbol{x}) : \boldsymbol{\tau}(\boldsymbol{x}), \tag{14.73}$$

where $\boldsymbol{C}^{-1}$ is the compliance tensor.

Ensemble averaging the expressions above for the microscopic energies yields the equivalent energy functionals

$$U[\varepsilon] \equiv \langle u_{\mathcal{E}}(\boldsymbol{x})\rangle = \frac{1}{2}\langle \varepsilon(\boldsymbol{x}) : \boldsymbol{C}(\boldsymbol{x}) : \varepsilon(\boldsymbol{x})\rangle, \tag{14.74}$$

$$U[\boldsymbol{\tau}] \equiv \langle u_{\mathcal{T}}(\boldsymbol{x})\rangle = \frac{1}{2}\langle \boldsymbol{\tau}(\boldsymbol{x}) : \boldsymbol{C}^{-1}(\boldsymbol{x}) : \boldsymbol{\tau}(\boldsymbol{x})\rangle. \tag{14.75}$$

By treating the heterogeneous material as a homogeneous material with uniform fields $\langle \varepsilon(\boldsymbol{x})\rangle$ and $\langle \boldsymbol{\tau}(\boldsymbol{x})\rangle$, we can define the "macroscopic energy" $\tilde{U}$ according to the relations

$$\begin{aligned}\tilde{U} &= \frac{1}{2}\langle \varepsilon(\boldsymbol{x})\rangle : \langle \boldsymbol{\tau}(\boldsymbol{x})\rangle \\ &= \frac{1}{2}\langle \varepsilon(\boldsymbol{x})\rangle : \boldsymbol{C}_e : \langle \varepsilon(\boldsymbol{x})\rangle \\ &= \frac{1}{2}\langle \boldsymbol{\tau}(\boldsymbol{x})\rangle : \boldsymbol{C}_e^{-1} : \langle \boldsymbol{\tau}(\boldsymbol{x})\rangle.\end{aligned} \tag{14.76}$$

It will now be shown that under rather general conditions, the macroscopic energy $\tilde{U}$ is equal to U, the average of the microscopic energy, thus providing an energy representation of the effective stiffness tensor $\boldsymbol{C}_e$ or effective compliance tensor $\boldsymbol{C}_e^{-1}$.

Theorem 14.6 *For ergodic macroscopically anisotropic elastic multiphase composites with anisotropic constituents, the macroscopic energy $\tilde{U}$ is equal to the average of the microscopic energy U, i.e., the effective stiffness tensor $\boldsymbol{C}_e$ has the energy representation*

$$\frac{1}{2}\langle \varepsilon\rangle : \boldsymbol{C}_e : \langle \varepsilon\rangle = \frac{1}{2}\langle \varepsilon : \boldsymbol{C} : \varepsilon\rangle, \tag{14.77}$$

or, equivalently,

$$\frac{1}{2}\langle \boldsymbol{\tau}\rangle : \boldsymbol{C}_e^{-1} : \langle \boldsymbol{\tau}\rangle = \frac{1}{2}\langle \boldsymbol{\tau} : \boldsymbol{C}^{-1} : \boldsymbol{\tau}\rangle. \tag{14.78}$$

Proof: Using the integration by parts formula (14.10) and relation (14.60), we can write the average on the right side of (14.77) as

$$\begin{aligned}\langle \varepsilon : \boldsymbol{\tau}\rangle &= \langle \nabla \cdot (\boldsymbol{u} \cdot \boldsymbol{\tau})\rangle - \langle \boldsymbol{u}(\nabla \cdot \boldsymbol{\tau})\rangle \\ &= \lim_{V\to\infty} \frac{1}{V}\int_S \boldsymbol{u} \cdot (\boldsymbol{\tau} \cdot \boldsymbol{n})dS \\ &= \langle \varepsilon\rangle : \langle \boldsymbol{\tau}\rangle + \lim_{V\to\infty} \frac{1}{V}\int_S \boldsymbol{u}' \cdot (\boldsymbol{\tau}' \cdot \boldsymbol{n})dS \\ &= \langle \varepsilon\rangle : \langle \boldsymbol{\tau}\rangle,\end{aligned} \tag{14.79}$$

which proves the theorem. The second line of (14.79) follows from (14.10) with $\boldsymbol{A} = \boldsymbol{u}$ and $\boldsymbol{B} = \boldsymbol{\tau}$ and the use of the divergence-free condition (14.58). The third line follows after substitution of the decompositions (14.63) and (14.65) into the second line and use of the divergence theorem (14.4). The fourth line of (14.79) follows from ergodicity. ∎

Remarks:

1. The result (14.79), valid for ergodic media, implies that the average of the product $\varepsilon' : \boldsymbol{\tau}'$ vanishes, i.e.,
$$\langle \varepsilon' : \boldsymbol{\tau}' \rangle = 0. \tag{14.80}$$
This means that the fluctuating quantities ε' and $\boldsymbol{\tau}'$ are orthogonal to one another on average. Thus, it also follows that for ergodic media
$$\langle \varepsilon' : \boldsymbol{\tau} \rangle = \langle \varepsilon : \boldsymbol{\tau}' \rangle = 0,$$
or, equivalently,
$$\langle \varepsilon' : \boldsymbol{C} : \varepsilon \rangle = \langle \boldsymbol{\tau}' : \boldsymbol{C}^{-1} : \boldsymbol{\tau} \rangle = 0. \tag{14.81}$$
This last identity will prove useful to us in subsequent derivations.
2. It follows from Theorem 14.6 that the effective stiffness tensor $\boldsymbol{C}_e$ is a monotonic function of the local stiffness tensor $\boldsymbol{C}(\boldsymbol{x})$, i.e., $\boldsymbol{C}_e^* > \boldsymbol{C}_e$ when $\boldsymbol{C}^*(\boldsymbol{x}) > \boldsymbol{C}(\boldsymbol{x})$ for all $\boldsymbol{x}$. The proof is simple. In light of the conditions on $\boldsymbol{C}_e^*$ and $\boldsymbol{C}(\boldsymbol{x})$, it follows that $\langle \varepsilon : \boldsymbol{C}^* : \varepsilon \rangle > \langle \varepsilon : \boldsymbol{C} : \varepsilon \rangle$ for all ε, which in turn implies that $\langle \varepsilon \rangle : \boldsymbol{C}_e^* : \langle \varepsilon \rangle > \langle \varepsilon \rangle : \boldsymbol{C}_e : \langle \varepsilon \rangle$ for all $\langle \varepsilon \rangle$. Therefore, $\boldsymbol{C}_e^* > \boldsymbol{C}_e$.
3. In the case of a macroscopically isotropic composite with isotropic constituents, the local and effective stiffness tensors are given in terms of the bulk and shear moduli (Section 13.3.3), i.e.,
$$\boldsymbol{C}(\boldsymbol{x}) = dK(\boldsymbol{x})\boldsymbol{\Lambda}_h + 2G(\boldsymbol{x})\boldsymbol{\Lambda}_s, \tag{14.82}$$
$$\boldsymbol{C}_e = dK_e\boldsymbol{\Lambda}_h + 2G_e\boldsymbol{\Lambda}_s, \tag{14.83}$$
where $\boldsymbol{\Lambda}_h$ and $\boldsymbol{\Lambda}_s$ are the hydrostatic and shear projection tensors defined by (13.96) and (13.97), $K(\boldsymbol{x})$ and $G(\boldsymbol{x})$ are the local bulk and shear moduli, and K_e and G_e are the effective bulk and shear moduli. Similarly, the local and effective compliance tensors are given by
$$\boldsymbol{C}^{-1}(\boldsymbol{x}) = \frac{1}{dK(\boldsymbol{x})}\boldsymbol{\Lambda}_h + \frac{1}{2G(\boldsymbol{x})}\boldsymbol{\Lambda}_s, \tag{14.84}$$
$$\boldsymbol{C}_e^{-1} = \frac{1}{dK_e}\boldsymbol{\Lambda}_h + \frac{1}{2G_e}\boldsymbol{\Lambda}_s. \tag{14.85}$$
Thus, for a macroscopically isotropic composite, the equivalent energy representations (14.77) and (14.78) of Theorem 14.6 reduce respectively to
$$K_e \langle \mathrm{Tr}\, \varepsilon \rangle^2 + 2G_e \langle \boldsymbol{e} \rangle : \langle \boldsymbol{e} \rangle = \langle K\, [\mathrm{Tr}\, \varepsilon]^2 \rangle + 2\langle G\, \boldsymbol{e} : \boldsymbol{e} \rangle, \tag{14.86}$$

$$\frac{K_e^{-1}\langle \text{Tr } \boldsymbol{\tau}\rangle^2}{d^2} + \frac{G_e^{-1}\langle \boldsymbol{s}\rangle : \langle \boldsymbol{s}\rangle}{2} = \frac{\langle K^{-1}\,[\text{Tr } \boldsymbol{\tau}]^2\rangle}{d^2} + \frac{\langle G^{-1}\,\boldsymbol{s} : \boldsymbol{s}\rangle}{2}, \tag{14.87}$$

where $\boldsymbol{e} = \boldsymbol{\varepsilon} - (\text{Tr } \boldsymbol{\varepsilon}/d)\boldsymbol{I}$ and $\boldsymbol{s} = \boldsymbol{\tau} - (\text{Tr } \boldsymbol{\tau}/d)\boldsymbol{I}$ are the deviatoric parts of the strain and stress, respectively.

Theorem 14.7 *The effective stiffness tensor $\boldsymbol{C}_e$ is symmetric and positive definite.*

Proof: To prove the symmetry of $\boldsymbol{C}_e$, we decompose the strain field $\boldsymbol{\varepsilon}(\boldsymbol{x})$ into its average and fluctuating parts according to (14.62). The fluctuating part can be written as the scalar product of a fourth-order tensor $\boldsymbol{e}$ and the average field, i.e., $\boldsymbol{\varepsilon}' = \boldsymbol{e} : \langle \boldsymbol{\varepsilon}\rangle$ (see Chapter 20). Using this decomposition and (14.77) yields that the effective stiffness tensor is given by

$$(C_e)_{ijkl} = \langle C_{ijkl}\rangle + \langle e_{ijmn} C_{mnpq} e_{klpq}\rangle. \tag{14.88}$$

This relation and the symmetry properties of the local stiffness tensor, i.e.,

$$C_{ijkl} = C_{jikl} = C_{ijlk} = C_{klij},$$

immediately lead to the result that the effective tensor is symmetric, i.e.,

$$(C_e)_{ijkl} = (C_e)_{jikl} = (C_e)_{ijlk} = (C_e)_{klij}.$$

Note that since the local stiffness tensor $\boldsymbol{C}$ is positive definite (see Section 13.3.1), the right side of the energy relation (14.77) is positive. Hence, the effective stiffness tensor $\boldsymbol{C}_e$ is also positive definite. ∎

14.2.3 Minimum Energy Principles

As noted in the previous section, the principles of minimum potential energy and minimum complementary energy are classical ones in the theory of elasticity of homogeneous bodies. The extension of these principles to heterogeneous materials with an *arbitrary number of phases* will be proved here. The first principle is concerned with the construction of *trial strain fields* $\hat{\boldsymbol{\varepsilon}}$ that are compatible (and satisfy a mild average condition) but whose associated stress fields are not necessarily divergence-free. By contrast, the second principle is concerned with the construction of *trial stress fields* $\hat{\boldsymbol{\sigma}}$ that are divergence-free (and satisfy a mild average condition) but whose associated strain fields are not necessarily compatible.

Beran (1968) was the first to demonstrate how to extend variational principles for the elastic moduli of homogeneous bodies with prescribed boundary conditions to variational principles for the effective elastic moduli of random media. The minimum principles given below follow Beran's approach, but we give more details and treat boundary terms differently.

Theorem 14.8 *Minimum Potential Energy:*
Let B_U be the class of trial strain fields $\hat{\boldsymbol{\varepsilon}}$ defined by the set

$$B_U = \{ergodic\ \hat{\varepsilon};\ \nabla \times [\nabla \times \hat{\varepsilon}]^{\mathrm{T}} = 0,\ \langle\hat{\varepsilon}\rangle = \langle\varepsilon\rangle\}, \tag{14.89}$$

and let

$$U[\hat{\varepsilon}] = \frac{1}{2}\langle\hat{\varepsilon}(\boldsymbol{x}) : \boldsymbol{C}(\boldsymbol{x}) : \hat{\varepsilon}(\boldsymbol{x})\rangle \tag{14.90}$$

be the trial energy functional, where $\boldsymbol{C}$ *is the local stiffness tensor. Then among all trial fields* $\hat{\varepsilon}$*, the field that makes the associated stress divergence-free and symmetric is the one that uniquely minimizes the trial energy functional* $U[\hat{\varepsilon}]$*. In other words,*

$$U[\varepsilon] \leq W[\hat{\varepsilon}] \qquad \forall\hat{\varepsilon} \in B_U, \tag{14.91}$$

or, equivalently,

$$\langle\varepsilon\rangle : \boldsymbol{C}_e : \langle\varepsilon\rangle \leq \langle\hat{\varepsilon} : \boldsymbol{C} : \hat{\varepsilon}\rangle \qquad \forall\hat{\varepsilon} \in B_U. \tag{14.92}$$

Proof: Let the "difference" field $\boldsymbol{f}$ be defined by

$$\boldsymbol{f} = \hat{\varepsilon} - \varepsilon$$

such that the trial field $\hat{\varepsilon}$ satisfies

$$\nabla \times [\nabla \times \hat{\varepsilon}]^{\mathrm{T}} = 0,\ \ \langle\hat{\varepsilon}\rangle = \langle\varepsilon\rangle. \tag{14.93}$$

Therefore, $\langle\boldsymbol{f}\rangle = 0$, and since both ε and $\hat{\varepsilon}$ are compatible, $\boldsymbol{f}$ is also compatible, implying that it can be written as a symmetrized gradient of a vector $\boldsymbol{w}$, i.e.,

$$\boldsymbol{f} = \frac{1}{2}\Big[\nabla\boldsymbol{w} + \nabla\boldsymbol{w}^{\mathrm{T}}\Big].$$

Now consider the identity

$$\langle\hat{\varepsilon} : \boldsymbol{C} : \hat{\varepsilon}\rangle = \langle\varepsilon : \boldsymbol{C} : \varepsilon\rangle + 2\langle\boldsymbol{f} : \boldsymbol{C} : \varepsilon\rangle + \langle\boldsymbol{f} : \boldsymbol{C} : \boldsymbol{f}\rangle, \tag{14.94}$$

where the symmetries of $\boldsymbol{f}$, $\boldsymbol{C}$, and ε have been used. We now prove that the middle term of (14.94) is identically zero. Using the symmetry of $\boldsymbol{\tau}$ and the formula (14.10) for integration by parts with $\boldsymbol{A} = \boldsymbol{w}$ and $\boldsymbol{B} = \boldsymbol{\tau}$, we have

$$\begin{aligned}\langle\boldsymbol{f} : \boldsymbol{C} : \varepsilon\rangle &= \langle\nabla\boldsymbol{w} : \boldsymbol{\tau}\rangle \\ &= \lim_{V\to\infty}\frac{1}{V}\int_S \boldsymbol{w}\cdot(\boldsymbol{\tau}\cdot\boldsymbol{n})dS - \lim_{V\to\infty}\frac{1}{V}\int_V \boldsymbol{w}\cdot(\nabla\cdot\boldsymbol{\tau})dV \\ &= \lim_{V\to\infty}\frac{1}{V}\int_S \boldsymbol{w}'\cdot(\boldsymbol{\tau}'\cdot\boldsymbol{n})dS = 0.\end{aligned}$$

The third line of (14.95) follows after decomposing $\boldsymbol{w}$ and $\boldsymbol{\tau}$ into their mean and fluctuating parts (as in the proof of Theorem 14.6), using the condition $\langle\boldsymbol{f}\rangle = 0$, and employing the fact that $\boldsymbol{\tau}$ is divergence-free. The integral in the third line involving the fluctuating fields is zero by ergodicity. Using this result in combination with the energy representation (14.77) gives

$$\langle\hat{\varepsilon} : \boldsymbol{C} : \hat{\varepsilon}\rangle = \langle\varepsilon\rangle : \boldsymbol{C}_e : \langle\varepsilon\rangle + \langle\boldsymbol{f} : \boldsymbol{C} : \boldsymbol{f}\rangle.$$

Now since the second term on the right side is greater than or equal to zero, the statement (14.92) is proved. The equality sign holds only when the difference field $\boldsymbol{f}$ is zero throughout $\mathcal{V}$, i.e., when $\hat{\varepsilon} = \varepsilon$.

If there were another trial field ε^* from the set B_U, different from ε but with a divergence-free and symmetric stress, then we could apply the above theorem first to ε^* and then to ε, yielding the contradictory results that $U[\varepsilon^*] > U[\varepsilon]$ and $U[\varepsilon] > U[\varepsilon^*]$; hence ε is unique. ∎

Remarks:

1. In words, the minimum potential energy principle states that the actual macroscopic energy is bounded from above by the trial macroscopic energy and thus leads to an upper bound $\boldsymbol{C}_e^U$ on the effective stiffness tensor $\boldsymbol{C}_e$ in the positive semidefinite sense [cf. (13.66)].
2. For the special case of a macroscopically isotropic composite with isotropic phases, the minimum potential energy principle becomes

$$K_e \langle \mathrm{Tr}\, \varepsilon \rangle^2 + 2G_e \langle \boldsymbol{e} \rangle : \langle \boldsymbol{e} \rangle \le \langle K\, [\mathrm{Tr}\, \hat{\varepsilon}]^2 \rangle + 2\langle G\, \hat{\boldsymbol{e}} : \hat{\boldsymbol{e}} \rangle, \tag{14.95}$$

   where $\hat{\boldsymbol{e}} = \hat{\varepsilon} - (\mathrm{Tr}\, \hat{\varepsilon}/d)\boldsymbol{I}$ is the deviatoric part of the trial strain field. This easily follows in the same way from the isotropic form of the energy representation (14.86).
3. Theorem 14.8 also applies to periodic media (Bensoussan et al. 1978). Boundary terms are now over the unit cell and vanish because of periodicity.

Theorem 14.9 *Minimum Complementary Energy:*
Let B_L be the class of trial stress fields $\hat{\boldsymbol{\tau}}$ defined by the set

$$B_L = \{\text{ergodic } \hat{\boldsymbol{\tau}};\ \nabla \cdot \hat{\boldsymbol{\tau}} = 0\ ,\ \hat{\boldsymbol{\tau}} = \hat{\boldsymbol{\tau}}^{\mathrm{T}}\ ,\ \langle \hat{\boldsymbol{\tau}} \rangle = \langle \boldsymbol{\tau} \rangle\}, \tag{14.96}$$

and let

$$U[\hat{\boldsymbol{\tau}}] = \frac{1}{2} \langle \hat{\boldsymbol{\tau}}(\boldsymbol{x}) : \boldsymbol{C}^{-1}(\boldsymbol{x}) : \hat{\boldsymbol{\tau}}(\boldsymbol{x}) \rangle \tag{14.97}$$

be the trial energy functional, where $\boldsymbol{C}$ *is the local stiffness tensor. Then among all trial stress fields $\hat{\boldsymbol{\tau}}$, the field that makes the associated strain compatible is the one that uniquely minimizes the trial energy functional $U[\hat{\boldsymbol{\tau}}]$. In other words,*

$$U[\boldsymbol{\tau}] \le U[\hat{\boldsymbol{\tau}}] \qquad \forall \hat{\boldsymbol{\tau}} \in B_L, \tag{14.98}$$

or, equivalently,

$$\langle \boldsymbol{\tau} \rangle : \boldsymbol{C}_e^{-1} : \langle \boldsymbol{\tau} \rangle \le \langle \hat{\boldsymbol{\tau}} : \boldsymbol{C}^{-1} : \hat{\boldsymbol{\tau}} \rangle \qquad \forall \hat{\boldsymbol{\tau}} \in B_L. \tag{14.99}$$

Proof: Let the "difference" field $\boldsymbol{g}$ be defined by

$$\boldsymbol{g} = \hat{\boldsymbol{\tau}} - \boldsymbol{\tau} \tag{14.100}$$

such that

$$\nabla \cdot \hat{\boldsymbol{\tau}} = 0, \quad \langle \hat{\boldsymbol{\tau}} \rangle = \langle \boldsymbol{\tau} \rangle.$$

Therefore, $\langle \boldsymbol{g} \rangle = 0$, and since both $\boldsymbol{\tau}$ and $\hat{\boldsymbol{\tau}}$ are divergence-free, $\boldsymbol{g}$ is also divergence-free, i.e.,

$$\nabla \cdot \boldsymbol{g} = 0.$$

Consider the identity

$$\langle \hat{\boldsymbol{\tau}} : \boldsymbol{C}^{-1} : \hat{\boldsymbol{\tau}} \rangle = \langle \boldsymbol{\tau} : \boldsymbol{C}^{-1} : \boldsymbol{\tau} \rangle + 2 \langle \boldsymbol{g} : \boldsymbol{C}^{-1} : \boldsymbol{\tau} \rangle + \langle \boldsymbol{g} : \boldsymbol{C}^{-1} : \boldsymbol{g} \rangle. \tag{14.101}$$

The middle term may be rewritten as

$$\begin{aligned} \langle \boldsymbol{g} : \boldsymbol{C}^{-1} : \boldsymbol{\tau} \rangle &= \langle \boldsymbol{g} : \nabla \boldsymbol{u} \rangle \\ &= \lim_{V \to \infty} \frac{1}{V} \int_S \boldsymbol{u} \cdot (\boldsymbol{g} \cdot \boldsymbol{n}) dS - \lim_{V \to \infty} \frac{1}{V} \int_V \boldsymbol{u} \cdot (\nabla \cdot \boldsymbol{g}) dV \\ &= \lim_{V \to \infty} \frac{1}{V} \int_S \boldsymbol{u}' \cdot (\boldsymbol{g}' \cdot \boldsymbol{n}) dS = 0. \end{aligned}$$

The second line of (14.102) follows from formula (14.10) with $\boldsymbol{A} = \boldsymbol{u}$ and $\boldsymbol{B} = \boldsymbol{g}$. The third line results after decomposing $\boldsymbol{u}$ and $\boldsymbol{g}$ into their mean and fluctuating parts (as in the proof of Theorem 14.6), using the condition $\langle \boldsymbol{g} \rangle = 0$, and utilizing the fact that $\nabla \cdot \boldsymbol{g} = 0$. The integral in the third line of (14.102) involving the fluctuating fields is zero by ergodicity. Using this result in combination with energy representation (14.78) gives

$$\langle \hat{\boldsymbol{\tau}} : \boldsymbol{C}^{-1} : \hat{\boldsymbol{\tau}} \rangle = \langle \boldsymbol{\tau} \rangle : \boldsymbol{C}_e^{-1} : \langle \boldsymbol{\tau} \rangle + \langle \boldsymbol{g} : \boldsymbol{C}^{-1} : \boldsymbol{g} \rangle,$$

which proves statement (14.99). The equality sign holds only when the difference field $\boldsymbol{g}$ is zero throughout $\mathcal{V}$, i.e., when $\hat{\boldsymbol{\tau}} = \boldsymbol{\tau}$. The uniqueness of $\boldsymbol{\tau}$ follows using the same arguments as for Theorem 14.8. ∎

Remarks:

1. In words, the minimum complementary energy principle, Theorem 14.9, states that the actual macroscopic energy is bounded from above by the trial macroscopic energy and thus leads to a lower bound $\boldsymbol{C}_e^L$ on the stiffness tensor $\boldsymbol{C}_e$ in the positive semidefinite sense.
2. For the special case of a macroscopically isotropic composite with isotropic phases, the minimum complementary energy principle becomes

$$\frac{K_e^{-1} \langle \mathrm{Tr}\, \boldsymbol{\tau} \rangle^2}{d^2} + \frac{G_e^{-1} \langle \boldsymbol{s} \rangle : \langle \boldsymbol{s} \rangle}{2} \le \frac{\langle K^{-1} \, [\mathrm{Tr}\, \hat{\boldsymbol{\tau}}]^2 \rangle}{d^2} + \frac{\langle G^{-1} \, \hat{\boldsymbol{s}} : \hat{\boldsymbol{s}} \rangle}{2}, \tag{14.102}$$

   where $\hat{\boldsymbol{s}} = \hat{\boldsymbol{\tau}} - (\mathrm{Tr}\, \hat{\boldsymbol{\tau}}/d)\boldsymbol{I}$ is the deviatoric part of the trial stress field. This easily follows from the isotropic form of the energy representation (14.87).
3. Theorem 14.9 applies as well to periodic media (Bensoussan et al. 1978). Again, boundary terms are over the unit cell and vanish because of periodicity.

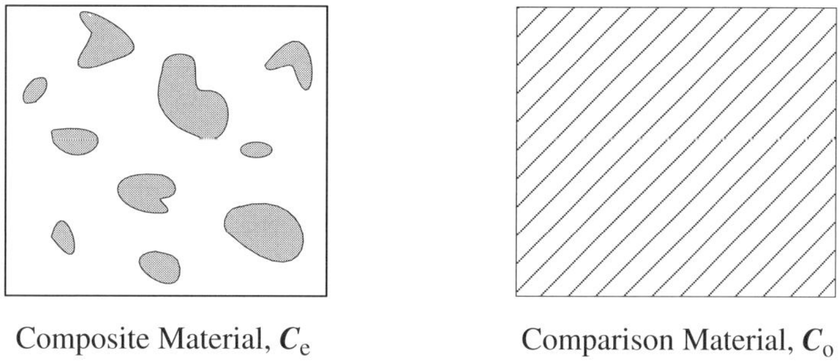

Figure 14.2 Portions of two materials: a composite material with effective stiffness tensor $\boldsymbol{C}_e$ and a homogeneous comparison material with stiffness tensor $\boldsymbol{C}_0$. The composite average strain field $\langle\varepsilon\rangle$ is equal to the uniform strain ε_0 in the comparison material.

14.2.4 Hashin–Shtrikman Principle

Consider an ergodic composite body with effective stiffness tensor $\boldsymbol{C}_e$ and a homogeneous *comparison* material (of the same shape and size) with a constant effective stiffness tensor $\boldsymbol{C}_0$ (see Figure 14.2). The strain field in the comparison material is the uniform field ε_0, and we require that the average strain field of the composite obey $\langle\varepsilon\rangle = \varepsilon_0$. The composite body is composed of two isotropic phases, and thus the local stiffness tensor $\boldsymbol{C}(\boldsymbol{x})$ is given by (14.82). The composite is characterized by the stiffness tensor $\boldsymbol{C}_e$. For simplicity, we will consider here macroscopically isotropic media, and hence $\boldsymbol{C}_e$ is given by (14.83) and

$$\boldsymbol{C}_0 = dK_0\boldsymbol{\Lambda}_h + 2G_0\boldsymbol{\Lambda}_s, \tag{14.103}$$

where K_0 and G_0 are the bulk and shear moduli, respectively, of the comparison material.

Let the local stress "polarization" field $\boldsymbol{p}$ be defined by

$$\boldsymbol{p} = \boldsymbol{\tau} - \boldsymbol{C}_0 : \varepsilon = (\boldsymbol{C} - \boldsymbol{C}_0) : \varepsilon. \tag{14.104}$$

Let us define also the difference between the local and comparison strain fields as

$$\boldsymbol{g} = \varepsilon - \varepsilon_0, \tag{14.105}$$

implying that

$$\nabla \times [\nabla \times \boldsymbol{g}]^{\mathrm{T}} = 0, \tag{14.106}$$

$$\langle \boldsymbol{g} \rangle = 0. \tag{14.107}$$

Substitution of (14.104) into (14.58) yields

$$\nabla \cdot (\boldsymbol{C}_0 : \boldsymbol{g}) + \nabla \cdot \boldsymbol{p} = 0, \tag{14.108}$$

where we have used (14.105) and the fact that $\nabla\varepsilon_0 = 0$.

We now state the Hashin–Shtrikman variational principle (Hashin and Shtrikman 1963). Again, the proof for ergodic media employs the same techniques used to derive the aforementioned classical minimum energy principles, and hence we do not give it here.

Theorem 14.10 *Hashin–Shtrikman Variational Principle:*
Consider any ergodic trial polarization field $\hat{\boldsymbol{p}}$ subject to the conditions

$$\nabla \cdot (\boldsymbol{C}_0 : \boldsymbol{g}) + \nabla \cdot \hat{\boldsymbol{p}} = 0, \tag{14.109}$$

$$\nabla \times [\nabla \times \boldsymbol{g}]^{\mathrm{T}} = 0, \tag{14.110}$$

$$\langle \hat{\boldsymbol{g}} \rangle = 0, \tag{14.111}$$

where $\hat{\boldsymbol{g}}$ is the associated trial difference strain field and $\boldsymbol{C}_0$ is the constant isotropic fourth-order tensor (14.103)*. If a hydrostatic boundary condition $(\varepsilon_0)_{ij} = \delta_{ij}/d$ is imposed, then for any two-phase isotropic composite,*

$$K_* = K_0 + \left\langle \hat{\boldsymbol{p}} : \hat{\boldsymbol{g}} + 2\hat{\boldsymbol{p}} : \varepsilon_0 - \hat{\boldsymbol{p}} : (\boldsymbol{C} - \boldsymbol{C}_0)^{-1} : \hat{\boldsymbol{p}} \right\rangle \tag{14.112}$$

is an upper bound on K_e when $G_0 \geq G_1$ and a lower bound on K_e when $G_0 \leq G_2$. However, if a deviatoric boundary condition $(\varepsilon_0)_{ij} = [\delta_{1i}\delta_{2j} + \delta_{2i}\delta_{1j}]/\sqrt{2}$ is imposed, then for any two-phase isotropic composite,

$$G_* = G_0 + \left\langle \hat{\boldsymbol{p}} : \hat{\boldsymbol{g}} + 2\hat{\boldsymbol{p}} : \varepsilon_0 - \hat{\boldsymbol{p}} : (\boldsymbol{C} - \boldsymbol{C}_0)^{-1} : \hat{\boldsymbol{p}} \right\rangle \tag{14.113}$$

is an upper bound on G_e when $K_0 \geq K_1$ and $G_0 \geq G_1$ and a lower bound on G_e when $K_0 \leq K_2$ and $G_0 \leq G_2$.

Remarks:

1. Using the stiffness relations (14.82) and (14.103) enables one to express the last term in the averaged relations (14.112) and (14.113) as

$$\hat{\boldsymbol{p}} : (\boldsymbol{C} - \boldsymbol{C}_0)^{-1} : \hat{\boldsymbol{p}} = \left[\frac{2(G - G_0) - d(K - K_0)}{2(K - K_0)(G - G_0)} \right] \frac{[\mathrm{Tr}\, \hat{\boldsymbol{p}}]^2}{d^2} + \frac{\hat{\boldsymbol{p}} : \hat{\boldsymbol{p}}}{2(G - G_0)}. \tag{14.114}$$

2. The original Hashin–Shtrikman variational principle was derived for a very large but finite composite body subjected to homogeneous displacement boundary conditions on the sample surface (rather than requiring $\langle \hat{\boldsymbol{g}} \rangle = 0$). However, their formulation in the infinite-volume limit is equivalent to Theorem 14.10, since boundary effects then vanish.
3. Theorem 14.10 also can be applied to periodic media (Milton and Kohn 1988, Nemat-Nasser and Hori 1993).
4. Theorem 14.10 has been generalized to treat anisotropic multiphase random composites with anisotropic constituents (Willis 1977).

14.3 Trapping Constant

We will derive variational principles that will enable us to bound the trapping constant γ (or, equivalently, mean survival time τ) from above and below. As in the case of the conduction and elasticity problems, these variational bounds are based on minimizing the "energy" of the system. First, it will be shown that the trapping constant has an energy representation. We then will prove the minimum energy principles that lead to rigorous upper and lower bounds on the trapping constant, or equivalently, the mean survival time.

We will consider *ergodic* porous media with a trap-free, or pore, region $\mathcal{V}_1$ and a trap region $\mathcal{V}_2$. Let the interface between the two regions be denoted by $\partial\mathcal{V}$. Referring to relations (13.148), (13.149), and (13.151), we see that the local equations for the *scaled* concentration field in the instance of perfectly absorbing traps are given by

$$\Delta u = -1 \qquad \text{in } \mathcal{V}_1, \tag{14.115}$$

$$u = 0 \qquad \text{on } \partial\mathcal{V}, \tag{14.116}$$

where we extend u into the trap region $\mathcal{V}_2$ to be zero. The averaged relation that defines the trapping constant γ is

$$\gamma^{-1} = \langle u\, \mathcal{I}\rangle = \langle u\rangle, \tag{14.117}$$

where $\mathcal{I} \equiv \mathcal{I}^{(1)}$ is the indicator function for the trap-free region. The second equality in (14.117) follows from the extension of u into $\mathcal{V}_2$.

14.3.1 Energy Representation

Recall that the trapping constant γ is the proportionality constant in the relation between the average concentration field and production rate per unit volume. We now derive an energy representation for the trapping constant.

Theorem 14.11 *For ergodic media, the trapping constant γ can be rewritten in terms of the energy functional:*

$$\gamma^{-1} = \langle \nabla u(\mathbf{x}) \cdot \nabla u(\mathbf{x})\, \mathcal{I}(\mathbf{x})\rangle. \tag{14.118}$$

Proof: Let $V_\mathcal{R}$ be a very large sphere of radius $\mathcal{R}$ centered at the origin in $\mathcal{V}$, and let $\partial V_\mathcal{R}$ be the surface of this sphere. Multiplying the diffusion equation (14.115) by u and ensemble averaging yields

$$\begin{aligned}
\gamma^{-1} &= -\langle u\Delta u\rangle \\
&= -\frac{1}{V_\mathcal{R}}\int_{V_\mathcal{R}} \langle u\Delta u\rangle dV = -\left\langle \frac{1}{V_\mathcal{R}}\int_{V_\mathcal{R}} u\Delta u dV\right\rangle \\
&= \left\langle \frac{1}{V_\mathcal{R}}\int_{V_\mathcal{R}} \nabla u\cdot\nabla u dV\right\rangle - \left\langle \frac{1}{V_\mathcal{R}}\int_{\partial V_\mathcal{R}} u\frac{\partial u}{\partial n} dS\right\rangle - \left\langle \frac{1}{V_\mathcal{R}}\int_{\partial\mathcal{V}} u\frac{\partial u}{\partial n} dS\right\rangle
\end{aligned}$$

$$= \langle \nabla u \cdot \nabla u \rangle - \frac{1}{V_{\mathcal{R}}} \int_{\partial V_{\mathcal{R}}} \left\langle u \frac{\partial u}{\partial n} \right\rangle dS.$$

The third line follows from the integration by parts formula (14.8) with $\boldsymbol{A} = u$ and $\boldsymbol{B} = \nabla u$. The fourth line follows because the surface integral over the pore-solid interface $\partial\mathcal{V}$ vanishes identically because of boundary condition (14.116). Since the surface integral over $\partial V_{\mathcal{R}}$ is $\mathcal{O}(\mathcal{R}^2)$, it vanishes in the limit $\mathcal{R} \to \infty$. Thus, letting $\mathcal{R} \to \infty$, we find that

$$\gamma^{-1} = \langle \nabla u \cdot \nabla u \rangle,$$

which, after extension of u into $\mathcal{V}_2$, proves the theorem. ∎

Remarks:

1. Again, surface integrals involving the boundary ∂V_R vanish by ergodicity.
2. The energy representation (14.118) clearly implies that the trapping constant γ is positive.

14.3.2 Minimum Energy Principles

Using heuristic arguments, Prager (1963a) and Reck and Prager (1965) stated variational principles that lead to upper and lower bounds, respectively, on the trapping constant γ for diffusion-controlled reactions among disordered, perfectly absorbing traps. Rubinstein and Torquato (1988) subsequently derived variational principles from which one can obtain rigorous upper and lower bounds on γ for ergodic media with perfectly absorbing traps. These variational principles are proved below. We also state variational principles for both γ and the principal relaxation time T_1 for partially absorbing traps.

First, we modify relations (14.115) and (14.116) slightly by introducing the function $f = \alpha u$, where α is some positive constant. Thus, the quantity f solves

$$\Delta f = -\alpha \quad \text{in } \mathcal{V}_1, \tag{14.119}$$

$$f = 0 \quad \text{on } \partial\mathcal{V} \text{ in } \mathcal{V}_2. \tag{14.120}$$

The trapping constant γ is then is also defined by

$$\gamma = \frac{\alpha}{\langle f \rangle} = \frac{\alpha}{\langle f\, \mathcal{I} \rangle}. \tag{14.121}$$

Theorem 14.12 *Variational Upper Bound:*
Let D_U be the class of trial concentration fields v defined by

$$D_U = \{ergodic\ v(\boldsymbol{x});\ v = 0 \ on\ \partial\mathcal{V},\ \langle v\, \mathcal{I} \rangle = \langle f\, \mathcal{I} \rangle\}. \tag{14.122}$$

Then the trapping constant is bounded from above by

$$\gamma \le \frac{\langle \nabla v \cdot \nabla v\, \mathcal{I} \rangle}{\langle f \rangle^2} \qquad \forall v \in D_U. \tag{14.123}$$

Proof: Eliminating α from (14.119) and (14.121) and applying Theorem 14.11 yields

$$\gamma = \frac{\langle \nabla f \cdot \nabla f \, \mathcal{I} \rangle}{\langle f \rangle^2}. \tag{14.124}$$

Let the "difference" field g be defined by

$$g = \upsilon - f,$$

where $\upsilon \in D_U$. Therefore, given the conditions on υ and f, we have that $\langle g \, \mathcal{I} \rangle = 0$ and $g = 0$ on $\partial\mathcal{V}$.

Now consider the identity

$$\langle \nabla \upsilon \cdot \nabla \upsilon \, \mathcal{I} \rangle = \langle \nabla f \cdot \nabla f \, \mathcal{I} \rangle + 2\langle \nabla f \cdot \nabla g \, \mathcal{I} \rangle + \langle \nabla g \cdot \nabla g \, \mathcal{I} \rangle. \tag{14.125}$$

We need only prove that the middle term in (14.125) is zero. The middle term can be rewritten, using the integration by parts formula (14.8) with $\boldsymbol{A} = g$ and $\boldsymbol{B} = \nabla f$ and ergodicity, as

$$\begin{aligned} \langle \nabla f \cdot \nabla g \, \mathcal{I} \rangle &= \langle \nabla \cdot (g \nabla f) \, \mathcal{I} \rangle - \langle g \Delta f \, \mathcal{I} \rangle \\ &= \alpha \langle g \, \mathcal{I} \rangle = 0. \end{aligned}$$

The first average on the right side of the first line can be written as surface integrals, all of which vanish identically because $g = 0$ on $\partial\mathcal{V}$ and by ergodicity (as in the proof of Theorem 14.11). The second line follows from $\Delta f = -\alpha$ and the average condition on g. Thus, $\langle \nabla \upsilon \cdot \nabla \upsilon \, \mathcal{I} \rangle \geq \langle \nabla f \cdot \nabla f \, \mathcal{I} \rangle$. This inequality combined with (14.124) proves Theorem 14.12. ∎

Theorem 14.13 *Variational Lower Bound:*
Let D_L be the class of trial concentration fields υ defined by

$$D_L = \{ergodic\ \upsilon(\boldsymbol{x});\quad \Delta \upsilon = -\alpha \ in\ \mathcal{V}_1\}. \tag{14.126}$$

Then the trapping constant is bounded from below by

$$\gamma \geq \frac{\alpha^2}{\langle \nabla \upsilon \cdot \nabla \upsilon \, \mathcal{I} \rangle} \qquad \forall \upsilon \in D_L. \tag{14.127}$$

Proof: From Theorem 14.11, it follows that

$$\gamma = \frac{\alpha^2}{\langle \nabla f \cdot \nabla f \, \mathcal{I} \rangle}. \tag{14.128}$$

Let the difference field g be defined by

$$g = \upsilon - f,$$

where $\upsilon \in D_L$. Therefore,

$$\Delta g = 0 \quad \text{in } \mathcal{V}_1.$$

Consider the identity

$$\langle \nabla v \cdot \nabla v \, \mathcal{I} \rangle = \langle \nabla f \cdot \nabla f \, \mathcal{I} \rangle + 2\langle \nabla f \cdot \nabla g \, \mathcal{I} \rangle + \langle \nabla g \cdot \nabla g \, \mathcal{I} \rangle.$$

However, use of the integration by parts formula (14.8) with $\boldsymbol{A} = f$ and $\boldsymbol{B} = \nabla g$, relation $\Delta g = 0$, boundary condition $f = 0$ on $\partial \mathcal{V}$, and ergodicity imply that the middle term vanishes, i.e.,

$$\langle \nabla f \cdot \nabla g \, \mathcal{I} \rangle = \langle f \Delta g \, \mathcal{I} \rangle = 0.$$

Therefore, we have that $\langle \nabla v \cdot \nabla v \, \mathcal{I} \rangle \geq \langle \nabla f \cdot \nabla f \, \mathcal{I} \rangle$. This last inequality in conjunction with (14.128) completes the proof. ∎

Remarks:

1. In light of the trivial inequality

$$\langle \nabla v \cdot \nabla v \rangle \geq \langle \nabla v \cdot \nabla v \, \mathcal{I} \rangle$$

and the inequalities (14.123) and (14.127), we have the cruder bounds

$$\gamma \leq \frac{\langle \nabla v \cdot \nabla v \rangle}{\langle f \rangle^2} \qquad \forall v \in D_U, \tag{14.129}$$

$$\gamma \geq \frac{\alpha^2}{\langle \nabla v \cdot \nabla v \rangle} \qquad \forall v \in D_L. \tag{14.130}$$

2. Rubinstein and Torquato (1988) also derived volume-averaged variational principles that are the analogues of (14.123) and (14.127).
3. Since the survival time is defined by $\tau = (\mathcal{D}\phi_1 \gamma)^{-1}$ (where $\mathcal{D}$ is the diffusion coefficient and ϕ_1 is the trap-free volume fraction), Theorem 14.12 also leads to a lower bound on τ. Likewise, Theorem 14.13 leads to an upper bound on τ.
4. Torquato and Avellaneda (1991) derived the following lower-bound variational principle on τ for the case of a finite surface rate constant κ (partially absorbing traps):

$$\tau \geq \frac{\langle f \rangle^2}{\mathcal{D}\phi_1 \langle \nabla v \cdot \nabla v \, \mathcal{I} \rangle + \kappa \phi_1 \langle v^2 \mathcal{M} \rangle}, \tag{14.131}$$

where $\mathcal{M}(\boldsymbol{x}) = |\nabla \mathcal{I}(\boldsymbol{x})|$ is the interface indicator function, v is a trial concentration field that satisfies $\langle v \, \mathcal{I} \rangle = \langle f \, \mathcal{I} \rangle$, and f solves

$$\Delta f = -\alpha \quad \text{in } \mathcal{V}_1,$$

$$\mathcal{D}\frac{\partial f}{\partial n} + \kappa f = 0 \quad \text{on } \partial \mathcal{V}.$$

5. Torquato and Avellaneda (1991) also derived the analogous lower-bound variational principle on the *principal relaxation time* T_1 (Section 13.7.1):

$$T_1 \geq \frac{\langle (\hat{\psi})^2 \rangle}{\mathcal{D}\langle \nabla \hat{\psi} \cdot \nabla \hat{\psi} \, \mathcal{I} \rangle + \kappa \langle (\hat{\psi})^2 \mathcal{M} \rangle}, \tag{14.132}$$

where $\hat{\psi}$ is a trial eigenfunction that need only be a scalar function in $\mathcal{V}_1$. The equality of (14.132) applies when $\hat{\psi}$ is the true eigenfunction, i.e., when it satisfies (13.190) and (13.191).

14.4 Fluid Permeability

We will derive variational principles that will enable us to bound the eigenvalues of the fluid permeability tensor from above and below. Consistent with all of the previous different physical problems, these variational bounds are based on minimizing the energy dissipation in the system. First, it will be shown that the fluid permeability tensor has an energy representation. We then will prove the minimum energy principles that lead to rigorous upper and lower bounds on the fluid permeability tensor.

We will consider *ergodic* porous media with a void or pore region $\mathcal{V}_1$ and a solid region $\mathcal{V}_2$. Let the interface between the two regions be denoted by $\partial\mathcal{V}$. Referring to expressions (13.171)–(13.173), we see that the local (Stokes) equations governing the components of the *tensor* velocity field $\boldsymbol{w}$ are given by

$$\Delta \boldsymbol{w} = \nabla \boldsymbol{\pi} - \boldsymbol{I} \qquad \text{in } \mathcal{V}_1, \tag{14.133}$$

$$\nabla \cdot \boldsymbol{w} = 0 \qquad \text{in } \mathcal{V}_1, \tag{14.134}$$

$$\boldsymbol{w} = 0 \qquad \text{on } \partial\mathcal{V}, \tag{14.135}$$

where we extend $\boldsymbol{w}$ into the solid region $\mathcal{V}_2$ to be zero. The fluid permeability tensor $\boldsymbol{k}$ has been shown to be the proportionality constant in Darcy's law. Thus, the average velocity field is related to the applied pressure gradient and is given by the average tensor velocity field, i.e.,

$$\boldsymbol{k} = \langle \boldsymbol{w}\, \mathcal{I} \rangle = \langle \boldsymbol{w} \rangle, \tag{14.136}$$

where $\mathcal{I} \equiv \mathcal{I}^{(1)}$ is the indicator function for the pore region. The second equality in (14.136) follows by extension of $\boldsymbol{w}$ into $\mathcal{V}_2$.

14.4.1 Energy Representation

We now derive an energy representation of the fluid permeability.

Theorem 14.14 *The fluid permeability tensor* $\boldsymbol{k}$ *of an ergodic porous medium can be rewritten in terms of the energy functional*

$$\boldsymbol{k} = \langle \nabla \boldsymbol{w} : \nabla \boldsymbol{w}\, \mathcal{I} \rangle. \tag{14.137}$$

Proof: Taking the dot product of the momentum equation (14.133) by $\boldsymbol{w}$ from the left and averaging gives

$$\boldsymbol{k} = \langle \boldsymbol{w} \cdot \nabla \boldsymbol{\pi} \rangle - \langle \boldsymbol{w} \cdot \Delta \boldsymbol{w} \rangle. \tag{14.138}$$

We will now show that the first term on the right side of (14.138) vanishes for an ergodic medium. Let $V_{\mathcal{R}}$ be a very large sphere of radius $\mathcal{R}$ centered at the origin in $\mathcal{V}$. Then integration by parts [formula (14.8) with $\boldsymbol{A} = \boldsymbol{\pi}$ and $\boldsymbol{B} = \boldsymbol{w}$] and the incompressibility condition (14.134) enables us to write

$$\begin{aligned}\langle \boldsymbol{w} \cdot \nabla \boldsymbol{\pi} \rangle &= \frac{1}{V_{\mathcal{R}}} \int_{V_{\mathcal{R}}} \langle \nabla \cdot (\boldsymbol{w}\boldsymbol{\pi}) \rangle dV = \left\langle \frac{1}{V_{\mathcal{R}}} \int_{V_{\mathcal{R}}} \nabla \cdot (\boldsymbol{w}\boldsymbol{\pi}) dV \right\rangle \\ &= \left\langle \frac{1}{V_{\mathcal{R}}} \int_{\partial \mathcal{V}} (\boldsymbol{w}\boldsymbol{\pi}) \cdot \boldsymbol{n} dS \right\rangle + \left\langle \frac{1}{V_{\mathcal{R}}} \int_{\partial V_{\mathcal{R}}} (\boldsymbol{w}\boldsymbol{\pi}) \cdot \boldsymbol{n} dS \right\rangle,\end{aligned}$$

where $\partial V_{\mathcal{R}}$ is the surface of the large sphere and $\boldsymbol{n}$ is the unit outward normal. Using the boundary condition $\boldsymbol{w} = 0$ on $\partial \mathcal{V}$ and letting $\mathcal{R} \to \infty$, we arrive at

$$\langle \boldsymbol{w} \cdot \nabla \boldsymbol{\pi} \rangle = 0.$$

We see again that surface integrals involving the boundary ∂V_R vanish by ergodicity. Integration by parts [formula (14.10) with $\boldsymbol{A} = \boldsymbol{w}$ and $\boldsymbol{B} = \nabla \boldsymbol{w}$], boundary condition (14.135), and ergodicity enable us to show that the second term in (14.138) is equivalent to

$$-\langle \boldsymbol{w} \cdot \Delta \boldsymbol{w} \rangle = \langle \nabla \boldsymbol{w} : \nabla \boldsymbol{w} \rangle.$$

Use of these identities in (14.138) yields

$$\boldsymbol{k} = \langle \nabla \boldsymbol{w} : \nabla \boldsymbol{w} \rangle,$$

which, after extension of $\boldsymbol{w}$ into $\mathcal{V}_2$, proves the theorem. ∎

Remark:

1. For isotropic media, recall that fluid permeability is a scalar given by $k = \langle \boldsymbol{w} \cdot \boldsymbol{e} \rangle$ [see relation (13.179)], where $\boldsymbol{w}$ is the scaled *vector* (not tensor) velocity. In this case, k has the energy representation

$$k = \langle \nabla \boldsymbol{w} : \nabla \boldsymbol{w} \, \mathcal{I} \rangle. \tag{14.139}$$

Theorem 14.15 *The permeability tensor* **k** *is symmetric and positive definite.*

Proof: The symmetry of $\boldsymbol{k}$ (i.e., $k_{ij} = k_{ji}$) immediately follows from (14.137). The positivity property is proved by considering, in component form, the product

$$a_i k_{ij} a_j = \left\langle a_i \frac{\partial w_{\ell i}}{\partial x_m} \frac{\partial w_{\ell j}}{\partial x_m} a_j \right\rangle,$$

where a_i is some arbitrary vector and k_{ij} is given by (14.137). Letting

$$\alpha_{\ell m} = a_i \frac{\partial w_{\ell i}}{\partial x_m}$$

gives that

$$a_i k_{ij} a_j = \langle \alpha_{\ell m} \alpha_{\ell m} \rangle \geq 0.$$

Since the equality only applies when $a_i = 0$, this proves that $\boldsymbol{k}$ is positive definite. ∎

14.4.2 Minimum Energy Principles

Prager (1961) and Weissberg and Prager (1970) were the first to derive upper bounds on the fluid permeability k of statistically isotropic porous media. It should be noted that their derivations were formulated through physical reasoning and that the Prager and Weissberg–Prager variational principles (and their respective admissible fields) were different. Berryman and Milton (1985) subsequently, using a volume-average approach, corrected a normalization constraint in the Prager variational principle. Torquato and Beasley (1987) reformulated the Weissberg–Prager variational upper bound on k in terms of ensemble averages. Rubinstein and Torquato (1989) developed new rigorous variational principles in terms of ensemble and volume averages that led to rigorous upper and lower bounds on k for ergodic porous media. The ensemble-average approach is particularly useful in obtaining upper bounds on the permeability; this avoids the difficulties encountered by Berryman and Milton (1985) and Caflisch and Rubinstein (1984) in handling boundary conditions for admissible fields in volume-average formulations. The variational principles that are proved below for isotropic media follow the derivation of Rubinstein and Torquato (1989). We also state variational principles for the anisotropic case.

Isotropic Media

To begin with, we will consider variational principles for macroscopically isotropic media, i.e., cases in which the fluid permeability tensor is isotropic, i.e., $\boldsymbol{k} = k\boldsymbol{I}$, where k is the scalar permeability. The isotropic formulation involving the scaled velocity field $\boldsymbol{w}$ is described by relations (13.180)–(13.182) with $\boldsymbol{w} = 0$ in $\mathcal{V}_2$. To derive the variational principles, we modify these equations slightly by introducing the functions $\boldsymbol{f} = \beta\boldsymbol{w}$ and $\zeta = \beta\pi$, where β is some positive constant. Then $\boldsymbol{f}$ and ζ solve

$$\Delta \boldsymbol{f} = \nabla\zeta - \beta\boldsymbol{e} \qquad \text{in } \mathcal{V}_1, \tag{14.140}$$

$$\nabla\cdot\boldsymbol{f} = 0 \qquad \text{in } \mathcal{V}_1, \tag{14.141}$$

$$\boldsymbol{f} = 0 \qquad \text{on } \partial\mathcal{V} \text{ and in } \mathcal{V}_2. \tag{14.142}$$

The scalar fluid permeability is then also defined by

$$k = \frac{1}{\beta}\langle \boldsymbol{f}\cdot\boldsymbol{e}\rangle = \frac{1}{\beta}\langle \boldsymbol{f}\cdot\boldsymbol{e}\,\mathcal{I}\rangle. \tag{14.143}$$

Theorem 14.16 *Variational Upper Bound:*
Let $\boldsymbol{q}$ be the class of trial velocity vector fields that are defined by

$$E_U = \{\text{ergodic } \boldsymbol{q}(\boldsymbol{x});\ \nabla\times(\Delta\boldsymbol{q} + \beta\boldsymbol{e}) = 0 \text{ in } \mathcal{V}_1\}. \tag{14.144}$$

Then the scalar fluid permeability k is bounded from above by

$$k \leq \frac{\langle \nabla \boldsymbol{q} : \nabla \boldsymbol{q}\, \mathcal{I} \rangle}{\beta^2} \qquad \forall \boldsymbol{q} \in E_U. \tag{14.145}$$

Proof: Using relation (14.139), we can write the energy representation of k as

$$k = \frac{\langle \nabla \boldsymbol{f} : \nabla \boldsymbol{f}\, \mathcal{I} \rangle}{\beta^2}. \tag{14.146}$$

Let the difference field $\boldsymbol{g}$ be defined by

$$\boldsymbol{g} = \boldsymbol{q} - \boldsymbol{f},$$

where $\boldsymbol{q} \in E_U$. Given the conditions on $\boldsymbol{f}$ and $\boldsymbol{q}$, we have therefore that

$$\nabla \times (\Delta \boldsymbol{q} + \beta \boldsymbol{e}) = 0 \qquad \text{in } \mathcal{V}_1.$$

Hence, there exists a function p^u such that

$$\Delta \boldsymbol{q} = \nabla p^u - \beta \boldsymbol{e}.$$

Use of the relations above and setting $p^u = \zeta + h$ yields

$$\Delta \boldsymbol{g} = \nabla h.$$

Consider the identity

$$\langle \nabla \boldsymbol{q} : \nabla \boldsymbol{q}\, \mathcal{I} \rangle = \langle \nabla \boldsymbol{f} : \nabla \boldsymbol{f}\, \mathcal{I} \rangle + 2\langle \nabla \boldsymbol{f} : \nabla \boldsymbol{g}\, \mathcal{I} \rangle + \langle \nabla \boldsymbol{g} : \nabla \boldsymbol{g}\, \mathcal{I} \rangle.$$

Use of the integration by parts formula (14.10) with $\boldsymbol{A} = \boldsymbol{f}$ and $\boldsymbol{B} = \nabla \boldsymbol{g}$, relation $\Delta \boldsymbol{g} = \nabla h$, boundary condition $\boldsymbol{f} = 0$ on $\partial \mathcal{V}$, and ergodicity enable us to show that the middle term vanishes, i.e.,

$$\begin{aligned} \langle \nabla \boldsymbol{f} : \nabla \boldsymbol{g}\, \mathcal{I} \rangle &= \langle \boldsymbol{f} \Delta \boldsymbol{g}\, \mathcal{I} \rangle \\ &= \langle h (\nabla \cdot \boldsymbol{f})\, \mathcal{I} \rangle = 0. \end{aligned}$$

The second line follows from relation $\Delta \boldsymbol{g} = \nabla h$, integration by parts formula (14.8) with $\boldsymbol{A} = h$ and $\boldsymbol{B} = \boldsymbol{f}$, and the solenoidal condition (14.141) on $\boldsymbol{f}$. This leads to the inequality $\langle \nabla \boldsymbol{q} : \nabla \boldsymbol{q}\, \mathcal{I} \rangle \geq \langle \nabla \boldsymbol{f} : \nabla \boldsymbol{f}\, \mathcal{I} \rangle$, which, when combined with (14.146), proves Theorem 14.16. ∎

Theorem 14.17 *Variational Lower Bound:*
Let $\boldsymbol{q}$ be the class of trial velocity vector fields that are defined by

$$E_L = \{ergodic\ \boldsymbol{q}(\boldsymbol{x});\ \boldsymbol{q} = 0 \text{ on } \partial \mathcal{V},\ \nabla \cdot \boldsymbol{q} = 0 \text{ in } \mathcal{V}_1,\ \langle \boldsymbol{q} \cdot \boldsymbol{e}\, \mathcal{I} \rangle = \langle \boldsymbol{f} \cdot \boldsymbol{e}\, \mathcal{I} \rangle\}.$$

Then the scalar fluid permeability k is bounded from below by

$$k \geq \frac{\langle \boldsymbol{f} \cdot \boldsymbol{e} \rangle^2}{\langle \nabla \boldsymbol{q} : \nabla \boldsymbol{q}\, \mathcal{I} \rangle} \qquad \forall \boldsymbol{q} \in E_L. \tag{14.147}$$

Proof: Using relation (14.139), we can write the energy representation of the fluid permeability as

$$k = \frac{\langle \boldsymbol{f} \cdot \boldsymbol{e}\, \mathcal{I} \rangle^2}{\langle \nabla \boldsymbol{f} : \nabla \boldsymbol{f}\, \mathcal{I} \rangle}. \tag{14.148}$$

Let the difference field $\boldsymbol{g}$ be defined by

$$\boldsymbol{g} = \boldsymbol{q} - \boldsymbol{f},$$

where $\boldsymbol{q} \in E_L$. The conditions on $\boldsymbol{q}$ and $\boldsymbol{f}$ imply the following conditions on $\boldsymbol{g}$:

$$\begin{aligned} \nabla \cdot \boldsymbol{g} &= 0 \quad \text{in } \mathcal{V}_1, \\ \boldsymbol{g} &= 0 \quad \text{on } \partial\mathcal{V}, \\ \langle \boldsymbol{g} \cdot \boldsymbol{e}\, \mathcal{I} \rangle &= 0. \end{aligned}$$

Consider the identity

$$\langle \nabla \boldsymbol{q} : \nabla \boldsymbol{q}\, \mathcal{I} \rangle = \langle \nabla \boldsymbol{f} : \nabla \boldsymbol{f}\, \mathcal{I}^{(1)} \rangle + 2\langle \nabla \boldsymbol{f} : \nabla \boldsymbol{g}\, \mathcal{I} \rangle + \langle \nabla \boldsymbol{g} : \nabla \boldsymbol{g}\, \mathcal{I} \rangle.$$

The integration by parts formula (14.10) with $\boldsymbol{A} = \boldsymbol{g}$ and $\boldsymbol{B} = \nabla \boldsymbol{f}$, ergodicity, and use of the above conditions on $\boldsymbol{g}$ enable us to show that the middle term vanishes, i.e.,

$$\begin{aligned} \langle \nabla \boldsymbol{f} : \nabla \boldsymbol{g}\, \mathcal{I} \rangle &= -\langle \boldsymbol{g} \cdot \Delta \boldsymbol{f}\, \mathcal{I} \rangle \\ &= -\langle \boldsymbol{g} \cdot \nabla \zeta\, \mathcal{I} \rangle + \beta \langle \boldsymbol{g} \cdot \boldsymbol{e}\, \mathcal{I} \rangle = 0. \end{aligned}$$

The second line follows directly from (14.140): the first term is shown to be zero after use of the integration by parts formula (14.8) with $\boldsymbol{A} = \zeta$ and $\boldsymbol{B} = \boldsymbol{g}$, and the second term is zero by the average condition on $\boldsymbol{g}$. Thus, we again have that $\langle \nabla \boldsymbol{q} : \nabla \boldsymbol{q}\, \mathcal{I} \rangle \geq \langle \nabla \boldsymbol{f} : \nabla \boldsymbol{f}\, \mathcal{I} \rangle$. This inequality and relation (14.148) complete the proof of the theorem. ∎

Remarks:

1. The upper-bound variational principle of Theorem 14.16 is different from those proposed by other investigators (Weissberg and Prager 1970, Berryman and Milton 1985). The normalization factor β^2 in (14.145) is deterministic, in contrast to the one used in the aforementioned works, which is an integral involving a random stress field. A formulation in which β is deterministic rather than stochastic is clearly preferred.
2. It is a challenge to construct trial velocity fields $\boldsymbol{q}$ for the lower-bound variational principle of Theorem 14.17 that do not lead to the trivial bound $k \geq 0$. The permeability is zero only when the pore phase does not percolate. By contrast, it is not as difficult to obtain corresponding lower bounds on the survival time τ, since it is generally nonzero, even for a nonpercolating pore phase. Specific lower bounds on k and γ are given in Chapters 21 and 22.
3. By virtue of the trivial inequality

$$\langle \nabla \boldsymbol{q} : \nabla \boldsymbol{q} \rangle \geq \langle \nabla \boldsymbol{q} : \nabla \boldsymbol{q}\, \mathcal{I} \rangle$$

and the inequalities (14.145) and (14.147), we have the cruder bounds

$$k \leq \frac{\langle \nabla \boldsymbol{q} : \nabla \boldsymbol{q} \rangle}{\beta^2} \qquad \forall \boldsymbol{q} \in E_U, \tag{14.149}$$

$$k \geq \frac{\langle \boldsymbol{f} \cdot \boldsymbol{e} \rangle^2}{\langle \nabla \boldsymbol{q} : \nabla \boldsymbol{q} \rangle} \qquad \forall \boldsymbol{q} \in E_L. \tag{14.150}$$

4. Rubinstein and Torquato (1989) also derived volume-averaged variational principles that are the analogues of (14.145) and (14.147).

Anisotropic Media

The variational principles given immediately above for isotropic media are very easily generalized to macroscopically anisotropic media with fluid permeability tensor $\boldsymbol{k}$. The anisotropic formulation is determined by the scaled *tensor* velocity $\boldsymbol{w}$ described by relations (14.133)–(14.135) with $\boldsymbol{w} = 0$ in $\mathcal{V}_2$. As before, we modify these equations slightly by introducing the functions $\boldsymbol{f} = \beta \boldsymbol{w}$ and $\boldsymbol{\zeta} = \beta \boldsymbol{\pi}$, where β is some positive constant. Here, of course, $\boldsymbol{f}$ is a second-order tensor and $\boldsymbol{\zeta}$ is a vector that solve

$$\Delta \boldsymbol{f} = \nabla \boldsymbol{\zeta} - \beta \boldsymbol{I} \qquad \text{in } \mathcal{V}_1, \tag{14.151}$$

$$\nabla \cdot \boldsymbol{f} = 0 \qquad \text{in } \mathcal{V}_1, \tag{14.152}$$

$$\boldsymbol{f} = 0 \qquad \text{on } \partial\mathcal{V} \text{ and in } \mathcal{V}_2. \tag{14.153}$$

The fluid permeability tensor is then also defined by

$$\boldsymbol{k} = \frac{1}{\beta}\langle \boldsymbol{f} \rangle = \frac{1}{\beta}\langle \boldsymbol{f}\, \mathcal{I} \rangle, \tag{14.154}$$

or, equivalently,

$$\boldsymbol{k} = \frac{1}{\beta^2}\langle \nabla \boldsymbol{f} : \nabla \boldsymbol{f}\, \mathcal{I} \rangle \tag{14.155}$$

$$= \langle \boldsymbol{f} \rangle \cdot \langle \nabla \boldsymbol{f} : \nabla \boldsymbol{f}\, \mathcal{I} \rangle^{-1} \cdot \langle \boldsymbol{f} \rangle. \tag{14.156}$$

Theorem 14.18 *Variational Upper Bound:*
Let $\boldsymbol{q}$ *be the class of trial tensor velocity fields that are defined by*

$$E_U = \{\text{ergodic } \boldsymbol{q}(\boldsymbol{x});\ \nabla \times (\Delta \boldsymbol{q} + \beta \boldsymbol{I}) = 0 \text{ in } \mathcal{V}_1\}. \tag{14.157}$$

Then the fluid permeability tensor $\boldsymbol{k}$ *is bounded from above (in the positive-semidefinite sense) by*

$$\boldsymbol{k} \leq \frac{\langle \nabla \boldsymbol{q} : \nabla \boldsymbol{q}\, \mathcal{I} \rangle}{\beta^2} \qquad \forall \boldsymbol{q} \in E_U. \tag{14.158}$$

Proof: The proof of this theorem follows in exactly the same fashion as that of Theorem 14.16 but using definition (14.155). ∎

Theorem 14.19 *Variational Lower Bound:*
Let $\boldsymbol{q}$ *be the class of trial tensor velocity fields that are defined by*

$$E_L = \{ergodic\ \boldsymbol{q}(\boldsymbol{x});\ \boldsymbol{q} = 0\ on\ \partial\mathcal{V},\ \nabla \cdot \boldsymbol{q} = 0\ in\ \mathcal{V}_1,\ \langle \boldsymbol{q}\,\mathcal{I}\rangle = \langle \boldsymbol{f}\,\mathcal{I}\rangle\}.$$

Then the fluid permeability tensor **k** *is bounded from below (in the positive-semidefinite sense) by*

$$\boldsymbol{k} \geq \langle \boldsymbol{f}\rangle \cdot \langle \nabla\boldsymbol{q} : \nabla\boldsymbol{q}\,\mathcal{I}\rangle^{-1} \cdot \langle \boldsymbol{f}\rangle \qquad \forall \boldsymbol{q} \in E_L. \tag{14.159}$$

Proof: The proof of this theorem follows in exactly the same fashion as that of Theorem 14.17 but using definition (14.156). ∎

CHAPTER 15

Phase-Interchange Relations

We refer to expressions that link the effective properties of a two-phase heterogeneous material to the effective properties of the same microstructure but with the phases interchanged as *phase-interchange relations*. When such rigorous relations apply to a wide class of heterogeneous materials, they can provide useful tests of analytical and numerical estimates of the effective properties. We focus here on the effective conductivity and elastic moduli of two-phase composites and, to a lesser extent, of polycrystals. The two-dimensional relations are shown to have interesting implications for *percolation* behavior. Finally, we remark on phase-interchange relations for the trapping constant and fluid permeability.

15.1 Conductivity

In the case of two dimensions, phase-interchange relations for the effective conductivity are equalities. For three-dimensional conduction problems, the phase-interchange relations are inequalities.

15.1.1 Duality for Two-Dimensional Media

Keller (1964) considered a heterogeneous material consisting of a rectangular lattice of identical parallel cylinders, each of which is symmetric in the x- and y-axes, with generators parallel to the z-axis. Thus, the problem is two-dimensional. He proved that the effective conductivity in the x-direction, σ_e^x, is related to the effective conductivity of the *phase-interchanged* composite in the y-direction, σ_e^y, by

$$\sigma_e^x(\sigma_1, \sigma_2)\sigma_e^y(\sigma_2, \sigma_1) = \sigma_1\sigma_2. \tag{15.1}$$

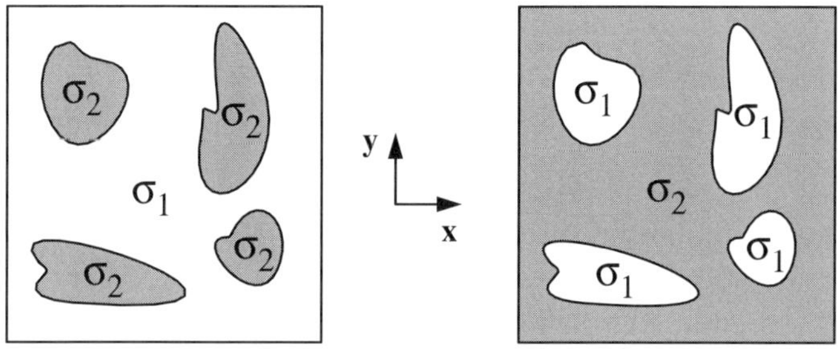

Figure 15.1 Left panel: A two-dimensional, two-phase composite with an effective conductivity in the x-direction given by $\sigma_e^x(\sigma_1, \sigma_2)$. Right panel: The phase-interchanged composite with an effective conductivity in the y-direction given by $\sigma_e^y(\sigma_2, \sigma_1)$ which is related to $\sigma_e^x(\sigma_1, \sigma_2)$ via (15.1).

In this relation, the first argument of σ_e^x (σ_e^y) denotes the conductivity of the matrix phase (included phase), and the second argument denotes the conductivity of the included phase (matrix phase). In related work, Dykhne (1971) showed that the effective conductivity of two-dimensional, two-phase composites possessing phase distributions that are statistically equivalent to one another is equal to the geometric mean $\sqrt{\sigma_1 \sigma_2}$. Subsequently, Mendelson (1975) generalized these results and, in particular, showed that relation (15.1) applies to *any* two-dimensional, two-phase composite (ordered or not) as long as x and y are the principal axes of the effective conductivity tensor, i.e., no geometrical symmetries are required (see Figure 15.1).

The proof of the duality theorem rests on the well-known result that a rotation of any two-dimensional curl-free vector field by 90° yields a divergence-free vector field and vice versa. That is, if we let $\boldsymbol{e}(\boldsymbol{x})$ and $\boldsymbol{j}(\boldsymbol{x})$ be curl-free and divergence-free two-dimensional vector fields, respectively, and denote by

$$\boldsymbol{R} = \begin{bmatrix} 0 & 1 \\ -1 & 0 \end{bmatrix} \tag{15.2}$$

the 90°-rotation tensor, then

$$\nabla \cdot (\boldsymbol{R} \cdot \boldsymbol{e}) = 0, \tag{15.3}$$

$$\nabla \times (\boldsymbol{R} \cdot \boldsymbol{j}) = 0. \tag{15.4}$$

Let us also note that for any second-order tensor $\boldsymbol{A}$, the quantity $\boldsymbol{R} \cdot \boldsymbol{A} \cdot \boldsymbol{R}^{\mathrm{T}}$ represents a 90° rotation of the tensor $\boldsymbol{A}$.

We are now in a position to state and prove the general theorem. The proof that we give below combines elements of the proofs given by Mendelson (1975) and Dykhne (1971). In words, the theorem states that a 90° rotation of the effective conductivity tensor of a two-dimensional heterogeneous material with local conductivity tensor $\boldsymbol{\sigma}$ determines the effective conductivity of the same material but one in which the local conductivity is the reciprocal tensor $\boldsymbol{\sigma}^{-1}$ rotated by 90°.

Theorem 15.1 *If $\sigma_e(\sigma)$ is the effective conductivity tensor of a two-dimensional heterogeneous material with local conductivity tensor $\sigma(x)$, then*

$$R \cdot \sigma_e(\sigma) \cdot R^T \cdot \sigma_e(\sigma') = \sigma_o^2 I, \tag{15.5}$$

where $\sigma'(x) = \sigma_o^2[R \cdot \sigma(x)^{-1} \cdot R^T]$ and σ_o is any convenient scalar normalization constant.

Proof: Consider in two dimensions the problem of conduction in either a two-phase composite or a polycrystal. Let us suppose that we can solve the local field equations (Section 13.2.1)

$$J(x) = \sigma(x) \cdot E(x), \tag{15.6}$$

$$\nabla \cdot J(x) = 0, \tag{15.7}$$

$$\nabla \times E(x) = 0, \tag{15.8}$$

where we generally take the local conductivity $\sigma(x)$ to be a second-order tensor. Now we introduce the new vector quantities

$$J'(x) = \sigma_o R \cdot E(x), \tag{15.9}$$

$$E'(x) = \sigma_o^{-1} R \cdot J(x), \tag{15.10}$$

where σ_o is a scalar conductivity to be specified later. Substitution of relation (15.6) into (15.10) yields

$$E'(x) = \sigma_o^{-1} R \cdot \sigma(x) \cdot E(x) = \sigma_o^{-2} R \cdot \sigma(x) \cdot R^T \cdot J'(x).$$

This relation can be solved for J' to give

$$J'(x) = \sigma'(x) \cdot E'(x), \tag{15.11}$$

where

$$\nabla \cdot J'(x) = 0, \tag{15.12}$$

$$\nabla \times E'(x) = 0, \tag{15.13}$$

and

$$\sigma'(x) = \sigma_o^2 \left[R \cdot \sigma(x) \cdot R^T\right]^{-1} = \sigma_o^2 \left[R \cdot \sigma(x)^{-1} \cdot R^T\right]. \tag{15.14}$$

Therefore, we see that relations (15.11)–(15.14) define a new conductivity problem.

The corresponding averaged relations that define the effective conductivities for the two different conductivity problems are given by

$$\langle J \rangle = \sigma_e \cdot \langle E \rangle, \tag{15.15}$$

$$\langle J' \rangle = \sigma_e' \cdot \langle E' \rangle, \tag{15.16}$$

where σ_e is the effective conductivity of the original material and σ_e' is the effective conductivity of the new material.

Averaging (15.9) and (15.10) gives

$$\langle \boldsymbol{J}' \rangle = \sigma_o \boldsymbol{R} \cdot \langle \boldsymbol{E} \rangle, \tag{15.17}$$

$$\langle \boldsymbol{E}' \rangle = \sigma_o^{-1} \boldsymbol{R} \cdot \langle \boldsymbol{J} \rangle. \tag{15.18}$$

Substitution of (15.15) into (15.18) and the use of (15.17) produces

$$\langle \boldsymbol{E}' \rangle = \sigma_o^{-2} \boldsymbol{R} \cdot \boldsymbol{\sigma}_e \cdot \boldsymbol{R}^{\mathrm{T}} \cdot \langle \boldsymbol{J}' \rangle,$$

and hence we have that

$$\boldsymbol{\sigma}_e' = \sigma_o^2 \left[\boldsymbol{R} \cdot \boldsymbol{\sigma}_e \cdot \boldsymbol{R}^{\mathrm{T}} \right]^{-1} = \sigma_o^2 \left[\boldsymbol{R} \cdot \boldsymbol{\sigma}_e^{-1} \cdot \boldsymbol{R}^{\mathrm{T}} \right]. \tag{15.19}$$

Relation (15.19) states that the effective conductivity $\boldsymbol{\sigma}_e'(\boldsymbol{\sigma}')$ of the new material with local conductivity $\boldsymbol{\sigma}'$ can be expressed in terms of the effective conductivity $\boldsymbol{\sigma}_e(\boldsymbol{\sigma})$ of the original material with local conductivity $\boldsymbol{\sigma}$. Since the effective conductivity function is the same for either problem, we can rewrite (15.19) as

$$\boldsymbol{R} \cdot \boldsymbol{\sigma}_e(\boldsymbol{\sigma}) \cdot \boldsymbol{R}^{\mathrm{T}} \cdot \boldsymbol{\sigma}_e(\boldsymbol{\sigma}') = \sigma_o^2 \boldsymbol{I},$$

which proves the duality relation. ∎

Remark:

1. Note that the duality relation of Theorem 15.1 is independent of the details of the microstructure: a rare finding. Since it is independent of the microstructure, it is a very useful test of theoretical and computer-simulation estimates of the effective conductivity of two-dimensional composites.

Isotropic Phases

Let us apply the duality relation (15.5) to the case of a composite consisting of two disjoint regions denoted by $\mathcal{V}_\alpha$ and $\mathcal{V}_\beta$ with isotropic materials with scalar conductivities σ_1 and σ_2, respectively. If we substitute $\sigma_o = \sqrt{\sigma_1 \sigma_2}$ into (15.5), then the local isotropic conductivity tensors are given by

$$\boldsymbol{\sigma}(\boldsymbol{x}) = \sigma(\boldsymbol{x}) \boldsymbol{I} = \left[\sigma_1 \mathcal{I}^{(1)}(\boldsymbol{x}) + \sigma_2 \mathcal{I}^{(2)}(\boldsymbol{x}) \right] \boldsymbol{I}, \tag{15.20}$$

$$\boldsymbol{\sigma}'(\boldsymbol{x}) = \frac{\sigma_1 \sigma_2}{\sigma(\boldsymbol{x})} \boldsymbol{I} = \left[\sigma_2 \mathcal{I}^{(1)}(\boldsymbol{x}) + \sigma_1 \mathcal{I}^{(2)}(\boldsymbol{x}) \right] \boldsymbol{I}, \tag{15.21}$$

where $\sigma(\boldsymbol{x})$ is the local scalar conductivity of the original material and $\mathcal{I}^{(i)}(\boldsymbol{x})$ is the indicator function for phase i given by (2.1). Note that the local conductivity of the new material is just the *phase-interchanged* conductivity of the original material; i.e., in the new composite, regions $\mathcal{V}_\alpha$ and $\mathcal{V}_\beta$ contain materials with conductivities σ_2 and σ_1, respectively. Thus, the duality relation of Theorem 15.1 yields

$$\boldsymbol{R} \cdot \boldsymbol{\sigma}_e(\sigma_1, \sigma_2) \cdot \boldsymbol{R}^{\mathrm{T}} \cdot \boldsymbol{\sigma}_e(\sigma_2, \sigma_1) = \sigma_1 \sigma_2 \boldsymbol{I}, \tag{15.22}$$

where $\boldsymbol{\sigma}_e(\sigma_1, \sigma_2)$ and $\boldsymbol{\sigma}_e(\sigma_2, \sigma_1)$ denote the effective conductivities of the original and phase-interchanged materials, respectively. Observe that the Keller–Mendelson phase-interchange relation (15.1) is easily recovered from the more general relation (15.22) by assuming that the coordinate system is oriented along the principal directions.

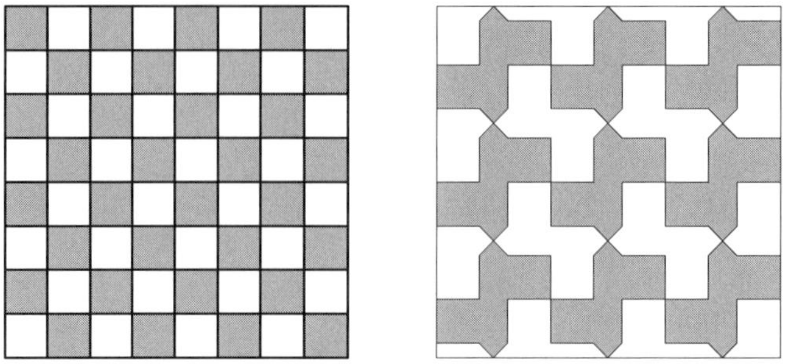

Figure 15.2 Two example geometries that satisfy the phase-interchange relation (15.24): Periodic checkerboard (left panel) and Schulgasser's proposal (right panel).

Isotropic Composites

If the composite is macroscopically isotropic (i.e., $\boldsymbol{\sigma}_e$ is rotationally invariant), then (15.22) reduces to

$$\sigma_e(\sigma_1, \sigma_2)\sigma_e(\sigma_2, \sigma_1) = \sigma_1\sigma_2. \tag{15.23}$$

Furthermore, if the two different composite materials possess phase distributions that are statistically equivalent to one another, then $\sigma_e(\sigma_1, \sigma_2) = \sigma_e(\sigma_2, \sigma_1) = \sigma_e$ and (15.23) gives that the effective conductivity is given by the *geometric mean* of the phase conductivities (Dykhne 1971), i.e.,

$$\sigma_e = \sqrt{\sigma_1\sigma_2}. \tag{15.24}$$

More precisely, such composites possess phase-inversion symmetry at $\phi_1 = \phi_2 = 1/2$ (see Section 2.2.2). Obvious examples of composites that satisfy (15.24) are a *periodic* checkerboard arrangement (see Figure 15.2), whose volume fraction equals 1/2, and all symmetric-cell materials with a volume fraction equal to 1/2, which includes the *random checkerboard* (see Figures 8.5 and 8.6). Schulgasser (1992) noted that the second microstructure shown in Figure 15.2 is a subtler example that satisfies (15.24), even though at first glance there is no reason to believe that this is a macroscopically isotropic system. This is easily seen by recognizing that for this geometry the effective conductivity under a 90° rotation is equal to the effective conductivity of the phase-interchanged composite, and hence from (15.22) we get (15.24).

Let us now examine two different isotropic composites but with identical microstructures. One composite is composed of a phase 1 material with conductivity σ_1 and a *perfectly conducting phase* 2, i.e., $\sigma_2 = \infty$, and has an effective conductivity σ_∞. The other composite, *with the same microstructure*, is composed of a phase 1 material with conductivity $\hat{\sigma}_1$ and a *perfectly insulating phase* 2, i.e., $\hat{\sigma}_2 = 0$, and has an effective conductivity σ_0. Provided that phase 2 in these composites is such that it does not percolate, the quantities σ_∞ and σ_0^{-1} remain finite. The phase-interchange relation (15.23) enables us to link σ_∞ to σ_0. To do so, consider a composite made of two phases with the conductivities σ_1 and $\lambda\hat{\sigma}_1$, where λ is a dimensionless parameter. Denote by σ_λ and $\hat{\sigma}_\lambda$

the effective conductivities of this composite and the corresponding *phase-interchanged* composite, respectively. Clearly,

$$\sigma_\infty = \lim_{\lambda\to\infty} \sigma_\lambda, \qquad \sigma_0 = \lim_{\lambda\to\infty} \hat{\sigma}_\lambda. \tag{15.25}$$

Theorem 15.2 *The effective conductivities σ_∞ and σ_0 satisfy the following equality:*

$$\sigma_\infty \sigma_0 = \sigma_1 \hat{\sigma}_1.$$

Proof: The equality follows immediately from the phase-interchange relation (15.23) in the limits specified by (15.25). ∎

An interesting corollary follows for the *continuum* system near the percolation threshold. Assuming the usual power-law behavior near the threshold $\phi_{2c} = 1 - \phi_{1c}$ (Section 9.2.3) and taking $\hat{\sigma}_1 = \sigma_1$ yields

$$\sigma_0/\sigma_1 = A_0(\phi_1 - \phi_{1c})^t, \qquad \phi_1 \to \phi_{1c}^+, \tag{15.26}$$

$$\sigma_\infty/\sigma_1 = A_\infty(\phi_1 - \phi_{1c})^{-s}, \qquad \phi_1 \to \phi_{1c}^+, \tag{15.27}$$

where t and s are the positive critical exponents (Sections 9.1.3 and 9.2.3), and A_0 and A_∞ are the corresponding dimensionless *amplitudes*. Implicit in the use of (15.26) and (15.27) is that only one phase can percolate at a time; i.e., the system is not bicontinuous.

Corollary 15.1 *The critical exponents s and t in the scaling relations* (15.26) *and* (15.27) *satisfy the equality*

$$s = t, \tag{15.28}$$

and the amplitudes A_0 and A_∞ satisfy the equality

$$A_0 A_\infty = 1. \tag{15.29}$$

Proof: The proof of this corollary follows directly from Theorem 15.2 (with $\hat{\sigma}_1 = \sigma_1$) and the scaling laws (15.26) and (15.27). ∎

Remark:

1. Corollary 15.1 applies not only to continuum but also to lattice percolation. The latter follows because continuum percolation contains bond percolation as a special case (Section 9.2.3). Site percolation is also included, since site and bond percolation are known to be in the same universality class.

Polycrystals

Now let us apply the duality relation to determine the effective conductivity of a two-dimensional polycrystal composed of identical single crystals (see Figure 15.3).

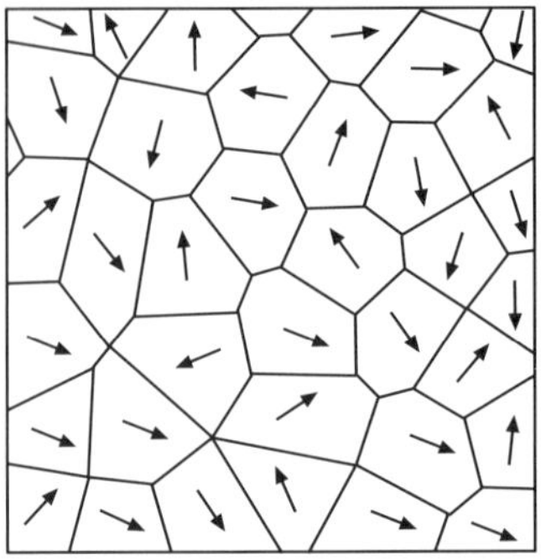

Figure 15.3 A schematic of a two-dimensional polycrystal. Arrows indicate the orientations of the grains.

Let σ_a and σ_b denote the principal conductivities (or eigenvalues) of the conductivity tensor. The local conductivity tensor is given by

$$\boldsymbol{\sigma}(\boldsymbol{x}) = \boldsymbol{R}(\boldsymbol{x}) \begin{bmatrix} \sigma_a & 0 \\ 0 & \sigma_b \end{bmatrix} \boldsymbol{R}^{\mathrm{T}}(\boldsymbol{x}), \tag{15.30}$$

where $\boldsymbol{R}(\boldsymbol{x})$ is any two-dimensional rotation matrix. It is easily shown that

$$\left[\boldsymbol{R} \cdot \boldsymbol{\sigma}(\boldsymbol{x}) \cdot \boldsymbol{R}^{\mathrm{T}}\right]^{-1} = \frac{\boldsymbol{\sigma}(\boldsymbol{x})}{\sigma_a \sigma_b} \tag{15.31}$$

and

$$\left[\boldsymbol{R} \cdot \boldsymbol{\sigma}_e \cdot \boldsymbol{R}^{\mathrm{T}}\right]^{-1} = \frac{\boldsymbol{\sigma}_e}{\det \boldsymbol{\sigma}_e}. \tag{15.32}$$

Letting $\sigma_o = \sqrt{\sigma_a \sigma_b}$ in the duality relation (15.5), we see upon use of (15.31) that the local conductivity $\boldsymbol{\sigma}'$ of the new material remains unchanged, i.e.,

$$\boldsymbol{\sigma}'(\boldsymbol{x}) = \boldsymbol{\sigma}(\boldsymbol{x}),$$

and hence $\boldsymbol{\sigma}_e' = \boldsymbol{\sigma}_e$. Therefore, the duality relation yields that

$$\det \boldsymbol{\sigma}_e = \sigma_a \sigma_b. \tag{15.33}$$

If $\boldsymbol{\sigma}_e$ is rotationally invariant, and thus isotropic, then (15.33) reduces to

$$\sigma_e = \sqrt{\sigma_a \sigma_b}, \tag{15.34}$$

which is independent of the details of the microstructure.

In three dimensions there is no corresponding equality. Instead, one has upper and lower bounds. The simplest bounds for macroscopically isotropic polycrystals are the arithmetic- and harmonic-mean bounds (Molyneux 1970):

$$\left[\frac{1}{3}\left(\frac{1}{\sigma_a} + \frac{1}{\sigma_b} + \frac{1}{\sigma_c}\right)\right]^{-1} \le \sigma_e \le \frac{1}{3}(\sigma_a + \sigma_b + \sigma_c). \tag{15.35}$$

Continuous Local Conductivity

Before closing this section, it is worth mentioning that Dykhne (1971) also demonstrated that the duality property of two dimensions enables one to obtain an exact relation for the case in which a statistically isotropic heterogeneous medium is described by a special random *continuous* (not discrete-valued) local scalar conductivity $\sigma(\boldsymbol{x})$. Specifically, he showed that if the probability density function f characterizing the random field $\ln \sigma(\boldsymbol{x})$ is an even function of $\ln \sigma(\boldsymbol{x})$, then the effective conductivity is given exactly by the *geometric mean*, i.e.,

$$\sigma_e = \exp(\langle \ln \sigma \rangle) = \left(\frac{\langle \sigma \rangle}{\langle \sigma^{-1} \rangle} \right)^{1/2}, \tag{15.36}$$

where the average of any function $g(\sigma)$ is

$$\langle g \rangle \equiv \int_{-\infty}^{\infty} g(\sigma) f(\ln \sigma)\, d \ln \sigma = \int_0^{\infty} \frac{g(\sigma)}{\sigma} f(\ln \sigma)\, d\sigma.$$

In the special case where $\ln \sigma(\boldsymbol{x})$ is Gaussian distributed [i.e., $\sigma(\boldsymbol{x})$ is log-normally distributed], then (15.36) takes the form

$$\sigma_e = \langle \sigma \rangle \exp(\beta^2/2), \tag{15.37}$$

where $\beta^2 = \langle (\ln \sigma)^2 \rangle - \langle \ln \sigma \rangle^2$.

For isotropic random media in any dimension d with a log-normally distributed conductivity, Matheron (1967) conjectured that the effective conductivity is given by

$$\sigma_e = \exp(\langle \ln \sigma \rangle)\, \exp\left[\beta^2 \left(\frac{1}{2} - \frac{1}{d} \right) \right]. \tag{15.38}$$

This expression is exact for $d = 1, 2$, and ∞, the latter being true since $\sigma_e = \langle \sigma \rangle$ for $d = \infty$ (see Chapter 20); for $d = 1$, $\sigma_e = \langle \sigma^{-1} \rangle^{-1}$. However, for other d, σ_e depends on the details of the microstructure as measured by correlation functions, and hence (15.38) cannot be exact. Indeed, for $d = 3$, Abramovich and Indelman (1995) showed that formula (15.38) deviates from the exact result at third order in the log-conductivity variance.

15.1.2 Three-Dimensional Media

For macroscopically isotropic three-dimensional two-phase media, the phase-interchange equality in the form of (15.23) is not valid, since 90° rotations of curl-free vector fields generally do not result in divergence-free vector fields and vice versa. Relation (15.23) is replaced by the system of inequalities given by

$$\frac{\phi_1 \phi_2}{\phi_1 \sigma_1 + \phi_2 \sigma_2 - \sigma_e} + \frac{\phi_1 \phi_2}{\phi_1 \sigma_2 + \phi_2 \sigma_1 - \hat{\sigma}_e} \le \frac{3(\sigma_1 + \sigma_2)}{(\sigma_1 - \sigma_2)^2}, \tag{15.39}$$

$$\frac{\sigma_e \hat{\sigma}_e}{\sigma_1 \sigma_2} + \frac{\sigma_e + \hat{\sigma}_e}{\sigma_1 + \sigma_2} \ge 2, \tag{15.40}$$

where $\hat{\sigma}_e = \sigma_e(\sigma_2, \sigma_1)$ is the phase-interchanged effective conductivity and ϕ_i is the volume fraction of phase i. Relation (15.39) is an obvious corollary of a bound due to Bergman (1978). Relation (15.40) was conjectured by Milton (1981a) and later proved by Avellaneda, Cherkaev, Lurie and Milton (1988); it is always sharper than the lower bound

$$\sigma_e \hat{\sigma}_e \geq \sigma_1 \sigma_2 \tag{15.41}$$

obtained by Schulgasser (1976a). The phase-interchange inequality (15.40) has been generalized for multiphase composites by Nesi (1991).

15.2 Elastic Moduli

Except for very special cases of isotropic two-dimensional, two-phase media, there are no phase-interchange equalities for the effective elastic moduli. The two exceptional cases have been studied by Berdichevsky (1983) and Helsing, Milton and Movchan (1997). Generally, phase-interchange relations for isotropic, two-phase media in two and three dimensions have been shown by Gibiansky and Torquato (1996a) to take the form of inequalities.

15.2.1 Two-Dimensional Media

All of the results given here apply to macroscopically isotropic two-dimensional composites consisting of two isotropic phases. Let the effective planar bulk and shear moduli be denoted by $k_e(k_1, G_1, k_2, G_2)$ and $G_e(k_1, G_1, k_2, G_2)$, respectively, and let $\hat{k}_e = k_e(k_2, G_2, k_1, G_1)$ and $\hat{G}_e = G_e(k_2, G_2, k_1, G_1)$ denote the corresponding phase-interchanged moduli. For the following two special cases one has phase-interchange equalities. Berdichevsky (1983) proved that for a two-dimensional, two-phase composite with *incompressible phases*, the following equality holds:

$$G_e \hat{G}_e = G_1 G_2. \tag{15.42}$$

For a two-dimensional composite consisting of two phases *with equal bulk moduli*, this relation was generalized by Helsing et al. (1997) as follows:

$$E_e \hat{E}_e = E_1 E_2, \tag{15.43}$$

where $E_1 = 4kG_1/(k + G_1)$, $E_2 = 4kG_2/(k + G_2)$ are the phase Young's moduli, and $E_e = 4kG_e/(k + G_e)$, $\hat{E}_e = 4k\hat{G}_e/(k + \hat{G}_e)$ are the Young's moduli of the composite and phase-interchanged composite, respectively.

For general two-phase media, Gibiansky and Torquato (1996a) derived the sharpest known phase-interchange inequalities for the effective bulk and shear moduli. These inequalities are stated below as theorems and contain as special cases the equalities (15.42) and (15.43). The reader is referred to Gibiansky and Torquato (1996a) for the proofs.

The following theorem specifies phase-interchange relations for the bulk modulus in two dimensions:

Theorem 15.3 *The bulk moduli k_e and $\hat{k}_e$ of a two-dimensional composite are restricted by the inequalities:*

$$\frac{\phi_1\phi_2 k_e}{k_e\langle\tilde{k}\rangle - k_1k_2} + \frac{\phi_1\phi_2\hat{k}_e}{\hat{k}_e\langle k\rangle - k_1k_2} \le \frac{k_1k_2(G_1+G_2) + G_1G_2(k_1+k_2)}{G_1G_2(k_1-k_2)^2}, \tag{15.44}$$

$$\frac{k_e\hat{k}_e - G_1G_2}{k_e\hat{k}_e(G_1+G_2) + G_1G_2(k_e+\hat{k}_e)} \le \frac{k_1k_2 - G_1G_2}{k_1k_2(G_1+G_2) + G_1G_2(k_1+k_2)}, \tag{15.45}$$

where $\langle\tilde{k}\rangle = k_2\phi_1 + k_1\phi_2$ and $\langle k\rangle = k_1\phi_1 + k_2\phi_2$. In the k_e-$\hat{k}_e$ plane, (15.44) *is a lower bound and* (15.45) *is an upper bound.*

The phase-interchange relations for the shear moduli in two dimensions are summarized by the following theorem:

Theorem 15.4 *(i) The shear modulus lower bound in the G_e-$\hat{G}_e$ plane is given by the sharpest of the following inequalities:*

$$\frac{G_1G_2 - G_e\hat{G}_e}{G_1+G_2-G_e-\hat{G}_e} \ge \frac{G_1G_2(G_+k_+ + G_-k_- + 2G_1G_2)}{k_1k_2(G_1+G_2) + k_+G_+^2 + k_-G_-^2 + 2G_1G_2(G_1+G_2+k_1+k_2)},$$

$$\frac{\phi_1\phi_2 G_e}{G_e\langle\tilde{G}\rangle - G_1G_2} + \frac{\phi_1\phi_2\hat{G}_e}{\hat{G}_e\langle G\rangle - G_1G_2} \le \frac{2k_1k_2(G_1+G_2) + 2G_1G_2(k_1+k_2)}{k_1k_2(G_1-G_2)^2},$$

where k_- and G_- (k_+ and G_+) are the minimal (maximal) bulk and shear moduli, respectively, $\langle\tilde{G}\rangle = G_2\phi_1 + G_1\phi_2$, and $\langle G\rangle = G_1\phi_1 + G_2\phi_2$.

(ii) The shear modulus upper bound in the G_e-$\hat{G}_e$ plane is given by the following inequality:

$$\frac{G_1G_2 - G_e\hat{G}_e}{G_1+G_2-G_e-\hat{G}_e} \le \frac{G_1G_2(G_+k_- + G_-k_+ + 2G_1G_2)}{k_1k_2(G_1+G_2) + k_-G_+^2 + k_+G_-^2 + 2G_1G_2(G_1+G_2+k_1+k_2)}.$$

Consider now two different composites with identical microstructures. One composite is composed of a phase 1 material with the moduli k_1 and G_1 and a *perfectly rigid phase* 2, i.e., $k_2 = \infty$, $G_2 = \infty$. We denote by k_∞ and G_∞ the effective moduli of this composite. The other composite, *with the same microstructure*, is compossed of a phase 1 material with the moduli $\hat{k}_1$ and $\hat{G}_1$ and a *void or cavity phase* 2, i.e., $\hat{k}_2 = 0$, $\hat{G}_2 = 0$. We denote by k_0 and G_0 the effective moduli of this composite. Provided that phase 2 in these composites is such that it does not elastically percolate, the quantities k_∞, G_∞, k_0^{-1}, and G_0^{-1} remain finite. We will now show that Theorems 15.3 and 15.4 provide links between the moduli of these two different composites.

To establish these relationships, consider a composite made of two phases with the moduli k_1, G_1 and $\lambda\hat{k}_1$, $\lambda\hat{G}_1$, respectively, where λ is a dimensionless parameter. Let us denote the effective moduli of this composite and the corresponding *phase-interchanged* composite by k_λ, G_λ and $\hat{k}_\lambda$, $\hat{G}_\lambda$, respectively. One can easily see that

$$k_\infty = \lim_{\lambda\to\infty} k_\lambda, \quad G_\infty = \lim_{\lambda\to\infty} G_\lambda, \quad k_0 = \lim_{\lambda\to\infty} \frac{\hat{k}_\lambda}{\lambda}, \quad G_0 = \lim_{\lambda\to\infty} \frac{\hat{G}_\lambda}{\lambda}. \tag{15.46}$$

Theorem 15.5 *The effective moduli k_∞, G_∞, k_0, and G_0 satisfy the following upper bounds:*

$$k_0 k_\infty(\hat{k}_1 + \hat{G}_1) + k_0(G_1\hat{G}_1 - k_1\hat{k}_1) - \hat{k}_1\hat{G}_1(k_1 + G_1) \le 0,$$
$$G_0 G_\infty(k_1 + 2G_1)(\hat{k}_1 + \hat{G}_1) - G_0 G_1(k_1\hat{G}_1 + \hat{k}_1 G_1 + 2G_1\hat{G}_1) - \hat{k}_1 G_1\hat{G}_1(k_1 + G_1) \le 0.$$

Proof: The above bounds follow immediately from the phase-interchange inequalities of Theorems 15.3 and 15.4 in the limits specified by (15.46). ∎

Remark:

1. The corresponding lower bounds of Theorems 15.3 and 15.4 degenerate to the appropriate limiting Hashin–Shtrikman lower bounds as $\lambda \to \infty$ and hence are uncoupled.

Gibiansky and Torquato (1996a) showed that Theorem 15.5 has interesting corollaries when the volume fraction of the cavities or rigid phase is small or when the *continuum* system is near the percolation threshold. Consider the latter case, take $\hat{k}_1 = k_1$ and $\hat{G}_1 = G_1$, and assume the usual power law behavior of the moduli near the threshold $\phi_{2c} = 1 - \phi_{1c}$, i.e.,

$$k_0/k_1 = A_0^k(\phi_1 - \phi_{1c})^f, \quad G_0/G_1 = A_0^G(\phi_1 - \phi_{1c})^g, \qquad \phi_1 \to \phi_{1c}^+, \tag{15.47}$$

$$k_\infty/k_1 = A_\infty^k(\phi_1 - \phi_{1c})^{-a}, \quad G_\infty/G_1 = A_\infty^G(\phi_1 - \phi_{1c})^{-b}, \qquad \phi_1 \to \phi_{1c}^+, \tag{15.48}$$

where f, g, a, and b are the positive critical exponents (Section 9.1.3), and A_0^k, A_0^G, A_∞^k, and A_∞^G are the corresponding dimensionless *amplitudes*. Implicit in the use of relations (15.47) and (15.48) is the assumption that only one phase can percolate at a time; i.e., the system is not bicontinuous.

Corollary 15.2 *The critical exponents a, b, f, and g in the scaling relations* (15.47) *and* (15.48) *satisfy the bounds*

$$f \ge a, \qquad g \ge b. \tag{15.49}$$

Moreover, if

$$f = a, \qquad g = b, \tag{15.50}$$

then the amplitudes A_0^k, A_0^G, A_∞^k, and A_∞^G satisfy the upper bounds

$$A_0^k A_\infty^k k_1(k_1 + G_1) - G_1(k_1 + G_1) \le 0, \tag{15.51}$$

$$A_0^G A_\infty^G(k_1 + 2G_1)(k_1 + G_1) - k_1(k_1 + G_1) \le 0. \tag{15.52}$$

Proof: The proof of this corollary follows directly from Theorem 15.5 (with $\hat{k}_1 = k_1$ and $\hat{G}_1 = G_1$) and the scaling laws (15.47) and (15.48). ∎

Remarks:

1. Corollary 15.2 applies not only to continuum percolation but also to lattice percolation for the same reasons given in Section 15.1.1.
2. Numerical and experimental measurements of the critical exponents (see Section 9.1.3) are consistent with the bounds (15.49) of Corollary 15.2.
3. Corollary 15.2 was used to derive the exact results (16.67) and (16.69) for ordered arrays of circular rigid inclusions given the corresponding results (16.66) and (16.68) for ordered arrays of circular holes (Gibiansky and Torquato 1996a).

15.2.2 Three-Dimensional Media

For macroscopically isotropic three-dimensional two-phase media, the existing phase-interchange inequalities for the effective bulk and shear moduli were derived by Gibiansky and Torquato (1996a) using the translation method and geometrical-parameter bounds. Interested readers are referred to this paper for the proofs. Let $K_e(K_1, G_1, K_2, G_2)$ and $G_e(K_1, G_1, K_2, G_2)$ denote the effective bulk and shear moduli, respectively, and let $\hat{K}_e = K_e(K_2, G_2, K_1, G_1)$ and $\hat{G}_e = G_e(K_2, G_2, K_1, G_1)$ denote the corresponding phase-interchanged moduli.

The following theorem specifies the phase-interchange relations for the bulk modulus in three dimensions:

Theorem 15.6 *The bulk moduli K_e and $\hat{K}_e$ of a three-dimensional composite are restricted by the inequalities*

$$\frac{\phi_1\phi_2 K_e}{K_e\langle\tilde{K}\rangle - K_1K_2} + \frac{\phi_1\phi_2\hat{K}_e}{\hat{K}_e\langle K\rangle - K_1K_2} \leq \frac{3K_1K_2(G_1+G_2)/4 + G_1G_2(K_1+K_2)}{G_1G_2(K_1-K_2)^2}, \tag{15.53}$$

$$\frac{\phi_1\phi_2}{\langle K\rangle - K_e} + \frac{\phi_1\phi_2}{\langle\tilde{K}\rangle - \hat{K}_e} \leq \frac{K_1+K_2+4(G_1+G_2)/3}{(K_1-K_2)^2}, \tag{15.54}$$

where $\langle\tilde{K}\rangle = K_2\phi_1 + K_1\phi_2$ and $\langle K\rangle = K_1\phi_1 + K_2\phi_2$. In the K_e-$\hat{K}_e$ plane, (15.53) *is a lower bound and* (15.54) *is an upper bound.*

Gibiansky and Torquato (1996a) gave corresponding phase-interchange relations for the effective shear modulus, but they are more complicated and hence will not be presented here in their full generality. In the special case of incompressible phases, these inequalities take simple forms.

Theorem 15.7 *The shear moduli G_e and $\hat{G}_e$ of a three-dimensional composite with incompressible phases are restricted by the inequalities*

$$\frac{\phi_1\phi_2 G_e}{G_e\langle\tilde{G}\rangle - G_1G_2} + \frac{\phi_1\phi_2\hat{G}_e}{\hat{G}_e\langle G\rangle - G_1G_2} \leq \frac{5(G_1+G_2)}{3(G_1-G_2)^2}, \tag{15.55}$$

$$\frac{\phi_1\phi_2}{\langle G\rangle - G_e} + \frac{\phi_1\phi_2}{\langle \tilde{G}\rangle - \hat{G}_e} \le \frac{5(G_1+G_2)}{2(G_1-G_2)^2}, \tag{15.56}$$

where $\langle \tilde{G}\rangle = G_2\phi_1 + G_1\phi_2$ *and* $\langle G\rangle = G_1\phi_1 + G_2\phi_2$. *In the* G_e-$\hat{G}_e$ *plane,* (15.55) *is a lower bound and* (15.56) *is an upper bound.*

15.3 Trapping Constant and Fluid Permeability

In contrast to the previous properties, microstructure-independent phase-interchange relations for the trapping constant γ or fluid permeability $\boldsymbol{k}$ currently do not exist. The difficulty in obtaining such relations is related to the fact that all of the known bounds on γ and $\boldsymbol{k}$ depend on the specific forms of two-point and higher-order correlation functions of the material (see Section 21.6). Microstructure-dependent phase-interchange inequalities for γ and $\boldsymbol{k}$ have been derived by Torquato and Rubinstein (1989) and Rubinstein and Torquato (1989).

CHAPTER 16

Exact Results

Due to the complexity of the microstructure, there are relatively few situations in which one can evaluate the effective properties of heterogeneous materials exactly. Such rare results are nonetheless quite valuable as benchmarks to test theories and computer simulations. Exact results also provide useful insights into the mechanisms responsible for the effective behavior, insights that extend beyond the specific microstructures for which they are derived. In Chapter 15 we discussed exact phase-interchange relations. In the ensuing sections we will derive and discuss some other known exact results that we will make use of in subsequent chapters.

In lieu of obtaining the effective properties exactly, some investigators have first idealized the microstructure in some fashion. A common idealization is to assume that the heterogeneous medium is periodic with a simple repeating cell, for example, spheres centered on the sites of a regular lattice. The symmetries associated with the periodicity enable one to find exact asymptotic expressions for the effective properties in the low- and high-concentration limits. We report such results for periodic arrays of spheres and aligned cylinders for the conduction, elasticity, trapping, and flow problems. In Chapter 19 we separately discuss exact results for random dispersions of inclusions at low concentrations. In Chapter 20 we derive exact contrast expansions.

In the cases of conduction and elasticity, we begin, respectively, with derivations of the effective conductivity and effective bulk modulus of special isotropic dispersions of spheres called the Hashin–Shtrikman construction. Such a construction is remarkable in that it is one of the very few instances in which one can compute the effective conductivity or bulk modulus of a dispersion of particles in a matrix exactly for arbitrary phase volume fractions and phase properties. We then proceed to examine an important class of macroscopically anisotropic composites whose effective conductivity or stiffness tensor can be evaluated exactly: certain hierarchically structured laminate composites. These hierarchical laminates have proven to be of immense value in the construction of composite structures that attain prescribed effective properties and thus can be used to identify *extremal* structures. Asymptotically exact results for the effective properties for periodic arrays of inclusions are subsequently given. We then report exact results for various systems and situations: effective properties of cellular solids, sheets with holes, and dispersions of particles or bubbles in a liquid; field fluctuations; the link between the tranverse conductivity and axial shear modulus of a fiber-reinforced material; and the link between certain thermoelastic constants and the bulk modulus of a two-phase system.

There are fewer exact results for the trapping constant (or survival time) and fluid permeability than for the conductivity and elastic moduli. For the former properties, besides reporting asymptotic expressions for transport exterior to periodic arrays of spheres and cylinders, we derive exact results for the survival time and fluid permeability when transport occurs inside certain simple domains such as layered systems and cylindrical tubes or, more generally, hyperspheres.

16.1 Conductivity

16.1.1 Coated-Spheres Model

There is a special isotropic dispersion of spheres for which one can obtain the effective conductivity σ_e exactly for any inclusion volume fraction ϕ_2 and phase contrast. This two-phase model, introduced by Hashin and Shtrikman (1962), consists of composite spheres that are composed of a spherical core, of conductivity σ_2 and radius a, which is surrounded by a concentric shell of conductivity σ_1 with an outer radius b. The ratio $(a/b)^3$ is fixed and equal to the inclusion volume fraction ϕ_2. The composite spheres fill all space, implying that there is a distribution in their sizes ranging to the infinitesimally small. Such a construction is depicted in two dimensions in Figure 16.1. Hashin (1970) also evaluated the effective conductivity for the analogous two-dimensional multisized dispersion, i.e., the coated-cylinders model in which, of course, $(a/b)^2 = \phi_2$.

We will derive the effective conductivity σ_e for the coated-spheres model in any space dimension d. To find σ_e, it will be shown that it is sufficient to obtain the local fields for a single d-dimensional composite sphere. Consider a homogeneous body with an unknown conductivity σ_e, in which we impose a uniform field $\boldsymbol{E}_0$ throughout the body. Now replace a d-dimensional sphere of radius b with a composite sphere. If we can find the constant σ_e such that the field outside the composite sphere remains unchanged, then there will be no change in the energy dissipation, and hence no change in the system conductivity due to the replacement. Repeating this process of replacement with composite spheres *ad infinitum* will result in a two-phase coated-spheres dispersion characterized by an effective conductivity σ_e.

Accordingly, consider a d-dimensional composite sphere in an infinite matrix of conductivity σ_e and let $\boldsymbol{r}$ be the radial vector emanating from the sphere center (see

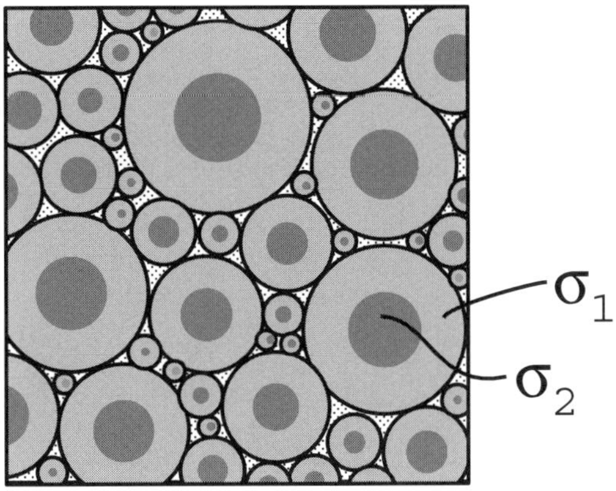

Figure 16.1 Coated-spheres model for effective conductivity for $d = 2$.

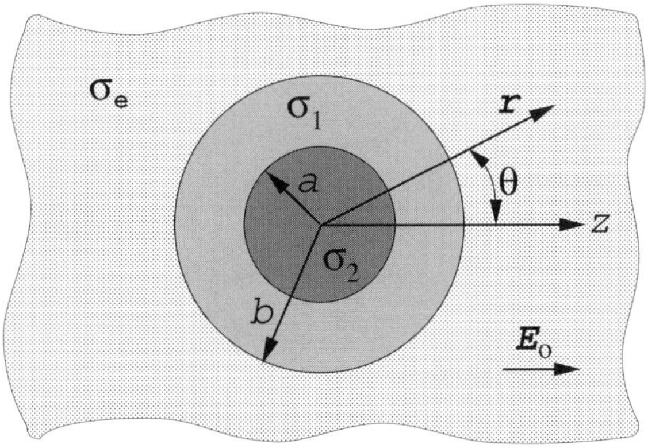

Figure 16.2 A single coated sphere for conductivity problem.

Figure 16.2). The potential (temperature) distribution T must satisfy Laplace's equation

$$\Delta T(\boldsymbol{r}) = 0 \tag{16.1}$$

everywhere subject to the continuity conditions (13.11) at the interfaces $r = a$ and $r = b$. Laplace's equation is arrived at by combining relations (13.3)–(13.5) and (13.6). The solution has the general form

$$T = \begin{cases} -E_0 r\cos\theta, & r \geq b, \\ AE_0 r\cos\theta + B\dfrac{E_0\cos\theta}{r^{d-1}}, & a \leq r \leq b, \\ CE_0 r\cos\theta, & r \leq a, \end{cases} \tag{16.2}$$

where $E_0 = |\boldsymbol{E}_0|$, $r = |\boldsymbol{r}|$, $\cos\theta = \boldsymbol{E}_0 \cdot \boldsymbol{r}/(E_0 r)$, and A, B, C are unknown parameters. The corresponding intensity field $\boldsymbol{E}$ is given by

$$\boldsymbol{E} = -\nabla T = \begin{cases} \boldsymbol{E}_0, & r > b, \\ -A\boldsymbol{E}_0 + B\Big[\dfrac{d\boldsymbol{nn} - \boldsymbol{I}}{r^d}\Big] \cdot \boldsymbol{E}_0, & a < r < b, \\ -C\boldsymbol{E}_0, & r < a, \end{cases} \tag{16.3}$$

where $\boldsymbol{n} = \boldsymbol{r}/r$ is the radial unit vector. The three unknown parameters A, B, C are determined from continuity of the potential at $r = a$ and $r = b$ and continuity of the radial component of the flux $(-\sigma\nabla T \cdot \boldsymbol{n})$ at $r = a$ [see (13.11)]. It is found that

$$A = \frac{(1-d)\sigma_1 - \sigma_2}{\sigma_2 + (d-1)\sigma_1 - \phi_2(\sigma_2 - \sigma_1)}, \tag{16.4}$$

$$B = \frac{a^d(\sigma_2 - \sigma_1)}{\sigma_2 + (d-1)\sigma_1 - \phi_2(\sigma_2 - \sigma_1)}, \tag{16.5}$$

$$C = \frac{-d\sigma_1}{\sigma_2 + (d-1)\sigma_1 - \phi_2(\sigma_2 - \sigma_1)}, \tag{16.6}$$

where

$$\phi_2 = \left(\frac{a}{b}\right)^d \tag{16.7}$$

is the volume fraction of the included phase.

Now, since continuity of the normal flux at $r = b$ implies that

$$\sigma_e = \frac{B\sigma_1(d-1)}{b^2} - A\sigma_1,$$

the conductivity of the homogeneous body must satisfy the relation

$$\sigma_e = \langle\sigma\rangle - \frac{(\sigma_2 - \sigma_1)^2\phi_1\phi_2}{\langle\tilde{\sigma}\rangle + (d-1)\sigma_1}, \tag{16.8}$$

where

$$\langle\sigma\rangle = \sigma_1\phi_1 + \sigma_2\phi_2, \qquad \langle\tilde{\sigma}\rangle = \sigma_1\phi_2 + \sigma_2\phi_1. \tag{16.9}$$

In summary, if we continue this process of replacement until all of the original material is replaced by composite spheres of all sizes down to the infinitesimally small, then a macroscopically isotropic and homogeneous material can be constructed whose effective conductivity is given exactly by (16.8).

We notice that phase 2 is always disconnected, i.e., does not percolate, except in the trivial instance when $\phi_2 = 1$. It will be shown in Chapter 21 that (16.8) is identical to the well-known Hashin–Shtrikman lower bound (Hashin and Shtrikman 1962) if $\sigma_2 \geq \sigma_1$. If the phase labels are reversed in (16.8), then it becomes identical to the Hashin–Shtrikman upper bound if $\sigma_2 \geq \sigma_1$. The consequences of the realizability of the bounds will be described in Chapter 21.

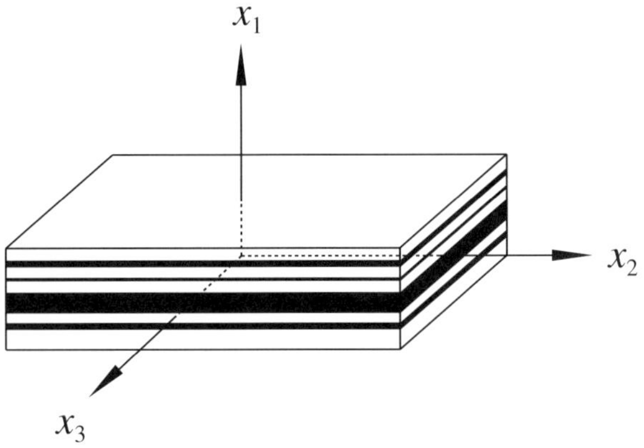

Figure 16.3 Portion of a rank 1 laminate composite consisting of alternating layers of phases of random thicknesses. The composite fills all of space.

It is clear that the Hashin–Shtrikman construction can be generalized by utilizing *multicoated* inclusions in the replacement process such that they do not change the field outside the inclusions. This idea was proposed by Schulgasser (1977) and fruitfully extended (even to ellipsoidal shapes) by Milton (1981a, 1981b), among others (see also Section 21.1.2).

In the ensuing sections we examine another important class of microstructures that yield to exact solutions of the effective conductivity: laminate composites.

16.1.2 Simple Laminates

Consider a simple laminate composite consisting of alternating layers of isotropic phases 1 and 2 of random thicknesses. In the language of modern composites theory, this is referred to as a laminate of rank 1, as will be described in Section 16.1.3. Phase i has conductivity σ_i and volume fraction ϕ_i. Assume that a Cartesian coordinate system of this laminate composite is oriented along the principal axes as depicted in Figure 16.3. The composite is assumed to fill all of space, so that we can neglect edge effects. The associated indicator function for phase 2, $\mathcal{I}^{(2)}(x_1)$, is illustrated in Figure 16.4. It is well known that the effective conductivity tensor of this macroscopically anisotropic composite is given by

$$\boldsymbol{\sigma}_e = \begin{bmatrix} \langle \sigma^{-1} \rangle^{-1} & 0 & 0 \\ 0 & \langle \sigma \rangle & 0 \\ 0 & 0 & \langle \sigma \rangle \end{bmatrix}, \tag{16.10}$$

where

$$\langle \sigma \rangle = \sigma_1 \phi_1 + \sigma_2 \phi_2 \tag{16.11}$$

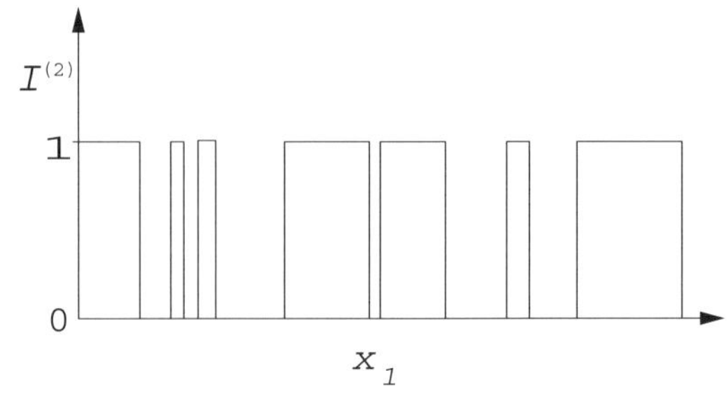

Figure 16.4 Indicator function for phase 2, $\mathcal{I}^{(2)}(x_1)$, of a rank 1 laminate composite.

is the *arithmetic* average of the phase conductivities and

$$\langle\sigma^{-1}\rangle^{-1} = \left(\frac{\phi_1}{\sigma_1} + \frac{\phi_2}{\sigma_2}\right)^{-1} = \frac{\sigma_1\sigma_2}{\sigma_1\phi_2 + \sigma_2\phi_1} \tag{16.12}$$

is the *harmonic average* of the phase conductivities.

We now prove (16.10) by reducing the task to one-dimensional conduction problems. The local fields will depend only on the variable x_1 when uniform applied fields are imposed. First consider a uniform applied field $\boldsymbol{E}_0$ in the x_1-direction. The local and averaged constitutive relations are given respectively by

$$J(x_1) = \sigma(x_1)E(x_1),$$
$$\langle J\rangle = (\sigma_e)_{11}\langle E\rangle.$$

Steady-state conduction requires that

$$\frac{\partial}{\partial x_1}[\sigma(x_1)E(x_1)] = 0,$$

implying that the local flux $J(x_1)$ is a constant everywhere and thus $J(x_1) = \langle J\rangle$. It immediately follows from the local constitutive law that the local intensity field is a *constant in each phase* (i.e., piecewise constant) and is given by

$$E(x_1) = \frac{\langle J\rangle}{\sigma(x_1)}.$$

Averaging this relation and comparing it to the averaged constitutive relation given above yields

$$(\sigma_e)_{11} = \langle\sigma^{-1}\rangle^{-1},$$

which agrees with (16.10). This harmonic-average result is equivalent to *resistance* (inverse conductivity) in series.

To complete the proof of (16.10), let us now consider the applied field in the x_2-direction. Here the local intensity $E = \langle E\rangle$ is the same constant in each phase, and hence the local constitutive relation becomes

$$J(x_2) = \sigma(x_2)\langle E\rangle.$$

Averaging this relation and comparing it to the averaged constitutive relation

$$\langle J\rangle = (\sigma_e)_{22}\langle E\rangle$$

yields

$$(\sigma_e)_{22} = \langle\sigma\rangle,$$

which again agrees with (16.10). This arithmetic-average result is equivalent to resistance in parallel. Similarly,

$$(\sigma_e)_{33} = \langle\sigma\rangle.$$

Observe that the effective conductivity tensor $\boldsymbol{\sigma}_e$ for the laminate given by (16.10) can be recast in the form

$$(\boldsymbol{\sigma}_e - \boldsymbol{\sigma}_1)^{-1} = \frac{1}{\phi_2}(\boldsymbol{\sigma}_2 - \boldsymbol{\sigma}_1)^{-1} + \frac{\phi_1}{\phi_2}\frac{1}{\sigma_1}\boldsymbol{A}, \tag{16.13}$$

where $\boldsymbol{\sigma}_i = \sigma_i \boldsymbol{I}$ and $\boldsymbol{A}$ is the 3×3 matrix

$$\boldsymbol{A} = \begin{bmatrix} 1 & 0 & 0 \\ 0 & 0 & 0 \\ 0 & 0 & 0 \end{bmatrix}. \tag{16.14}$$

The representation (16.13) of the effective conductivity tensor of a first rank laminate facilitates the determination of the effective conductivity of more general laminates, as will be discussed in the next section.

Remarks:

1. It is important to note that the local fields in the simple laminate are piecewise constant (under the uniform applied fields that we imposed) and, of course, satisfy the continuity conditions (13.11) at the two-phase interface. The piecewise-constant nature of the local fields has vital consequences for constructing composites with desired effective conductivities, as we will see in the next section.
2. There has been a temptation in the literature to apply the simple mixture rules (16.11) and (16.12) to *macroscopically isotropic* composites. However, such estimates are almost always poor, except when the phase conductivities are nearly equal to one another. Generally, the arithmetic-average formula (16.11) overestimates the effective conductivity of isotropic composites (especially when the phase contrast ratio is large). The reason for this is that this formula corresponds to conduction along the slabs (x_2-x_3 plane) for the laminate composite of Figure 16.3. In the direction of the slabs, the phases are connected from one end of the sample to the opposite end, and thus the current (or heat flow) will be relatively unimpeded even if one phase is a poor conductor. Indeed, we will show that (16.11) is a rigorous upper bound on the effective conductivity of isotropic composites. On the other hand, the

harmonic-average formula (16.12) generally underestimates the effective conductivity of isotropic composites, especially when the phase conductivities widely differ. This turns out to be the case, since the formula (16.12) corresponds to conduction perpendicular to the slabs (x_1-direction), and hence the current (heat flow) will be relatively impeded in this direction and can be impeded entirely if one phase is perfectly insulating. Formula (16.12) will be shown to be a rigorous lower bound on the effective conductivity of isotropic composites.

3. Note that the effective conductivity tensor of a first-rank laminate but with M different phases is still given by (16.10) but with

$$\langle\sigma\rangle = \sum_{i=1}^{M} \sigma_i \phi_i, \tag{16.15}$$

$$\langle\sigma^{-1}\rangle^{-1} = \left(\sum_{i=1}^{M} \frac{\phi_i}{\sigma_i}\right)^{-1}. \tag{16.16}$$

These relations are simply the generalizations of the arithmetic and harmonic mean formulas for an M-phase composite.

16.1.3 Higher-Order Laminates and Attainability

It was recognized by Schulgasser (1976b) that the effective conductivity tensor of a wide class of composites can be evaluated exactly via a lamination construction. These laminates have structural hierarchy; i.e., they possess structure on multiple length scales. The number of different levels of hierarchy is referred to as the *rank* of the laminate. A rank 1 (or first-rank) laminate is made simply by alternating layers of phase 1 and phase 2 materials, as shown in Figure 16.3, to form a multilayered sandwich, where the vector $\boldsymbol{n}^{(1)}$ is taken to be the normal to the slabs. To make a laminate of rank 2, we take a large piece of the rank 1 laminate composite (at some arbitrary orientation) and layer it with, say, phase 1 material to form the multilayered sandwich depicted in Figure 16.5. The normal vector $\boldsymbol{n}^{(2)}$ specifies the orientation of the slabs in the second stage. The heterogeneity length scale of the composite layers must be much smaller than the slab lengths of the pure phase 1 material. We note that phase 1 is a connected (or continuous) *matrix* phase. Continuing in this fashion produces laminates of arbitrarily high rank, and hence a broad class of composites can be constructed. Hierarchical laminates are topologically similar to the coated-spheres models in that for each class, one phase is a continuous matrix phase and the other is a disconnected inclusion phase.

As we have observed, the effective conductivity tensor of laminates of rank one is simple to calculate, since the local fields are piecewise constant under uniform applied fields. The effective properties of laminates of higher rank are also easy to evaluate analytically because the local fields are again piecewise constant under the assumption of infinite separation of the length scales between the different structural levels. Thus, if there is an infinite separation of the length scales, one can repeatedly use the effective

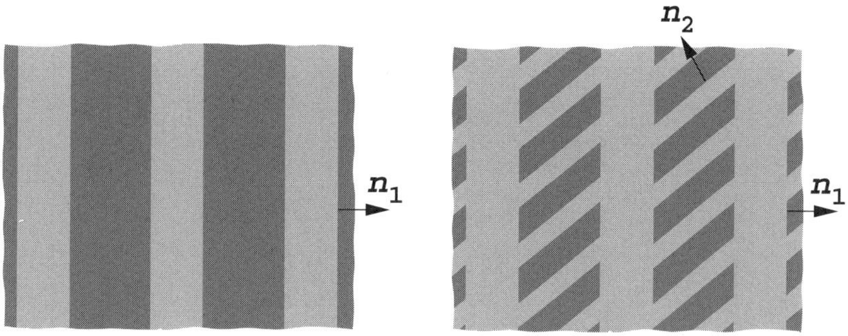

Figure 16.5 Examples of a first-rank laminate (left) and a second-rank laminate (right).

conductivity tensor result for a first-rank laminate successively at each level of the hierarchy to get the effective conductivity of the entire laminate. This point will be illustrated below via an example for a second-rank laminate.

Schulgasser (1976b) was interested in finding the maximum range of the effective conductivity that an isotropic polycrystal composed of grains having the principal conductivities σ_a, σ_b, and σ_c can take for all possible orientations, shapes, and configurations of the grains. He found that the well-known upper bound (15.35) on the effective conductivity σ_e of a three-dimensional polycrystal is attained by a laminate of rank 3.

Subsequently, it was discovered that a class of laminates realize rigorous bounds on the effective conductivity of two-phase composites. In particular, Milton (1981c) found that many bounds on the complex dielectric constant of two-phase composites can be realized by laminates. In independent work, Lurie and Cherkaev (1984) showed that second-rank laminates with orthogonal layering directions ($\boldsymbol{n}^{(1)} \cdot \boldsymbol{n}^{(2)} = 0$) attain the two-dimensional Hashin–Shtrikman bounds on the effective conductivity (see Figure 16.6). This result was subsequently given for any dimension d by Murat and Tartar (1985) and by Lurie and Cherkaev (1986). Laminates of rank d achieve the d-dimensional Hashin–Shtrikman bounds (21.20) for the effective conductivity. It should be noted that Murat and Tartar give an explicit formula to compute the effective conductivity of laminates of arbitrary rank and layering orientations.

We will not attempt to discuss the large subject of attainability of effective properties via lamination in detail. Instead, to illustrate the idea, we will show that formula (16.8) for the coated-spheres model with $d = 2$ (also equal to the Hashin–Shtrikman lower bound for $\sigma_2 \geq \sigma_1$) is attained by a laminate of rank 2. The ensuing derivation is not limited to periodic second-rank laminates, such as the one shown in Figure 16.6, but includes random second-rank laminates, such as the one depicted in Figure 7.4.

Under the assumption of an infinite separation of length scales ($D_1/D_2 \to \infty$), the effective conductivity tensor $\boldsymbol{\sigma}_e^{(2)}$ for the material between the slabs of phase 1 at the first stage of construction is a diagonal tensor with components

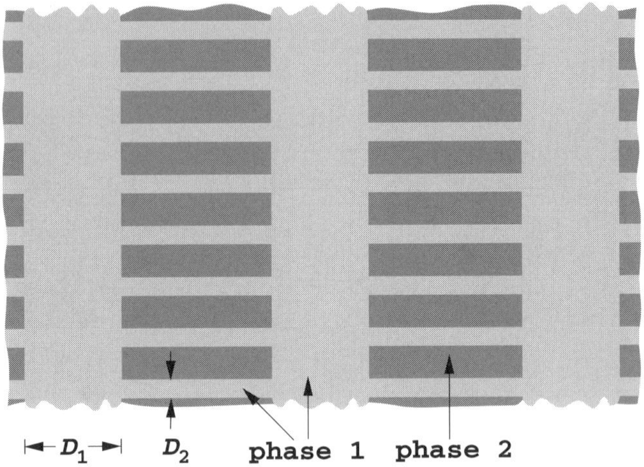

Figure 16.6 An example of a two-dimensional second-rank laminate whose effective conductivity is exactly given by (16.8) in the limit $D_1/D_2 \to \infty$. Here $\boldsymbol{n}^{(1)} \cdot \boldsymbol{n}^{(2)} = 0$.

$$\left(\sigma_e^{(2)}\right)_{11} = \left[\frac{\phi_1^{(2)}}{\sigma_1} + \frac{\phi_2^{(2)}}{\sigma_2}\right]^{-1}, \qquad \left(\sigma_e^{(2)}\right)_{22} = \sigma_1\phi_1^{(2)} + \sigma_2\phi_2^{(2)},$$

where $\phi_i^{(j)}$ is the volume fraction of phase i in the jth stage with $\phi_2 = \phi_2^{(1)}\phi_2^{(2)}$. This result immediately follows from our result (16.10) for the first-rank laminate but for $d = 2$. The effective conductivity tensor of the entire laminate composite is therefore given by

$$\boldsymbol{\sigma}_e = \begin{bmatrix} (\sigma_e)_{11} & 0 \\ 0 & (\sigma_e)_{22} \end{bmatrix}, \tag{16.17}$$

where

$$(\sigma_e)_{11} = \sigma_1\phi_1^{(1)} + (\sigma_e^{(2)})_{11}\phi_2^{(1)}, \qquad (\sigma_e)_{22} = \left[\frac{\phi_1^{(1)}}{\sigma_1} + \frac{\phi_2^{(1)}}{(\sigma_e^{(2)})_{22}}\right]^{-1}.$$

In order for the tensor (16.17) to be isotropic, we must have $(\sigma_e)_{11} = (\sigma_e)_{22}$, implying the volume fraction requirement

$$\phi_1^{(1)} = \frac{\phi_1^{(2)}}{\phi_2^{(2)}}. \tag{16.18}$$

It follows, after some algebra, that the corresponding isotropic effective conductivity is given by

$$\sigma_e = \langle\sigma\rangle - \frac{(\sigma_2 - \sigma_1)^2\phi_1\phi_2}{\langle\tilde{\sigma}\rangle + \sigma_1}, \tag{16.19}$$

where $\langle\sigma\rangle$ and $\langle\tilde{\sigma}\rangle$ are defined by the relations in (16.9). We recognize result (16.19) to be identical to the exact expression for the coated-cylinder model given by (16.8) with

$d = 2$. We also observe that the same expression is obtained when the widths D_1 and D_2 are random variables, provided that there is still an infinite separation of length scales. In Chapter 21 we will show that (16.19) is an extremal result.

The procedure outlined above to obtain the effective conductivity tensor for a second-rank laminate becomes unwieldy algebraically for arbitrary layering orientations and as the rank increases. Murat and Tartar (1985) have given the following general formula for the effective conductivity tensor for laminates of arbitrary rank N and layering orientations $\boldsymbol{n}^{(1)}, \boldsymbol{n}^{(2)}, \ldots, \boldsymbol{n}^{(N)}$ (under the assumption of an infinite separation of length scales):

$$(\boldsymbol{\sigma}_e - \sigma_1)^{-1} = \frac{1}{\phi_2}(\sigma_2 - \sigma_1)^{-1} + \frac{\phi_1}{\phi_2}\frac{1}{\sigma_1}\sum_{i=1}^{N} \alpha_i \boldsymbol{n}^{(i)}\boldsymbol{n}^{(i)}, \tag{16.20}$$

where α_i is a known parameter that generally depends on the volume fractions $\phi_1^{(i)}, \phi_2^{(i)}$ of stage i, obeying the conditions

$$\sum_{i=1}^{N} \alpha_i = 1, \qquad \alpha_i \geq 0.$$

Formula (16.20) is obtained by making repeated use of expression (16.13) for the first-rank laminate. Clearly, relation (16.20) is identical to (16.13) when $N = 1$ and the normal $\boldsymbol{n}^{(1)}$ is oriented along the x_1 direction.

16.1.4 Fiber-Reinforced Materials

Consider transversely isotropic fiber-reinforced composites whose phase boundaries are cylindrical surfaces of arbitrary shape with generators parallel to one axis (see Figure 13.6). The effective principal conductivities are determined by two parameters: the effective longitudinal (or axial) conductivity σ_e^L and effective transverse conductivity σ_e^T. The former is trivially given by the arithmetic average of the phases, i.e., $\sigma_e^L = \sigma_1\phi_1 + \sigma_2\phi_2$, since the electric field in the longitudinal direction is the same constant in each phase. By contrast, σ_e^T depends on the microstructure.

16.1.5 Periodic Arrays of Inclusions

Lord Rayleigh (1892) was the first to consider the effective conductivity of a periodic lattice of spheres, focusing on the simple cubic lattice. This problem was revisited by a number of investigators using Rayleigh's method, but the definitive work for arbitrary sphere volume fractions was carried out by McPhedran and McKenzie (1978) for simple cubic lattices and by McKenzie, McPhedran and Derrick (1978) for body-centered and face-centered cubic lattices. The asymptotic formula in the form presented by Lord Rayleigh for these three cubic lattices is given by

$$\frac{\sigma_e}{\sigma_1} = 1 + \frac{3\beta\phi_2}{1 - \beta\phi_2 - a_1\frac{\beta^2}{1+2\beta/7}\phi_2^{10/3}}, \tag{16.21}$$

where, for simple cubic lattices, $a_1 = 1.6772$; for body-centered cubic lattices, $a_1 = 0.073886$; for face-centered cubic lattices, $a_1 = 0.0060503$; and β in any dimension is given by

$$\beta = \frac{\sigma_2 - \sigma_1}{\sigma_2 + (d-1)\sigma_1}. \tag{16.22}$$

Higher-order asymptotic expressions have been given by McPhedran and McKenzie (1978) for simple cubic lattices and by McKenzie et al. (1978) for the other two lattices. Expanding (16.21) in powers of the volume fraction ϕ_2 yields the exact asymptotic expansion

$$\frac{\sigma_e}{\sigma_1} = 1 + 3\beta\phi_2 + 3\beta^2\phi_2^2 + 3\beta^3\phi_2^3 + \frac{3a_1\beta^3}{1+2\beta/7}\phi_2^{10/3} + \mathcal{O}(\phi_2^4). \tag{16.23}$$

The coated-spheres formula (16.8) agrees with the periodic formula (16.23) through third order in ϕ_2, since the spheres in both models are always "well separated" from one another.

Corresponding asymptotic formulas have been developed for two-dimensional lattices of disks (oriented infinitely long cylinders) by Perrins, McKenzie and McPhedran (1979). They found that for disks centered on the sites of a square lattice (square array),

$$\frac{\sigma_e}{\sigma_1} = 1 + \frac{2\beta\phi_2}{1 - \beta\phi_2 - 0.305827\beta^2\phi_2^4}, \tag{16.24}$$

and for disks centered on the sites of a triangular lattice (also know as the hexagonal array),

$$\frac{\sigma_e}{\sigma_1} = 1 + \frac{2\beta\phi_2}{1 - \beta\phi_2 - 0.075422\beta^2\phi_2^6}. \tag{16.25}$$

They actually obtained asymptotic expressions that are of higher order than those reported here. Note that relations (16.24) and (16.25) are exact through fourth and sixth order in ϕ_2, respectively.

As observed by Keller (1963) and Batchelor and O'Brien (1977), for volume fractions near the maximum close-packing values, the interaction between spheres is concentrated in the regions near the points of contact. A local analysis of the region between two nearly touching spheres, in conjunction with knowledge of nearest-neighbor locations, produces the dominant contribution of the interactions for the entire system.

For a simple cubic array of identical perfectly conducting spheres in a matrix of conductivity σ_1 near the particle-phase threshold $\phi_{2c} = \pi/6$, Keller (1963) found that the effective conductivity σ_e is given by

$$\frac{\sigma_e}{\sigma_1} = -\frac{\pi}{2}\ln(\phi_{2c} - \phi_2). \tag{16.26}$$

Keller (1963) obtained a corresponding asymptotic expression for the effective conductivity of a square array of identical perfectly conducting circular cylinders in a matrix of conductivity σ_1 near the particle-phase threshold $\phi_{2c} = \pi/4$:

$$\frac{\sigma_e}{\sigma_1} = \frac{\pi^{3/2}}{2}(\phi_{2c} - \phi_2)^{-1/2}. \tag{16.27}$$

He also showed that the effective conductivity for a square array of perfectly insulating inclusions near ϕ_{2c} is given by

$$\frac{\sigma_e}{\sigma_1} = \frac{2}{\pi^{3/2}}(\phi_{2c} - \phi_2)^{1/2}, \tag{16.28}$$

which we see is the reciprocal of the result for perfectly conducting cylinders. It is seen that (16.28) follows immediately from the use of the phase-interchange Theorem 15.2 and knowledge of (16.27).

16.1.6 Low-Density Cellular Solids

Cellular solids abound in nature and are fabricated by man on a large scale. Examples include wood, cancellous bone, cork, foams for insulation and packaging, sandwich panels in aircraft, and filters, to mention but a few (Gibson and Ashby 1997). Cellular solids may be regarded as porous materials with a very high porosity or very low solid volume fraction. We take phase 1 to be the solid phase with conductivity σ_1 and phase 2 to be the void phase ($\sigma_2 = 0$). Moreover, we consider the limit $\phi_1 \ll 1$; i.e., the solid volume fraction is very small.

Consider two-dimensional cellular solids consisting of regular triangular, square, or regular hexagonal cells. Such cells are the Voronoi cells associated with the sites of a honeycomb, square, or triangular lattice (see Figure 8.2). The effective conductivities of these three different structures are easily obtained analytically (Torquato, Gibiansky, Silva and Gibson 1998), since the behavior is governed by transport along individual cell-wall segments (effectively one-dimensional transport), and are all equal to each other according to

$$\frac{\sigma_e}{\sigma_1} = \frac{\phi_1}{2}. \tag{16.29}$$

Observe that the Hashin–Shtrikman upper bound on the effective conductivity of two-dimensional porous solids in which $\phi_1 \ll 1$ is given by $\sigma_e/\sigma_1 \le \phi_1/2$. Therefore, we see that two-dimensional cellular solids with either triangular, square, or hexagonal cells maximize the effective conductivity (Torquato et al. 1998). It is not as well known that the Kagomé network (see Figure 9.4) is also optimal, i.e, $\sigma_e/\sigma_1 = \phi_1/2$.

We note that the d-dimensional Hashin–Shtrikman upper bound in the low-density asymptotic limit is given by

$$\frac{\sigma_e}{\sigma_1} = \frac{d-1}{d}\phi_1. \tag{16.30}$$

This expression is, of course, realized for the special *foam* obtained by considering the Hashin–Shtrikman composite-sphere assemblages in the low-density asymptotic limit. It is also realized by space-filling d-dimensional cubical cells arranged on a d-dimensional simple cubic lattice in the same limit.

16.1.7 Field Fluctuations

In the study of heterogeneous materials, the preponderance of work has been devoted to finding the effective properties of the material, which amounts to knowing only the first moment of the local field. When heterogeneous materials are subjected to constant applied fields, the associated local fields exhibit strong spatial fluctuations. The distribution (probability density) of the local field (or, equivalently, all moments of the field when the probability density is continuous) is of great fundamental and practical importance in understanding many crucial material properties, such as breakdown phenomena (Li and Duxbury 1989) and the nonlinear behavior of composites (Levy and Bergman 1994).

Here we state a known exact result for the scalar second moment of the intensity field $\boldsymbol{E}$ within phase i for a macroscopically isotropic composite with M isotropic phases. This moment is proportional to the partial derivative of the effective conductivity σ_e with respect to the conductivity of phase i, namely,

$$\frac{\partial \sigma_e}{\partial \sigma_i} = \phi_i \frac{\langle \mathcal{I}^{(i)}(\boldsymbol{x})\boldsymbol{E}(\boldsymbol{x}) \cdot \boldsymbol{E}(\boldsymbol{x})\rangle}{\langle \boldsymbol{E}\rangle \cdot \langle \boldsymbol{E}\rangle}, \tag{16.31}$$

where $\mathcal{I}^{(i)}(\boldsymbol{x})$ is the indicator function for phase i. This relation was first noted by Bergman (1978) and later rediscovered by Bobeth and Diener (1986). We see that the second moment can be computed given an explicit expression for the effective conductivity as a function of the phase conductivities. In particular, one can utilize the exact expressions given in the present chapter as well as in Chapters 19 and 20, approximate expressions given in Chapters 18 and 20, and bounds given in Chapter 21 to find corresponding expressions for the second moment.

Relation (16.31) is easily derived by considering the first variation of the energy representation (14.31) of the effective conductivity for fixed average field $\langle \boldsymbol{E}\rangle$. This process yields

$$\delta\sigma_e \langle \boldsymbol{E}\rangle \cdot \langle \boldsymbol{E}\rangle = \langle \delta\sigma(\boldsymbol{x})\boldsymbol{E} \cdot \boldsymbol{E}\rangle + \langle \sigma(\boldsymbol{x})\delta(\boldsymbol{E} \cdot \boldsymbol{E})\rangle. \tag{16.32}$$

The second term on the right-hand side vanishes, since we know from the variational principle of Theorem 14.3 that field $\boldsymbol{E}(\boldsymbol{x})$, for fixed $\sigma(\boldsymbol{x})$, minimizes the right-hand side of (14.31). Substitution of

$$\delta\sigma(\boldsymbol{x}) = \sum_{i=1}^{M} \mathcal{I}^{(i)}(\boldsymbol{x})\delta\sigma_i$$

into (16.32) yields (16.31).

We note in passing that in the case of the Hashin–Shtrikman coated-sphere assemblages, the full distribution of the local electric field (all of the moments) has been evaluated exactly (Cule and Torquato 1998).

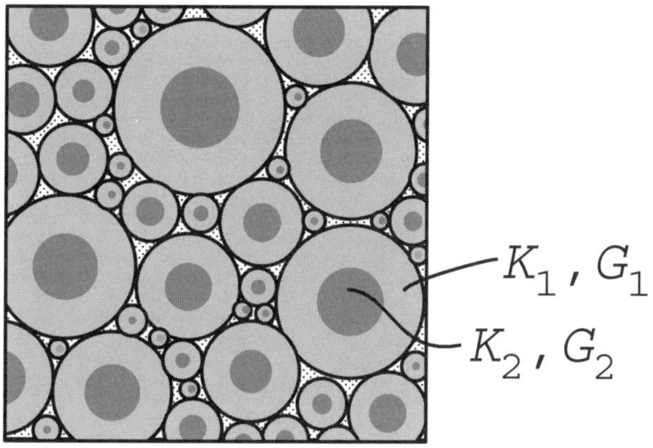

Figure 16.7 Coated-spheres model for effective bulk modulus for $d = 2$.

16.2 Elastic Moduli

16.2.1 Coated-Spheres Model

The effective bulk modulus K_e also can be found exactly for the coated-spheres model (Section 16.1.1) for arbitrary inclusion volume fraction ϕ_2, as first discovered by Hashin (1962). Here each composite sphere is composed of a spherical core of bulk and shear moduli K_2 and G_2 and radius a, surrounded by a concentric shell of bulk and shear moduli K_1 and G_1 with outer radius b (Figure 16.7). The effective bulk modulus of the analogous two-dimensional multisized construction, i.e., coated cylinders, was given by Hashin and Rosen (1964).

In what follows we will derive the effective bulk modulus K_e for the coated-spheres model in any space dimension d. Let us consider a homogeneous body with an unknown bulk modulus K_e in which we impose a uniform hydrostatic strain field ε_0 throughout the body. Let us now replace a d-dimensional sphere of radius b with a composite sphere. If we can find the constant K_e such that the field outside the composite sphere remains unchanged, then there will be no change in the energy dissipation, and hence there is no change of the bulk modulus of the system due to the replacement. Repeating this process of replacement with composite spheres *ad infinitum* will result in a two-phase coated-spheres dispersion characterized by an effective bulk modulus K_e.

To find K_e, it will be sufficient to obtain the local elastostatic fields for a single d-dimensional composite sphere in an infinite matrix of bulk modulus K_e in which the strain is a uniform hydrostatic field, i.e., $\boldsymbol{\varepsilon}_0 = \varepsilon_0 \boldsymbol{I}/d$. Let $\boldsymbol{r}$ be the radial vector emanating from the sphere center (see Figure 16.8). The displacement field $\boldsymbol{u}$ must satisfy the equilibrium (or Navier) equations

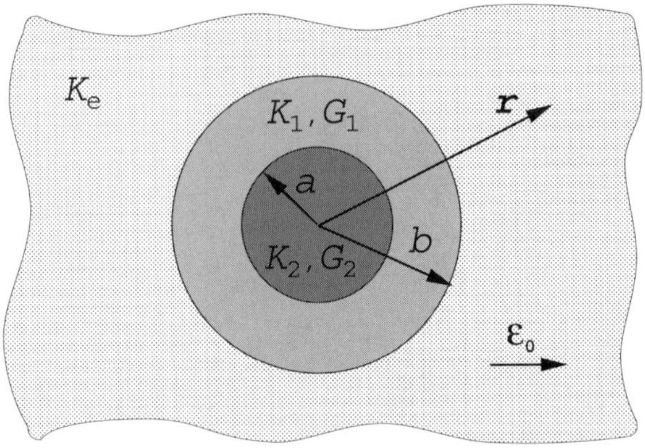

Figure 16.8 A single coated sphere for bulk modulus problem.

$$(\lambda_e + G_e)\nabla(\nabla \cdot \boldsymbol{u}) + G_e \Delta \boldsymbol{u} = 0, \qquad r \geq b, \tag{16.33}$$

$$(\lambda_1 + G_1)\nabla(\nabla \cdot \boldsymbol{u}) + G_1 \Delta \boldsymbol{u} = 0, \qquad a \leq r \leq b, \tag{16.34}$$

$$(\lambda_2 + G_2)\nabla(\nabla \cdot \boldsymbol{u}) + G_2 \Delta \boldsymbol{u} = 0, \qquad r \leq a, \tag{16.35}$$

subject to the continuity conditions (13.59) at $r = a$ and $r = b$, where we recall that $\lambda_i = K_i - 2G_i/d$ and $\lambda_e = K_e - 2G_e/d$. The Navier equations for $\boldsymbol{u}$ are arrived at by combining relations (13.50)–(13.52) and (13.89). The general form of the solution is

$$\boldsymbol{u} = \begin{cases} \varepsilon_0 \cdot \boldsymbol{r}, & r \geq b, \\ A\varepsilon_0 \cdot \boldsymbol{r} + B\dfrac{\varepsilon_0 \cdot \boldsymbol{n}}{r^{d-1}}, & a \leq r \leq b, \\ C\varepsilon_0 \cdot \boldsymbol{r}, & r \leq a, \end{cases} \tag{16.36}$$

where $r = |\boldsymbol{r}|$, $\boldsymbol{n} = \boldsymbol{r}/r$ is the radial unit vector, and A, B, C are unknown parameters. The corresponding strain field ε is given by

$$\varepsilon = \frac{1}{2}\left[\nabla \boldsymbol{u} + (\nabla \boldsymbol{u})^{\mathrm{T}}\right] = \begin{cases} \varepsilon_0, & r > b, \\ A\varepsilon_0 - B\left[\dfrac{d\boldsymbol{nn} - \boldsymbol{I}}{r^d}\right] \cdot \varepsilon_0, & a < r < b, \\ C\varepsilon_0, & r < a. \end{cases} \tag{16.37}$$

The three unknown parameters A, B, C are determined from the continuity of displacement at $r = a$ and $r = b$ and the continuity of the radial component of the stress $\boldsymbol{\tau} \cdot \boldsymbol{n}$, [cf. (13.59)], where

$$\boldsymbol{\tau} = \left(K - \frac{2G}{d}\right)(\mathrm{Tr}\ \varepsilon)\boldsymbol{I} + 2G\varepsilon.$$

Employing these conditions and that $\phi_2 = (a/b)^d$ is the inclusion volume fraction, we obtain that

$$A = \frac{K_2 + 2(d-1)G_1/d}{K_2 + 2(d-1)G_1/d - \phi_2(K_2 - K_1)}, \tag{16.38}$$

$$B = \frac{-a^d(K_2 - K_1)}{K_2 + 2(d-1)G_1/d - \phi_2(K_2 - K_1)}, \tag{16.39}$$

$$C = \frac{K_1 + 2(d-1)G_1/d}{K_2 + 2(d-1)G_1/d - \phi_2(K_2 - K_1)}. \tag{16.40}$$

Continuity of the radial component of the stress at $r = b$ implies that

$$K_e = AK_1 - \frac{BG_1 2(d-1)}{b^d d}.$$

The bulk modulus of the homogeneous body must therefore satisfy

$$K_e = \langle K \rangle - \frac{(K_2 - K_1)^2 \phi_1 \phi_2}{\langle \tilde{K} \rangle + \dfrac{2(d-1)}{d} G_1}, \tag{16.41}$$

where

$$\langle K \rangle = K_1\phi_1 + K_2\phi_2, \qquad \langle \tilde{K} \rangle = K_1\phi_2 + K_2\phi_1. \tag{16.42}$$

To summarize, if this replacement process is repeated *ad infinitum* until all of the original material is replaced by composite spheres (implying sizes down to the infinitesimally small), then a macroscopically isotropic and homogeneous material can be constructed whose effective bulk modulus is exactly given by (16.41).

As observed earlier, the included phase 2 is always disconnected (i.e., does not percolate) except in the trivial instance when $\phi_2 = 1$. In Chapter 21 we will see that formula (16.41) is identical to the well-known Hashin–Shtrikman lower bound if $G_2 \geq G_1$. If the phase labels are reversed in relation (16.41), then it becomes identical to the Hashin–Shtrikman upper bound if $G_2 \geq G_1$. The consequences of the realizability of the bounds will be discussed in Chapter 21.

It is important to note that the coated-spheres construction does not permit exact evaluation of the effective shear modulus. Hashin and Rosen (1964) showed that the arguments used to obtain the effective conductivity and effective bulk modulus for the coated-spheres model break down, since the shear field outside a composite sphere is not uniform (in contrast to the pressure field or electric field). Interestingly, Christensen and Lo (1979) showed that a variant of the coated-spheres model, a "three-phase model," does permit an exact solution of G_e.

16.2.2 Simple Laminates

We want to derive the effective stiffness tensor $\boldsymbol{C}_e$ for two-phase laminates consisting of isotropic phases. Although the determination of the elastic response is more involved algebraically than the corresponding electrical (thermal) response described earlier, it is still an elementary calculation, since the local fields are piecewise constant under uniform applied fields.

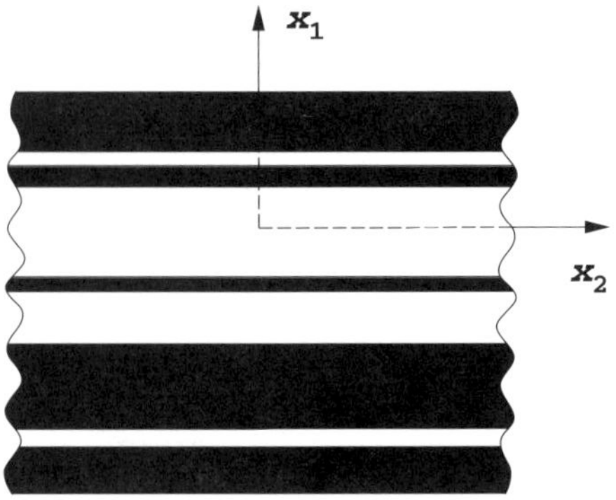

Figure 16.9 Portion of a two-dimensional laminate of rank one.

It is simpler and instructive to consider first the calculation of $\boldsymbol{C}_e$ for the two-dimensional laminate of rank 1 depicted in Figure 16.9. Let E_i and ν_i be the two-dimensional (planar) Young modulus and Poisson ratio of phase i, respectively. The local fields can depend at most upon the x_1 coordinate when uniform fields are imposed at the boundaries. The effective stiffness tensor for this laminate in plane elasticity is most conveniently represented by the 3×3 matrix

$$\boldsymbol{C}_e = \begin{bmatrix} (C_e)_{11} & (C_e)_{12} & 0 \\ (C_e)_{12} & (C_e)_{22} & 0 \\ 0 & 0 & (C_e)_{66} \end{bmatrix}, \tag{16.43}$$

where

$$(C_e)_{11} = \left\langle \frac{1-\nu^2}{E} \right\rangle^{-1}, \quad (C_e)_{12} = \langle \nu \rangle \left\langle \frac{1-\nu^2}{E} \right\rangle^{-1}, \tag{16.44}$$

$$(C_e)_{22} = \langle \nu^2 \rangle \left\langle \frac{1-\nu^2}{E} \right\rangle^{-1} + \langle E \rangle, \qquad (C_e)_{66} = \left\langle G^{-1} \right\rangle^{-1}, \tag{16.45}$$

and for any property c, $\langle c \rangle = c_1\phi_1 + c_2\phi_2$. Since the effective Young moduli in the x_1- and x_2-directions are given by

$$(E_e)_{11} = (C_e)_{11} - \frac{(C_e)_{12}^2}{(C_e)_{22}}, \qquad (E_e)_{22} = (C_e)_{22} - \frac{(C_e)_{12}^2}{(C_e)_{11}},$$

we have that

$$(E_e)_{11} = \left\langle \frac{1-\nu^2}{E} \right\rangle^{-1} \left[\frac{\langle \nu^2 \rangle - \langle \nu \rangle^2 + \langle E \rangle \left\langle \frac{1-\nu^2}{E} \right\rangle}{\langle \nu^2 \rangle + \langle E \rangle \left\langle \frac{1-\nu^2}{E} \right\rangle} \right], \tag{16.46}$$

$$(E_e)_{22} = \langle E \rangle + (\langle \nu^2 \rangle - \langle \nu \rangle^2) \left\langle \frac{1-\nu^2}{E} \right\rangle^{-1}. \tag{16.47}$$

Moreover, since the effective shear modulus $(G_e)_{12}$ is equal to $(C_e)_{66}$, we have that

$$(G_e)_{12} = \langle G^{-1} \rangle^{-1}. \tag{16.48}$$

We see that the effective Young modulus $(E_e)_{22}$ is simply given by the arithmetic average of the phase Young moduli when $\nu_1 = \nu_2$, and the effective shear modulus $(G_e)_{12}$ is given by the harmonic average of the phase shear moduli. Note that it is only when the phase Poisson ratios are equal to zero ($\nu_1 = \nu_2 = 0$), that the effective Young modulus in the x_1-direction, (16.46), reduces to the harmonic average of the Young moduli, i.e.,

$$(E_e)_{11} = \langle E^{-1} \rangle^{-1} . \tag{16.49}$$

We now derive result (16.43). Let us consider applying three independent uniform strain fields: a uniaxial strain ε^0_{11} in the x_1-direction, a uniaxial strain ε^0_{22} in the x_2-direction, and a uniform shear strain ε^0_{12}. The local fields must satisfy the equilibrium equations in each phase, i.e.,

$$\frac{\partial \tau_{11}}{\partial x_1} + \frac{\partial \tau_{12}}{\partial x_2} = 0, \tag{16.50}$$

$$\frac{\partial \tau_{12}}{\partial x_1} + \frac{\partial \tau_{22}}{\partial x_2} = 0, \tag{16.51}$$

and also obey Hooke's law in each phase:

$$\tau_{11}(x_1) = \frac{E(x_1)}{1-\nu^2(x_1)} \left[\varepsilon_{11}(x_1) + \nu(x_1)\varepsilon_{22}(x_1) \right], \tag{16.52}$$

$$\tau_{22}(x_1) = \frac{E(x_1)}{1-\nu^2(x_1)} \left[\nu(x_1)\varepsilon_{11}(x_1) + \varepsilon_{22}(x_1) \right], \tag{16.53}$$

$$\tau_{12}(x_1) = 2G(x_1)\varepsilon_{12}(x_{12}). \tag{16.54}$$

Since the local fields are independent of x_2, it follows from (16.50) and (16.51) that $\tau_{11}(x_1)$ and $\tau_{12}(x_1)$ are constants throughout the composite:

$$\tau_{11}(x_1) = \langle \tau_{11} \rangle, \qquad \tau_{12}(x_1) = \langle \tau_{12} \rangle.$$

Moreover, it follows that the local strain in the x_2-direction must be a constant throughout the composite:

$$\varepsilon_{22}(x_1) = \langle \varepsilon_{22} \rangle = \varepsilon^0_{22}.$$

The average of the strain component $\varepsilon_{11}(x_1)$ must satisfy the boundary condition

$$\langle \varepsilon_{11} \rangle = \varepsilon^0_{11}.$$

Therefore, from expression (16.52), we see that the 11-component of strain must be a piecewise constant function according to the relation

$$\varepsilon_{11}(x_1) = \frac{1-\nu^2(x_1)}{E(x_1)} \langle \tau_{11} \rangle - \nu(x_1) \langle \varepsilon_{22} \rangle. \tag{16.55}$$

Averaging (16.53) [after use of (16.55)], (16.54), and (16.55) yields the averaged constitutive relations

$$\begin{aligned}
\langle\tau_{11}\rangle &= \left\langle\frac{1-\nu^2}{E}\right\rangle^{-1}\langle\varepsilon_{11}\rangle + \langle\nu\rangle\left\langle\frac{1-\nu^2}{E}\right\rangle^{-1}\langle\varepsilon_{22}\rangle,\\
\langle\tau_{22}\rangle &= \langle\nu\rangle\left\langle\frac{1-\nu^2}{E}\right\rangle^{-1}\langle\varepsilon_{11}\rangle + \left(\langle\nu^2\rangle\left\langle\frac{1-\nu^2}{E}\right\rangle^{-1} + \langle E\rangle\right)\langle\varepsilon_{22}\rangle,\\
\langle\tau_{12}\rangle &= 2\langle G^{-1}\rangle^{-1}\langle\varepsilon_{12}\rangle.
\end{aligned}$$

Comparison of the coefficients of the average strain components with the matrix (16.43) yields the desired expressions (16.44) and (16.45).

It is important to note that expression (16.43) for the effective stiffness tensor $\boldsymbol{C}_e$ of a simple laminate can be rewritten in the form

$$(\boldsymbol{C}_e - \boldsymbol{C}_1)^{-1} = \frac{1}{\phi_2}(\boldsymbol{C}_2 - \boldsymbol{C}_1)^{-1} + \frac{2\phi_1}{\phi_2}\frac{1}{G_1}\boldsymbol{B}, \tag{16.56}$$

where, in 3×3 matrix form,

$$\boldsymbol{C}_i = \frac{2G_i}{1-\nu_i}\begin{bmatrix} 1 & \nu_i & 0\\ \nu_i & 1 & 0\\ 0 & 0 & 1-\nu_i\end{bmatrix}, \qquad \boldsymbol{B} = \begin{bmatrix} 1-\nu_1 & 0 & 0\\ 0 & 0 & 0\\ 0 & 0 & 1\end{bmatrix}. \tag{16.57}$$

The formula (16.56) is invaluable in obtaining corresponding expressions for the effective stiffness tensor of more general laminates, as will be discussed in the next section.

We can obtain the elastic response of the simple three-dimensional laminate shown in Figure 16.3, using the same procedure. The effective stiffness matrix of this transversely isotropic composite is given by

$$\boldsymbol{C}_e = \begin{bmatrix}
(C_e)_{11} & (C_e)_{12} & (C_e)_{12} & 0 & 0 & 0\\
(C_e)_{12} & (C_e)_{22} & (C_e)_{23} & 0 & 0 & 0\\
(C_e)_{12} & (C_e)_{23} & (C_e)_{22} & 0 & 0 & 0\\
0 & 0 & 0 & \frac{1}{2}[(C_e)_{22} - (C_e)_{23}] & 0 & 0\\
0 & 0 & 0 & 0 & (C_e)_{66} & 0\\
0 & 0 & 0 & 0 & 0 & (C_e)_{66}
\end{bmatrix}, \tag{16.58}$$

where

$$\begin{aligned}
(C_e)_{11} &= \langle(K+4G/3)^{-1}\rangle^{-1}, \quad (C_e)_{12} = \left\langle\frac{K-2G/3}{K+4G/3}\right\rangle(C_e)_{11}, \quad (C_e)_{66} = \langle G^{-1}\rangle^{-1},\\
(C_e)_{22} &= \left\langle\frac{K-2G/3}{K+4G/3}\right\rangle^2(C_e)_{11} + \langle K+4G/3\rangle - \left\langle\frac{(K-2G/3)^2}{K+4G/3}\right\rangle,\\
(C_e)_{23} &= \left\langle\frac{K-2G/3}{K+4G/3}\right\rangle^2(C_e)_{11} + \langle K-2G/3\rangle - \left\langle\frac{(K-2G/3)^2}{K+4G/3}\right\rangle.
\end{aligned}$$

This result was first given by Postma (1955).

The effective Young moduli along the coordinate axes are given by

$$(E_e)_{11} = (C_e)_{11} - \frac{2(C_e)_{12}^2}{(C_e)_{22} + (C_e)_{23}},$$

$$(E_e)_{22} = (E_e)_{33} = (C_e)_{22} - \frac{(C_e)_{12}^2[(C_e)_{23} - (C_e)_{22}] + (C_e)_{23}[(C_e)_{12}^2 - (C_e)_{11}(C_e)_{23}]}{(C_e)_{11}(C_e)_{22} - (C_e)_{12}^2}.$$

If Poisson's ratios of the phases are equal ($\nu_1 = \nu_2$), then the effective Young moduli in the transverse plane reduce to the arithmetic average:

$$(E_e)_{22} = (E_e)_{33} = \langle E \rangle.$$

However, Young's modulus along the axis of symmetry reduces to the harmonic average

$$(E_e)_{11} = \langle E^{-1} \rangle^{-1}$$

only when Poisson's ratios of both phases vanish ($\nu_1 = \nu_2 = 0$).

Remarks:

1. Under uniform applied fields, we see that the local fields in the laminate are piecewise constant and satisfy the continuity conditions (13.59) at the two-phase interface. This simple but important property of the local fields has crucial implications for constructing composites with desired effective stiffness tensors, as we will see in the next section.
2. We have seen that under certain situations for the simple laminate, the effective Young modulus in the plane of the slabs may reduce to the arithmetic average $\langle E \rangle$, and in the direction normal to the slabs, it may reduce to the harmonic average $\langle E^{-1} \rangle^{-1}$. Generally, for arbitrary phase Poisson's ratios, the effective Young moduli in these directions do not differ appreciably from $\langle E \rangle$ and $\langle E^{-1} \rangle^{-1}$, respectively.
3. There has been a tendency in the literature to apply the aforementioned simple mixture rules to macroscopically isotropic media. However, such estimates are almost always poor, except when the phase moduli are nearly the same. Generally, the arithmetic average $\langle E \rangle$ overestimates the effective Young modulus E_e of an isotropic composite, especially when the phase contrast ratio is large. The reason for this is that the estimate $\langle E \rangle$ corresponds, to a very good approximation, to the effective Young modulus of the simple laminate in the direction in which the slabs are connected from one end of the composite to the other. Indeed, for $\nu_1 = \nu_2$, $\langle E \rangle$ will be shown in Chapter 21 to be an upper bound on the effective Young modulus E_e of isotropic composites. On the other hand, the harmonic average $\langle E^{-1} \rangle^{-1}$ generally underestimates E_e, especially when the phase moduli differ widely. The estimate $\langle E^{-1} \rangle^{-1}$ corresponds, to a very good approximation, to the effective Young modulus of the simple laminate in the direction normal to the slabs (roughly speaking, the most "disconnected" direction). The average $\langle E^{-1} \rangle^{-1}$ will be shown in Chapter 21 to be a lower bound on E_e for isotropic composites, irrespective of the phase Poisson ratios.

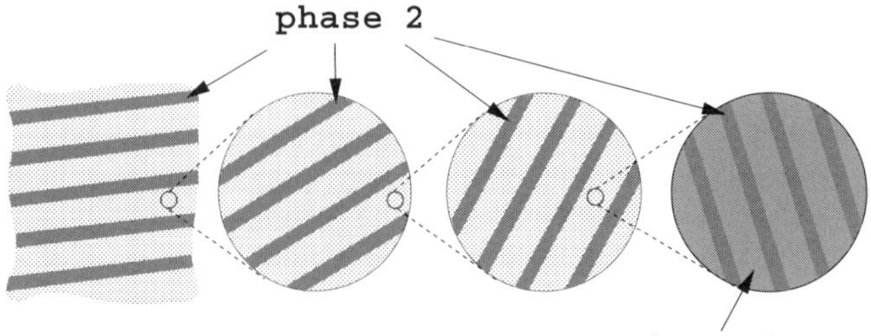

Figure 16.10 An example of a hierarchical laminate composite of infinite rank that achieves the Hashin–Shtrikman bounds (21.73) for the effective shear modulus of two-phase isotropic composites.

16.2.3 Higher-Order Laminates and Attainability

We recall from Section 16.1.3 that hierarchical nth-rank laminates with an infinite separation of length scales serve as very useful theoretical tools, since their effective conductivities can be calculated exactly (because the local fields are piecewise constant) and also because they can be used to attain the effective conductivities of a wide class of composites. Precisely the same is true for the effective stiffness tensor.

For years, an outstanding problem in composites theory had been whether the Hashin–Shtrikman bounds on the effective shear modulus of two-phase isotropic composites (derived in Section 21.2.2) could be attained by any microstructure. About fifteen years ago, a number of different investigators showed that certain types of laminates attained these bounds. Milton (1986) demonstrated that hierarchical laminates of infinite rank (with random orientations at different levels of the hierarchy) achieve the Hashin–Shtrikman shear modulus bounds (see Figure 16.10). Norris (1985) and Lurie and Cherkaev (1985) found that these bounds were attained by structures that are essentially similar to the one depicted in Figure 16.10. Francfort and Murat (1986) showed that finite-rank laminates can achieve the Hashin–Shtrikman shear modulus bounds. For $d = 3$, certain sixth-rank laminates attain the bounds. On the other hand, the two-dimensional versions of these isotropic bounds are achieved by third-rank laminates.

Francfort and Murat (1986) gave a general formula for the effective stiffness tensor $\boldsymbol{C}_e$ for arbitrary rank N and layering orientations $\boldsymbol{n}^{(1)}, \boldsymbol{n}^{(2)}, \ldots, \boldsymbol{n}^{(N)}$. For $d = 2$, the formula can be expressed as follows:

$$(\boldsymbol{C}_e - \boldsymbol{C}_1)^{-1} = \frac{1}{\phi_2}(\boldsymbol{C}_2 - \boldsymbol{C}_1)^{-1} + \frac{2\phi_1}{\phi_2}\frac{1}{G_1}\sum_{i=1}^{N} \alpha_i \boldsymbol{B}^{(i)}, \tag{16.59}$$

where $\boldsymbol{B}^{(i)}$ is the fourth-order tensor for the ith stage given by

$$\boldsymbol{B}^{(i)} = (1+\nu_1)\boldsymbol{n}^{(i)}\boldsymbol{n}^{(i)}\boldsymbol{n}^{(i)}\boldsymbol{n}^{(i)} + \boldsymbol{n}^{(i)}\boldsymbol{m}^{(i)}[\boldsymbol{n}^{(i)}\boldsymbol{m}^{(i)} + \boldsymbol{m}^{(i)}\boldsymbol{n}^{(i)}] + \boldsymbol{m}^{(i)}\boldsymbol{n}^{(i)}[\boldsymbol{m}^{(i)}\boldsymbol{n}^{(i)} + \boldsymbol{n}^{(i)}\boldsymbol{m}^{(i)}], \tag{16.60}$$

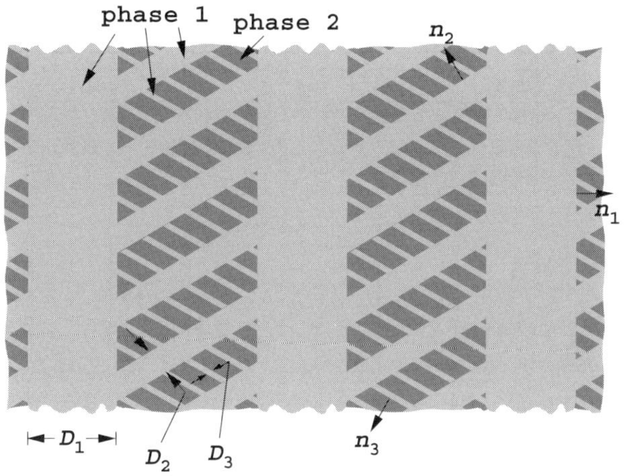

Figure 16.11 An example of a two-dimensional third-rank laminate whose effective bulk and shear moduli are exactly given by (16.62) and (16.63) in the limit $D_1/D_2 \to \infty$ and $D_2/D_3 \to \infty$. Here $\boldsymbol{n}^{(1)} \cdot \boldsymbol{n}^{(2)} = \boldsymbol{n}^{(2)} \cdot \boldsymbol{n}^{(3)} = -0.5$.

$\boldsymbol{m}^{(i)}$ is the 90° rotation of the vector $\boldsymbol{n}^{(i)}$, and α_i are known parameters that generally depend on the volume fractions of stage i and satisfy the condition

$$\sum_{i=1}^{N} \alpha_i = 1, \qquad \alpha_i \geq 0. \tag{16.61}$$

Formula (16.59) is obtained by making repeated use of relation (16.56) for the laminate of rank 1. Upon representing the fourth-order tensors in the 3×3 matrix form of (16.43), one can see that (16.59) reduces to (16.56) when $N = 1$ and the normal $\boldsymbol{n}^{(1)}$ is oriented along the x_1 direction. For general space dimension d, the laminate formula corresponding to (16.59) is more complicated but is still explicitly known (Francfort and Murat 1986).

The two-dimensional Hashin–Shtrikman expressions are achieved for the third-rank laminates in which the layering direction at each stage differs by $2\pi/3$ ($\boldsymbol{n}^{(1)} \cdot \boldsymbol{n}^{(2)} = \boldsymbol{n}^{(2)} \cdot \boldsymbol{n}^{(3)} = -0.5$), as illustrated in Figure 16.11. By applying the laminate formula (16.59) for the effective stiffness tensor for such a third-rank laminate and requiring isotropy, one can show that the isotropic effective planar bulk modulus k_e and shear modulus G_e are given respectively by

$$k_e = \langle k \rangle - \frac{(k_2 - k_1)^2 \phi_1 \phi_2}{\langle \tilde{k} \rangle + G_1}, \tag{16.62}$$

$$G_e = \langle G \rangle - \frac{(G_2 - G_1)^2 \phi_1 \phi_2}{\langle \tilde{G} \rangle + \dfrac{k_1 G_1}{k_1 + 2G_1}}, \tag{16.63}$$

where for any property c,

$$\langle \tilde{c} \rangle = c_1 \phi_2 + c_2 \phi_1.$$

Table 16.1 Coefficients in the low-concentration asymptotic expressions.

Lattice	a_1	a_2	a_3	a_4
SC	−1.396	1.714	2.889	−3.077
BCC	0.430	−0.520	−6.163	3.373
FCC	0.388	−0.411	−5.928	2.750

Notice that result (16.62) for the effective bulk modulus is identical to the exact expression for the coated-cylinder model given by (16.41) with $d = 2$. Unlike the effective shear modulus expression (16.63), the effective bulk modulus expression (16.62) is also exact for a *square symmetric* second-rank laminate, an example of which is depicted in Figure 16.6.

16.2.4 Periodic Arrays of Inclusions

Nunan and Keller (1984) studied the effective moduli of the three cubic arrays of *rigid* spheres in a matrix (phase 1). Such a composite has cubic elastic symmetry, and therefore the effective stiffness tensor is specified by three elastic moduli (see Section 13.3.2) according to the relation

$$(C_e)_{ijkl} = (\lambda_1 + G_1\gamma)\delta_{ij}\delta_{kl} + G_1(1+\beta)(\delta_{ik}\delta_{jl} + \delta_{il}\delta_{jk}) + 2G_1(\alpha - \beta)\delta_{ijkl}. \tag{16.64}$$

This expression corrects a misprint contained in Nunan and Keller (1984). Here $\lambda_1 = K_1 - 2G_1/3$ is the matrix Lamé constant, and δ_{ijkl} is equal to one if all the subscripts are equal and zero otherwise. They found low-density asymptotic relations for the parameters α, β, and γ as follows:

$$\alpha = \frac{15}{2}\frac{(1-\nu_1)\phi_2}{4-5\nu_1}\left[1 - \left(1 - \frac{3a_1}{4-5\nu_1}\right)\phi_2 + \frac{3a_2}{4-5\nu_1}\phi_2^{5/3} + \mathcal{O}\left(\phi_2^{7/3}\right)\right]^{-1},$$
$$\beta = \frac{15}{2}\frac{(1-\nu_1)\phi_2}{4-5\nu_1}\left[1 - \left(1 + \frac{2a_1}{4-5\nu_1}\right)\phi_2 - \frac{2a_2}{4-5\nu_1}\phi_2^{5/3} + \mathcal{O}\left(\phi_2^{7/3}\right)\right]^{-1},$$
$$\gamma = \frac{3(1-\nu_1)\phi_2}{(1-2\nu_1)(4-5\nu_1)}\left[1 - \left(\frac{3\nu_1 + (1-2\nu_1)a_3}{4-5\nu_1}\right)\phi_2 + \frac{(1-2\nu_1)a_4}{4-5\nu_1}\phi_2^{5/3} + \mathcal{O}\left(\phi_2^{7/3}\right)\right]^{-1}.$$

The coefficients a_i in these formulas depend upon the lattice geometry, but not the matrix Poisson ratio ν_1, and are summarized in Table 16.1.

Consider the opposite asymptotic regime of nearly close-packed spheres. The elastic interaction between spheres near the maximum close-packing density is concentrated in the regions near the points of contact, as observed by Flaherty and Keller (1973) and Nunan and Keller (1984). A local analysis of the region between two nearly touching spheres, in conjunction with knowledge of nearest-neighbor locations, produces the dominant contribution of the interaction for the entire sphere.

Table 16.2 Coefficients in the high-concentration asymptotic expressions.

	SC	BCC	FCC
B_α	$\frac{\pi}{4}[1+(1-2\nu_1)^{-1}]$	$\frac{\pi\sqrt{3}}{4}$	$\frac{\pi\sqrt{2}}{8}[4+(1-2\nu_1)^{-1}]$
B_β	$\frac{\pi}{4}$	$\frac{\pi\sqrt{3}}{6}[\frac{3}{2}+(1-2\nu_1)^{-1}]$	$\frac{\pi\sqrt{2}}{4}[2+(1-2\nu_1)^{-1}]$
B_γ	0	$\frac{\pi\sqrt{3}}{6}(1-2\nu_1)^{-1}$	$\frac{\pi\sqrt{2}}{4}(1-2\nu_1)^{-1}$

For cubic lattices, the high-concentration asymptotic formulas for α, β, and γ have been obtained by Nunan and Keller. For α, they obtained

$$\alpha = -B_\alpha(\nu_1)\ln\left[1-\left(\frac{\phi_2}{\phi_{2c}}\right)^{1/3}\right]+\mathcal{O}(1), \tag{16.65}$$

where ϕ_{2c} is the close-packing sphere volume fraction ($\phi_{2c} = \pi/6, \pi\sqrt{3}/8$, and $\pi\sqrt{2}/6$ for simple cubic, body-centered cubic, and face-centered cubic lattices, respectively). Notice that the coefficient $B_\alpha(\nu_1)$ generally depends on the matrix Poisson ratio ν_1. The expressions for β and γ have the same form but $B_\alpha(\nu_1)$ is replaced by $B_\beta(\nu_1)$ and $B_\gamma(\nu_1)$. The coefficients B_α, B_β, and B_γ for the three cubic lattices are given in Table 16.2.

In two dimensions, asymptotic expressions for the effective elastic moduli of periodic lattices of circular holes and of perfectly rigid, circular inclusions that apply near the percolation thresholds have been developed. Let k_1, G_1, and $\nu_1 = (k_1-G_1)/(k_1+G_1)$ be the plane bulk modulus, plane shear modulus, and plane Poisson's ratio of the matrix phase, respectively. Day, Snyder, Garboczi and Thorpe (1992) showed that the elastic moduli of a composite with circular holes arranged on the sites of a honeycomb, or hexagonal, lattice near the threshold $\phi_{2c} = \pi/(3\sqrt{3})$ are given by

$$\frac{k_e}{k_1} = \frac{\sqrt{3}(1-\nu_1)}{\pi}\left(\frac{\phi_{2c}-\phi_2}{\phi_{2c}}\right)^{1/2}, \quad \frac{G_e}{G_1} = \frac{\sqrt{3}(1+\nu_1)}{2\pi}\left(\frac{\phi_{2c}-\phi_2}{\phi_{2c}}\right)^{1/2}. \tag{16.66}$$

For composites with perfectly rigid circular inclusions arranged on the sites of a hexagonal lattice, Chen, Thorpe and Davis (1995) have found that

$$\frac{k_e}{k_1} = \frac{\pi}{\sqrt{3}(1+\nu_1)}\left(\frac{\phi_{2c}-\phi_2}{\phi_{2c}}\right)^{-1/2}, \quad \frac{G_e}{G_1} = \frac{2\pi}{\sqrt{3}(3-\nu_1)}\left(\frac{\phi_{2c}-\phi_2}{\phi_{2c}}\right)^{-1/2}. \tag{16.67}$$

For the case of circular holes on the sites of a triangular lattice, the moduli near the threshold $\phi_{2c} = \pi/(2\sqrt{3})$ (Day et al. 1992) are given by

$$\frac{k_e}{k_1} = \frac{1-\nu_1}{\pi\sqrt{3}}\left(\frac{\phi_{2c}-\phi_2}{\phi_{2c}}\right)^{1/2}, \quad \frac{G_e}{G_1} = \frac{4(1+\nu_1)}{3\pi\sqrt{3}}\left(\frac{\phi_{2c}-\phi_2}{\phi_{2c}}\right)^{3/2}. \tag{16.68}$$

For the corresponding composite with perfectly rigid circular inclusions (Chen et al. 1995), the moduli are described by

$$\frac{k_e}{k_1} = \frac{\pi\sqrt{3}}{1+\nu_1}\left(\frac{\phi_{2c}-\phi_2}{\phi_{2c}}\right)^{-1/2}, \quad \frac{G_e}{G_1} = \frac{\sqrt{3}(3-\nu_1)\pi}{4(1-\nu_1)}\left(\frac{\phi_{2c}-\phi_2}{\phi_{2c}}\right)^{-1/2}. \tag{16.69}$$

Gibiansky and Torquato (1996a) showed that knowledge of the results (16.66) for circular holes can lead to the results (16.67) for rigid inclusions (and vica versa) upon use of the phase-interchange Theorem 15.5. Similarly, knowledge of the bulk-modulus result in (16.68) leads to the bulk-modulus result in (16.69) (and vice versa) upon use of Theorem 15.5.

16.2.5 Low-Density Cellular Solids

Here we present results for the effective elastic moduli of cellular solids that correspond to the conductivity results given in Section 16.1.6. Phases 1 and 2 are again taken to be the solid and pore (void) phases, respectively, and we consider the limit $\phi_1 \ll 1$, i.e., the solid volume fraction very small.

The effective elastic moduli of two-dimensional cellular solids are easily found by considering the deformation mechanisms in individual cell-wall segments and using simple *beam* theory. For two-dimensional cellular solids consisting of *regular triangular cells*, the effective elastic moduli (Gibson and Ashby 1997) are given by

$$\frac{k_e}{E_1} = \frac{\phi_1}{4}, \qquad \frac{G_e}{E_1} = \frac{\phi_1}{8}, \qquad \frac{E_e}{E_1} = \frac{\phi_1}{3}, \qquad \nu_e = \frac{1}{3}. \tag{16.70}$$

Christensen (2000) observed that these results apply also to the Kagomé network (see Figure 9.4). In the case of *regular hexagonal cells*, the corresponding results (Christensen 1995) are given by

$$\frac{k_e}{E_1} = \frac{\phi_1}{4}, \qquad \frac{G_e}{E_1} = \frac{3\phi_1^3}{8}, \qquad \frac{E_e}{E_1} = \frac{3\phi_1^3}{2}, \qquad \nu_e = 1. \tag{16.71}$$

Effective moduli that are proportional to ϕ_1 are due to the axial extension or contraction of the cell walls, whereas the effective moduli that are proportional to ϕ_1^3 (substantially more compliant behavior) are due to bending of the cell walls. The reader is referred to the work of Christensen (2000) for results on the moduli of other two-dimensional structures.

It is of interest to note that the Hashin–Shtrikman upper bounds on the effective elastic moduli of two-dimensional porous solids in which $\phi_1 \ll 1$ are as follows: $k_e/E_1 \leq \phi_1/4$, $G_e/E_1 \leq \phi_1/8$, and $E_e/E_1 \leq \phi_1/3$. Thus, we see that two-dimensional cellular solids with triangular or Kagomé cells offer *optimal performance*; i.e., they are the stiffest structures under both bulk and shear deformation. Indeed, civil engineering structures, such as bridges and towers, are often based on triangular cell structures.

The study of the mechanical behavior of three-dimensional cellular solids is a rich and challenging area of research. In contrast to the two-dimensional case, one can have open-cell or closed-cell structures in three dimensions. This interesting subject is amply reviewed by Gibson and Ashby (1997), Kraynik, Nielsen, Reinelt and Warren (1999), Christensen (2000), and Weaire and Hutzler (2000).

16.2.6 Equal Phase Shear Moduli

Consider a macroscopically isotropic two-phase composite of arbitrary microstructure in which the phase shear moduli are equal ($G_1 = G_2 = G$). Hill (1963) showed that in three dimensions there is an exact solution for the effective elastic moduli. Specifically, he found that the effective shear modulus is trivially $G_e = G$ and the effective bulk modulus is given exactly by formula (16.41) with $d = 3$. Thus, formula (16.41) represents the d-dimensional generalization of Hill's result for equal shear moduli.

16.2.7 Sheets with Holes

Consider a two-dimensional sheet containing a statistically isotropic distribution of holes. Denote by E_1 and ν_1 the Young modulus and Poisson ratio, respectively, of the solid material phase. Cherkaev, Lurie and Milton (1992) have rigorously shown that the dimensionless effective Young modulus E_e/E_1 is independent of the Poisson ratio of the sheet ν_1, thus explaining the numerical findings of Day et al. (1992). (This is actually a consequence of a more general principle concerning the invariance of the stress field under a certain shift of the compliance tensor.) We note that this result does not generalize to porous materials in three dimensions but is approximately true (Christensen 1993).

16.2.8 Dispersions of Particles in a Liquid

It is well known that the effective bulk modulus of a dispersion of isotropic elastic particles (phase 2) with moduli K_2 and G_2 in a liquid (phase 1) with moduli $K_1 \neq 0$ and $G_1 = 0$ subjected to hydrostatic compression is given by

$$\frac{1}{K_e} = \left\langle \frac{1}{K} \right\rangle. \tag{16.72}$$

This follows from the fact that local stress is uniform everywhere [i.e., $\boldsymbol{\tau}(\boldsymbol{x}) = -p\boldsymbol{I}$, where p is the pressure], and as a result, the trace of the local strain is piecewise uniform [i.e., Tr $\varepsilon(\boldsymbol{x}) = -p/K_1$ and Tr $\varepsilon(\boldsymbol{x}) = -p/K_2$ in the liquid and particles, respectively]. (This assumes that the particle phase does not percolate.) Therefore, the trace of the average strain is given by Tr $\langle\varepsilon\rangle = -p(\phi_1/K_1 + \phi_2/K_2)$. Since the effective stiffness tensor is $\boldsymbol{C}_e = dK_e\boldsymbol{\Lambda}_h$ (the dispersion cannot resist shear), then the averaged Hooke's law (13.53) yields (16.72). Formula (16.72) also applies to dispersions with $M-1$ different particles ($M \geq 2$) but where $\langle 1/K\rangle = \sum_{j=1}^{M} \phi_j/K_j$.

16.2.9 Cavities (Bubbles) in an Incompressible Matrix (Liquid)

Interestingly, the problem of determining the effective Lamé constant $\lambda_e = K_e - 2G_e/d$ of a composite consisting of spherical cavities in an incompressible matrix ($K_1/G_1 = \infty$) of shear modulus G_1 is mathematically equivalent to finding the effective expansion viscosity of an incompressible liquid of shear viscosity μ_1 containing spherical air bubbles. The analogues of the shear modulus, bulk modulus, and Lamé constant in the liquid problem are the shear viscosity, bulk viscosity, and expansion viscosity, respectively. Thus, interpreted in this fashion, the quantities K_e/G_1, G_e/G_1, and λ_e/G_1 are also the dimensionless effective bulk, shear, and expansion viscosities of a liquid containing air bubbles. Neglecting interactions between three-dimensional spherical bubbles ($\phi_2 \ll 1$), Taylor (1954) found the effective expansion viscosity. In d dimensions, his result is easily generalized; it is given by

$$\frac{\lambda_e}{G_1} = \frac{2(d-1)}{d\phi_2}. \tag{16.73}$$

16.2.10 Field Fluctuations

In Section 16.1.7 we discussed the importance of quantifying the fluctuations of the local fields and, in particular, presented exact results for the scalar second moment of the intensity field $\boldsymbol{E}$ within phase i for a macroscopically isotropic composite with M isotropic phases. Bobeth and Diener (1986) found analogous results for the scalar second moments of the strain field in three dimensions. These moments are simply related to the partial derivative of the effective elastic moduli with respect to the elastic moduli of phase i. Here we generalize and state their results in d dimensions. When the average strain field is hydrostatic, i.e., $\langle \varepsilon \rangle = (\text{Tr}\,\langle \varepsilon \rangle / d)\boldsymbol{I}$, then

$$\frac{\partial K_e}{\partial K_i} = \frac{\phi_i}{d} \frac{\langle \mathcal{I}^{(i)}(\boldsymbol{x})\text{Tr}[\varepsilon(\boldsymbol{x})]^2 \rangle}{\langle \varepsilon \rangle : \langle \varepsilon \rangle}, \qquad \frac{\partial K_e}{\partial G_i} = \frac{2\phi_i}{d} \frac{\langle \mathcal{I}^{(i)}(\boldsymbol{x})\varepsilon(\boldsymbol{x}) : \varepsilon(\boldsymbol{x}) \rangle}{\langle \varepsilon \rangle : \langle \varepsilon \rangle}, \tag{16.74}$$

where $\mathcal{I}^{(i)}(\boldsymbol{x})$ is the indicator function for phase i. When the average strain field is deviatoric, i.e., $\text{Tr}\,\langle \varepsilon \rangle = 0$, then

$$\frac{\partial G_e}{\partial K_i} = \frac{\phi_i}{2} \frac{\langle \mathcal{I}^{(i)}(\boldsymbol{x})\text{Tr}[\varepsilon(\boldsymbol{x})]^2 \rangle}{\langle \varepsilon \rangle : \langle \varepsilon \rangle}, \qquad \frac{\partial G_e}{\partial G_i} = \phi_i \frac{\langle \mathcal{I}^{(i)}(\boldsymbol{x})\varepsilon(\boldsymbol{x}) : \varepsilon(\boldsymbol{x}) \rangle}{\langle \varepsilon \rangle : \langle \varepsilon \rangle}. \tag{16.75}$$

The above relations are easily obtained by taking the first variation of the energy representation (14.82) and using the isotropic form of the minimum potential principle given by (14.95) (see Section 16.1.7 for the details in the similar conductivity problem). The scalar second moments of the strains can be found from relations for the effective moduli in terms of the phase moduli (see this chapter and Chapter 20 for exact relations, Chapters 18 and 20 for approximate relations, and Chapter 21 for bounds).

16.2.11 Link to Two-Dimensional Conductivity

Consider transversely isotropic fiber-reinforced composites whose phase boundaries are cylindrical surfaces of arbitrary shape with generators parallel to one axis (see Figure 13.6). Hashin (1970) has observed that the governing equation for conduction in the transverse plane (Laplace's equation) is identical to the governing equation for out-of-plane shear. Thus, the determination of the effective *transverse* conductivity σ_e is mathematically identical to the determination of the effective *longitudinal* (*axial*) shear modulus μ_e.

16.2.12 Link to Thermoelastic Constants

A less obvious *cross-property* relation is one due to Levin (1967) that links the effective thermal expansion coefficient α_e to the effective bulk modulus K_e. The thermal expansion of a material is a measure of the volumetric strain due to a change in temperature under traction-free conditions. Levin showed that for a macroscopically isotropic two-phase composite,

$$\alpha_e = \langle \alpha \rangle + \frac{\alpha_2 - \alpha_1}{1/K_2 - 1/K_1} \left(\frac{1}{K_e} - \left\langle \frac{1}{K} \right\rangle \right), \tag{16.76}$$

where $\langle \alpha \rangle = \alpha_1 \phi_1 + \alpha_2 \phi_2$ and α_i is the thermal expansion coefficient of phase i. Thus, given the effective bulk modulus and phase properties of the composite, one can compute the effective thermal expansion via (16.76). This formula shows how α_e deviates from the simple mixture rule $\alpha_e = \langle \alpha \rangle$ as a result of elastic interactions embodied in the factor $K_e^{-1} - \langle K^{-1} \rangle$, which is generally negative because of the inequality $K_e \geq \langle K^{-1} \rangle^{-1}$ (see Section 21.2). It is clear that when K_e is not known, one can use the rigorous bounds on K_e given in Section 21.2 to find corresponding bounds on the effective thermal expansion α_e using (16.76).

Levin's formula (16.76) has been generalized to treat macroscopically anisotropic two-phase media (Rosen and Hashin 1970) (in which case the effective thermal expansion tensor is related to the effective compliance tensor), polycrystals (Schulgasser 1987, Benveniste 1996), and elastic-plastic two-phase composites (Dvorak 1986), among other instances.

There is a variety of ways of deriving (16.76). One elegant way of deriving it is to utilize the notion of uniform fields in heterogeneous media (Dvorak 1990). Although a full treatment of uniform fields is beyond the scope of this book, it is useful to describe the salient idea behind this technique and the types of problems that can be attacked with it. It is not fully appreciated that uniform strain fields may exist in multiphase heterogeneous media of any phase geometry and elastic symmetry (Dvorak and Benveniste 1992). In uniformly strained two-phase materials, for example, such fields result from a linear combination of the actual mechanical strains and so-called "eigenstrains" (e.g., strain due to a temperature change). Since the strain in the composite will have the same uniform value, the resulting effective properties of the composite must be connected to each other. For example, for the isotropic two-phase thermal expansion problem, a direct consequence is Levin's formula (16.76). The method of uniform fields can be used to establish that only three of the five effective elastic moduli of a transversely isotropic two-phase fiber-reinforced material are independent (Hill 1964); see Section 13.3.4. It has also been employed by Benveniste and Dvorak (1992) to establish relations among the effective coefficients of piezoelectric composites. Recently, Grabovsky, Milton and Sage (2000) have developed a systematic procedure to find exact *microstructure-independent* relations for the effective properties such as Levin-type formulas, as well as others.

For composites with more than two phases, the effective compliance tensor generally does not determine the effective thermal expansion coefficients. Schapery (1968) and Rosen and Hashin (1970) have used variational principles to find the simplest Reuss–Voigt-type bounds on the thermal expansion coefficients of such general multiphase composites. Gibiansky and Torquato (1997) used the translation method to derive tight Hashin–Shtrikman-type bounds on the effective thermal expansion coefficient of isotropic multiphase composites as well as polycrystals. Application of these bounds for the case of three-phase composites reveals that one may achieve exotic thermal expansion behavior (not possible with just two phases), such as a negative effective thermal expansion with two positive thermal expansion phases and one void phase. Indeed, composite structures that possess such exotic behavior have been discovered (Sigmund and Torquato 1996).

Rosen and Hashin (1970) have shown that the effective specific heat at constant pressure $(c_p)_e$ of a two-phase isotropic composite is also determined by the effective bulk modulus K_e according to the relation

$$(c_p)_e = \langle c_p \rangle + 9T\left(\frac{\alpha_2 - \alpha_1}{1/K_2 - 1/K_1}\right)^2 \left(\frac{1}{K_e} - \left\langle \frac{1}{K} \right\rangle\right), \tag{16.77}$$

where T is absolute temperature. It follows from thermodynamics that the effective specific heat at constant volume $(c_v)_e$ is given by

$$(c_v)_e = (c_p)_e - 3K_e\alpha_e^2 T. \tag{16.78}$$

We also see that $(c_p)_e \leq \langle c_p \rangle$, since $K_e \geq \langle K^{-1} \rangle^{-1}$.

16.3 Trapping Constant

Here we describe exact results for the mean survival time τ for a few special cases. In particular, we obtain τ when transport occurs inside certain simple domains such as layered systems and cylindrical tubes or, more generally, hyperspheres. We also report asymptotic results for τ for transport exterior to periodic arrays of spherical and cylindrical traps.

16.3.1 Diffusion Inside Hyperspheres

Consider the steady-state problem of diffusion and reaction inside a hyperspherical pore of dimension d and radius a with a constant surface reaction rate κ. In this case, the diffusion equation and boundary conditions [cf. (13.154) and (13.155)] are

$$\frac{1}{r^{d-1}}\frac{\partial}{\partial r}\left(r^{d-1}\frac{\partial u}{\partial r}\right) = -1, \qquad r < a, \tag{16.79}$$

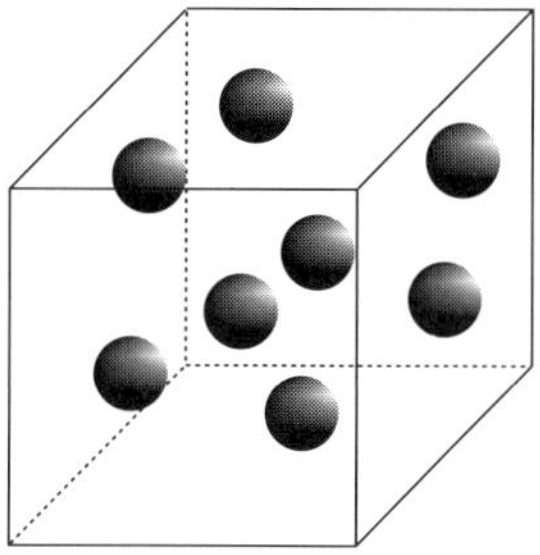

Figure 16.12 Diffusion inside arrays of spheres.

$$\mathcal{D}\frac{\partial u}{\partial r} + \kappa u = 0, \qquad r = a, \tag{16.80}$$

$$u < \infty, \qquad r = 0, \tag{16.81}$$

where u is the scaled concentration field. The cases $d = 1, 2,$ and 3 correspond to transport within a slab, cylinder, and sphere, respectively.

The solution of this problem is

$$u = \frac{1}{2d}(a^2 - r^2) + \frac{1}{d}\frac{a\mathcal{D}}{\kappa}. \tag{16.82}$$

Since the mean survival time is given by $\langle u \rangle / (\mathcal{D}\phi_1)$, we have for a single hypersphere that

$$\tau = \frac{\int_0^a u(r) r^{d-1} dr}{\mathcal{D}\int_0^a r^{d-1} dr} = \frac{1}{d(d+2)}\frac{a^2}{\mathcal{D}} + \frac{1}{d}\frac{a}{\kappa}. \tag{16.83}$$

This result also applies to a random array of such hyperspheres (Figure 16.12), since τ will be the same in each sphere. For diffusion-controlled reactions ($\kappa = \infty$), this expression gives

$$\tau = \frac{1}{d(d+2)}\frac{a^2}{\mathcal{D}}. \tag{16.84}$$

Unlike the mean survival time, the trapping constant $\gamma = (\mathcal{D}\phi_1\tau)^{-1}$ will depend on the porosity (volume fraction of the trap-free region). In particular, in the diffusion-controlled limit, we have that

$$\gamma = \frac{d(d+2)}{\phi_1 a^2}. \tag{16.85}$$

The determination of the trapping constant for diffusion outside of dilute arrays of traps is treated in Chapter 17.

16.3.2 Periodic Arrays of Traps

Felderhof (1985) has evaluated the "trapping rate," proportional to the trapping constant γ (Section 13.4), for diffusion exterior to cubic lattices of spherical traps of radius

R in the diffusion-controlled limit ($\kappa = \infty$). He found analytical expressions for this quantity in powers of the trap volume fraction ϕ_2 that are valid for arbitrary ϕ_2. In terms of γ, the first few terms of this expansion are

$$\frac{\gamma}{\gamma_s} = \frac{1}{1 - m_0\phi_2^{1/3} + \phi_2 - \frac{\phi_2^2}{5} + \mathcal{O}\left(\phi_2^{10/3}\right)}, \tag{16.86}$$

where $\gamma_s = 3\phi_2/R^2$ is the Smoluchowski trapping constant in the infinitely dilute limit (see the discussion in Chapter 19), and for the three cubic lattices, the numerical coefficient m_0 is given by

$$\begin{aligned} m_0 &= 1.760119, && \text{(SC)},\\ m_0 &= 1.791860, && \text{(BCC)},\\ m_0 &= 1.791753, && \text{(FCC)}. \end{aligned} \tag{16.87}$$

The corresponding expansion for the mean survival time $\tau = (\mathcal{D}\phi_1\gamma)^{-1}$ is

$$\frac{\tau}{\tau_s} = 1 - m_0\phi_2^{1/3} + 2\phi_2 - m_0\phi_2^{4/3} + \frac{9}{5}\phi_2^2 + \mathcal{O}\left(\phi_2^{7/3}\right), \tag{16.88}$$

where $\tau_s = R^2/(3\mathcal{D}\phi_2)$.

Low-concentration asymptotic expressions have also been obtained for two-dimensional lattices of circular traps of radius R by Torney and Goldstein (1987). For the square array, they found that

$$\frac{\gamma}{\gamma_0} = -\frac{1}{\ln\phi_2}\left[1 - \frac{1.4763}{\ln\phi_2} + \left(\frac{1.4763}{\ln\phi_2}\right)^2 + \mathcal{O}([\ln\phi_2]^{-3})\right], \tag{16.89}$$

where $\gamma_0 = 4\phi_2/R^2$. The corresponding expression for survival time is

$$\frac{\tau}{\tau_0} = -\ln\phi_2 - 1.4763 - \mathcal{O}([\ln\phi_2]^{-2}), \tag{16.90}$$

where $\tau_0 = R^2/(4\mathcal{D}\phi_2)$.

16.4 Fluid Permeability

In this section we describe exact results for the fluid permeability k for a few special cases. Specifically, we derive k when transport occurs inside certain simple domains such as layered media and cylindrical tubes. We also report asymptotic results for k when transport occurs exterior to periodic arrays of spheres and cylinders.

16.4.1 Flow Between Plates and Inside Tubes

The simplest models of porous media consist of flow within bundles of parallel cylindrical circular tubes, i.e., Poiseuille flow. Consider the steady-state flow of a fluid within a

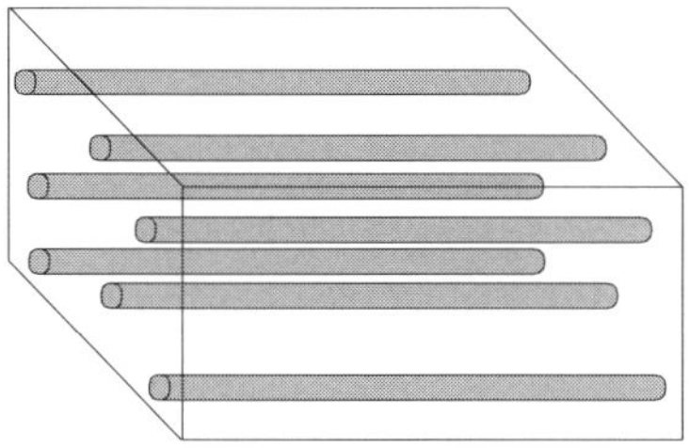

Figure 16.13 Slow viscous flow inside cylindrical circular tubes of radius a.

hyperspherical pore of dimension d and radius a. The problems of physical interest are the cases $d = 1$ (flow in layered media) and $d = 2$ (flow in a cylindrical tube). Assuming the no-slip condition on the hypersphere surface, the Stokes equations reduce to

$$\frac{1}{r^{d-1}}\frac{\partial}{\partial r}\left(r^{d-1}\frac{\partial w}{\partial r}\right) = -1, \qquad r < a, \tag{16.91}$$

$$w = 0, \qquad r = a, \tag{16.92}$$

$$w < \infty, \qquad r = 0, \tag{16.93}$$

where w represents the radial component of the scaled velocity. The right side of (16.91) represents the constant scaled pressure gradient.

The solution of this problem is

$$w = \frac{1}{2d}(a^2 - r^2). \tag{16.94}$$

Therefore, since the fluid permeability k is the average of w, we find that

$$k = \frac{a^2}{d(d+2)}. \tag{16.95}$$

If we consider an infinite number of hyperspherical pores distributed throughout a solid such that the porosity is ϕ_1, then the fluid permeability is given by

$$k = \gamma^{-1} = \frac{\phi_1 a^2}{d(d+2)}. \tag{16.96}$$

Notice that fluid permeability is exactly equal to the inverse trapping constant for this special geometry. This observation was first made by Torquato (1990) for the physically interesting case of circular cylindrical tubes (see Figure 16.13). In this instance, $d = 2$ and expression (16.96) becomes

$$k = \gamma^{-1} = \frac{\phi_1 a^2}{8}. \tag{16.97}$$

Since the *specific surface* s, defined to be the interfacial area per unit total volume, for circular cylindrical pores is given by $s = 2\phi_1/a$, we can rewrite relation (16.97) as

$$k = \gamma^{-1} = \frac{\phi_1^3}{2s^2}. \tag{16.98}$$

For cylinders of arbitrary shape, we can write

$$k = \gamma^{-1} = \frac{\phi_1^3}{cs^2}, \tag{16.99}$$

where c is a shape-dependent constant (e.g., $c = 2$ for circles, $c = 5/3$ for equilateral triangles, and $c = 1.78\ldots$ for squares). Equation (16.99) has been applied as an empirical relation for isotropic porous media of arbitrary microstructure; c then is some adjustable parameter ($c = 5$ models many porous media well). This empirical equation has a special status in the literature and is referred to as the *Kozeny–Carman* relation.

The determination of the permeability for flow past dilute arrays of obstacles is treated in Chapter 17.

16.4.2 Periodic Arrays of Obstacles

Hasimoto (1959) obtained low-concentration asymptotic formulas for the fluid permeability k associated with slow viscous flow past spheres of radius R arranged on the sites of the three cubic lattices. Specifically, he found that the dimensionless fluid permeabilities are given by

$$\frac{k}{k_s} = 1 - 1.76011\phi_2^{1/3} + \phi_2 - 1.5593\phi_2^2 + \mathcal{O}\left(\phi_2^{8/3}\right), \qquad \text{(SC)}, \tag{16.100}$$

$$\frac{k}{k_s} = 1 - 1.79186\phi_2^{1/3} + \phi_2 - 0.329\phi_2^2 + \mathcal{O}\left(\phi_2^{8/3}\right), \qquad \text{(BCC)}, \tag{16.101}$$

$$\frac{k}{k_s} = 1 - 1.79175\phi_2^{1/3} + \phi_2 - 0.302\phi_2^2 + \mathcal{O}\left(\phi_2^{8/3}\right), \qquad \text{(FCC)}, \tag{16.102}$$

where $k_s = (2R^2)/(9\phi_2)$ is the Stokes permeability in the infinitely dilute limit (see Chapter 19). The coefficients multiplying $\phi_2^{1/3}$ in the above expressions are identical to those in (16.88) up to the significant figures indicated. Sangani and Acrivos (1982) have obtained asymptotic relations for the fluid permeabilities for all three lattices to very high order in ϕ_2.

Hasimoto also obtained low-concentration asymptotic expressions for the fluid permeability k associated with flow past a square array of circular cylinders of radius R. He found that

$$\frac{k}{k_0} = -\ln\phi_2 - 1.4763 + 2\phi_2 + \mathcal{O}(\phi_2^2), \tag{16.103}$$

where $k_0 = R^2/(8\phi_2)$. For the hexagonal array, Sangani and Acrivos obtained

$$\frac{k}{k_o} = -\ln\phi_2 - 1.4975 + 2\phi_2 + \mathcal{O}(\phi_2^2). \tag{16.104}$$

Sangani and Acrivos (1982) have also found higher-order asymptotic expressions for these two-dimensional lattices.

CHAPTER 17

Single-Inclusion Solutions

A variety of estimates of the effective properties of heterogeneous media utilize the solution of the boundary value problem of the relevant field for a single inclusion of one material in a matrix of another material. Such estimates include effective properties of dispersions in the dilute-concentration limit (Chapter 19), a variety of effective-medium type approximations (Chapter 18), and rigorous bounds on the effective properties (Chapter 21). In this chapter we will derive single-inclusion solutions for all four classes of problems, focusing on spherical and ellipsoidal inclusions.

17.1 Conduction Problem

17.1.1 Spherical Inclusion

Here we obtain the potential (temperature) field T for a spherical inclusion of radius R and conductivity σ_2 in an infinite matrix of conductivity σ_1 in which the applied intensity field is the constant vector $\boldsymbol{E}_0$. Let $\boldsymbol{r}$ be a position vector emanating from the center of the sphere (see Figure 17.1). Within and outside the sphere, T must satisfy Laplace's equation (Section 16.1.1)

$$\Delta T(\boldsymbol{r}) = 0 \tag{17.1}$$

subject to the boundary conditions

$$T_+ = T_-, \qquad r = R, \tag{17.2}$$

$$\sigma_1 \boldsymbol{n} \cdot (\nabla T)_+ = \sigma_2 \boldsymbol{n} \cdot (\nabla T)_-, \qquad r = R, \tag{17.3}$$

$$T = -\boldsymbol{E}_0 \cdot \boldsymbol{r}, \qquad r \to \infty, \tag{17.4}$$

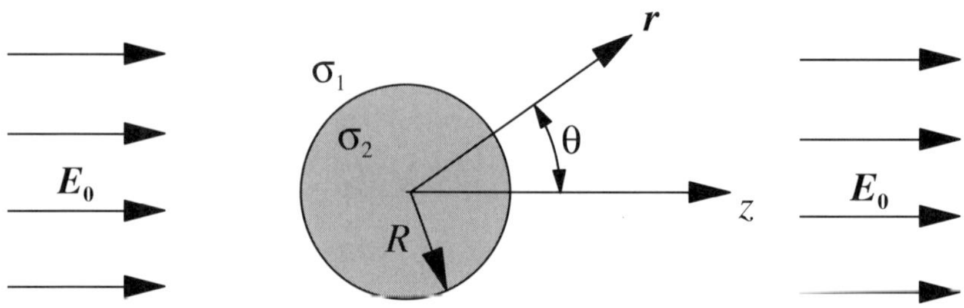

Figure 17.1 Coordinate system for sphere of conductivity σ_2 and radius R in an infinite matrix of conductivity σ_1. Here $\boldsymbol{E}_0$ is the intensity field at infinity.

where T_+ and T_- are the limits of T on the sphere surface $r \equiv |\boldsymbol{r}| = R$ taken from the outside and inside, respectively. Similarly, $(\nabla T)_+$ and $(\nabla T)_-$ are the limits of ∇T at $r = R$ from the outside and inside, respectively, where $\boldsymbol{n} = \boldsymbol{r}/|\boldsymbol{r}|$ is the radial unit normal.

The solution of equations (17.1)–(17.4) in three dimensions has been known for well over a century [see, for example, the treatise by Maxwell (1873)]. One can solve for T using the separation of variables technique (Jackson 1990). The general form of the solution in the matrix is given by

$$\begin{aligned} T &= -\boldsymbol{E}_0 \cdot \boldsymbol{r} + A\, \boldsymbol{E}_0 \cdot \nabla \left(\frac{1}{r}\right), \qquad r \geq R, \\ &= -E_0 r \cos\theta - A\, E_0 \frac{\cos\theta}{r^2}, \qquad r \geq R, \end{aligned} \tag{17.5}$$

where A is an arbitrary constant, θ is the angle between vectors $\boldsymbol{E}_0$ and $\boldsymbol{r}$, and $E_0 \equiv |\boldsymbol{E}_0|$. It is seen that the potential field is axisymmetric and is the superposition of two elementary solutions of Laplace's equation: the field due to a uniform intensity and the field due to a dipole located at the origin. As $r \to \infty$, relation (17.5) satisfies boundary condition (17.4). The general form of the solution inside the sphere is

$$\begin{aligned} T &= -\boldsymbol{E}_0 \cdot \boldsymbol{r} + B\, \boldsymbol{E}_0 \cdot \boldsymbol{r}, \qquad r \leq R, \\ &= -E_0 r \cos\theta + B\, E_0 r \cos\theta, \qquad r \leq R, \end{aligned} \tag{17.6}$$

where B is an arbitrary constant.

Application of the interface continuity conditions (17.2) and (17.3) in relations (17.5) and (17.6) yields two linear relations for the unknown constants A and B. This system is easily solved, yielding

$$A = -R^3 \frac{\sigma_2 - \sigma_1}{\sigma_2 + 2\sigma_1}, \qquad B = \frac{\sigma_2 - \sigma_1}{\sigma_2 + 2\sigma_1}.$$

Substitution of these relations into (17.5) and (17.6) gives

$$T = \begin{cases} -\boldsymbol{E}_0 \cdot \boldsymbol{r} + \beta_{21} \boldsymbol{E}_0 \cdot \boldsymbol{r} \left(\dfrac{R}{r}\right)^3, & r \geq R, \\ -\boldsymbol{E}_0 \cdot \boldsymbol{r} + \beta_{21} \boldsymbol{E}_0 \cdot \boldsymbol{r}, & r \leq R, \end{cases} \tag{17.7}$$

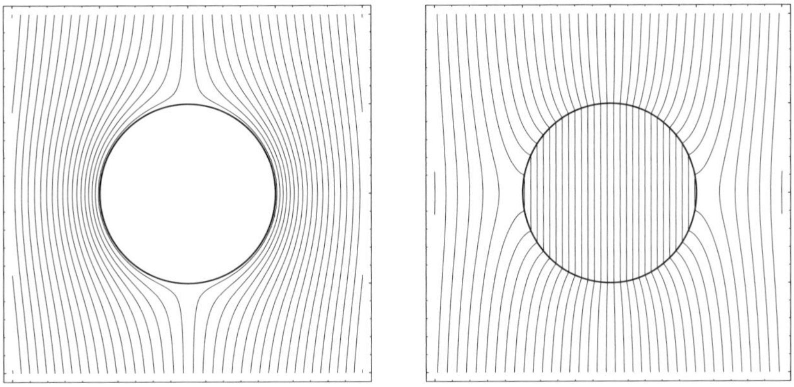

Figure 17.2 Equipotential lines for single inclusions. The field is applied horizontally. Left panel: Superconducting inclusion ($\sigma_2/\sigma_1 = \infty$). Right panel: Perfectly insulating inclusion ($\sigma_2/\sigma_1 = 0$).

where

$$\beta_{21} = \frac{\sigma_2 - \sigma_1}{\sigma_2 + 2\sigma_1}. \tag{17.8}$$

The quantity β_{21} is termed the "polarizability," since the potential outside the sphere is equivalent to the applied field plus the field of a dipole at the origin with dipole moment $\beta_{21}R^3E_0$. The associated expression for the intensity field $\boldsymbol{E} = -\nabla T$ is

$$\boldsymbol{E} = \begin{cases} \boldsymbol{E}_0 + \beta_{21}R^3\, \boldsymbol{t}(\boldsymbol{r}) \cdot \boldsymbol{E}_0, & r > R, \\ \boldsymbol{E}_0 - \beta_{21}\boldsymbol{E}_0, & r < R, \end{cases} \tag{17.9}$$

where

$$\boldsymbol{t}(\boldsymbol{r}) = \nabla\nabla\left(\frac{1}{r}\right) = \frac{3\boldsymbol{nn} - \boldsymbol{I}}{r^3} \tag{17.10}$$

is the dipole tensor. Figure 17.2 shows equipotential lines for superconducting and perfectly insulating inclusions, respectively.

For d-dimensional spheres with $d \geq 2$, this procedure is easily generalized. For $d \geq 3$, the potential outside the sphere has the form

$$T = -\boldsymbol{E}_0 \cdot \boldsymbol{r} + A\boldsymbol{E}_0 \cdot \nabla\left(\frac{1}{r^{d-2}}\right), \qquad r \geq R, \tag{17.11}$$

where the second term is the d-dimensional dipole field for $d \geq 3$. The form of the potential inside the sphere remains the same as (17.6). For $d = 2$ (circular disks), one replaces $1/r^{d-2}$ in (17.11) with $-\ln r$. Therefore, employing the aforementioned procedure yields the following solution of equations (17.1)–(17.4) for $d \geq 2$:

$$T = \begin{cases} -\boldsymbol{E}_0 \cdot \boldsymbol{r} + \beta_{21}\boldsymbol{E}_0 \cdot \boldsymbol{r}\left(\dfrac{R}{r}\right)^d, & r \geq R, \\ -\boldsymbol{E}_0 \cdot \boldsymbol{r} + \beta_{21}\boldsymbol{E}_0 \cdot \boldsymbol{r}, & r \leq R, \end{cases} \tag{17.12}$$

where the d-dimensional polarizability is given by

$$\beta_{21} = \frac{\sigma_2 - \sigma_1}{\sigma_2 + (d-1)\sigma_1}. \tag{17.13}$$

The associated intensity field $\boldsymbol{E} = -\nabla T$ for $d \geq 2$ is given by

$$\boldsymbol{E} = \begin{cases} \boldsymbol{E}_0 + \beta_{21} R^d\, \boldsymbol{t}(\boldsymbol{r}) \cdot \boldsymbol{E}_0, & r > R, \\ \boldsymbol{E}_0 - \beta_{21} \boldsymbol{E}_0, & r < R, \end{cases} \tag{17.14}$$

where the d-dimensional dipole tensor is

$$\boldsymbol{t}(\boldsymbol{r}) = \frac{d\boldsymbol{nn} - \boldsymbol{I}}{r^d} \tag{17.15}$$

and $\boldsymbol{n} = \boldsymbol{r}/|\boldsymbol{r}|$ is the radial unit vector.

It is noteworthy that the local field $\boldsymbol{E}$ is uniform *within the inclusion* [cf. (17.14)] and is linearly related to the applied field $\boldsymbol{E}_0$ by

$$\boldsymbol{E} = \boldsymbol{R} \cdot \boldsymbol{E}_0, \qquad r < R, \tag{17.16}$$

where $\boldsymbol{R}$ is the isotropic second-order tensor

$$\begin{aligned} \boldsymbol{R} &= (1 - \beta_{21})\boldsymbol{I} \\ &= \frac{d\sigma_1}{\sigma_2 + (d-1)\sigma_1} \boldsymbol{I}. \end{aligned} \tag{17.17}$$

Similarly, the polarization $\boldsymbol{P}$, defined to be

$$\boldsymbol{P} \equiv (\sigma_2 - \sigma_1)\boldsymbol{E}, \qquad r < R, \tag{17.18}$$

within the inclusion and zero otherwise, is also uniform within the inclusion according to

$$\boldsymbol{P} = \boldsymbol{M} \cdot \boldsymbol{E}_0, \qquad r < R, \tag{17.19}$$

where $\boldsymbol{M}$ is the isotropic second-order tensor

$$\begin{aligned} \boldsymbol{M} &= (\sigma_2 - \sigma_1)\boldsymbol{R} \\ &= d\sigma_1 \beta_{21} \boldsymbol{I}. \end{aligned} \tag{17.20}$$

The polarization measures the induced flux in the inclusion relative to the case in the absence of the inclusion (see Chapter 20 for a more general definition). Importantly, the polarization is the key ingredient in determining the effective conductivity of dilute dispersions (see Chapter 19) and approximations for the effective conductivity (see Chapter 18).

We emphasize that the prefactors multiplying $\boldsymbol{E}_0$ in both expressions (17.16) and (17.19) depend on the inclusion shape (in this case, spherical). This is due to the fact that we have expressed local fields (such as $\boldsymbol{E}$ or $\boldsymbol{P}$) in terms of the *applied* field $\boldsymbol{E}_0$. This shape dependence is clearly seen in the ensuing discussion concerning ellipsoidal inclusions. Of course, relations between local fields, as in relation (17.18), are shape-independent. This simple observation will be of use to us in Chapter 20, where we will derive exact expressions for the effective conductivity of composites.

17.1.2 Polarization Within an Ellipsoid

More generally, it is well known that the local polarization field $\boldsymbol{P}$ and intensity field $\boldsymbol{E}$ are uniform *within* an ellipsoidal inclusion of conductivity σ_2 immersed in an infinite matrix of conductivity σ_1 and subjected to a constant electric field $\boldsymbol{E}_0$ at infinity (Stratton 1941). The uniform fields are linearly related to the applied field $\boldsymbol{E}_0$ via the expressions

$$\boldsymbol{P} = \boldsymbol{M} \cdot \boldsymbol{E}_0, \tag{17.21}$$

$$\boldsymbol{E} = \boldsymbol{R} \cdot \boldsymbol{E}_0, \tag{17.22}$$

where $\boldsymbol{M}$ and $\boldsymbol{R}$ are constant shape-dependent second-order tensors given by

$$\boldsymbol{M} = (\sigma_2 - \sigma_1)\boldsymbol{R}, \tag{17.23}$$

$$\boldsymbol{R} = \left[\boldsymbol{I} + \boldsymbol{A}^* \frac{\sigma_2 - \sigma_1}{\sigma_1}\right]^{-1}. \tag{17.24}$$

The quantities $\boldsymbol{M}$ and $\boldsymbol{R}$ are referred to as the polarization and electric field *concentration* tensors, respectively. In relation (17.24), $\boldsymbol{A}^*$ is the symmetric *depolarization* tensor of the d-dimensional ellipsoid, which in the principal axes frame has diagonal components or eigenvalues (denoted by A_i^*, $i = 1, \ldots, d$) given by the elliptic integrals

$$A_i^* = \left(\prod_{j=1}^{d} \frac{a_j}{2}\right) \int_0^\infty \frac{dt}{(t + a_i^2)\sqrt{\prod_{j=1}^{d}(t + a_j^2)}}, \qquad i = 1, \ldots, d, \tag{17.25}$$

where a_i is the semiaxis of the ellipsoid along the x_i direction. Of course, the eigenvalues A_i^*, referred to as the *depolarization factors*, are positive. The depolarization tensor has the property that its trace is unity, i.e.,

$$\operatorname{Tr} \boldsymbol{A}^* = \sum_{i=1}^{d} A_i^* = 1. \tag{17.26}$$

Combining these properties we see that

$$0 \le A_i^* \le 1. \tag{17.27}$$

For a d-dimensional sphere, we have that

$$A_i^* = \frac{1}{d}, \qquad i = 1, 2, \ldots, d \tag{17.28}$$

in light of the trace condition and the symmetry of a sphere.

For subsequent calculations, we present results for a three-dimensional spheroidal inclusion whose symmetry axis is aligned along the x_3-axis, where $a_1 = a_2 = a$ and $a_3 = b$. In this special case, (17.25) yields exactly

$$\boldsymbol{A}^* = \begin{bmatrix} Q & 0 & 0 \\ 0 & Q & 0 \\ 0 & 0 & 1 - 2Q \end{bmatrix}. \tag{17.29}$$

For prolate spheroids,

$$Q = \frac{1}{2}\left\{1 + \frac{1}{(b/a)^2 - 1}\left[1 - \frac{1}{2\chi_b}\ln\left(\frac{1+\chi_b}{1-\chi_b}\right)\right]\right\}, \qquad \frac{b}{a} \geq 1, \tag{17.30}$$

and for oblate spheroids,

$$Q = \frac{1}{2}\left\{1 + \frac{1}{(b/a)^2 - 1}\left[1 - \frac{1}{\chi_a}\tan^{-1}(\chi_a)\right]\right\}, \qquad \frac{b}{a} \leq 1, \tag{17.31}$$

with

$$\chi_a^2 = -\chi_b^2 = (a^2/b^2) - 1. \tag{17.32}$$

From these results, we see that for spheres ($b/a = 1, Q = 1/3$)

$$A_1^* = A_2^* = A_3^* = \frac{1}{3}, \qquad \text{(spheres)} \tag{17.33}$$

for needle-shaped inclusions ($b/a = \infty, Q = 1/2$)

$$A_1^* = A_2^* = \frac{1}{2}, \quad A_3^* = 0, \qquad \text{(needles)} \tag{17.34}$$

and for disk-shaped inclusions ($b/a = 0, Q = 0$)

$$A_1^* = A_2^* = 0, \quad A_3^* = 1. \qquad \text{(disks)} \tag{17.35}$$

The polarization and intensity fields within the ellipsoid, averaged over all orientations of the ellipsoid, are easily obtained from the isotropic averages of the tensors $\boldsymbol{M}$ and $\boldsymbol{R}$, i.e.,

$$\boldsymbol{M} = M\boldsymbol{I}, \tag{17.36}$$

$$\boldsymbol{R} = R\boldsymbol{I}, \tag{17.37}$$

where the scalars M and R are given by

$$M = (\sigma_2 - \sigma_1)R, \tag{17.38}$$

$$R = \frac{R_{ii}}{d} = \frac{1}{d}\sum_{i=1}^{d}\frac{1}{1 + A_i^*\dfrac{\sigma_2 - \sigma_1}{\sigma_1}}. \tag{17.39}$$

17.2 Elasticity Problem

17.2.1 Spherical Inclusion

Consider obtaining the displacement field $\boldsymbol{u}$ for a spherical inclusion of radius R having Lamé constants λ_2, G_2 in an infinite matrix with Lamé constants λ_1, G_1 that is subjected to an applied strain field ε_0 at infinity. Let $\boldsymbol{r}$ be the position vector emanating from the

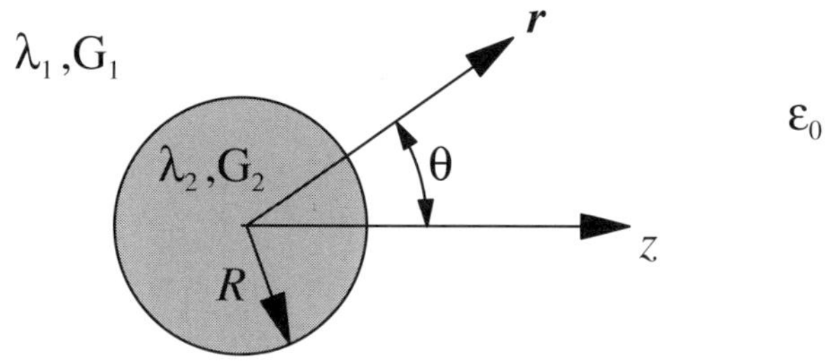

Figure 17.3 Coordinate system for sphere of Lamé constants λ_2, G_2 and radius R in an infinite matrix of Lamé constants λ_1, G_1. Here ε_0 is the strain at infinity.

center of the sphere (see Figure 17.3). Within and outside the sphere, one must solve the Navier equations for $\boldsymbol{u}$ (Section 16.2.1), i.e.,

$$(\lambda_1 + G_1)\nabla(\nabla \cdot \boldsymbol{u}) + G_1 \Delta \boldsymbol{u} = 0, \qquad r \geq R, \tag{17.40}$$

$$(\lambda_2 + G_2)\nabla(\nabla \cdot \boldsymbol{u}) + G_2 \Delta \boldsymbol{u} = 0, \qquad r \leq R, \tag{17.41}$$

subject to the boundary conditions

$$\boldsymbol{u}_+ = \boldsymbol{u}_-, \qquad r = R, \tag{17.42}$$

$$\boldsymbol{n} \cdot \boldsymbol{\tau}_+ = \boldsymbol{n} \cdot \boldsymbol{\tau}_-, \qquad r = R, \tag{17.43}$$

$$\boldsymbol{u} = \boldsymbol{\varepsilon}_0 \cdot \boldsymbol{r}, \qquad r \to \infty, \tag{17.44}$$

where + and − subscripts indicate that the limits of the functions at the sphere–matrix interface are to be taken from the outside and inside, respectively, and $\boldsymbol{n} = \boldsymbol{r}/|\boldsymbol{r}|$. The solution of equations (17.40)–(17.44) in three dimensions is well known (Goodier 1933) for hydrostatic and deviatoric applied fields.

Hydrostatic Applied Field

Let us first consider a hydrostatic applied field, i.e.,

$$\boldsymbol{\varepsilon}_0 = \frac{\varepsilon_0}{d} \boldsymbol{I},$$

where ε_0 is a scalar quantity and $\boldsymbol{I}$ is the second-order identity tensor. This problem bears a close resemblance to the single-sphere solution for the conduction problem, since both possess spherical symmetry. Exploiting the similarity between the conduction problem and the elastic problem for an isotropic applied strain field, we assume that the displacement field outside the sphere is given by the uniform field plus a dipole perturbation, i.e.,

$$\boldsymbol{u} = \boldsymbol{\varepsilon}_0 \cdot \boldsymbol{r} + A\, \boldsymbol{\varepsilon}_0 \cdot \nabla \left(\frac{1}{r}\right), \qquad r \geq R,$$

$$= \varepsilon_0 \boldsymbol{r} + A\, \varepsilon_0 \nabla \left(\frac{1}{r}\right), \qquad r \geq R, \tag{17.45}$$

where A is an arbitrary constant. Inside the sphere, we take the field to be given by

$$\boldsymbol{u} = \varepsilon_0 \cdot \boldsymbol{r} + B\, \varepsilon_0 \cdot \boldsymbol{r}, \qquad r \leq R. \tag{17.46}$$

Application of the interface conditions (17.42) and (17.43) in the relations (17.45) and (17.46) yields two linear relations for the unknown constants A and B, which have the solutions

$$A = \frac{3R^3(K_2 - K_1)}{3K_2 + 4G_1}, \qquad B = \frac{-3(K_2 - K_1)}{3K_2 + 4G_1}.$$

Here we have used the fact that the bulk modulus K_i can be expressed as $K_i = \lambda_i + 2G_i/3$ (Chapter 13). Substitution of the above relations for A and B into (17.45) and (17.46) gives

$$\boldsymbol{u} = \begin{cases} \varepsilon_0 \cdot \boldsymbol{r} + \kappa_{21}\varepsilon_0 \cdot \boldsymbol{r} \left(\dfrac{R}{r}\right)^3, & r \geq R, \\ \varepsilon_0 \cdot \boldsymbol{r} - \kappa_{21}\varepsilon_0 \cdot \boldsymbol{r}, & r \leq R, \end{cases} \tag{17.47}$$

where

$$\kappa_{21} = -B = \frac{K_2 - K_1}{K_2 + \frac{4}{3}G_1} \tag{17.48}$$

is the "bulk modulus polarizability." The associated strain field $\varepsilon = [\nabla \boldsymbol{u} + (\nabla \boldsymbol{u})^{\mathrm{T}}]/2$ is given by

$$\varepsilon = \begin{cases} \varepsilon_0 + \kappa_{21} R^3\, \boldsymbol{t}(\boldsymbol{r}) \cdot \varepsilon_0, & r > R, \\ \varepsilon_0 - \kappa_{21}\varepsilon_0, & r < R, \end{cases} \tag{17.49}$$

where $\boldsymbol{t}(\boldsymbol{r})$ is the dipole tensor (17.10).

This procedure is easily generalized to obtain the solution of (17.40)–(17.44) for d-dimensional spheres when $d \geq 2$. For $d \geq 3$, the displacement outside the sphere has the form

$$\boldsymbol{u} = \varepsilon_0 \cdot \boldsymbol{r} + A\, \varepsilon_0 \cdot \nabla \left(\frac{1}{r^{d-2}}\right), \qquad r \geq R. \tag{17.50}$$

The form of the displacement within the sphere does not change [cf. (17.46)]. For $d = 2$ (circular disks), the function $1/r^{d-2}$ in (17.50) is replaced by $-\ln r$. Application of the interface conditions (17.42) and (17.43) yields the unknown d-dimensional constants A and B with the result that for $d \geq 2$,

$$\boldsymbol{u} = \begin{cases} \varepsilon_0 \cdot \boldsymbol{r} + \kappa_{21}\varepsilon_0 \cdot \boldsymbol{r} \left(\dfrac{R}{r}\right)^d, & r \geq R, \\ \varepsilon_0 \cdot \boldsymbol{r} - \kappa_{21}\varepsilon_0 \cdot \boldsymbol{r}, & r \leq R, \end{cases} \tag{17.51}$$

where

$$\kappa_{21} = \frac{K_2 - K_1}{K_2 + \frac{2(d-1)}{d}G_1} \tag{17.52}$$

is the d-dimensional bulk modulus polarizability. The associated strain field for $d \geq 2$ is given by

$$\varepsilon = \begin{cases} \varepsilon_0 + \kappa_{21} R^d\, \boldsymbol{t}(\boldsymbol{r}) \cdot \varepsilon_0, & r > R, \\ \varepsilon_0 - \kappa_{21}\varepsilon_0, & r < R, \end{cases} \tag{17.53}$$

where $\boldsymbol{t}(\boldsymbol{r})$ is the d-dimensional dipole tensor, given by (17.15).

Note that the local strain ε_0 is uniform within the inclusion and is linearly related to the applied strain ε_0 by

$$\varepsilon = (1 - \kappa_{21})\varepsilon_0, \qquad r < R. \tag{17.54}$$

Similarly, the stress polarization tensor $\boldsymbol{p}$ is defined to be

$$\boldsymbol{p} \equiv (\boldsymbol{C}_2 - \boldsymbol{C}_1) : \varepsilon, \qquad r < R, \tag{17.55}$$

within the inclusion and zero otherwise, where $\boldsymbol{C}_i$ is the isotropic stiffness tensor [cf. (13.95)]

$$\boldsymbol{C}_i = dK_i \boldsymbol{\Lambda}_h + 2G_i \boldsymbol{\Lambda}_s, \qquad i = 1, 2. \tag{17.56}$$

Thus, $\boldsymbol{p}$ is also uniform within the inclusion according to

$$\boldsymbol{p} = d[K_1 + 2(d-1)G_1/d]\kappa_{21}\varepsilon_0. \tag{17.57}$$

The polarization measures the induced stress in the inclusion relative to the case in the absence of the inclusion (see Chapter 20 for a more general definition).

The prefactors multiplying the local fields in the relations (17.54) and (17.57) depend on the inclusion shape. This shape dependence is due to the fact that we have expressed local fields in terms of the applied field ε_0.

Deviatoric Applied Field

Consider determining the displacement field for a spherical inclusion of radius R having Lamé constants λ_2, G_2 in an infinite matrix with Lamé constants λ_1, G_1 in which the applied strain field ε_0 at infinity is *deviatoric*, i.e., $\text{Tr}\ \varepsilon_0 = 0$. We will obtain the local fields for this problem by extending an approach used by Landau and Lifshitz (1970) to obtain the solution for the special case of a spherical cavity, i.e., $\lambda_2 = G_2 = 0$. First, we observe that any solution of the *biharmonic* equations (17.40) and (17.41) can be written as a linear combination of axially symmetrical solutions (e.g., $r^2, r, 1, 1/r, 1/r^2$) and their spatial gradients. The most general form of the biharmonic vector $\boldsymbol{u}$ in the matrix satisfying relations (17.40) and (17.44) is given by

$$\boldsymbol{u} = \varepsilon_0 \cdot \boldsymbol{r} + A\, \varepsilon_0 \cdot \nabla\left(\frac{1}{r}\right) + B\, \varepsilon_0 : \nabla\nabla\nabla\left(\frac{1}{r}\right) + C\, \varepsilon_0 : \nabla\nabla\nabla\,(r), \qquad r \geq R. \tag{17.58}$$

Note that the second term in (17.58) is just a dipole contribution, whereas the third term is the gradient of the dipole tensor $\boldsymbol{t}(\boldsymbol{r})$, given by (17.10). Using the fact that ε_0 is traceless, we can simplify and rewrite (17.58) in component form as follows:

$$u_i = \left(1 - \frac{A}{r^3} + \frac{6B}{r^5} - \frac{2C}{r^3}\right)(\varepsilon_0)_{ij} r_j + 3\left(C - \frac{5B}{r^2}\right)\frac{r_i r_j r_k}{r^5}(\varepsilon_0)_{jk}, \qquad r \geq R. \quad (17.59)$$

Within the inclusion, we assume that

$$\boldsymbol{u} = \varepsilon_0 \cdot \boldsymbol{r} + D\varepsilon_0 \cdot \boldsymbol{r}, \qquad r \leq R. \quad (17.60)$$

Applying the interface conditions yields four linear equations for the four unknowns A, B, C, and D, which are satisfied when

$$A = -\frac{10R^3(G_1 - G_2)(3K_1 + 4G_1)}{3G_1(9K_1 + 8G_1) + 18G_2(K_1 + 2G_1)}, \quad (17.61)$$

$$B = \frac{R^5(G_1 - G_2)(3K_1 + G_1)}{3G_1(9K_1 + 8G_1) + 18G_2(K_1 + 2G_1)}, \quad (17.62)$$

$$C = \frac{5R^3(G_1 - G_2)(3K_1 + G_1)}{3G_1(9K_1 + 8G_1) + 18G_2(K_1 + 2G_1)}, \quad (17.63)$$

$$D = \frac{G_1 - G_2}{G_2 + H_1}, \quad (17.64)$$

where

$$H_1 = G_1 \frac{3K_1/2 + 4G_1/3}{K_1 + 2G_1}. \quad (17.65)$$

The bulk modulus K_i appears in the above relations by virtue of the interrelation $K_i = \lambda_i + 2G_i/3$.

We define the "shear modulus polarizability" μ_{21} via the relation

$$\mu_{21} \equiv -D = \frac{G_2 - G_1}{G_2 + H_1}. \quad (17.66)$$

Therefore, from relation (17.60), the displacement within the sphere is given by

$$\boldsymbol{u} = \varepsilon_0 \cdot \boldsymbol{r} - \mu_{21}\varepsilon_0 \cdot \boldsymbol{r}, \qquad r \leq R. \quad (17.67)$$

The associated strain field in the matrix in indicial notation is given by

$$\varepsilon_{ij} = (\varepsilon_0)_{ij} + \frac{3A}{2r^3}\left[\mathcal{G}_{ijkl}(\boldsymbol{n}) - \frac{2}{3}\delta_{ik}\delta_{jl}\right]\varepsilon^0_{kl} - \frac{15B}{r^5}\left[\mathcal{H}_{ijkl}(\boldsymbol{n}) - 7n_i n_j n_k n_l\right]\varepsilon^0_{kl}$$
$$+ \frac{3C}{r^3}\left[\mathcal{H}_{ijkl}(\boldsymbol{n}) - 5n_i n_j n_k n_l\right]\varepsilon^0_{kl}, \qquad r > R, \quad (17.68)$$

where

$$\mathcal{G}_{ijkl}(\boldsymbol{r}) = n_i n_l \delta_{jk} + n_j n_l \delta_{ik}, \quad (17.69)$$

$$\mathcal{H}_{ijkl}(\boldsymbol{r}) = n_j n_l \delta_{ik} + n_i n_l \delta_{jk} + n_k n_l \delta_{ij} + n_i n_k \delta_{jl} + n_j n_k \delta_{il}, \quad (17.70)$$

and δ_{ij} is the Kronecker delta given by (13.10). The strain field in the inclusion is uniform and is given by

$$\varepsilon_{ij} = \varepsilon_0 - \mu_{21}\varepsilon_0, \qquad r < R. \quad (17.71)$$

This procedure is generalizable to d-dimensional spheres for $d \ge 2$. For $d \ge 3$, the displacement in the matrix has the form

$$\boldsymbol{u} = \varepsilon_0 \cdot \boldsymbol{r} + A\, \varepsilon_0 : \nabla \left(\frac{1}{r^{d-2}}\right) + B\, \varepsilon_0 : \nabla\nabla\nabla \left(\frac{1}{r^{d-2}}\right) + C\, \varepsilon_0 : \nabla\nabla\nabla \left(r^{d-2}\right), \quad r \ge R. \tag{17.72}$$

The form of the displacement within the sphere is still given by relation (17.60). For $d = 2$ (circular disks), the function $1/r^{d-2}$ in expression (17.72) is replaced by $-\ln r$. Application of the interface conditions (17.42) and (17.43) yields the unknown d-dimensional constants A, B, C, and D for $d \ge 2$:

$$A = -\frac{2R^d(G_1 - G_2)(d+2)[dK_1 + 2(d-1)G_1]}{G_1[d^3K_1 + 2d(d+1)(d-2)G_1] + G_2[2d^2(K_1 + 2G_1)]}, \tag{17.73}$$

$$B = \frac{R^{d+2}(G_1 - G_2)[dK_1 + (d-2)G_1]}{G_1[d^3K_1 + 2d(d+1)(d-2)G_1] + G_2[2d^2(K_1 + 2G_1)]}, \tag{17.74}$$

$$C = \frac{R^d(G_1 - G_2)(d+2)[dK_1 + (d-2)G_1]}{G_1[d^3K_1 + 2d(d+1)(d-2)G_1] + G_2[2d^2(K_1 + 2G_1)]}, \tag{17.75}$$

$$D = \frac{G_1 - G_2}{G_2 + H_1}, \tag{17.76}$$

where

$$H_1 = \frac{G_1[dK_1/2 + (d+1)(d-2)G_1/d]}{K_1 + 2G_1}. \tag{17.77}$$

Here we have eliminated the constant λ_i in favor of the bulk modulus K_i via the interrelation $K_i = \lambda_i + 2G_i/d$.

The d-dimensional shear modulus polarizability μ_{21} is defined via

$$\mu_{21} \equiv -D = \frac{G_2 - G_1}{G_2 + H_1}. \tag{17.78}$$

Hence, from expression (17.60) for the displacement within the sphere, we have

$$\boldsymbol{u} = \varepsilon_0 \cdot \boldsymbol{r} - \mu_{21}\varepsilon_0 \cdot \boldsymbol{r}, \qquad r \le R, \tag{17.79}$$

where H_1 is given by (17.77).

The associated strain field in the matrix for $d \ge 2$ in component form is given by

$$\varepsilon_{ij} = (\varepsilon_0)_{ij} + \frac{d}{2}\frac{A}{r^d}\left[\mathcal{G}_{ijkl}(\boldsymbol{n}) - \frac{2}{d}\delta_{ik}\delta_{jl}\right](\varepsilon_0)_{kl} - \frac{d(d+2)B}{r^{d+2}}\left[\mathcal{H}_{ijkl}(\boldsymbol{n}) - (d+4)n_in_jn_kn_l\right](\varepsilon_0)_{kl} + \frac{dC}{r^d}\left[\mathcal{H}_{ijkl}(\boldsymbol{n}) - (d+2)n_in_jn_kn_l\right](\varepsilon_0)_{kl}, \quad r > R, \tag{17.80}$$

where $\mathcal{G}_{ijkl}$ and $\mathcal{H}_{ijkl}$ are given by (17.69) and (17.70), respectively, and A, B, and C are given by (17.73), (17.74), and (17.75), respectively. The strain field in the inclusion is uniform and equal to

$$\varepsilon = \varepsilon_0 - \mu_{21}\varepsilon_0, \qquad r < R, \tag{17.81}$$

where μ_{21} is given by (17.78).

Note that the local strain ε is uniform *within the inclusion* and is linearly related to the applied strain ε_0 by

$$\varepsilon = (1 - \mu_{21})\varepsilon_0, \qquad r < R. \tag{17.82}$$

Similarly, the stress polarization $\boldsymbol{p}$ within the spherical inclusion, given by (17.55), is the uniform field

$$\boldsymbol{p} = 2(G_1 + H_1)\mu_{21}\varepsilon_0, \qquad r < R. \tag{17.83}$$

The polarization within the d-dimensional sphere is also uniform for either hydrostatic or deviatoric applied strain fields [see (17.57) and (17.83)].

We can combine the above results for hydrostatic and deviatoric applied fields to obtain the local fields for an arbitrary applied strain field, i.e.,

$$\varepsilon = \boldsymbol{T} : \varepsilon_0, \qquad r < R, \tag{17.84}$$

where $\boldsymbol{T}$ is the isotropic fourth-order tensor

$$\boldsymbol{T} = (1 - \kappa_{21})\boldsymbol{\Lambda}_h + (1 - \mu_{21})\boldsymbol{\Lambda}_s, \tag{17.85}$$

and

$$\boldsymbol{p} = \boldsymbol{N} : \varepsilon_0, \qquad r < R, \tag{17.86}$$

where $\boldsymbol{N}$ is the isotropic fourth-order tensor

$$\begin{aligned} \boldsymbol{N} &= (\boldsymbol{C}_2 - \boldsymbol{C}_1) : \boldsymbol{T} \\ &= d[K_1 + 2(d-1)G_1/d]\kappa_{21}\boldsymbol{\Lambda}_h + 2[G_1 + H_1]\mu_{21}\boldsymbol{\Lambda}_s. \end{aligned} \tag{17.87}$$

Note that the polarization within an inclusion is the key ingredient in determining the effective elastic moduli of dilute dispersions (see Chapter 19) and of approximations for the effective elastic moduli (see Chapter 18).

Again, we emphasize that the prefactors multiplying the applied fields in the expressions above depend on the inclusion shape. This shape dependence is due to the fact that we have expressed local fields in terms of the applied field ε_0 and is clearly seen in the ensuing discussion concerning ellipsoidal inclusions. Obviously, relations between local fields, as in expression (17.55), are shape-independent. This basic observation will aid us later in deriving exact expressions for the effective moduli of composites that are free of conditionally convergent integrals (see Chapter 20).

17.2.2 Polarization Within an Ellipsoid

More generally, Eshelby (1957) showed that the polarization tensor $\boldsymbol{p}$ and strain tensor ε are uniform *within* an isotropic ellipsoidal inclusion of stiffness tensor $\boldsymbol{C}_2$ immersed in an infinite isotropic matrix of stiffness tensor $\boldsymbol{C}_1$ and subjected to a constant strain field ε_0 at infinity, where $\boldsymbol{C}_i$ is given by (17.56). The uniform fields are linearly related to the applied field ε_0 via the relations

$$\boldsymbol{p} = \boldsymbol{N} : \varepsilon_0, \tag{17.88}$$

$$\varepsilon = \boldsymbol{T} : \varepsilon_0, \tag{17.89}$$

where $\boldsymbol{N}$ and $\boldsymbol{T}$ are constant shape-dependent fourth-order tensors given by

$$\boldsymbol{N} = (\boldsymbol{C}_2 - \boldsymbol{C}_1) : \boldsymbol{T}, \tag{17.90}$$

$$\boldsymbol{T} = [\boldsymbol{I} + \boldsymbol{S} : \boldsymbol{C}_1^{-1} : (\boldsymbol{C}_2 - \boldsymbol{C}_1)]^{-1}. \tag{17.91}$$

It is common to refer to $\boldsymbol{N}$ and $\boldsymbol{T}$ as the polarization and strain *concentration* tensors, respectively. The dimensionless fourth-order tensor $\boldsymbol{S}$ in (17.91) is called the Eshelby tensor of the ellipsoid and is seen to be the analogue of the depolarization tensor $\boldsymbol{A}^*$ in the conduction problem (see Section 17.1.2). It is symmetric with respect to the first two indices and second two indices, i.e.,

$$S_{ijkl} = S_{jikl} = S_{ijlk}, \tag{17.92}$$

but generally it is not symmetric with respect to interchange of ij and kl, i.e.,

$$S_{ijkl} \neq S_{klij}. \tag{17.93}$$

Eshelby's tensor is generally given in terms of elliptic integrals of the first and second kinds (Mura 1987).

In the case of a three-dimensional spheroidal inclusion with semiaxes $a_1 = a_2 = a$ and $a_3 = b$ and whose symmetry axis is aligned in the x_3-direction, the elliptic integrals can be evaluated analytically (Mura 1987). The components of S_{ijkl} are given by

$$\begin{aligned}
S_{1111} = S_{2222} &= \frac{3}{8(1-\nu_1)}\frac{\alpha^2}{\alpha^2-1} + \frac{1}{4(1-\nu_1)}\left[1 - 2\nu_1 - \frac{9}{4(\alpha^2-1)}\right]q, \\
S_{3333} &= \frac{1}{2(1-\nu_1)}\left\{1 - 2\nu_1 + \frac{3\alpha^2-1}{\alpha^2-1} - \left[1 - 2\nu_1 + \frac{3\alpha^2}{\alpha^2-1}\right]q\right\}, \\
S_{1122} = S_{2211} &= \frac{1}{4(1-\nu_1)}\left\{\frac{\alpha^2}{2(\alpha^2-1)} - \left[1 - 2\nu_1 + \frac{3}{4(\alpha^2-1)}\right]q\right\}, \\
S_{1133} = S_{2233} &= \frac{1}{2(1-\nu_1)}\left\{\frac{-\alpha^2}{\alpha^2-1} + \frac{1}{2}\left[\frac{3\alpha^2}{\alpha^2-1} - (1-2\nu_1)\right]q\right\}, \\
S_{3311} = S_{3322} &= \frac{1}{2(1-\nu_1)}\left\{2\nu_1 - 1 - \frac{1}{\alpha^2-1} + \left[1 - 2\nu_1 + \frac{3}{2(\alpha^2-1)}\right]q\right\}, \\
S_{1212} &= \frac{1}{4(1-\nu_1)}\left\{\frac{\alpha^2}{2(\alpha^2-1)} + \left[1 - 2\nu_1 - \frac{3}{4(\alpha^2-1)}\right]q\right\}, \\
S_{1313} = S_{2323} &= \frac{1}{4(1-\nu_1)}\left\{1 - 2\nu_1 - \frac{\alpha^2+1}{\alpha^2-1} - \frac{1}{2}\left[1 - 2\nu_1 - \frac{3(\alpha^2+1)}{\alpha^2-1}\right]q\right\},
\end{aligned} \tag{17.94}$$

where ν_1 is the Poisson ratio of the matrix, $\alpha = b/a$ is the aspect ratio of the spheroid, and q is given by

$$q = \begin{cases} \dfrac{\alpha}{(\alpha^2-1)^{3/2}}\left[\alpha(\alpha^2-1)^{1/2} - \cosh^{-1}\alpha\right], & \alpha \geq 1, \\ \dfrac{\alpha}{(1-\alpha^2)^{3/2}}\left[\cos^{-1}\alpha - \alpha(1-\alpha^2)^{1/2}\right], & \alpha \leq 1. \end{cases} \tag{17.95}$$

From these results, we see that for spheres ($\alpha = 1, q = 2/3$),

$$S_{ijkl} = \frac{5\nu_1 - 1}{15(1-\nu_1)}\delta_{ij}\delta_{kl} + \frac{4-5\nu_1}{15(1-\nu_1)}(\delta_{ik}\delta_{jl} + \delta_{il}\delta_{jk}). \tag{17.96}$$

For a needle-shaped inclusion ($\alpha = \infty, q = 1$), one has

$$\begin{aligned} S_{1111} &= S_{2222} = \frac{5-4\nu_1}{8(1-\nu_1)}, \quad S_{3333} = 0, \\ S_{1122} &= S_{2211} = \frac{4\nu_1 - 1}{8(1-\nu_1)}, \quad S_{1133} = S_{2233} = \frac{\nu_1}{2(1-\nu_1)}, \\ S_{3311} &= S_{3322} = 0, \quad S_{1212} = \frac{3-4\nu_1}{8(1-\nu_1)}, \quad S_{1313} = S_{2323} = \frac{1}{4}. \end{aligned} \tag{17.97}$$

For a disk-shaped inclusion ($\alpha = 0, q = 0$), the only nonvanishing components are

$$S_{3333} = 1, \quad S_{3311} = S_{3322} = \frac{\nu_1}{1-\nu_1}, \quad S_{1313} = S_{2323} = \frac{1}{2}. \tag{17.98}$$

The polarization and intensity fields within the ellipsoid, averaged over all orientations of the ellipsoid, are obtained from the isotropic averages of the tensors $\boldsymbol{N}$ and $\boldsymbol{T}$, i.e.,

$$\boldsymbol{N} = N_h \boldsymbol{\Lambda}_h + N_s \boldsymbol{\Lambda}_s, \tag{17.99}$$

$$\boldsymbol{T} = T_h \boldsymbol{\Lambda}_h + T_s \boldsymbol{\Lambda}_s, \tag{17.100}$$

where the scalars N_h, N_s, T_h, and T_s, for any d, are given by

$$N_h = d(K_2 - K_1)T_h, \qquad N_s = 2(G_2 - G_1)T_s, \tag{17.101}$$

$$T_h = \frac{T_{iijj}}{d}, \qquad T_s = \frac{2}{(d+2)(d-1)}\left(T_{ijij} - \frac{T_{iijj}}{d}\right). \tag{17.102}$$

Therefore, the corresponding local polarization and strain fields within the ellipsoid are still given by relations (17.88) and (17.89) but where $\boldsymbol{N}$ and $\boldsymbol{T}$ are given by (17.99) and (17.100), respectively.

One can obtain the coefficients N_h, N_s, T_h, and T_s for three-dimensional spheroids using the relations (17.90), (17.91), (17.94), (17.101), and (17.102). Values of T_h and T_s for the limiting cases of spheres, needles, and disks are given in Table 17.1 in terms of the moduli K_i and G_i. The corresponding values of N_h and N_s are easily obtained from the expressions (17.101). The formulas for general spheroidal inclusions were first given by Wu (1966); see also Berryman (1980a).

Table 17.1 Values of the coefficients T_h and T_s defined by (17.102) for the limiting cases of spheres, needles, and disks in three dimensions. For needles, $E_1 \equiv G_1(3K_1+G_1)/(3K_1+7G_1)$, and for disks, $H_2 = G_2(3K_2/2+4G_2/3)/(K_2+2G_2)$. Corresponding values of N_h and N_s are easily obtained from the formulas (17.101). Note that the results for spheres agree with terms of (17.84) for $d=3$.

Inclusion Shape	T_h	T_s
Spheres	$\frac{K_1+4G_1/3}{K_2+4G_1/3}$	$\frac{G_1+H_1}{G_2+H_1}$
Needles	$\frac{K_1+G_1+G_2/3}{K_2+G_1+G_2/3}$	$\frac{1}{5}\left[\frac{4G_1}{G_1+G_2}+\frac{2G_1(G_1+E_1)}{(G_2+E_1)}+\frac{K_2+4G_1/3}{K_2+G_1+G_2/3}\right]$
Disks	$\frac{K_1+4G_2/3}{K_2+4G_2/3}$	$\frac{G_1+H_2}{G_2+H_2}$

17.3 Trapping Problem

17.3.1 Spherical Trap

We are interested in obtaining the concentration field exterior to a spherical trap of radius R. This is a classical problem dating back to the work of Smoluchowski (1917). According to the local equation (13.136), the steady-state concentration field satisfies a Poisson equation subject to the appropriate boundary conditions. This system of equations can be replaced by a Laplace equation for the concentration field, with the production term in the Poisson equation replaced by a uniform concentration field at infinity.

Consider inserting a *partially absorbing* three-dimensional spherical trap with radius R and surface rate constant κ into an infinite medium in which the unperturbed concentration field is the uniform value C_0. Let $\boldsymbol{r}$ be the position vector emanating from the trap center (see Figure 17.4). Under steady-state conditions, the local concentration field $c(\boldsymbol{r})$ depends only on the radial distance $r \equiv |\boldsymbol{r}|$ and is governed by

$$\Delta c(\boldsymbol{r}) = \frac{1}{r^2}\frac{\partial}{\partial r}\left(r^2\frac{\partial c}{\partial r}\right) = 0, \qquad r \geq R, \tag{17.103}$$

$$\mathcal{D}\frac{\partial c}{\partial r} = \kappa c, \qquad r = R, \tag{17.104}$$

$$c = C_0, \qquad r \to \infty. \tag{17.105}$$

The general form of the solution of this diffusion equation is given by

$$c(r) = A + \frac{B}{r}, \qquad r \geq R. \tag{17.106}$$

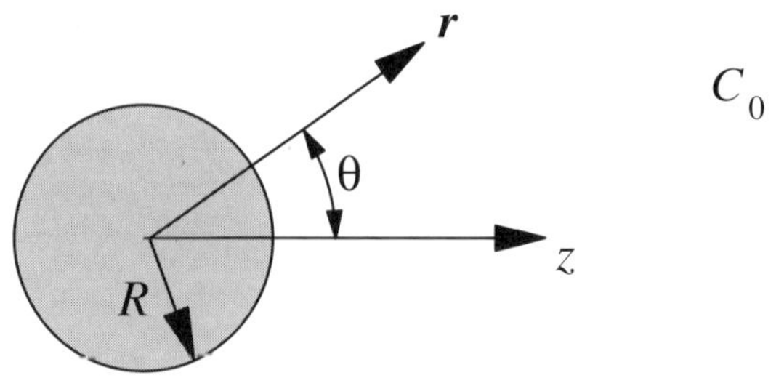

Figure 17.4 Coordinate system for a spherical trap of radius R in an infinite medium with constant concentration field C_0 at infinity.

Application of the boundary conditions reveals that

$$A = C_0, \qquad B = \frac{C_0 \kappa R^2}{\mathcal{D} + \kappa R},$$

and therefore the desired solution is

$$c(r) = C_0 \left[1 + \frac{\kappa R}{\mathcal{D} + \kappa R} \left(\frac{R}{r} \right) \right], \qquad r \geq R. \tag{17.107}$$

The total flux J_T into the sphere can be calculated by integrating the normal flux over the surface of the sphere, i.e.,

$$\begin{aligned} J_T &= \int_{r=R} -\mathcal{D}\, \nabla c \cdot \boldsymbol{n}\, R^2 d\Omega \\ &= \frac{4\pi\, \kappa R^2 \mathcal{D} C_0}{\kappa R + \mathcal{D}}, \end{aligned} \tag{17.108}$$

where $d\Omega$ is the differential solid angle contained in a three-dimensional sphere.

In the *diffusion-controlled* case ($\kappa R/\mathcal{D} = \infty$), the solution (17.107) gives

$$c(r) = C_0 \left[1 - \left(\frac{R}{r} \right) \right], \qquad r \geq R, \tag{17.109}$$

and the total flux (17.108) becomes

$$J_T = 4\pi R \mathcal{D} C_0. \tag{17.110}$$

Of course, in the *reaction-controlled* case ($\kappa R/\mathcal{D} = 0$), the solution (17.107) is trivial, i.e., $c = C_0$ and $J_T = 0$, since there is no absorption.

The solution of the concentration field obtained above for a spherical trap in three dimensions has no analogue in two dimensions. Specifically, there is no solution of the Laplace equation for a circular trap in an infinite medium with an unperturbed concentration field C_0 that simultaneously satisfies the condition $c = 0$ on the trap surface and the boundary condition at infinity. The nonexistence of such a solution is the analogue of Stokes's paradox, described in the next section in the case of uniform flow past a circular disk.

The concentration field for a d-dimensional spherical trap of radius R for any $d \geq 3$ can be obtained from the boundary value problem

$$\frac{1}{r^{d-1}}\frac{\partial}{\partial r}\left(r^{d-1}\frac{\partial c}{\partial r}\right) = 0, \qquad r \geq R, \tag{17.111}$$

$$\mathcal{D}\frac{\partial c}{\partial r} = \kappa c, \qquad r = R, \tag{17.112}$$

$$c = C_0, \qquad r \to \infty. \tag{17.113}$$

The solution of this boundary value problem is

$$c(r) = C_0\left[1 + \frac{\kappa R}{(d-2)\mathcal{D} + \kappa R}\left(\frac{R}{r}\right)^{d-2}\right], \qquad r \geq R. \tag{17.114}$$

The total flux J_T into the sphere is then

$$\begin{aligned} J_T &= \int_{r=R} -\mathcal{D}\, \nabla c \cdot \boldsymbol{n}\, R^{d-1} d\Omega \\ &= \frac{(d-2)\Omega\, \kappa R^{d-1}\mathcal{D}C_0}{\kappa R + (d-2)\mathcal{D}}, \qquad d \geq 3 \end{aligned} \tag{17.115}$$

where $\Omega(d)$ is the total solid angle contained in a d-dimensional sphere, given by (2.56).

In the *diffusion-controlled* case ($\kappa R/\mathcal{D} = \infty$), the solution (17.114) becomes

$$c(r) = C_0\left[1 - \left(\frac{R}{r}\right)^{d-2}\right], \qquad r \geq R. \tag{17.116}$$

From (17.115), we find that the total flux in this instance becomes

$$J_T = (d-2)\Omega\, R^{d-2}\mathcal{D}C_0. \tag{17.117}$$

17.3.2 Spheroidal Trap

Consider a perfectly absorbing spheroidal trap in three dimensions with semiaxes $a_1 = a_2 = a$ and $a_3 = b$ in which the concentration at infinity is C_0. The concentration field c outside of this trap is obtained by solving $\Delta c = 0$ in spheroidal coordinates (η, θ, ψ) (Abramowitz and Stegun 1972), with $c(\mathcal{S}) = 0$ on the trap surface $\mathcal{S}$ and $c = C_0$ at infinity (Miller, Kim and Torquato 1991). The level surfaces of η (η = constant) are a confocal family of spheroids having their common center at the origin. The trap surface $\mathcal{S}$, defined by $\eta = \eta_0$, is related to the semiaxes a and b through

$$a = \sqrt{b^2 - a^2}\sinh\eta_0, \qquad b = \sqrt{b^2 - a^2}\cosh\eta_0, \qquad b \geq a, \tag{17.118}$$

$$a = \sqrt{a^2 - b^2}\cosh\eta_0, \qquad b = \sqrt{a^2 - b^2}\sinh\eta_0, \qquad b \leq a. \tag{17.119}$$

The coordinates θ and ψ are angles.

By symmetry, c is a function only of η and in prolate spheroidal coordinates (Miller et al. 1991) the governing equations simplify as follows:

$$\frac{d^2c}{d\eta^2} + (\coth\eta)\frac{dc}{d\eta} = 0, \tag{17.120}$$

$$c(\eta = \eta_0) = 0, \tag{17.121}$$

$$c(\eta \to \infty) = C_0. \tag{17.122}$$

The concentration outside the prolate spheroidal trap ($b \geq a$) is therefore

$$c(\eta) = C_0 \left\{1 - \frac{\ln[\coth(\eta/2)]}{\ln[\coth(\eta_0/2)]}\right\}. \tag{17.123}$$

The total flux J_T into a single isolated prolate spheroidal trap is given by

$$J_T = \int -\nabla c(\eta = \eta_0) \cdot \boldsymbol{n}\, d\mathcal{S}, \tag{17.124}$$

where $\boldsymbol{n}$ is the unit vector normal to the trap surface. Carrying out the indicated operations, Miller et al. (1991) obtained

$$J_T = 4\pi b \mathcal{D} C_0 \frac{2\chi_b}{\ln\left(\dfrac{1+\chi_b}{1-\chi_b}\right)}, \tag{17.125}$$

where χ_b is given by (17.32).

For oblate spheroidal ($b \leq a$) traps, the appropriate differential equation (Miller et al. 1991) is given by

$$\frac{d^2c}{d\eta^2} + (\tanh\eta)\frac{dc}{d\eta} = 0. \tag{17.126}$$

The concentration field that satisfies this equation along with the boundary conditions (17.121) and (17.122) is

$$c(\eta) = C_0 \left[1 - \frac{\cot^{-1}(\sinh\eta)}{\cot^{-1}(\sinh\eta_0)}\right]. \tag{17.127}$$

Following the same procedure as above, Miller et al. (1991) found that the total flux into an isolated oblate spheroidal trap is given by

$$J_T = 4\pi b \mathcal{D} C_0 \frac{\chi_a}{\tan^{-1}\chi_a}, \tag{17.128}$$

where χ_a is given by (17.32).

Specific limiting cases for the flux due to needle-like and disk-like traps can be trivially extracted from the corresponding dilute trapping-constant results (19.100) and (19.101), respectively, and relations (19.90) and (19.91).

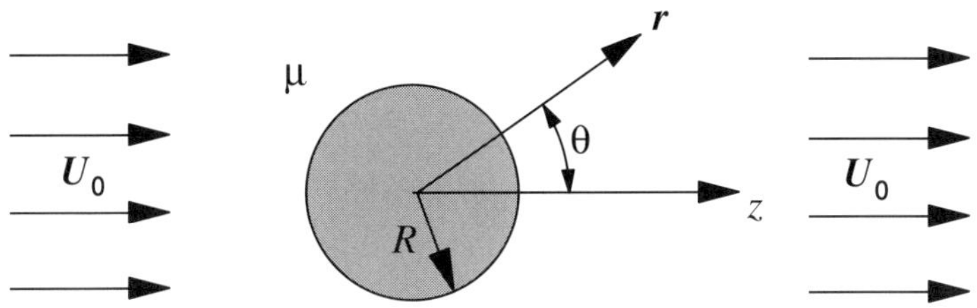

Figure 17.5 Coordinate system for sphere of radius R in which fluid velocity at infinity is the constant vector $\boldsymbol{U}_0 = U_0\boldsymbol{e}$. Here μ is the dynamic viscosity.

17.4 Flow Problem

17.4.1 Spherical Obstacle

It is desired to find the velocity field $\boldsymbol{v}$ and pressure field p for a flow past a fixed sphere of radius R under creeping flow conditions, i.e., when the Reynold's number is much less than unity. This is a very old problem, dating back to the work of Stokes (1851). According to the local equations given in Section 13.5.2, the steady-state velocity field must satisfy the Stokes equations with an *applied* pressure gradient term. This system of equations can be replaced by the Stokes equations without an applied pressure gradient, but with an imposed uniform velocity at infinity.

Let $\boldsymbol{r}$ be the position vector emanating from the sphere center (see Figure 17.5). When the Reynold's number is negligibly small, the flow exterior to the sphere is governed by the Stokes equations

$$\mu\Delta\boldsymbol{v} = \nabla p, \qquad r \geq R, \tag{17.129}$$

$$\nabla\cdot\boldsymbol{v} = 0, \qquad r \geq R, \tag{17.130}$$

$$\boldsymbol{v} = 0, \qquad r = R, \tag{17.131}$$

$$\boldsymbol{v} = \boldsymbol{U}_0, \qquad r \to \infty, \tag{17.132}$$

where μ is the dynamic viscosity. Taking the divergence of the momentum equation (17.129) reveals that the pressure is a harmonic function, i.e.,

$$\Delta p = 0. \tag{17.133}$$

The velocity field corresponding to uniform flow past a stationary sphere may be obtained by a superposition of the solutions for uniform flow, a doublet (i.e., dipole), and a *Stokeslet* (defined below). Thus, we assume that the velocity and pressure fields are respectively

$$\boldsymbol{v} = U_0\boldsymbol{e} + A\,\boldsymbol{t}(\boldsymbol{r})\cdot\boldsymbol{e} + B\,\boldsymbol{O}(\boldsymbol{r})\cdot\boldsymbol{e}, \tag{17.134}$$

$$p = p_0 + 2B\,\mu\frac{\boldsymbol{n}\cdot\boldsymbol{e}}{r^2}, \tag{17.135}$$

where $\boldsymbol{t}(\boldsymbol{r})$ is the dipole tensor (17.10),

$$O(r) = \frac{I}{r} + \frac{nn}{r} \tag{17.136}$$

is the Oseen tensor, p_0 is the constant pressure at infinity, and e is the unit vector in the direction of the velocity at infinity. The contribution to the velocity field (17.134) involving the Oseen tensor plus the pressure field (17.135) constitute what is referred to as a Stokeslet. Applying the no-slip boundary condition $v = 0$ on the surface of the sphere ($r = R$) in (17.134) enables us to determine the constants A and B as follows:

$$A = \frac{R^3}{4} U_0, \qquad B = -\frac{3R}{4} U_0.$$

Substituting these expressions into relations (17.134) and (17.135) yields the velocity and pressure distributions to be

$$v = U_0 + \frac{U_0}{4}\left(\frac{R}{r}\right)^3 [3nn - I] \cdot e - \frac{3U_0}{4}\left(\frac{R}{r}\right)[I + nn] \cdot e, \tag{17.137}$$

$$p = p_0 - \frac{3R}{2} U_0 \mu \frac{n \cdot e}{r^2}. \tag{17.138}$$

The uniform-flow and dipole terms in (17.137) do not exert forces or torques on the surrounding fluid. Although the Stokeslet does not exert a torque on the fluid, it does exert a force or stress on it. The stress tensor τ is generally given by

$$\tau = -pI + \mu\left[\nabla v + (\nabla v)^{\mathrm{T}}\right]. \tag{17.139}$$

Therefore, the contribution to the stress from the Stokeslet is given by

$$\tau = -p_0 I + \frac{9\mu R U_0}{2} \frac{nn(n \cdot e)}{r^2}. \tag{17.140}$$

The drag force F acting on the sphere can be computed by integrating the traction $\tau \cdot n$ over the surface of the sphere, i.e.,

$$F = \int_{r=R} \tau \cdot n R^2 d\Omega, \tag{17.141}$$

and $d\Omega$ is the differential solid angle contained in a three-dimensional sphere. Substitution of (17.140) into (17.141) finally yields

$$F = 6\pi\mu R\, U_0, \tag{17.142}$$

which is the well-known Stokes drag law for a sphere in a uniform viscous flow.

The solution obtained above for a sphere in three dimensions has no counterpart in two dimensions. Specifically, there is no solution of the Stokes equations for uniform flow past a circular disk that simultaneously satisfies the no-slip condition on the disk surface and the boundary condition at infinity. The lack of such a solution is known as *Stokes's paradox*. Mathematically, the Stokes approximation is an asymptotic limit of the Navier–Stokes equations in the limit that the Reynolds number (Re) goes to zero. In three dimensions, the zeroth-order term in an asymptotic expansion of the velocity in powers of Re exists, but in two dimensions it does not exist, resulting in a singular perturbation problem. Physically, the difficulty encountered is associated with the neglect of the convection of momentum of the fluid, an assumption that is not valid far from the body.

The velocity and pressure fields for flow past a d-dimensional sphere of radius R for any $d \geq 3$ that satisfies the Stokes equations (17.129)–(17.131) and the boundary condition (17.132) is easily obtained as

$$v = U_0 \boldsymbol{e} + \frac{U_0}{d+1}\left(\frac{R}{r}\right)^d [d\boldsymbol{nn} - \boldsymbol{I}] \cdot \boldsymbol{e} - \frac{dU_0}{d+1}\left(\frac{R}{r}\right)^{d-2} [\boldsymbol{I} + \boldsymbol{nn}] \cdot \boldsymbol{e},$$

$$p = p_0 - \frac{2dR^{d-2}}{d+1} U_0 \mu \frac{\boldsymbol{n} \cdot \boldsymbol{e}}{r^{d-1}}.$$

These results in combination with (17.139) give the Stokeslet contribution to the stress to be

$$\boldsymbol{\tau} = -p_0 \boldsymbol{I} + \frac{\mu R^{d-2} U_0}{d+1} \frac{[2d^2 \boldsymbol{nn}(\boldsymbol{n} \cdot \boldsymbol{e}) + d(d-3)(\boldsymbol{ne} + \boldsymbol{en})]}{r^{d-1}}, \qquad d \geq 3. \tag{17.143}$$

The drag force $\boldsymbol{F}$ acting on the sphere is obtained via the integral

$$\boldsymbol{F} = \int_{r=R} \boldsymbol{\tau} \cdot \boldsymbol{n} R^{d-1} d\Omega, \tag{17.144}$$

where $d\Omega$ is the differential solid angle contained in a d-dimensional sphere. Substitution of (17.143) into (17.144) then gives the drag force as

$$\boldsymbol{F} = \frac{d^2 - 3}{d+1} \Omega \mu R^{d-2} \, \boldsymbol{U}_0, \qquad d \geq 3, \tag{17.145}$$

where $\Omega(d)$ is the total solid angle contained in a d-dimensional sphere, given by (2.56). Here we have used the identity

$$\int_\Omega \boldsymbol{nn} d\Omega = \frac{\Omega(d)}{d} \boldsymbol{I}.$$

17.4.2 Spheroidal Obstacle

Consider now computing the drag force $\boldsymbol{F}$ on an aligned but stationary three-dimensional spheroid in three dimensions with semiaxes $a_1 = a_2 = a$ and $a_3 = b$ in which a constant velocity $\boldsymbol{U}_0$ is imposed at infinity. This solution can be found by solving the Stokes equations in spheroidal coordinates and dates back to the work of Oberbeck (1876); see also Lamb (1932). Here we will just summarize the results.

For a spheroidal obstacle, the drag force can be expressed as

$$\boldsymbol{F} = 6\pi\mu b[X(b/a)\,(\boldsymbol{I} - \boldsymbol{nn}) + Y(b/a)\,\boldsymbol{nn}] \cdot \boldsymbol{U}_0, \tag{17.146}$$

where

$$X(b/a) = \begin{cases} \dfrac{16\chi_b^3}{3}\left[(3\chi_b^2 - 1)\ln\left(\dfrac{1+\chi_b}{1-\chi_b}\right) + 2\chi_b\right]^{-1}, & b/a \geq 1, \\ \dfrac{8\chi_a^3}{3}\left[(1 + 3\chi_a^2)\tan^{-1}\chi_a - \chi_a\right]^{-1}, & b/a \leq 1, \end{cases} \tag{17.147}$$

$$Y(b/a) = \begin{cases} \dfrac{8\chi_b^3}{3}\left[(1 + \chi_b^2)\ln\left(\dfrac{1+\chi_b}{1-\chi_b}\right) - 2\chi_b\right]^{-1}, & b/a \geq 1, \\ \dfrac{4\chi_a^3}{3}\left[(\chi_a^2 - 1)\tan^{-1}\chi_a + \chi_a\right]^{-1}, & b/a \leq 1. \end{cases} \tag{17.148}$$

In these equations, $\boldsymbol{n}$ denotes the unit vector along the symmetry axis, and χ_a and χ_b are given by (17.32).

Specific limiting cases for the drag on needle-like and disk-like obstacles can be trivially extracted from the corresponding dilute permeability results (19.117) and (19.118), respectively, and relations (19.104) and (19.105).

CHAPTER 18

Effective-Medium Approximations

In this chapter we will show that single-inclusion solutions obtained in Chapter 17 can also be utilized to derive popular *effective-medium approximations*, which investigators have employed to estimate effective properties for a wide range of volume fractions, as well as phase properties (when appropriate). However, these approximations necessarily can account only for simple microstructural information, such as volume fraction and inclusion shape. Thus, although effective-medium approximations can provide qualitative trends on the behavior of the effective properties of dispersions, they cannot be quantitatively predictive for general situations. In applying the variety of different effective-medium approximations that have been proposed, it is important to understand the class of microstructures and conditions under which they are valid. Moreover, any reasonable approximation should satisfy existing property bounds.

18.1 Conductivity

Here we will derive and describe several types of effective-medium approximations, including Maxwell-type approximations, self-consistent approximations, and differential effective-medium approximations for the effective conductivity of multiphase media. All of these approximations rely on knowing the effective conductivity of suspensions in the infinitely dilute limit and therefore the solutions of the single-inclusion boundary value problems described in Section 17.1.

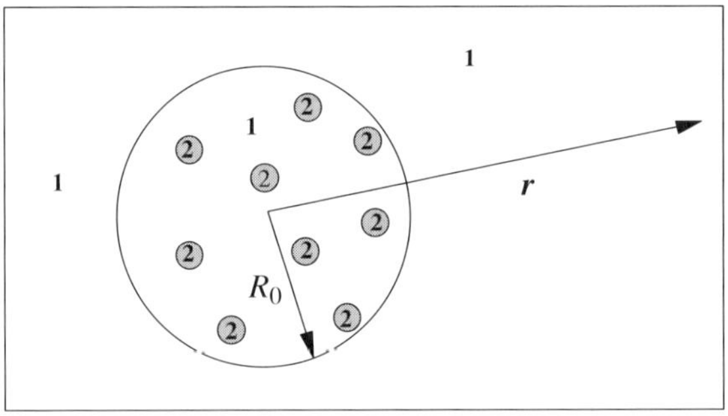

Figure 18.1 A schematic corresponding to Maxwell's approximation. A composite sphere of radius R_o, composed of spherical particles (phase 2) in a matrix (phase 1), is immersed in an infinite matrix of phase 1.

18.1.1 Maxwell Approximations

Spherical Inclusions

The ensuing derivation is the d-dimensional generalization of the approximation scheme given by Maxwell (1873) for the case $d = 3$. Consider a large sphere of radius R_o composed of N smaller spheres of radius R and conductivity σ_2 in a matrix of conductivity σ_1 in d spatial dimensions (see Figure 18.1). A constant intensity field $\boldsymbol{E}_0$ is applied at infinity. The volume fraction of the small spheres $\phi_2 = N(R/R_o)^d$ is sufficiently small that interactions between them can be neglected. At radial distances that are large compared to R_o, the field $\boldsymbol{E}(\boldsymbol{r})$ at the radial position $\boldsymbol{r}$ is simply the superposition of the field due to each small sphere, given by (17.14), and hence

$$\boldsymbol{E}(\boldsymbol{r}) = \boldsymbol{E}_0 + \frac{\phi_2 R_o^d\,\beta_{21}}{r^d}[d\boldsymbol{nn} - \boldsymbol{I}]\cdot\boldsymbol{E}_0, \tag{18.1}$$

where β_{21} is the polarizability given by (17.13). The large sphere can be treated as a homogeneous sphere with effective conductivity σ_e, and hence the intensity field that it produces at large distances is

$$\boldsymbol{E}(\boldsymbol{r}) = \boldsymbol{E}_0 + \frac{R_o^d\,\beta_{e1}}{r^d}[d\boldsymbol{nn} - \boldsymbol{I}]\cdot\boldsymbol{E}_0, \tag{18.2}$$

where

$$\beta_{e1} = \frac{\sigma_e - \sigma_1}{\sigma_e + (d-1)\sigma_1}. \tag{18.3}$$

Assuming that expressions (18.1) and (18.2) are equal to each other implies

$$\beta_{e1} = \phi_2\beta_{21}. \tag{18.4}$$

Using expression (17.13) for the polarizability β_{21}, this relation can be rewritten more explicitly as

$$\frac{\sigma_e - \sigma_1}{\sigma_e + (d-1)\sigma_1} = \phi_2 \left[\frac{\sigma_2 - \sigma_1}{\sigma_2 + (d-1)\sigma_1} \right]. \tag{18.5}$$

We will refer to formula (18.5) as Maxwell's approximation. It is sometimes called the Maxwell–Garnett approximation (Landauer 1978), and it is related to the Clausius–Mossotti formula (Hansen and McDonald 1986) for the dielectric constant of a fluid. The Maxwell approximation is an *explicit* scheme: It yields the effective properties by direct substitution.

Given our use of the single-inclusion field solution to derive (18.5), one might surmise that Maxwell's approximation should yield useful estimates of the effective conductivity only for small inclusion concentrations. Indeed, Maxwell's approximation will give good estimates of σ_e at *nondilute* conditions, provided that the spheres are well separated from each other, because it is identical to result (16.8) for the coated-spheres model. Moreover, relation (18.5) coincides with the well-known *optimal* Hashin–Shtrikman bounds, i.e., it equals the lower bound (21.21) when $\sigma_2 \geq \sigma_1$ and the corresponding upper bound (21.22) when $\sigma_2 \leq \sigma_1$. Note that the mean-field argument leading to (18.5) is applicable to spheres with a polydispersivity in size and hence (18.5) is insensitive to polydispersivity. In general, the effective conductivity of dispersions actually depends on the degree of polydispersivity.

When the spheres are superconducting relative to the matrix ($\sigma_2/\sigma_1 = \infty$), Maxwell's approximation (18.5) simplifies as follows:

$$\frac{\sigma_e}{\sigma_1} = \frac{1 + (d-1)\phi_2}{1 - \phi_2} \qquad \text{if} \quad \sigma_2/\sigma_1 = \infty. \tag{18.6}$$

Thus, the composite becomes superconducting at the trivial percolation threshold $\phi_{2c} = 1$. When the spheres are perfectly insulating relative to the matrix ($\sigma_2/\sigma_1 = 0$), Maxwell's approximation (18.5) reduces to

$$\frac{\sigma_e}{\sigma_1} = \frac{(d-1)(1-\phi_2)}{d - \phi_1} \qquad \text{if} \quad \sigma_2/\sigma_1 = 0. \tag{18.7}$$

The composite becomes nonconducting at the trivial point $\phi_{2c} = 1$.

We note that Maxwell's expression (18.5) is easily generalized to suspensions with $M - 1$ different types of spheres ($M \geq 2$) with volume fractions $\phi_2, \ldots, \phi_M$ and conductivities $\sigma_2, \ldots, \sigma_M$. For such a multiphase composite, it is found that

$$\frac{\sigma_e - \sigma_1}{\sigma_e + (d-1)\sigma_1} = \sum_{j=1}^{M} \phi_j \left[\frac{\sigma_j - \sigma_1}{\sigma_j + (d-1)\sigma_1} \right]. \tag{18.8}$$

When all the inclusions are more (less) conducting than the matrix phase, relation (18.8) coincides with the multiphase Hashin–Shtrikman lower (upper) bound [cf. (21.23)]. Benveniste (1986) derived this formula for $d = 3$ using the Mori–Tanaka method.

Ellipsoidal Inclusions

A generalization of the Maxwell approximation (18.8) for a macroscopically anisotropic composite consisting of $M-1$ different types of unidirectionally aligned isotropic ellipsoidal inclusions of the same shape is given by

$$\sum_{j=1}^{M} \phi_j(\boldsymbol{\sigma}_e - \boldsymbol{\sigma}_j) \cdot \boldsymbol{R}^{(j1)} = 0. \tag{18.9}$$

Here $\boldsymbol{\sigma}_e$ is the effective conductivity tensor, $\boldsymbol{\sigma}_j = \sigma_j \boldsymbol{I}$,

$$\boldsymbol{R}^{(j1)} = \left[\boldsymbol{I} + \boldsymbol{A}^* \frac{\sigma_j - \sigma_1}{\sigma_1}\right]^{-1} \tag{18.10}$$

is the *electric field concentration tensor* for an inclusion of conductivity σ_j in a matrix of conductivity σ_1 [cf. (17.22)], and $\boldsymbol{A}^*$ is the corresponding depolarization tensor. It can easily be shown that formula (18.9) reduces to relation (18.8) in the limit of spherical inclusions. Formula (18.9) is the conductivity analogue of the Mori–Tanaka (1973) approximation for the effective stiffness tensor of dispersions of aligned ellipsoids of the same shape [cf. (18.46)]. Moreover, (18.9) coincides with one of the optimal bounds (21.29), due to Willis (1977), for the same microstructure, depending on whether the matrix is more or less conducting than all of the inclusions.

If instead the ellipsoids are randomly oriented, then one need only replace $\boldsymbol{R}^{(j1)}$ in (18.9) with its isotropic form, i.e., $R^{(j1)} = \text{Tr}\, \boldsymbol{R}^{(j1)}/d$ [cf. (17.39)]:

$$\sum_{j=1}^{M} \phi_j(\sigma_e - \sigma_j) R^{(j1)} = 0. \tag{18.11}$$

Although formula (18.11) always lies between the isotropic Hashin–Shtrikman bounds (21.23) for two phases ($M = 2$), it can violate these bounds for $M \geq 3$ for certain nonspherical inclusion shapes (Norris 1989). Whenever a formula violates optimal Hashin–Shtrikman-type bounds, it implies that the formula does not correspond to realizable microstructures. Realizability is an important criterion that any reasonable approximation scheme should obey. The drawbacks of Mori–Tanaka schemes are more fully discussed in Section 18.2.1.

18.1.2 Self-Consistent Approximations

Spherical Inclusions

Consider a macroscopically isotropic composite aggregate composed of M different types of spherical inclusions with volume fractions $\phi_1, \ldots, \phi_M$ and conductivities $\sigma_1, \ldots, \sigma_M$. In the self-consistent approximation, originally due to Bruggeman (1935) and elaborated upon by Landauer (1952, 1978), the effect of all of the material outside a type j inclusion is to produce a homogeneous medium (matrix) whose effective conductivity σ_e is the unknown to be calculated. One now requires that the perturbations to a uniform field be zero *on average*. According to relation (17.14) for the intensity field, a single spherical inclusion produces a dipole perturbation. Thus, we choose the effective conductivity σ_e by the self-consistency requirement that perturbations due to each type of "test" sphere vanish on average, i.e.,

$$\sum_{j=1}^{M} \phi_j \beta_{je} = 0. \tag{18.12}$$

More explicitly, this formula yields

$$\sum_{j=1}^{M} \phi_j \frac{\sigma_j - \sigma_e}{\sigma_j + (d-1)\sigma_e} = 0. \tag{18.13}$$

Relation (18.13) is referred to as the self-consistent (SC) approximation and can be seen to be a quadratic equation for the effective conductivity. For the two-phase case ($M = 2$), it has the solution

$$\sigma_e = \frac{\alpha + \sqrt{\alpha^2 + 4(d-1)\sigma_1\sigma_2}}{2(d-1)}, \quad \text{with} \quad \alpha = \sigma_1(d\phi_1 - 1) + \sigma_2(d\phi_2 - 1), \tag{18.14}$$

when σ_1 and σ_2 are both real and positive. In contrast to Maxwell-type approximations, the SC approximation is *implicit*: It requires solving implicit equations for the effective properties. Because formula (18.13) treats each phase symmetrically in that it is invariant to simultaneous interchange of the phase conductivities and volume fractions, it is often referred to as the *symmetric* SC approximation. For example, for $M = 2$, it is invariant under the simultaneous interchange $\sigma_1 \leftrightarrow \sigma_2$ and $\phi_1 \leftrightarrow \phi_2$. The SC approximation is analogous to the *coherent potential approximation* successfully used in the study of electronic structure in solid state physics (Ziman 1979).

Because of its simplicity, the SC approximation is a popular way of estimating not only the effective conductivity but also, as described in the following section, the effective elastic moduli of dispersions. However, the SC approximation is generally a poor predictor of the effective properties of dispersions encountered in practice. As will become apparent, the main problem is not the SC approximation itself but rather a lack of understanding of the microstructures and conditions under which it provides a valid estimate of the effective property.

What are the deficiencies of the SC approximation when it is naively applied to estimate the conductivities of typical dispersions? First, it does not include information about the spatial distribution of the inclusions or possible correlations between the properties of neighboring inclusions. Clearly, applying it to composites that do not possess phase-inversion symmetry (e.g., dispersions of identical spheres; see Section 2.2.2) is highly questionable. Second, the SC approximation assumes that the effective medium exists just outside the test sphere. This is clearly impossible in the case of packings of identical spheres, since gaps always exist between the spheres. Acrivos and Chang (1986) have addressed this problem for sphere packings by first surrounding the test sphere with a shell of matrix material and then surrounding this coated inclusion with the effective medium.

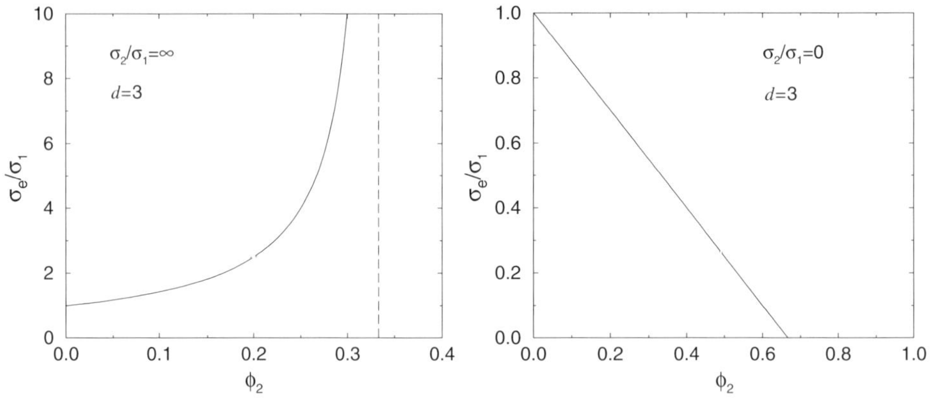

Figure 18.2 Scaled effective conductivity σ_e/σ_1 versus ϕ_2 in the SC approximation for infinite-contrast situations in three dimensions. Left panel: Formula (18.15) for a superconducting phase 2 predicts a percolation threshold $\phi_{2c} = 1/3$. Right panel: Formula (18.16) for a perfectly insulating phase 2 predicts that the composite can no longer conduct at $\phi_{2c} = 2/3$.

Third, a well-known failure of the SC approximation occurs when applied to dispersions with widely different phase conductivities. In the infinite-contrast cases, it predicts nontrivial but spurious percolation thresholds. For example, for the case of two phases, (18.14) yields the simple expression

$$\frac{\sigma_e}{\sigma_1} = \frac{1}{1 - d\,\phi_2} \qquad \text{if} \quad \sigma_2/\sigma_1 = \infty, \tag{18.15}$$

in the limiting case of superconducting spherical inclusions relative to phase 2 ($\sigma_2/\sigma_1 = \infty$). Thus, according to (18.15), phase 2 percolates (i.e., composite becomes superconducting) at the critical volume fraction of $\phi_2^c = 1/d$, independent of the microstructural details (see Figure 18.2). However, we know from Chapter 9 that the percolation threshold ϕ_{2c} varies widely from composite to composite, being *very sensitive* to the microstructure. For example, we have seen that identical three-dimensional overlapping spheres percolate at $\phi_{2c} \approx 0.29$, while three-dimensional hard spheres in equilibrium along the metastable branch percolate at the maximally-random-jammed value of $\phi_{2c} \approx 0.64$; see Section 10.6.3. Furthermore, in the opposite limit of perfectly insulating spherical inclusions relative to phase 1 ($\sigma_2/\sigma_1 = 0$), the SC approximation (18.14) gives

$$\frac{\sigma_e}{\sigma_1} = 1 - \frac{d}{d-1}\phi_2 \qquad \text{if} \quad \sigma_2/\sigma_1 = 0. \tag{18.16}$$

This predicts that the point at which the composite no longer conducts is $\phi_{2c} = (d-1)/d$, independent of the microstructure (see Figure 18.2).

Finally, note that the SC approximation will generally violate bounds that improve upon the Hashin–Shtrikman bounds for realistic models of dispersions. An example of such a violation is depicted in Figure 20.3.

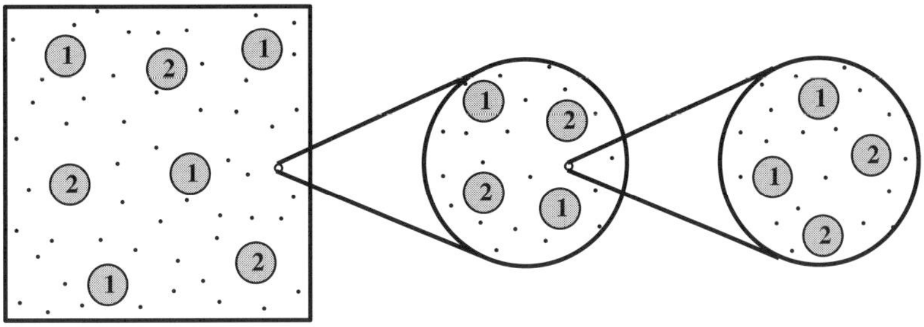

Figure 18.3 Hierarchical geometry corresponding to the SC approximation (18.14) at different levels of magnification (Milton 1984, Milton 1985).

However, the foregoing critique does not mean that SC approximations are useless. Indeed, Milton (1984, 1985) has provided a rigorous basis for them by showing that there is a particular class of microstructures that correspond exactly to the SC relations for the conductivity, given by (18.13), and the elastic moduli, given by (18.51) and (18.53). For two phases, this class consists of granular aggregates such that spherical grains of the two phases of comparable size are well-separated with self-similarity on many length scales. At a particular length scale, well-separated large spheres of the two phases (i.e., in dilute proportions) are surrounded by a matrix consisting of much smaller well-separated spheres of the two phases, which in turn are surrounded by a matrix consisting of even much smaller well-separated spheres of the two phases, and so on, *ad infinitum* (see Figure 18.3). At the end of the process, the different-sized spheres of the two phases fill up all of space in accordance with the overall phase volume fractions. At any stage of the construction, the effective properties are calculated using the dilute formulas, i.e., relations (19.23), (19.67), and (19.68) for σ_e, K_e, and G_e, respectively. The reader is referred to Milton (1985) for a detailed proof that this fractal-like construction leads to the SC formulas. Since SC approximations are realizable, *they are always within the Hashin–Shtrikman bounds* and can be used as a benchmark to test new theories.

Torquato and Hyun (2001) have found single-scale periodic dispersions that realize the two-phase SC formula (18.14) when $d = 2$ for all volume fractions at a given phase contrast. Moreover, to an excellent approximation (but not exactly), the same structures realize the SC formula for almost the entire range of phase conductivities and volume fractions. Unlike Milton's hierarchical structures, the Torquato–Hyun structures involve a simple inclusion shape at a single length scale, namely, the *generalized hypocylcoid*, defined in the x-y plane by the equation

$$x^{2/b} + y^{2/b} = a^{2/b}, \tag{18.17}$$

where a and b are dimensionless volume-fraction-dependent parameters and all distances (x, y, and a) are given in units of the cell length. Except for $b = 1$, this interface is nonsmooth. The special cases $b = 1, 2$, and 3 specify the circle, square, and four-

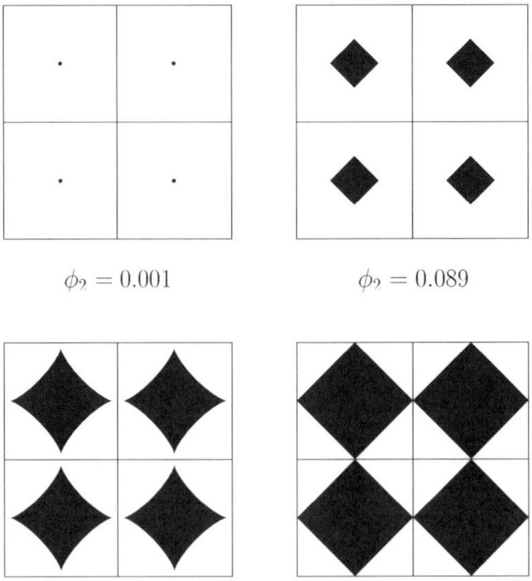

Figure 18.4 A class of single-scale dispersions that achieves the SC approximation (18.14) for $d = 2$ at infinite contrast for several volume fractions (Torquato and Hyun 2001): $\phi_2 = 0.001$ ($a \approx 0$, $b \approx 1$), $\phi_2 = 0.089$ ($a = 0.42137$, $b = 2$), $\phi_2 = 0.3$ ($a = 0.89740$, $b = 2.56810$), and $\phi_2 = 0.5$ ($a = 1$, $b = 2$).

cusped hypocycloid. For example, for $\phi_2 \ll 1$ ($a \to 0$, $b \to 1$), the structure consists of identical circular inclusions of phase 2 on the sites of a square lattice in a matrix of phase 1, and for $\phi_2 = 1/2$ ($a = 1$, $b = 2$), the structure is the periodic checkerboard (Figure 18.4). However, like the Milton constructions, the Torquato–Hyun structures possess phase-inversion symmetry (Section 2.2.2). For this reason, we show structures only in the range $0 \le \phi_2 \le 1/2$ in Figure 18.4. Note that at $\phi_2 = 1/2$, the SC formula (18.14) yields $\sigma_e = \sqrt{\sigma_1\sigma_2}$, which we have seen is exact for composites with statistically equivalent phase distributions [cf. (15.24)], the regular checkboard being one such example.

The restrictions as to the class of microstructures to which one can apply the SC formula are now quite clear, a key feature being that all phases are treated symmetrically.

It is useful now to revisit the percolation characteristics of the two-phase SC formula (18.14), since it is realizable exactly by the aforementioned fractal-like microstructures. In three dimensions, relation (18.15) states that the SC composite becomes superconducting (when phase 2 is superconducting) at the percolation threshold $\phi_2^c = 1/3$, whereas relation (18.16) states that the SC composite no longer conducts (when phase 2 is a perfect insulator) at $\phi_2^c = 2/3$. Why are the thresholds different for these two cases? The difference in the thresholds is just a reflection of the fact that phase 2 changes from a *geometrically* disconnected phase (for $0 \le \phi_2 < 1/3$) to a geometrically connected phase for $\phi_2 \geq 1/3$, and phase 1 changes from a connected phase (for $0 \leq \phi_2 < 2/3$) to a disconnected phase for $\phi_2 \geq 2/3$. Thus, it is bicontinuous for $1/3 \leq \phi_2 < 2/3$. Bicontinuity is a realistic feature of a variety of two-phase materials, including sedimentary rocks and cermets (see Figures 1.2 and 1.3). On the other hand, in two dimensions, statistically isotropic two-phase media that are bicontinuous are *rare* in practice, although such media can be devised (Chapter 9). Consistent with this observation, we see that formulas (18.15) and (18.16) for $d = 2$ predict the same percolation threshold $\phi_2^c = 0.5$; i.e., the composite becomes superconducting or ceases to conduct at the same volume fraction.

Ellipsoidal Inclusions

A generalization of the SC formula (18.13) for a macroscopically anisotropic composite consisting of M different types of unidirectionally aligned isotropic ellipsoidal inclusions of the same shape is given by

$$\sum_{j=1}^{M} \phi_j(\boldsymbol{\sigma}_j - \boldsymbol{\sigma}_e) \cdot \boldsymbol{R}^{(je)} = 0, \tag{18.18}$$

where

$$\boldsymbol{R}^{(je)} = [\boldsymbol{I} + \boldsymbol{A}^* \cdot \boldsymbol{\sigma}_e^{-1} \cdot (\boldsymbol{\sigma}_j - \boldsymbol{\sigma}_e)]^{-1} \tag{18.19}$$

and $\boldsymbol{A}^*$ is the depolarization tensor for an ellipsoid in a matrix with an effective conductivity tensor $\boldsymbol{\sigma}_e$. When the embedding medium is anisotropic, the intensity field within the ellipsoid is still a constant, given by (17.22) (Mura 1987), but the corresponding depolarization tensor $\boldsymbol{A}^*$ is not as simple as it is in the isotropic case discussed in Section 17.1.2. Formula (18.18) reduces to relation (18.13) in the limit of spherical inclusions. When the ellipsoids are randomly oriented, then (18.18) becomes

$$\sum_{j=1}^{M} \phi_j(\sigma_j - \sigma_e) R^{(je)} = 0, \tag{18.20}$$

where $R^{(je)} = \text{Tr}\, \boldsymbol{R}^{(je)}/d$. As in the case of spherical inclusions, the above SC approximations are realizable by hierarchical microstructures (Milton 1985, Avellaneda 1987a) and hence lie between Willis's optimal bounds (21.29) on $\boldsymbol{\sigma}_e$.

18.1.3 Differential Effective-Medium Approximations

Another popular implicit approximation scheme is the differential effective-medium (DEM) estimate first introduced by Bruggeman (1935). Unlike the SC approximation, the DEM approximation does not treat each phase symmetrically. However, both SC and DEM approximations share the idea of incremental homogenization. For a two-phase composite, phase 1 is treated as a matrix phase with volume fraction ϕ_1, and phase 2 is taken to be an included phase with volume fraction ϕ_2. Let us suppose that the effective conductivity $\sigma_e(\phi_2)$ at one value of ϕ_2 is known. Now treat $\sigma_e(\phi_2)$ as the composite host conductivity and let $\sigma_e(\phi_2+\Delta\phi_2)$ represent the effective conductivity of the composite after a small fraction $\Delta\phi_2/(1-\phi_2)$ of composite host has been replaced by inclusions of phase 2. On average, a fraction $\Delta\phi_2/(1-\phi_2)$ of the composite host must be replaced by phase 2 material in order to change the overall fraction of phase 2 to $\phi_2+\Delta\phi_2$. In the case of d-dimensional spherical inclusions, we can use the dilute-limit result (19.20) for the effective conductivity to obtain

$$\sigma_e(\phi_2+\Delta\phi_2)-\sigma_e(\phi_2)=\sigma_e(\phi_2)\left[\frac{\sigma_2-\sigma_e(\phi_2)}{\sigma_2+(d-1)\sigma_e(\phi_2)}\right]\frac{\Delta\phi_2}{1-\phi_2}d. \tag{18.21}$$

In the limit $\Delta\phi_2\to 0$, this expression becomes the differential equation

$$(1-\phi_2)\,\frac{d\sigma_e}{d\phi_2}=\sigma_e\left[\frac{\sigma_2-\sigma_e}{\sigma_2+(d-1)\sigma_e}\right]d, \tag{18.22}$$

with the initial condition $\sigma_e(\phi_2=0)=\sigma_1$. The differential equation (18.22) can be integrated analytically to yield

$$\left(\frac{\sigma_2-\sigma_e}{\sigma_2-\sigma_1}\right)\left(\frac{\sigma_1}{\sigma_e}\right)^{1/d}=1-\phi_2. \tag{18.23}$$

The embedding process in the DEM approximation ensures that the initial matrix material remains *connected* in the resulting composite (Yonezawa and Cohen 1983). This feature enables the DEM scheme to successfully reproduce an empirical law due to Archie (1942) for the effective conductivity of sedimentary rock containing a conducting fluid of conductivity σ_1 (Sen, Scala and Cohen 1981). With the conductivity of the rock (phase 2) taken to be zero (a perfect insulator), Archie's law is given by

$$\frac{\sigma_e}{\sigma_1}=\phi_1^m, \tag{18.24}$$

where ϕ_1 is the porosity and m is an exponent that varies between 1.5 and 4. Using the DEM expression (18.23) yields

$$\frac{\sigma_e}{\sigma_1}=\phi_1^{d/(d-1)}\qquad \text{if}\quad \sigma_2/\sigma_1=0, \tag{18.25}$$

and hence for $d=3$ produces

$$\frac{\sigma_e}{\sigma_1}=\phi_1^{3/2}\qquad \text{if}\quad \sigma_2/\sigma_1=0. \tag{18.26}$$

This is to be contrasted with the three-dimensional estimates $\sigma_e=\sigma_1\cdot 2\phi_1(3-\phi_1)^{-1}$ and $\sigma_e=\sigma_1(3\phi_1-1)/2$ given by Maxwell's and SC approximations, respectively, which are in poor agreement with Archie's law for the low porosities typical of sedimentary rock.

Note that when phase 2 is superconducting relative to phase 1, the DEM expression (18.23) simplifies as follows:

$$\frac{\sigma_e}{\sigma_1}=\frac{1}{(1-\phi_2)^d}\qquad \text{if}\quad \sigma_2/\sigma_1=\infty. \tag{18.27}$$

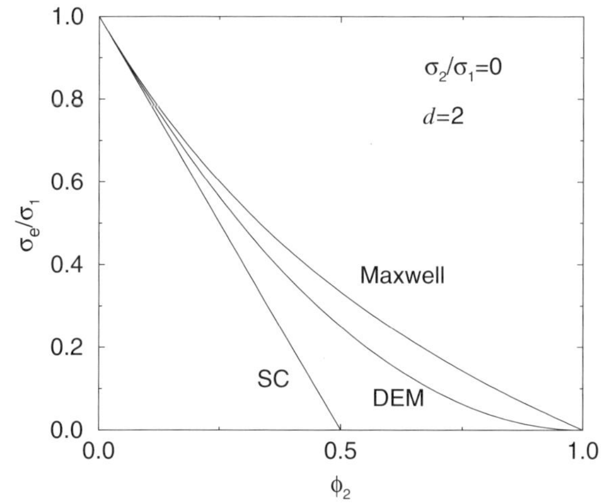

Figure 18.5 Comparison of the predictions of the three different effective-medium approximations for the scaled effective conductivity σ_e/σ_1 versus ϕ_2 of a two-phase, two-dimensional composite in which $\sigma_2/\sigma_1 = 0$. These results correspond to formulas (18.7), (18.16), and (18.25) with $d = 2$.

The DEM approximation (18.23) was shown by Milton (1985) to be realizable for hierarchical models similar to the class that realizes the SC approximation, and hence the DEM expression also must lie between the Hashin–Shtrikman bounds. Norris, Callegari and Sheng (1985) have generalized the DEM scheme for multiphase composites that by design is realizable via an incremental homogenization construction process. The final properties depend on the construction path taken and contain both the SC and DEM approximations as special cases. Avellaneda (1987a) extended these results by proving the realizability of the generalized DEM scheme for the effective conductivity and stiffness tensors for multiphase composites consisting of anisotropic grains of arbitrary shape.

Figure 18.5 compares the Maxwell, SC, and DEM approximations for a two-phase two-dimensional medium in the extreme case where phase 2 is nonconducting relative to phase 1 ($\sigma_2/\sigma_1 = 0$). Whereas the SC formula predicts a nontrivial percolation threshold, both the Maxwell and DEM approximations imply that phase 1 remains connected until the point $\phi_2 = 1$ is reached. Similar qualitative trends apply in higher dimensions as well. The behavior of these three effective-medium approximations in the opposite extreme of a superconducting phase 2 ($\sigma_2/\sigma_1 = \infty$) is qualitatively similar to the behavior of the corresponding approximations for the effective elastic moduli when phase 2 is perfectly rigid (see Figure 18.7). Clearly, both the Maxwell and DEM approximations are not appropriate estimates of the conductivities of dispersions in which the inclusions form large clusters.

Ellipsoidal Inclusions

A generalization of the DEM formula (18.22) for a macroscopically anisotropic composite consisting of identical unidirectionally aligned isotropic ellipsoidal inclusions in a matrix is given by the differential equation

$$(1-\phi_2)\frac{d\boldsymbol{\sigma}_e}{d\phi_2} = (\boldsymbol{\sigma}_2 - \boldsymbol{\sigma}_e)\cdot \boldsymbol{R}^{(2e)}, \tag{18.28}$$

where $\boldsymbol{R}^{(2e)}$ is given by (18.19) with $j = 2$ and $\boldsymbol{\sigma}_e(\phi_2 = 0) = \boldsymbol{\sigma}_1$. We see that formula (18.28) with $j = 2$ reduces to relation (18.23) in the limit of spherical inclusions. The realizability of the corresponding DEM approximation (18.66) for the effective stiffness tensor of general multiphase anisotropic composites has been proven by Avellaneda (1987a). Using the same arguments, expression (18.28) for the effective conductivity tensor can be shown to be realizable by the same class of microstructures. When the ellipsoids are randomly oriented, then we have

$$(1-\phi_2)\frac{d\sigma_e}{d\phi_2} = (\sigma_2 - \sigma_e)R^{(2e)}, \tag{18.29}$$

where $R^{(2e)} = \mathrm{Tr}\,\boldsymbol{R}^{(2e)}/d$.

18.2 Elastic Moduli

Effective-medium approximations based on the solutions of the single-inclusion boundary value problems described in Section 17.2, completely analogous to those described above for the effective conductivity, have been derived for the effective elastic moduli. Here we will describe explicit Maxwell-type approximations, implicit self-consistent approximations, and implicit differential effective-medium approximations for the effective moduli of multiphase media.

18.2.1 Maxwell Approximations

Spherical Inclusions

Consider a large d-dimensional sphere of radius R_o containing N smaller spheres of radius R in a matrix of bulk modulus K_1 and shear modulus G_1 (see Figure 18.1). The spheres have a bulk modulus K_2 and shear modulus G_2. The volume fraction of the small spheres $\phi_2 = N(R/R_o)^d$ is sufficiently small that interactions between them can be neglected. To find the effective bulk modulus K_e, we apply a hydrostatic strain at infinity, i.e., $\boldsymbol{\varepsilon}_0 = \varepsilon_0 \boldsymbol{I}/d$. At radial distances that are large compared to R_o, the strain field $\boldsymbol{\varepsilon}(\boldsymbol{r})$ at the radial position $\boldsymbol{r}$ is simply the superposition of the field due to each small sphere, given by (17.53), and thus

$$\boldsymbol{\varepsilon}(\boldsymbol{r}) = \boldsymbol{\varepsilon}_0 + \frac{\phi_2 R_o^d \kappa_{21}}{r^d}[d\boldsymbol{nn} - \boldsymbol{I}]\cdot \boldsymbol{\varepsilon}_0, \tag{18.30}$$

where κ_{21} is the bulk modulus polarizability given by (17.52). The large sphere can be treated as a homogeneous sphere with effective bulk modulus K_e, and hence the strain that it produces at large distances is given by

$$\varepsilon(\boldsymbol{r}) = \varepsilon_0 + \frac{R_o^d \kappa_{e1}}{r^d}[d\boldsymbol{nn} - \boldsymbol{I}] \cdot \varepsilon_0, \tag{18.31}$$

where

$$\kappa_{e1} = \frac{K_e - K_1}{K_e + \frac{2(d-1)}{d} G_1}. \tag{18.32}$$

Assuming that expressions (18.30) and (18.31) are equal to each other gives

$$\kappa_{e1} = \phi_2 \kappa_{21}. \tag{18.33}$$

Relation (18.33) can be rewritten more explicitly as

$$\frac{K_e - K_1}{K_e + \frac{2(d-1)}{d} G_1} = \phi_2 \left[\frac{K_2 - K_1}{K_2 + \frac{2(d-1)}{d} G_1} \right], \tag{18.34}$$

where we have used (17.52) for the bulk modulus polarizability κ_{21}.

We refer to (18.34) as Maxwell's approximation for the effective bulk modulus of suspensions of identical d-dimensional spheres. As in the case of the conductivity, Maxwell's approximation will yield good estimates of K_e at nondilute conditions, provided that the spheres are well separated from each other, because (18.34) is identical to result (16.41) for the coated-spheres model. Furthermore, Maxwell's approximation (18.34) coincides with the optimal Hashin–Shtrikman lower bound (21.71) when $G_2 \geq G_1$, and the corresponding optimal upper bound (21.72) when $G_2 \leq G_1$.

The determination of the effective shear modulus G_e of suspensions of identical spheres follows in the same way, except, of course, that one applies a uniform shear strain field far from the large sphere. Utilizing the single-sphere solution (17.80) for inclusions of type 2 in a matrix of type 1 and Maxwell's approach, as described above, leads to the relation

$$\mu_{e1} = \phi_2 \mu_{21}, \tag{18.35}$$

where $\mu_{21} = (G_2 - G_1)/(G_2 + H_1)$ is the shear modulus polarizability for an inclusion of type 2 in a matrix of type 1 [cf. (17.78)],

$$\mu_{e1} = \frac{G_e - G_1}{G_e + H_1}, \qquad H_i \equiv \frac{G_i[dK_i/2 + (d+1)(d-2)G_i/d]}{K_i + 2G_i}. \tag{18.36}$$

Relation (18.35) can be rewritten more explicitly as

$$\frac{G_e - G_1}{G_e + H_1} = \phi_2 \left[\frac{G_2 - G_1}{G_2 + H_1} \right]. \tag{18.37}$$

We refer to (18.37) as Maxwell's approximation for the effective shear modulus of suspensions of identical d-dimensional spheres. This formula does not correspond to the coated-spheres model [in contrast to approximation (18.34)], but it is identical to relation (21.74), which is exact for the hierarchical laminate constructions that realize the Hashin–Shtrikman lower bound (21.74) when $K_2 \geq K_1$ and $G_2 \geq G_1$. This class of laminates is actually a special class of dispersions, i.e., one phase is an included disconnected phase and the other is a connected matrix phase. Note that the mean-field arguments leading to both (18.34) and (18.37) are applicable to spheres with a polydispersivity in size, and hence these expressions cannot capture the effect of polydispersivity.

When the spheres are perfectly rigid relative to the matrix ($G_2/G_1 = K_2/K_1 = \infty$), the Maxwell approximations (18.34) and (18.37) reduce to

$$\frac{K_e}{K_1} = \frac{1 + \dfrac{2(d-1)G_1}{dK_1}\phi_2}{1-\phi_2} \quad \text{if} \quad G_2/G_1 = K_2/K_1 = \infty, \tag{18.38}$$

$$\frac{G_e}{G_1} = \frac{1 + \dfrac{H_1}{G_1}\phi_2}{1-\phi_2} \quad \text{if} \quad G_2/G_1 = K_2/K_1 = \infty. \tag{18.39}$$

We again see that percolation occurs at the trivial point $\phi_{2c} = 1$.

If both phases are incompressible phases ($K_1/G_1 = \infty, K_2/G_2 = \infty$), relation (18.34) correctly predicts that the composite itself is incompressible, and (18.37) yields

$$\frac{G_e}{G_1} = \frac{1 + \dfrac{d}{2}\mu_{21}\phi_2}{1-\mu_{21}\phi_2} \quad \text{if} \quad K_1/G_1 = K_2/G_2 = \infty, \tag{18.40}$$

where $\mu_{21} = (G_2 - G_1)/(G_2 + dG_1/2)$. If we further let the spheres be perfectly rigid relative to phase 1 ($G_2/G_1 = \infty$), then (18.40) becomes

$$\frac{G_e}{G_1} = \frac{1 + \dfrac{d}{2}\phi_2}{1-\phi_2} \quad \text{if} \quad K_1/G_1 = K_2/G_2 = \infty,\ G_2/G_1 = \infty. \tag{18.41}$$

We see that through first order in ϕ_2, relation (18.41) yields $G_e/G_1 = 1 + (d+2)\phi_2/2$, which is the d-dimensional version of Einstein's famous dilute result (Einstein 1906); see also Section 13.3.6 and formula (19.77).

When the spheres are cavities ($K_2 = G_2 = 0$) and the matrix is assumed to be incompressible ($K_1/G_1 = \infty$), then the Maxwell approximations yield

$$\frac{K_e}{G_1} = \frac{2(d-1)}{d}\frac{\phi_1}{\phi_2} \quad \text{if} \quad K_2 = G_2 = 0,\ K_1/G_1 = \infty, \tag{18.42}$$

$$\frac{G_e}{G_1} = \frac{d(1-\phi_2)}{d+2\phi_2} \quad \text{if} \quad K_2 = G_2 = 0,\ K_1/G_1 = \infty. \tag{18.43}$$

These formulas predict that the composite ceases to support hydrostatic or shear loads at the trivial point $\phi_{2c} = 1$. They can be combined to yield the Lamé constant $\lambda_e = K_e - 2G_e/d$. In particular, for $\phi_2 \ll 1$, this gives $\lambda_e = 4G_1/(3\phi_2)$ for $d = 3$, which agrees with the corresponding exact formula (16.73) for the effective expansion viscosity of a dilute concentration of spherical bubbles in an incompressible liquid derived by Taylor (1954). Thus, the aforementioned combination of (18.42) and (18.43) that yields λ_e approximately extends Taylor's result to arbitrary order in ϕ_2.

Maxwell's formulas (18.34) and (18.37) for $M-1$ different types of spheres ($M \geq 2$) with volume fractions $\phi_2, \ldots, \phi_M$ and moduli $K_2, G_2, \ldots, K_M, G_M$ generalize as

$$\frac{K_e - K_1}{K_e + \frac{2(d-1)}{d} G_1} = \sum_{j=1}^{M} \phi_j \frac{K_j - K_1}{K_j + \frac{2(d-1)}{d} G_1}, \tag{18.44}$$

$$\frac{G_e - G_1}{G_e + H_1} = \sum_{j=1}^{M} \phi_j \frac{G_j - G_1}{G_j + H_1}. \tag{18.45}$$

When all of the inclusions are stiffer (more compliant) than the matrix phase, formulas (18.44) and (18.45) coincide with the Hashin–Shtrikman–Walpole lower (upper) bounds [cf. (21.77) and (21.78)]. For $d = 3$, these formulas are equivalent to those derived by Mori and Tanaka (1973) using a different approach; see also Weng (1984).

Ellipsoidal Inclusions

A generalization of the Maxwell approximations (18.44) and (18.45) for a macroscopically anisotropic composite consisting of $M-1$ different types of unidirectionally aligned isotropic ellipsoidal inclusions of the same shape is given by

$$\sum_{j=1}^{M} \phi_j (\boldsymbol{C}_e - \boldsymbol{C}_j) \cdot \boldsymbol{T}^{(j1)} = 0, \tag{18.46}$$

where $\boldsymbol{C}_e$ is the effective stiffness tensor,

$$\boldsymbol{T}^{(j1)} = [\boldsymbol{I} + \boldsymbol{S}_1 : \boldsymbol{C}_1{}^{-1} : (\boldsymbol{C}_j - \boldsymbol{C}_1)]^{-1} \tag{18.47}$$

is the *strain concentration tensor* for an inclusion of stiffness $\boldsymbol{C}_j$ in a matrix of stiffness $\boldsymbol{C}_1$ [cf. (17.89)], and $\boldsymbol{S}_1$ is Eshelby's tensor for an ellipsoid in an isotropic matrix of stiffness $\boldsymbol{C}_1$ (see Section 17.2.2). Simple algebra reveals that formula (18.46) implies (18.44) and (18.45) in the limit of spherical inclusions. Formula (18.46) is identical to the approximation due to Mori and Tanaka (1973) for the effective stiffness tensor of dispersions of aligned ellipsoids of the same shape in a matrix. Moreover, (18.46) coincides with one of the bounds (21.82), due to Willis (1977), for the same microstructure, depending on whether the matrix is stiffer or more compliant than all of the inclusions (Weng 1992).

The case of randomly oriented ellipsoidal inclusions is easily obtained from (18.46) by replacing the tensor $\boldsymbol{T}^{(j1)}$ with its isotropic form. This yields the following expressions for the effective bulk and shear moduli:

$$\sum_{j=1}^{M} \phi_j (K_e - K_j) T_h^{(j1)} = 0, \tag{18.48}$$

$$\sum_{j=1}^{M} \phi_j (G_e - G_j) T_s^{(j1)} = 0, \tag{18.49}$$

where $T_h^{(j1)}$ and $T_s^{(j1)}$ are the hydrostatic and shear contributions to $\boldsymbol{T}^{(j1)}$ [cf. (17.100)].

Mori–Tanaka schemes work best when the inclusions are spheres or unidirectionally aligned ellipsoids for $M \geq 2$ (Benveniste 1987, Weng 1992) or for randomly oriented ellipsoids for $M = 2$ (Benveniste 1987). In other instances, Mori–Tanaka approximations are more problematic and may even violate fundamental conditions. Benveniste (1987) has shown that the Mori–Tanaka scheme can yield an effective tensor $\boldsymbol{C}_e$ that is not symmetric for multiphase composites with different alignment and/or shapes of the inclusions. Norris (1989) demonstrated that Mori–Tanaka estimates for randomly oriented ellipsoids can give results that violate the Hashin–Shtrikman bounds for $M \geq 3$. Moreover, Berryman and Berge (1996) have compared expressions (18.48) and (18.49) to experimental data and concluded that these estimates should not be used when the inclusion volume fraction is greater than about 20–30%. Ferrari and Filipponi (1991) has shown that the Mori–Tanaka scheme yields the absurd result, when the composite is anisotropic, that the effective tensor $\boldsymbol{C}_e$ can still depend on the matrix elastic properties even in the limit of vanishing matrix volume fraction.

18.2.2 Self-Consistent Approximations

Spherical Inclusions

Implicit self-consistent (SC) approximations completely analogous to those derived for the effective conductivity have been obtained for the effective elastic moduli. Such approximations were first derived by Budiansky (1965) and Hill (1965a,1965b) using a different approach from the one leading to the self-consistent approximations for the effective conductivity outlined in Section 18.1.2. Also, Gubernatis and Krumhansl (1975) and Berryman (1980b) used a scattering approach to obtain self-consistent approximations. The derivation presented below is the analogue of the conductivity formulation.

Consider a macroscopically isotropic composite composed of M different types of spherical inclusions with volume fractions $\phi_1, \ldots, \phi_M$, bulk moduli $K_1, \ldots, K_M$, and shear moduli $G_1, \ldots, G_M$. The effect of all the material outside a type j inclusion is to produce a homogeneous medium (matrix) whose effective bulk modulus K_e is the unknown to be calculated. One now requires that the perturbations to a uniform field be zero *on average*. According to relation (17.53) for the intensity field, a single spherical inclusion produces a dipole perturbation. Thus, we choose the effective bulk modulus K_e by the self-consistency requirement that perturbations due to each type of inclusion vanish on average, i.e.,

$$\sum_{j=1}^{M} \phi_j \kappa_{je} = 0. \tag{18.50}$$

More explicitly, this is equivalent to

$$\sum_{j=1}^{M} \phi_j \frac{K_j - K_e}{K_j + \frac{2(d-1)}{d} G_e} = 0. \tag{18.51}$$

This is the SC estimate of the effective bulk modulus K_e, which depends upon the unknown effective shear modulus G_e.

Following the same procedure as for the bulk modulus, we find the SC approximation for the effective shear modulus to be given by

$$\sum_{j=1}^{M} \phi_j \mu_{je} = 0, \tag{18.52}$$

or, more explicitly, by

$$\sum_{j=1}^{M} \phi_j \frac{G_j - G_e}{G_j + H_e} = 0, \tag{18.53}$$

where

$$H_e \equiv \frac{G_e[dK_e/2 + (d+1)(d-2)G_e/d]}{K_e + 2G_e}. \tag{18.54}$$

Note that the SC relations (18.51) and (18.53) are coupled and hence provide the effective moduli K_e and G_e only implicitly. These values are found by iterating to the fixed point, which is known to be stable and unique for positive values of the moduli. Relations (18.51) and (18.53) are the direct analogues of formula (18.13) for the effective conductivity and are symmetric in the phases in the same sense as (18.13).

The problems with applying SC approximations to general dispersions have already been discussed in Section 18.1.2. Accordingly, most of these criticisms will not be repeated here. It is useful to examine the behavior of the SC approximations (18.51) and (18.53) for infinite-contrast situations. In the case of two incompressible phases, relation (18.51) correctly predicts that the composite itself is incompressible, whereas (18.53) simplifies as

$$\phi_1 \frac{G_e - G_1}{dG_e/2 + G_1} + \phi_2 \frac{G_e - G_2}{dG_e/2 + G_2} = 0 \qquad \text{if} \quad K_1/G_1 = K_2/G_2 = \infty. \tag{18.55}$$

Letting phase 2 be perfectly rigid relative to phase 1 allows this relation to be further simplified as

$$\frac{G_e}{G_1} = \frac{1}{1 - \frac{d+2}{2}\phi_2} \qquad \text{if} \quad K_1/G_1 = K_2/G_2 = \infty,\ G_2/G_1 = \infty. \tag{18.56}$$

Thus, the SC approximation predicts that an incompressible dispersion with a perfectly rigid included phase forms a percolating cluster (i.e., composite becomes perfectly rigid) at the threshold of $\phi_{2c} = 2/(d+2)$, independent of the microstructure (see Figure 18.6). Since percolation thresholds are very sensitive to microstructural details and vary over a wide range of volume fractions, this threshold prediction must be regarded as spurious for general dispersions.

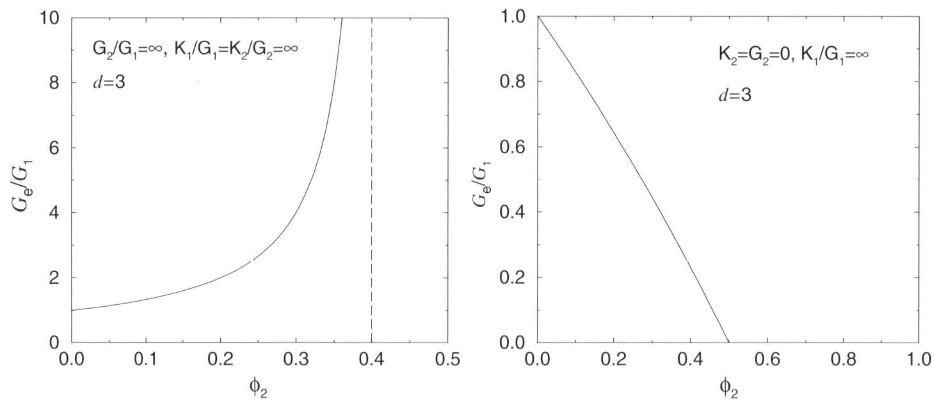

Figure 18.6 Scaled shear modulus G_e/G_1 versus ϕ_2 in the SC approximation for infinite-contrast situations in three dimensions. Left panel: Formula (18.56) for a rigid phase 2 predicts a percolation threshold $\phi_{2c} = 0.4$. Right panel: Formula (18.58) for cavities (phase 2) in an incompressible matrix predicts that at $\phi_{2c} = 0.5$, the composite can no longer carry any shear load.

If phase 2 is taken to be a cavity or void phase and phase 1 is assumed to be incompressible, then the SC approximations (18.51) and (18.53) simplify as

$$K_e = \frac{2(d-1)\phi_1}{d\,\phi_2} G_e \qquad \text{if} \quad K_2 = G_2 = 0, \quad K_1/G_1 = \infty, \tag{18.57}$$

$$\frac{G_e}{G_1} = \frac{d[(d-1)-(d+1)\phi_2]}{d(d-1)-2\phi_2} \qquad \text{if} \quad K_2 = G_2 = 0, \quad K_1/G_1 = \infty. \tag{18.58}$$

Therefore, for such a porous medium, both the effective bulk and shear moduli are predicted to vanish for $\phi_2 \geq (d-1)/(d+1)$ (see Figure 18.6).

As in the case of the conductivity, the SC approximations for the elastic moduli will generally violate bounds that improve upon the Hashin–Shtrikman bounds for realistic models of dispersions. An example of such a violation is depicted in Figure 20.5.

Recall that Milton (1984, 1985) showed that the SC relations (18.51) and (18.53) are realizable by the fractal-like granular aggregates discussed earlier (see Figure 18.3). Thus, these relations always lie between the Hashin–Shtrikman bounds and can be used as benchmarks to test theories.

Let us now reexamine the percolation behavior of the SC expressions in the light that they are realizable. Consider the case $d = 3$. Formula (18.56) states that the SC composite becomes perfectly rigid at the threshold $\phi_2^c = 2/5$, which is above the superconducting threshold $\phi_2^c = 1/3$ [cf. (18.15)]. Thus, although geometric and conduction connectivity coincide, *rigidity* connectivity requires a higher volume fraction for the SC structures. Similarly, relations (18.57) and (18.58) state that the SC composites with voids can no longer carry any hydrostatic or shear loads at $\phi_2^c = 1/2$, which is lower than the nonconducting threshold $\phi_2^c = 2/3$ [cf. (18.16)]. Unlike the superconducting and nonconducting SC formulas for the effective conductivity for $d = 2$, formulas (18.56) and (18.58) give different thresholds for $d = 2$. Expression (18.56) for $d = 2$ yields $\phi_2^c = 1/2$, which happens to equal the superconducting or connectivity threshold. However, relation (18.58) for $d = 2$ yields $\phi_2^c = 1/3$.

Ellipsoidal Inclusions

A generalization of the SC formulas (18.51) and (18.53) for a macroscopically anisotropic composite consisting of M different types of identical unidirectionally aligned isotropic ellipsoidal inclusions is given by

$$\sum_{j=1}^{M} \phi_j (\boldsymbol{C}_j - \boldsymbol{C}_e) \cdot \boldsymbol{T}^{(je)} = 0, \tag{18.59}$$

where

$$\boldsymbol{T}^{(je)} = [\boldsymbol{I} + \boldsymbol{S}_e : \boldsymbol{C}_e^{-1} : (\boldsymbol{C}_j - \boldsymbol{C}_e)]^{-1} \tag{18.60}$$

and $\boldsymbol{S}_e$ is Eshelby's tensor for an ellipsoid in a matrix with a stiffness tensor $\boldsymbol{C}_e$ (Berryman 1980a). When the embedding medium is anisotropic, the intensity field within the ellipsoid is still a constant given by (17.89) (Eshelby 1957, Mura 1987), but the corresponding Eshelby tensor is more intricate than it is in the isotropic case discussed in Section 17.2.2. Formula (18.59) reduces to expressions (18.51) and (18.53) in the limit of spherical inclusions.

When the ellipsoids are randomly oriented, we need only consider the isotropic form of $\boldsymbol{T}^{(je)}$ in (18.18), and we obtain

$$\sum_{j=1}^{M} \phi_j (K_j - K_e) T_h^{(je)} = 0, \tag{18.61}$$

$$\sum_{j=1}^{M} \phi_j (G_j - G_e) T_s^{(je)} = 0, \tag{18.62}$$

where $T_h^{(je)}$ and $T_s^{(je)}$ are the hydrostatic and shear contributions to the tensor $\boldsymbol{T}^{(je)}$ [cf. (17.100)]. The indices j and e in $T_h^{(je)}$ and $T_s^{(je)}$ indicate that a type j inclusion is embedded in a matrix with effective moduli K_e and G_e.

Another class of approximations developed by Wu (1966) and Boucher (1974), also referred to as SC estimates, do not yield the same results as (18.61) and (18.62), except for spherical inclusions. Relations (18.61) and (18.62) are preferred, since unlike this other class of SC approximations, they always correspond to realizable microstructures (Milton 1985, Avellaneda 1987a).

18.2.3 Differential Effective-Medium Approximations

The ideas of Bruggeman (1935) have been extended to obtain analogues of the conductivity DEM approximation for the effective elastic moduli K_e and G_e; see, for example,

Boucher (1974) and McLaughlin (1977). In contrast to the SC approximation, the DEM approximation does not treat each phase symmetrically. For a two-phase composite, phase 1 is treated as a matrix phase with volume fraction ϕ_1, and phase 2 is taken to be an included phase with volume fraction ϕ_2. Let us suppose that the effective bulk modulus $K_e(\phi_2)$ at one value of ϕ_2 is known. Now treat $K_e(\phi_2)$ as the composite host bulk modulus and let $K_e(\phi_2 + \Delta\phi_2)$ represent the effective bulk modulus of the composite after a small fraction $\Delta\phi_2/(1 - \phi_2)$ of composite host has been replaced by inclusions of phase 2 (see Section 18.1.3). In the case of d-dimensional spherical inclusions, we can use the dilute-limit result (19.67) to get the difference $K_e(\phi_2 + \Delta\phi_2) - K_e(\phi_2)$ and, in the limit $\Delta\phi_2 \to 0$, we obtain the differential equation

$$(1 - \phi_2)\,\frac{dK_e}{d\phi_2} = \left[K_e + \frac{2(d-1)}{d}G_e\right]\frac{K_2 - K_e}{K_2 + \frac{2(d-1)}{d}G_e}, \tag{18.63}$$

with $K_e\ (\phi_2 = 0) = K_1$. The corresponding formula for the effective shear modulus is given by

$$(1 - \phi_2)\,\frac{dG_e}{d\phi_2} = [G_e + H_e]\frac{G_2 - G_e}{G_2 + H_e}, \tag{18.64}$$

with $G_e\ (\phi_2 = 0) = G_1$ and where H_e is given by (18.54). The embedding process ensures that the initial matrix material remains *connected* in the resulting composite.

The DEM approximations (18.63) and (18.64) are realizable, since they are special cases of the generalized DEM scheme devised by Norris (1985), and hence fall between the Hashin–Shtrikman bounds. The latter is the analogue of the aforementioned generalized DEM procedure for the conductivity (Norris et al. 1985) which, by design, corresponds to realizable hierarchical models.

The DEM approximations admit simple solutions in infinite-contrast cases. For example, when both phases are incompressible and phase 2 is perfectly rigid relative to phase 1, then the DEM approximation (18.64) yields

$$\frac{G_e}{G_1} = \frac{1}{(1 - \phi_2)^{(d+2)/2}} \qquad \text{if} \quad K_1/G_1 = K_2/G_2 = \infty,\ G_2/G_1 = \infty. \tag{18.65}$$

Figure 18.7 compares the Maxwell, SC, and DEM approximations for a two-phase two-dimensional medium in this extreme case, but where, as before, $k_i \equiv K_i$. Unlike the prediction of the SC formula, the Maxwell and DEM approximations imply that the rigid phase does not become connected until the point $\phi_2 = 1$ is reached. The behavior of these three approximations in the opposite extreme, when phase 2 consists of cavities, is qualitatively similar to the behavior of the corresponding approximations for the effective conductivity in this limit (see Figure 18.5).

Ellipsoidal Inclusions

A generalization of the DEM formulas (18.63) and (18.64) for a macroscopically anisotropic composite consisting of isotropic ellipsoidal inclusions (of the same shape and unidirectionally aligned) in a matrix is given by

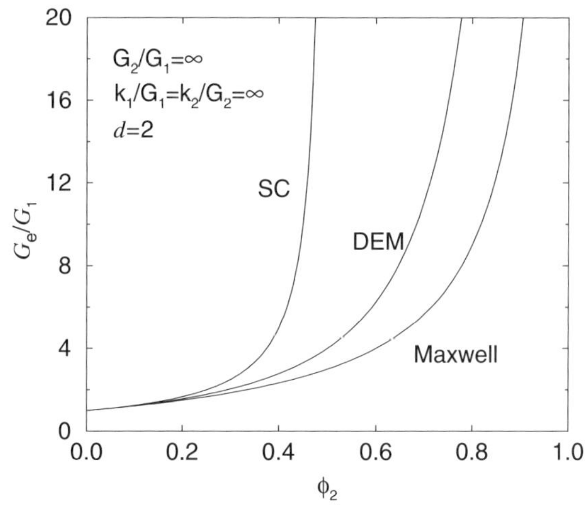

Figure 18.7 Comparison of the predictions of the three different effective-medium approximations for the scaled shear modulus G_e/G_1 versus ϕ_2 of a two-phase composite in which phase 2 is rigid and $d = 2$. These formulas correspond to (18.41), (18.56), and (18.65).

$$(1-\phi_2)\frac{d\boldsymbol{C}_e}{d\phi_2} = (\boldsymbol{C}_2 - \boldsymbol{C}_e) : \boldsymbol{T}^{(2e)}, \tag{18.66}$$

where $\boldsymbol{T}^{(2e)}$ is given by (18.60) with $j = 2$ and $\boldsymbol{C}_e(\phi_2 = 0) = \boldsymbol{C}_1$ (McLaughlin 1977). We see that formula (18.66) reduces to expressions (18.63) and (18.64) in the limit of spherical inclusions. Avellaneda (1987a) has proved the realizability of the DEM approximation (18.66) for multiphase anisotropic composites. When the ellipsoids are randomly oriented, then we have

$$(1-\phi_2)\frac{dK_e}{d\phi_2} = (K_2 - K_e)T_h^{(2e)}, \tag{18.67}$$

$$(1-\phi_2)\frac{dG_e}{d\phi_2} = (G_2 - G_e)T_s^{(2e)}, \tag{18.68}$$

where $T_h^{(2e)}$ and $T_s^{(2e)}$ are the hydrostatic and shear contributions to the tensor $\boldsymbol{T}^{(2e)}$ [cf. (17.100)].

18.3 Trapping Constant

Numerous sophisticated approximations have been developed to estimate the trapping constant γ for diffusive processes among traps; see, for example, Muthukumar and Cukier (1981), Calef and Deutch (1983), Richards (1987) and references therein. Here we derive a simple self-consistent approximation that is based on the approach that Brinkman (1947) employed to estimate the fluid permeability of porous media.

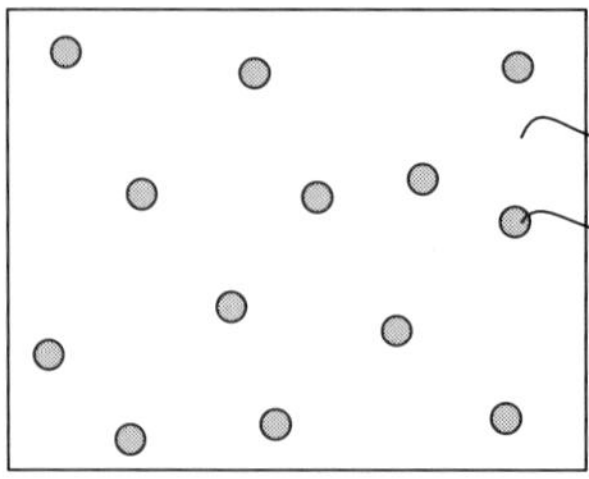

Figure 18.8 Distribution of spherical traps (phase 2).

Consider inserting a spherical trap of radius R into an effective medium containing N traps in which the production rate per unit trap-free volume is G. Let the concentration field infinitely far from this sphere be the constant C_0 and take the diffusion coefficient $\mathcal{D}$ to be unity. What is the microscopic equation for this problem? Employing the ideas of Brinkman (1947) for the fluid permeability, one could propose the following governing equations for the concentration field C outside the test sphere:

$$\Delta C(r) - \gamma C(r) = -G, \qquad r > R, \tag{18.69}$$

$$C = 0, \qquad r = R, \tag{18.70}$$

$$C = C_0, \qquad r = \infty. \tag{18.71}$$

In the limit that the trap volume fraction tends to zero, $\gamma \to 0$ and the microscopic steady-state diffusion equation $\Delta C = -G$ is recovered from (18.69). On the other hand, for nondilute situations, the first term on the left side of (18.69) is $\mathcal{O}(\tilde{C}/L^2)$, where $\tilde{C}$ is a typical concentration and L is a typical length scale for macroscopic variation in the diffusion process. The second term is $\mathcal{O}(\gamma\tilde{C})$ or $\mathcal{O}(\tilde{C}/\ell^2)$, where ℓ is a typical pore length scale. Therefore, since $L \gg \ell$, the first term in (18.69) is negligibly small compared to the second term, and we effectively recover the macroscopic relation $G = \gamma C_0$. Thus, the relation (18.69) interpolates between the usual microscopic diffusion equation (13.140) and the usual macroscopic relation (13.142).

The solution of (18.69)–(18.71) is given by

$$C(r) = C_0\left[1 - \frac{R}{r}e^{-\sqrt{\gamma}(r-R)}\right]. \tag{18.72}$$

Comparing this to the single-sphere solution (17.107) reveals that the $1/r$ interaction is *screened* by virtue of the exponentially decaying term that multiplies it. Physically, screening effects arise because the test sphere sees its surroundings as an effective medium with trapping constant γ. When the "screening length" $1/\sqrt{\gamma}$ tends to infinity, $C(r)$ tends to the unscreened single-sphere solution (17.107). The flux into one sink, j, is given by

$$j = \int -\nabla C(r) \cdot \boldsymbol{n}\, dS = 4\pi R(1 + \sqrt{\gamma}R)C_0. \tag{18.73}$$

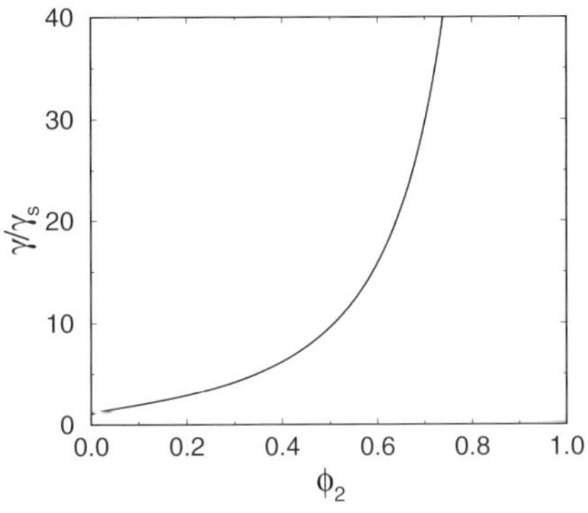

Figure 18.9 Scaled trapping constant γ/γ_s versus trap volume fraction ϕ_2 in the SC approximation (18.75).

Therefore, the total flux for N traps, $J = Nj$, is given by

$$J = N \cdot 4\pi R(1 + \sqrt{\gamma}R)C_0. \tag{18.74}$$

Now we require that the flux due to an assembly of traps be equal to the macroscopic flux, i.e.,

$$N \cdot 4\pi R(1 + \sqrt{\gamma}R)C_0 = GV_1 = \gamma C_0 V_1,$$

where we have used relation (13.150), and V_1 is the total volume of the pore space. In the thermodynamic limit, this yields

$$\gamma = \frac{\gamma_s}{\phi_1}(1 + \sqrt{\gamma}\, R), \tag{18.75}$$

where γ_s is the Smoluchowski dilute-limit result (19.89). Indeed, (18.75) gives the proper volume-fraction correction to the Smoluchowski result:

$$\frac{\gamma}{\gamma_s} = 1 + \sqrt{3}\phi_2^{1/2} + \cdots, \tag{18.76}$$

which agrees with the exact dilute-concentration result (19.102) through the same order in ϕ_2. However, at arbitrary values of ϕ_2, formula (18.75) cannot be expected to be quantitatively accurate, since it is independent of the spatial arrangement of the traps. Figure 18.9 depicts the scaled trapping constant γ/γ_s versus the trap volume fraction ϕ_2 as predicted by (18.75).

18.4 Fluid Permeability

Brinkman (1947) obtained a self-consistent approximation for the fluid permeability k of a porous medium. Consider a cylindrical porous medium composed of an isotropic

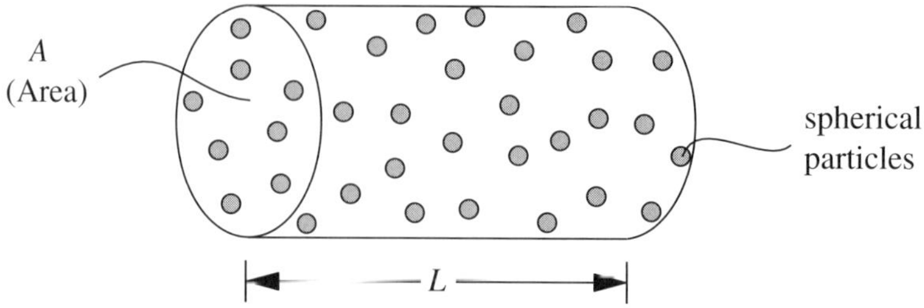

Figure 18.10 Cylindrical porous medium composed of spherical obstacles.

distribution of identical spherical obstacles of radius R (see Figure 18.10) whose generator is parallel to the z-direction. Let ΔP denote the macroscopic pressure drop along the the length L of the cylinder, let A be the cross-sectional area of the medium transverse to the average flow direction, and set the dynamic viscosity μ equal to unity. Now embed in the porous medium a test sphere of radius R and let the velocity infinitely far from this sphere be the constant $\boldsymbol{U} = U_0\, \boldsymbol{i}_z$, where $\boldsymbol{i}_z$ is the unit vector in the positive z-direction.

What are the microscopic equations for flow around the test sphere of radius R that is embedded in this porous medium? Brinkman proposed the equations

$$\Delta \boldsymbol{U} - \frac{1}{k}\boldsymbol{U} = \nabla p, \qquad r > R, \tag{18.77}$$

$$\nabla \cdot \boldsymbol{U} = 0, \qquad r > R, \tag{18.78}$$

$$\boldsymbol{U} = 0, \qquad r = R, \tag{18.79}$$

$$\boldsymbol{U} = \boldsymbol{U}_0, \qquad r = \infty, \tag{18.80}$$

where $\boldsymbol{U}$ is the velocity field and p is the pressure field. In the limit that the solid concentration tends to zero, $k \to \infty$, and thus the microscopic Stokes momentum equation is recovered from (18.77). On the other hand, for realistic porous media, the first term on the left side of (18.77) is $\mathcal{O}(\tilde{U}/L^2)$, where $\tilde{U}$ and L are typical velocity and length scales, respectively, for macroscopic variation in the flow. The second term is $\mathcal{O}(\tilde{U}/k)$ or $\mathcal{O}(\tilde{U}/\ell^2)$, where ℓ is a typical pore length scale. Therefore, since $L \gg \ell$, the first term in (18.77) can be dropped, and we effectively recover Darcy's law. Thus, Brinkman's momentum equation (18.77) interpolates between the usual microscopic momentum equation (13.157) and and the macroscopic Darcy's law (13.156).

The solution of (18.77)–(18.80) is given by

$$\boldsymbol{U}(\boldsymbol{r}) = \nabla\left\{\left[U_0 r + a\frac{e^{-\lambda r}(1+\lambda r) - 1}{\lambda^2 r^2} - \frac{b}{r^2}\right]\cos\theta\right\} + a\frac{e^{-\lambda r}}{r}\boldsymbol{i}_z, \tag{18.81}$$

$$p = \left(-\lambda^2 U_0 r + \frac{a}{r^2} + \frac{b\lambda^2}{r^2}\right)\cos\theta, \tag{18.82}$$

where r, θ, and ϕ are spherical polar coordinates with a polar axis in the positive z-direction, and

$$\lambda = k^{-1/2}, \quad a = -\frac{3}{2}U_0 R e^{\lambda R}, \quad b = \frac{1}{2}U_0 R\left[-R^2 + \frac{3}{\lambda^2}(e^{\lambda R} - 1 - \lambda R)\right].$$

As in the trapping problem, the fact that the test sphere sees its surroundings as an effective medium results in terms that screen the interactions that would result if the test sphere were surrounded by pure fluid [cf. (17.137)]. As expected, when the "screening length" $\sqrt{k}$ tends to infinity, $\boldsymbol{U}(\boldsymbol{r})$ tends to the unscreened single-sphere solution (17.137). The stress components σ_{rr} and $\sigma_{r\theta}$ are obtained from the relations

$$\sigma_{rr} = -p + 2\frac{\partial u_r}{\partial r}, \tag{18.83}$$

$$\sigma_{r\theta} = r\frac{\partial(u_\theta/r)}{\partial r} + \frac{1}{r}\frac{\partial u_r}{\partial \theta}. \tag{18.84}$$

The magnitude of the *drag force* f on the embedded sphere is found by integrating the z-component of (18.83) and (18.84) over the test sphere surface:

$$\begin{aligned} f &= \int (\sigma_{rr}\cos\theta + \sigma_{r\theta}\sin\theta)\, dS \\ &= 6\pi R U_0\left(1 + \frac{R}{\sqrt{k}} + \frac{R^2}{3k}\right). \end{aligned} \tag{18.85}$$

The magnitude of the drag force due to all of the spheres is $F = Nf$, and thus

$$F = Nf = 6\pi R U_0 N\left(1 + \frac{R}{\sqrt{k}} + \frac{R^2}{3k}\right). \tag{18.86}$$

The permeability is calculated self-consistently by requiring that (18.86) be equal to the drag computed from the macroscopic relation, i.e.,

$$F = \Delta P \cdot A = \frac{U_0 L}{k} \cdot A, \tag{18.87}$$

and thus

$$k = k_s\left(1 + \frac{R}{\sqrt{k}} + \frac{R^2}{3k}\right)^{-1}, \tag{18.88}$$

where k_s is the Stokes dilute-limit result (19.103). Indeed, we see that (18.88) gives the proper volume-fraction correction to the Stokes result; i.e., it yields the *inverse* permeability

$$\frac{k_s}{k} = 1 + \frac{3}{\sqrt{2}}\phi_2^{1/2} + \cdots, \tag{18.89}$$

which agrees with the exact dilute-concentration result (19.119) through the same order in ϕ_2. However, relation (18.88) cannot be expected to be quantitatively accurate for

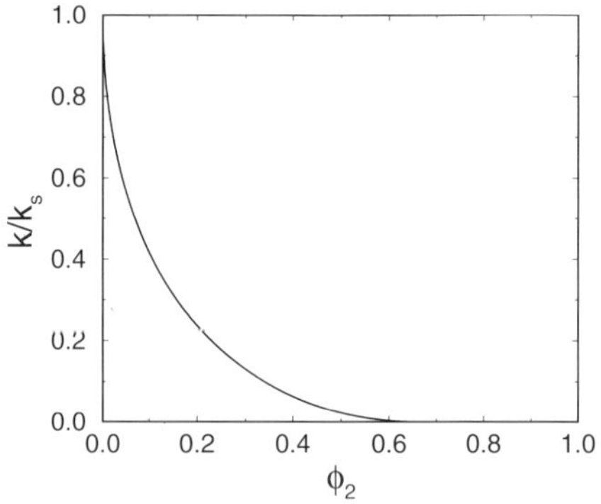

Figure 18.11 Scaled fluid permeability k/k_s versus sphere volume fraction ϕ_2 in the Brinkman approximation (18.88).

arbitrary volume fractions, since it does not incorporate nontrivial information about the microstructure.

In Figure 18.11, the dimensionless permeability k/k_s predicted by Brinkman's expression (18.88) is plotted versus the sphere volume fraction ϕ_2. Note that k vanishes when $\phi_2 = 2/3$ as in the three-dimensional SC approximation (18.16) for the effective conductivity when phase 2 is perfectly insulating. As in the case of the SC approximations for the conductivity and elastic moduli, such a percolation threshold must be regarded as spurious, since percolation thresholds vary widely, depending on the microstructure.

Finally, we remark that Brinkman allowed for an effective viscosity μ' multiplying the Laplacian term in (18.77) in his original derivation. In most applications of Brinkman's equations, it is assumed that $\mu' = 1$, i.e., it is equal to the pure fluid viscosity. However, it can be argued that μ' will be altered by the porous medium at nonnegligible solid concentration. For a discussion of this point, the reader is referred to Koplik, Levine and Zee (1983) and to Martys, Bentz and Garboczi (1994a), who showed (via simulations) that $\mu' > 1$.

CHAPTER 19

Cluster Expansions

Heterogeneous materials composed of well-defined inclusions (e.g., spheres, cylinders, ellipsoids) distributed randomly throughout a matrix material have served as excellent starting points for modeling the complex field interactions in random composites. The celebrated formulas of Maxwell (1873) and Einstein (1906) for the effective conductivity and effective viscosity of dispersions of spheres, respectively, assume that the particles do not interact with another and therefore are valid through first order in the sphere volume fraction ϕ_2. Similar formulas for the trapping constant of dilute distributions of traps or the fluid permeability of dilute beds of spheres can easily be obtained from the classical results of Smoluchowski (1917) and Stokes (1851), respectively, given in Chapter 17.

One can systematically correct such dilute-limit formulas by taking into account interactions between pairs of spheres, triplets of spheres, and so on. Following the language of statistical mechanics, we refer to such schemes as *cluster-expansion* techniques. Although it is generally impossible to compute the effective properties of random dispersions exactly through all orders in ϕ_2, low-density expansions provide benchmark results to test approximate theories and experiments.

In this chapter we will derive exact expressions for the effective properties of dispersions in the dilute (noninteracting) limit for spherical or ellipsoidal inclusions. Then we will describe the full cluster-expansion procedure for each effective property and state the first few terms in the density expansion in the case of dispersions of spheres.

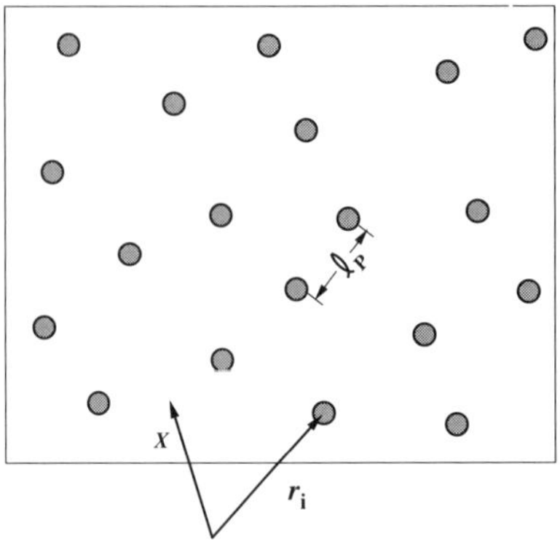

Figure 19.1 A dispersion of identical spheres of radius R. Here ℓ_P is the mean nearest neighbor distance, r_i is the position of the ith sphere, and x is an arbitrary position in the system. For a dilute dispersion, $\ell_P \gg R$ and the sphere volume fraction satisfies $\phi_2 \ll 1$.

19.1 Conductivity

Consider a dispersion of identical spheres of radius R and conductivity σ_2 in a matrix of conductivity σ_1, and let the position of the ith particle be denoted by $\boldsymbol{r}_i$ (see Figure 19.1). The local conductivity $\sigma(\boldsymbol{x})$ can be expressed as

$$\sigma(\boldsymbol{x}) = \sigma_1 + (\sigma_2 - \sigma_1)\mathcal{I}^{(2)}(\boldsymbol{x}), \tag{19.1}$$

where

$$\begin{aligned} \mathcal{I}^{(2)}(\boldsymbol{x}; \boldsymbol{r}^N) &= 1 - \prod_{i=1}^{N}[1 - m(|\boldsymbol{x} - \boldsymbol{r}_i|)] \\ &= \sum_{i=1}^{N} m(|\boldsymbol{x} - \boldsymbol{r}_i|) - \sum_{i<j}^{N} m(|\boldsymbol{x} - \boldsymbol{r}_i|)m(|\boldsymbol{x} - \boldsymbol{r}_j|) + \cdots \end{aligned} \tag{19.2}$$

is the indicator function for N possibly overlapping spheres (phase 2), as obtained from the matrix-phase relation (4.3), and $m(r)$ is the inclusion indicator function, given by (4.4). The flux $\boldsymbol{J}$ is related to the electric field $\boldsymbol{E}$ by Ohm's law (13.6) and, by using (19.1), can be written as

$$\boldsymbol{J}(\boldsymbol{x}) = \sigma_1 \boldsymbol{E}(\boldsymbol{x}) + \boldsymbol{P}(\boldsymbol{x}), \tag{19.3}$$

where $\boldsymbol{P}$ is the "flux polarization field," given by

$$\boldsymbol{P}(\boldsymbol{x}) = (\sigma_2 - \sigma_1)\mathcal{I}^{(2)}(\boldsymbol{x})\boldsymbol{E}(\boldsymbol{x}). \tag{19.4}$$

The polarization field $\boldsymbol{P}$ is the flux induced within the inclusions over and above the flux in the absence of inclusions and thus is zero in the matrix (see Chapter 20 for a more general definition). The effective conductivity tensor $\boldsymbol{\sigma}_e$ can also be found from the alternative averaged relation

$$\langle \boldsymbol{P} \rangle = (\boldsymbol{\sigma}_e - \sigma_1) \cdot \langle \boldsymbol{E} \rangle. \tag{19.5}$$

To indicate that the local field depends on the positions of the particles $\boldsymbol{r}^N$, we write it as $\boldsymbol{E}(\boldsymbol{x}; \boldsymbol{r}^N)$. We can decompose this many-body function into interactions arising from no spheres, single spheres, pairs of spheres, and so on. To show this decomposition, let us impose an applied field $\boldsymbol{E}_0$ on a matrix without inclusions and consider the contributions to the local field $\boldsymbol{E}$ as we sequentially add inclusions to the system. Clearly, when there are no inclusions, the local and applied fields are equal, i.e.,

$$\boldsymbol{E}(\boldsymbol{x}) = \boldsymbol{E}_0.$$

After the addition of a single inclusion at $\boldsymbol{r}_1$, the local field is given by

$$\boldsymbol{E}(\boldsymbol{x}; \boldsymbol{r}_1) = \boldsymbol{E}_0 + \boldsymbol{K}_1(\boldsymbol{x}; \boldsymbol{r}_1) \cdot \boldsymbol{E}_0,$$

where $\boldsymbol{K}_1(\boldsymbol{r})$ is the single-body operator due to interactions over and above the applied field $\boldsymbol{E}_0$, and $\boldsymbol{r} = \boldsymbol{x} - \boldsymbol{r}_1$. This second-order tensor is found from the solution of the single-sphere boundary value problem (17.14) as

$$\boldsymbol{K}_1(\boldsymbol{r}) = \begin{cases} \beta_{21}\left(\dfrac{R}{r}\right)^d [d\hat{\boldsymbol{r}}\hat{\boldsymbol{r}} - \boldsymbol{I}], & r > R, \\ -\beta_{21}\boldsymbol{I}, & r < R, \end{cases} \tag{19.6}$$

where

$$\beta_{21} = \frac{\sigma_2 - \sigma_1}{\sigma_2 + (d-1)\sigma_1} \tag{19.7}$$

is the *polarizability*, given previously by (17.13) and defined more generally in Section 20.1.1. After the addition of another inclusion at $\boldsymbol{r}_2$, the field is

$$\boldsymbol{E}(\boldsymbol{x}; \boldsymbol{r}_1, \boldsymbol{r}_2) = \boldsymbol{E}_0 + \boldsymbol{K}_1(\boldsymbol{x}; \boldsymbol{r}_1) \cdot \boldsymbol{E}_0 + \boldsymbol{K}_1(\boldsymbol{x}; \boldsymbol{r}_2) \cdot \boldsymbol{E}_0 + \boldsymbol{K}_2(\boldsymbol{x}; \boldsymbol{r}_1, \boldsymbol{r}_2) \cdot \boldsymbol{E}_0,$$

where $\boldsymbol{K}_2(\boldsymbol{x}; \boldsymbol{r}_1, \boldsymbol{r}_2)$ is the two-body operator that accounts for interactions over and above the applied field $\boldsymbol{E}_0$ and single-body interactions; it is obtainable from the two-sphere boundary value problem (Jeffrey 1973).

More generally, for an N-body system, we have

$$\begin{aligned} \boldsymbol{E}(\boldsymbol{x}; \boldsymbol{r}^N) = \boldsymbol{E}_0 &+ \sum_{i=1}^{N} \boldsymbol{K}_1(\boldsymbol{x}; \boldsymbol{r}_i) \cdot \boldsymbol{E}_0 + \sum_{i<j}^{N} \boldsymbol{K}_2(\boldsymbol{x}; \boldsymbol{r}_i, \boldsymbol{r}_j) \cdot \boldsymbol{E}_0 \\ &+ \sum_{i<j<k}^{N} \boldsymbol{K}_3(\boldsymbol{x}; \boldsymbol{r}_i, \boldsymbol{r}_j, \boldsymbol{r}_k) \cdot \boldsymbol{E}_0 + \cdots, \end{aligned} \tag{19.8}$$

where $\boldsymbol{K}_n(\boldsymbol{x}; \boldsymbol{r}^n)$ is a second-order tensor operator that accounts for intrinsically n-body interactions. Equation (19.8) represents a *cluster* expansion of the many-body function $\boldsymbol{E}(\boldsymbol{x}; \boldsymbol{r}^N)$ into a zero-inclusion term, singlet terms, doublet terms, etc. The nth sum in (19.8) is over all distinguishable n-tuplets of particles and thus contains $N!/[n!(N-n)!]$ terms.

Ensemble averaging (19.8) yields

$$\langle \boldsymbol{E} \rangle \equiv \int \boldsymbol{E}(\boldsymbol{r}^N) P_N(\boldsymbol{r}^N) d\boldsymbol{r}^N$$
$$= \boldsymbol{E}_0 + \left\langle \sum_{i=1}^{N} \boldsymbol{K}_1 \cdot \boldsymbol{E}_0 \right\rangle + \left\langle \sum_{i<j}^{N} \boldsymbol{K}_2 \cdot \boldsymbol{E}_0 \right\rangle + \cdots, \tag{19.9}$$

where we recall that $P_N(\boldsymbol{r}^N)$ is the *specific* probability density function (3.4) associated with finding a configuration of spheres with positions $\boldsymbol{r}^N$ and $d\boldsymbol{r}^N \equiv d\boldsymbol{r}_1 \cdots d\boldsymbol{r}_N$. The averages on the right side of (19.9) *are not necessarily zero* and in fact depend upon the *shape* of the *macroscopic volume*! Thus, generally

$$\langle \boldsymbol{E} \rangle \neq \boldsymbol{E}_0. \tag{19.10}$$

This *conditional convergence* arises because for large separations ($|\boldsymbol{x} - \boldsymbol{r}_i| \to \infty$), the $\boldsymbol{K}_n$ go to zero as $|\boldsymbol{x} - \boldsymbol{r}_i|^{-d}$.

19.1.1 Dilute Dispersions of Spheres

We seek to obtain an expression for the effective conductivity of a dispersion of spheres through first order in the sphere volume fraction ϕ_2. To do so, we will ultimately use the constitutive relation (19.5) for the average polarization $\langle \boldsymbol{P} \rangle$. We begin by substituting relation (19.8) for the local field $\boldsymbol{E}$ into expression (19.4) for the local polarization $\boldsymbol{P}$, retaining lowest-order terms, and averaging to get

$$\langle \boldsymbol{P} \rangle = (\sigma_2 - \sigma_1)[\phi_2 \boldsymbol{I} + \langle \mathcal{I}^{(2)} \Sigma \boldsymbol{K}_1 \rangle + \cdots] \cdot \boldsymbol{E}_0. \tag{19.11}$$

Since we are interested in the dilute limit ($\phi_2 \ll 1$), we can neglect the overlap contributions to relation (19.2) for the indicator function, obtaining

$$\mathcal{I}^{(2)}(\boldsymbol{x}; \boldsymbol{r}^N) = \sum_{i=1}^{N} m(|\boldsymbol{x} - \boldsymbol{r}_i|). \tag{19.12}$$

Therefore, we write the ensemble average appearing in (19.11) as

$$\left\langle \mathcal{I}^{(2)} \sum_{i=1}^{N} \boldsymbol{K}_1 \right\rangle = \int \sum_{i=1}^{N} \sum_{j=1}^{N} m(|\boldsymbol{x} - \boldsymbol{r}_i|) \boldsymbol{K}_1(\boldsymbol{x} - \boldsymbol{r}_j) P_N(\boldsymbol{r}^N) d\boldsymbol{r}^N$$
$$= \int \sum_{i=1}^{N} m(|\boldsymbol{x} - \boldsymbol{r}_i|) \boldsymbol{K}_1(\boldsymbol{x} - \boldsymbol{r}_i) P_N(\boldsymbol{r}^N) d\boldsymbol{r}^N$$
$$+ \int \sum_{i \neq j}^{N} m(|\boldsymbol{x} - \boldsymbol{r}_i|) \boldsymbol{K}_1(\boldsymbol{x} - \boldsymbol{r}_j) P_N(\boldsymbol{r}^N) d\boldsymbol{r}^N$$
$$= \int m(|\boldsymbol{x} - \boldsymbol{r}_1|) \boldsymbol{K}_1(\boldsymbol{x} - \boldsymbol{r}_1) \rho_1(\boldsymbol{r}_1) d\boldsymbol{r}_1$$
$$+ \int m(|\boldsymbol{x} - \boldsymbol{r}_1|) \boldsymbol{K}_1(\boldsymbol{x} - \boldsymbol{r}_2) \rho_2(\boldsymbol{r}_1, \boldsymbol{r}_2) d\boldsymbol{r}_1 d\boldsymbol{r}_2. \tag{19.13}$$

Recall that $\rho_n(\mathbf{r}^n)$ is the *generic* n-particle probability density function defined by (3.7). For statistically homogeneous suspensions, $\rho_1(\mathbf{r}_1)$ equals the number density $\rho = N/V$ (where $N \to \infty$, $V \to \infty$, such that ρ is a fixed number).

Assuming statistical homogeneity and noting that ρ_2 is $\mathcal{O}(\rho^2)$, we obtain

$$\langle \mathcal{I}^{(2)} \Sigma \boldsymbol{K}_1 \rangle = \rho \int m(y) \boldsymbol{K}_1(\boldsymbol{y}) d\boldsymbol{y} + \mathcal{O}(\rho^2), \tag{19.14}$$

where we have made a change of variables to $\boldsymbol{y} = \boldsymbol{x} - \boldsymbol{r}_1$, with $y = |\boldsymbol{y}|$. Inverting (19.8), retaining lowest-order terms, and averaging yield

$$\boldsymbol{E}_0 = \langle \boldsymbol{E} \rangle - \rho \int \boldsymbol{K}_1(\boldsymbol{y}) \cdot \langle \boldsymbol{E} \rangle d\boldsymbol{y} + \mathcal{O}(\rho^2). \tag{19.15}$$

Substitution of (19.14) and (19.15) into (19.11) yields that $\langle \boldsymbol{P} \rangle$ and $\langle \boldsymbol{E} \rangle$, through order ρ (or ϕ_2), are related by

$$\langle \boldsymbol{P} \rangle = (\sigma_2 - \sigma_1) \left[\phi_2 \boldsymbol{I} + \rho \int m(y) \boldsymbol{K}_1(\boldsymbol{y}) d\boldsymbol{y} \right] \cdot \langle \boldsymbol{E} \rangle + \mathcal{O}(\phi_2^2). \tag{19.16}$$

Since the step function $m(y)$ appears in the integrand, we need $\boldsymbol{K}_1$ [cf. (19.6)] for only $y < R$, which is the constant isotropic tensor

$$\boldsymbol{K}_1(\boldsymbol{y}) = \boldsymbol{R} - \boldsymbol{I} = -\beta_{21} \boldsymbol{I}, \qquad y < R, \tag{19.17}$$

where $\boldsymbol{R}$ is defined by (17.17). Carrying out the integration in (19.16) gives

$$\langle \boldsymbol{P} \rangle = \boldsymbol{M} \cdot \langle \boldsymbol{E} \rangle \, \phi_2 + \mathcal{O}(\phi_2^2), \tag{19.18}$$

where $\boldsymbol{M} = (\sigma_2 - \sigma_1)\boldsymbol{R}$ is the constant, isotropic second-order polarization concentration tensor defined by (17.20), i.e.,

$$\boldsymbol{M} = d\sigma_1 \beta_{21} \boldsymbol{I}. \tag{19.19}$$

Comparing (19.5) to (19.18) and recognizing that the effective conductivity tensor is isotropic (i.e., $\boldsymbol{\sigma}_e = \sigma_e \boldsymbol{I}$) lead to the desired dilute-limit relation for the scalar effective conductivity σ_e in any spatial dimension d:

$$\sigma_e = \sigma_1 + d\sigma_1 \beta_{21} \phi_2 + \mathcal{O}(\phi_2^2). \tag{19.20}$$

This result for $d = 3$ is usually attributed to Maxwell (1873), since his approximate formula (18.5) is exact through first order in ϕ_2. One can easily verify that the "polarization (concentration) coefficient" $d\sigma_1\beta_{21}$ multiplying ϕ_2 in (19.20) is *not changed* if there is *polydispersivity in the sphere sizes*. Indeed, this independence of the size distribution is true for inclusions of *arbitrary shape*. Observe also that since expression (19.20) coincides with one of the Hashin–Shtrikman bounds (21.20) when the latter is expanded through first order in ϕ_2, the polarization coefficient for spheres provides bounds on the corresponding coefficient for isotropic dispersions of inclusions of *arbitrary* shape. Specifically, if the corresponding expansion for the general situation has the form $\sigma_e = \sigma_1 + \sigma_1 f \beta_{21} \phi_2 + \mathcal{O}(\phi_2^2)$, then the coefficient f, which generally depends on the phase conductivities, obeys the following inequality:

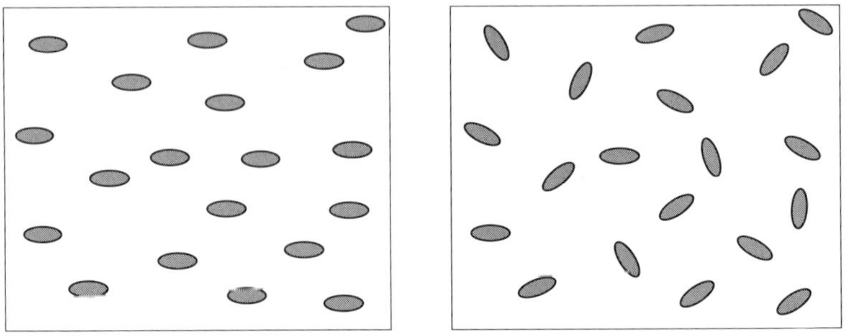

Figure 19.2 Dispersions of spheroids. Left panel: Aligned ellipsoids with the structure of an idealized nematic liquid crystal (Section 7.1). Right panel: Randomly oriented ellipsoids.

$$f \geq d. \tag{19.21}$$

Note that result (19.29) for ellipsoidal inclusions correctly obeys the bound (19.21), as do results for polygonal inclusions (Hetherington and Thorpe 1992). Finally, we note that in the limit that the dimension d goes to infinity, the conductivity expression (19.20) becomes the arithmetic average $\langle \sigma \rangle$, as it should according to the general Theorem 20.2.

For a dispersion of perfectly insulating spheres ($\sigma_2 = 0$), (19.20) reduces to

$$\frac{\sigma_e}{\sigma_1} = 1 - \frac{d}{d-1}\phi_2 + \mathcal{O}(\phi_2^2). \tag{19.22}$$

For a dispersion of superconducting spheres relative to the matrix ($\sigma_2/\sigma_1 = \infty$), (19.20) becomes

$$\frac{\sigma_e}{\sigma_1} = 1 + d\phi_2 + \mathcal{O}(\phi_2^2). \tag{19.23}$$

19.1.2 Dilute Dispersions of Ellipsoids

Consider a dilute suspension of d-dimensional ellipsoidal inclusions with semiaxes $a_1, a_2, \ldots, a_d$ with specified orientations. The conductivities of the matrix and inclusions are σ_1 and σ_2, respectively. We treat both aligned ellipsoids and randomly oriented ellipsoids (see Figure 19.2).

First, let the ellipsoids be aligned along the x_d-axis. Following the same analysis given above for spheres reveals that the average polarization for aligned ellipsoids through first order in ϕ_2 is still given by (19.18), but where $\boldsymbol{M}$ is the polarization concentration tensor for an ellipsoid given by (17.21). Hence, comparison to (19.5) gives

$$\boldsymbol{\sigma}_e = \boldsymbol{\sigma}_1 + \boldsymbol{M}\phi_2 + \mathcal{O}(\phi_2^2), \tag{19.24}$$

where

$$\boldsymbol{M} = (\sigma_2 - \sigma_1)\left[\boldsymbol{I} + \boldsymbol{A}^* \frac{(\sigma_2 - \sigma_1)}{\sigma_1}\right]^{-1} \tag{19.25}$$

and $\boldsymbol{A}^*$ is the depolarization tensor for an ellipsoidal inclusion defined in Section 17.1.2.

The case of aligned three-dimensional spheroids in which $a = a_1 = a_2$ and $b = a_3$ follows immediately from (17.29) and (19.24) with $d = 3$. We find that the effective conductivity tensor for this *transversely isotropic* composite is given by

$$\boldsymbol{\sigma}_e = \begin{bmatrix} (\sigma_e)_{11} & 0 & 0 \\ 0 & (\sigma_e)_{22} & 0 \\ 0 & 0 & (\sigma_e)_{33} \end{bmatrix}, \tag{19.26}$$

where

$$(\sigma_e)_{11} = (\sigma_e)_{22} = \sigma_1 + \frac{(\sigma_2 - \sigma_1)\phi_2}{1 + Q(\sigma_2 - \sigma_1)/\sigma_1} + \mathcal{O}(\phi_2^2), \tag{19.27}$$

$$(\sigma_e)_{33} = \sigma_1 + \frac{(\sigma_2 - \sigma_1)\phi_2}{1 + (1 - 2Q)(\sigma_2 - \sigma_1)/\sigma_1} + \mathcal{O}(\phi_2^2), \tag{19.28}$$

Q is the shape factor given by (17.30) and (17.31) for prolate and oblate spheroids, respectively, and $\phi_2 = \rho 4\pi a^2 b/3$ [cf. (7.22)]. We see that for needle-shaped inclusions ($Q = 1/2$), $(\sigma_e)_{33}$ reduces to the arithmetic average, as expected. For disk-shaped inclusions ($Q = 0$), $(\sigma_e)_{33} = \sigma_1[1 + (\sigma_2 - \sigma_1)\phi_2] + \mathcal{O}(\phi_2^2)$, equal to the harmonic average through first order in ϕ_2. We note that Douglas and Garboczi (1995) give the tensor coefficient $\boldsymbol{M}$ for other particle shapes.

The scalar effective conductivity for the case of a dilute isotropic suspension of randomly oriented ellipsoidal inclusions is obtained by forming the scalar product of (19.24) with the identity tensor $\boldsymbol{I}$, yielding

$$\sigma_e = \sigma_1 + M\phi_2 + \mathcal{O}(\phi_2^2), \tag{19.29}$$

where M is just the *scalar* polarization coefficient

$$M = \frac{\sigma_2 - \sigma_1}{d} \sum_{k=1}^{d} \frac{1}{1 + A_k^*(\sigma_2 - \sigma_1)/\sigma_1} \tag{19.30}$$

and the A_k^* are the "depolarization" factors defined by (17.25). This result in three dimensions was given by Polder and Van Santen (1946). From the results (17.34), (17.35), and (19.29), it is seen that the effective conductivities for three-dimensional needles and disks are given by

$$\frac{\sigma_e}{\sigma_1} = 1 + \frac{(\sigma_2 + 5\sigma_1)(\sigma_2 - \sigma_1)}{3\sigma_1(\sigma_1 + \sigma_2)}\phi_2 + \mathcal{O}(\phi_2^2) \qquad \text{(needles)}, \tag{19.31}$$

$$\frac{\sigma_e}{\sigma_1} = 1 + \frac{(2\sigma_2 + \sigma_1)(\sigma_2 - \sigma_1)}{3\sigma_1\sigma_2}\phi_2 + \mathcal{O}(\phi_2^2) \qquad \text{(disks)}. \tag{19.32}$$

19.1.3 Nondilute Concentrations

We seek a "cluster expansion" for the effective conductivity σ_e of a suspension of particles, i.e., an expansion that accounts for the effect of a successively larger number of particles. However, one wants to pass to the infinite-volume limit *without shape-dependent (conditionally convergent) integrals* appearing in the expression for σ_e. The history of the determination of the effective properties of heterogeneous media is replete with studies that either ignored or were oblivious to this subtle obstacle. This even includes the renowned Lord Rayleigh (1892), who obtained a result for σ_e that depended on the shape of the macroscopic sample.

Jeffrey (1973) evaluated the effective conductivity of dispersions of spheres through second order in the volume fraction and used a method due to Batchelor (1972) to make the two-body integral absolutely convergent. By extending this "renormalization" technique to bypass conditionally convergent integrals to higher-order terms, Jeffrey (1974) later obtained σ_e to all orders. However, this renormalization procedure is not systematic.

A systematic procedure to obtain cluster expansions (*or any other expansion*) for the effective conductivity that are free of conditional-convergence problems was first laid out by Brown (1955), who obtained a contrast expansion (see Chapter 20). The basic idea is as follows: One first considers a large but finite sample of *arbitrary shape* in an arbitrary applied field $\boldsymbol{E}_0(\boldsymbol{x})$. The average polarization $\langle \boldsymbol{P} \rangle$ is then expressed as a formal operator acting on the applied field $\boldsymbol{E}_0$. The formal operator, an ensemble-averaged quantity, is then expanded in the desired fashion. As is well known from continuum theories, however, relations between average fields and $\boldsymbol{E}_0$ are dependent on the shape of the sample (see, e.g., Section 17.1.2), and hence the integrals involved here must necessarily be *conditionally convergent*. Accordingly, one then seeks the appropriate series expression for $\boldsymbol{E}_0$ in terms of the average field $\langle \boldsymbol{E} \rangle$. Using this expression, the applied field $\boldsymbol{E}_0$ is eliminated in favor of $\langle \boldsymbol{E} \rangle$ in the aforementioned expression for $\langle \boldsymbol{P} \rangle$. The resulting expression between $\langle \boldsymbol{P} \rangle$ and $\langle \boldsymbol{E} \rangle$ is localized, i.e., independent of the shape of the sample, and hence involves *absolutely convergent* integrals. One may now pass to the limit of an infinite volume without ambiguity and obtain from this "localized" relation the particular expansion for σ_e of statistically homogeneous media.

Felderhof, Ford and Cohen (1982) used this procedure to obtain a cluster expansion of σ_e in the case of *identical mutually impenetrable* spherical inclusions. Torquato (1984) generalized this result to the instance of identical spheres distributed with an arbitrary degree of impenetrability. In the thermodynamic limit, it is found that

$$\sigma_e = \sigma_1 + \sum_{n=1}^{\infty} \frac{1}{n!} \int W_n(\boldsymbol{r}^n) d\boldsymbol{r}^n, \tag{19.33}$$

where $W_n(\boldsymbol{r}^n)$ is a complicated functional of the n-body cluster operators $\boldsymbol{K}_n$ and the set of n-particle probability density functions $\rho_1, \ldots, \rho_n$. Clearly, W_n depends upon the density ρ of the spheres. If W_n is analytic in ρ, we can obtain a density or volume-fraction expansion from (19.33):

$$\frac{\sigma_e}{\sigma_1} = 1 + \sum_{n=1}^{\infty} B_n \phi_2^n. \tag{19.34}$$

The coefficient B_n is a multidimensional integral over the solutions $\boldsymbol{K}_1, \ldots, \boldsymbol{K}_n$ weighted with the n-particle functions $\rho_1, \ldots, \rho_n$.

Following Torquato (1984), the effective conductivity for possibly overlapping spheres in d dimensions through $\mathcal{O}(\phi_2^2)$ is

$$\frac{\sigma_e}{\sigma_1} = 1 + B_1\phi_2 + B_2\phi_2^2, \tag{19.35}$$

where

$$B_1 = d\beta_{21}, \tag{19.36}$$

$$B_2 = d\beta_{21}^2 + \int f[\boldsymbol{K}_1, \boldsymbol{K}_2] g_2^{(0)}(r)\, d\boldsymbol{r}, \tag{19.37}$$

$g_2^{(0)}(r)$ is the zero-density limit of the radial distribution function, and $f[\boldsymbol{K}_1, \boldsymbol{K}_2]$ denotes a scalar functional of $\boldsymbol{K}_1$ and $\boldsymbol{K}_2$. Torquato (1984) divided up the region of integration into parts: one for $r > 2R$, which gives the contribution to B_2 for a *reference* system of *totally* impenetrable spheres, and one for $r < 2R$, which gives the contribution to B_2 (in excess of the reference part) due entirely to *overlap* or *clustering* effects:

$$B_2 = B_2^* + B_2^+, \tag{19.38}$$

$$B_2^* = \int f[\boldsymbol{K}_1, \boldsymbol{K}_2] B_2^{(0)}(r)\, d\boldsymbol{r}, \tag{19.39}$$

$$B_2^+ = \int f[\boldsymbol{K}_1, \boldsymbol{K}_2] P_2^{(0)}(r)\, d\boldsymbol{r}, \tag{19.40}$$

where

$$g_2^{(0)}(r) = B_2^{(0)}(r) + P_2^{(0)}(r). \tag{19.41}$$

Here $B_2^{(0)}(r)$ and $P_2^{(0)}(r)$ are the zero-density limits of the pair-blocking and pair-connectedness functions, respectively, defined in Chapter 9.

Impenetrable Spheres

Consider three-dimensional dispersions of impenetrable spheres. Jeffrey (1973) computed $B_2 = B_2^*$ ($B_2^+ = 0$) for such a system using

$$g_2^{(0)}(r) = B_2^{(0)}(r) = \begin{cases} 0, & r \le 2R, \\ 1, & r > 2R, \end{cases} \tag{19.42}$$

which corresponds to the most random distribution of dilute spheres subject to the impenetrability constraint. (Note that $P_2^{(0)} = 0$.) In the extreme contrast situations, he found that

$$B_2(\sigma_2/\sigma_1 = 0) \approx 0.588, \qquad B_2(\sigma_2/\sigma_1 = \infty) \approx 4.51.$$

Therefore, for perfectly insulating inclusions ($\sigma_2/\sigma_1 = 0$), Jeffrey obtained

$$\frac{\sigma_e}{\sigma_1} = 1 - \frac{3}{2}\phi_2 + 0.588\phi_2^2 + \mathcal{O}(\phi_2^3). \tag{19.43}$$

If instead we use a radial distribution function corresponding to a "well-separated" dispersion in d dimensions,

$$g_2^{(0)}(r) = B_2^{(0)}(r) = \begin{cases} 0, & r \le R\phi_2^{-1/d}, \\ 1, & r > R\phi_2^{-1/d}, \end{cases} \tag{19.44}$$

then the integral of (19.39) vanishes (for any phase contrast and dimension d), and we instead get for $d = 3$

$$\frac{\sigma_e}{\sigma_1} = 1 - \frac{3}{2}\phi_2 + \frac{3}{4}\phi_2^2 + \mathcal{O}(\phi_2^3). \tag{19.45}$$

This formula can be obtained by expanding Maxwell's approximation (18.5) or the Hashin–Shtrikman *upper* bound (21.22) through second order in ϕ_2. This last point is not surprising, since the Hashin–Shtrikman bounds are realized by coated inclusions that are well separated from each other (see Chapter 16).

For superconducting inclusions ($\sigma_2/\sigma_1 = \infty$) obeying (19.42),

$$\frac{\sigma_e}{\sigma_1} = 1 + 3\phi_2 + 4.51\phi_2^2 + \mathcal{O}(\phi_2^3). \tag{19.46}$$

This is to be contrasted with the expression that results when we instead use (19.44),

$$\frac{\sigma_e}{\sigma_1} = 1 + 3\phi_2 + 3\phi_2^2 + \mathcal{O}(\phi_2^3), \tag{19.47}$$

which again can be obtained by expanding Maxwell's approximation (18.5) or the Hashin–Shtrikman *lower* bound (21.21) through second order in ϕ_2. Not surprisingly, it is seen that a dispersion in which the spheres are allowed to get arbitrarily close to each other has a higher effective conductivity than one in which the spheres are well separated from each other.

The analogous two-dimensional calculation for impenetrable disks was performed by Peterson and Hermans (1969). We note that in the case of superconducting disks, they found, using (19.42), that the effective conductivity is approximately given by

$$\frac{\sigma_e}{\sigma_1} = 1 + 2\phi_2 + 2.74\phi_2^2 + \mathcal{O}(\phi_2^3). \tag{19.48}$$

This is to be contrasted with the result using (19.44), in which the second-order coefficient 2.74 is replaced with 2. Choy, Alexopoulos and Thorpe (1998) obtained corresponding second-order expansions for σ_e for d-dimensional spheres at any phase contrast.

Interpenetrable Spheres

For interpenetrable-sphere models in three dimensions, Torquato (1985b) estimated B_2, or equivalently B_2^+ [given by (19.40)], as a function of the contrast ratio σ_2/σ_1. For example, in the fully-penetrable-sphere limit of the cherry-pit model (i.e., when the impenetrability parameter λ is equal to 0), he found that for perfectly insulating inclusions

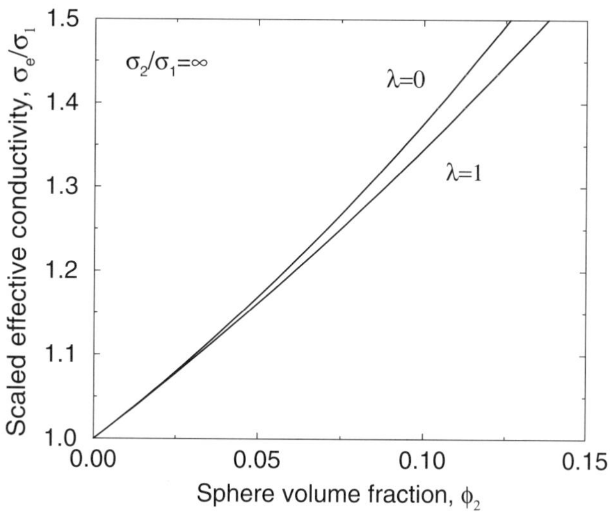

Figure 19.3 Comparison of the effective conductivities for systems of impenetrable ($\lambda = 1$) and fully penetrable ($\lambda = 0$) superconducting spheres ($d = 3$) as computed from the second-order expressions (19.46) and (19.51).

$$B_2(\sigma_2/\sigma_1 = 0) \approx 0.345, \qquad B_2^+(\sigma_2/\sigma_1 = 0) \approx -0.243,$$

and for superconducting inclusions

$$B_2(\sigma_2/\sigma_1 = \infty) \approx 7.56, \qquad B_2^+(\sigma_2/\sigma_1 = \infty) \approx 3.05.$$

These results are obtained using (19.42) and

$$P_2^{(0)}(r) = \begin{cases} 1, & r \le 2R, \\ 0, & r > 2R. \end{cases} \tag{19.49}$$

Therefore, for $\sigma_2/\sigma_1 = 0$ and $\lambda = 0$, it was found that

$$\frac{\sigma_e}{\sigma_1} \approx 1 - \frac{3}{2}\phi_2 + 0.345\phi_2^2 + \mathcal{O}(\phi_2^3). \tag{19.50}$$

This is to be contrasted with the corresponding impenetrable-sphere result (19.45). For $\sigma_2/\sigma_1 \approx \infty$ and $\lambda = 0$,

$$\frac{\sigma_e}{\sigma_1} \approx 1 + 3\phi_2 + 7.56\phi_2^2 + \mathcal{O}(\phi_2^3). \tag{19.51}$$

This expression for fully penetrable spheres is compared to (19.46) for totally impenetrable spheres in Figure 19.3.

To summarize, through second order in the sphere volume fraction, the effect of clustering is to *increase* the effective conductivity σ_e when the inclusions are superconducting ($\sigma_2/\sigma_1 = \infty$) at fixed ϕ_2. On the other hand, when the inclusions are perfectly insulating ($\sigma_2/\sigma_1 = 0$) at fixed ϕ_2, the effect of clustering is to *decrease* the effective conductivity σ_e. These are rather intuitively obvious results.

An exact evaluation of the cluster expansion for arbitrary densities is generally out of the question. Cluster-expansion techniques become difficult to implement in practice because they require information about the n-body problems (n-body boundary value problem as well as n-body statistics), which are generally intractable analytically. The solution of the three-body boundary value problem is already very difficult to obtain. Some investigators have attempted to sum the series approximately by approximating the n-body contributions with lower-body contributions. For examples of such studies, the reader is referred to Cichocki and Felderhof (1988), Ju and Chen (1994), and Buryachenko (2000).

19.2 Elastic Moduli

Consider a dispersion of identical spheres of radius R and elastic moduli K_2, G_2 in a matrix of elastic moduli K_1, G_1 and denote by $\boldsymbol{r}_i$ the position of the ith particle (see Figure 19.1). The local stiffness tensor $\boldsymbol{C}(\boldsymbol{x})$ can be expressed as

$$\boldsymbol{C}(\boldsymbol{x}) = \boldsymbol{C}_1 + (\boldsymbol{C}_2 - \boldsymbol{C}_1)\mathcal{I}^{(2)}(\boldsymbol{x}), \tag{19.52}$$

where $\mathcal{I}^{(2)}(\boldsymbol{x}; \boldsymbol{r}^N)$ is the indicator function for N possibly overlapping spheres, given by (19.2), and

$$\boldsymbol{C}_i = dK_i\boldsymbol{\Lambda}_h + 2G_i\boldsymbol{\Lambda}_s, \tag{19.53}$$

where $\boldsymbol{\Lambda}_h$ and $\boldsymbol{\Lambda}_s$ are the hydrostatic and shear projection tensors defined by (13.96) and (13.97), respectively. The stress $\boldsymbol{\tau}$ is related to the strain field $\boldsymbol{\varepsilon}$ by Hooke's law (13.53), and by using (19.52), it can be written as

$$\boldsymbol{\tau}(\boldsymbol{x}) = \boldsymbol{C}_1\boldsymbol{\varepsilon}(\boldsymbol{x}) + \boldsymbol{p}(\boldsymbol{x}), \tag{19.54}$$

where $\boldsymbol{p}$ is the "stress polarization field" given by

$$\boldsymbol{p}(\boldsymbol{x}) = (\boldsymbol{C}_2 - \boldsymbol{C}_1)\mathcal{I}^{(2)}(\boldsymbol{x}) : \boldsymbol{\varepsilon}(\boldsymbol{x}). \tag{19.55}$$

The polarization field $\boldsymbol{p}$ is the stress induced within the inclusions over and above the stress in the absence of inclusions and therefore is zero in the matrix (see Chapter 20 for a more general definition). The effective stiffness tensor $\boldsymbol{C}_e$ can then be obtained from the alternative averaged relation

$$\langle \boldsymbol{p} \rangle = (\boldsymbol{C}_e - \boldsymbol{C}_1) : \langle \boldsymbol{\varepsilon} \rangle. \tag{19.56}$$

In light of the fact that the local strain field depends on the positions of the particles $\boldsymbol{r}^N$, we write it as $\boldsymbol{\varepsilon}(\boldsymbol{x}; \boldsymbol{r}^N)$. We can decompose this many-body function into a cluster expansion, as we did for the electric field earlier. Therefore, for an applied strain field $\boldsymbol{\varepsilon}_0$, we have

$$\varepsilon(\boldsymbol{x};\boldsymbol{r}^N) = \varepsilon_0 + \sum_{i=1}^{N} \boldsymbol{L}_1(\boldsymbol{x};\boldsymbol{r}_i)\cdot\varepsilon_0 + \sum_{i<j}^{N} \boldsymbol{L}_2(\boldsymbol{x};\boldsymbol{r}_i,\boldsymbol{r}_j)\cdot\varepsilon_0 + \sum_{i<j<k}^{N} \boldsymbol{L}_3(\boldsymbol{x};\boldsymbol{r}_i,\boldsymbol{r}_j,\boldsymbol{r}_k)\cdot\varepsilon_0 + \cdots, \tag{19.57}$$

where $\boldsymbol{L}_n(\boldsymbol{x};\boldsymbol{r}^n)$ is a fourth-order tensor operator that accounts for intrinsically n-body interactions, i.e., over and above zero-body, single-body,..., and $(n-1)$-body interactions. The operator $\boldsymbol{L}_n(\boldsymbol{x};\boldsymbol{r}^n)$ is obtained by solving the n-sphere elasticity boundary value problem.

Ensemble averaging (19.57) yields

$$\begin{aligned}\langle\varepsilon\rangle &\equiv \int \varepsilon(\boldsymbol{r}^N)P_N(\boldsymbol{r}^N)d\boldsymbol{r}^N \\ &= \varepsilon_0 + \left\langle \sum_{i=1}^{N} \boldsymbol{L}_1\cdot\varepsilon_0 \right\rangle + \left\langle \sum_{i<j}^{N} \boldsymbol{L}_2\cdot\varepsilon_0 \right\rangle + \cdots.\end{aligned} \tag{19.58}$$

Once again, the averages on the right-hand side of (19.58) *are not necessarily zero* and in fact depend upon the *shape* of the *macroscopic volume*, so that generally

$$\langle\varepsilon\rangle \neq \varepsilon_0. \tag{19.59}$$

This *conditional convergence* arises because for large separations ($|\boldsymbol{x}-\boldsymbol{r}_i| \to \infty$), the $\boldsymbol{L}_n$ go to zero as $|\boldsymbol{x}-\boldsymbol{r}_i|^{-d}$.

19.2.1 Dilute Dispersions of Spheres

Following the procedure outlined in Section 19.1.1, we derive expressions for the effective elastic moduli of dispersions of spheres through first order in the sphere volume fraction ϕ_2. This is accomplished via the constitutive relation (19.56) for the average polarization $\langle\boldsymbol{p}\rangle$. We start by substituting relation (19.57) for the local strain ε into relation (19.55) for the local polarization $\boldsymbol{p}$, retaining lowest-order terms, and averaging to obtain

$$\langle\boldsymbol{p}\rangle = (\boldsymbol{C}_2 - \boldsymbol{C}_1) : [\phi_2\boldsymbol{I} + \langle\mathcal{I}^{(2)}\boldsymbol{L}_1\rangle + \cdots] : \varepsilon_0, \tag{19.60}$$

where $\boldsymbol{I}$ represents the fourth-order identity tensor. In the dilute limit, the indicator function $\mathcal{I}^{(2)}$ for the sphere phase is given by (19.12).

Following the derivation of (19.14), we write the ensemble average appearing in (19.60) as

$$\langle\mathcal{I}^{(2)}\boldsymbol{L}_1\rangle = \rho\int m(y)\boldsymbol{L}_1(\boldsymbol{y})d\boldsymbol{y} + \mathcal{O}(\rho^2), \tag{19.61}$$

where we have assumed statistical homogeneity and have retained lowest-order terms. Here ρ is the number density, $\boldsymbol{y} = \boldsymbol{x}-\boldsymbol{r}_1$, $y = |\boldsymbol{x}-\boldsymbol{r}_1|$, and we have made a simple change of variables. Inverting (19.57), retaining lowest order terms, and averaging yields

$$\varepsilon_0 = \langle \varepsilon \rangle - \rho \int \boldsymbol{L}_1(\boldsymbol{y}) : \langle \varepsilon \rangle d\boldsymbol{y} + \mathcal{O}(\rho^2). \tag{19.62}$$

Substitution of (19.61) and (19.62) into (19.60) yields that $\langle \boldsymbol{p} \rangle$ and $\langle \varepsilon \rangle$, through order ρ (or ϕ_2), are related by

$$\langle \boldsymbol{p} \rangle = (\boldsymbol{C}_2 - \boldsymbol{C}_1) : \left[\phi_2 \boldsymbol{I} + \rho \int m(y) \boldsymbol{L}_1(\boldsymbol{y}) d\boldsymbol{y} \right] : \langle \varepsilon \rangle + \mathcal{O}(\phi_2^2). \tag{19.63}$$

Given the presence of the step function $m(y)$ in the integrand, we need the tensor $\boldsymbol{L}_1$ for only $y < R$, which is the constant isotropic tensor

$$\boldsymbol{L}_1(\boldsymbol{y}) = \boldsymbol{T} - \boldsymbol{I} = -\kappa_{21} \boldsymbol{\Lambda}_h - \mu_{21} \boldsymbol{\Lambda}_s, \qquad y < R, \tag{19.64}$$

where $\boldsymbol{T}$ is given by (17.85), and κ_{21} and μ_{21} are the *bulk modulus polarizability* and *shear modulus polarizability*, respectively, given by (17.52) and (17.78). (These polarizabilities are defined more generally in Section 20.2.1.) Carrying out the integration in (19.63) yields

$$\langle \boldsymbol{p} \rangle = \boldsymbol{N} : \langle \varepsilon \rangle \, \phi_2 + \mathcal{O}(\phi_2^2), \tag{19.65}$$

where $\boldsymbol{N} = (\boldsymbol{C}_2 - \boldsymbol{C}_1) : \boldsymbol{T}$ is the constant isotropic fourth-order polarization concentration tensor defined by (17.87), i.e.,

$$\boldsymbol{N} = d[K_1 + 2(d-1)G_1/d]\kappa_{21} \boldsymbol{\Lambda}_h + 2[G_1 + H_1]\mu_{21} \boldsymbol{\Lambda}_s. \tag{19.66}$$

Comparison of (19.55) and (19.65) reveals that the effective stiffness tensor $\boldsymbol{C}_e$ is isotropic (i.e., $\boldsymbol{C}_e = dK_e \boldsymbol{\Lambda}_h + 2G_e \boldsymbol{\Lambda}_s$), and therefore the d-dimensional effective bulk and shear moduli, through first order in the sphere-volume fraction, are given respectively by

$$K_e = K_1 + [K_1 + 2(d-1)G_1/d]\kappa_{21}\phi_2 + \mathcal{O}(\phi_2^2), \tag{19.67}$$

$$G_e = G_1 + [G_1 + H_1]\mu_{21}\phi_2 + \mathcal{O}(\phi_2^2). \tag{19.68}$$

One can easily check that the "polarization coefficients" multiplying ϕ_2 in (19.67) and (19.68) are *not changed* if there is a *polydispersivity in the sphere sizes*. As in the conductivity expansion, the polarization coefficients provide bounds on the corresponding coefficients for isotropic dispersions of inclusions of *arbitrary* shape, since the expressions (19.67) and (19.68) coincide with the Hashin–Shtrikman bounds (21.70) and (21.73) when the latter are expanded through first order in ϕ_2. Specifically, if the corresponding expansions for the general instance have the forms $K_e = K_1 + K_1 g \kappa_{21}\phi_2 + \mathcal{O}(\phi_2^2)$ and $G_e = G_1 + G_1 h \mu_{21}\phi_2 + \mathcal{O}(\phi_2^2)$, then the coefficients g and h, which generally depend on the phase moduli, obey the following inequalities:

$$g \geq 1 + \frac{2(d-1)G_1}{K_1 d}, \qquad h \geq 1 + \frac{H_1}{G_1}. \tag{19.69}$$

Observe that results (19.80) and (19.81) for ellipsoidal inclusions correctly obey the bounds of (19.69), as do results for polygonal holes (Jasiuk, Chen and Thorpe 1994).

Note that in the infinite-dimensional limit, the shear-modulus expression (19.68) becomes the arithmetic average $\langle G \rangle$, as it should according to the general Theorem 20.2. The bulk modulus relation is not equal to the arithmetic average in this limit, consistent with Theorem 20.3.

For a dispersion of spherical cavities ($K_2 = G_2 = 0$), (19.67) and (19.68) reduce to

$$\frac{K_e}{K_1} = 1 - \frac{dK_1 + 2(d-1)G_1}{2(d-1)G_1}\phi_2 + \mathcal{O}(\phi_2^2), \tag{19.70}$$

$$\frac{G_e}{G_1} = 1 - \frac{G_1 + H_1}{H_1}\phi_2 + \mathcal{O}(\phi_2^2). \tag{19.71}$$

For the special cases of $d = 2$ and $d = 3$, it is useful to recast these into corresponding expressions for the effective Young modulus E_e and effective Poisson ratio ν_e using the interrelations (13.103) and (13.104). Thus, for cavities when $d = 2$, we have

$$\frac{E_e}{E_1} = 1 - 3\phi_2 + \mathcal{O}(\phi_2^2), \tag{19.72}$$

$$\nu_e = \nu_1 + (1 - 3\nu_1)\phi_2 + \mathcal{O}(\phi_2^2), \tag{19.73}$$

and when $d = 3$, we have

$$\frac{E_e}{E_1} = 1 - \frac{3(9 + 5\nu_1)(1 - \nu_1)}{2(7 - 5\nu_1)}\phi_2 + \mathcal{O}(\phi_2^2), \tag{19.74}$$

$$\nu_e = \nu_1 + \frac{3(1 - 5\nu_1)(1 - \nu_1^2)}{2(7 - 5\nu_1)}\phi_2 + \mathcal{O}(\phi_2^2). \tag{19.75}$$

First, observe that the effective Young modulus E_e for $d = 2$ (19.72) is independent of the Poisson ratio of the matrix ν_1, consistent with the general theorem stated in Section 16.2.7. For $d = 3$, E_e depends on ν_1. Second, in the dilute limit, increasing the porosity ϕ_2 can actually increase the effective Poisson's ratio ν_e over and above the solid matrix value ν_1 for the range $-1 \le \nu_1 < 1/3$ when $d = 2$ or for $-1 \le \nu_1 < 0.2$ when $d = 3$. This, of course, applies in the dilute limit.

For a dispersion of incompressible particles ($K_2/G_2 = \infty$) in an incompressible matrix ($K_1/G_1 = \infty$), the effective shear modulus obtained from (19.68) is

$$\frac{G_e}{G_1} = 1 + \frac{(d+2)(G_2 - G_1)}{2G_2 + dG_1}\phi_2 + \mathcal{O}(\phi_2^2). \tag{19.76}$$

If in addition to incompressible phases, we allow the particles to be perfectly rigid ($G_2/G_1 = \infty$), then the above formula yields

$$\frac{G_e}{G_1} = 1 + \frac{d+2}{2}\phi_2 + \mathcal{O}(\phi_2^2). \tag{19.77}$$

We see that formula (19.77) gives the celebrated Einstein formula (Einstein 1906) for $d = 3$; i.e., the coefficient of the first-order term is exactly 5/2.

19.2.2 Dilute Dispersions of Ellipsoids

Consider a dilute suspension of three-dimensional ellipsoidal inclusions with semi-axes a_1, a_2, a_3. The elastic moduli of the matrix and inclusions are K_1, G_1 and K_2, G_2, respectively. Both aligned and randomly oriented ellipsoids are examined (see Figure 19.2).

Let the ellipsoids be aligned along the x_3-axis. The same analysis used for spheres shows that relation (19.65) still holds, but where $\boldsymbol{N}$ is the polarization concentration tensor for an ellipsoid, given by (17.90). Comparing this expression to (19.56) gives the effective stiffness tensor of the dilute suspension as

$$\boldsymbol{C}_e = \boldsymbol{C}_1 + \boldsymbol{N}\phi_2 + \mathcal{O}(\phi_2^2), \tag{19.78}$$

where

$$\boldsymbol{N} = (\boldsymbol{C}_2 - \boldsymbol{C}_1) : [\boldsymbol{I} + (\boldsymbol{C}_2 - \boldsymbol{C}_1) : \boldsymbol{C}_1^{-1} : \boldsymbol{S}]^{-1} \tag{19.79}$$

and $\boldsymbol{S}$ is the Eshelby tensor for an ellipsoidal inclusion. Using the components of the Eshelby tensor, given in Section 17.2.2 for a spheroidal inclusion ($a = a_1 = a_2$ and $b = a_3$), one can immediately apply (19.78) to compute the effective stiffness tensor of a dilute dispersion of oriented spheroids.

The effective moduli for a dilute isotropic dispersion of randomly oriented ellipsoids are easily obtained from relation (19.78) by utilizing the isotropic form of $\boldsymbol{N}$ for ellipsoids, given by (17.99). We find that the effective moduli are given by

$$\frac{K_e}{K_1} = 1 + T_h(K_2 - K_1)\phi_2 + \mathcal{O}(\phi_2^2), \tag{19.80}$$

$$\frac{G_e}{G_1} = 1 + T_s(G_2 - G_1)\phi_2 + \mathcal{O}(\phi_2^2), \tag{19.81}$$

where T_h and T_s are the hydrostatic and shear contributions to the orientationally averaged $\boldsymbol{T}$ tensor defined by (17.100).

Using the results given in Table 17.1, we see that for randomly oriented needle-like inclusions in three dimensions,

$$\frac{K_e}{K_1} = 1 + \left[\frac{K_1 + G_1 + G_2/3}{K_2 + G_1 + G_2/3}\right](K_2 - K_1)\phi_2 + \mathcal{O}(\phi_2^2), \tag{19.82}$$

$$\frac{G_e}{G_1} = 1 + \frac{1}{5}\left[\frac{4G_1}{G_1 + G_2} + \frac{2G_1(6K_1 + 8G_1)}{G_1(3K_1 + G_1) + G_2(3K_1 + 7G_1)} + \frac{K_2 + 4G_1/3}{K_2 + G_1 + G_2/3}\right] \times (G_2 - G_1)\phi_2 + \mathcal{O}(\phi_2^2). \tag{19.83}$$

Similarly, for three-dimensional randomly oriented disk-like inclusions,

$$\frac{K_e}{K_1} = 1 + \left[\frac{K_1 + 4G_2/3}{K_2 + 4G_2/3}\right](K_2 - K_1)\phi_2 + \mathcal{O}(\phi_2^2), \tag{19.84}$$

$$\frac{G_e}{G_1} = 1 + \left[\frac{G_1 + H_2}{G_2 + H_2}\right](G_2 - G_1)\phi_2 + \mathcal{O}(\phi_2^2). \tag{19.85}$$

19.2.3 Nondilute Concentrations

As in the conductivity problem, we seek a "cluster expansion" for the effective stiffness tensor $\boldsymbol{C}_e$ of a suspension of particles, i.e., an expansion that accounts for the effect of successively larger number of particles. However, one wants to pass to the infinite-volume limit *without shape-dependent (conditionally convergent) integrals* appearing in the expression for $\boldsymbol{C}_e$. A systematic procedure for achieving this in any expansion technique was already described for the conductivity problem. Without repeating the details, it suffices here to say that expressing the average polarization $\langle \boldsymbol{p} \rangle$ as an operator acting on the average field $\langle \varepsilon \rangle$ results in a localized relation having *absolutely convergent* integrals.

The full cluster expansion for the effective elastic moduli of isotropic dispersions of identical mutually impenetrable spheres was derived by Jones and Schmitz (1984). This expression is completely analogous to the cluster expansion of the effective conductivity derived by Felderhof et al. (1982). In the thermodynamic limit, the results of Jones and Schmitz can be formally expressed as

$$\boldsymbol{C}_e = \boldsymbol{C}_1 + \sum_{n=1}^{\infty} \frac{1}{n!} \int \boldsymbol{V}_n(\boldsymbol{r}^n) d\boldsymbol{r}^n, \tag{19.86}$$

where $\boldsymbol{V}_n(\boldsymbol{r}^n)$ is a functional of the n-body cluster operators $\boldsymbol{L}_n$ and the set of n-particle probability density functions $\rho_1, \ldots, \rho_n$. Obviously, $\boldsymbol{V}_n$ depends upon the density of spheres ρ. If $\boldsymbol{V}_n$ is analytic in ρ, we can obtain a density or volume-fraction expansion from (19.86):

$$\boldsymbol{C}_e = \boldsymbol{C}_1 + \sum_{n=1}^{\infty} \boldsymbol{D}_n \phi_2^n. \tag{19.87}$$

The coefficient $\boldsymbol{D}_n$ is a multidimensional integral over the operators $\boldsymbol{L}_1, \ldots, \boldsymbol{L}_n$ weighted with the statistical functions $\rho_1, \ldots, \rho_n$.

Chen and Acrivos (1978a) solved the two-sphere elasticity problem for certain phase moduli and evaluated K_e and G_e through second order in ϕ_2 (Chen and Acrivos 1978b). In the case of perfectly rigid spherical inclusions ($d = 3$) in an incompressible matrix, they found that the effective shear modulus is given by

$$\frac{G_e}{G_1} = 1 + \frac{5}{2}\phi_2 + 5.01\phi_2^2 + \mathcal{O}(\phi_2^3) \tag{19.88}$$

when the radial distribution function is given by (19.42).

As already noted, the cluster-expansion technique becomes difficult to implement in practice because it requires information about the n-body problems (n-body boundary value problem as well as n-body statistics), which are generally intractable analytically. Thus, an exact evaluation of the cluster expansion for arbitrary densities is generally not possible. Consequently, one may attempt to sum the series approximately by estimating the n-body contributions in terms of lower-body contributions (Ersfeld 1997).

19.3 Trapping Constant

One of the classical problems in the theory of diffusion-controlled reactions is to determine the effective steady-state trapping constant γ for species diffusing among a distribution of stationary but perfectly absorbing ($\kappa = \infty$) spherical traps (Smoluchowski 1917). When a dispersion of identical spherical traps is dilute enough so that concentration fields from each trap do not interfere with each other, the trapping constant (as shown below) is given by the well-known Smoluchowski result

$$\gamma_s = \frac{3\phi_2}{R^2}, \tag{19.89}$$

where ϕ_2 and R are the volume fraction and radius of the spheres, respectively. For nondilute concentrations there is a competition between the traps for the diffusing species. A number of investigators have attempted to find density expansions of γ that correct Smoluchowski's result for nondilute concentrations (Felderhof and Deutch 1976, Lebenhaft and Kapral 1979, Muthukumar and Cukier 1981, Mattern and Felderhof 1987).

In what follows, we derive the dilute limit of the trapping constant for partially absorbing spherical traps ($\kappa < \infty$) and for perfectly absorbing spheroidal traps. We then discuss cluster-type expansions for nondilute concentrations.

19.3.1 Dilute Dispersions of Spherical Traps

We recall that the trapping constant γ for a system of traps with surface rate constant κ is generally defined by the constitutive relation

$$G = \gamma \mathcal{D} C, \tag{19.90}$$

where G is the production rate per unit volume, $\mathcal{D}$ is the diffusion coefficient, and C is the average concentration field [cf. (13.142)]. Consider now a dilute distribution of d-dimensional spherical traps of radius R with a finite surface rate constant κ. If the concentration fields from each trap do not interfere with each other, then we need to know the flux J_T into only one such trap in order to compute γ for the entire array. If ρ is the number density of traps, then the production rate per unit volume is

$$G = \rho J_T. \tag{19.91}$$

In Chapter 17 we found that the total flux into a single trap in an otherwise quiescent medium with uniform concentration field C_0 is given by (17.115). Substitution of this result into (19.91) yields

$$G = \frac{d(d-2)\kappa R \mathcal{D}}{\kappa R - (d-2)\mathcal{D} C_0} \frac{\phi_2}{R^2}, \qquad d \geq 3, \tag{19.92}$$

where we have used the fact that $v_1 = \Omega R^d/d$ is the volume of a trap and $\phi_2 = \rho v_1$ is the trap volume fraction. Use of (19.90) and (19.92) and the fact that in the infinite-volume limit $C_0 = C$ yields γ to be

$$\gamma = \frac{d(d-2)}{1 - \dfrac{(d-2)\mathcal{D}}{\kappa R}} \frac{\phi_2}{R^2}, \qquad d \geq 3. \tag{19.93}$$

For nonzero κ, it is seen that as the space dimension d tends to infinity (while the remaining parameters are held fixed), γ also tends to infinity. This result is not surprising since the surface-to-volume ratio of the spheres varies as d. In Section 21.3.1 we prove that $\gamma \to \infty$ as $d \to \infty$ for arbitrary microstructures in the diffusion-controlled limit. In the diffusion-controlled limit, we see that (19.93) gives

$$\gamma = d(d-2)\frac{\phi_2}{R^2}, \qquad d \geq 3. \tag{19.94}$$

When $d = 3$, we recover the well-known Smoluchowski result (19.89) for three-dimensional perfectly absorbing spherical traps.

What is the effect of polydispersivity in trap size on the dilute limit of the trapping constant γ? In the diffusion-controlled limit ($\kappa R/\mathcal{D} \to \infty$), one can easily verify, using similar arguments as given above, that γ for dilute arrays of polydisperse d-dimensional spherical traps is given by

$$\gamma = d(d-2)\frac{\langle R \rangle \phi_2}{\langle R^3 \rangle}, \tag{19.95}$$

where

$$\langle R^n \rangle = \int_0^\infty R^n f(R) dR \tag{19.96}$$

is the nth moment of the probability density function $f(R)$, which for the discrete-size case is given by (4.80). Thus, unlike the monodisperse case, the polydisperse result (19.95) depends on two length scales: $\langle R \rangle$ and $\langle R^3 \rangle^{1/3}$.

19.3.2 Dilute Dispersions of Spheroidal Traps

Consider a dilute dispersion of three-dimensional spheroidal traps with semiaxes a and b at number density ρ. In this case, the total flux J_T into a single prolate or oblate spheroidal trap is given by (17.125) or (17.128), respectively, independent of the orientation. Using these relations for J_T, (19.90) and (19.91) give the trapping constant for dilute spheroidal traps (Miller et al. 1991):

$$\gamma = \frac{3\phi_2}{a^2} f(b/a), \tag{19.97}$$

where

$$f(b/a) = \begin{cases} \dfrac{2\chi_b}{\ln\left(\dfrac{1+\chi_b}{1-\chi_b}\right)}, & b/a \geq 1, \\[2ex] \dfrac{\chi_a}{\tan^{-1}\chi_a}, & b/a \leq 1, \end{cases} \tag{19.98}$$

$$\chi_a^2 = -\chi_b^2 = \frac{a^2}{b^2} - 1. \tag{19.99}$$

We recall that the inclusion volume fraction for a spheroidal dispersion is $\phi_2 = \rho 4\pi a^2 b/3$ [cf. (7.22)].

We see that for spheres ($b/a = 1$), (19.97) reduces to the Smoluchowski result (19.89), as expected. For very slender inclusions ($b \gg a$), $f(b/a) \sim [\ln(2b/a)]^{-1}$, and so

$$\gamma = \frac{3\phi_2}{a^2} \frac{1}{\ln(2b/a)} \qquad \text{(needles)}. \tag{19.100}$$

For disk-like inclusions ($b \ll a$), $f(b/a) \sim 2a/\pi b$, and so

$$\gamma = \frac{3\phi_2}{a^2} \frac{2a}{\pi b} \qquad \text{(disks)}. \tag{19.101}$$

19.3.3 Nondilute Concentrations

The cluster-expansion procedure outlined above for the conductivity and elasticity problems cannot be used for the trapping and flow problems because the relevant fields decay to zero more slowly for large sphere separations than in the former instances. Thus, a naive cluster expansion of the flux into a successively larger number of sphere interactions actually results in divergent integrals. Different procedures must be found to incorporate interparticle interactions in such a way as to lead to absolutely convergent integrals. As in the case of Coulombic *screening* in the Debye-Hückel theory of ionic solutions, the divergence difficulties in the trapping problem are expected to be screened out by the long-range interactions between the traps. Thus, a correct cluster-expansion technique involves properly ordering the multiple-sphere contributions. Noteworthy studies that have attacked this problem include those of Felderhof and Deutch (1976), Lebenhaft and Kapral (1979), Muthukumar and Cukier (1981), and Mattern and Felderhof (1987).

Mattern and Felderhof (1987) obtained the following expansion for dispersions of identical spherical traps:

$$\frac{\gamma}{\gamma_s} = 1 + \sqrt{3}\phi_2^{1/2} + \frac{3\phi_2}{2} \ln \phi_2 + 12.839\phi_2 + o(\phi_2), \tag{19.102}$$

where γ_s is the Smoluchowski dilute result (19.89) and $o(\phi_2)$ indicates terms of higher order than ϕ_2. Note that the expansion is nonanalytic in the trap volume fraction ϕ_2. The nonanalyticity is due to screening effects. Observe also that the self-consistent approximation (18.75) agrees with (19.102) through order $\sqrt{\phi_2}$ [cf. (18.76)]. The reader should also note the striking similarity between this result and the corresponding one for the inverse fluid permeability (19.119).

19.4 Fluid Permeability

Idealized models of porous media that have been well investigated consist of random beds of spheres. It is desired to find the fluid permeability k associated with the slow viscous flow of a fluid in the space between the spheres. When a bed of identical spheres is sufficiently dilute so that the velocity fields from each sphere do not interfere with each other, the fluid permeability (as shown below) is given by the well-known Stokes result

$$k_s = \frac{2R^2}{9\phi_2}, \tag{19.103}$$

where ϕ_2 and R are the volume fraction and radius of the spheres, respectively. For nondilute concentrations, the flow field can be quite complicated. Considerable effort has been expended to find density expansions of k that correct Stokes's result for nondilute concentrations (Childress 1972, Howells 1974, Hinch 1977, Freed and Muthukumar 1978).

In the ensuing discussion, we derive the Stokes dilute limit of the fluid permeability for identical spheres. This is followed by a description of cluster-type expansions for nondilute concentrations.

19.4.1 Dilute Beds of Spheres

Recall that the fluid permeability k of any isotropic porous medium is generally defined by Darcy's law:

$$\boldsymbol{U} = -\frac{k}{\mu}\nabla p_0, \tag{19.104}$$

where $\boldsymbol{U}$ is the average velocity, μ is the dynamic viscosity, and ∇p_0 is the applied pressure gradient [cf. (13.156)]. Let us consider a dilute bed of d-dimensional spheres of radius R. If the velocity fields from each sphere do not interfere with each other, then we need to know the drag force $\boldsymbol{F}$ only on one such sphere in order to compute k for the entire array. If ρ is the number density of spheres, then the applied pressure gradient (drag force per unit volume) is

$$\nabla p_0 = -\rho \boldsymbol{F}. \tag{19.105}$$

The drag force $\boldsymbol{F}$ on an isolated fixed sphere in d dimensions ($d \geq 3$) was found to be given by (17.145). This result combined with (19.105) gives

$$\nabla p_0 = -\left[\frac{d(d^2-3)}{d+1}\right]\frac{\mu\phi_2}{R^2}\boldsymbol{U}_0, \qquad d \geq 3, \tag{19.106}$$

where $\boldsymbol{U}_0$ is the uniform velocity at infinity. Substitution of (19.106) into (19.104) yields the permeability to be

$$k = \left[\frac{d+1}{d(d^2-3)}\right]\frac{R^2}{\phi_2}, \qquad d \geq 3, \tag{19.107}$$

where $\phi_2 = \rho v_1$ is the sphere volume fraction and $v_1 = \Omega R^d/d$.

We see that as the space dimension d tends to infinity (while the remaining parameters are held fixed), k tends to zero. The reason is that the surface-to-volume ratio of the spheres varies as d. In Section 21.4.1, we prove that $k \to 0$ as $d \to \infty$ for arbitrary microstructures. When $d = 3$, we recover the well-known Stokes result (19.103) for three-dimensional spheres.

Does polydispersivity in sphere size change the Stokes dilute-limit expression for the permeability? One can easily verify, using the same reasoning as given above, that the permeability for a dilute bed of polydisperse d-dimensional spheres is given by

$$k = \left[\frac{d+1}{d(d^2-3)}\right]\frac{\langle R^3\rangle}{\langle R\rangle\phi_2}, \qquad d \geq 3, \tag{19.108}$$

where $\langle R^n\rangle$ is the nth moment of the probability density function $f(R)$, given by (19.96). The density function for a discrete set of sizes is given by (4.80). Thus, we see that unlike the monodisperse case, the polydisperse result (19.108) depends on two length scales: $\langle R\rangle$ and $\langle R^3\rangle^{1/3}$.

19.4.2 Dilute Beds of Spheroids

Consider flow past a dilute dispersion of spheroids with semiaxes $a = a_1 = a_2$ and $b = a_3$. We will consider obtaining expressions for the fluid permeability in two instances: (1) spheroids whose symmetry axis is aligned with the x_3-axis, and (2) randomly oriented spheroids. In the first case, the permeability is a second-order tensor $\boldsymbol{k}$ defined by the statistically homogeneous but anisotropic version of Darcy's law:

$$\boldsymbol{U} = -\frac{\boldsymbol{k}}{\mu}\cdot\nabla p_0. \tag{19.109}$$

We can determine the permeability tensor $\boldsymbol{k}$ in the principal axes frame of reference by finding the drag force $\boldsymbol{F}$ on an isolated oriented spheroid in each of the principal directions due to an average velocity $\boldsymbol{U}$ in each of these directions. Thus, use of (19.105), (19.109), and expression (17.146) for the drag force on an oriented spheroid yields

$$\boldsymbol{k} = \frac{2a^2}{9\phi_2}\begin{bmatrix} f(b/a) & 0 & 0 \\ 0 & f(b/a) & 0 \\ 0 & 0 & g(b/a)\end{bmatrix}, \tag{19.110}$$

where $\phi_2 = \rho 4\pi a^2 b/3$,

$$f(b/a) = \begin{cases} \dfrac{3}{16\chi_b^3}\left[(3\chi_b^2 - 1)\ln\left(\dfrac{1+\chi_b}{1-\chi_b}\right) + 2\chi_b\right], & b/a \geq 1, \\[2ex] \dfrac{3}{8\chi_a^3}\left[(1+3\chi_a^2)\tan^{-1}\chi_a - \chi_a\right], & b/a \leq 1, \end{cases} \tag{19.111}$$

$$g(b/a) = \begin{cases} \dfrac{3}{8\chi_b^3}\left[(1+\chi_b^2)\ln\left(\dfrac{1+\chi_b}{1-\chi_b}\right) - 2\chi_b\right], & b/a \geq 1, \\ \dfrac{3}{4\chi_a^3}\left[(\chi_a^2 - 1)\tan^{-1}\chi_a + \chi_a\right], & b/a \leq 1, \end{cases} \tag{19.112}$$

and χ_a and χ_b are given by (19.99). Note that for spheres ($b/a = 1$), (19.110) is an isotropic tensor with components equal to the Stokes result (19.103). For needles ($b \gg a$), the shape factors in (19.110) simplify as

$$f(b/a) = \frac{3}{4}\ln(2b/a), \qquad g(b/a) = \frac{3}{2}\ln(2b/a) \qquad \text{(needles)}. \tag{19.113}$$

For disks ($b \ll a$), the shape factors in (19.110) are given by

$$f(b/a) = \frac{9\pi b}{16a}, \qquad g(b/a) = \frac{3\pi b}{8a} \qquad \text{(disks)}. \tag{19.114}$$

The scalar permeability k for a dilute bed of randomly oriented spheroids is found from the relation $k = \text{Tr}\, \boldsymbol{k}/d$, where $\boldsymbol{k}$ is given by (19.110), i.e.,

$$k = \frac{2a^2}{9\phi_2} h(b/a), \tag{19.115}$$

where

$$h(b/a) = \begin{cases} \dfrac{\ln\left(\dfrac{1+\chi_b}{1-\chi_b}\right)}{2\chi_b}, & b/a \geq 1, \\ \dfrac{\tan^{-1}\chi_a}{\chi_a}, & b/a \leq 1. \end{cases} \tag{19.116}$$

Note that the shape factor $h(b/a)$ is exactly the inverse of the shape factor $f(b/a)$ given in relation (19.97) for the trapping constant $\gamma(b/a)$ of a dilute dispersion of spheroidal traps. The reason for this connection is discussed in Section 23.2. For very slender bodies ($b \gg a$), (19.116) reduces to

$$k = \frac{2a^2}{9\phi_2}\ln(2b/a) \qquad \text{(needles)}. \tag{19.117}$$

For disks ($b \ll a$), (19.116) yields

$$k = \frac{2a^2}{9\phi_2}\frac{\pi b}{2a} \qquad \text{(disks)}. \tag{19.118}$$

19.4.3 Nondilute Concentrations

The long-range field interactions in the flow problem require a different cluster-expansion procedure from those outlined above for the conductivity and elasticity problems. The same divergence difficulty also arose in the trapping problem, and so we refer the reader to Section 19.3.3 for a brief explanation of the difficulty one encounters. Childress (1972), Howells (1974), and Hinch (1977) arrived at absolutely convergent cluster-type expansions for the fluid permeability of random beds of identical spheres.

Howells (1974) and Hinch (1977) found the first few terms in the following expansion for the inverse permeability:

$$\frac{k_s}{k} = 1 + \frac{3}{\sqrt{2}}\phi_2^{1/2} + \frac{135}{64}\phi_2 \ln \phi_2 + 16.456\phi_2 + o(\phi_2), \tag{19.119}$$

where k_s is the Stokes dilute result (19.103). As in the trapping problem, the nonanalyticity in the sphere volume fraction ϕ_2 is due to *screening* effects, except that here it is *hydrodynamic* in nature. (Note that there are computational errors in Hinch's calculation in the order ϕ_2 term, and when corrected his result agrees with Howells's.) Observe that the self-consistent approximation (18.88) agrees with (19.119) through order $\sqrt{\phi_2}$ [cf. (18.89)]. We choose to expand the inverse permeability in order to compare this expansion with the corresponding expression for the trapping constant (19.102). Note the strong similarities between the two different expressions. We will see in Chapter 23 that this is more than a coincidence, since the permeability (inverse permeability) is rigorously related to the mean survival time (trapping constant).

CHAPTER 20

Exact Contrast Expansions

For two-phase media in which variations in the phase properties are small, formally exact perturbation series for both the effective conductivity (Beran 1968, Phan-Thien and Milton 1982) and effective elastic moduli (Beran 1968, Dederichs and Zeller 1973, Willis 1981) have been developed. Such *weak-contrast* expansions are found by first obtaining corresponding expansions of either the local electric field $\boldsymbol{E}(\boldsymbol{x})$ or the local strain field $\boldsymbol{\varepsilon}(\boldsymbol{x})$ via integral equations. For specificity, it is useful to state the weak-contrast form of the effective conductivity σ_e of a macroscopically isotropic medium, keeping in mind that analogous results exist for anisotropic media and for the effective elastic tensor. The weak-contrast expansion of σ_e may generally be written as

$$\sigma_e = \langle \sigma \rangle + \sigma_0 \sum_{n=2}^{\infty} a_n \left(\frac{\sigma_2 - \sigma_1}{\sigma_0} \right)^n, \tag{20.1}$$

where σ_0 denotes some reference conductivity and the a_n are coefficients that depend on the microstructure. The convergence of (20.1) depends on the value of σ_0, often taken to be either σ_1, σ_2, or $\langle \sigma \rangle$. Weak-contrast expansions are fundamentally important, since such trial fields lead to rigorous bounds on the effective conductivity or elastic moduli (see Chapter 21). Although they are applicable at all volume fractions, a drawback of such classical perturbation expansions is that they are valid only when the two phase properties are nearly the same.

Due to the nature of the integral operator in the conduction and elasticity problems, one must contend with conditionally convergent integrals, i.e., integrals that depend on the shape of the macroscopic sample owing to the long-range behavior of the fields. Of course, a true bulk property cannot depend on the shape of the macroscopic specimen and therefore must ultimately involve absolutely convergent integrals. One approach to resolve this problem has been to carry out a "renormalization" analysis, which amounts to identifying physically what the conditionally convergent terms ought to contribute and replacing them by convergent terms that make this contribution vanish (McCoy 1979). However, it is clearly more desirable to achieve absolute convergence systematically.

Brown (1955) devised a procedure to find a perturbation expansion of a *rational function* of the effective conductivity σ_e of three-dimensional, two-phase isotropic media in powers of rational functions of the phase conductivities. We refer to such expansions as *strong-contrast* expansions. In general, for the same reference conductivity σ_0, strong-contrast expansions possess a larger radius of convergence than their weak-contrast counterparts. Torquato (1985a) extended Brown's results for isotropic media to any space dimension d. Sen and Torquato (1989) later derived strong-contrast expansions for the effective conductivity tensor $\boldsymbol{\sigma}_e$ of macroscopically anisotropic media for any d. Torquato (1997) generalized this procedure to find strong-contrast expansions for the effective stiffness matrix $\boldsymbol{C}_e$ of macroscopically anisotropic two-phase media for any d.

There are other advantages that the Brown procedure enjoys. First, the resulting strong-contrast expansions are not formal, but rather the nth-order coefficients are given explicitly in terms of integrals over certain tensor fields weighted by n-point microstructural correlation functions that render the integrals absolutely convergent. Thus, no renormalization analysis is required. Second, by appropriately choosing the free parameters in the procedure, one can obtain strong-contrast expansions that perturb about the structures that either maximize or minimize the effective property, as shown below. Other choices may be explored that will lead to expansions that perturb about other known structures. [Indeed, the weak-contrast expansion (20.1) may be regarded as one that perturbs about a simple laminate, since truncation of the series in (20.1) results in the arithmetic-average estimate.] The ability to perturb about a specific structure can be practically useful if one wants to estimate the effective property of an actual composite that resembles the reference structure. In the case of the effective conductivity, we will show that one can find strong-contrast expansions that perturb about the phase-symmetric-type microstructures that realize the self-consistent approximations described in Chapter 18.

We begin with a treatment of strong-contrast expansions for the effective conductivity tensor $\boldsymbol{\sigma}_e$ of macroscopically anisotropic two-phase media in any space dimension d. This is followed by an analogous treatment of the effective stiffness tensor $\boldsymbol{C}_e$ for such media.

20.1 Conductivity Tensor

We derive the strong-contrast expansions of Sen and Torquato (1989) for effective conductivity tensor $\boldsymbol{\sigma}_e$ of macroscopically anisotropic two-phase d-dimensional media in detail heretofore not given. The derivation follows the one given for the effective stiffness tensor (Torquato 1997). Various applications of the expansions are considered.

We also discuss the behavior of the effective conductivity of isotropic media as d tends to infinity.

20.1.1 Integral Equation for Cavity Electric Field

Consider a *large* but finite-sized ellipsoidal macroscopically anisotropic composite specimen in arbitrary space dimension d composed of two isotropic phases with conductivities σ_1 and σ_2. The shape of the composite specimen is purposely chosen to be nonspherical, since any rigorously correct expression for the effective conductivity tensor must ultimately be independent of the shape of the composite specimen in the infinite-volume limit. The microstructure is perfectly general and possesses a characteristic microscopic length scale that is much smaller than the semiaxes of the ellipsoid. Thus, the specimen is virtually statistically homogeneous. Ultimately, we will take the infinite-volume limit and hence consider statistically homogeneous media. The local scalar conductivity $\sigma(\mathbf{x})$ is expressible as

$$\sigma(\mathbf{x}) = \sigma_1 \mathcal{I}^{(1)}(\mathbf{x}) + \sigma_2 \mathcal{I}^{(2)}(\mathbf{x}), \tag{20.2}$$

where

$$\mathcal{I}^{(p)}(\mathbf{x}) = \begin{cases} 1, & \mathbf{x} \text{ in phase } p, \\ 0, & \text{otherwise}, \end{cases} \tag{20.3}$$

is the indicator function for phase p ($p = 1, 2$).

Now let us embed this d-dimensional ellipsoidal composite specimen in an infinite *reference* phase q, which is subjected to an applied electric field $\mathbf{E}_0(\mathbf{x})$ at infinity (see Figure 20.1). The reference phase can be chosen to be arbitrary, but for our purposes, we will take it to be either phase 1 or phase 2, i.e., $q = 1$ or $q = 2$. Introducing the *polarization* field defined by

$$\mathbf{P}(\mathbf{x}) = [\sigma(\mathbf{x}) - \sigma_q]\mathbf{E}(\mathbf{x}) \tag{20.4}$$

enables us to reexpress the flux $\mathbf{J}$, defined by Ohm's law (13.6), as follows:

$$\mathbf{J}(\mathbf{x}) = \sigma_q \mathbf{E}(\mathbf{x}) + \mathbf{P}(\mathbf{x}). \tag{20.5}$$

The vector $\mathbf{P}(\mathbf{x})$ is the *induced flux polarization field* relative to the medium in the absence of phase p and hence is zero in the reference phase q and nonzero in the "polarized" phase p ($p \neq q$). Throughout the chapter, the indices p and q will be reserved only for the polarized and reference phases, respectively. The choice of which is the reference or polarized phase is arbitrary; all of the results are valid for any $p \neq q$, i.e., $p = 1$ and $q = 2$ or $p = 2$ and $q = 1$.

Under steady-state conditions without sources, the flux is divergence-free [cf. (13.3)], and hence, with the aid of (20.5), this solenoidal condition can be rewritten as

$$\sigma_q \Delta \hat{\varphi}(\mathbf{x}) = \nabla \cdot \mathbf{P}(\mathbf{x}), \tag{20.6}$$

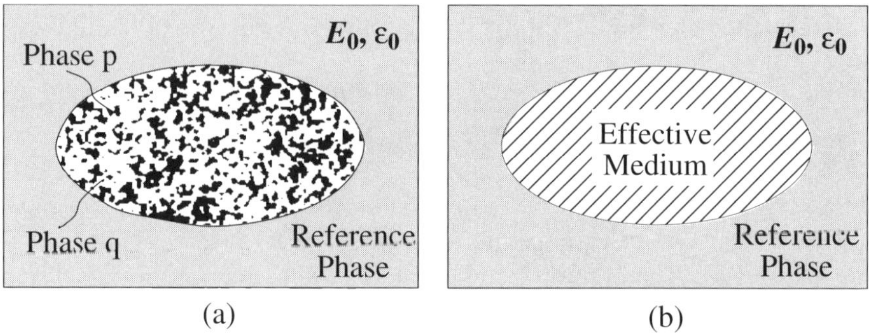

Figure 20.1 (a) A schematic of a large d-dimensional ellipsoidal, macroscopically anisotropic two-phase composite specimen embedded in an infinite reference phase under either an applied electric field E_0 or applied strain field ε_0 at infinity. The composite consists of two phases: phase p (black region) and phase q (white region). (b) After homogenization, the same ellipsoid can be viewed as having an effective conductivity tensor σ_e or effective stiffness tensor C_e.

$$\hat{\varphi}(\boldsymbol{x}) \to 0, \qquad |\boldsymbol{x}| \to \infty, \tag{20.7}$$

where $\hat{\varphi}(\boldsymbol{x})$ is the potential in excess of the potential at infinity $\varphi_0(\boldsymbol{x})$, i.e., $\hat{\varphi}(\boldsymbol{x}) = \varphi(\boldsymbol{x}) - \varphi_0(\boldsymbol{x})$. The infinite-space Green's function $g^{(q)}$ is defined by requiring it to satisfy

$$\sigma_q \Delta g^{(q)}(\boldsymbol{x}, \boldsymbol{x}') = -\delta(\boldsymbol{x} - \boldsymbol{x}'), \tag{20.8}$$

$$g^{(q)}(\boldsymbol{x}, \boldsymbol{x}') \to 0, \quad |\boldsymbol{x}| \to \infty. \tag{20.9}$$

Multiplying (20.6) by the Green's function and integrating by parts yields the integral relation

$$\varphi(\boldsymbol{x}) = \varphi_0(\boldsymbol{x}) - \int d\boldsymbol{x}' \; \nabla g^{(q)}(\boldsymbol{x}, \boldsymbol{x}') \cdot \boldsymbol{P}(\boldsymbol{x}'), \tag{20.10}$$

where φ_0 is the potential field at infinity. Note that the presence of the polarization $\boldsymbol{P}$ in (20.10) implies that the integration volume extends only over the region of space occupied by the finite-sized ellipsoidal composite specimen.

It is a simple matter to show that the d-dimensional Green's function that satisfies (20.8) and (20.9) is given by

$$g^{(q)}(\boldsymbol{r}) = \begin{cases} \dfrac{1}{\Omega \sigma_q} \ln\left(\dfrac{1}{r}\right), & d = 2, \\[2ex] \dfrac{1}{(d-2)\Omega \sigma_q} \dfrac{1}{r^{d-2}}, & d \geq 3, \end{cases} \tag{20.11}$$

where $\Omega(d)$ is the total solid angle contained in a d-dimensional sphere, given by (2.56), $\boldsymbol{r} = \boldsymbol{x} - \boldsymbol{x}'$, and $\boldsymbol{n} = \boldsymbol{r}/|\boldsymbol{r}|$.

The integral of (20.10) is *improper* because the Green's function possesses a singularity at the point $\boldsymbol{x}' = \boldsymbol{x}$. To determine whether any improper integral converges, or exists, one must exclude a small region containing the point $\boldsymbol{x}' = \boldsymbol{x}$. The integral exists in the limit that the excluded region shrinks to zero, independent of the shape of the excluded region (Kellogg 1953). If the limit exists, then any convenient excluded cavity shape may be used. A d-dimensional sphere is often the most convenient shape. If the excluded region is chosen to be a sphere centered at $\boldsymbol{x}' = \boldsymbol{x}$, the limit, if it exists, is the d-dimensional analogue of the Cauchy principal value of a one-dimensional improper integral. According to this criterion, the integral of (20.10) is convergent.

To obtain the field $\boldsymbol{E}$, one must differentiate (20.10); however, because of the singular nature of the integral, one cannot simply differentiate under the integral sign. Excluding a spherical region or "cavity" from the point $\boldsymbol{x}' = \boldsymbol{x}$ in (20.10), integrating by parts, and using the divergence theorem, we find that the electric field satisfies the integral relation

$$\boldsymbol{E}(\boldsymbol{x}) = \boldsymbol{E}_0(\boldsymbol{x}) + \int d\boldsymbol{x}'\, \boldsymbol{G}^{(q)}(\boldsymbol{r}) \cdot \boldsymbol{P}(\boldsymbol{x}'), \tag{20.12}$$

where

$$\boldsymbol{G}^{(q)}(\boldsymbol{r}) = -\boldsymbol{D}^{(q)}\delta(\boldsymbol{r}) + \boldsymbol{H}^{(q)}(\boldsymbol{r}). \tag{20.13}$$

In relation (20.13), the constant second-order tensor $\boldsymbol{D}^{(q)}$ that arises because of the exclusion of the spherical cavity in (20.10) is given by

$$\boldsymbol{D}^{(q)} = \frac{1}{d\sigma_q}\boldsymbol{I}, \tag{20.14}$$

where $\boldsymbol{I}$ is the second-order identity tensor and $\boldsymbol{H}^{(q)}(\boldsymbol{r})$ is the double gradient $\nabla\nabla g^{(q)}(r)$ given by

$$\boldsymbol{H}^{(q)}(\boldsymbol{r}) = \frac{1}{\Omega\sigma_q}\frac{d\boldsymbol{nn} - \boldsymbol{I}}{r^d}. \tag{20.15}$$

It is understood that integrals involving the second-order tensor $\boldsymbol{H}^{(q)}$ are to be carried out by excluding at $\boldsymbol{x}' = \boldsymbol{x}$ an infinitesimal sphere in the limit that the sphere radius shrinks to zero. Moreover, the integral of $\boldsymbol{H}^{(q)}(\boldsymbol{r})$ over the surface of a sphere of radius $R > 0$ is identically zero, i.e.,

$$\int_{r=R} \boldsymbol{H}^{(q)}(\boldsymbol{r}) d\Omega = 0. \tag{20.16}$$

A detailed proof of the analogue of (20.13) for the elasticity problem is given by Torquato (1997). The infinite-space Green's function for an *elastic* isotropic medium contains, among other contributions, the *Coulombic* term (20.11).

Substitution of (20.13) into expression (20.12) yields an integral equation for the *cavity* intensity field $\boldsymbol{F}(\boldsymbol{x})$:

$$\boldsymbol{F}(\boldsymbol{x}) = \boldsymbol{E}_0(\boldsymbol{x}) + \int_\epsilon d\boldsymbol{x}'\boldsymbol{H}^{(q)}(\boldsymbol{x} - \boldsymbol{x}') \cdot \boldsymbol{P}(\boldsymbol{x}'). \tag{20.17}$$

The cavity intensity field $\boldsymbol{F}(\boldsymbol{x})$ is related to $\boldsymbol{E}(\boldsymbol{x})$ by the expression

$$F(x) = \{I + D^{(q)}[\sigma(x) - \sigma_q]\} \cdot E(x), \tag{20.18}$$

where the constant tensor $D^{(q)}$ is given by (20.14). We also define

$$\int_\epsilon f(x,x')\,dx' \equiv \lim_{\epsilon\to 0} \int_{|x-x'|>\epsilon} f(x,x')\,dx'. \tag{20.19}$$

We refer to $F(x)$ as the *cavity* intensity field because, as can be seen from (20.18), it is a modified intensity field equal to the usual field plus a contribution involving the constant tensor $D^{(q)}$, which arises as a result of excluding a spherical cavity from the point $x' = x$ in (20.10). In dielectric theory, the cavity strain field is referred to as the Lorentz field.

Combination of the expressions (20.4) and (20.18) gives a relation between the polarization and cavity intensity field:

$$P(x) = L^{(q)}(x) \cdot F(x), \tag{20.20}$$

where

$$L^{(q)}(x) = \{\sigma(x) - \sigma_q\}\{I + D^{(q)}[\sigma(x) - \sigma_q]\}^{-1}. \tag{20.21}$$

Notice that $L^{(q)}(x)$ is a fractional linear transformation of the local conductivity tensor $\sigma(x) = \sigma(x)I$ and, in light of (20.14), is an isotropic tensor, i.e., $L^{(q)}(x) = L^{(q)}(x)I$, where

$$L^{(q)}(x) = \sigma_q d\beta_{pq}\mathcal{I}^{(p)}(x), \tag{20.22}$$

$$\beta_{pq} = \frac{\sigma_p - \sigma_q}{\sigma_p + (d-1)\sigma_q}. \tag{20.23}$$

The *polarizability* β_{pq} is a scalar (not a tensor) parameter that depends on the conductivities of the polarized and reference phases p and q, respectively, and is bounded according to

$$-(d-1)^{-1} \le \beta_{pq} \le 1. \tag{20.24}$$

Also, since $L^{(q)}(x)$ a scalar quantity, we have

$$P(x) = L^{(q)}(x)F(x). \tag{20.25}$$

20.1.2 Strong-Contrast Expansions

The effective conductivity tensor σ_e can also be defined through an expression relating the average polarization to the average Lorentz field, i.e.,

$$\langle P(x)\rangle = L_e^{(q)} \cdot \langle F(x)\rangle, \tag{20.26}$$

where the second-order tensor $L_e^{(q)}$ is the following fractional linear transformation of σ_e:

$$
\begin{aligned}
\boldsymbol{L}_e^{(q)} &= \{\boldsymbol{\sigma}_e - \sigma_q \boldsymbol{I}\} \cdot \{\boldsymbol{I} + \boldsymbol{D}^{(q)} \cdot [\boldsymbol{\sigma}(\boldsymbol{x}) - \boldsymbol{\sigma}_q]\}^{-1} \\
&= \sigma_q d \{\boldsymbol{\sigma}_e - \sigma_q \boldsymbol{I}\} \cdot \{\boldsymbol{\sigma}_e + (d-1)\sigma_q \boldsymbol{I}\}^{-1}. \qquad (20.27)
\end{aligned}
$$

The constitutive relation (20.26) is localized; i.e., it is independent of the shape of the ellipsoidal composite specimen in the infinite-volume limit. This relation is completely equivalent to the averaged Ohm's law (13.46) that defines the effective conductivity tensor. In what follows, we will temporarily drop the superscript q in the quantities $L^{(q)}$, $\boldsymbol{L}_e^{(q)}$, and $\boldsymbol{H}^{(q)}$.

We want to obtain an expression for the effective conductivity tensor $\boldsymbol{\sigma}_e$ from relation (20.26) using the solution of the integral equation (20.17), which is recast as

$$
\boldsymbol{F}(1) = \boldsymbol{E}_0(1) + \int_\epsilon d2\, \boldsymbol{H}(1,2) \cdot \boldsymbol{P}(2),
$$

where we have adopted the shorthand notation of representing $\boldsymbol{x}$ and $\boldsymbol{x}'$ by 1 and 2, respectively. In schematic operator form, this integral equation can be tersely rewritten as

$$
\boldsymbol{F} = \boldsymbol{E}_0 + \boldsymbol{H}\boldsymbol{P}. \qquad (20.28)
$$

Multiplying this relation by the scalar $L(\boldsymbol{x})$, defined by (20.22), yields

$$
\boldsymbol{P} = L\boldsymbol{E}_0 + L\boldsymbol{H}\boldsymbol{P}. \qquad (20.29)
$$

A solution for the polarization $\boldsymbol{P}$ in terms of the applied field $\boldsymbol{E}_0$ can be obtained by successive substitutions using (20.29), with the result

$$
\begin{aligned}
\boldsymbol{P} &= L\boldsymbol{E}_0 + L\boldsymbol{H}L\boldsymbol{E}_0 + L\boldsymbol{H}L\boldsymbol{H}L\boldsymbol{E}_0 + \cdots \\
&= \boldsymbol{S}\boldsymbol{E}_0, \qquad (20.30)
\end{aligned}
$$

where the second-order tensor operator $\boldsymbol{S}$ is given by

$$
\boldsymbol{S} = L[\boldsymbol{I} - L\boldsymbol{H}]^{-1}.
$$

For concreteness, we write out (20.30) more explicitly as

$$
\begin{aligned}
\boldsymbol{P}(1) &= \int d2\, \boldsymbol{S}(1,2) \cdot \boldsymbol{E}_0(2) \\
&= L(1)\boldsymbol{E}_0(1) + \int d2\, L(1)\, L(2) \boldsymbol{H}(1,2) \cdot \boldsymbol{E}_0(2) \\
&\quad + \int d2\, d3\, L(1)\, L(2)\, L(3) \boldsymbol{H}(1,2) \cdot \boldsymbol{H}(2,3) \cdot \boldsymbol{E}_0(3) + \cdots . \qquad (20.31)
\end{aligned}
$$

Ensemble averaging (20.30) gives

$$
\langle \boldsymbol{P} \rangle = \langle \boldsymbol{S} \rangle \boldsymbol{E}_0. \qquad (20.32)
$$

The operator $\langle \boldsymbol{S} \rangle$ involves products of the tensor $\boldsymbol{H}$, which decays to zero like r^{-d} for large r, and hence $\langle \boldsymbol{S} \rangle$ at best involves conditionally convergent integrals. In other words, $\langle \boldsymbol{S} \rangle$ is dependent upon the shape of the ellipsoidal composite specimen. Indeed, this nonlocal nature of the relation between the polarization and applied intensity fields is completely consistent with the well-known electrostatic results described in Section 17.1. That is, when an ellipsoidal inclusion in an infinite matrix of another material is subjected to a constant field $\boldsymbol{E}_0$ at infinity, the polarization field within the ellipsoid is uniform and, when expressed in terms of $\boldsymbol{E}_0$, depends upon the shape of the ellipsoid.

Thus, in order to obtain a local (shape-independent) relation between average polarization $\langle \boldsymbol{P} \rangle$ and average Lorentz field $\langle \boldsymbol{F} \rangle$ as prescribed by (20.26), we must eliminate the applied field $\boldsymbol{E}_0$ in favor of the appropriate average field. Inverting (20.32) yields

$$\boldsymbol{E}_0 = \langle \boldsymbol{S} \rangle^{-1} \langle \boldsymbol{P} \rangle, \tag{20.33}$$

and averaging (20.28) gives

$$\langle \boldsymbol{F} \rangle = \boldsymbol{E}_0 + \boldsymbol{H} \langle \boldsymbol{P} \rangle. \tag{20.34}$$

Eliminating the applied field $\boldsymbol{E}_0$ in (20.34) using (20.33) yields

$$\langle \boldsymbol{F} \rangle = \boldsymbol{Q} \langle \boldsymbol{P} \rangle, \tag{20.35}$$

where

$$\boldsymbol{Q} = \langle \boldsymbol{S} \rangle^{-1} + \boldsymbol{H}.$$

Explicitly, (20.35) reads

$$\begin{aligned}
\langle \boldsymbol{F}(1) \rangle &= \int d2 \boldsymbol{Q}(1,2) \cdot \langle \boldsymbol{P}(2) \rangle \\
&= \frac{\langle \boldsymbol{P}(1) \rangle}{\langle L(1) \rangle} - \int d2 \left[\frac{\langle L(1) L(2) \rangle - \langle L(1) \rangle \langle L(2) \rangle}{\langle L(1) \rangle \langle L(2) \rangle} \right] \boldsymbol{H}(1,2) \cdot \langle \boldsymbol{P}(2) \rangle \\
&\quad - \int d2\, d3 \left[\frac{\langle L(1) L(2) L(3) \rangle}{\langle L(1) \rangle \langle L(2) \rangle} - \frac{\langle L(1) L(2) \rangle \langle L(2) L(3) \rangle}{\langle L(1) \rangle \langle L(2) \rangle \langle L(3) \rangle} \right] \boldsymbol{H}(1,2) \cdot \boldsymbol{H}(2,3) \cdot \langle \boldsymbol{P}(3) \rangle \\
&\quad - \cdots .
\end{aligned}$$

The n-point correlation function $\langle L(1) \cdots L(n) \rangle$ which arises here is directly related to the n-point probability function $S_n^{(p)}$ of phase p, which, we recall from Section 2.2, is defined by the following ensemble average:

$$S_n^{(p)}(\boldsymbol{x}_1, \ldots, \boldsymbol{x}_n) = \langle \mathcal{I}^{(p)}(\boldsymbol{x}_1) \mathcal{I}^{(p)}(\boldsymbol{x}_2) \cdots \mathcal{I}^{(p)}(\boldsymbol{x}_n) \rangle. \tag{20.36}$$

For example, returning to the notation of explicitly indicating that the reference medium is phase q, we have

$$\begin{gathered}
\langle L^{(q)}(1) \rangle = \sigma_q d\, \beta_{pq} S_1^{(p)}(1), \\
\langle L^{(q)}(1) L^{(q)}(2) \rangle - \langle L^{(q)}(1) \rangle \langle L^{(q)}(2) \rangle = (\sigma_q d \beta_{pq})^2 [S_2^{(p)}(1,2) - S_1^{(p)}(1) S_1^{(p)}(2)].
\end{gathered}$$

Comparing expressions (20.26) and (20.35) yields the desired result for the effective tensor $\boldsymbol{L}_e^{(q)}$, i.e.,

$$\boldsymbol{L}_e^{-1} = \boldsymbol{Q}.$$

Multiplying this equation by the scalar $\sigma_q \beta_{pq}^2 \phi_p^2 d$ gives the general expansion as

$$\beta_{pq}^2\phi_p^2\{\boldsymbol{\sigma}_e - \sigma_q \boldsymbol{I}\}^{-1}\cdot\{\boldsymbol{\sigma}_e + (d-1)\sigma_q\boldsymbol{I}\} = \phi_p\beta_{pq}\boldsymbol{I} - \sum_{n=2}^{\infty} \boldsymbol{A}_n^{(p)}\beta_{pq}^n, \quad p \neq q. \tag{20.37}$$

In this expression, the n-point tensor coefficients $\boldsymbol{A}_n^{(p)}$ are certain integrals over the $S_n^{(p)}$ associated with phase p: for $n = 2$,

$$\boldsymbol{A}_2^{(p)} = \frac{d}{\Omega}\int_\epsilon d2\ \boldsymbol{t}(1,2)\left[S_2^{(p)}(1,2) - \phi_p^2\right], \tag{20.38}$$

and for $n \geq 3$,

$$\boldsymbol{A}_n^{(p)} = \left(\frac{-1}{\phi_p}\right)^{n-2}\left(\frac{d}{\Omega}\right)^{n-1}\int d2\cdots\int dn\ \boldsymbol{t}(1,2)\cdot\boldsymbol{t}(2,3)\cdots\boldsymbol{t}(n-1,n)\Delta_n^{(p)}(1,\ldots,n). \tag{20.39}$$

Here $\boldsymbol{t}(\boldsymbol{r})$ is the property-independent dipole tensor [cf. (17.15)] trivially related to the tensor $\boldsymbol{H}(\boldsymbol{r})$ according to

$$\boldsymbol{t}(\boldsymbol{r}) = \Omega\sigma_q\boldsymbol{H}(\boldsymbol{r}) = \frac{d\boldsymbol{nn} - \boldsymbol{I}}{r^d}, \tag{20.40}$$

and $\Delta_n^{(p)}$ is a position-dependent determinant associated with phase p:

$$\Delta_n^{(p)} = \begin{vmatrix} S_2^{(p)}(1,2) & S_1^{(p)}(2) & \cdots & 0 \\ S_3^{(p)}(1,2,3) & S_2^{(p)}(2,3) & \cdots & 0 \\ \vdots & \vdots & \ddots & \vdots \\ S_n^{(p)}(1,\ldots,n) & S_{n-1}^{(p)}(2,\ldots,n) & \cdots & S_2^{(p)}(n-1,n) \end{vmatrix}. \tag{20.41}$$

Remarks:

1. Result (20.37) actually represents two different series expansions: one for $p = 1$ and $q = 2$ and the other for $p = 2$ and $q = 1$.
2. Note that the n-point tensor coefficients $\boldsymbol{A}_n^{(p)}$ are purely microstructural parameters, independent of the phase properties. They are functionals of the correlation functions $S_1^{(p)}, \ldots, S_n^{(p)}$. This is in contrast to other approaches in which the coefficients depend on derivatives of the correlation functions (Beran 1968, Hori 1973). The former description is clearly preferred.
3. At first glance, one might be concerned that the integrals of (20.38) and (20.39) are conditionally convergent [as in the work of Hori (1973)] because of the appearance in the integrand of the tensor $\boldsymbol{t}(\boldsymbol{r})$, which decays as r^{-d} for large r. However, the quantity within the brackets of (20.38) and the determinant $\Delta_n^{(p)}$ of (20.39) identically vanish at the boundary of the ellipsoidal sample because of the asymptotic properties of the $S_n^{(p)}$ (see Section 2.2.2). The integrals in (20.37), therefore, are absolutely convergent (i.e., independent of the shape of the macroscopic ellipsoid), and hence any convenient shape (such as a d-dimensional sphere) may be used in the infinite-volume limit. Moreover, when $n \geq 3$, the limiting process of excluding an in-

finitesimally small cavity about $r_{ij} = 0$ in the integrals (20.39) is no longer necessary, since $\Delta_n^{(p)}$ again is identically zero for such values.

4. Classical perturbation expansions involve parameters of smallness that are simple differences in the phase conductivities (Beran 1968, Hori 1973). By contrast, the series expansions represented by (20.37) are *nonclassical* in the sense that the expansion parameter is the polarizability β_{pq}, a rational function, and therefore have convergence properties that are superior to those of the classical expansions. Moreover, for certain microstructures (described below), the first few terms of the nonclassical series provide excellent approximations of $\boldsymbol{\sigma}_e$ for any phase contrast.
5. Observe that the expansion parameter β_{pq} arises because of our choice of excluding a spherical cavity at $\boldsymbol{x}' = \boldsymbol{x}$ in the integral (20.10) and integrating by parts. By choosing a nonspherical cavity shape, we would have obtained a different cavity intensity field and hence different expansion parameters. The implications of excluding *nonspherical* cavities has not been fully explored. Another free parameter is the reference medium, which is not restricted to phase 1 or 2. The convergence of the series can be accelerated by judiciously choosing the exclusion-cavity shape and reference medium (Ramshaw 1984, Eyre and Milton 1999).
6. The n-point tensors $\boldsymbol{A}_n^{(p)}$ generally will not possess common principal axes. This implies that for general media, the principal axes of the effective conductivity tensor $\boldsymbol{\sigma}_e$ will rotate as the phase contrast changes, as will be the case for composites with *chirality* (Milton 1986). However, there exists a large class of media that has the symmetry required for all the $\boldsymbol{A}_n^{(p)}$ to possess common principal axes (e.g., arrays of unidirectionally oriented ellipsoids or unidirectionally oriented cylinders in a matrix); in such instances n-point tensor multiplication is commutative.
7. For macroscopically anisotropic media, the series expressions (20.37) may be regarded as expansions that perturb about the optimal structures that realize the generalized Hashin–Shtrikman bounds derived by Willis (1977) and by Sen and Torquato (1989). This point will be demonstrated below. For macroscopically isotropic media, it is shown in Section 20.1.6 that the series expressions (20.37) may be regarded as expansions that perturb about the optimal structures that realize the bounds of Hashin and Shtrikman (1962).

Before showing this last point, we first observe that the lowest-order parameter $\boldsymbol{A}_2^{(p)}$ in the series (20.37) generally does not vanish for statistically anisotropic media, since the two-point function $S_2^{(p)}(\boldsymbol{r})$ depends on the distance $r = |\boldsymbol{r}|$ as well as the orientation of the vector $\boldsymbol{r}$. For macroscopically isotropic media,

$$\boldsymbol{A}_2^{(p)} = \boldsymbol{0}, \tag{20.42}$$

since $\boldsymbol{A}_2^{(p)}$ is traceless. This easily follows from the fact that the tensor $\boldsymbol{t}$, defined by (20.40) and appearing in (20.38), is itself traceless, i.e., $\text{Tr}\,\boldsymbol{t} = 0$. However, statistical anisotropy does not necessarily imply macroscopic anisotropy. For example, cubic lattices of spheres are statistically anisotropic but are macroscopically isotropic with respect to the effective conductivity (see Section 13.2.5).

Now consider microstructures for which the n-point tensor coefficients $\boldsymbol{A}_n^{(p)} = \boldsymbol{0}$ for all $n \geq 3$. For such structures, (20.37) reduces exactly to

$$\phi_p^2\{\boldsymbol{\sigma}_e - \sigma_q \boldsymbol{I}\}^{-1} \cdot \{\boldsymbol{\sigma}_e + (d-1)\sigma_q \boldsymbol{I}\} = \phi_p \beta_{pq}^{-1} \boldsymbol{I} - \boldsymbol{A}_2^{(p)}, \quad p \neq q. \tag{20.43}$$

Relations (20.43) are equivalent to the generalized Hashin–Shtrikman bounds (21.24) for d-dimensional anisotropic composites, derived by Sen and Torquato (1989) using the method of Padé approximants. Willis (1977) was the first to derive (20.43) for $d = 3$ using variational principles. (The reader is referred to Chapter 21 for a full discussion of bounds and their realizable structures.) The bounds embodied in (20.43) are optimal, since they are achieved by certain oriented singly-coated ellipsoidal assemblages (Bergman 1980, Milton 1981a, Tartar 1985) as well as finite-rank laminates (Tartar 1985). Importantly, for such structures, one of the phases is always a disconnected, dispersed phase in a connected matrix phase (except in the trivial instance when the generally dispersed phase fills all of space). Thus, the general expansion (20.37) for macroscopically anisotropic media may be regarded as one that perturbs about such optimal composites. Consequently, the first few terms of the expansion (20.37) will provide an *excellent estimate* of $\boldsymbol{\sigma}_e$ for composites that are similar in structure to the optimal ones.

In the special case of anisotropic composite microstructures in which $S_2^{(p)}(\boldsymbol{r})$ possesses ellipsoidal symmetry (e.g., arrays of oriented similar ellipsoidal inclusions with a nematic-liquid-crystal structure in a matrix) , Sen and Torquato (1989) showed that $\boldsymbol{A}_2^{(p)}$ can be written in the form

$$\boldsymbol{A}_2^{(p)} = [\boldsymbol{I} - d\boldsymbol{A}^*]\phi_1\phi_2, \tag{20.44}$$

where $\boldsymbol{A}^*$ is the d-dimensional *depolarization* tensor (17.25) for an ellipsoid.

20.1.3 Some Tensor Properties

The n-point microstructural parameters associated with phases 1 and 2, $\boldsymbol{A}_n^{(1)}$ and $\boldsymbol{A}_n^{(2)}$, are related to one another, since $\boldsymbol{\sigma}_e$ is invariant under different labeling of the reference phase. To obtain such relationships, we compare two different series expansions for $\boldsymbol{\sigma}_e$, using (20.37) or the weak-contrast expansion (20.48), one with phase 1 as the reference material ($q = 1$) and one with phase 2 as the reference material ($q = 2$). For $n = 2$, one immediately obtains that

$$\boldsymbol{A}_2^{(1)} = \boldsymbol{A}_2^{(2)} \equiv \boldsymbol{A}_2. \tag{20.45}$$

After some algebra, one also finds that

$$\boldsymbol{A}_3^{(1)} + \boldsymbol{A}_3^{(2)} = (d-1)\phi_1\phi_2 \boldsymbol{I} - (d-2)\boldsymbol{A}_2 - \frac{\boldsymbol{A}_2 \cdot \boldsymbol{A}_2}{\phi_1\phi_2}, \tag{20.46}$$

$$\boldsymbol{A}_4^{(1)} - \boldsymbol{A}_4^{(2)} = (d-2)[\boldsymbol{A}_3^{(2)} - \boldsymbol{A}_3^{(1)}] + \frac{\boldsymbol{A}_2 \cdot \boldsymbol{A}_3^{(2)}}{\phi_1\phi_2} - \frac{\boldsymbol{A}_3^{(1)} \cdot \boldsymbol{A}_2}{\phi_1\phi_2}. \tag{20.47}$$

By employing the method described above, one may obtain similar relations for higher-order microstructural parameters. However, the evaluation of n-point probability functions, and hence the $\boldsymbol{A}_n^{(p)}$, for $n \geq 5$ is presently very difficult (either theoretically or experimentally) for general microstructures. Thus, the three relations given above are the ones of practical value.

Milton (1987) has employed the mathematically elegant formalism of continued fractions to obtain bounds on $\boldsymbol{\sigma}_e$ for multiphase composites. His treatment leads to a sequence of matrices that he refers to as weights $\boldsymbol{W}_n^{(p)}$ and normalization factors $\boldsymbol{N}_n$. The weights and normalization factors formally depend upon the microstructure. However, his results are formal in the sense that he does not express these matrices explicitly in terms of integrals over the n-point correlation functions. The $\boldsymbol{W}_n^{(p)}$ and $\boldsymbol{N}_n$ are, in fact, related to the $\boldsymbol{A}_n^{(p)}$ given here. The reader is referred to Sen and Torquato (1989) for a discussion of these relationships. Milton's formalism readily leads to important properties of the matrices, which are much more difficult to prove generally using the approach of Sen and Torquato (1989).

20.1.4 Weak-Contrast Expansions

For some subsequent calculations it will be convenient to use the expansion of the effective conductivity tensor in powers of the difference in the phase conductivities, i.e.,

$$\boldsymbol{\sigma}_e = \sigma_q \boldsymbol{I} + \sigma_q \sum_{n=1}^{\infty} \boldsymbol{a}_n^{(p)} \left(\frac{\sigma_p - \sigma_q}{\sigma_q} \right)^n, \qquad p \neq q. \tag{20.48}$$

Employing (20.37), we find through fourth order that

$$\boldsymbol{a}_1^{(p)} = \phi_p \boldsymbol{I}, \tag{20.49}$$

$$\boldsymbol{a}_2^{(p)} = \frac{1}{d}\left[\boldsymbol{A}_2^{(p)} - \phi_p \phi_q \boldsymbol{I}\right], \tag{20.50}$$

$$\boldsymbol{a}_3^{(p)} = \frac{1}{d^2}\left\{ \phi_p \phi_q^2 \left[\frac{\boldsymbol{A}_2^{(p)}}{\phi_p \phi_q} - \boldsymbol{I} \right]^2 + \boldsymbol{A}_3^{(p)} \right\}, \tag{20.51}$$

$$\begin{aligned} \boldsymbol{a}_4^{(p)} = \frac{1}{d^3}\Bigg\{ & \phi_p \phi_q^3 \left[\frac{\boldsymbol{A}_2^{(p)}}{\phi_p \phi_q} - \boldsymbol{I} \right]^3 + \phi_q \left[\frac{\boldsymbol{A}_2^{(p)}}{\phi_p \phi_q} - \boldsymbol{I} \right] \cdot \boldsymbol{A}_3^{(p)} \\ & + \phi_q \boldsymbol{A}_3^{(p)} \cdot \left[\frac{\boldsymbol{A}_2^{(p)}}{\phi_p \phi_q} - \boldsymbol{I} \right] - \boldsymbol{A}_3^{(p)} + \boldsymbol{A}_4^{(p)} \Bigg\}. \end{aligned} \tag{20.52}$$

Here, for any tensor $\boldsymbol{W}$, $\boldsymbol{W}^2$ and $\boldsymbol{W}^3$ are abbreviations for $\boldsymbol{W} \cdot \boldsymbol{W}$ and $\boldsymbol{W} \cdot \boldsymbol{W} \cdot \boldsymbol{W}$, respectively.

20.1.5 Expansion of Local Electric Field

The local field $\boldsymbol{E}(\boldsymbol{x})$ itself can be expressed as a contrast expansion. We choose to expand the field in a weak-contrast form, i.e.,

$$\boldsymbol{E}(\boldsymbol{x}) = \langle \boldsymbol{E} \rangle + \sum_{k=1}^{\infty} \boldsymbol{E}_k(\boldsymbol{x}), \qquad \langle \boldsymbol{E}_k \rangle = 0 \quad \forall k, \tag{20.53}$$

where $\boldsymbol{E}_k$ is the zero-mean fluctuating contribution to the field of kth order in the difference in the phase conductivities, i.e., $[(\sigma_p - \sigma_q)/\sigma_q)]^k$. The k-point microstructural parameter $\boldsymbol{a}_k^{(p)}$ can be written in terms of certain averages of the local field $\boldsymbol{E}_k$. To show this, we use (20.53) and the definition $\boldsymbol{\sigma}_e \cdot \langle \boldsymbol{E} \rangle = \langle \boldsymbol{\sigma}(\boldsymbol{x}) \cdot \boldsymbol{E}(\boldsymbol{x}) \rangle$ to obtain

$$\boldsymbol{\sigma}_e \cdot \langle \boldsymbol{E} \rangle = \langle \boldsymbol{\sigma} \rangle \cdot \langle \boldsymbol{E} \rangle + \langle \delta\boldsymbol{\sigma} \cdot \boldsymbol{E}_1 \rangle + \langle \delta\boldsymbol{\sigma} \cdot \boldsymbol{E}_2 \rangle + \cdots,$$

where $\delta\boldsymbol{\sigma}(\boldsymbol{x}) = \boldsymbol{\sigma}(\boldsymbol{x}) - \boldsymbol{\sigma}_q$. Comparing this to (20.48) reveals that

$$\langle \delta\boldsymbol{\sigma} \cdot \boldsymbol{E}_k \rangle = \frac{(\sigma_p - \sigma_q)^{k+1}}{\sigma_q^k} \boldsymbol{a}_{k+1}^{(p)} \cdot \langle \boldsymbol{E} \rangle, \qquad k \geq 1. \tag{20.54}$$

This weak-contrast formulation will prove valuable in constructing trial fields for variational bounds on the effective conductivity in Chapter 21. There we will also employ the following useful identities:

$$\frac{(\sigma_p - \sigma_q)^{k+1}}{\sigma_q^k} \langle \boldsymbol{E} \rangle \cdot \boldsymbol{a}_{k+1}^{(p)} \cdot \langle \boldsymbol{E} \rangle = \begin{cases} -\langle \boldsymbol{E}_k \cdot \boldsymbol{\sigma}_q \cdot \boldsymbol{E}_k \rangle, & k = 1, 3, 5, \ldots, \\ \langle \boldsymbol{E}_{k-1} \cdot \delta\boldsymbol{\sigma} \cdot \boldsymbol{E}_{k-1} \rangle, & k = 2, 4, 6, \ldots, \end{cases} \tag{20.55}$$

where *no summation* is implied on the right side of (20.55) involving index k. These identities follow immediately from (20.54) and the identities

$$\langle \boldsymbol{E}_k \cdot \boldsymbol{\sigma}_q \cdot \boldsymbol{E}_i \rangle = -\langle \boldsymbol{E}_k \cdot \delta\boldsymbol{\sigma} \cdot \boldsymbol{E}_{i-1} \rangle, \qquad k \geq 1,\ i \geq 1, \tag{20.56}$$

where $\boldsymbol{E}_0 \equiv \langle \boldsymbol{E} \rangle$. Relations (20.56) follow easily after $\boldsymbol{E}' = \boldsymbol{E}_k$ and (20.53) are substituted into identity (14.35) and terms of like order collected.

20.1.6 Isotropic Media

Here we specialize to the case of macroscopically isotropic media, i.e., $\boldsymbol{\sigma}_e = \sigma_e \boldsymbol{I}$. Thus, we take the trace of both sides of relation (20.37) to obtain

$$\beta_{pq}^2 \phi_p^2 \beta_{eq}^{-1} = \phi_p \beta_{pq} - \sum_{n=3}^{\infty} A_n^{(p)} \beta_{pq}^n, \quad p \neq q, \tag{20.57}$$

where $A_n^{(p)} = \text{Tr}\, \boldsymbol{A}_n^{(p)}/d$ and β_{eq} is an *effective polarizability* defined by

$$\beta_{eq} = \frac{\sigma_e - \sigma_q}{\sigma_e + (d-1)\sigma_q}. \tag{20.58}$$

Here we have employed the result that $\boldsymbol{A}_2^{(p)} = \mathbf{0}$ for macroscopically isotropic media [cf. (20.42)]. Expression (20.57) was first given by Brown (1955) for $d = 3$ and by Torquato (1985a) for arbitrary space dimension d.

The three- and four-point tensor parameters can be written in the forms

$$\boldsymbol{A}_3^{(p)} = (d-1)\phi_1\phi_2\zeta_p\boldsymbol{I}, \tag{20.59}$$

$$\boldsymbol{A}_4^{(p)} = (d-1)\phi_1\phi_2\gamma_p\boldsymbol{I}. \tag{20.60}$$

Therefore, relations (20.46) and (20.47) reduce to

$$\zeta_1 + \zeta_2 = 1, \tag{20.61}$$

$$\gamma_1 - \gamma_2 = (d-2)(\zeta_2 - \zeta_1). \tag{20.62}$$

These results were obtained by Torquato (1985a) for macroscopically isotropic composites. Relation (20.61) was proven earlier by Torquato (1980) and by Milton (1981b). Note that for $d = 2$, relation (20.62) gives $\gamma_1 = \gamma_2$. Moreover, Torquato (1985a) showed, using the phase-interchange theorem, that the four-point parameters are exactly zero ($\gamma_1 = \gamma_2 = 0$) for $d = 2$.

By using (20.39) with $n = 3$ and (20.59), the three-point parameter ζ_p can be expressed in any space dimension d (Torquato 1985a, Torquato 1997) as

$$\zeta_p = \frac{d^2}{(d-1)\phi_1\phi_2} I[\bar{S}_3^{(p)}], \tag{20.63}$$

where I is the integral operator

$$I[f] = \frac{1}{\Omega^2}\int\int \frac{d\boldsymbol{r}}{r^d}\frac{d\boldsymbol{s}}{s^d}[d(\boldsymbol{n}\cdot\boldsymbol{m})^2 - 1]f(r,s,t), \tag{20.64}$$

$$\bar{S}_3^{(p)}(r,s,t) = S_3^{(p)}(r,s,t) - \frac{S_2^{(p)}(r)S_2^{(p)}(s)}{\phi_p}, \tag{20.65}$$

and $\boldsymbol{n} = \boldsymbol{r}/|\boldsymbol{r}|$ and $\boldsymbol{m} = \boldsymbol{s}/|\boldsymbol{s}|$ are unit vectors. The parameter ζ_p lies in the closed interval $[0, 1]$ (Torquato 1980, Milton 1981a, Milton 1981b).

Observe that the quantity $[d(\boldsymbol{n}\cdot\boldsymbol{m})^2 - 1]$ can be written in terms of d-dimensional spherical orthogonal polynomials. The d^2-fold integral of the operator (20.64) can be reduced (by integrating over angles) to a three-fold integral over the lengths r, s, and t of the sides of the triangle or, equivalently, over the two lengths r and s and the angle θ opposite the side of length t, where $t^2 = r^2 + s^2 - 2rs\cos\theta$, with $\cos\theta = \boldsymbol{n}\cdot\boldsymbol{m}$. For example, for $d = 2$,

$$\zeta_p = \frac{4}{\pi\phi_1\phi_2}\int_0^\infty \frac{dr}{r}\int_0^\infty \frac{ds}{s}\int_0^\pi d\theta\cos(2\theta)\left[S_3^{(p)}(r,s,\theta) - \frac{S_2^{(p)}(r)S_2^{(p)}(s)}{S_1^{(p)}}\right], \tag{20.66}$$

and for $d = 3$,

$$\zeta_p = \frac{9}{2\phi_1\phi_2}\int_0^\infty \frac{dr}{r}\int_0^\infty \frac{ds}{s}\int_{-1}^1 d(\cos\theta)P_2(\cos\theta)\left[S_3^{(p)}(r,s,\theta) - \frac{S_2^{(p)}(r)S_2^{(p)}(s)}{S_1^{(p)}}\right], \tag{20.67}$$

Table 20.1 The three-point parameter ζ_2 versus the particle volume fraction ϕ_2 for simple cubic (SC), body-centered cubic (BCC), and face-centered cubic (FCC) arrays of spheres (McPhedran and Milton 1981).

	Three-point parameter ζ_2		
ϕ_2	SC Array	BCC Array	FCC Array
0.10	0.0003	0.0000	0.0000
0.20	0.0050	0.0007	0.0004
0.30	0.0220	0.0031	0.0021
0.40	0.0678	0.0107	0.0078
0.45	0.1104	0.0184	0.0136
0.50	0.1738	0.0307	0.0232
0.60		0.0796	0.0619
0.66		0.1381	0.1095
0.70			0.1596
0.71			0.1756

where P_2 is the Legendre polynomial of order 2.

We also note two simple properties of the operator (20.64): (i) If the function f does not depend on t, then

$$I[f(r,s)] = 0 \tag{20.68}$$

by virtue of the orthogonality of the spherical harmonics; (ii) less obviously, if f depends only on t, then

$$I[f(t)] = \frac{d-1}{d^2}[f(0) - f(\infty)], \tag{20.69}$$

provided that $f(\infty)$ is finite and the Fourier transform of $f(t) - f(\infty)$ is convergent. The second property states that the integral $I[f(t)]$ depends only on the extreme values of the function $f(t)$, not on intermediate values. Property (20.69) is the d-dimensional generalization of the ones proved for $d = 2$ and $d = 3$ using integration by parts (Torquato 1980, Milton 1982) or Fourier transform techniques (Lado and Torquato 1986).

The evaluation of ζ_p for a variety of two- and three-dimensional models of *random* media is detailed in Section 22.1.1. For periodic lattices of particles, the associated three-point parameter ζ_p can be ascertained without the integral (20.63) having to be evaluated directly, since the effective conductivity is often known essentially exactly. Tables 20.1 and 20.2 give ζ_2 for three- and two-dimensional lattices as found by McPhedran and Milton (1981). The parameter ζ_2 has been also determined numerically for periodic arrays of triangular and hexagonal inclusions (Hyun and Torquato 2000).

Table 20.2 The three-point parameter ζ_2 versus the particle volume fraction ϕ_2 for square and hexagonal arrays of aligned, infinitely long circular cylinders (McPhedran and Milton 1981).

	Three-point parameter ζ_2	
ϕ_2	Square Array	Hexagonal Array
0.10	3.398×10^{-5}	8.380×10^{-8}
0.20	6.117×10^{-4}	6.034×10^{-6}
0.30	3.540×10^{-3}	7.855×10^{-5}
0.40	1.306×10^{-2}	5.149×10^{-4}
0.50	3.833×10^{-2}	2.357×10^{-3}
0.60	9.965×10^{-2}	8.798×10^{-3}
0.70	0.2473	2.958×10^{-2}
0.76	0.4314	6.057×10^{-2}
0.78	0.5229	7.722×10^{-2}
0.80		9.888×10^{-2}
0.89		0.3409
0.90		0.4010
0.905		0.4364

In order to better understand the physical meaning of the expansion (20.57), it is helpful to consider the microstructures for which the microstructural parameters $A_n^{(p)}$ vanish for all values of the phase conductivities. That is, we may ask, for which class of composites is the relation

$$\beta_{eq} = \phi_p \beta_{pq} \tag{20.70}$$

or, equivalently,

$$\frac{\sigma_e - \sigma_q}{\sigma_e + (d-1)\sigma_q} = \frac{\sigma_p - \sigma_q}{\sigma_p + (d-1)\sigma_q}\phi_p \tag{20.71}$$

exact? For $d = 2$ and $d = 3$, (20.71) coincides with the Hashin–Shtrikman bounds on the effective conductivity for any isotropic two-phase composite (Hashin and Shtrikman 1962, Hashin 1965b) and hence is exact for the assemblages of coated spheres and finite-rank laminates discussed in Chapters 16 and 21. Importantly, for either the coated-inclusion assemblages or finite-rank laminates, one of the phases is always a disconnected, dispersed phase in a connected matrix phase (except in the trivial instance when the generally dispersed phase fills all of space). Result (20.71) is the d-dimensional generalization of the Hashin–Shtrikman bounds on σ_e for any $d \geq 2$; for $\sigma_2 \geq \sigma_1$, it gives a lower bound for $q = 1$ and $p = 2$ and an upper bound for $q = 2$ and $p = 1$ (see Chapter 21).

In light of this discussion, series (20.57) can be viewed as an expansion that perturbs around the Hashin–Shtrikman structures. Therefore, it is expected that the first few terms of the expansion (20.57) will yield an excellent approximation of σ_e for any values of the phase conductivities for dispersions in which the inclusions, taken to be the polarized phase, are prevented from forming large clusters. This is indeed the case for a variety of ordered and disordered dispersions when (20.57) is truncated after third-order terms, as shown at the end of the present section and in Section 22.1.1. However, we note that this truncated expansion may also be applied to bicontinuous porous media, as discussed shortly below.

Now we show that by utilizing (20.37) we can easily obtain a strong-contrast expansion that perturbs about another known exact class of structures, namely, the phase-inversion-symmetric structures that realize the self-consistent (SC) formula (18.13) for two phases ($M = 2$) and arbitrary d. All we need to do is add the expansion (20.37) with $p = 2$ and $q = 1$ to the one with $p = 1$ and $q = 2$ to get

$$\phi_2 \frac{\sigma_e + (d-1)\sigma_1}{\sigma_e - \sigma_1} + \phi_1 \frac{\sigma_e + (d-1)\sigma_2}{\sigma_e - \sigma_2} = 2 - d - \sum_{n=3}^{\infty} \left[\frac{A_n^{(2)}}{\phi_2} \beta_{21}^{n-2} + \frac{A_n^{(1)}}{\phi_1} \beta_{12}^{n-2} \right]. \tag{20.72}$$

If the sum on the right-hand side of (20.72) is set equal to zero, then we recognize the remaining expression to be equivalent to the self-consistent (SC) formula (15.14) for *spherical grains* with $M = 2$. This observation, combined with the fact that β_{12} can be expressed in terms of β_{21} via

$$\beta_{12} = -\frac{\beta_{21}}{1 + (d-2)\beta_{21}},$$

enables us to conclude that (20.72) is a strong-contrast expansion of σ_e in powers of β_{21} that perturbs about the structures that realize the two-phase self-consistent (SC) formula (15.14). The n-point parameters for the SC geometries follow immediately from (20.72), since the coefficients of order β_{21}^{n-2} must all be zero; for example, the three-, four-, five-, and six-point parameters are

$$A_3^{(p)} = (d-1)\phi_q \phi_p^2, \qquad A_4^{(p)} = (d-1)(d-2)\phi_q \phi_p^3, \tag{20.73}$$

$$A_5^{(p)} = (d-1)\phi_q \phi_p^3 [(d^2 - 5d + 5)\phi_p + d + 1], \tag{20.74}$$

$$A_6^{(p)} = (d-1)(d-2)\phi_q \phi_p^4 [(d^2 - 7d + 7)\phi_p + 3(d-3)]. \tag{20.75}$$

Note that combination of (20.59), (20.60), and (20.73) gives the scaled three- and four-point parameters as

$$\zeta_p = \phi_p, \qquad \gamma_p = (d-2)\phi_p^2 \tag{20.76}$$

for the SC geometries. The result $\zeta_p = \phi_p$ was first given by Milton (1984). The use of a truncated form of (20.72) as an approximation formula is discussed at the end of the present section.

Weak-Contrast Expansions

From the anisotropic weak-contrast expansion (20.48) and relations (20.59) and (20.60), we find the corresponding expansion for macroscopically isotropic media:

$$\sigma_e = \sigma_q + a_1^{(p)}(\sigma_p - \sigma_q) + a_2^{(p)}\frac{(\sigma_p - \sigma_q)^2}{\sigma_q} + a_3^{(p)}\frac{(\sigma_p - \sigma_q)^3}{\sigma_q^2}$$
$$+ a_4^{(p)}\frac{(\sigma_p - \sigma_q)^4}{\sigma_q^3} + \mathcal{O}\left[\frac{(\sigma_p - \sigma_q)^5}{\sigma_q^4}\right], \qquad p \neq q, \tag{20.77}$$

where $a_k^{(p)} = \text{Tr}\,\boldsymbol{a}_k^{(p)}/d$. The first four parameters are

$$a_1^{(p)} = \phi_p, \tag{20.78}$$

$$a_2^{(p)} = \frac{-\phi_p\phi_q}{d}, \tag{20.79}$$

$$a_3^{(p)} = \frac{\phi_p\phi_q}{d^2}[\phi_q + (d-1)\zeta_p], \tag{20.80}$$

$$a_4^{(p)} = \frac{\phi_p\phi_q}{d^3}[-\phi_q^2 - (1+2\phi_q)(d-1)\zeta_p + (d-1)\gamma_p]. \tag{20.81}$$

Effective Behavior in the Limit of Infinite Dimension

Let us consider the effective conductivity in the limit that the space dimension tends to infinity ($d \to \infty$). In this limit, Torquato (1997) showed that the effective conductivity σ_e for any isotropic multiphase composite is given exactly by the *arithmetic average*, provided that the phase conductivities are nonzero. This result applies to *nonlinear* composites as well.

The first hint of this interesting behavior as $d \to \infty$ can be gleaned by examining the exact fourth-order expansion (20.77) in this limit. We find from this expression that as $d \to \infty$,

$$\sigma_e = \phi_1\sigma_1 + \phi_2\sigma_2,$$

which is just the arithmetic average of the phase conductivities.

The general proof for arbitrary conditions and for any number of phases requires the d-dimensional Hashin–Shtrikman bounds for M-phase composites given by (21.23).

Theorem 20.1 *For any d-dimensional macroscopically isotropic M-phase composite possessing nonzero phase conductivities, the effective conductivity σ_e is independent of the microstructure and is exactly given by the arithmetic average*

$$\sigma_e = \sum_{i=1}^{M} \phi_i\sigma_i \tag{20.82}$$

in the limit that the space dimension becomes infinite ($d \to \infty$).

Proof: Under the stated conditions, the d-dimensional, M-phase bounds (21.23) on the effective conductivity coincide and equal expression (20.82). ∎

Remarks:

1. Why does the effective conductivity tend to the arithmetic mean as $d \to \infty$ (implying the same constant electric field in each phase)? This can be explained physically by considering the phase-interface continuity conditions on the electric or strain fields and the fact that the energy associated with an *isotropic composite* must be the same in any direction. For example, in the case of conduction, the components of the electric field vector are continuous in all of the d directions, except the direction normal to the interface. Thus, for an isotropic composite in the limit $d \to \infty$, the electric fields in each phase approach the same constant value. Note that this limiting behavior applies also to composites possessing nonlinear flux-intensity constitutive laws.
2. As we will show, the effective shear modulus G_e of isotropic composites also tends to the arithmetic average in the limit $d \to \infty$.

Approximation Formulas

Let us assume that phases 1 and 2 are the matrix and dispersed phases, respectively, of a dispersion. Truncation of the series (20.57) after third-order terms yields the expression

$$\frac{\sigma_e}{\sigma_1} = \frac{1 + (d-1)\phi_2\beta_{21} - (d-1)\phi_1\zeta_2\beta_{21}^2}{1 - \phi_2\beta_{21} - (d-1)\phi_1\zeta_2\beta_{21}^2}, \tag{20.83}$$

which was shown by Torquato (1985a) to give an *accurate approximation* of the effective conductivity of *dispersions in which the particles (polarized phase), generally, do not form large clusters*. However, for the special case of bicontinuous porous media in which the void (polarized) phase is perfectly insulating ($\sigma_2 = 0$), the approximation (20.83) may still provide a good estimate of σ_e. The reason for this is that the flux is carried by the *connected* solid (i.e., nonpolarized) phase, regardless of whether the void phase is connected or disconnected.

We refer to (20.83) as a *three-point approximation*, since for $d \geq 3$, it incorporates information of up to three-point correlation functions. For $d = 2$, relation (20.83) has been shown (Torquato 1985a) to coincide with the four-point bounds (21.42) derived by Milton (1981a). It turns out that for $d = 2$, the four-point parameter γ_2, defined by (20.60), vanishes identically (Torquato 1985a). For $d \geq 3$, the three-point approximation (20.83) always lies between the best available rigorous three-point bounds [cf. (21.35) and (21.37)], provided that $\zeta_2 \leq 1/(d-1)$. Thus, formula (20.83) should be applied for $d = 3$ only when $\zeta_2 \leq 0.5$. It is noteworthy that for all of the realistic models of random media discussed in Section 22.1.1, the parameter ζ_2 always meets this condition, and hence estimates based on (20.83) always lie between the tightest three-point bounds.

To test the three-point approximation (20.83) for three-dimensional dispersions, Torquato (1985a) evaluated it for the benchmark model of periodic arrays of hard spheres for the extreme case of superconducting particles ($\sigma_2/\sigma_1 = \infty$), the most difficult case to treat theoretically (see Figure 20.2). Figure 20.2 shows that (20.83) provides

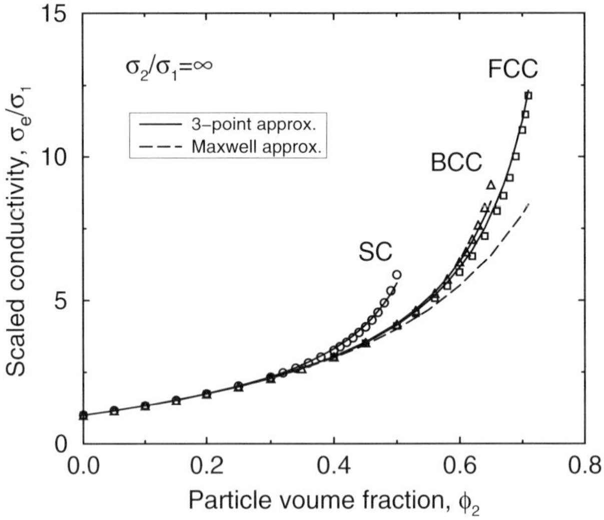

Figure 20.2 The scaled effective conductivity σ_e/σ_1 versus ϕ_2 for systems of superconducting spheres arranged on three cubic lattices. The solid curves are the predictions of (20.83) for $d = 3$, and the circles, triangles, and squares denote numerical data (McPhedran and McKenzie 1978, McKenzie et al. 1978). Maxwell's formula (18.5) or, equivalently, the Hashin–Shtrikman lower bound is included.

excellent agreement with the exact results up to the maximum volume fractions reported, i.e., up to about 95% of the respective close-packing volume fractions. Thus, (20.83) does indeed capture the sensitivity of σ_e to microstructure for three-dimensional dispersions. This is in contrast to Maxwell's approximation (18.5) (equivalent to the Hashin–Shtrikman lower bound), which is insensitive to the lattice arrangement (see Figure 20.2). Note that the accuracy of (20.83) for periodic arrays in which $\sigma_2/\sigma_1 < \infty$ will be even better than in the superconducting case. This example demonstrates that (20.83) will provide accurate estimates of σ_e for ordered as well as random dispersions, for all phase contrast ratios, provided that the particles generally do not form large clusters. In Section 22.1.1 formula (20.83) is used to assess the accuracy of rigorous bounds on the effective conductivity of *random* dispersions.

In Figure 20.3 we compare formula (20.83) for an FCC lattice of spheres in which the contrast ratio is $\sigma_2/\sigma_1 = 50$ to the popular self-consistent (SC) expression (18.13). Not surprisingly, the SC formula does not provide a good estimate of the effective conductivity for this dispersion, for reasons stated in Section 18.1.2. However, it is not well known that the SC formula actually *violates* the rigorous three-point upper bound (21.35).

Consider truncating the "self-consistent" strong-contrast expansion (20.72) after third-order terms. Accordingly, we find that

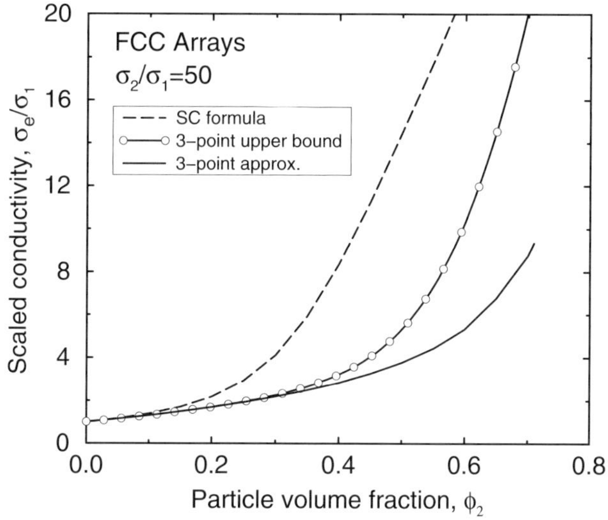

Figure 20.3 The scaled effective conductivity σ_e/σ_1 versus ϕ_2 for a system of spheres arranged on FCC lattices with $\sigma_2/\sigma_1 = 50$. Comparison of the three-point approximation (20.83), self-consistent (SC) formula (18.13), and rigorous three-point upper bound (21.35) for $d = 3$.

$$\phi_2 \frac{\sigma_e + (d-1)\sigma_1}{\sigma_e - \sigma_1} + \phi_1 \frac{\sigma_e + (d-1)\sigma_2}{\sigma_e - \sigma_2} = 2 - d + (d-1)(\phi_2 - \zeta_2)\beta_{21}. \tag{20.84}$$

The three-point formula (20.84) should yield good approximations of the effective conductivity of composite structures that resemble those that realize the SC formula (18.13) (Section 18.1.2), a key feature being phase-inversion symmetry. Thus, it should provide a good estimate of σ_e for symmetric-cell materials with spherical cells (see Chapter 8). In Section 22.1.1 it is shown that ζ_2 for symmetric-cell materials is a linear function of the volume fraction whose slope depends on the shape of the cell. For example, for d-dimensional spherical cells, $\zeta_2 = \phi_2$, which is exactly its value for the SC spherical-grain structures [cf. (20.76)]. Thus, (20.84) predicts that such symmetric-cell materials will have an effective conductivity that is identical to the two-phase SC formula (18.13). As the cell shape of a symmetric-cell material deviates from that of a sphere, the accuracy of (20.84) is expected to diminish. An example of an actual microstructure for which (20.84) may provide a good estimate of σ_e is the bicontinuous boron carbide/aluminum composite shown in Figure 1.2. Formula (20.84) should not be applied to composites that do not possess this type of phase symmetry (e.g., particulate media). However, the utility of (20.84) has yet to be fully explored.

At any rate, we require that the approximation (20.84) always lie within the tightest three-point bounds (21.35) and (21.37). This places possible restrictions on ζ_2 that generally are complicated functions of the phase conductivities. However, when phase 2 is superconducting relative to phase 1 ($\sigma_2/\sigma_1 = \infty$), then the three-point approximation (20.84) yields

$$\frac{\sigma_e}{\sigma_1} = \frac{1 + (d-1)(\phi_2 - \zeta_2)}{\phi_1 - (d-1)\zeta_2} \quad \text{if} \quad \sigma_2/\sigma_1 = \infty, \tag{20.85}$$

with the restriction that $\zeta_2 \le [d/(d-1)] - \phi_1$. This restriction is satisfied for $d = 2$ for all allowable values of ζ_2. Note that (20.85) predicts that phase 2 percolates (composite becomes superconducting) when $\phi_1 = (d-1)\zeta_2$. This percolation-threshold prediction, unlike the one from the SC formula (18.13), is dependent on the microstructure (as it should be) via the three-point parameter ζ_2. For a symmetric-cell material with spherical cells, (20.85) predicts the same threshold as the SC formula, i.e., $\phi_{2c} = 1/d$.

When phase 2 is perfectly insulating relative to phase 1 ($\sigma_2/\sigma_1 = 0$), then the approximation (20.85) gives

$$\frac{\sigma_e}{\sigma_1} = 1 - \frac{d\phi_2}{d - \phi_1 - \zeta_2} \quad \text{if} \quad \sigma_2/\sigma_1 = 0, \tag{20.86}$$

with *no restrictions* on ζ_2. Expression (20.86) predicts a microstructure-dependent percolation threshold when $\phi_1 = \zeta_2/(d-1)$, i.e., the point that the composite no longer conducts. For a symmetric-cell material with spherical cells, (20.86) predicts the same threshold as the SC formula, i.e., $\phi_{2c} = (d-1)/d$.

20.2 Stiffness Tensor

Torquato (1997) generalized the approach of the previous section to derive contrast expansions for the effective stiffness tensor. The key to the derivation is the formulation of an integral equation for the local "cavity" strain field $\boldsymbol{f}(\boldsymbol{x})$ in d spatial dimensions. A relationship between $\boldsymbol{f}(\boldsymbol{x})$ and the local "polarization" stress field $\boldsymbol{p}(\boldsymbol{x})$ is then established. By carefully manipulating integral equations for $\boldsymbol{p}(\boldsymbol{x})$ and $\boldsymbol{f}(\boldsymbol{x})$ and averaging, he then found series expansions for the effective tensor in terms of powers of the *elastic polarizabilities*. This formulation is detailed below and various applications of the expansions are examined. The behavior of the effective elastic moduli of isotropic media as d tends to infinity is also described.

20.2.1 Integral Equation for the Cavity Strain Field

As before, consider a large but finite-sized ellipsoidal, macroscopically anisotropic composite specimen in arbitrary space dimension d composed of two isotropic phases with fourth-order stiffness tensors $\boldsymbol{C}_1$ and $\boldsymbol{C}_2$. The microstructure is perfectly general and possesses a characteristic microscopic length scale that is much smaller than the smallest semiaxes of the ellipsoid. Thus, the specimen is virtually statistically homogeneous. Ultimately, we will take the infinite-volume limit and hence consider statistically homogeneous media. In this distinguished limit, the effective stiffness tensor should be independent of the shape of the ellipsoid. The local stiffness tensor $\boldsymbol{C}(\boldsymbol{x})$ can be expressed in terms of $\mathcal{I}^{(p)}$, the indicator function for phase p ($p = 1, 2$) defined by (20.3), as follows:

$$\boldsymbol{C}(\boldsymbol{x}) = \boldsymbol{C}_1 \mathcal{I}^{(1)}(\boldsymbol{x}) + \boldsymbol{C}_2 \mathcal{I}^{(2)}(\boldsymbol{x}). \tag{20.87}$$

The stiffness of phase p can in turn be written in terms of the bulk modulus K_p and shear modulus G_p of phase p according to the relation

$$\boldsymbol{C}_p = dK_p \boldsymbol{\Lambda}_h + 2G_p \boldsymbol{\Lambda}_s, \tag{20.88}$$

where $\boldsymbol{\Lambda}_h$ and $\boldsymbol{\Lambda}_s$ are the fourth-order hydrostatic and shear projection tensors, respectively, defined by (13.96) and (13.97).

Now let us embed this d-dimensional ellipsoidal composite specimen in an infinite *reference* phase q that is subjected to an applied strain field $\varepsilon_0(\boldsymbol{x})$ at infinity (see Figure 20.1). The reference phase can be chosen to be arbitrary, but here we will take it to be either phase 1 or phase 2, i.e., $q = 1$ or $q = 2$. Introducing the *polarization* field, defined by

$$\boldsymbol{p}(\boldsymbol{x}) = [\boldsymbol{C}(\boldsymbol{x}) - \boldsymbol{C}_q] : \varepsilon(\boldsymbol{x}), \tag{20.89}$$

enables us to reexpress the symmetric stress tensor $\boldsymbol{\tau}$, defined by Hooke's law (13.53), as follows:

$$\boldsymbol{\tau}(\boldsymbol{x}) = \boldsymbol{C}_q \varepsilon(\boldsymbol{x}) + \boldsymbol{p}(\boldsymbol{x}). \tag{20.90}$$

The symmetric second-order tensor $\boldsymbol{p}(\boldsymbol{x})$ is the *induced stress polarization field* relative to the medium in the absence of phase p and hence is zero in the reference phase q and nonzero in the "polarized" phase p ($p \neq q$) (see also Section 20.1.1).

With the aid of (20.90), the divergence-free (equilibrium) condition (13.50) on the stress can be rewritten in component form as

$$(C_q)_{ijkl} \frac{\partial^2 \hat{u}_k(\boldsymbol{x})}{\partial x_j \partial x_l} = -\frac{\partial p_{ij}(\boldsymbol{x})}{\partial x_j}, \tag{20.91}$$

$$\hat{u}_k(\boldsymbol{x}) \to 0, \quad |\boldsymbol{x}| \to \infty, \tag{20.92}$$

where $\hat{\boldsymbol{u}}(\boldsymbol{x})$ is the displacement field in excess of the displacement field $\boldsymbol{u}_0(\boldsymbol{x})$ at infinity, i.e., $\hat{\boldsymbol{u}}(\boldsymbol{x}) = \boldsymbol{u}(\boldsymbol{x}) - \boldsymbol{u}_0(\boldsymbol{x})$. The infinite-space Green's function $g_{ij}^{(q)}$ is the solution of

$$(C_q)_{ijkl} \frac{\partial^2 g_{im}^{(q)}(\boldsymbol{x}, \boldsymbol{x}')}{\partial x_j \partial x_l} = -\delta_{km} \delta(\boldsymbol{x} - \boldsymbol{x}'), \tag{20.93}$$

$$g_{km}^{(q)}(\boldsymbol{x}, \boldsymbol{x}') \to 0, \quad |\boldsymbol{x}| \to \infty, \tag{20.94}$$

where it is to be recalled that we are not summing over the phase index q. Multiplying (20.91) by the Green's function and integrating by parts yields the integral relation

$$\boldsymbol{u}(\boldsymbol{x}) = \boldsymbol{u}_0(\boldsymbol{x}) + \int d\boldsymbol{x}' \, \nabla \boldsymbol{g}^{(q)}(\boldsymbol{x}, \boldsymbol{x}') : \boldsymbol{p}(\boldsymbol{x}'), \tag{20.95}$$

where $\boldsymbol{u}_0$ is the displacement field at infinity. Note that the presence of the polarization $\boldsymbol{p}$ in (20.95) implies that the integration volume extends only over the region of space occupied by the finite-sized ellipsoidal composite specimen. Integral relations of the form (20.95) have been derived previously by various investigators (Dederichs and Zeller 1973, Gubernatis and Krumhansl 1975, Willis 1981) for the case of three dimensions ($d = 3$).

It is a simple matter to show (Torquato 1997) that the d-dimensional Green's function that satisfies (20.93) and (20.94) is given by

$$g_{ij}^{(q)}(\boldsymbol{r}) = \begin{cases} \dfrac{1}{2\Omega G_q} \ln\left(\dfrac{1}{r}\right)\delta_{ij} + b_q n_i n_j, & d = 2, \\ a_q \dfrac{\delta_{ij}}{r^{d-2}} + b_q \dfrac{n_i n_j}{r^{d-2}}, & d \geq 3, \end{cases} \tag{20.96}$$

where

$$a_q = \frac{1}{2(d-2)\Omega G_q} \frac{dK_q + (3d-2)G_q}{dK_q + 2(d-1)G_q}, \tag{20.97}$$

$$b_q = \frac{1}{2\Omega G_q} \frac{dK_q + (d-2)G_q}{dK_q + 2(d-1)G_q}, \tag{20.98}$$

$\Omega(d)$ is given by (2.56), $\boldsymbol{r} = \boldsymbol{x} - \boldsymbol{x}'$, and $\boldsymbol{n} = \boldsymbol{r}/|\boldsymbol{r}|$.

The fact that the Green's function possesses a singularity at the point $\boldsymbol{x}' = \boldsymbol{x}$ requires one to exclude a small region containing the singularity for reasons already elaborated on for the conductivity problem (see the discussion in Section 20.1.1). Briefly, an improper integral is convergent if this integral exists in the limit that the excluded region shrinks to zero, independent of the shape of the excluded region [see Torquato (1997)]. According to this criterion, the integral of (20.95) is convergent.

To obtain the strain, one must differentiate (20.95); however, because of the singular nature of the integral, one cannot simply differentiate under the integral sign. Excluding a spherical region or "cavity" from the point $\boldsymbol{x}' = \boldsymbol{x}$ in (20.95), integrating by parts, and using the divergence theorem, Torquato (1997) proved that the strain field is given by the integral relation

$$\boldsymbol{\varepsilon}(\boldsymbol{x}) = \boldsymbol{\varepsilon}_0(\boldsymbol{x}) + \int d\boldsymbol{x}'\, \boldsymbol{G}^{(q)}(\boldsymbol{r}) : \boldsymbol{p}(\boldsymbol{x}'), \tag{20.99}$$

where

$$\boldsymbol{G}^{(q)}(\boldsymbol{r}) = -\boldsymbol{D}^{(q)}\delta(\boldsymbol{r}) + \boldsymbol{H}^{(q)}(\boldsymbol{r}). \tag{20.100}$$

In relation (20.100), the constant fourth-order tensor $\boldsymbol{D}^{(q)}$ [which arises because of the exclusion of the spherical cavity in (20.95)] is given by

$$\boldsymbol{D}^{(q)} = \frac{\boldsymbol{\Lambda}_h}{dK_q + 2(d-1)G_q} + \frac{d(K_q + 2G_q)\boldsymbol{\Lambda}_s}{G_1(d+2)[dK_q + 2(d-1)G_q]}, \tag{20.101}$$

and $\boldsymbol{H}^{(q)}(\boldsymbol{r})$ is the symmetrized double gradient $\nabla\nabla \boldsymbol{g}^{(q)}(\boldsymbol{r})$ given by

$$H_{ijkl}^{(q)}(\boldsymbol{r}) = \frac{1}{2\Omega[dK_q + 2(d-1)G_q]}\frac{1}{r^d}\Big[\alpha_q\delta_{ij}\delta_{kl} - d(\delta_{ik}\delta_{jl} + \delta_{il}\delta_{jk}) - d\alpha_q(\delta_{ij}n_kn_l + \delta_{kl}n_in_j)$$
$$+ \frac{d(d-\alpha_q)}{2}(\delta_{ik}n_jn_l + \delta_{il}n_jn_k + \delta_{jk}n_in_l + \delta_{jl}n_in_k) + d(d+2)\alpha_q n_in_jn_kn_l\Big], \quad (20.102)$$

where

$$\alpha_q = dK_q/G_q + (d-2) \quad (20.103)$$

is a dimensionless parameter. It is understood that integrals involving the fourth-order tensor $\boldsymbol{H}^{(q)}$ are to be carried out by excluding at $\boldsymbol{x}' = \boldsymbol{x}$ an infinitesimal sphere in the limit that the sphere radius shrinks to zero. The tensor $H_{ijkl}^{(q)}$ is symmetric with respect to the first two indices, to the second two indices, and to interchange of ij and kl, i.e.,

$$H_{ijkl}^{(q)} = H_{jikl}^{(q)} = H_{ijlk}^{(q)} = H_{klij}^{(q)}.$$

Moreover, the integral of $\boldsymbol{H}^{(q)}(\boldsymbol{r})$ over the surface of a sphere of radius $R > 0$ is identically zero, i.e.,

$$\int_{r=R} \boldsymbol{H}^{(q)}(\boldsymbol{r})d\Omega = 0. \quad (20.104)$$

Some contractions of the tensor $\boldsymbol{H}^{(q)}$ that will be of use to us in the subsequent analysis are as follows:

$$H_{ijkk}^{(q)}(\boldsymbol{r}) = \frac{d}{\Omega[dK_q + 2(d-1)G_q]}\frac{1}{r^d}(dn_in_j - \delta_{ij}), \quad (20.105)$$

$$H_{iikl}^{(q)}(\boldsymbol{r}) = \frac{d}{\Omega[dK_q + 2(d-1)G_q]}\frac{1}{r^d}(dn_kn_l - \delta_{kl}), \quad (20.106)$$

$$H_{iikk}^{(q)}(\boldsymbol{r}) = H_{ikik}^{(q)}(\boldsymbol{r}) = 0. \quad (20.107)$$

We will also utilize the following scalar identities:

$$H_{iikl}^{(q)}(\boldsymbol{r})H_{kljj}^{(q)}(\boldsymbol{s}) = \frac{d^3}{\Omega^2[dK_q + 2(d-1)G_q]^2}\frac{1}{r^d}\frac{1}{s^d}[d(\boldsymbol{n}\cdot\boldsymbol{m})^2 - 1], \quad (20.108)$$

$$H_{ijkl}^{(q)}(\boldsymbol{r})H_{klij}^{(q)}(\boldsymbol{s}) = \frac{1}{4\Omega^2[dK_q + 2(d-1)G_q]^2}\frac{1}{r^d}\frac{1}{s^d}\Big\{d(d+2)\alpha_q^2[d(d+2)(\boldsymbol{n}\cdot\boldsymbol{m})^4 - 3]$$
$$- d(5d+6)\alpha_q^2[d(\boldsymbol{n}\cdot\boldsymbol{m})^2 - 1] + 2d^2(d-2)\alpha_q[d(\boldsymbol{n}\cdot\boldsymbol{m})^2 - 1]$$
$$+ d^3(d+2)[d(\boldsymbol{n}\cdot\boldsymbol{m})^2 - 1]\Big\}, \quad (20.109)$$

where $\boldsymbol{m} = \boldsymbol{s}/|\boldsymbol{s}|$ is a unit vector in the direction of $\boldsymbol{s}$.

At this stage of the analysis, Torquato (1997) departed from previous treatments by introducing an integral equation for the "cavity" strain field $\boldsymbol{f}$. Specifically, upon substitution of (20.100) into expression (20.99), we obtain the integral equation

$$\boldsymbol{f}(\boldsymbol{x}) = \boldsymbol{\varepsilon}_0(\boldsymbol{x}) + \int_\epsilon d\boldsymbol{x}'\boldsymbol{H}^{(q)}(\boldsymbol{x}-\boldsymbol{x}') : \boldsymbol{p}(\boldsymbol{x}), \quad (20.110)$$

which is related to the usual strain by the expression

$$f(x) = \left\{ I + D^{(q)} : [C(x) - C_q] \right\} : \varepsilon(x), \tag{20.111}$$

where the constant tensor $D^{(q)}$ is given by (20.101) and the subscript ϵ on an integral is defined by (20.19). We refer to $f(x)$ as the *cavity strain field* because, as can be seen from (20.111), it is a modified strain field equal to the usual strain plus a contribution involving the constant tensor $D^{(q)}$, which arises as a result of excluding a spherical cavity from the point $x' = x$ in (20.95). The cavity strain field is the elasticity analogue of the Lorentz electric field used in dielectric theory (Brown 1955, Torquato 1985a, Sen and Torquato 1989).

Combination of the expressions (20.89) and (20.111) gives a relation between the stress polarization and cavity strain field:

$$p(x) = \mathcal{L}^{(q)}(x) : f(x), \tag{20.112}$$

where the fourth-order tensor $\mathcal{L}^{(q)}(x)$ is the fractional linear transformation of $C(x)$ given by

$$\mathcal{L}^{(q)}(x) = \{C(x) - C_q\}\{I + D^{(q)} : [C(x) - C_q]\}^{-1}. \tag{20.113}$$

Clearly, $\mathcal{L}^{(q)}(x)$ has the same symmetry properties as the stiffness tensor $C(x)$. If the phases are isotropic as specified by (20.88), then $\mathcal{L}^{(q)}(x)$ can be written as a constant tensor $L^{(q)}$ multiplied by the indicator function $\mathcal{I}^{(p)}(x)$, i.e.,

$$\mathcal{L}^{(q)}(x) = L^{(q)}\mathcal{I}^{(p)}(x), \tag{20.114}$$

where

$$L^{(q)} = [dK_q + 2(d-1)G_q][\kappa_{pq}\Lambda_h + \frac{(d+2)G_q}{d(K_q + 2G_q)}\mu_{pq}\Lambda_s], \tag{20.115}$$

$$\kappa_{pq} = \frac{K_p - K_q}{K_p + \dfrac{2(d-1)}{d}G_q}, \tag{20.116}$$

$$\mu_{pq} = \frac{G_p - G_q}{G_p + \dfrac{G_q[dK_q/2 + (d+1)(d-2)G_q/d]}{K_q + 2G_q}}. \tag{20.117}$$

Note that the coefficients κ_{pq} and μ_{pq} are *scalar* (not tensor) parameters that depend on the moduli of the polarized and reference phases p and q, respectively. In analogy with dielectric theory, we refer to the scalar parameters κ_{pq} and μ_{pq} as the *bulk modulus polarizability* and the *shear modulus polarizability*, respectively.

20.2.2 Strong-Contrast Expansions

The effective tensor $L_e^{(q)}$ is defined via the relation linking the average polarization to the average cavity strain field, i.e.,

$$\langle \boldsymbol{p}(\boldsymbol{x}) \rangle = \boldsymbol{L}_e^{(q)} : \langle \boldsymbol{f}(\boldsymbol{x}) \rangle, \tag{20.118}$$

where $\boldsymbol{L}_e^{(q)}$ is the following fractional linear transformation of $\boldsymbol{C}_e$:

$$\boldsymbol{L}_e^{(q)} = \{\boldsymbol{C}_e - \boldsymbol{C}_q\} \cdot \{\boldsymbol{I} + \boldsymbol{D}^{(q)} : [\boldsymbol{C}_e - \boldsymbol{C}_q]\}^{-1}. \tag{20.119}$$

The constitutive relation (20.118) is localized, i.e., it is independent of the shape of the ellipsoidal composite specimen in the infinite-volume limit. In light of (20.119), we see that the effective tensor $\boldsymbol{L}_e^{(q)}$ has the same symmetry properties as the effective stiffness tensor $\boldsymbol{C}_e$. Note that the constitutive relation (20.118) that defines the effective tensor $\boldsymbol{L}_e^{(q)}$ is entirely equivalent to the averaged Hooke's law (13.113) that defines the effective stiffness tensor $\boldsymbol{C}_e$. Keeping in mind that the tensors $\mathcal{L}^{(q)}$, $\boldsymbol{L}_e^{(q)}$, and $\boldsymbol{H}^{(q)}$ are associated with the reference phase q, we will temporarily drop the superscript q when referring to these tensors in the subsequent discussion.

It is desired to find an explicit expression for the effective moduli $\boldsymbol{L}_e^{(q)}$ using the solution of the integral equation (20.110), which we rewrite as

$$\boldsymbol{f}(1) = \boldsymbol{\varepsilon}_0(1) + \int_\epsilon d2\boldsymbol{H}(1,2) : \boldsymbol{p}(2),$$

where 1 and 2 denote $\boldsymbol{x}$ and $\boldsymbol{x}'$, respectively. In schematic operator form, this equation can be tersely rewritten as

$$\boldsymbol{f} = \boldsymbol{\varepsilon}_0 + \boldsymbol{H}\boldsymbol{p}. \tag{20.120}$$

Multiplying the integral equation for the cavity field from the left by $\mathcal{L}(\boldsymbol{x})$ [defined by (20.115)] yields the equation

$$\boldsymbol{p} = \mathcal{L}\boldsymbol{\varepsilon}_0 + \mathcal{L}\boldsymbol{H}\boldsymbol{p}. \tag{20.121}$$

A solution for the polarization $\boldsymbol{p}$ in terms of an operator acting on the applied strain field $\boldsymbol{\varepsilon}_0$ is found by successive substitutions using relation (20.121), giving

$$\begin{aligned} \boldsymbol{p} &= \mathcal{L}\boldsymbol{\varepsilon}_0 + \mathcal{L}\boldsymbol{H}\mathcal{L}\boldsymbol{\varepsilon}_0 + \mathcal{L}\boldsymbol{H}\mathcal{L}\boldsymbol{H}\mathcal{L}\boldsymbol{\varepsilon}_0 + \cdots, \\ &= \boldsymbol{T}\boldsymbol{\varepsilon}_0, \end{aligned} \tag{20.122}$$

where the fourth-order tensor operator $\boldsymbol{T}$ is given by

$$\boldsymbol{T} = \mathcal{L}[\boldsymbol{I} - \mathcal{L}\boldsymbol{H}]^{-1}.$$

Ensemble averaging (20.122) yields

$$\langle \boldsymbol{p} \rangle = \langle \boldsymbol{T} \rangle \boldsymbol{\varepsilon}_0. \tag{20.123}$$

The operator $\langle \boldsymbol{T} \rangle$ involves products of the tensor $\boldsymbol{H}$, which decays to zero like r^{-d} for large r. Thus, $\langle \boldsymbol{T} \rangle$ at best involves conditionally convergent integrals and hence is dependent upon the shape of the ellipsoidal composite specimen. Indeed, this nonlocal nature of the relation between the polarization and applied strain is completely consistent with a well-known continuum result: When an ellipsoidal inclusion in an infinite matrix of another material is subjected to a constant strain field ε_0 at infinity, the polarization stress field within the ellipsoid is uniform and, when expressed in terms of ε_0, depends upon the shape of the ellipsoid (Section 17.2.2).

Given the nonlocal nature of the relation (20.123), the remaining strategy is clear. To obtain a relation between the average polarization $\langle \boldsymbol{p} \rangle$ and average cavity field $\langle \boldsymbol{f} \rangle$ as prescribed by (20.118), the applied field ε_0 must be eliminated in favor of the appropriate average field. Thus, inverting (20.123) gives

$$\varepsilon_0 = \langle \boldsymbol{T} \rangle^{-1} \langle \boldsymbol{p} \rangle, \tag{20.124}$$

and averaging (20.120) yields

$$\langle \boldsymbol{f} \rangle = \varepsilon_0 + \boldsymbol{H} \langle \boldsymbol{p} \rangle. \tag{20.125}$$

We can now eliminate the applied field in (20.125) using (20.124) to obtain

$$\langle \boldsymbol{f} \rangle = \boldsymbol{X} \langle \boldsymbol{p} \rangle, \tag{20.126}$$

where

$$\boldsymbol{X} = \langle \boldsymbol{T} \rangle^{-1} + \boldsymbol{H}.$$

Comparing expressions (20.118) and (20.126) (and returning to the notation of explicitly indicating that the reference medium is phase q) yields the desired result for the inverse of the effective tensor $\boldsymbol{L}_e^{(q)}$, i.e.,

$$[\boldsymbol{L}_e^{(q)}]^{-1} = \boldsymbol{X}.$$

It is convenient to multiply this equation by the constant fourth-order tensor $\boldsymbol{L}^{(q)}$ [defined by (20.115)] from the left to obtain

$$\boldsymbol{L}^{(q)} \left[\boldsymbol{L}_e^{(q)} \right]^{-1} = \boldsymbol{L}^{(q)} \boldsymbol{X},$$

or, more explicitly,

$$\begin{aligned} \boldsymbol{L}^{(q)} : \left[\boldsymbol{L}_e^{(q)} \right]^{-1} = \frac{\boldsymbol{I}}{S_1^{(p)}(1)} &- \int d2 \left[\frac{S_2^{(p)}(1,2) - S_1^{(p)}(1) S_1^{(p)}(2)}{S_1^{(p)}(1) S_1^{(p)}(2)} \right] \boldsymbol{U}^{(q)}(1,2) \\ &- \int d2 d3 \left[\frac{S_3^{(p)}(1,2,3)}{S_1^{(p)}(1) S_1^{(p)}(2)} - \frac{S_2^{(p)}(1,2) S_2^{(p)}(2,3)}{S_1^{(p)}(1) S_1^{(p)}(2) S_1^{(p)}(3)} \right] \boldsymbol{U}^{(q)}(1,2) : \boldsymbol{U}^{(q)}(2,3) - \cdots, \end{aligned}$$

where

$$\begin{aligned} U_{ijkl}^{(q)}(\boldsymbol{r}) &= L_{ijmn}^{(q)} H_{mnkl}^{(q)}(\boldsymbol{r}) \\ &= [dK_q + 2(d-1)G_q] \left\{ \left[\kappa_{pq} - \frac{(d+2)G_q}{d(K_q + 2G_q)} \mu_{pq} \right] \frac{\delta_{ij}}{d} H_{mmkl}^{(q)}(\boldsymbol{r}) \right. \\ &\qquad \left. + \frac{(d+2)G_q}{d(K_q + 2G_q)} \mu_{pq} H_{ijkl}^{(q)}(\boldsymbol{r}) \right\}. \end{aligned} \tag{20.127}$$

Here the n-point probability function $S_n^{(p)}$ for the polarized phase p is defined by (20.36) and arises in the corresponding expansion for the effective conductivity (20.37). The reason why $S_n^{(p)}$ arises in the expansion (20.127) is that the operator $\boldsymbol{X}$ contains averages over products of the position-dependent tensor $\mathcal{L}^{(q)}(\boldsymbol{x})$, which in turn depends on the indicator function $\mathcal{I}^{(p)}(\boldsymbol{x})$ [cf. (20.115)]. Note that at this stage of the analysis we have not passed to the statistically homogeneous infinite-volume limit.

The general term of the expansion (20.127) is given by

$$\phi_p^2 \boldsymbol{L}^{(q)} : \left[\boldsymbol{L}_e^{(q)}\right]^{-1} = \phi_p \boldsymbol{I} - \sum_{n=2}^{\infty} \boldsymbol{B}_n^{(p)}, \tag{20.128}$$

where the fourth-order tensor coefficients $\boldsymbol{B}_n^{(p)}$ are the following integrals over products of the $\boldsymbol{U}^{(q)}$ tensors and the $S_n^{(p)}$ associated with phase p:

$$\boldsymbol{B}_2^{(p)} = \int_\epsilon d2 \boldsymbol{U}^{(q)}(1,2)[S_2^{(p)}(1,2) - \phi_p^2], \tag{20.129}$$

$$\boldsymbol{B}_n^{(p)} = (-1)^n \left(\frac{1}{\phi_p}\right)^{n-2} \int d2 \cdots \int dn\, \boldsymbol{U}^{(q)}(1,2){:}\boldsymbol{U}^{(q)}(2,3){:}\cdots{:}\boldsymbol{U}^{(q)}(n-1,n) \times \Delta_n^{(p)}(1,\ldots,n), \qquad n \geq 3, \tag{20.130}$$

and $\Delta_n^{(p)}$ is exactly the same position-dependent determinant (20.41) associated with phase p that arises in the conductivity problem.

Remarks:

1. Result (20.128) actually represents two different series expansions: one for $p = 1$ and $q = 2$ and the other for $p = 2$ and $q = 1$.
2. The n-point tensor coefficients $\boldsymbol{B}_n^{(p)}$ are functionals of the correlation functions $S_1^{(p)}, \ldots, S_n^{(p)}$ but unlike their conductivity counterparts $\boldsymbol{A}_n^{(p)}$, they also depend on the phase properties through the polarizabilities κ_{pq} and μ_{pq}. To see this, one can write the product of the tensors $\boldsymbol{U}^{(q)}$ appearing in (20.130) in terms of products of the tensor $\boldsymbol{H}^{(q)}$ via relation (20.127). Such products of $\boldsymbol{H}^{(q)}$ will involve the powers $\kappa_{pq}^m \mu_{pq}^{n-m}$, where m takes on integer values from 0 to n. Depending on the value of n, some of these terms will vanish identically. For example, it is easily seen that for any n, all terms involving κ_{pq}^n will vanish because of the contraction property (20.107).
3. Initially, one might surmise that the integrals of (20.129) and (20.130) are conditionally convergent (dependent on the shape of the macroscopic ellipsoid) because of the appearance in the integrand of the tensor $\boldsymbol{U}^{(q)}(\boldsymbol{r})$, which decays as r^{-d} for large r. Indeed, the integrals in (20.128) are absolutely convergent, since the quantity within the brackets of (20.129) and the determinant $\Delta_n^{(p)}$ in (20.130) identically vanish at the boundary of the ellipsoidal sample because of the asymptotic properties of the $S_n^{(p)}$ (Section 2.2.4). Thus, any convenient shape (such as a d-dimensional sphere) may be used in the infinite-volume limit. Also, when $n \geq 3$, the limiting process of excluding a small cavity about $r_{ij} = 0$ in the integrals (20.130) is no longer necessary, since $\Delta_n^{(p)}$ again vanishes for such values.

4. Classical perturbation expansions involve parameters of smallness that are simple differences in the phase moduli (Beran 1968, Dederichs and Zeller 1973, Gubernatis and Krumhansl 1975, Willis 1981). By contrast, the series expansions represented by (20.128) are *nonclassical* in that the expansion parameters are the polarizabilities κ_{pq} and μ_{pq}, rational functions, and thus have convergence properties that are superior to those of the classical expansions. Additionally, for certain structures (described below), the first few terms of the nonclassical series yield excellent approximations of $\boldsymbol{C}_e$ for any phase contrast.
5. Note that the expansion parameters κ_{pq} and μ_{pq} arise because of our choice of excluding a spherical cavity from the point $\boldsymbol{x}' = \boldsymbol{x}$ in the integral (20.95) and integrating by parts. By choosing a nonspherical cavity shape we would have obtained a different cavity strain field and hence different expansion parameters. The implications of excluding *nonspherical* cavities have heretofore not been explored. Another free parameter is the reference medium, which is not restricted to phase 1 or 2. The convergence of the series can be accelerated by judiciously choosing the exclusion-cavity shape and reference medium (Ramshaw 1984, Eyre and Milton 1999).
6. The n-point tensors $\boldsymbol{B}_n^{(p)}$ for all n generally will not possess common principal axes. This implies that for general media, the principal axes of the macroscopic stiffness tensor $\boldsymbol{C}_e$ (which has the same symmetry properties as $\boldsymbol{L}_e^{(q)}$) will rotate as the phase moduli ratio changes, as will be the case for composites with chirality. Nonetheless, there exists a large class of media that has the symmetry required for all the $\boldsymbol{B}_n^{(p)}$ to possess common principal axes (e.g., distributions of unidirectionally oriented ellipsoids or unidirectionally oriented cylinders in a matrix); in such instances n-point tensor multiplication is commutative.
7. For macroscopically anisotropic media, the series expressions (20.128) may be regarded as expansions that perturb about the structures that realize Willis's (1977) bounds. This point is demonstrated below. For macroscopically isotropic media, it is shown in Section 20.2.5 that the series expressions (20.128) may be regarded as expansions that perturb about the structures that realize the Hashin–Shtrikman bounds.

Before demonstrating this last point, we note that the two-point tensor coefficient $\boldsymbol{B}_2^{(p)}$ [cf. (20.129)] does not vanish for statistically anisotropic media, since the two-point probability function $S_2^{(p)}(\boldsymbol{r})$ depends on the distance $r = |\boldsymbol{r}|$ as well as the orientation of the vector $\boldsymbol{r}$. It is important to observe that for macroscopically isotropic media,

$$\boldsymbol{B}_2^{(p)} = \boldsymbol{0}, \tag{20.131}$$

since

$$\boldsymbol{\Lambda}_h \,\vdots\, \boldsymbol{B}_n^{(p)} = \boldsymbol{\Lambda}_s \,\vdots\, \boldsymbol{B}_n^{(p)} = 0,$$

where $\vdots$ denotes the quadruple dot product. This easily follows from the fact that the scalar invariants $H_{iijj}^{(q)}$ and $H_{ijij}^{(q)}$ vanish identically [cf. (20.107)]. Unlike in the conduction case, statistical anisotropy always implies macroscopic anisotropy in so far as the effective stiffness is concerned.

Let us now examine microstructures for which $\boldsymbol{B}_n^{(p)} = \mathbf{0}$ for all $n \geq 3$. For such composites, (20.128) reduces exactly to

$$\phi_p^2 \boldsymbol{L}^{(q)} : \left[\boldsymbol{L}_e^{(q)}\right]^{-1} = \phi_p \boldsymbol{I} - \boldsymbol{B}_2^{(p)}. \tag{20.132}$$

Multiplying this relation by $\left[\boldsymbol{L}^{(q)}\right]^{-1}$ from the left gives

$$\phi_p^2 \left[\boldsymbol{L}_e^{(q)}\right]^{-1} = \phi_p \left[\boldsymbol{L}^{(q)}\right]^{-1} + \phi_p \phi_q [\boldsymbol{P}^{(q)} - \boldsymbol{D}^{(q)}], \tag{20.133}$$

where the fourth-order tensor $\boldsymbol{P}^{(q)}$ is defined by

$$\begin{aligned} \boldsymbol{P}^{(q)} &= \boldsymbol{D}^{(q)} - \frac{1}{\phi_p \phi_q} \left[\boldsymbol{L}^{(q)}\right]^{-1} : \boldsymbol{B}_2^{(p)} \\ &= \boldsymbol{D}^{(q)} - \frac{1}{\phi_p \phi_q} \int_\epsilon d2 \boldsymbol{H}^{(q)}(1,2) \left[S_2^{(p)}(1,2) - \phi_p^2\right], \end{aligned} \tag{20.134}$$

where $\boldsymbol{D}^{(q)}$ is given by (20.101). Expressions (20.133) indeed are the generalized Hashin–Shtrikman bounds (21.79) for anisotropic composites derived by Willis (1977), albeit expressed in a different form than given originally by Willis. (See Chapter 21 for a detailed discussion of bounds and the structures that attain them.)

In the special case of anisotropic composites consisting of unidirectionally oriented similar ellipsoidal inclusions (i.e., idealized nematic-liquid-crystal structure) in a matrix or microstructures in which $S_2^{(p)}(\boldsymbol{r})$ possesses ellipsoidal symmetry, it can be shown (Willis 1977, Weng 1992) that $\boldsymbol{P}^{(q)}$ is related to Eshelby's tensor $\boldsymbol{S}^{(q)}$ for an ellipsoidal inclusion in a matrix of stiffness $\boldsymbol{C}_q$ (discussed in Section 17.2.2) via

$$\boldsymbol{P}^{(q)} = \boldsymbol{S}^{(q)} : \boldsymbol{C}_q^{-1}. \tag{20.135}$$

20.2.3 Weak-Contrast Expansions

From the strong-contrast expansion (20.128) we can obtain the corresponding weak-contrast expansion in the difference in the phase stiffnesses $(\boldsymbol{C}_p - \boldsymbol{C}_q)$, i.e.,

$$\boldsymbol{C}_e = \boldsymbol{C}_q + \sum_{n=1}^{\infty} \boldsymbol{b}_n^{(p)}. \tag{20.136}$$

The n-point fourth-order tensors $\boldsymbol{b}_n^{(p)}$ are related to the n-point microstructural parameters $\boldsymbol{B}_n^{(p)}$ defined by relations (20.129) and (20.130). Through third order, we obtain

$$\boldsymbol{b}_1^{(p)} = (\boldsymbol{C}_p - \boldsymbol{C}_q)\phi_p, \tag{20.137}$$

$$\boldsymbol{b}_2^{(p)} = -\phi_p \phi_q (\boldsymbol{C}_p - \boldsymbol{C}_q) : \boldsymbol{P}^{(q)} : (\boldsymbol{C}_p - \boldsymbol{C}_q), \tag{20.138}$$

$$\boldsymbol{b}_3^{(p)} = \phi_p\phi_q^2(\boldsymbol{C}_p - \boldsymbol{C}_q) : \boldsymbol{P}^{(q)} : (\boldsymbol{C}_p - \boldsymbol{C}_q) : \boldsymbol{P}^{(q)} : (\boldsymbol{C}_p - \boldsymbol{C}_q) + (\boldsymbol{C}_p - \boldsymbol{C}_q) : \boldsymbol{Q}^{(q)} : (\boldsymbol{C}_p - \boldsymbol{C}_q), \tag{20.139}$$

where $\boldsymbol{Q}^{(q)}$ is the three-point parameter given by $\left[\boldsymbol{L}^{(q)}\right]^{-1} : \boldsymbol{B}_3^{(p)}/(\phi_p\phi_q)$ expanded through third order, i.e.,

$$\boldsymbol{Q}^{(q)} = \frac{1}{\phi_p\phi_q}\int d2 \int d3 \boldsymbol{H}^{(q)}(1,2) : (\boldsymbol{C}_p - \boldsymbol{C}_q) : \boldsymbol{H}^{(q)}(2,3) \times \left[S_3^{(p)}(1,2,3) - \frac{S_2^{(p)}(1,2)S_2^{(p)}(2,3)}{\phi_p}\right], \tag{20.140}$$

and $\boldsymbol{P}^{(q)}$ is given by (20.134). The parameters $\boldsymbol{b}_1^{(p)}, \boldsymbol{b}_2^{(p)}, \boldsymbol{b}_3^{(p)}$ are explicitly given below (Section 20.2.5) for macroscopically isotropic media.

20.2.4 Expansion of Local Strain Field

The local strain field $\varepsilon(\boldsymbol{x})$ itself can be expressed as a contrast expansion, i.e.,

$$\varepsilon(\boldsymbol{x}) = \langle\varepsilon\rangle + \sum_{k=1}^{\infty} \varepsilon_k(\boldsymbol{x}), \qquad \langle\varepsilon_k\rangle = 0, \tag{20.141}$$

where ε_k is the zero-mean fluctuating contribution to the field of kth order in the particular contrast parameter. Combining (20.141) and the Hooke's law definition $\boldsymbol{C}_e : \langle\varepsilon\rangle = \langle\boldsymbol{C}(\boldsymbol{x}) : \varepsilon(\boldsymbol{x})\rangle$ gives

$$\boldsymbol{C}_e : \langle\varepsilon\rangle = \langle\boldsymbol{C}\rangle : \langle\varepsilon\rangle + \langle\delta\boldsymbol{C} : \varepsilon_1\rangle + \langle\delta\boldsymbol{C} : \varepsilon_2\rangle + \cdots,$$

where $\delta\boldsymbol{C}(\boldsymbol{x}) = \boldsymbol{C}(\boldsymbol{x}) - \boldsymbol{C}_q$. Comparing this relation to the weak-contrast expansion (20.136) yields that $\boldsymbol{b}_1^{(p)} \equiv \langle\delta\boldsymbol{C}\rangle = (\boldsymbol{C}_p - \boldsymbol{C}_q)\phi_p$ and

$$\boldsymbol{b}_{k+1}^{(p)} : \langle\varepsilon\rangle = \langle\delta\boldsymbol{C} : \varepsilon_k\rangle, \qquad k \geq 1. \tag{20.142}$$

We see that the weak-contrast n-point tensors $\boldsymbol{b}_k^{(p)}$ are related to the averages $\langle\delta\boldsymbol{C} : \varepsilon_k\rangle$. This weak-contrast expansion will prove valuable in Chapter 21 in formulating trial fields in variational bounds on the effective stiffness. There we will also make use of the following identities:

$$\langle\varepsilon\rangle : \boldsymbol{b}_{k+1}^{(p)} : \langle\varepsilon\rangle = \begin{cases} -\langle\varepsilon_k : \boldsymbol{C}_q : \varepsilon_k\rangle, & k = 1,3,5,\ldots, \\ \langle\varepsilon_{k-1} : \delta\boldsymbol{C} : \varepsilon_{k-1}\rangle, & k = 2,4,6,\ldots, \end{cases} \tag{20.143}$$

where *no summation* is implied on the right side of (20.143) involving index k. These identities follow immediately from (20.142) and the identities

$$\langle\varepsilon_k : \boldsymbol{C}_q : \varepsilon_i\rangle = -\langle\varepsilon_k : \delta\boldsymbol{C} : \varepsilon_{i-1}\rangle, \qquad k \geq 1,\ i \geq 1, \tag{20.144}$$

where $\varepsilon_0 \equiv \langle\varepsilon\rangle$. Relations (20.144) follow easily after we substitute $\varepsilon' = \varepsilon_k$ and (20.136) into identity (14.81) and collect terms of like order.

20.2.5 Isotropic Media

In this section we specialize the previous results to macroscopically isotropic media. For such composites, it is seen from formula (20.119) that

$$\boldsymbol{L}_e^{(q)} = [dK_q + 2(d-1)G_q]\left[\kappa_{eq}\boldsymbol{\Lambda}_h + \frac{(d+2)G_q}{d(K_q+2G_q)}\mu_{eq}\boldsymbol{\Lambda}_s\right], \tag{20.145}$$

where the *scalar effective polarizabilities* κ_{eq} and μ_{eq} are defined by

$$\kappa_{eq} = \frac{K_e - K_q}{K_e + \dfrac{2(d-1)}{d}G_q}, \quad q = 1, 2, \tag{20.146}$$

$$\mu_{eq} = \frac{G_e - G_q}{G_e + \dfrac{G_q[dK_q/2 + (d+1)(d-2)G_q/d]}{K_q + 2G_q}}, \quad q = 1, 2. \tag{20.147}$$

Therefore, series (20.128) becomes

$$\phi_p^2\left[\frac{\kappa_{pq}}{\kappa_{eq}}\boldsymbol{\Lambda}_h + \frac{\mu_{pq}}{\mu_{eq}}\boldsymbol{\Lambda}_s\right] = \phi_p\boldsymbol{I} - \sum_{n=3}^{\infty}\boldsymbol{B}_n^{(p)}, \tag{20.148}$$

where we have utilized the result that $\boldsymbol{B}_2^{(p)} = \boldsymbol{0}$ for macroscopically isotropic media, as discussed earlier [cf. (20.131)].

Bulk Modulus

In order to obtain an explicit expression for the effective bulk modulus K_e, we take the quadruple dot product of the hydrostatic projection tensor $\boldsymbol{\Lambda}_h$ with relation (20.148) and use the identities of (13.98) to obtain

$$\phi_p^2\frac{\kappa_{pq}}{\kappa_{eq}} = \phi_p - \sum_{n=3}^{\infty} C_n^{(p)}, \tag{20.149}$$

where the scalar microstructural coefficients are given by

$$C_n^{(p)} = \boldsymbol{\Lambda}_h \,\vdots\, \boldsymbol{B}_n^{(p)}, \tag{20.150}$$

and $\boldsymbol{B}_n^{(p)}$ is given by (20.130).

Note that when the shear moduli of the phases are equal ($G_1 = G_2$), the well-known exact result

$$\kappa_{eq} = \phi_p\kappa_{pq} \tag{20.151}$$

duc to Hill (1963) immediately follows from (20.149). This is because

$$U_{ijkl}^{(q)}(\boldsymbol{r}) = [dK_q + 2(d-1)G_q]\kappa_{pq}\frac{\delta_{ij}}{d}H_{mmkl}^{(q)}(\boldsymbol{r}), \tag{20.152}$$

and therefore each coefficient $C_n^{(p)}$ possesses the invariant $\boldsymbol{\Lambda}_h \,\vdots\, \boldsymbol{H}^{(q)}$, which vanishes. We see that for such a composite, the effective bulk modulus K_e is independent of thc microstructure in any space dimension d.

In order to better understand the physical meaning of the expansion (20.149), it is helpful to consider the microstructures for which the microstructural parameters $C_n^{(p)}$ vanish for all values of the phase moduli. That is, we may ask, for which class of composites is the relation

$$\kappa_{eq} = \phi_p \kappa_{pq}, \tag{20.153}$$

or, equivalently,

$$\frac{K_e - K_q}{K_e + \dfrac{2(d-1)}{d}G_q} = \frac{K_p - K_q}{K_p + \dfrac{2(d-1)}{d}G_q}\phi_p, \tag{20.154}$$

exact? [Note that this is the same as the Hill relation (20.151), but here we are not restricting the phase shear moduli.] For $d = 2$ and $d = 3$, expression (20.154) is recognized to coincide with the Hashin–Shtrikman bounds on the effective bulk modulus for any isotropic two-phase composite (Hashin and Shtrikman 1963, Hashin 1965b) and hence is exact for the assemblages of coated circles ($d = 2$) and coated spheres ($d = 3$) that realize the bounds. The Hashin–Shtrikman bounds are also realized for certain finite-rank laminates in both two and three dimensions (Francfort and Murat 1986). For either the coated-inclusion assemblages or finite-rank laminates, one of the phases is always a disconnected dispersed phase in a connected matrix phase (except in the trivial instance when the generally dispersed phase fills all of space). Result (20.154) is the d-dimensional generalization of the Hashin–Shtrikman bounds on K_e for any $d \geq 2$; for $K_2 \geq K_1$ and $G_2 \geq G_1$, it gives a lower bound for $q = 1$ and $p = 2$ and an upper bound for $q = 2$ and $p = 1$. The reader should consult Chapter 21 for a complete discussion of bounds and realizable microstructures.

In light of this discussion, series (20.149) can be viewed as an expansion that perturbs around the Hashin–Shtrikman structures. Consequently, the *first few terms* of the expansion (20.149) should provide an *excellent approximation* of the effective bulk modulus K_e of *dispersions in which the inclusions, taken to be the polarized phase, are generally prevented from forming large clusters.* This is indeed the case for a variety of ordered and disordered dispersions when (20.149) is truncated after third-order terms, as shown at the end of the present section and in Section 22.2.1. As in the case of the conductivity, this truncated expansion may also be applied to bicontinuous porous media, as discussed shortly below.

Accordingly, let us write out (20.149) through third-order terms and simplify. We find that

$$\begin{aligned}\phi_p \frac{\kappa_{pq}}{\kappa_{eq}} &= 1 - \frac{C_3^{(p)}}{\phi_p} \\ &= 1 - \frac{(d+2)G_q \kappa_{pq} \mu_{pq}}{d(K_q + 2G_q)} \frac{M_p}{\phi_p},\end{aligned} \tag{20.155}$$

where M_p is a three-point microstructural parameter, independent of the phase moduli, given by

$$M_p = \frac{d^2}{\Omega^2} \int\int \frac{d\boldsymbol{r}}{r^d} \frac{d\boldsymbol{s}}{s^d} [d(\boldsymbol{n} \cdot \boldsymbol{m})^2 - 1] \left[S_3^{(p)}(\boldsymbol{r}, \boldsymbol{s}) - \frac{S_2^{(p)}(\boldsymbol{r}) S_2^{(p)}(\boldsymbol{s})}{\phi_p} \right], \tag{20.156}$$

and $\boldsymbol{n} = \boldsymbol{r}/|\boldsymbol{r}|$ and $\boldsymbol{m} = \boldsymbol{s}/|\boldsymbol{s}|$ are unit vectors. Here we have used the identity (20.108). The three-point microstructural parameters M_1 and M_2 are not independent of one another; specifically, one has that

$$M_1 + M_2 = (d-1)\phi_1\phi_2. \tag{20.157}$$

Relation (20.157) easily follows from the fact that (20.155) yields exact results for the effective bulk modulus K_e through third order in the difference in the moduli and that K_e remains invariant under different labeling of the reference phase. It is seen that the three-point parameter M_p is identical to the three-point parameter $A_3^{(p)}$ [defined by (20.39) with $n = 3$] that arises in the conductivity expansion (20.37). Thus, for any space dimension d, we have that M_p is related to ζ_p [cf. (20.59)] via

$$\zeta_p = \frac{M_p}{(d-1)\phi_1\phi_2}, \tag{20.158}$$

where ζ_2 is given by (20.63). Equations (20.66) and (20.67) give the simplified forms of ζ_p for $d = 2$ and $d = 3$, respectively. The parameter ζ_p as a function of ϕ_2 is given for certain periodic media in Section 20.1.6 and for random media in Section 22.1.1.

Shear Modulus

In order to obtain an explicit expression for the effective shear modulus G_e, we take a quadruple dot product of the shear tensor $\boldsymbol{\Lambda}_s$ with equation (20.148) and use the identities of (13.98) to get

$$\phi_p^2 \frac{\mu_{pq}}{\mu_{eq}} = \phi_p - \sum_{n=3}^{\infty} D_n^{(p)}, \tag{20.159}$$

where the scalar coefficients $D_n^{(p)}$ are given by

$$D_n^{(p)} = \frac{2}{(d+2)(d-1)} \boldsymbol{\Lambda}_s \vdots \boldsymbol{B}_n^{(p)} \tag{20.160}$$

and $\boldsymbol{B}_n^{(p)}$ is given by (20.130). Note that this series begins with the third-order term, i.e., $D_2^{(p)}$ is zero because the invariant $H_{ikik}^{(q)}$ vanishes [cf. (20.107)].

Truncating the series (20.159) after the first term yields

$$\mu_{eq} = \phi_p \mu_{pq}, \tag{20.161}$$

or, equivalently,

$$\frac{G_e - G_q}{G_e + \frac{G_q[dK_q/2 + (d+1)(d-2)G_q/d]}{K_q + 2G_q}} = \frac{G_p - G_q}{G_p + \frac{G_q[dK_q/2 + (d+1)(d-2)G_q/d]}{K_q + 2G_q}} \phi_p. \tag{20.162}$$

Following the discussion on the bulk modulus, it is helpful to consider the class of microstructures for which the formula (20.161) is exact. For $d = 2$ and $d = 3$, expression (20.161) is recognized to coincide with the Hashin–Shtrikman bounds on the effective shear modulus for any isotropic two-phase composite and hence is exact for the finite-rank hierarchical laminate composites that realize the bounds (Francfort and Murat 1986). Again, for such hierarchical laminates one of the phases is always a disconnected, dispersed phase in a connected matrix phase (except in the trivial instance when the generally dispersed phase fills all of space). Result (20.161) is the d-dimensional generalization of the Hashin–Shtrikman bounds on G_e for any $d \geq 2$; for the "well-ordered" case $(K_2 - K_1)(G_2 - G_1) \geq 0$ and $G_2 \geq G_1$, it gives a lower bound for $q = 1$ and $p = 2$ and an upper bound for $q = 2$ and $p = 1$. The reader is referred to Chapter 21 for a complete discussion of bounds and realizable microstructures.

Consequently, series (20.159) can be viewed as an expansion that perturbs around the Hashin–Shtrikman hierarchical laminates. Therefore, it is expected that the *truncation* of expansion (20.159) after the first few terms will provide an *excellent approximation* of G_e for any values of the phase moduli *for dispersions in which the inclusions, taken to be the polarized phase, are generally prevented from forming large clusters*. Indeed, truncation of (20.159) after third-order terms does give an excellent approximation to the effective shear modulus of a variety of ordered and disordered dispersions, as shown at the end of the present section and in Section 22.2.1. Again, this truncated expansion may also be applied to bicontinuous porous media, as discussed shortly below.

Writing out (20.159) through third-order terms and simplifying yields

$$\phi_p \frac{\mu_{pq}}{\mu_{eq}} = 1 - \frac{D_3^{(p)}}{\phi_p}, \tag{20.163}$$

where

$$\frac{D_3^{(p)}}{\phi_p} = \frac{2G_q \kappa_{pq} \mu_{pq}}{d(d-1)(K_q + 2G_q)} \frac{M_p}{\phi_p} + \frac{(d^2-4)G_q(2K_q + 3G_q)\mu_{pq}^2}{2d(d-1)(K_q + 2G_q)^2} \frac{M_p}{\phi_p}$$
$$+ \frac{1}{2d(d-1)} \left[\frac{dK_q + (d-2)G_q}{K_q + 2G_q} \right]^2 \mu_{pq}^2 \frac{N_p}{\phi_p}. \tag{20.164}$$

In obtaining (20.163), we used the identity (20.109) and definition (20.103). The quantity N_p is a three-point parameter independent of the phase moduli given by

$$N_p = -\frac{(d+2)(5d+6)}{d^2} M_p + (d+2)^2 J[\bar{S}_3^{(p)}], \tag{20.165}$$

where J is the integral operator

$$J[f] = \frac{1}{\Omega^2} \int \int \frac{d\boldsymbol{r}}{r^d} \frac{d\boldsymbol{s}}{s^d} [d(d+2)(\boldsymbol{n} \cdot \boldsymbol{m})^4 - 3] f(r, s, t), \tag{20.166}$$

$\boldsymbol{n} = \boldsymbol{r}/|\boldsymbol{r}|$ and $\boldsymbol{m} = \boldsymbol{s}/|\boldsymbol{s}|$ are unit vectors, and $\bar{S}_3^{(p)}$ is given by (20.65). The quantity $[d(d+2)(\boldsymbol{n} \cdot \boldsymbol{m})^4 - 3]$ can be expressed in terms of d-dimensional spherical orthogonal polynomials. As in the case of the microstructural parameters M_1 and M_2, the three-point parameters N_1 and N_2 are not independent of one another; specifically, one has that

$$N_1 + N_2 = (d-1)\phi_1\phi_2. \tag{20.167}$$

This is shown using the fact that relation (20.163) yields the effective shear modulus G_e exactly through third order in the difference in the moduli and that G_e remains invariant under different labeling of the reference phase.

For any space dimension d, Torquato (1997) has shown that N_p is related to a parameter η_p that lies in the interval $[0, 1]$ via the relation

$$\begin{aligned}\eta_p &= \frac{N_p}{(d-1)\phi_1\phi_2} \\ &= -\frac{(d+2)(5d+6)}{d^2}\zeta_p + \frac{(d+2)^2}{(d-1)\phi_1\phi_2}J[\bar{S}_3^{(p)}],\end{aligned} \tag{20.168}$$

where ζ_p is given by (20.63). Thus, combining (20.167) and (20.168) gives

$$\eta_1 + \eta_2 = 1. \tag{20.169}$$

As in the case of ζ_p, the d^2-fold integral of the operator (20.166) can be reduced (by integrating over angles) to a threefold integral over the lengths r, s, and t of the sides of the triangle or, equivalently, over the two lengths r and s and the angle θ opposite the side of length t, where $t^2 = r^2 + s^2 - 2rs\cos\theta$, with $\cos\theta = \boldsymbol{n}\cdot\boldsymbol{m}$. For $d = 2$, we recover the result

$$\eta_p = \frac{16}{\pi\phi_q\phi_p}\int_0^\infty \frac{dr}{r}\int_0^\infty \frac{ds}{s}\int_0^\pi d\theta\cos(4\theta)\left[S_3^{(p)}(r,s,t) - \frac{S_2^{(p)}(r)S_2^{(p)}(s)}{\phi_p}\right], \tag{20.170}$$

first given by Milton (1982), and for $d = 3$, we recover the result

$$\eta_p = \frac{5\zeta_p}{21} + \frac{150}{7\phi_q\phi_p}\int_0^\infty \frac{dr}{r}\int_0^\infty \frac{ds}{s}\int_{-1}^1 d(\cos\theta)P_4(\cos\theta)\left[S_3^{(p)}(r,s,t) - \frac{S_2^{(p)}(r)S_2^{(p)}(s)}{\phi_p}\right], \tag{20.171}$$

first given by Milton (1981b). Here P_4 is the Legendre polynomial of order four.

As in the case of the operator (20.64) that defines ζ_2, the operator (20.166) has two simple properties: (i) If the function f does not depend on t, then

$$J[f(r,s)] = 0 \tag{20.172}$$

by virtue of the orthogonality of the spherical harmonics; (ii) less obviously, if f depends only on t, then

$$J[f(t)] = \frac{2(3d^2+8d+6)}{d^2}[f(0) - f(\infty)], \tag{20.173}$$

provided that $f(\infty)$ is finite and the Fourier transform of $f(t) - f(\infty)$ is convergent. Property (20.173) is the d-dimensional generalization of the ones proved for $d = 2$ and $d = 3$ using integration by parts (Milton 1982).

The parameter η_p has been computed for a variety of two- and three-dimensional models of *random* media, as detailed in Section 22.2.1. For periodic arrays of particles (phase 2) that are macroscopically isotropic, the three-point parameter η_2 (as well as ζ_2) can be determined without the integrals above having to be evaluated directly, since the effective shear modulus is often known essentially exactly. The first such calculation of η_2 was performed by Eischen and Torquato (1993) to determine bounds on the effective transverse shear modulus of hexagonal arrays of unidirectionally aligned infinitely long circular cylinders (circular disks in two dimensions). This evaluation was subsequently refined by Helsing (1994a). The parameter η_2 has also been found numerically for periodic arrays of triangular and hexagonal inclusions (Hyun and Torquato 2000).

Weak-Contrast Expansions

The weak-contrast expansion (20.136) for $\boldsymbol{C}_e$ in the macroscopically isotropic case, through third order in the difference in the phase moduli, reduces to

$$\boldsymbol{C}_e = \boldsymbol{C}_q + \boldsymbol{b}_1^{(p)} + \boldsymbol{b}_2^{(p)} + \boldsymbol{b}_3^{(p)}, \tag{20.174}$$

where

$$\boldsymbol{b}_1^{(p)} = d\phi_p(K_p - K_q)\boldsymbol{\Lambda}_h + 2\phi_p(G_p - G_q)\boldsymbol{\Lambda}_s, \tag{20.175}$$

$$\boldsymbol{b}_2^{(p)} = db_{2h}^{(p)}(K_p - K_q)^2\boldsymbol{\Lambda}_h + 2b_{2s}^{(p)}(G_p - G_q)^2\boldsymbol{\Lambda}_s, \tag{20.176}$$

$$\boldsymbol{b}_3^{(p)} = db_{3h}^{(p)}(K_p - K_q)^2\boldsymbol{\Lambda}_h + 2b_{3s}^{(p)}(G_p - G_q)^2\boldsymbol{\Lambda}_s, \tag{20.177}$$

and

$$b_{2h}^{(p)} = -\frac{d\phi_p\phi_q}{[dK_q + 2(d-1)G_q]}, \quad b_{2s}^{(p)} = -\frac{2d\phi_p\phi_q(K_q + 2G_q)}{(d+2)G_q[dK_q + 2(d-1)G_q]}, \tag{20.178}$$

$$b_{3h}^{(p)} = \frac{d\phi_p\phi_q}{[dK_q + 2(d-1)G_q]^2}[d\phi_q(K_p - K_q) + 2(d-1)\zeta_p(G_p - G_q)], \tag{20.179}$$

$$\begin{aligned} b_{3s}^{(p)} = \frac{2d\phi_p\phi_q}{[dK_q + 2(d-1)G_q]^2}\Bigg\{\frac{2\zeta_p}{(d+2)}(K_p - K_q) + \Bigg[\frac{2d(K_q + 2G_q)^2\phi_q}{(d+2)^2G_q^2} \\ + \frac{(d-2)(2K_q + 3G_q)\zeta_p}{(d+2)G_q} + \left[\frac{dK_q + (d-2)G_q}{(d+2)G_q}\right]^2\eta_p\Bigg](G_p - G_q)\Bigg\}. \end{aligned} \tag{20.180}$$

Forming the quadruple dot products of $\boldsymbol{\Lambda}_h$ and $\boldsymbol{\Lambda}_s$ with relation (20.174) leads to the following third-order weak-contrast expansion for the effective bulk and shear moduli:

$$K_e = K_q + \phi_p(K_p - K_q) + b_{2h}^{(p)}(K_p - K_q)^2 + b_{3h}^{(p)}(K_p - K_q)^2, \tag{20.181}$$

$$G_e = G_q + \phi_p(G_p - G_q) + b_{2s}^{(p)}(G_p - G_q)^2 + b_{3s}^{(p)}(G_p - G_q)^2. \tag{20.182}$$

Note that for $d = 2$ and $d = 3$, expansions (20.181) agree with the corresponding expansions of the Silnutzer (1972) and Beran and Molyneux (1966) bounds on K_e, respectively. For $d = 2$ and $d = 3$, expansions (20.182) agree with the corresponding expansions of the Silnutzer (1972) and McCoy (1970) bounds on G_e, respectively. The reader is also referred to the work of Milton (1981c, 1982a).

Effective Behavior in the Limit of Infinite Dimension

It is of interest to study the effective elastic behavior in the limit that the space dimension tends to infinity ($d \to \infty$). In this limit, Torquato (1997) demonstrated that the effective shear modulus G_e, for any isotropic composite, is given exactly by the *arithmetic average*, provided that the phase moduli are nonzero. As discussed earlier, the effective electrical (thermal) conductivity of an isotropic composite also tends to the arithmetic average as $d \to \infty$. All of the aforementioned results turn out to apply not only to two-phase composites but also to M-phase composites with linear or *nonlinear* stress-strain laws. By contrast, we will demonstrate that the effective bulk modulus K_e depends on the microstructure as $d \to \infty$.

The first hint of this interesting behavior as $d \to \infty$ can be gleaned by examining the exact third-order expansions (20.181) and (20.182) in this limit. We find from these relations that as $d \to \infty$,

$$
\begin{aligned}
K_e &= K_1 + \phi_2(K_2 - K_1) - \frac{\phi_1\phi_2}{K_1 + 2G_1}(K_2 - K_1)^2 - \frac{\phi_1^2\phi_2}{(K_1 + 2G_1)^2}(K_2 - K_1)^3 \\
&\quad + \frac{2\phi_1\phi_2\zeta_2}{(K_1 + 2G_1)^2}(K_2 - K_1)^2(G_2 - G_1), \\
G_e &= G_1 + \phi_2(G_2 - G_1) = \phi_1 G_1 + \phi_2 G_2,
\end{aligned}
$$

where we have taken $p = 2$ and $q = 1$. Here we used the fact that N_p is asymptotic to d as $d \to \infty$. Thus, we see that in contrast to the effective bulk modulus, the second- and third-order terms vanish in the case of the effective shear modulus.

The general proof for arbitrary conditions and for any number of phases requires the d-dimensional Hashin–Shtrikman–Walpole bounds for M-phase composites given by (21.77) and (21.78).

Theorem 20.2 *For any d-dimensional macroscopically isotropic M-phase composite possessing nonzero phase moduli, the effective shear modulus G_e is independent of the microstructure and is exactly given by the arithmetic average*

$$G_e = \sum_{i=1}^{n} \phi_i G_i \tag{20.183}$$

in the limit that the space dimension becomes infinite ($d \to \infty$).

Proof: The proof is immediate given the d-dimensional bounds for M phases ($M \geq 2$) on the effective shear modulus given by relation (21.78). Under the stated conditions, the bounds coincide and equal expression (20.183). ∎

Theorem 20.3 *For any d-dimensional macroscopically isotropic M-phase composite, the effective bulk modulus K_e generally depends on the microstructure in the limit that the space dimension becomes infinite ($d \to \infty$).*

Proof: The proof follows from the fact that the d-dimensional M-phase bounds on the effective bulk modulus given by (21.77) are realizable (see Section 21.2.2) and generally do not coincide in the limit $d \to \infty$. ∎

Remarks:

1. Interestingly, in lower dimensions ($d = 2$ or $d = 3$), it has been shown that the effective conductivity of two-phase composites is more closely related to the effective bulk modulus rather than the effective shear modulus (Milton 1984, Gibiansky and Torquato 1995a, Gibiansky and Torquato 1996b).
2. Why does the shear modulus tend to the arithmetic mean as $d \to \infty$ (implying the same constant strain field in each phase), in contrast to the behavior of the effective bulk modulus in this limit? In analogy with the conductivity, this can be explained by considering the phase-interface continuity conditions on the strain field and the fact that the energy associated with an *isotropic composite* must be the same in any direction. The components of the strain are continuous in all directions, except that normal to the interface, and hence for an isotropic composite in the limit $d \to \infty$, the strains in each phase approach the same constant value. The bulk modulus behaves differently, since the quantity dK_e, rather than K_e, is an eigenvalue of the stiffness tensor, and thus the energy associated with the former is unbounded for isotropic composites as $d \to \infty$. Note that all of these conclusions apply also to materials possessing *nonlinear* stress–strain laws.

Approximation Formulas

Let us assume that phases 1 and 2 are the matrix and dispersed phases, respectively, of a dispersion. Then the third-order truncated formulas (20.155) and (20.163) become

$$\phi_2 \frac{\kappa_{21}}{\kappa_{e1}} = 1 - \frac{(d+2)(d-1)G_q \kappa_{21}\mu_{21}}{d(K_1 + 2G_1)} \phi_1 \zeta_2 \,, \tag{20.184}$$

$$\begin{aligned} \phi_2 \frac{\mu_{21}}{\mu_{e1}} = 1 &- \frac{2G_1 \kappa_{21}\mu_{21}}{d(K_1 + 2G_1)} \phi_1 \zeta_2 - \frac{(d^2-4)G_1(2K_1 + 3G_1)\mu_{21}^2}{2d(K_1 + 2G_1)^2} \phi_1 \zeta_2 \\ &- \frac{1}{2d} \left[\frac{dK_1 + (d-2)G_1}{K_1 + 2G_1} \right]^2 \mu_{21}^2 \phi_1 \eta_2 \,. \end{aligned} \tag{20.185}$$

Torquato (1998) has shown that (20.184) and (20.185) give excellent approximations of the effective moduli of dispersions for all phase contrasts, provided that the particles (polarized phase) generally do not form large clusters. However, for the special case of bicontinuous porous media in which the void (polarized) phase possesses vanishing moduli ($K_2 = G_2 = 0$), the approximations (20.184) and (20.185) may still provide good estimates of K_e and G_e (see, for example, Figure 22.14). The reason for this is that the stress is carried by the *connected*, solid (i.e., nonpolarized) phase, regardless of whether the void phase is connected or disconnected.

We will require that the three-point approximations (20.184) and (20.185) always lie within the most restrictive three-point bounds (see Section 21.2.2). This generally will place restrictions on the range of values of the geometrical parameters ζ_2 and η_2 beyond the normal condition that they must lie in the interval [0, 1], as discussed by Torquato (1998). In some instances, these restrictions can be determined only numerically, but in other cases, one can obtain them analytically. For example, for $d = 2$, the effective bulk modulus formula (20.184) will be within the tightest three-point bounds when $\zeta_2 \leq 0.5$. The parameter ζ_2 is less restricted in higher dimensions. For the shear modulus relation (20.185), it is usually difficult to obtain similar analytical conditions. However, in the instance of a two-dimensional composite in which both phases are incompressible, it can be shown that there is no additional restriction on η_2 beyond it lying in the interval [0, 1]. For three-dimensional incompressible media, (20.185) always lies below the best three-point upper bound (20.185) for allowable η_2, but (20.185) obeys the best three-point lower bound, provided that

$$\frac{1+\eta_2}{2-3\eta_2} \geq \frac{1-\frac{11}{16}\zeta_2-\frac{5}{16}\eta_2}{2(1-\zeta_2)(1+\frac{5}{16}\zeta_2-\frac{21}{16}\eta_2)}.$$

It is noteworthy that for all of the realistic models of two- and three-dimensional dispersions detailed in Sections 22.1.1 and 22.2.1, the parameters ζ_2 and η_2 are such that the estimates from relations (20.184) and (20.185) always lie within the tightest three-point bounds.

Torquato (1998) demonstrated that (20.184) and (20.185) provide very sharp estimates of the effective elastic moduli of two- and three-dimensional dispersions whose particles are arranged on the sites of periodic lattices. Such examples are excellent test cases, since the effective moduli are often known exactly, either analytically or numerically. To illustrate the utility of the three-point approximations, we begin by estimating the effective bulk modulus K_e of face-centered cubic arrays of spheres using result (20.184) for $d = 3$, which applies not only to isotropic composites but to composites with cubic symmetry as well. The parameter ζ_2 is taken from Table 20.1. Figure 20.4 compares the three-point approximation (20.184) and the self-consistent (SC) formula (18.51) for $d = 3$ to the numerical data of Nunan and Keller (1984) for FCC arrays of perfectly rigid spheres in a compressible matrix. The prediction of (20.184) is remarkably accurate, whereas the SC approximation begins to diverge increasingly from the data for values of the particle volume fraction ϕ_2 larger than 0.3.

In Figure 20.5 we compare the prediction of (20.185) for the effective transverse shear modulus G_e of hexagonal arrays of aligned infinitely long circular glass fibers in an epoxy matrix ($d = 2$) to the corresponding simulation data of Eischen and Torquato (1993). The parameter η_2 is taken from Eischen and Torquato (1993), and ζ_2 is taken from Table 20.1. It is seen that formula (20.185) provides an excellent estimate of the shear modulus for the entire range of volume fractions. [The prediction of the effective bulk modulus from relation (20.184) is equally accurate but is not shown

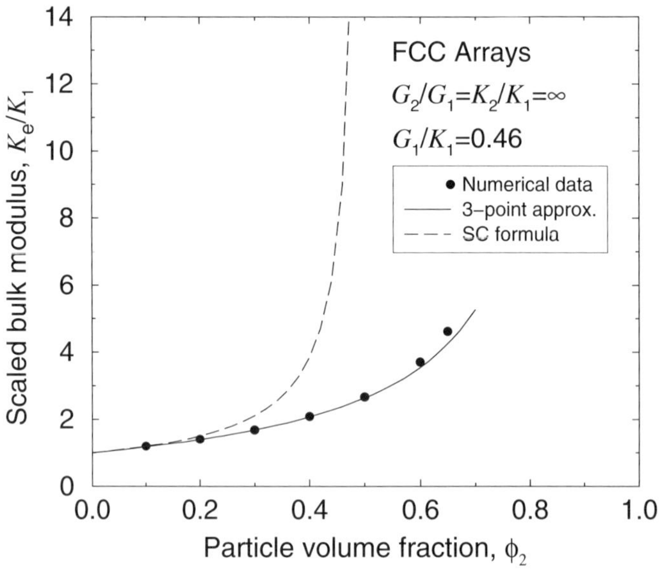

Figure 20.4 The scaled effective bulk modulus K_e/K_1 versus ϕ_2 for face-centered cubic arrays of rigid spheres in a compressible matrix. Filled circles are numerical data (Nunan and Keller 1984), solid curve is the three-point approximation (20.184), and dashed curve is the self-consistent formula (18.51) for $d = 3$.

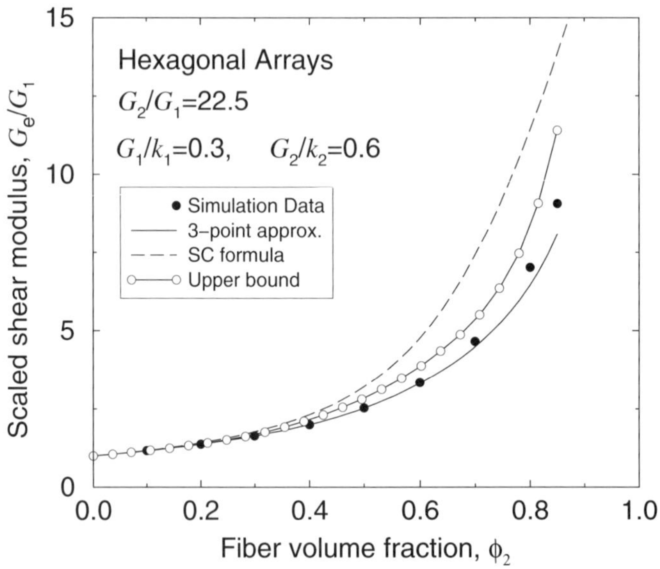

Figure 20.5 Scaled effective transverse shear modulus G_e/G_1 versus ϕ_2 for hexagonal arrays of aligned infinitely long circular glass fibers in an epoxy matrix. Shown are the three-point approximation (20.185) for $d = 2$, SC formula (18.53) for $d = 2$, three-point upper bound (21.100), and simulation data (Eischen and Torquato 1993).

here.] The two-dimensional SC approximation (18.53) and the best three-point upper bound (21.100) are also included in the figure. It is seen that the SC approximation not only overestimates the effective shear modulus of the glass–epoxy composite for moderate to high fiber volume fractions, but it also begins to violate the upper bound at small values of ϕ_2.

In Section 22.2.1, formulas (20.184) and (20.185) are employed to assess the sharpness of rigorous bounds on the effective elastic moduli of *random* dispersions.

CHAPTER 21

Rigorous Bounds

Since it is generally impossible to determine exactly the effective properties of random heterogeneous media, any rigorous statements about effective properties must take the form of bounds. In Chapter 14 we formulated variational (extremum) principles for the effective properties that are given in terms of trial fields. Here we will derive specific existing bounds on the effective conductivity, effective elastic moduli, trapping constant, and fluid permeability by constructing specific trial fields that conform with the variational principles.

It will be demonstrated that bounds are useful because:

- they rigorously incorporate nontrivial information about the microstructure via statistical correlation functions and consequently serve as a guide in identifying appropriate statistical descriptors;
- as successively more microstructural information is included, the bounds become progressively narrower;
- one of the bounds can provide a relatively sharp estimate of the property for a wide range of conditions, even when the reciprocal bound diverges from it;
- they are usually exact under certain conditions;
- they can be utilized to test the merits of a theory or computer experiment;
- they provide a unified framework to study a variety of different effective properties.

The bounds derived here on the effective properties will generally incorporate certain n-point correlation functions that contain microstructural information beyond volume fractions. We refer to bounds that depend *nontrivially* upon two-point and higher-order correlation functions as *improved bounds*. In the instances of the conductivity and elastic moduli of isotropic two-phase media, for example, improved bounds are those that are more stringent than the well-known Hashin–Shtrikman bounds. A variety of different correlation functions will be shown to arise in the bounds on the effective properties, including:

- n-point probability functions, $S_n^{(i)}$;
- surface correlation functions, F_{sv}, F_{ss}, F_{sss}, etc.;
- pore-size probability density function, P;
- point/q-particle correlation functions, G_n $(n = 1 + q)$;
- nearest-neighbor probability density function, H_P.

All of these correlation functions have been defined and discussed in Chapter 2 and quantified, for various models of random media, in Chapters 5–7 and 12. Whereas the first three types of correlation functions are applicable to random media of arbitrary microstructure, the last two apply only to dispersions of particles.

Techniques other than variational principles can be used to find bounds on the effective conductivity tensor $\boldsymbol{\sigma}_e$ and the effective stiffness tensor $\boldsymbol{C}_e$. The analytic method, initiated by Bergman (1978) for $\boldsymbol{\sigma}_e$ and by Kantor and Bergman (1984) for $\boldsymbol{C}_e$, exploits the analytic properties of these effective properties as a function of the phase properties and has been applied by others (Milton 1981c, Milton 1981a, Golden and Papanicolaou 1983, Felderhof 1984). One of the advantages of the analytic method is that it is easy to extend to produce bounds on $\boldsymbol{\sigma}_e$ when σ_1 and σ_2 are complex (Milton 1981c, Milton 1981a, Bergman 1982, Felderhof 1984). The method of Padé approximants developed by Milton and Golden (1985) is closely related to the analytic method because the conductivity tensor $\boldsymbol{\sigma}_e$ is a Stieltjes function. This method was used by Torquato (1985a) and by Sen and Torquato (1989) to obtain higher-order bounds on the effective conductivity of isotropic and anisotropic composites, respectively. Another technique, called the translation (or compensated compactness) method, was initiated by Murat and Tartar (Tartar 1979, Tartar 1980, Murat and Tartar 1985) and by Lurie and Cherkaev (1984, 1986) to bound $\boldsymbol{\sigma}_e$ and $\boldsymbol{C}_e$. This method was subsequently applied by Francfort and Murat (1986), Kohn and Milton (1986), Milton (1990), and others. Finally, the field equation recursion method was developed by Milton (1987).

Only variational principles are available to derive bounds on the trapping constant γ and fluid permeability tensor $\boldsymbol{k}$. Since variational principles can be applied to all four classes of problems considered in this book, this is the method that we will employ to derive bounds on all of the effective properties.

A word on notation is in order here. In general, we will be concerned with bounds on effective property tensors in the *positive semidefinite* sense, i.e., in the sense of (13.16) and (13.66) in the case of symmetric second-order and fourth-order tensors, respectively. More specifically, for any two symmetric tensors $\boldsymbol{C}$ and $\boldsymbol{D}$ of the same tensorial order, the condition

$$\boldsymbol{C} \leq \boldsymbol{D}$$

means that the difference $(\boldsymbol{D} - \boldsymbol{C})$ is positive semidefinite.

21.1 Conductivity

The minimum potential energy principle requires a trial intensity field $\hat{\boldsymbol{E}}(\boldsymbol{x})$ that is curl-free and whose average equals the actual average, i.e., $\langle\hat{\boldsymbol{E}}(\boldsymbol{x})\rangle = \langle\boldsymbol{E}\rangle$. The minimum complementary energy principle requires a trial current field $\hat{\boldsymbol{J}}(\boldsymbol{x})$ that is divergence-free and whose average equals the actual average, i.e., $\langle\hat{\boldsymbol{J}}(\boldsymbol{x})\rangle = \langle\boldsymbol{J}\rangle$. In the ensuing discussion we consider specific trial fields that lead to progressively tighter bounds.

21.1.1 General Considerations

To understand the basic idea behind generating bounds on the effective conductivity $\boldsymbol{\sigma}_e$, let us consider a simple application of the minimum principles of Theorems 14.3 and 14.4 for a macroscopically anisotropic composite consisting of M anisotropic phases. The constant fields $\langle\boldsymbol{E}\rangle$ and $\langle\boldsymbol{J}\rangle$ trivially satisfy the admissibility conditions on the trial intensity and current fields, respectively. Use of these trial fields in the variational principles results in the following simple bounds on the effective conductivity tensor $\boldsymbol{\sigma}_e$:

$$\langle\boldsymbol{\sigma}^{-1}\rangle^{-1} \leq \boldsymbol{\sigma}_e \leq \langle\boldsymbol{\sigma}\rangle. \tag{21.1}$$

Thus, the effective conductivity tensor is bounded from above by the arithmetic mean of the phase conductivities and from below by the harmonic mean of the phase conductivities. We refer to these results as *one-point bounds*, since they involve information of up to the level of the volume fraction, which is a one-point correlation function. As already noted in Chapters 1 and 16, these bounds are generally far apart from one another. In order to improve upon them, one must construct *nonuniform* trial fields that better reflect the field interactions between the phases.

One such allowable nontrivial trial intensity field is the following:

$$\hat{\boldsymbol{E}}(\boldsymbol{x}) = \langle\boldsymbol{E}\rangle + \sum_{k=1}^{n} \xi_k \boldsymbol{E}_k(\boldsymbol{x}), \qquad \langle\boldsymbol{E}_k\rangle = 0 \quad \forall k, \tag{21.2}$$

where ξ_k are constant multipliers that are to be optimized. The field $\boldsymbol{E}_k$ can be based on the exact kth-order solution of the intensity field from the contrast expansion (Chapter 20) or on the exact solution of the intensity field for k interacting spheres (Chapter 19) . Substituting relation (21.2) into the upper bound (14.40) and setting the partial derivative of the right-hand side with respect to ξ_k equal to zero gives

$$\langle\boldsymbol{E}\rangle \cdot \boldsymbol{\sigma}_e \cdot \langle\boldsymbol{E}\rangle \leq \langle\boldsymbol{E}\rangle \cdot \langle\boldsymbol{\sigma}\rangle \cdot \langle\boldsymbol{E}\rangle - \boldsymbol{V}^{\mathrm{T}}\boldsymbol{W}^{-1}\boldsymbol{V}. \tag{21.3}$$

Here $\boldsymbol{V}^{\mathrm{T}}$ denotes the transpose of the column vector $\boldsymbol{V}$ with scalar elements

$$V_k = \langle\boldsymbol{E}\rangle \cdot \langle\boldsymbol{\sigma} \cdot \boldsymbol{E}_k\rangle, \tag{21.4}$$

and $\boldsymbol{W}^{-1}$ denotes the inverse of the matrix $\boldsymbol{W}$ with scalar elements

$$W_{kl} = \langle\boldsymbol{E}_k \cdot \boldsymbol{\sigma} \cdot \boldsymbol{E}_l\rangle. \tag{21.5}$$

Whether the trial fields $\boldsymbol{E}_i$ $(i = 1, \ldots, n)$ are of the contrast type or cluster type, the upper bound (21.3) contains microstructural information up to the level of the relevant $(2n+1)$-point correlation function. Accordingly, we refer to such bounds as $(2n+1)$-*point bounds*.

A corresponding allowable nontrivial trial current field is the following:

$$\hat{\boldsymbol{J}}(\boldsymbol{x}) = \langle \boldsymbol{J} \rangle + \sum_{k=1}^{n} \omega_k \boldsymbol{J}_k(\boldsymbol{x}), \qquad \langle \boldsymbol{J}_k \rangle = 0 \quad \forall k, \tag{21.6}$$

where ω_k are constant multipliers. The field $\boldsymbol{J}_k$ can be based on the exact kth-order solution of the current from the contrast expansion or on the exact solution of the current for k interacting spheres. Substituting (21.6) into bound (14.45) and setting the partial derivative of the right-hand side with respect to the ω_k equal to zero gives

$$\langle \boldsymbol{J} \rangle \cdot \boldsymbol{\sigma}_e^{-1} \cdot \langle \boldsymbol{J} \rangle \leq \langle \boldsymbol{J} \rangle \cdot \langle \boldsymbol{\sigma}^{-1} \rangle \cdot \langle \boldsymbol{J} \rangle - \boldsymbol{X}^{\mathrm{T}} \boldsymbol{Y}^{-1} \boldsymbol{X}. \tag{21.7}$$

Here $\boldsymbol{X}^{\mathrm{T}}$ denotes the transpose of the column vector with scalar elements

$$X_k = \langle \boldsymbol{J} \rangle \cdot \langle \boldsymbol{\sigma}^{-1} \cdot \boldsymbol{J}_k \rangle, \tag{21.8}$$

and $\boldsymbol{Y}^{-1}$ denotes the inverse of the matrix $\boldsymbol{Y}$ with scalar elements

$$Y_{kl} = \langle \boldsymbol{J}_k \cdot \boldsymbol{\sigma}^{-1} \cdot \boldsymbol{J}_l \rangle. \tag{21.9}$$

Expression (21.7) leads to a $(2n+1)$-point lower bound on the effective conductivity tensor $\boldsymbol{\sigma}_e$.

Of particular interest are the optimized three-points bounds $(n = 1)$:

$$\langle \boldsymbol{E} \rangle \cdot \boldsymbol{\sigma}_e \cdot \langle \boldsymbol{E} \rangle \leq \langle \boldsymbol{E} \rangle \cdot \langle \boldsymbol{\sigma} \rangle \cdot \langle \boldsymbol{E} \rangle - \frac{|\langle \boldsymbol{E} \rangle \cdot \langle \boldsymbol{\sigma} \cdot \boldsymbol{E}_1 \rangle|^2}{\langle \boldsymbol{E}_1 \cdot \boldsymbol{\sigma} \cdot \boldsymbol{E}_1 \rangle}, \tag{21.10}$$

$$\langle \boldsymbol{J} \rangle \cdot \boldsymbol{\sigma}_e^{-1} \cdot \langle \boldsymbol{J} \rangle \leq \langle \boldsymbol{J} \rangle \cdot \langle \boldsymbol{\sigma}^{-1} \rangle \cdot \langle \boldsymbol{J} \rangle - \frac{|\langle \boldsymbol{J} \rangle \cdot \langle \boldsymbol{\sigma}^{-1} \cdot \boldsymbol{J}_1 \rangle|^2}{\langle \boldsymbol{J}_1 \cdot \boldsymbol{\sigma}^{-1} \cdot \boldsymbol{J}_1 \rangle}. \tag{21.11}$$

21.1.2 Contrast Bounds

We will derive many of the existing bounds on the effective conductivity whose trial fields are based on the contrast expansion. We will focus primarily on one-, two-, three-, and four-point bounds. By construction, such n-point bounds are exact through nth-order in the difference in the phase conductivities. We will mainly treat composites consisting of isotropic phases; i.e., the conductivity tensor of phase i is given by

$$\boldsymbol{\sigma}_i = \sigma_i \boldsymbol{I}, \tag{21.12}$$

and hence is characterized by the scalar conductivity σ_i. Both macroscopically isotropic and anisotropic composites will be considered.

One-Point Bounds

For macroscopically anisotropic composites with isotropic phases, simple one-point bounds are trivially obtained from (21.1) using (21.12) as

$$\langle\sigma^{-1}\rangle^{-1}\boldsymbol{I} \le \boldsymbol{\sigma}_e \le \langle\sigma\rangle\boldsymbol{I}. \tag{21.13}$$

Recall that the upper and lower bounds are realized by laminates of rank one in which the fields are oriented parallel and perpendicular to the slabs, respectively (see Section 16.1.2). If we further specify that the medium is macroscopically isotropic, i.e., $\boldsymbol{\sigma}_e = \sigma_e\boldsymbol{I}$, then the bounds above reduce to

$$\langle\sigma^{-1}\rangle^{-1} \le \sigma_e \le \langle\sigma\rangle. \tag{21.14}$$

The bounds (21.14) are attributed to Wiener (1912), who obtained them via certain algebraic arguments.

Two-Point Bounds

To begin, we consider macroscopically isotropic composites. Hashin and Shtrikman (1962) derived the best possible bounds on the effective conductivity of two-phase isotropic three-dimensional composites given just volume-fraction information. These are actually two-point bounds (i.e., they incorporate the two-point probability function $S_2^{(i)}$), even though they explicitly involve only the volume fractions. It turns out that for isotropic media, the key integral involving $S_2^{(i)}$ just depends on the end points of the function, i.e., ϕ_i and ϕ_i^2. Hashin (1970) obtained the two-dimensional version of the Hashin–Shtrikman bounds.

We will derive the Hashin–Shtrikman bounds for any space dimension d using the Hashin–Shtrikman variational principle of Theorem 14.5. The key quantities are the trial difference field $\hat{\boldsymbol{G}}$, trial polarization field $\hat{\boldsymbol{P}}$, and the conductivity σ_0 of the comparison medium. We take the trial difference field to be given by the *fluctuating* part of the trial field (21.2), i.e.,

$$\hat{\boldsymbol{G}}(\boldsymbol{x}) = \sum_{k=1}^{n} \xi_k \boldsymbol{E}_k(\boldsymbol{x}), \tag{21.15}$$

where the field $\boldsymbol{E}_k$ is the exact kth-order solution of the field in the exact contrast expansion discussed in Section 20.1.4. In a similar spirit, we take the trial polarization to be

$$\hat{\boldsymbol{P}}(\boldsymbol{x}) = [\sigma(\boldsymbol{x}) - \sigma_0]\sum_{k=1}^{n} \xi_k \boldsymbol{E}_{k-1}, \tag{21.16}$$

where $\boldsymbol{E}_0 \equiv \langle\boldsymbol{E}\rangle$. Substitution of these trial fields in the Hashin–Shtrikman variational principle and optimizing over the vector $\boldsymbol{\xi}$ yields a $2n$-point lower bound on the effective conductivity when $\sigma_0 = \sigma_1$. Interchanging the phase subscripts gives the corresponding $2n$-point upper bound.

Our specific concerns here are two-point bounds. Taking $n = 1$ and $\sigma_0 = \sigma_1$, we see from (21.15) and (21.16) that

$$\hat{\boldsymbol{G}}(\boldsymbol{x}) = \xi_1 \boldsymbol{E}_1(\boldsymbol{x}), \qquad \hat{\boldsymbol{P}}(\boldsymbol{x}) = \xi_1(\sigma_2 - \sigma_1)\mathcal{I}^{(2)}\langle\boldsymbol{E}\rangle. \tag{21.17}$$

The Hashin–Shtrikman principle requires us to compute the following averages:

$$\left\langle \frac{\hat{P}\cdot\hat{P}}{\sigma-\sigma_1}\right\rangle = (\sigma_2-\sigma_1)\phi_2\xi_1^2, \qquad \langle \hat{P}\cdot e\rangle = (\sigma_2-\sigma_1)\phi_2\xi_1,$$

$$\langle \hat{P}\cdot\hat{G}\rangle = (\sigma_2-\sigma_1)\langle \mathcal{I}^{(2)}E_1\rangle\cdot e\xi_1^2 = a_2\frac{(\sigma_2-\sigma_1)^2}{\sigma_1}\xi_1^2 = -\frac{(\sigma_2-\sigma_1)^2}{d\sigma_1}\phi_1\phi_2\xi_1^2,$$

where we have used identity (20.54), $a_2 \equiv a_2^{(1)} = a_2^{(2)}$ is the second-order coefficient (20.79) in the contrast expansion (20.77) or (20.48), and, without loss of generality, we have taken $\langle E\rangle = e$, a unit vector. If we substitute the averages above into the Hashin–Shtrikman principle (14.56) and optimize over ξ_1, we obtain the lower bound

$$\sigma_* = \langle\sigma\rangle - \frac{\phi_1\phi_2(\sigma_2-\sigma_1)^2}{\langle\tilde{\sigma}\rangle + (d-1)\sigma_1}, \tag{21.18}$$

where for any property c,

$$\langle\tilde{c}\rangle = c_1\phi_2 + c_2\phi_1. \tag{21.19}$$

The upper bound is obtained by interchanging the subscripts 1 and 2.

To summarize, the d-dimensional Hashin–Shtrikman bounds on σ_e for two-phase isotropic media in which $\sigma_2 \geq \sigma_1$ are

$$\sigma_L^{(2)} \leq \sigma_e \leq \sigma_U^{(2)}, \tag{21.20}$$

where

$$\sigma_L^{(2)} = \langle\sigma\rangle - \frac{\phi_1\phi_2(\sigma_2-\sigma_1)^2}{\langle\tilde{\sigma}\rangle + (d-1)\sigma_1}, \tag{21.21}$$

$$\sigma_U^{(2)} = \langle\sigma\rangle - \frac{\phi_1\phi_2(\sigma_2-\sigma_1)^2}{\langle\tilde{\sigma}\rangle + (d-1)\sigma_2}. \tag{21.22}$$

As expected, the bounds (21.20) are exact through second order in the difference in the phase conductivities [cf. (20.77)].

The Hashin–Shtrikman bounds are realized by the singly-coated d-dimensional sphere assemblages and the second-rank laminates described in Chapter 16. Accordingly, because the bounds are attainable by certain microstructures, they are the best possible bounds on the effective conductivity of isotropic two-phase composites given volume-fraction information only. For $d = 2$, the single-scale Vigdergauz (1989, 1994) constructions that realize the Hashin–Shtrikman bounds on the effective planar bulk modulus also realize the two-point bounds (21.20).

The corresponding two-point bounds on the effective conductivity σ_e of d-dimensional, macroscopically isotropic composites consisting of M isotropic phases are also known. Let σ_i be the conductivity of phase i, and denote by $\sigma_{\max}$ and $\sigma_{\min}$ the largest and smallest phase conductivities, respectively. We have

$$\left[\sum_{i=1}^{M}\phi_i(\sigma_{\min}^*+\sigma_i)^{-1}\right]^{-1} - \sigma_{\min}^* \leq \sigma_e \leq \left[\sum_{i=1}^{M}\phi_i(\sigma_{\max}^*+\sigma_i)^{-1}\right]^{-1} - \sigma_{\max}^*, \tag{21.23}$$

where

$$\sigma^*_{\min} = (d-1)\sigma_{\min}, \qquad \sigma^*_{\max} = (d-1)\sigma_{\max}.$$

These multiphase bounds on the effective conductivity were shown by Milton (1981d) to be realizable by certain multicoated spheres ($d = 3$) for a wide range of cases. The bounds for any dimension $d \geq 2$ are realizable by the corresponding d-dimensional multicoated spheres.

Now let us consider a macroscopically anisotropic composite medium consisting of two isotropic phases that is characterized by an effective conductivity tensor $\boldsymbol{\sigma}_e$. The two-point anisotropic generalizations of the Hashin–Shtrikman bounds on $\boldsymbol{\sigma}_e$ when $\sigma_2 \geq \sigma_1$ are given by

$$\boldsymbol{\sigma}_L^{(2)} \leq \boldsymbol{\sigma}_e \leq \boldsymbol{\sigma}_U^{(2)}, \tag{21.24}$$

where

$$\boldsymbol{\sigma}_L^{(2)} = \langle \boldsymbol{\sigma} \rangle + (\sigma_1 - \sigma_2)^2 \boldsymbol{a}_2 \cdot \left[\sigma_1 \boldsymbol{I} + \frac{(\sigma_1 - \sigma_2)}{\phi_2} \boldsymbol{a}_2 \right]^{-1}, \tag{21.25}$$

$$\boldsymbol{\sigma}_U^{(2)} = \langle \boldsymbol{\sigma} \rangle + (\sigma_2 - \sigma_1)^2 \boldsymbol{a}_2 \cdot \left[\sigma_2 \boldsymbol{I} + \frac{(\sigma_2 - \sigma_1)}{\phi_1} \boldsymbol{a}_2 \right]^{-1}. \tag{21.26}$$

Here $\boldsymbol{a}_2 \equiv \boldsymbol{a}_2^{(1)} = \boldsymbol{a}_2^{(2)}$ is the two-point tensor parameter that arose in the contrast expansion (20.48) defined by (20.50). The above two-point bounds have been derived by a variety of methods. Willis (1977) first derived them for $d = 3$ using the anisotropic generalizations of the Hashin–Shtrikman principles. Sen and Torquato (1989) obtained them in arbitrary dimension d using the method of Padé approximants (Milton and Golden 1985). The bounds (21.24) are obtained by forming the $[1,1]$ Padé approximant of the contrast expansion (20.48); when $\sigma_2 \geq \sigma_1$, the lower bound is obtained for $j = 1$ and $i = 2$, and the upper bound is obtained when $j = 2$ and $i = 1$. The $[M,N]$ Padé approximant of a power series of a function $f(z) = \sum_{n=0}^{\infty} c_n z^n$ is defined [see Baker (1975)] as a rational function

$$[M,N] = \frac{a_0 + a_1 z + \cdots + a_M z^M}{1 + b_1 z + \cdots + b_N z^N}$$

whose Taylor series agrees with that of $f(z)$ up to order $M + N$. We will show in the ensuing discussion how to derive the two-point bounds (21.24) from the corresponding three-point bounds.

It is noteworthy that the two-point anisotropic bounds (21.25) and (21.26) are realized by a variety of models, including oriented, d-dimensional singly coated ellipsoids (Milton 1980, Bergman 1980, Milton 1981a, Tartar 1985) and the class of finite-rank laminates (Tartar 1985) discussed in Section 16.1.3. Notice that the d-dimensional Hashin–Shtrikman bounds (21.20) are recovered for isotropic media from (21.25) and (21.26), since (20.50) simplifies as

$$\boldsymbol{a}_2 = -\frac{\phi_1\phi_2}{d}\boldsymbol{I}. \tag{21.27}$$

For statistically anisotropic microstructures in which $S_2^{(p)}(\boldsymbol{r})$ possesses ellipsoidal symmetry (e.g., oriented similar ellipsoidal inclusions in a matrix with nematic-liquid-crystal structure), the two-point parameter is given by

$$\boldsymbol{a}_2 = -\phi_1\phi_2\boldsymbol{A}^*, \tag{21.28}$$

where $\boldsymbol{A}^*$ is the polarization tensor (17.25) for an ellipsoid.

We note in passing that one can eliminate the parameter $\boldsymbol{a}_2$ by utilizing the property that $\mathrm{Tr}\,\boldsymbol{a}_2 = -\phi_1\phi_2$ to yield the simpler bounds on the eigenvalues λ_i $(i = 1, \ldots, d)$ of $\boldsymbol{\sigma}_e$ obtained by Lurie and Cherkaev (1984, 1986) and by Tartar (1985). Thus, these bounds incorporate only volume-fraction information.

Willis (1977) actually found two-point bounds for three-dimensional multiphase anisotropic composites with M anisotropic phases of conductivities $\boldsymbol{\sigma}_1, \boldsymbol{\sigma}_2, \ldots, \boldsymbol{\sigma}_M$ and a comparison phase 0 of conductivity $\boldsymbol{\sigma}_0$. These bounds are easily generalized to arbitrary d using the above considerations and are given by

$$\boldsymbol{\sigma}_e \begin{cases} \geq \bar{\boldsymbol{\sigma}}, & \boldsymbol{\sigma}_p \geq \boldsymbol{\sigma}_0, \\ \leq \bar{\boldsymbol{\sigma}}, & \boldsymbol{\sigma}_p \leq \boldsymbol{\sigma}_0, \end{cases} \qquad p = 1, 2, \ldots, M, \tag{21.29}$$

where

$$\bar{\boldsymbol{\sigma}} = \left[\sum_{p=1}^{M} \phi_p \boldsymbol{\sigma}_p \boldsymbol{R}^{(p0)}\right] \cdot \left[\sum_{p=1}^{M} \phi_p \boldsymbol{R}^{(p0)}\right]^{-1}, \tag{21.30}$$

$$\boldsymbol{R}^{(p0)} = [\boldsymbol{I} + \boldsymbol{P}^{(0)} \cdot (\boldsymbol{\sigma}_p - \boldsymbol{\sigma}_0)]^{-1}, \tag{21.31}$$

$$\boldsymbol{P}^{(0)}[S_2^{(pq)}(0) - \phi_p\phi_q] = -\int \boldsymbol{G}^{(0)}(\boldsymbol{x} - \boldsymbol{x}')[S_2^{(pq)}(\boldsymbol{x} - \boldsymbol{x}') - \phi_p\phi_q]d\boldsymbol{x}'. \tag{21.32}$$

Here $S_2^{(pq)}(\boldsymbol{r})$ is the probability of finding one point in phase p and another point in phase q with a relative separation of $\boldsymbol{r}$. In the absence of long-range order, it has the following asymptotic properties:

$$S_2^{(pq)}(0) = \begin{cases} \phi_p, & \text{if } p = q, \\ 0, & \text{if } p \neq q, \end{cases} \qquad S_2^{(pq)}(\infty) = \phi_p\phi_q.$$

Moreover, the second order tensor $\boldsymbol{G}^{(0)}(\boldsymbol{r})$ is the double gradient of the infinite-space Green's function for a medium with conductivity tensor $\boldsymbol{\sigma}_0$; for an isotropic medium, it is given by (20.13) with phase $q = 0$. For the two-phase case ($M = 2$), we note that $\boldsymbol{P}^{(0)} = -\boldsymbol{a}_2/(\phi_1\phi_2\sigma_0)$, and therefore (21.29) reduces to (21.24) when the comparison media are appropriately chosen.

Three-Point Bounds

The determination of three-point bounds on diffusion-type properties of heterogeneous media began with the pioneering works by Prager (1963b) and Beran (1965). The approach that we discuss here is that of Beran, who utilized trial fields based on the contrast expansion of the effective conductivity.

Beran derived three-point bounds on σ_e for three-dimensional two-phase isotropic media. Torquato (1980) and Milton (1981a) independently showed that the three point Beran bounds, which involve sixfold integrals of certain three-point correlation functions, can be expressed in terms of the volume fraction ϕ_i of phase i and a single threefold integral involving the three-point probability function $S_3^{(i)}$ of phase i described below. Silnutzer (1972) obtained two-dimensional analogues of the Beran bounds for transversely isotropic fiber-reinforced materials, which Schulgasser (1976c) and Milton (1982) simplified in a similar fashion.

The d-dimensional three-point contrast bounds on the effective conductivity σ_e for two-phase isotropic composites of *any* microstructure are given by

$$\sigma_L^{(3)} \le \sigma_e \le \sigma_U^{(3)}, \tag{21.33}$$

where

$$\sigma_L^{(3)} = \langle \sigma \rangle - \frac{\phi_1 \phi_2 (\sigma_2 - \sigma_1)^2}{\langle \tilde{\sigma} \rangle + (d-1)\langle \sigma^{-1} \rangle_\zeta^{-1}}, \tag{21.34}$$

$$\sigma_U^{(3)} = \langle \sigma \rangle - \frac{\phi_1 \phi_2 (\sigma_2 - \sigma_1)^2}{\langle \tilde{\sigma} \rangle + (d-1)\langle \sigma \rangle_\zeta}, \tag{21.35}$$

and for any property c we have

$$\langle c \rangle_\zeta = c_1 \zeta_1 + c_2 \zeta_2, \quad \text{with} \quad \zeta_1 + \zeta_2 = 1. \tag{21.36}$$

The bounds (21.33), which we derive below, depend not only on ϕ_i but also on the three-point microstructural parameter ζ_i: a multidimensional integral involving $S_3^{(i)}$ defined by (20.63). Since ζ_i must lie in the closed interval [0,1], the bounds (21.33) always improve upon the two-point Hashin–Shtrikman bounds. When $\zeta_2 = 0$, the bounds (21.33) coincide and equal the two-point Hashin–Shtrikman lower bound (for $\sigma_2 \ge \sigma_1$). When $\zeta_2 = 1$, the bounds (21.33) coincide and equal the aforementioned two-point bound for $\sigma_2 \ge \sigma_1$. The bounds (21.33) are exact through third order in the difference in the phase conductivities [cf. (20.77)].

Milton (1984) has derived a three-point lower bound on σ_e for the case $d = 3$ and $\sigma_2 \ge \sigma_1$ that improves upon the Beran lower bound (21.34) for isotropic media. These bounds can be written in the form (Torquato 1985a)

$$\frac{\sigma_L^{(3)}}{\sigma_1} = \frac{1 + (1 + 2\phi_2)\beta_{21} - 2(\phi_1 \zeta_2 - \phi_2)\beta_{21}^2}{1 + \phi_1 \beta_{21} - (2\phi_1 \zeta_2 + \phi_2)\beta_{21}^2}, \tag{21.37}$$

where $\beta_{21} = (\sigma_2 - \sigma_1)/(\sigma_2 + 2\sigma_1)$. This is actually the best possible lower bound on σ_e, given ϕ_i and ζ_2, since it is exactly realized for space-filling doubly-coated composite spheres (see Figure 21.1 for a two-dimensional analogue).

Sen and Torquato (1989) derived corresponding three-point bounds for d-dimensional macroscopically anisotropic two-phase media consisting of isotropic phases using the Padé approach. Using the general variational approach, we will derive their three-point upper bound, which reduces to (21.35) when the composite is isotropic. Taking the fluctuating trial field $\boldsymbol{E}_1$ in (21.10) to be the exact first-order solution in the weak-contrast expansion (20.53), we have from (20.54) with $k = 1$, $q = 1$, and $p = 2$ that

$$\langle \boldsymbol{E}\rangle \cdot \langle \boldsymbol{\sigma} \cdot \boldsymbol{E}_1\rangle = \langle \boldsymbol{E}\rangle \cdot \langle \delta\boldsymbol{\sigma} \cdot \boldsymbol{E}_1\rangle = \frac{(\sigma_2-\sigma_1)^2}{\sigma_1}\langle \boldsymbol{E}\rangle \cdot \boldsymbol{a}_2^{(2)} \cdot \langle \boldsymbol{E}\rangle, \tag{21.38}$$

where $\delta\boldsymbol{\sigma}(\boldsymbol{x}) = \boldsymbol{\sigma}(\boldsymbol{x}) - \boldsymbol{\sigma}_1$ and $\boldsymbol{a}_2^{(2)}$ is the two-point tensor parameter for phase 2 explicitly given by (20.50). Moreover, use of the identities (20.55) gives

$$\begin{aligned}\langle \boldsymbol{E}_1 \cdot \boldsymbol{\sigma} \cdot \boldsymbol{E}_1\rangle &= \langle \boldsymbol{E}_1 \cdot \boldsymbol{\sigma}_1 \cdot \boldsymbol{E}_1\rangle + \langle \boldsymbol{E}_1 \cdot \delta\boldsymbol{\sigma} \cdot \boldsymbol{E}_1\rangle \\ &= \frac{(\sigma_2-\sigma_1)^2}{\sigma_1}\langle \boldsymbol{E}\rangle \cdot \left[\frac{(\sigma_2-\sigma_1)}{\sigma_1}\boldsymbol{a}_3^{(2)} - \boldsymbol{a}_2^{(2)}\right] \cdot \langle \boldsymbol{E}\rangle,\end{aligned} \tag{21.39}$$

where $\boldsymbol{a}_3^{(2)}$ is the three-point tensor parameter for phase 2 explicitly given by (20.51). Substitution of relations (21.38) and (21.39) into the general three-point upper bound (21.10) finally yields

$$\boldsymbol{\sigma}_e \le \langle \boldsymbol{\sigma}\rangle + (\sigma_2-\sigma_1)^2\boldsymbol{a}_2 \cdot [\sigma_1\boldsymbol{I} + (\sigma_1-\sigma_2)\boldsymbol{a}_3^{(2)} \cdot \boldsymbol{a}_2^{-1}]^{-1}, \tag{21.40}$$

where $\boldsymbol{a}_2 \equiv \boldsymbol{a}_2^{(1)} = \boldsymbol{a}_2^{(2)}$. This upper bound was derived by Sen and Torquato not by the variational approach used here but by forming the [2, 1] Padé approximant of the contrast expansion (20.48). The corresponding Sen–Torquato lower bound ([1, 2] Padé approximant) can be obtained in the same manner using (21.11). Notice that the three-point upper bound (21.35) for isotropic media is recovered from (21.40), since for such composites, $\boldsymbol{a}_2$ is given by (21.27) and

$$\boldsymbol{a}_3^{(2)} = \frac{\phi_1\phi_2}{d^2}[\phi_1 + (d-1)\zeta_2]\boldsymbol{I},$$

as obtained from (20.80).

The use of Schwarz's inequality in the three-point bounds enables us to obtain weaker but optimal two-point bounds. For example, the two-point anisotropic upper bound (21.26) can be obtained from the general three-point upper bound (21.10). The contribution to the latter containing three-point information is the average $\langle \mathcal{I}^{(2)}\boldsymbol{E}_1\boldsymbol{E}_1\rangle$, a dyadic, and this average is bounded from above via Schwarz's inequality:

$$\langle \mathcal{I}^{(2)}\boldsymbol{E}_1\boldsymbol{E}_1\rangle \le \langle \boldsymbol{E}_1\boldsymbol{E}_1\rangle - \frac{1}{\phi_1}\langle \mathcal{I}^{(2)}\boldsymbol{E}_1\rangle\langle \mathcal{I}^{(2)}\boldsymbol{E}_1\rangle. \tag{21.41}$$

This inequality, in conjunction with the identities (20.54) and (20.55), implies that we can bound the three-point tensor $\boldsymbol{a}_3^{(2)}$, given by (20.51), according to

$$\boldsymbol{a}_3^{(2)} \le -\boldsymbol{a}_2 \cdot \left[\boldsymbol{I} + \frac{\boldsymbol{a}_2}{\phi_1}\right].$$

Substitution of this last inequality into the three-point upper bound (21.40), under the condition that $\boldsymbol{\sigma}_2 \geq \boldsymbol{\sigma}_1$, yields the two-point upper bound (21.26). The corresponding two-point lower bound can be obtained from the three-point lower bound in the same fashion.

Using an elegant continued-fractions formalism, Milton (1987) derived formal n-point bounds that, for $n = 3$, are equivalent to the Sen–Torquato anisotropic bounds. The bounds are formal in that the microstructural parameters are not explicitly given in terms of the n-point probability functions. However, important tensor properties of the parameters follow readily from this treatment, which is not the case using the approach of Sen and Torquato.

Four-Point Bounds

Four-point bounds on the effective conductivity can be derived using the aforementioned procedures. We will not derive such bounds here but will instead state four-point bounds for isotropic media. Milton (1981a) was the first to derive such bounds. Sen and Torquato (1989) found anisotropic analogues of these bounds, which in the isotropic case and for $\sigma_2 \geq \sigma_1$ reduce to the following form:

$$\sigma_L^{(4)} \leq \sigma_e \leq \sigma_U^{(4)}, \tag{21.42}$$

where

$$\frac{\sigma_L^{(4)}}{\sigma_1} = \frac{1 + [(d-1)\phi_2 - \gamma_2/\zeta_2]\beta_{21} + (1-d)[\phi_1\zeta_2 + \phi_2\gamma_2/\zeta_2]\beta_{21}^2}{1 - [\phi_2 + \gamma_2/\zeta_2]\beta_{21} + [\phi_1(1-d)\zeta_2 + \phi_2\gamma_2/\zeta_2]\beta_{21}^2}, \tag{21.43}$$

$$\frac{\sigma_U^{(4)}}{\sigma_2} = \frac{1 + [(d-1)\phi_1 - \gamma_1/\zeta_1]\beta_{12} + (1-d)[\phi_2\zeta_1 + \phi_1\gamma_1/\zeta_1]\beta_{12}^2}{1 - [\phi_1 + \gamma_1/\zeta_1]\beta_{12} + [\phi_2(1-d)\zeta_1 + \phi_1\gamma_1/\zeta_1]\beta_{12}^2}, \tag{21.44}$$

$$\gamma_1 - \gamma_2 = (d-2)(\zeta_2 - \zeta_1), \tag{21.45}$$

and

$$\beta_{ij} = \frac{\sigma_i - \sigma_j}{\sigma_i + (d-1)\sigma_j}, \qquad i \neq j. \tag{21.46}$$

The four-point bounds (21.42) depend upon ϕ_i, ζ_i, and the four-point parameters γ_i [defined by (20.60) and related to $\boldsymbol{a}_4^{(i)}$ of (20.52)], which depend upon $S_1^{(i)}$, $S_2^{(i)}$, $S_3^{(i)}$, and $S_4^{(i)}$. Note that for $d = 2$, the four-point parameters are exactly zero ($\gamma_1 = \gamma_2 = 0$), and hence the four-point bounds depend only on the phase volume fractions and parameter $\zeta_2 = 1 - \zeta_1$.

For $d = 2$, Milton (1981a) showed that the four-point bounds (21.42) are realized by space-filling doubly-coated cylinders (circular disks in two dimensions). Each cylinder (disk) is similar, within a scale factor, to any other multicoated cylinder in the composite, and there is a distribution in their sizes ranging to the infinitesimally small (see Figure 21.1). Note that when the radius of the inner cylinder goes to zero, i.e., $\zeta_i \to 0$ ($i = 1, 2$), one recovers the singly coated cylinder assemblages corresponding to the Hashin–Shtrikman two-point bounds.

Figure 21.1 The doubly coated cylinders (circles) geometry associated with the four-point lower bound (21.43) (Milton 1981a) for $\sigma_2 \geq \sigma_1$. Here phases 1 and 2 are white and black phases, respectively. The ratio of core volume to the core plus inner shell is $\phi_1\zeta_2$. The three-dimensional analogue of this geometry corresponds to the three-point lower bound (21.37) of Milton (1981a). The four-point upper bound (21.44) for $d = 2$ and $\sigma_2 \geq \sigma_1$ is realized by this geometry but with the phases interchanged.

21.1.3 Cluster Bounds

In cases in which one of the phases is composed of well-defined inclusions, it is desired to derive bounds that incorporate such specific microstructural information. The so-called nth-order "cluster" bounds of Torquato (1986b) accomplish this by using admissible fields that exactly account for interactions between n inclusions. (This class of bounds is sometimes referred to as "multiple-scattering" bounds; see the discussion below on the trapping constant and fluid permeability.) Such bounds are thus exact through nth order in the inclusion volume fraction ϕ_2. Weissberg (1963) actually was the first to derive first-order cluster bounds for a system of perfectly insulating ($\sigma_2/\sigma_1 = 0$), *fully penetrable spheres* of equal radius. (This model is discussed in Chapter 5.) DeVera and Strieder (1977) extended Weissberg's results for fully penetrable spheres for all σ_2/σ_1. All of these bounds are special cases of the general first-order (or three-point) cluster bounds derived by Torquato (1986b) for distributions of equisized spheres with an *arbitrary* degree of impenetrability (see Chapter 3).

The three-point Torquato bounds can be written in compact notation as

$$\sigma_L^{(3)}[\sigma_1, \sigma_2, \phi_2, \xi] \leq \sigma_e \leq \sigma_U^{(3)}[\sigma_1, \sigma_2, \phi_2, \xi], \tag{21.47}$$

where ξ represents certain integrals involving the microstrutural functions G_2 and G_3. Recall from Section 2.9 that the point/q-particle function $G_n(\boldsymbol{x}; \boldsymbol{r}^q)$ ($n = 1 + q$) gives the probability associated with finding a point at $\boldsymbol{x}$ exterior to the spheres and a configuration of q spheres with centers at $\boldsymbol{r}^q \equiv \boldsymbol{r}_1, \ldots, \boldsymbol{r}_q$. [Note that Torquato (1986b) used slightly different notation for G_n.] The first-order cluster bounds (21.47) are clearly three-point bounds, since they involve information up to the level of G_3. In general, nth-order cluster bounds are $(n+2)$-point bounds.

The upper bound of (21.47) is generated for dispersions of N spheres of radius R in d dimensions by using the trial field

$$\boldsymbol{E}_1(\boldsymbol{x}; \boldsymbol{r}^N) = \sum_{i=1}^{N} \boldsymbol{K}_1(\boldsymbol{x} - \boldsymbol{r}_i) \cdot \langle \boldsymbol{E} \rangle - \rho \int d\boldsymbol{r}_1 \boldsymbol{K}_1(\boldsymbol{x} - \boldsymbol{r}_1) \cdot \langle \boldsymbol{E} \rangle, \tag{21.48}$$

where $\boldsymbol{K}_1$ is the single-body operator (19.6) derivable from the solution of the field for the single-sphere boundary value problem (17.14):

$$\boldsymbol{K}_1(\boldsymbol{r}) = \begin{cases} \dfrac{\beta_{21} R^d}{r^d}[d\boldsymbol{nn} - \boldsymbol{I}], & r > R, \\ -\beta_{21} \boldsymbol{I}, & r < R. \end{cases}$$

Here β_{21} is given by (21.46), $\boldsymbol{n} = \boldsymbol{r}/|\boldsymbol{r}|$ is a unit normal vector, and ρ is the inclusion number density; ultimately we consider the *thermodynamic limit*, i.e., $N \to \infty$, $V \to \infty$, such that $\rho = N/V$ is fixed. Substitution of the trial field (21.48) into the general upper bound (21.10) yields integrals, represented by ξ, that involve the functions G_2 and G_3. The form of the trial field (21.48) ensures the absolute convergence of the integrals resulting from the ensemble averages. The corresponding lower bound is obtained by using the analogue of (21.48) for the trial current field.

Markov (1998) generalized Torquato's cluster bounds to the case of spheroids and, in particular, examined the limit in which the spheroids become cracks. Interestingly, the bounds coincide with the so-called "noninteracting" cracks approximation, thus rigorously proving that crack interactions always decrease the effective conductivity of a microcracked solid.

21.1.4 Security-Spheres Bounds

Conventional bounds, such as contrast and cluster bounds, generally diverge from one another in the limit of infinite phase contrast. For example, when phase 2 is superconducting relative to phase 1 ($\sigma_2/\sigma_1 = \infty$), such finite-order upper bounds diverge to infinity, since the bounds take into account realizations in which phase 2 is percolating, even if phase 2 is actually not percolating. This is due to the fact that the limited microstructural information that the bounds incorporate is insufficient to distinguish phase connectivity. Similarly, when phase 2 is perfectly insulating ($\sigma_2/\sigma_1 = 0$), finite-order conventional lower bounds vanish identically in order to take into account realizations in which only phase 2 is percolating.

Torquato (1985a) has noted, however, that the remaining bound (in either case $\sigma_2/\sigma_1 = \infty$ or $\sigma_2/\sigma_1 = 0$) can yield a useful estimate of the effective conductivity σ_e, provided that certain conditions are met (see Section 21.6). Nonetheless, it is useful to derive bounds that account for connectedness information in such a way that the bounds do not diverge in the limit of infinite phase contrast. A first step in this direction was taken by Keller et al. (1967), who used the concept of a "security" sphere to construct trial fields to bound the effective viscosity of suspensions. Torquato and Rubinstein (1991) applied the same idea to find bounds on the effective conductivity, which we describe below.

Consider a statistically isotropic system of identical spheres of radius R of conductivity σ_2 in a matrix of conductivity σ_1. Surround the ith sphere in the system by a larger "security" sphere such that the concentric-shell region between the actual sphere and the surface of the security sphere contains only phase 1. In order to find an upper bound on the effective conductivity σ_e, Torquato and Rubinstein used a trial intensity field $\hat{\boldsymbol{E}}$ taken to be the uniform field $\boldsymbol{E}_0$ in the matrix region outside the security spheres. The fields within the security spheres are evaluated exactly by applying a potential equal to $-\boldsymbol{E}_0 \cdot \boldsymbol{x}$ on the boundary of each security sphere and requiring continuity of potential and normal flux at the two-phase interface $r = R$, where r is measured from the center of a typical sphere. The resulting trial energy then provides an upper bound on the effective conductivity via the minimum energy principle. By constructing a trial flux in the same manner, Torquato and Rubinstein found lower bounds on σ_e using the minimum complementary energy principle.

Here we report their results but generalized to arbitrary space dimension d for two limiting cases. For superconducting inclusions ($\sigma_2/\sigma_1 = \infty$), the d-dimensional generalization of their upper bound is given by

$$\frac{\sigma_e}{\sigma_1} \leq 1 + d\phi_2 \int_0^\infty \frac{x^d}{x^d - 1} D H_P(x) dx, \tag{21.49}$$

where $D = 2R$ is the sphere diameter, $x = r/D$ is a dimensionless distance, ϕ_2 is the sphere volume fraction, and $H_P(r)$ is the nearest-neighbor probability density function defined by (2.89). We do not give the corresponding lower bound, since conventional three-point lower bounds are superior. Notice that the function multiplying H_P in (21.49) contains a pole at $x = 1$, i.e., when the spheres touch. Thus, unlike conventional upper bounds, the security-sphere upper bound (21.49) will remain finite in this superconducting limit, provided that $H_P(x)$ vanishes sufficiently rapidly as $x \to 1$ to make the integral convergent.

For perfectly insulating inclusions ($\sigma_2/\sigma_1 = 0$), the d-dimensional generalization of their lower bound is given by

$$\frac{\sigma_e}{\sigma_1} \geq \left[1 + \frac{d\phi_2}{d-1} \int_0^\infty \frac{x^d}{x^d - 1} D H_P(x) dx\right]^{-1}. \tag{21.50}$$

We see that unlike conventional lower bounds, the security-sphere lower bound (21.50) will remain nonzero in this zero-conductivity limit, provided that $H_P(x)$ vanishes sufficiently rapidly as $x \to 1$ to make the integral convergent. The corresponding upper bound is not given, since conventional three-point lower bounds are sharper.

Note that the bounds (21.49) and (21.50) will deliver nontrivial results in the case of *nontouching* spheres that are centered on a regular lattice with a minimum interparticle spacing of bD, in which case

$$DH_P(x) = \delta(x - b). \tag{21.51}$$

Using the results for H_P for random arrays discussed in Chapter 5, one can find nontrivial security-spheres bounds provided that the spheres do not touch (Torquato and Rubinstein 1991).

Bruno (1991) found security-spheres type bounds on σ_e using the complex variable method. However, these bounds are given in terms of the minimum interparticle distance among all of the particles (not H_P).

21.2 Elastic Moduli

The minimum potential energy principle requires a trial strain field $\hat{\varepsilon}(\boldsymbol{r})$ that is compatible and whose average equals the actual average, i.e., $\langle\hat{\varepsilon}(\boldsymbol{r})\rangle = \langle\varepsilon\rangle$. The minimum complementary energy principle requires a trial stress field $\hat{\tau}(\boldsymbol{r})$ that is divergence-free and whose average equals the actual average, i.e., $\langle\hat{\tau}\rangle = \langle\tau\rangle$. In what follows, we consider specific trial fields that lead to progressively tighter bounds. The reader will note the strong similarities between the derivations presented here and those given in Section 21.1 for the conductivity.

21.2.1 General Considerations

To grasp the essential idea behind generating bounds on the effective stiffness $\boldsymbol{C}_e$, let us consider a simple application of the minimum principles of Theorems 14.8 and 14.9 for a macroscopically anisotropic composite consisting of M anisotropic phases. The constant fields $\langle\varepsilon\rangle$ and $\langle\tau\rangle$ trivially satisfy the admissibility conditions on the trial strain and stress fields, respectively. Use of such trial fields in the variational principles results in the so-called Reuss–Voigt bounds on the effective stiffness tensor $\boldsymbol{C}_e$:

$$\langle \boldsymbol{C}^{-1}\rangle^{-1} \le \boldsymbol{C}_e \le \langle \boldsymbol{C}\rangle. \tag{21.52}$$

Thus, the effective stiffness tensor is bounded from above by the arithmetic mean of the phase stiffnesses and from below by the harmonic mean of the phase stiffnesses. As before, these are one-point bounds and are generally far apart from one another. In order to improve upon them, one must use nonuniform trial fields that better reflect the interactions between the phases.

One such allowable nontrivial trial strain field is the following:

$$\hat{\varepsilon}(\boldsymbol{x}) = \langle\varepsilon\rangle + \sum_{k=1}^{n} \xi_k \varepsilon_k(\boldsymbol{x}), \qquad \langle\varepsilon_k\rangle = 0 \quad \forall k, \tag{21.53}$$

where ξ_k are constant multipliers to be optimized. The field ε_k can be based on the exact kth-order solution of the strain from the contrast expansion (Chapter 20) or on the exact solution of the strain for k interacting spheres (Chapter 19). Substituting expression (21.53) into upper bound (14.92) and setting the partial derivative of the right-hand side with respect to ξ_k equal to zero yields

$$\langle\varepsilon\rangle : \boldsymbol{C}_e : \langle\varepsilon\rangle \leq \langle\varepsilon\rangle : \langle\boldsymbol{C}\rangle : \langle\varepsilon\rangle - \boldsymbol{V}^{\mathrm{T}}\boldsymbol{W}^{-1}\boldsymbol{V}. \tag{21.54}$$

Here $\boldsymbol{V}^{\mathrm{T}}$ denotes the transpose of the column vector $\boldsymbol{V}$ with scalar elements

$$V_k = \langle\varepsilon\rangle : \langle\boldsymbol{C} : \varepsilon_k\rangle, \tag{21.55}$$

and $\boldsymbol{W}^{-1}$ denotes the inverse of the matrix $\boldsymbol{W}$ with scalar elements

$$W_{kl} = \langle\varepsilon_k : \boldsymbol{C} : \varepsilon_l\rangle. \tag{21.56}$$

Whether the trial fields ε_i $(i = 1, \ldots, n)$ are of the contrast type or cluster type, the upper bound (21.54) is a $(2n+1)$-point bound, since it contains microstructural information up to the level of the relevant $(2n+1)$-point correlation function.

A corresponding allowable nontrivial trial stress field is

$$\hat{\boldsymbol{\tau}}(\boldsymbol{x}) = \langle\boldsymbol{\tau}\rangle + \sum_{k=1}^{n} \omega_k \boldsymbol{\tau}_k(\boldsymbol{x}), \qquad \langle\boldsymbol{\tau}_k\rangle = 0 \quad \forall k, \tag{21.57}$$

where ω_k are constant multipliers. The field $\boldsymbol{\tau}_k$ can be based on the exact kth-order solution of the stress from the contrast expansion or on the exact solution of the stress for k interacting spheres. Substituting (21.57) into the bound (14.99) and setting the partial derivative of the right-hand side with respect to ω_k equal to zero gives

$$\langle\boldsymbol{\tau}\rangle : \boldsymbol{C}_e^{-1} : \langle\boldsymbol{\tau}\rangle \leq \langle\boldsymbol{\tau}\rangle : \langle\boldsymbol{C}^{-1}\rangle : \langle\boldsymbol{\tau}\rangle - \boldsymbol{X}^{\mathrm{T}}\boldsymbol{Y}^{-1}\boldsymbol{X}. \tag{21.58}$$

Here $\boldsymbol{X}^{\mathrm{T}}$ denotes the transpose of the vector $\boldsymbol{X}$ with scalar elements

$$X_k = \langle\boldsymbol{\tau}\rangle : \langle\boldsymbol{C}^{-1} : \boldsymbol{\tau}_k\rangle, \tag{21.59}$$

and $\boldsymbol{Y}^{-1}$ denotes the inverse of the matrix $\boldsymbol{Y}$ with scalar elements

$$Y_{kl} = \langle\boldsymbol{\tau}_k : \boldsymbol{C}^{-1} : \boldsymbol{\tau}_l\rangle. \tag{21.60}$$

Expression (21.58) leads to a $(2n+1)$-point lower bound on the effective stiffness tensor $\boldsymbol{C}_e$.

Of particular interest are the optimized three-points bounds when $n = 1$:

$$\langle\varepsilon\rangle : \boldsymbol{C}_e : \langle\varepsilon\rangle \leq \langle\varepsilon\rangle : \langle\boldsymbol{C}\rangle : \langle\varepsilon\rangle - \frac{|\langle\varepsilon\rangle : \langle\boldsymbol{C} : \varepsilon_1\rangle|^2}{\langle\varepsilon_1 : \boldsymbol{C} : \varepsilon_1\rangle},$$

$$\langle\boldsymbol{\tau}\rangle : \boldsymbol{C}_e^{-1} : \langle\boldsymbol{\tau}\rangle \leq \langle\boldsymbol{\tau}\rangle : \langle\boldsymbol{C}^{-1}\rangle : \langle\boldsymbol{\tau}\rangle - \frac{|\langle\boldsymbol{\tau}\rangle : \langle\boldsymbol{C}^{-1} : \boldsymbol{\tau}_1\rangle|^2}{\langle\boldsymbol{\tau}_1 : \boldsymbol{C}^{-1} : \boldsymbol{\tau}_1\rangle}.$$

21.2.2 Contrast Bounds

We will derive many of the existing bounds on the effective elastic moduli whose trial fields are based on the contrast expansion. We will focus primarily on one-, two-, three-, and four-point bounds. Again, by construction, such n-point bounds are exact through nth order in the difference in the phase moduli. We will mainly treat composites consisting of isotropic phases; i.e., the stiffness tensor for phase i is given by

$$\boldsymbol{C}_i = dK_i \boldsymbol{\Lambda}_h + 2G_i \boldsymbol{\Lambda}_s, \tag{21.61}$$

where K_i and G_i are the bulk and shear moduli of phase i, respectively, and $\boldsymbol{\Lambda}_h$ and $\boldsymbol{\Lambda}_s$ are the fourth-order hydrostatic and shear projection tensors, respectively, given by (13.96) and (13.97). Both macroscopically isotropic and anisotropic composites will be considered.

One-Point Bounds

For a macroscopically anisotropic composite with isotropic constituents, the one-point bounds (21.52) simplify as follows:

$$d\langle K^{-1}\rangle^{-1}\boldsymbol{\Lambda}_h + 2\langle G^{-1}\rangle^{-1}\boldsymbol{\Lambda}_s \le \boldsymbol{C}_e \le d\langle K\rangle\boldsymbol{\Lambda}_h + 2\langle G\rangle\boldsymbol{\Lambda}_s. \tag{21.62}$$

If we further specify that the composite is macroscopically isotropic, i.e., $\boldsymbol{C}_e = dK_e\boldsymbol{\Lambda}_h + 2G_e\boldsymbol{\Lambda}_s$, then the bounds above reduce to the following simple one-point bounds on the effective elastic moduli K_e and G_e:

$$\langle K^{-1}\rangle^{-1} \le K_e \le \langle K\rangle, \tag{21.63}$$

$$\langle G^{-1}\rangle^{-1} \le G_e \le \langle G\rangle. \tag{21.64}$$

We remind the reader that these bounds are valid for M phases. The effective Young modulus is also bounded from below by the harmonic average, i.e.,

$$E_e \ge \langle E^{-1}\rangle^{-1}.$$

This is easily obtained using the expressions relating Young's moduli to the bulk and shear moduli [cf. (13.103) and (13.130)] and the lower bounds of both (21.63) and (21.64). The use of these interrelations and the upper bounds of (21.63) and (21.64) do not generally lead to an arithmetic upper bound on E_e. However, this expression reduces to the upper bound

$$E_e \le \langle E\rangle$$

when all the phases have the same Poisson's ratio. Paul (1960) derived such one-point bounds on the effective shear and Young's moduli for the case of three-dimensional composites.

Two-Point Bounds

Consider first macroscopically isotropic composites. Hashin and Shtrikman (1963) derived the best possible bounds on the effective elastic moduli of two-phase isotropic three-dimensional composites given just volume-fraction information. As before, these are actually two-point bounds, but the key integral involving $S_2^{(i)}$ just depends on the end points of the function, i.e., ϕ_i and ϕ_i^2. Hashin (1965b) derived the two-dimensional version of the Hashin–Shtrikman bounds.

We will derive the Hashin–Shtrikman bounds for any space dimension d using the Hashin–Shtrikman variational principle of Theorem 14.10. The quantities of central concern are the trial difference field $\hat{\boldsymbol{g}}$, trial polarization field $\hat{\boldsymbol{p}}$, and the stiffness tensor C_0 of the comparison medium. We take the trial difference field to be given by the *fluctuating* part of the trial strain field (21.53), i.e.,

$$\hat{\boldsymbol{g}}(\boldsymbol{x}) = \sum_{k=1}^{n} \xi_k \varepsilon_k(\boldsymbol{x}), \tag{21.65}$$

where the field ε_k is the exact kth-order solution of the field in the exact contrast expansion discussed in Section 20.2.4. In a similar spirit, we take the trial polarization tensor to be

$$\hat{\boldsymbol{p}}(\boldsymbol{x}) = [\boldsymbol{C}(\boldsymbol{x}) - \boldsymbol{C}_0] : \sum_{k=1}^{n} \xi_k \varepsilon_{k-1}, \tag{21.66}$$

where $\varepsilon_0 \equiv \langle \varepsilon \rangle$. Substitution of these trial fields in the Hashin–Shtrikman variational principle and optimizing over the vector $\boldsymbol{\xi}$ yields a $2n$-point lower bound on the effective moduli when $\boldsymbol{C}_0 = \boldsymbol{C}_1$. Interchanging the phase subscripts gives the corresponding $2n$-point upper bound.

The specific case of two-point bounds is obtained from (21.65) and (21.66) by taking $n = 1$ and $\boldsymbol{C}_0 = \boldsymbol{C}_1$, i.e.,

$$\hat{\boldsymbol{g}}(\boldsymbol{x}) = \xi_1 \varepsilon_1(\boldsymbol{x}), \qquad \hat{\boldsymbol{p}}(\boldsymbol{x}) = \xi_1 (\boldsymbol{C}_2 - \boldsymbol{C}_1)\mathcal{I}^{(2)}(\boldsymbol{x}) : \langle \varepsilon \rangle. \tag{21.67}$$

The Hashin–Shtrikman principle requires us to compute the following averages for the bounds on the effective bulk modulus K_e:

$$\langle \hat{\boldsymbol{p}} : (\boldsymbol{C} - \boldsymbol{C}_1)^{-1} : \hat{\boldsymbol{p}} \rangle = (K_2 - K_1)\phi_2 \xi_1^2, \qquad \langle \hat{\boldsymbol{p}} \rangle : \langle \varepsilon \rangle = (K_2 - K_1)\phi_2 \xi_1,$$

$$\langle \hat{\boldsymbol{p}} : \hat{\boldsymbol{g}} \rangle = d(K_2 - K_1)\langle \mathcal{I}^{(2)} \varepsilon_1 \rangle : \langle \varepsilon \rangle \xi_1^2 = (K_2 - K_1)^2 b_{2h}^{(1)} \xi_1^2 = -\frac{d(K_2 - K_1)^2}{dK_1 + 2(d-1)G_1} \phi_1 \phi_2 \xi_1^2,$$

where we have used identity (20.142), $b_{2h}^{(1)}$ is the second-order coefficient (20.178) in the weak-contrast expansion given by (20.174), and $\langle \varepsilon \rangle = \boldsymbol{I}/d$. If we substitute the averages above into the Hashin–Shtrikman principle (14.112) and optimize over ξ_1, we obtain the lower bound

$$K_* = \langle K \rangle - \frac{\phi_1 \phi_2 (K_2 - K_1)^2}{\langle \tilde{K} \rangle + \dfrac{2(d-1)G_1}{d}}, \tag{21.68}$$

where for any property c, $\langle \tilde{c} \rangle$ is given by (21.19). The upper bound is obtained by interchanging the subscripts 1 and 2.

For the corresponding two-point lower bound on the effective shear modulus G_e, use of the trial fields (21.67) results in the following necessary averages:

$$\langle \hat{\boldsymbol{p}} : (\boldsymbol{C} - \boldsymbol{C}_1)^{-1} : \hat{\boldsymbol{p}} \rangle = 2(G_2 - G_1)\phi_2 \xi_1^2, \qquad \langle \hat{\boldsymbol{p}} \rangle : \langle \varepsilon \rangle = 2(G_2 - G_1)\phi_2 \xi_1,$$

$$\begin{aligned}\langle \hat{\boldsymbol{p}} : \hat{\boldsymbol{g}} \rangle &= 2(G_2 - G_1)\langle \mathcal{I}^{(2)} \varepsilon_1 \rangle : \langle \varepsilon \rangle \xi_1^2 = (G_2 - G_1)^2 b_{2s}^{(1)} \xi_1^2 \\ &= -\frac{2d(G_2 - G_1)^2 (K_1 + 2G_1)}{(d+2)G_1[dK_1 + 2(d-1)G_1]} \phi_1 \phi_2 \xi_1^2,\end{aligned}$$

where we have used identity (20.142), $b_{2s}^{(1)}$ is the second-order coefficient (20.178) in the weak-contrast expansion given by (20.174), and $\langle \varepsilon_{ij} \rangle = (\delta_{i1}\delta_{j2} + \delta_{i2}\delta_{j1})/2$. Substituting the averages above into the Hashin–Shtrikman principle (14.113) and optimizing over ξ_1 yields the lower bound

$$G_* = \langle G \rangle - \frac{\phi_1 \phi_2 (G_2 - G_1)^2}{\langle \tilde{G} \rangle + H_1}, \tag{21.69}$$

where H_1 is given below by (21.76), and for any property c, $\langle \tilde{c} \rangle$ is given by (21.19). The upper bound is obtained by interchanging the subscripts 1 and 2.

To summarize, the d-dimensional Hashin–Shtrikman bounds on the effective bulk modulus K_e for two-phase isotropic composites in which $G_2 \geq G_1$ are

$$K_L^{(2)} \leq K_e \leq K_U^{(2)}, \tag{21.70}$$

where

$$K_L^{(2)} = \langle K \rangle - \frac{\phi_1 \phi_2 (K_2 - K_1)^2}{\langle \tilde{K} \rangle + \dfrac{2(d-1)}{d} G_1}, \tag{21.71}$$

$$K_U^{(2)} = \langle K \rangle - \frac{\phi_1 \phi_2 (K_2 - K_1)^2}{\langle \tilde{K} \rangle + \dfrac{2(d-1)}{d} G_2}. \tag{21.72}$$

The d-dimensional Hashin–Shtrikman bounds on the effective shear modulus G_e for two-phase isotropic composites in which $K_2 \geq K_1$ and $G_2 \geq G_1$ are

$$G_L^{(2)} \leq G_e \leq G_U^{(2)}, \tag{21.73}$$

where

$$G_L^{(2)} = \langle G \rangle - \frac{\phi_1 \phi_2 (G_2 - G_1)^2}{\langle \tilde{G} \rangle + H_1}, \tag{21.74}$$

$$G_U^{(2)} = \langle G \rangle - \frac{\phi_1 \phi_2 (G_2 - G_1)^2}{\langle \tilde{G} \rangle + H_2}, \tag{21.75}$$

and

$$H_i = G_i \left[\frac{dK_i/2 + (d+1)(d-2)G_i/d}{K_i + 2G_i} \right]. \tag{21.76}$$

One can verify that the Hashin–Shtrikman bounds are exact through second order in the difference in the phase moduli [cf. (20.181) and (20.182)].

Given the set of all two-phase isotropic composites with prescribed volume fractions, the Hashin–Shtrikman bounds on K_e are optimal, since they are realizable by a number of different models. For example, as indicated in Chapter 16, they are attained by singly coated sphere assemblages (Hashin and Shtrikman 1963). They are also realized by certain finite-rank laminates (Francfort and Murat 1986) (see Figure 16.11) as well as the single-scale Vigdergauz constructions for $d = 2$ (Vigdergauz 1989, Vigdergauz 1994, Grabovsky and Kohn 1995). The Hashin–Shtrikman bounds on the effective shear modulus G_e are not achieved by the aforementioned singly coated sphere assemblages for reasons already noted in Chapter 16. The shear modulus bounds (21.73) are optimal, however, since they are attained by the various laminate constructions (Norris 1985, Lurie and Cherkaev 1985, Francfort and Murat 1986, Milton 1986) discussed in Chapter 16.

Milton (1986) found that the Hashin–Shtrikman bounds on either K_e or G_e are attained if and only if the field in phase 1 (or phase 2) is uniform. Thus, any laminate material that is constructed by starting with phase 1 (or phase 2) and approximately laminating it with the other phase, until a material of some desired rank is achieved, must necessarily attain the Hashin–Shtrikman bounds, since the field in the starting matrix material is always uniform when the applied field is uniform.

For the case of three dimensions, Walpole (1966a) generalized the Hashin–Shtrikman bounds by removing the restrictions on the phase moduli; i.e., the bounds apply not only in the "well ordered" case $(K_2 - K_1)(G_2 - G_1) \geq 0$ (as considered by Hashin and Shtrikman) but also in the "badly ordered" case $(K_2 - K_1)(G_1 - G_2) \geq 0$. To date, no composite structures have been found that realize the bounds in the badly ordered case.

Here we state the Hashin–Shtrikman–Walpole bounds for a d-dimensional macroscopically isotropic composite consisting of M isotropic phases. These more general results are a simple extension of the two-phase bounds discussed in the text, and therefore the details of the derivation are omitted. Let K_i and G_i be the bulk and shear moduli, respectively, of the ith phase, and let ϕ_i be the corresponding volume fraction. Let the largest and smallest phase bulk moduli be denoted by $K_{\max}$ and $K_{\min}$, respectively, and the largest and smallest phase shear moduli be denoted by $G_{\max}$ and $G_{\min}$, respectively. Then the effective bulk modulus K_e and effective shear modulus G_e are bounded according to the relations

$$\left[\sum_{i=1}^{M} \phi_i (K^*_{\min} + K_i)^{-1}\right]^{-1} - K^*_{\min} \leq K_e \leq \left[\sum_{i=1}^{M} \phi_i (K^*_{\max} + K_i)^{-1}\right]^{-1} - K^*_{\max}, \quad (21.77)$$

$$\left[\sum_{i=1}^{M} \phi_i (G^*_{\min} + G_i)^{-1}\right]^{-1} - G^*_{\min} \leq G_e \leq \left[\sum_{i=1}^{M} \phi_i (G^*_{\max} + G_i)^{-1}\right]^{-1} - G^*_{\max}, \quad (21.78)$$

where

$$K_{\min}^* = \frac{2(d-1)}{d} G_{\min}, \quad K_{\max}^* = \frac{2(d-1)}{d} G_{\max}, \quad G_{\min}^* = H_{\min}, \quad G_{\max}^* = H_{\max},$$

and $H_{\min}$ and $H_{\max}$ are the smallest and largest moduli H_i given by (21.76), respectively. Note that the multiphase bounds on the effective bulk modulus were shown by Milton (1981d) to be realizable by certain multicoated spheres ($d = 3$) for a wide range of cases. The bounds on K_e for any dimension $d \geq 2$ are realizable by the corresponding d-dimensional multicoated spheres in these cases.

Consider now macroscopically anisotropic composite media consisting of two isotropic phases characterized by an effective stiffness tensor $\boldsymbol{C}_e$. The two-point anisotropic generalizations of the Hashin–Shtrikman bounds on $\boldsymbol{C}_e$ when $\boldsymbol{C}_2 \geq \boldsymbol{C}_1$ are given, for any d, by

$$\boldsymbol{C}_L^{(2)} \leq \boldsymbol{C}_e \leq \boldsymbol{C}_U^{(2)}, \tag{21.79}$$

where

$$\boldsymbol{C}_L^{(2)} = \langle \boldsymbol{C} \rangle - \phi_1\phi_2(\boldsymbol{C}_1 - \boldsymbol{C}_2){:}\boldsymbol{P}^{(1)}{:}(\boldsymbol{C}_1 - \boldsymbol{C}_2){:} \left[\boldsymbol{I} + \phi_1 \boldsymbol{P}^{(1)}{:}(\boldsymbol{C}_2 - \boldsymbol{C}_1)\right]^{-1}, \tag{21.80}$$

$$\boldsymbol{C}_U^{(2)} = \langle \boldsymbol{C} \rangle - \phi_1\phi_2(\boldsymbol{C}_2 - \boldsymbol{C}_1){:}\boldsymbol{P}^{(2)}{:}(\boldsymbol{C}_2 - \boldsymbol{C}_1){:} \left[\boldsymbol{I} + \phi_2 \boldsymbol{P}^{(2)}{:}(\boldsymbol{C}_1 - \boldsymbol{C}_2)\right]^{-1}. \tag{21.81}$$

Here $\boldsymbol{P}^{(q)}$ is the two-point tensor parameter, given by (20.134), that arose in the weak-contrast expansion (20.136). The bounds (21.79) have been shown by Avellaneda (1987b) to be attainable by the class of finite-rank laminates discussed in Section 16.2.3.

For $d = 3$, the bounds (21.79) are special cases of multiphase bounds due to Willis (1977), who derived them using the anisotropic generalizations of the Hashin–Shtrikman principles. Here we state the d-dimensional generalizations of Willis's bounds for M phases with stiffness tensors $\boldsymbol{C}_1, \boldsymbol{C}_2, \ldots, \boldsymbol{C}_M$ and a comparison phase 0 with stiffness tensor $\boldsymbol{C}_0$:

$$\boldsymbol{C}_e \begin{cases} \geq \bar{\boldsymbol{C}}, & \boldsymbol{C}_p \geq \boldsymbol{C}_0, \\ \leq \bar{\boldsymbol{C}}, & \boldsymbol{C}_p \leq \boldsymbol{C}_0, \end{cases} \qquad p = 1, 2, \ldots, M, \tag{21.82}$$

where

$$\bar{\boldsymbol{C}} = \left[\sum_{p=1}^{M} \phi_p \boldsymbol{C}_p \boldsymbol{T}^{(p0)}\right] \cdot \left[\sum_{p=1}^{M} \phi_p \boldsymbol{T}^{(p0)}\right]^{-1}, \tag{21.83}$$

$$\boldsymbol{T}^{(p0)} = [\boldsymbol{I} + \boldsymbol{P}^{(0)} : (\boldsymbol{C}_p - \boldsymbol{C}_0)]^{-1}, \tag{21.84}$$

$$\boldsymbol{P}^{(0)}[S_2^{(pq)}(0) - \phi_p\phi_q] = -\int \boldsymbol{G}^{(0)}(\boldsymbol{x} - \boldsymbol{x}')[S_2^{(pq)}(\boldsymbol{x} - \boldsymbol{x}') - \phi_p\phi_q] d\boldsymbol{x}'. \tag{21.85}$$

Here $S_2^{(pq)}(\boldsymbol{r})$ is the two-point probability function defined in Section 21.1.2 and the fourth-order tensor $\boldsymbol{G}^{(0)}(\boldsymbol{r})$ is the double gradient of the infinite-space Green's function for a medium with stiffness tensor $\boldsymbol{C}_0$; for an isotropic medium, it is given by (20.100) with phase $q = 0$. For the two-phase case ($M = 2$), we note that (21.82) reduces to (21.79) when the comparison media are appropriately chosen.

Milton and Kohn (1988) derived Hashin–Shtrikman-type bounds on the effective stiffness tensor $\boldsymbol{C}_e$ for anisotropic composites that involve only volume-fraction information. It is interesting to note that these bounds can be derived from the two-point bounds (21.82) by performing trace operations. The reader is also referred to the monograph of Nemat-Nasser and Hori (1993) for a discussion of Hashin–Shtrikman bounds for anisotropic composites.

Three-Point Bounds

Employing the minimum principles of Theorems 14.8 and 14.9, and the admissible fields based upon the first few terms of the contrast expansion of the fields, Beran and Molyneux (1966) and McCoy (1970) derived three-point bounds on K_e and G_e for $d = 3$. Silnutzer (1972) obtained two-dimensional analogues of these bounds on K_e and G_e. Milton (1981c, 1982a) subsequently simplified each of the above three-point bounds, showing that the bounds on the effective bulk moduli for $d = 2$ or 3 can be expressed in terms of ϕ_2 and the three-point parameter ζ_2 defined by (20.66) and (20.67) and that the bounds on the effective shear moduli for $d = 2$ or 3 can be expressed in terms of ϕ_2, ζ_2, and another three-point parameter η_2, defined by (20.170) and (20.171). Some of these bounds were improved upon by Milton and Phan-Thien (1982) and by Gibiansky and Torquato (1995b).

To illustrate the basic idea behind the derivation of three-point bounds on the effective elastic moduli, we will obtain the d-dimensional generalizations of the Beran-type bounds on the effective bulk modulus K_e. The other bounds are stated and discussed.

The d-dimensional three-point contrast bounds on the effective bulk modulus K_e for two-phase isotropic composites are given by

$$K_L^{(3)} \leq K_e \leq K_U^{(3)}, \tag{21.86}$$

where

$$K_L^{(3)} = \langle K \rangle - \frac{\phi_1\phi_2(K_2 - K_1)^2}{\langle \tilde{K} \rangle + \frac{2(d-1)}{d}\langle G^{-1} \rangle_\zeta^{-1}}, \tag{21.87}$$

$$K_U^{(3)} = \langle K \rangle - \frac{\phi_1\phi_2(K_2 - K_1)^2}{\langle \tilde{K} \rangle + \frac{2(d-1)}{d}\langle G \rangle_\zeta}. \tag{21.88}$$

We will now derive the three-point upper bound (21.88) on the effective bulk modulus. Taking the fluctuating trial field ε_1 in (21.61) to be the exact first-order solution in the weak-contrast expansion (20.141), we have from (20.142) with $k = 1, q = 1$, and $p = 2$ that

$$\langle \varepsilon \rangle : \langle \boldsymbol{C} : \varepsilon_1 \rangle = \langle \varepsilon \rangle : \langle \delta \boldsymbol{C} : \varepsilon_1 \rangle = \langle \varepsilon \rangle : \boldsymbol{b}_2^{(2)} : \langle \varepsilon \rangle, \tag{21.89}$$

where $\delta\boldsymbol{C}(\boldsymbol{x}) = \boldsymbol{C}(\boldsymbol{x}) - \boldsymbol{C}_1$ and $\boldsymbol{b}_2^{(2)}$ is the two-point tensor parameter for phase 2 defined by (20.138). Furthermore, use of the identities (20.143) yields

$$\langle \varepsilon_1 : \boldsymbol{C} : \varepsilon_1 \rangle = \langle \varepsilon_1 : \boldsymbol{C}_1 : \varepsilon_1 \rangle + \langle \varepsilon_1 : \delta \boldsymbol{C} : \varepsilon_1 \rangle$$
$$= \langle \varepsilon \rangle : [\boldsymbol{b}_3^{(2)} - \boldsymbol{b}_2^{(2)}] : \langle \varepsilon \rangle, \tag{21.90}$$

where $\boldsymbol{b}_3^{(2)}$ is the three-point tensor parameter for phase 2 defined by (20.139). Substitution of relations (21.89) and (21.90) into the general three-point upper bound (21.61) finally yields

$$\boldsymbol{C}_e \leq \langle \boldsymbol{C} \rangle + \boldsymbol{b}_2^{(2)} : \left\{ \boldsymbol{I} - \boldsymbol{b}_3^{(2)} : [\boldsymbol{b}_2^{(2)}]^{-1} \right\}^{-1}. \tag{21.91}$$

The bound (21.91) is valid for macroscopically anisotropic media.

Recall from Section 20.2.5 that for isotropic media, the microstructural tensor parameters in the weak-contrast form can be written as follows:

$$\boldsymbol{b}_2^{(2)} = d b_{2h}^{(2)} (K_2 - K_1)^2 \boldsymbol{\Lambda}_h + 2 b_{2s}^{(2)} (G_2 - G_1)^2 \boldsymbol{\Lambda}_s,$$
$$\boldsymbol{b}_3^{(2)} = d b_{3h}^{(2)} (K_2 - K_1)^2 \boldsymbol{\Lambda}_h + 2 b_{3s}^{(2)} (G_2 - G_1)^2 \boldsymbol{\Lambda}_s,$$

where $\boldsymbol{\Lambda}_h$ and $\boldsymbol{\Lambda}_s$ are the hydrostatic and shear projection tensors, respectively. Taking the scalar product of $\boldsymbol{\Lambda}_h$ with bound (21.91) gives the following upper bound on the effective bulk modulus:

$$K_e \leq \langle K \rangle + \frac{b_{2h}^{(2)} (K_2 - K_1)^2}{1 - b_{3h}^{(2)} \big/ b_{2h}^{(2)}}. \tag{21.92}$$

Now we know from the weak-contrast expansion (20.174) that

$$b_{2h}^{(2)} = -\frac{d\phi_1\phi_2}{dK_1 + 2(d-1)G_1},$$
$$b_{3h}^{(2)} = \frac{d\phi_1\phi_2}{[dK_1 + 2(d-1)G_1]^2} [d\phi_1 (K_2 - K_1) + 2(d-1)\zeta_2 (G_2 - G_1)].$$

Substitution of the these relations into (21.92) yields the upper bound (21.88). The corresponding lower bound (21.87) can be obtained in the same manner from (21.61).

In the two-dimensional case, Gibiansky and Torquato (1995b) improved upon Silnutzer's upper bound on the effective planar bulk modulus k_e using the so-called *translation method*. (Recall that for $d = 2$, we define $k \equiv K$.) The Gibiansky–Torquato upper bound is given by

$$k_e \leq \langle k \rangle - \frac{\phi_1\phi_2 (k_2 - k_1)^2}{\langle \tilde{k} \rangle + y}, \tag{21.93}$$

where

$$y = \frac{G_1 G_2 + k_{\max} \langle G \rangle_\zeta}{k_{\max} + \langle \tilde{G} \rangle_\zeta}, \qquad \langle \tilde{G} \rangle_\zeta = G_1 \zeta_2 + G_2 \zeta_1.$$

The derivation of three-point bounds on the effective shear modulus can be carried out in the same way as outlined above for the effective bulk modulus. However, here one must be more judicious about the choice of the reference medium and hence the expansion parameter. For example, the choice of phase 1 or 2 as the reference medium [as in (20.174)] does not produce the best bounds. We will simply state the best available bounds on the effective shear modulus for $d = 2$ and $d = 3$ below.

The simplified forms of the Silnutzer three-point bounds on the effective transverse shear modulus for fiber-reinforced materials ($d = 2$) are given by

$$G_L^{(3)} \le G_e \le G_U^{(3)}, \tag{21.94}$$

where

$$G_L^{(3)} = \langle G \rangle - \frac{\phi_1\phi_2(G_2 - G_1)^2}{\langle \tilde{G} \rangle + \Xi}, \tag{21.95}$$

$$G_U^{(3)} = \langle G \rangle - \frac{\phi_1\phi_2(G_2 - G_1)^2}{\langle \tilde{G} \rangle + \Theta}, \tag{21.96}$$

$$\Xi = \left(2\langle 1/k \rangle_\zeta + \langle 1/G \rangle_\eta\right)^{-1}, \tag{21.97}$$

$$\Theta = \frac{2\langle k \rangle_\zeta \langle G \rangle^2 + \langle k \rangle^2 \langle G \rangle_\eta}{\langle k + 2G \rangle^2}, \tag{21.98}$$

and for any property c,

$$\langle c \rangle_\eta = c_1\eta_1 + c_2\eta_2, \qquad \text{with} \qquad \eta_1 = 1 - \eta_2. \tag{21.99}$$

It is seen that the three-point bounds on G_e depend not only on ϕ_i and ζ_i, but also on another three-point parameter η_i, defined by (20.170).

Using the translation method, Gibiansky and Torquato (1995b) improved upon Silnutzer's upper bound on the effective transverse shear modulus G_e of fiber-reinforced composites ($d = 2$). The Gibiansky-Torquato upper bound incorporates ϕ_i, ζ_i, and η_i, and is given by

$$G_e \le \langle G \rangle - \frac{\phi_1\phi_2(G_2 - G_1)^2}{\langle \tilde{G} \rangle + A^{-1}}, \tag{21.100}$$

where

$$A = \begin{cases} y_{1*}, & \text{if } t \in [-k_{\max}^{-1}, G_{\max}^{-1}], \\ y_{2*}, & \text{if } t \le -k_{\max}^{-1}, \\ y_{3*}, & \text{if } t \ge G_{\max}^{-1}, \end{cases}$$

$$y_{1*} = \langle 2k^{-1} \rangle_\zeta + \langle G^{-1} \rangle_\eta - \frac{\left[\sqrt{2\zeta_1\zeta_2(k_1^{-1} - k_2^{-1})^2} + \sqrt{\eta_1\eta_2(G_1^{-1} - G_2^{-1})^2}\,\right]^2}{\eta_1 G_2^{-1} + \eta_2 G_1^{-1} + 2\zeta_1 k_2^{-1} + 2\zeta_2 k_1^{-1}},$$

$$y_{2*} = \left\langle \frac{1}{G^{-1} + k_{\max}^{-1}} \right\rangle_\eta^{-1} + \frac{1}{k_{\max}}, \qquad y_{3*} = 2\left\langle \frac{1}{k^{-1} + G_{\max}^{-1}} \right\rangle_\eta^{-1} - \frac{1}{G_{\max}},$$

$$t = \frac{\dfrac{\eta_1 G_2^{-1} + \eta_2 G_1^{-1}}{\sqrt{\eta_1\eta_2(G_1^{-1} - G_2^{-1})^2}} - \dfrac{\zeta_1 k_2^{-1} + \zeta_2 k_1^{-1}}{\sqrt{2\zeta_1\zeta_2(k_1^{-1} - k_2^{-1})^2}}}{\dfrac{1}{\sqrt{\eta_1\eta_2(G_1^{-1} - G_2^{-1})^2}} + \dfrac{1}{\sqrt{2\zeta_1\zeta_2(k_1^{-1} - k_2^{-1})^2}}}.$$

Kublanov and Milton (1991) also improved upon Silnutzer's upper bound, but it is not as restrictive as (21.100). We also note that Gibiansky and Torquato (1995b) found the sharpest upper and lower bounds on G_e that involve only ϕ_i and ζ_i.

McCoy (1970) was the first to find three-point bounds on the effective shear modulus of three-dimensional two-phase isotropic composites. Subsequently, Milton and Phan-Thien (1982) improved upon these bounds. The three-point Milton–Phan-Thien bounds on the effective shear modulus G_e for $d = 3$ are still given by the aforementioned expressions (21.95) and (21.96), except that here

$$\Xi = \frac{\left\langle \frac{128}{K} + \frac{99}{G} \right\rangle_\zeta + 45\left\langle \frac{1}{G} \right\rangle_\eta}{30\left\langle \frac{1}{G} \right\rangle_\zeta \left\langle \frac{6}{K} - \frac{1}{G} \right\rangle_\zeta + 6\left\langle \frac{1}{G} \right\rangle_\eta \left\langle \frac{2}{K} + \frac{21}{G} \right\rangle_\zeta}, \tag{21.101}$$

$$\Theta = \frac{3\langle G\rangle_\eta \langle 6K + 7G\rangle_\zeta - 5\langle G\rangle_\zeta^2}{6\langle 2K - G\rangle_\zeta + 30\langle G\rangle_\eta}. \tag{21.102}$$

Milton and Phan-Thien also obtained the sharpest upper and lower bounds on G_e that incorporate only ϕ_i and ζ_i.

Four-Point Bounds

Milton and Phan-Thien (1982) also derived four-point bounds on the elastic moduli of three-dimensional two-phase composites. These results are not explicitly given here, however. We note that whereas the bulk-modulus bounds involve three different microstructural parameters, the shear-modulus bounds involve eight parameters. The determination of such a large number of parameters is not practically useful, however.

21.2.3 Cluster Bounds

When one of the phases is composed of well-defined inclusions, it is desired to derive bounds that incorporate such specific microstructural information. Quintanilla and Torquato (1995) have derived three-point "cluster" bounds on the effective moduli of isotropic dispersions of spheres in matrix that are the analogues of the conductivity bounds of Torquato (1986b). The three-point Quintanilla–Torquato bounds on the effective moduli depend on the microstructure via the volume fraction and integrals involving the microstructural functions G_2 and G_3. In general, the point/q-particle function $G_n(\boldsymbol{x}; \boldsymbol{r}^q)$ gives the probability associated with finding a point at $\boldsymbol{x}$ exterior to the spheres and a configuration of q spheres with centers at $\boldsymbol{r}^q \equiv \boldsymbol{r}_1, \ldots, \boldsymbol{r}_q$ (Section 2.9).

The Quintanilla–Torquato upper bound is obtained for dispersions of N spheres of radius R in d dimensions by using the trial field

$$\varepsilon_1(\boldsymbol{x}; \boldsymbol{r}^N) = \sum_{i=1}^{N} \boldsymbol{L}_1(\boldsymbol{x} - \boldsymbol{r}_i) \cdot \langle \varepsilon \rangle - \int d\boldsymbol{r}_1 \rho_1(\boldsymbol{r}_1) \boldsymbol{L}_1(\boldsymbol{x} - \boldsymbol{r}_1) \cdot \langle \varepsilon \rangle, \tag{21.103}$$

where $\boldsymbol{L}_1$ is the single-body fourth-order tensor operator derivable from the solution of the field for the single-sphere boundary value problem (17.53), i.e.,

$$L_1(\boldsymbol{r}) = \begin{cases} \dfrac{\kappa_{21} R^d}{r^d}[d\boldsymbol{nn} - \boldsymbol{I}]\boldsymbol{I}, & r > R, \\ \\ -\kappa_{21}\boldsymbol{\Lambda}_h, & r < R. \end{cases}$$

Here $\boldsymbol{\Lambda}_h$ is given by (13.96), κ_{21} is given by (17.52), and $\boldsymbol{n} = \boldsymbol{r}/|\boldsymbol{r}|$ is a unit normal. Moreover, ρ is the inclusion number density; ultimately, we consider the *thermodynamic limit*, i.e., $N \to \infty$, $V \to \infty$, such that $\rho = N/V$ is fixed. Substitution of the trial field (21.103) into the general upper bound (21.61) yields integrals that involve G_2 and G_3. The form of the trial field (21.103) ensures the absolute convergence of the integrals resulting from the ensemble averages. The corresponding lower bound is found using the analogue of (21.103) for the trial stress field.

Markov (1998) generalized the Quintanilla–Torquato cluster bounds to the case of spheroids and, in particular, examined the limit in which the spheroids become cracks. As in the conductivity problem, the bounds coincide with the so-called "noninteracting" cracks approximation, thus rigorously proving that crack interactions always decrease the elastic moduli of a microcracked solid.

21.2.4 Security-Spheres Bounds

Conventional bounds on the effective moduli, such as contrast and cluster bounds, generally diverge from one another in the limit of infinite phase contrast, as we found for conventional bounds on the effective conductivity. For example, when phase 2 is perfectly rigid relative to phase 1, such finite-order upper bounds diverge to infinity (even if phase 2 is actually not percolating), since the bounds take into account realizations in which phase 2 is percolating. As noted earlier, the microstructural information contained in the bounds is incapable of accounting for phase connectivity. Similarly, when phase 2 is a cavity phase, finite-order conventional lower bounds vanish identically in order to take into account realizations in which only phase 2 is percolating.

In any of these limits, Torquato (1991b) has observed that the remaining bound can yield a useful estimate of the effective modulus, provided that certain conditions are met (see the discussion at the end of this chapter). The security-sphere concept introduced by Keller et al. (1967) provides a way to derive bounds that account for connectedness information in such a way that the bounds do not diverge in the limit of infinite phase contrast. Rubenfeld and Keller (1969) applied this idea to obtain bounds on the effective elastic moduli in three dimensions, which are described below.

Consider a statistically isotropic system of identical spheres of radius R of moduli K_2, G_2 in a matrix of moduli K_1, G_1. Surround the ith sphere in the system by a generally larger "security" sphere such that the concentric-shell region between the actual sphere and the surface of the security sphere contains only phase 1. In order to find an upper bound on the effective moduli K_e or G_e, Rubenfeld and Keller used a trial strain field $\hat{\varepsilon}$ taken to be the uniform strain ε_0 in the matrix region outside the security spheres. The fields within the security spheres are evaluated exactly by applying a displacement equal to $\varepsilon_0 \cdot \mathbf{x}$ on the boundary of each security sphere and requiring continuity of displacements and tractions at the two-phase interface $r = R$, where r is measured from the center of a typical sphere. The resulting trial energy then provides an upper bound on the effective moduli via the minimum energy principle. Taking ε_0 to be hydrostatic and deviatoric yields bounds on K_e and G_e, respectively. By constructing a trial stress in the same manner, Rubenfeld and Keller derived lower bounds on K_e and G_e using the minimum complementary energy principle.

Here we report their bulk-modulus results but generalized to arbitrary space dimension d for two limiting cases. For perfectly rigid inclusions ($K_2/K_1 = \infty$), the d-dimensional generalization of their upper bound is given by

$$\frac{K_e}{K_1} \leq 1 + \frac{K_1 + 2(d-1)G_1/d}{K_1}\phi_2 \int_0^\infty \frac{x^d}{x^d - 1} D H_P(x) dx, \tag{21.104}$$

where $D = 2R$ is the sphere diameter, $x = r/D$ is a dimensionless distance, ϕ_2 is the sphere volume fraction, and $H_P(r)$ is the nearest-neighbor probability density function. We do not give the corresponding lower bound, since conventional three-point lower bounds are sharper. Note that the function multiplying H_P in (21.104) contains a pole at $x = 1$, i.e., when the spheres touch. Therefore, unlike conventional upper bounds, the security-sphere upper bound (21.104) will remain finite in this perfectly rigid limit, provided that $H_P(x)$ vanishes sufficiently rapidly as $x \to 1$ to make the integral convergent. This will be true for nontouching periodic arrays [cf. (21.51)] as well as nontouching random arrays (see Chapter 3).

For spherical cavities ($K_2 = G_2 = 0$), the d-dimensional generalization of the Rubenfeld-Keller lower bound is given by

$$\frac{K_e}{K_1} \geq \left[1 + \frac{dK_1 + 2(d-1)G_1}{2(d-1)G_1}\phi_2 \int_0^\infty \frac{x^d}{x^d - 1} D H_P(x) dx\right]^{-1}. \tag{21.105}$$

The corresponding upper bound is not given, since conventional three-point upper bounds are superior. It is seen that unlike conventional lower bounds, the security-sphere lower bound (21.105) will remain nonzero in this cavity limit, provided that $H_P(x)$ vanishes sufficiently rapidly as $x \to 1$ to make the integral convergent.

21.3 Trapping Constant

In order to apply the variational bounds on the trapping constant γ derived in Chapter 14 for models of random media, we must choose admissible trial fields, substitute such trials fields into the bounds, and then perform the necessary averaging. In what follows, we derive five different types of bounds: interfacial-surface, void, cluster, security-spheres, and pore-size bounds. All of the bounds, except for the pore-size bounds (Prager 1963a, Torquato and Avellaneda 1991), were derived by Rubinstein and Torquato (1988) and Torquato and Rubinstein (1989). As in Chapter 14, we take phase 1 to be the trap-free, or "void," phase.

21.3.1 Interfacial-Surface Lower Bound

Rubinstein and Torquato (1988) rederived a lower bound on γ, obtained by Doi (1976), using the variational lower bound of Theorem 14.13. This derivation, presented below and generalized to arbitrary dimension d, is different from his, and in fact, his bound corresponds to a special choice of a trial field in the set D_L, (14.126), and not to a new variational principle as Doi stated. Specifically, we choose the trial concentration field

$$v_1(\mathbf{x}) = \alpha \int g(\mathbf{x}-\mathbf{y}) \cdot [\mathcal{I}(\mathbf{y}) - \xi(\mathbf{y})\mathcal{M}(\mathbf{y})]d\mathbf{y}, \tag{21.106}$$

where

$$g(\mathbf{r}) = \frac{1}{(d-2)\Omega r^{d-2}}, \qquad d \geq 3, \tag{21.107}$$

is the d-dimensional Green's function of the Laplacian operator Δ with $r = |\mathbf{r}|$ [cf. (20.11)], Ω is given by (2.56), $\mathcal{I}(\mathbf{x}) \equiv \mathcal{I}^{(1)}(\mathbf{x})$ is the void (pore) phase indicator function defined by (2.1), and $\mathcal{M}(\mathbf{x})$ is the interface indicator function defined by (2.3). The quantity $\xi(\mathbf{x})$ is an arbitrary function that is defined on the interfacial surface $\partial\mathcal{V}$. Accordingly, we refer to (21.106) as an *interfacial-surface* trial concentration field.

We now substitute (21.106) into the weaker variational bound (14.130) and compute the ensemble average $\langle \nabla v_1 \cdot \nabla v_1 \rangle$. In carrying out the averages, we will make use of the following two-point correlation functions:

$$F_{vv}(\mathbf{r}) = \langle \mathcal{I}(\mathbf{x})\mathcal{I}(\mathbf{x}+\mathbf{r}) \rangle, \tag{21.108}$$

$$F_{sv}(\mathbf{r}) = \langle \mathcal{M}(\mathbf{x})\mathcal{I}(\mathbf{x}+\mathbf{r}) \rangle, \tag{21.109}$$

$$F_{ss}(\mathbf{r}) = \langle \mathcal{M}(\mathbf{x})\mathcal{M}(\mathbf{x}+\mathbf{r}) \rangle. \tag{21.110}$$

These functions are called void–void, surface–void, and surface–surface correlation functions, respectively. Such correlation functions and their generalizations (e.g., F_{ssv}, F_{sss}) have been extensively studied (Torquato 1986a, Torquato 1986c) and are discussed in Chapters 2, 4, 5, and 6. Note that the void–void function F_{vv} is just the two-point probability function S_2 for the void phase defined in Chapter 2.

We are free to choose $\xi(\mathbf{x})$ subject to the constraint that $\langle \nabla v_1 \cdot \nabla v_1 \rangle$ is finite. For simplicity, we choose ξ to be a constant ξ_0 and obtain the bound

$$\gamma \geq \left\{ \int g(\mathbf{r}) \left[\xi_0^2 F_{ss}(\mathbf{r}) - 2\xi_0 F_{sv}(\mathbf{r}) + F_{vv}(\mathbf{r}) \right] d\mathbf{r} \right\}^{-1}. \tag{21.111}$$

Given the asymptotic behavior as $|\mathbf{r}| \to \infty$ of the correlation functions in relation (21.111) as described by (2.21) and (2.63), the only choice of ξ_0 for which the integral of (21.111) exists is

$$\xi_0 = \phi_1 / s. \tag{21.112}$$

Therefore, for statistically homogeneous media, we have the lower bound

$$\gamma \geq \left\{ \int \frac{1}{(d-2)\Omega r^{d-2}} \left[\frac{\phi_1^2}{s^2} F_{ss}(\mathbf{r}) - \frac{2\phi_1}{s} F_{sv}(\mathbf{r}) + F_{vv}(\mathbf{r}) \right] d\mathbf{r} \right\}^{-1}, \qquad d \geq 3. \tag{21.113}$$

For statistically isotropic media, the correlation functions depend only on the magnitude $r = |\boldsymbol{r}|$, and thus we get the lower bound

$$\gamma \geq \left\{\frac{1}{d-2}\int_0^\infty \left[\frac{\phi_1^2}{s^2}F_{ss}(r) - \frac{2\phi_1}{s}F_{sv}(r) + F_{vv}(r)\right] r\,dr\right\}^{-1}, \qquad d \geq 3. \tag{21.114}$$

This lower bound enables us to conclude that in the limit $d \to \infty$, $\gamma \to \infty$ (or, equivalently, the survival time τ vanishes) for arbitrary microstructures.

The interfacial-surface lower bound (21.113), originally derived by Doi for $d = 3$ using a different approach, is valid for any statistically homogeneous medium. Doi made the choice $\xi_0 = \phi_1/s$ after "optimizing" bound (21.111) over all possible ξ_0. However, as we have noted, any other choice for ξ_0 will provide a trivial bound (namely, $\gamma \geq 0$), so there is actually no room for optimization.

Nevertheless, (21.113) is the best possible lower bound on γ given void–void, surface–void, and surface–surface correlation functions for the medium. However, unlike bounds for conductivity and elastic moduli, it is not known what microstructures achieve the bounds (21.113) and (21.114).

We note that higher-order interfacial-surface bounds on γ can be derived by allowing the quantity ξ in (21.106) to have spatial dependence; see, for example, Given et al. (1990a).

21.3.2 Void Lower Bound

Torquato and Rubinstein (1989) considered the following admissible field in the set D_L, (14.126):

$$v_2(\boldsymbol{x}) = \frac{\alpha}{\phi_2}\int g(\boldsymbol{x}-\boldsymbol{y})[\mathcal{I}(\boldsymbol{y}) - \phi_1]d\boldsymbol{y}. \tag{21.115}$$

This is to be contrasted with the interfacial-surface trial field (21.106) with $\xi = \phi_1/s$, i.e.,

$$v_1(\boldsymbol{x}) = \alpha\int g(\boldsymbol{x}-\boldsymbol{y})\left[\mathcal{I}(\boldsymbol{y}) - \frac{\phi_1}{s}\mathcal{M}(\boldsymbol{y})\right]d\boldsymbol{y}. \tag{21.116}$$

Consider the terms within the brackets of (21.115) and (21.116). Although the first terms are the same to within a factor of ϕ_2, the second term of (21.115), unlike that of (21.116), does not involve interfacial information. For statistically homogeneous but anisotropic media, substitution of (21.115) into (14.127) yields

$$\gamma \geq \left\{\frac{1}{\phi_2^2}\int \frac{1}{(d-2)\Omega r^{d-2}}[F_{vv}(\boldsymbol{r}) - \phi_1^2]d\boldsymbol{r}\right\}^{-1}, \qquad d \geq 3, \tag{21.117}$$

which is the d-dimensional generalization of the Torquato–Rubinstein void lower bound. Therefore, for general isotropic media, we have

$$\gamma \geq \left\{\frac{1}{(d-2)\phi_2^2}\int_0^\infty [F_{vv}(r) - \phi_1^2]r\,dr\right\}^{-1}, \qquad d \geq 3. \tag{21.118}$$

21.3.3 Cluster Lower Bounds

If the medium is composed of a distribution of inclusions, we can construct trial fields that are based on the field solutions for a single inclusion, pairs of inclusions, etc. We refer to them here as cluster bounds (as in previous sections), but they have also been referred to as *multiple-scattering* bounds. For the case of a distribution of N identical spheres of radius R, Rubinstein and Torquato (1988) constructed the following single-scatterer trial field:

$$v_3(\boldsymbol{x}) = \alpha\left[\int g(\boldsymbol{x}-\boldsymbol{y})d\boldsymbol{y} - a\sum_{i=1}^{N} g(\boldsymbol{x}-\boldsymbol{r}_1)\chi(|\boldsymbol{x}-\boldsymbol{r}_i|)\right], \qquad (21.119)$$

where

$$\chi(r) = \begin{cases} 0, & r < R, \\ 1, & r \geq R, \end{cases} \qquad (21.120)$$

is the indicator function for the exterior of a single sphere, $\boldsymbol{r}_i$ is the position of the ith sphere, and a is a parameter to be optimized. The trial field (21.119) in combination with (21.107) for g enables us to generalize the results of Rubinstein and Torquato to any dimension $d \geq 3$.

It turns out that the only choice for which the energy $\langle \nabla v_3 \cdot \nabla v_3 \mathcal{I}\rangle$ is finite is $a = 1/\rho$, where ρ is the number density of the spheres. Trial functions of this type have been employed in variational bounds on the effective conductivity (Section 21.1.3), effective elastic moduli (Section 21.2.3), and the fluid permeability (Section 21.4.3).

Substitution of (21.119) into the weaker variational bound (14.130) then yields the optimized two-point lower bound

$$\gamma \geq \left[\frac{1}{\rho}\int \chi(z_1)\nabla g(z_1)\cdot\nabla g(z_1)d\boldsymbol{r}_1 + \int \chi(z_1)\chi(z_2)h(\boldsymbol{r}_{12})\nabla g(z_1)\cdot\nabla g(z_2)d\boldsymbol{r}_1 d\boldsymbol{r}_2\right]^{-1},$$

where $h(\boldsymbol{r})$ is the total (pair) correlation function defined by (3.30), $\boldsymbol{z}_i = \boldsymbol{x}-\boldsymbol{r}_i, z_i = |\boldsymbol{z}_i|$, and $\boldsymbol{r}_{ij} = \boldsymbol{r}_j - \boldsymbol{r}_i$. The first integral in this inequality can be evaluated explicitly, yielding the bound

$$\gamma \geq \left[\frac{R^2}{d(d-2)\phi_2} + \int \chi(z_1)\chi(z_2)h(\boldsymbol{r}_{12})\nabla g(z_1)\cdot\nabla g(z_2)d\boldsymbol{r}_1 d\boldsymbol{r}_2\right]^{-1}, \qquad d \geq 3. \qquad (21.121)$$

The term $d(d-2)\phi_2/R^2$ in (21.121) is the exact Smoluchowski dilute-limit result (19.94). The integral in (21.121) is absolutely convergent, provided that $h(\boldsymbol{r})$ tends to zero more rapidly than r^{-2} as $r \to \infty$ for any $d \geq 3$ (a very mild condition).

Combination of (21.119) and (14.127) yields the following optimized three-point lower bound:

$$\gamma \geq \left[\frac{1}{\rho^2}\int G_2(z_1)|\nabla g(z_1)|^2 d\boldsymbol{r}_1 + \frac{1}{\rho^2}\int\int Q_3(z_1,z_2)\nabla g(z_1)\cdot\nabla g(z_2)d\boldsymbol{r}_1 d\boldsymbol{r}_2\right]^{-1}, \qquad d \geq 3, \qquad (21.122)$$

where

$$Q_3(z_1, z_2) = G_3(\boldsymbol{x}; \boldsymbol{r}_1, \boldsymbol{r}_2) - \rho G_2(\boldsymbol{x}; \boldsymbol{r}_1) - \rho G_2(\boldsymbol{x}; \boldsymbol{r}_2) + \rho^2 \phi_1, \tag{21.123}$$

and $G_n(\boldsymbol{x}; \boldsymbol{r}^q)$ is the point/q-particle correlation function (Torquato 1986b) defined in Chapter 2. For virtually all realistic media without long-range order, the asymptotic behavior of Q_3 for large separation of the points ensures the absolute convergence of the second integral of (21.122). An explicit representation of the G_n for equisized spheres is given in Section 4.4.

There is an interesting relation between the interfacial-surface trial field (21.116) and the cluster trial field (21.119). Consider a distribution of N totally impenetrable spheres ($\lambda = 1$), and restrict $v_1(\boldsymbol{x})$ to $\boldsymbol{x} \in \mathcal{V}_1$. Then (21.116) yields

$$\begin{aligned} v_1(\boldsymbol{x}) &= \alpha \left[\int_{\mathcal{V}_1} g(\boldsymbol{x} - \boldsymbol{y}) d\boldsymbol{y} - \frac{\phi_1}{\Omega R^{d-1} \rho} \int_{\partial \mathcal{V}} g(\boldsymbol{x} - \boldsymbol{y}) dS_y \right] \\ &= \alpha \left[\int_{\mathcal{V}_1} g(\boldsymbol{x} - \boldsymbol{y}) d\boldsymbol{y} - \frac{\phi_1}{\rho} \sum_{i=1}^{N} g(\boldsymbol{x} - \boldsymbol{r}_i) \right], \end{aligned} \tag{21.124}$$

where we have used the mean value theorem for harmonic functions. Equation (21.124) is strikingly similar (but *not* identical) to (21.119). The advantage of using (21.124) and not (21.119) is not obvious when there is a large degree of penetrability of the spheres (e.g., small λ in the cherry-pit model discussed in Chapter 3), since the union of N nonoverlapping sphere surfaces does not represent the surface of the medium $\partial \mathcal{V}$ well in such instances. This explains why, for example, the three-point multiple scattering bound (21.122) is poorer than the two-point interfacial bound (21.114) for $\lambda = 0$ [see relation (22.72)]. Therefore, bounds that incorporate extra microstructural information on the medium are not necessarily the sharpest. Nonetheless, within a certain class of trial fields (interfacial-surface fields, etc.), increasing the level of information leads to improved bounds. As the degree of impenetrability increases, however, the union of N nonoverlapping sphere surfaces better represents the surface $\partial \mathcal{V}$. For large λ, therefore, the three-point cluster bound is superior to the two-point interfacial-surface bound because the former contains a greater amount of statistical information (see Section 22.3.3).

The interfacial-surface and cluster bounds each have their own merits. The main advantage of the former is that it can be applied to media of arbitrary geometry (not just models such as distributions of spheres). On the other hand, the cluster bounds can be applied to any system composed of distributions of inclusions and can be systematically upgraded to include higher-order multiple-inclusion solutions.

21.3.4 Security-Spheres Upper Bound

The construction of a trial concentration field v for the upper bound (14.123) is more problematic than that for the lower bound because of the condition that $u = 0$ on the interface $\partial \mathcal{V}$. Rubinstein and Torquato (1988) considered constructing a trial field for a distribution of N identical spheres of radius R. Let the distance between the ith sphere and its nearest neighbor be denoted by $2\tilde{R}_i$. In addition, we assume $\tilde{R}_i > R$ for all i. It is more convenient here to use a volume-average formulation. By invoking an ergodic hypothesis, we can equate volume and ensemble averaging. A trial field $v \in D_U$ [where D_U is specified by (14.122)] is chosen as follows: For every sphere i, we consider the domain defined by itself and a concentric "security" sphere of radius $\tilde{R}_i$. In that domain we solve

$$\begin{aligned} \Delta v_i(\boldsymbol{x}) = 0 \quad & \text{in} \quad R < |\boldsymbol{x} - \boldsymbol{r}_i| < \tilde{R}_i, \\ v_i = 0 \quad & \text{on} \quad |\boldsymbol{x} - \boldsymbol{r}_i| = R, \\ v_i = \xi \quad & \text{on} \quad |\boldsymbol{x} - \boldsymbol{r}_i| = \tilde{R}_i. \end{aligned} \tag{21.125}$$

The trial field v is chosen to be equal to v_i in the ith security shell and to be ξ elsewhere. Finally, we choose ξ such that $\langle v \rangle = \langle f \rangle$. The security-spheres method has been employed to find bounds on the viscosity of suspensions (Keller et al. 1967), elastic moduli (Rubenfeld and Keller 1969) and conductivity (Torquato and Rubinstein 1991) of composites, and permeability of porous media (Rubinstein and Torquato 1989).

Substitution of the solution of (21.125) into the inequality (14.123) and elimination of ξ yields in three dimensions

$$\frac{\gamma}{\gamma_s} \le \frac{\dfrac{1}{N}\displaystyle\sum_{i=1}^{N} e(b_i)}{\left[1 - \dfrac{\phi_2}{2}\dfrac{1}{N}\displaystyle\sum_{i=1}^{N} f(b_i)\right]^2}, \tag{21.126}$$

$$e(x) = \frac{x}{x-1}, \qquad f(x) = x(1+x),$$

where $b_i = \tilde{R}_i/R$, $\phi_2 = 4\pi R^3\rho/3$, ρ is the sphere number density, and $\gamma_s = 3\phi_2/R^2$ is the Smoluchowski dilute limit. Using the law of large numbers, we can write (21.126) as

$$\frac{\gamma}{\gamma_s} \le \frac{\displaystyle\int_1^\infty e(x) D H_P(x)\,dx}{\left[1 - \dfrac{\phi_2}{2}\displaystyle\int_1^\infty f(x) D H_P(x)\,dx\right]^2}, \tag{21.127}$$

where $D = 2R$ is the sphere diameter, $x = r/D$ is a dimensionless distance, ϕ_2 is the sphere volume fraction, and $H_P(r)$ is the nearest-neighbor probability density function (having dimensions of inverse length). Observe that the function $e(x)$ in the integrand of (21.127) contains a pole at $x = 1$, i.e., when the spheres touch. The upper bound (21.127) is nontrivially finite, provided that $H_P(x)$ vanishes sufficiently rapidly as $x \to 1$ to make the integral in the numerator convergent. This will be true for nontouching periodic arrays [cf. (21.51)] as well as nontouching random arrays (see Chapter 3).

Torquato and Rubinstein (1989) obtained an improved security-spheres bound by satisfying the Poisson equation $\Delta v_i(\boldsymbol{x}) = \alpha$, instead of the Laplace equation of (21.125).

By optimizing the constant α, they derived a bound that improves on the choice $\alpha = 0$, i.e., on (21.127).

21.3.5 Pore-Size Upper Bound

Prager (1963a) found an upper bound on the trapping constant (or, equivalently, a lower bound on the mean survival time τ) for perfectly absorbing traps in terms of the first moment of the pore-size density function $P(\delta)$, i.e.,

$$\langle \delta \rangle = \int_0^\infty \delta P(\delta) d\delta. \tag{21.128}$$

Torquato and Avellaneda (1991) generalized Prager's result for partially absorbing traps, i.e., finite surface reaction rate κ, as described below.

Consider a trial concentration field of the form

$$v(\boldsymbol{x}) = \frac{\langle f \rangle \psi(\delta)}{\phi_1 \int_0^\infty \psi(\delta) P(\delta) d\delta}, \tag{21.129}$$

where $\delta(\boldsymbol{x})$ is the minimum distance to the pore–solid interface from a point $\boldsymbol{x}$ in the pore space (not a Dirac delta function) and $\psi(\delta)$ is some deterministic function. We emphasize, however, that δ is a random function of $\boldsymbol{x}$, since it varies from point to point in a stochastic fashion. Insertion of the trial field (21.129) into the variational bound of Theorem 14.12 yields

$$\tau \mathcal{D} \geq \frac{[\int_0^\infty \psi(\delta) P(\delta) d\delta]^2}{\int_0^\infty (d\psi/d\delta)^2 P(\delta) d\delta + (\kappa s/\mathcal{D}\phi_1)\psi^2(0)}, \tag{21.130}$$

where s is the specific surface and $\mathcal{D}$ is the diffusion coefficient.

Let the deterministic function ψ be given by

$$\psi(\delta) = a\delta + b, \tag{21.131}$$

where a and b are constants to be optimized. Substitution of (21.131) into (21.130) yields the *optimized lower bound* on τ:

$$\tau \geq \frac{\langle \delta \rangle^2}{\mathcal{D}} + \frac{\phi_1}{\kappa s}, \tag{21.132}$$

which is valid in any dimension d. For $\kappa \to \infty$, (21.132) reduces to the diffusion-controlled-limit bound obtained by Prager (1963a). As expected, a finite value of κ yields a survival time bound that is larger than the one for the diffusion-controlled limit.

Remark:

1. Torquato and Avellaneda (1991) found a lower bound on the principal diffusion relaxation time T_1 for arbitrary κ using the variational principle (14.132) that is the analogue of (21.132). In the reaction-controlled and diffusion-controlled regimes, the bound asymptotically yields

$$T_1 \geq \frac{\phi_1}{\kappa s} + \frac{2(2\langle\delta\rangle^2 - \langle\delta^2\rangle)}{\mathcal{D}}, \qquad \frac{\kappa}{\mathcal{D}s} \ll 1, \tag{21.133}$$

and

$$T_1 \geq \frac{\langle\delta^2\rangle}{\mathcal{D}} + \frac{3\phi_1\langle\delta\rangle^2}{4\kappa s\langle\delta^2\rangle}, \qquad \frac{\kappa}{\mathcal{D}s} \gg 1, \tag{21.134}$$

respectively, where $\langle\delta^2\rangle$ is the second moment of the pore-size density function $P(\delta)$.

21.4 Fluid Permeability

To derive explicit rigorous bounds on the fluid permeability k for models of random porous media, we must construct admissible trial fields that are contained in the sets (14.144) or (14.147), substitute such trial fields into the appropriate variational bound, and finally perform the necessary averaging. In what follows, we derive four different types of bounds: interfacial-surface, void, cluster, and security-spheres bounds. All of the bounds were derived by Rubinstein and Torquato (1989), and it is their approach that we follow here. As in Chapter 14, we take phase 1 to be the fluid, or "void," phase.

21.4.1 Interfacial-Surface Upper Bound

Rubinstein and Torquato (1989) rederived an upper bound on k first obtained by Doi (1976) using the variational principle of Theorem 14.16. This derivation, given below and generalized to any dimension $d \geq 3$, is different from his, corresponding to a special choice of a trial field in the set E_U, (14.144), and not to a new variational principle as Doi stated. In particular, we use the trial velocity field

$$\boldsymbol{q}_1(\boldsymbol{x}) = \beta \int \boldsymbol{\Psi}(\boldsymbol{x}-\boldsymbol{y}) \cdot [\boldsymbol{e}\mathcal{I}(\boldsymbol{y}) - \boldsymbol{\xi}(\boldsymbol{y})\mathcal{M}(\boldsymbol{y})]d\boldsymbol{y}, \tag{21.135}$$

where $\boldsymbol{\Psi}$ is the d-dimensional Green's function (second-order tensor) associated with the velocity for Stokes flow given by

$$\boldsymbol{\Psi}(\boldsymbol{r}) = \frac{d}{(d^2-3)\Omega}\left[\frac{\boldsymbol{I}}{r^{d-2}} + \frac{\boldsymbol{nn}}{r^{d-2}}\right], \qquad d \geq 3, \tag{21.136}$$

Ω is given by (2.56), $r = |\boldsymbol{r}|$, $\boldsymbol{n} = \boldsymbol{r}/r$, $\mathcal{I}(\boldsymbol{x}) \equiv \mathcal{I}^{(1)}(\boldsymbol{x})$ is the void phase indicator function defined by (2.1), and $\mathcal{M}(\boldsymbol{x})$ is the interface indicator function defined by (2.3). We see that $\boldsymbol{\Psi}$ is proportional to the Oseen tensor (Section 17.4.1). The vector field $\boldsymbol{\xi}(\boldsymbol{x})$ defined on the interfacial surface $\partial\mathcal{V}$ is arbitrary. Accordingly, in analogy with the trapping problem, we refer to (21.135) as an *interfacial-surface* trial velocity field.

We are free to choose $\boldsymbol{\xi}$ subject to the constraint that $\langle\nabla\boldsymbol{q}_1{:}\nabla\boldsymbol{q}_1\rangle$ is finite. For simplicity, we choose $\boldsymbol{\xi}$ to be a constant vector $\boldsymbol{\xi}_0$. Substituting (21.135) into the weaker variational bound (14.149) and computing the ensemble average $\langle\nabla\boldsymbol{q}_1 : \nabla\boldsymbol{q}_1\rangle$ gives

$$k \leq \int [\boldsymbol{e}\cdot\boldsymbol{\Psi}(\boldsymbol{r})\cdot\boldsymbol{e}F_{vv}(\boldsymbol{r}) - 2\boldsymbol{\xi}_0\cdot\boldsymbol{\Psi}(\boldsymbol{r})\cdot\boldsymbol{e}F_{sv}(\boldsymbol{r}) + \boldsymbol{\xi}_0\cdot\boldsymbol{\Psi}(\boldsymbol{r})\cdot\boldsymbol{\xi}_0F_{ss}(\boldsymbol{r})]d\boldsymbol{r}, \tag{21.137}$$

where F_{vv}, F_{sv}, and F_{ss} are the void–void, surface–void, and surface–surface correlation functions, defined by relations (21.108)–(21.110), that arose in the interfacial-surface bound on the trapping constant and were discussed in Chapters 2, 4, 5, and 6. Given the asymptotic behavior of the two-point correlation functions as $|\boldsymbol{r}| \to \infty$ [cf. (2.21) and (2.63)], it is seen that the only choice of $\boldsymbol{\xi}_0$ for which the integral of (21.137) exists is

$$\boldsymbol{\xi}_0 = \frac{\phi_1^2}{s}\boldsymbol{e}. \tag{21.138}$$

Therefore, substitution of (21.138) into (21.137) yields for statistically homogeneous media the upper bound

$$k \le \int \boldsymbol{e} \cdot \boldsymbol{\Psi}(\boldsymbol{r}) \cdot \boldsymbol{e} \left[\frac{\phi_1^2}{s^2} F_{ss}(\boldsymbol{r}) - \frac{2\phi_1}{s} F_{sv}(\boldsymbol{r}) + F_{vv}(\boldsymbol{r}) \right] d\boldsymbol{r}, \qquad d \ge 3. \tag{21.139}$$

For statistically isotropic media, the correlation functions depend only on the magnitude $r = |\boldsymbol{r}|$, and we have the upper bound

$$k \le \frac{d+1}{d^2-3} \int_0^\infty \left[\frac{\phi_1^2}{s^2} F_{ss}(r) - \frac{2\phi_1}{s} F_{sv}(r) + F_{vv}(r) \right] r\, dr, \qquad d \ge 3. \tag{21.140}$$

This upper bound enables us to conclude that in the limit $d \to \infty$, $k \to 0$ for arbitrary microstructures.

The two-point upper bound (21.139), first derived by Doi using a different approach, applies to any microstructure that is statistically homogeneous. Doi made the choice $\xi_0 = \phi_1/s$ after "optimizing" over all possible $\boldsymbol{\xi}_0$. As we have seen, however, any other choice for $\boldsymbol{\xi}_0$ provides the trivial bound $k \le \infty$, so there is actually no room for optimization. Note that apart from a trivial constant factor, the isotropic upper bound (21.140) on k is identical to the upper bound on the inverse trapping constant γ^{-1} obtained from (21.114). The relationship between the permeability and trapping constant is discussed in Chapter 23.

As in the interfacial-surface bound on the trapping constant, (21.139) is the best possible lower bound on k given void–void, surface–void, and surface–surface correlation functions for the medium. However, in contrast to bounds on the conductivity and elastic moduli, it is not known what microstructures achieve the bounds (21.139) and (21.140).

Observe that higher-order interfacial-surface bounds on k can be derived by allowing the quantity $\boldsymbol{\xi}$ in (21.135) to have spatial dependence.

21.4.2 Void Upper Bound

Rubinstein and Torquato (1989) considered the following admissible field in the set E_U, (14.144):

$$\boldsymbol{q}_2(\boldsymbol{x}) = \frac{\beta}{\phi_2} \int \boldsymbol{\Psi}(\boldsymbol{x}-\boldsymbol{y}) \cdot \boldsymbol{e}[\mathcal{I}(\boldsymbol{y}) - \phi_1] d\boldsymbol{y}. \tag{21.141}$$

This is to be contrasted with the interfacial-surface trial field (21.135) with $\boldsymbol{\xi} = (\phi_1/s)\boldsymbol{e}$, i.e.,

$$\boldsymbol{q}_1(\boldsymbol{x}) = \beta \int \boldsymbol{\Psi}(\boldsymbol{x}-\boldsymbol{y})\cdot \boldsymbol{e}\left[\mathcal{I}(\boldsymbol{y}) - \frac{\phi_1}{s}\mathcal{M}(\boldsymbol{y})\right] d\boldsymbol{y}. \tag{21.142}$$

Consider the terms within the brackets of (21.141) and (21.142). Although the first terms are the same to within a factor of ϕ_2, the second term of (21.141), unlike that of (21.142), does not involve interfacial information. The surface integration indicated in (21.142) is required in order to obtain the correct dilute-limit result for spheres. Substitution of (21.141) into (14.145) yields

$$k \le \frac{d+1}{(d^2-3)\phi_2^2}\int_0^\infty [F_{vv}(r) - \phi_1^2] r\, dr, \qquad d \ge 3, \tag{21.143}$$

for general statistically isotropic media. This result generalizes the Rubinstein–Torquato void lower bound to any dimension $d \ge 3$.

For $d = 3$, the two-point void bound (21.143) is equal to the corrected form of a bound due to Prager (1961). Berryman and Milton (1985) corrected a coefficient in Prager's bound using a variational principle, different from (14.145), in which the normalization factor β is a stochastic rather than a deterministic quantity. Moreover, they used a volume-averaged approach. It is difficult to construct admissible fields that satisfy the required boundary conditions in the volume-average approach (Rubinstein and Torquato 1989). Nevertheless, the above derivation demonstrates that (21.143) is a rigorous upper bound on k.

21.4.3 Cluster Upper Bounds

If the medium is composed of a distribution of inclusions and flow is considered to occur in the region exterior to the inclusions (phase 1), we can construct trial fields that are based on the solutions for scattering from a single inclusion, pairs of inclusions, etc. Consistent with our previous terminology, we refer to bounds so obtained as cluster bounds, although they have also been referred to as "multiple-scattering" bounds. The following analysis generalizes the results of Rubinstein and Torquato (1989) for arrays of three-dimensional identical spheres with radius R to any dimension $d \ge 3$. The single-sphere scattering function $\boldsymbol{S}(\boldsymbol{r})$ in d dimensions is given by

$$\boldsymbol{S}(\boldsymbol{r}) = \left(1 + \frac{1}{2d}R^2\Delta\right)\boldsymbol{\Psi}(\boldsymbol{r}), \qquad d \ge 3, \tag{21.144}$$

where $\boldsymbol{\Psi}$ is the Stokeslet given by (21.136) and Δ is the Laplacian differential operator. Note that the velocity field outside a single sphere moving with velocity $\boldsymbol{U}$ in an otherwise quiescent viscous fluid is $\boldsymbol{S}\cdot\boldsymbol{U}$. Rubinstein and Torquato (1989) then constructed the following single-scatterer trial field:

$$\boldsymbol{q}_3(\boldsymbol{x}) = \beta\left[\int \boldsymbol{S}(\boldsymbol{x}-\boldsymbol{y})\cdot\boldsymbol{e}\, d\boldsymbol{y} - a\sum_{i=1}^{\infty}\boldsymbol{S}(\boldsymbol{x}-\boldsymbol{r}_i)\cdot\boldsymbol{e}\,\chi(|\boldsymbol{x}-\boldsymbol{r}_i|)\right], \tag{21.145}$$

where χ is the indicator function for the exterior of a single sphere given by (21.120), $\boldsymbol{r}_i$ denotes the position of the ith sphere, and a is a parameter to be optimized. It turns out that the only choice for which the energy $\langle \nabla \boldsymbol{q}_3 : \nabla \boldsymbol{q}_3 \mathcal{I} \rangle$ is finite is $a = 1/\rho$, where ρ is the number density of spheres. Trial fields of this kind have been utilized in variational bounds on the conductivity (Torquato 1986b) and elastic moduli (Quintanilla and Torquato 1995) of composite media and on the trapping constant of heterogeneous media (Rubinstein and Torquato 1988). Next we define the tensor $\boldsymbol{T}$ through

$$\boldsymbol{T} = \nabla(\boldsymbol{S} \cdot \boldsymbol{e}) + \nabla(\boldsymbol{S} \cdot \boldsymbol{e})^{\mathrm{T}}, \tag{21.146}$$

where the second term on the right-hand side is the transpose of the first term.

Substitution of (21.145) into the weaker variational bound (14.149) then yields the two-point upper bound

$$k \le \frac{1}{2\rho} \int \chi(z_1) \boldsymbol{T}(z_1) : \boldsymbol{T}(z_1) d\boldsymbol{r}_1 + \frac{1}{2} \int \int \chi(z_1)\chi(z_2) h(\boldsymbol{r}_{12}) \boldsymbol{T}(z_1) : \boldsymbol{T}(z_2) d\boldsymbol{r}_1 d\boldsymbol{r}_2,$$

where $h(\boldsymbol{r})$ is the total pair correlation function defined by (3.30), $z_i = \boldsymbol{x} - \boldsymbol{r}_i, z_i = |z_i|$, and $\boldsymbol{r}_{ij} = \boldsymbol{r}_j - \boldsymbol{r}_i$. The first integral in this inequality can be calculated explicitly, giving the bound

$$k \le \frac{(d+1)R^2}{d(d^2-3)\phi_2} + \frac{1}{2} \int \int \chi(y_1)\chi(y_2) h(\boldsymbol{r}_{12}) \boldsymbol{T}(z_1) : \boldsymbol{T}(z_2) d\boldsymbol{r}_1 d\boldsymbol{r}_2, \qquad d \ge 3. \tag{21.147}$$

The first term on the right side of (21.147) is the exact Stokes dilute-limit result (19.107). The second integral in (21.147) is absolutely convergent, provided that $h(\boldsymbol{r})$ tends to zero more rapidly than r^{-2} as $r \to \infty$ for any $d \ge 3$ (a very mild condition).

Substituting (21.145) into (14.145) gives the following three-point upper bound:

$$k \le \frac{1}{2\rho^2} \int G_2(z_1) \boldsymbol{T}(z_1) : \boldsymbol{T}(z_1) d\boldsymbol{r}_1 + \frac{1}{2\rho^2} \int \int Q_3(z_1, z_2) \boldsymbol{T}(z_1) : \boldsymbol{T}(z_2) d\boldsymbol{r}_1 d\boldsymbol{r}_2 \tag{21.148}$$

for $d \ge 3$, where Q_3 is a three-point function given by (21.123). Again, the asymptotic properties of Q_3 ensure the absolute convergence of the six-dimensional integral of (21.148) for virutally all realistic media without long-range order.

It should be noted that the cluster upper bound on k derived by Torquato and Beasley (1987) [using a probabilistic formulation of the Weissberg and Prager (1970) variational principle] is equivalent to (21.148), albeit expressed in a slightly different form. Torquato and Beasley found that

$$k \le \frac{\int G_2(z_1) \boldsymbol{T}(z_1) : \boldsymbol{T}(z_1) d\boldsymbol{r}_1 + \int \int Q_3(z_1, z_2) \boldsymbol{T}(z_1) : \boldsymbol{T}(z_2) d\boldsymbol{r}_1 d\boldsymbol{r}_2}{\Lambda^2},$$

where Λ is a multidimensional integral involving the trial stress field and the surface/particle correlation function F_{sp} (Section 2.10). The advantage of the variational principle leading to (21.148) is that the normalization factor is deterministic and does not have to be explicitly evaluated.

Finally, we mention that Beasley and Torquato (1989) showed that the bound (21.148) can be improved upon by using the same trial fields but with undetermined coefficients that were subsequently optimized. We refer to this result as the *optimized* three-point cluster bound.

21.4.4 Security-Spheres Lower Bound

It is a challenge to obtain lower bounds on the permeability k that are not trivially equal to zero, corresponding to a *nonpercolating fluid phase*, because of the no-slip and incompressibility conditions on the trial velocity field. Rubinstein and Torquato (1989) employed the security-spheres method to construct trial fields in the set E_L, specified by (14.147), to find nonzero lower bounds. This method goes back to Keller et al. (1967), who used it to estimate the effective viscosity of suspensions. It has been used by Rubinstein and Keller (1987) to derive upper bounds on the effective drag coefficient.

Consider a distribution of N identical spheres with radius R. Let the distance between the ith sphere and its nearest neighbor be denoted by $2\tilde{R}_i$. For every sphere i, we consider the domain defined by itself and a concentric security sphere of radius $\tilde{R}_i$. In that domain we solve

$$\begin{aligned}
\Delta \boldsymbol{q}^i &= \nabla p^i \quad \text{in} \quad R < |\boldsymbol{x} - \boldsymbol{r}_i| < \tilde{R}_i, \\
\nabla \cdot \boldsymbol{q}^i &= 0 \quad \text{in} \quad R < |\boldsymbol{x} - \boldsymbol{r}_i| < \tilde{R}_i, \\
\boldsymbol{q}^i &= 0 \quad \text{on} \quad |\boldsymbol{x} - \boldsymbol{r}_i| = R, \\
\boldsymbol{q}^i &= \boldsymbol{U} \quad \text{on} \quad |\boldsymbol{x} - \boldsymbol{r}_i| = \tilde{R}_i,
\end{aligned} \tag{21.149}$$

where $\boldsymbol{U}$ is an arbitrary constant vector. A trial field $\boldsymbol{q}(\boldsymbol{x})$ belonging to the set (14.147) is defined to be equal to $\boldsymbol{q}^i$ for $R < |\boldsymbol{x} - \boldsymbol{r}_i| < \tilde{R}$ and equal to $\boldsymbol{U}$ when x is outside all the security spheres. Without loss of generality, we choose $\boldsymbol{e}$ and $\boldsymbol{U}$ in the positive z-direction. The incompressibility of the flow in *three dimensions* implies

$$\int_{R<|\boldsymbol{x}-\boldsymbol{r}_i|<\tilde{R}_i} q_z \boldsymbol{e}\, d\boldsymbol{x} = \frac{4}{3}\pi \tilde{R}_i^3 \boldsymbol{U}. \tag{21.150}$$

The volume-average formulation is convenient here. By invoking an ergodic hypothesis, we can equate volume and ensemble averaging.

By utilizing the solution of (21.149), the numerator in (14.147) is easily shown to be equal to $|\boldsymbol{U}|^2$. The only contributions to the rate of energy dissipation are from the security spheres. In the kth security shell, we find in three dimensions that

$$\int_{R<|\boldsymbol{x}-\boldsymbol{r}_i|<\tilde{R}_i} \nabla \boldsymbol{q}^i : \nabla \boldsymbol{q}^i d\boldsymbol{x} = 6\pi R |\boldsymbol{U}|^2 f(b_i), \tag{21.151}$$

where $b_i = R/\tilde{R}_i$ is a dimensionless distance and

$$f(x) = (1 - x^5)\left(1 - \frac{9}{4}x + \frac{5}{2}x^3 - \frac{9}{4}x^5 + x^6\right)^{-1}. \tag{21.152}$$

Hence, we obtain from the variational lower bound (14.147) that

$$k \geq \frac{1}{6\pi R\rho}\left[\lim_{N\to\infty}\frac{1}{N}\sum_{i=1}^{N} f(b_i)\right]^{-1}, \tag{21.153}$$

where ρ is the sphere number density. Scaling k with the Stokes permeability $k_s = (6\pi R\rho)^{-1}$ and using the law of large numbers, we have in three dimensions that

$$\frac{k}{k_s} \geq \left[\int_0^\infty f(x) D H_P(x) dx\right]^{-1}, \tag{21.154}$$

where $D = 2R$ is the sphere diameter, $x = r/D$ is a dimensionless distance, ϕ_2 is the sphere volume fraction, and $H_P(r)$ is the nearest-neighbor probability density function (having dimensions of inverse length). Note that the function $f(x)$ in the integrand of (21.154) contains a pole at $x = 1$, i.e., when the spheres touch. The lower bound (21.154) will not be trivially equal to zero, provided that $H_P(x)$ vanishes sufficiently rapidly as $x \to 1$ to make the integral convergent. This will be true for nontouching periodic arrays [cf. (21.51)] as well as nontouching random arrays (see Chapter 3).

The bound (21.154) coincides with the Rubinstein–Keller bound on the effective drag coefficient. This is, however, merely a coincidence and results from our particular choice of $\boldsymbol{q}^i$ (21.149) and the identity (21.151). Other trial fields [e.g., in the spirit of Torquato and Rubinstein (1989)] would yield improved bounds.

21.5 Structural Optimization

In the study of random media, it is useful not only to seek the range of possible values that the effective property can take given limited morphological information (i.e., rigorous bounds) but also to identify the microstructures that correspond to the extreme values, i.e., to determine whether the bounds are optimal. Such investigations have important implications for structural optimization (Kohn 1988, Kohn 1989, Cherkaev 2000). This topic was touched upon in this chapter, but an in-depth discussion is beyond the scope of this monograph. It is useful to make a few comments, however. As we have already discussed, an extensive literature has developed that describes the realizability of bounds on $\boldsymbol{\sigma}_e$ and $\boldsymbol{C}_e$ by laminates, indicating the importance of laminates in modeling composites; see the review of Milton (1986). The class of sequential laminates that achieves the bounds is larger than the aforementioned coated-inclusion assemblages that realize bounds.

Heretofore, microstructures that realize the aforementioned bounds on the trapping constant γ and fluid permeability tensor $\boldsymbol{k}$ have not been identified. The difficulty in identifying optimal microstructures for these classes of problems lies in the fact that γ and $\boldsymbol{k}$ are length-scale dependent quantities and known bounds on γ and $\boldsymbol{k}$ depend nontrivially on the specific forms of two-point and higher-order correlation functions.

The so-called G-closures are of fundamental importance to solving optimization problems. Given a set U of conductivity (or stiffness) tensors corresponding to various

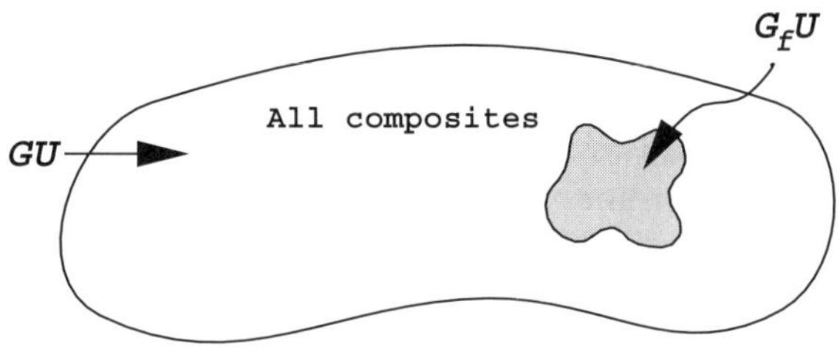

Figure 21.2 A schematic depiction of the G-closure and G_f-closure.

material phases, the G-closure of U, GU, is defined as the complete set of effective conductivity (or stiffness) tensors associated with composites formed from these phases in any orientation (see Figure 21.2). The G-closures of a variety of sets have been determined (Tartar 1978, Lurie and Cherkaev 1984, Murat and Tartar 1985, Lurie and Cherkaev 1986).

When the phase volume fractions are also specified, the effective tensors are restricted to a smaller set G_fU (see Figure 21.2). Accordingly, G_fU is defined as the complete set of effective conductivity (or stiffness) tensors of all composites formed from these phases in prescribed proportions, i.e., for fixed volume fractions. Whereas the G_f-closure for the effective conductivity tensor $\boldsymbol{\sigma}_e$ of two-phase composites consisting of isotropic phases is known (Tartar 1978, Lurie and Cherkaev 1984, Murat and Tartar 1985, Lurie and Cherkaev 1986), the same is not true for multiphase composites. Much less is known about the G_f-closure for the stiffness tensor $\boldsymbol{C}_e$. It is known for macroscopically isotropic two-dimensional composites composed of two incompressible isotropic phases (Lipton 1988). However, for general macroscopically isotropic two-dimensional composites made of two isotropic phases, the G_f-closure is not completely known (Cherkaev and Gibiansky 1993). The reader is referred to Milton and Nesi (1999), Cherkaev (2000), and references therein for more recent developments.

A promising numerical technique for the systematic design of composites as well as smart material systems is the *topology optimization method*. This numerical method was initiated by Bendsøe and Kikuchi (1988) originally for the design of mechanical structures. The basic topology optimization problem can be stated as follows: Distribute a given amount of material in a design domain such that an objective function is extremized, subject to certain constraints (Bendsøe and Kikuchi 1988, Bendsøe 1995). The objective function can be any combination of the individual components of the relevant effective property tensor. The topology optimization method has been used to design composites with exotic properties, such as negative Poisson's ratio (Sigmund 1994, Sigmund et al. 1998) and negative thermal expansion (Sigmund and Torquato 1997). More recently, it has been adapted to find composite microstructures with targeted properties (Hyun and Torquato 2001, Torquato and Hyun 2001): one of the Holy Grails of materials science.

21.6 Utility of Bounds

How useful are bounds when they diverge from one another? One of the bounds may still provide useful estimates in such cases. To fix ideas, let us first consider the isotropic conduction problem. It has been observed (Torquato 1985a, Torquato 1991b) that because of the aforementioned correspondence between the *conventional* n-point bounds on σ_e and certain realizable geometries, lower-order *lower* bounds are expected to yield good estimates of σ_e/σ_1 for $\sigma_2 \gg \sigma_1$, provided that the highly conducting phase is not percolating (i.e., $\phi_2 < \phi_{2c}$, where ϕ_{2c} is the percolation threshold) and the characteristic *cluster* size of phase 2, Λ_2, is much smaller than the macroscopic length scale L. For dispersions of particles, Λ_2 can be estimated from the *average cluster volume* defined in Section 9.2.1. Note that the condition $\Lambda_2 \ll L$ alone implies $\phi_2 < \phi_{2c}$. For spatially periodic or equilibrium arrays of d-dimensional spheres, the condition $\Lambda_2 \ll L$ is satisfied for all ϕ_2 except at the close-packing or percolation threshold value ϕ_{2c}.

In summary, for general media, even though conventional upper bounds on σ_e are much larger than the lower bounds for $\sigma_2 \gg \sigma_1$ (and, in fact, go to infinity in the limit $\sigma_2/\sigma_1 \to \infty$), the lower-order lower bounds should give good estimates of σ_e/σ_1, provided that $\Lambda_2 \ll L$. The accuracy of the lower-order lower bounds increases as n increases. Similarly, conventional lower-order *upper* bounds are expected to yield useful estimates of σ_e/σ_1 for $\sigma_2 \gg \sigma_1$ given that $\phi_2 > \phi_{2c}$ and $\Lambda_1 \ll L$.

Very similar arguments apply to conventional lower-order elastic moduli bounds when one phase is much stiffer than the other (Torquato and Lado 1986, Sen, Lado and Torquato 1987a). The claim that one of the improved bounds can provide relatively sharp estimates of the effective conductivities and elastic moduli for a wide range of conditions is corroborated by specific calculations given in Chapter 22.

Statements regarding the utility of lower-order bounds on the trapping constant γ and fluid permeability tensor $\boldsymbol{k}$ cannot be made as strong as in the conduction and elasticity problems because geometries corresponding to the bounds have yet to be identified. However, with some restrictions, one of the bounds on γ or k will provide a useful estimate of the property for a range of volume fractions, as will be shown in Chapter 22.

In the subsequent chapter we will compute bounds on the effective properties for a variety of different microstructures. Examples will be presented that support the aforementioned statements concerning the utility of bounds.

CHAPTER 22

Evaluation of Bounds

Although some improved bounds have been in existence for nearly four decades, they have, until the recent past, lain dormant and untested because of the difficulty associated with ascertaining the various types of correlation functions involved, even for simple models (e.g., random arrays of spherical particles). Advances in the quantitative characterization of the microstructure, described in Part I of this book, have paved the way for the computation of improved bounds on the effective conductivity, effective elastic moduli, trapping constant, and fluid permeability of nontrivial models of two-phase random heterogeneous materials.

Improved bounds have not only been shown to yield accurate estimates of effective properties for simple isotropic microstructures, where some investigators may prefer to use a simpler approximation scheme, but also of more complex microstructures, such as suspensions of polydisperse particles, anisotropic arrays of inclusions, bicontinuous media, and statistically inhomogeneous media. For these and other complex microstructures, the few available approximate formulas are of dubious value, since they typically account only for the barest of microstructural information. It is for complex microstructures that bounds have begun to make and will continue to make a significant practical impact.

In this chapter we will discuss the evaluation of improved bounds. It will be demonstrated that improved bounds, such as three- and four-point bounds, can provide relatively sharp estimates of the effective properties for a wide range of conditions. Even in cases when improved bounds are not tight, one of the bounds can provide a good estimate of the effective property. From a practical point of view, such predictive power is encouraging, since the measurement of five-point and higher-order correlation functions of real heterogeneous materials is beyond presently available technology.

The bounds on the properties will be compared to available experimental and computer-simulation results. No attempt will be made to cover the extensive literature concerning the direct calculation of effective properties of random media via computer simulations. However, it is useful to point out that there is a variety of simulation methods that have been utilized, including:

- Brownian motion methods (Schwartz and Banavar 1989, Torquato and Kim 1989, Kim and Torquato 1990b, Torquato, Kim and Cule 1999b)
- finite element methods (Ghosh and Mukhopadhyay 1993, Garboczi and Day 1995, Cruz, Ghaddar and Patera 1995, Ostoja-Starzewski and Wang 1999, Oden, Vemaganti and Moes 1999)
- finite difference methods (Adler 1992, Martys and Garboczi 1992, Riley, Muzzio and Reyes 1998)
- boundary element methods (Eischen and Torquato 1993, Helsing 1994a)
- multipole methods (Sangani and Yao 1988, Bonnecaze and Brady 1991, Rodin 1993, Greengard and Moura 1994, Cheng and Greengard 1997)
- Fourier-integral equation methods (Tao, Chen and Sheng 1990, Moulinec and Suquet 1998, Eyre and Milton 1999)
- lattice-Boltzmann methods (Cancelliere, Chang, Foti, Rothman and Succi 1990, Martys, Hagedorn, Goujon and Devaney 2000)

22.1 Conductivity

Since the early work on improved bounds on the effective conductivity (Prager 1961, Beran 1968, Miller 1969), significant progress has been made in the evaluation of such bounds for nontrivial model microstructures as well as real heterogeneous materials. In particular, we describe below specific evaluations of contrast, cluster, and security-spheres bounds on the effective conductivity of heterogeneous materials. Results are given for model microstructures as well as real heterogeneous media.

22.1.1 Contrast Bounds

As shown in Section 21.1.2, the quantity ζ_2, defined by (20.63) in any dimension d, is the key microstructural parameter that arises in the three-point contrast bounds (21.33) and (21.37) and, for $d = 2$, the four-point contrast bounds (21.42) on the effective conductivity of macroscopically isotropic two-phase media. For $d = 2$ and $d = 3$, this parameter simplifies according to (20.66) and (20.67), respectively.

Most of this section is devoted to the evaluation of ζ_2 and the associated bounds on σ_e for isotropic microstructures. We close with a brief discussion on the calculation of contrast bounds on the conductivity tensor for anisotropic microstructures.

ζ_2 for Dilute Dispersions

It is useful first to outline procedures for obtaining ζ_2 analytically or at least reducing the problem to quadratures. To begin, consider a random array of inclusions of fixed arbitrary shape with a size distribution (phase 2) in a matrix (phase 1). Let us suppose that we are given the Taylor series expansion for the effective conductivity about $\phi_2 = 0$, i.e.,

$$\frac{\sigma_e}{\sigma_1} = 1 + B_1\phi_2 + B_2\phi_2^2 + \cdots. \tag{22.1}$$

The coefficients B_i are independent of the volume fraction but depend on the ratio σ_2/σ_1. Torquato (1985b) used the low-density expansion (22.1) to find the corresponding low-density expansion of the parameter ζ_2, i.e.,

$$\zeta_2 = \zeta_2^{(0)} + \zeta_2^{(1)}\phi_2 + \cdots. \tag{22.2}$$

To illustrate this methodology, we will show how to obtain the leading-order term $\zeta_2^{(0)}$. Since the coefficient B_1 in (22.1) is analytic in σ_2/σ_1, we can expand it in powers of $\delta \equiv \sigma_2/\sigma_1 - 1$, i.e.,

$$B_1 = B_1^{(1)}\delta + B_1^{(2)}\delta^2 + B_1^{(3)}\delta^3 + \mathcal{O}(\delta^4), \tag{22.3}$$

and thus, through first order in ϕ_2,

$$\frac{\sigma_e}{\sigma_1} = 1 + B_1^{(1)}\phi_2\delta + B_1^{(2)}\phi_2\delta^2 + B_1^{(3)}\phi_2\delta^3 + \mathcal{O}(\delta^4). \tag{22.4}$$

Evaluating the weak-contrast relation (20.77) through first order in ϕ_2 gives

$$\frac{\sigma_e}{\sigma_1} = 1 + \phi_2\delta - \frac{\phi_2}{d}\delta^2 + \frac{\phi_2}{d^2}[1 + (d-1)\zeta_2^{(0)}]\delta^3 + \mathcal{O}(\delta^4). \tag{22.5}$$

Comparing (22.4) and (22.5), we obtain

$$B_1^{(1)} = 1, \qquad B_1^{(2)} = -\frac{1}{d}, \qquad B_1^{(3)} = \frac{1 + (d-1)\zeta_2^{(0)}}{d^2}, \tag{22.6}$$

and solving for the leading-order term in (22.2) gives

$$\zeta_2^{(0)} = \frac{d^2 B_1^{(3)} - 1}{d - 1}. \tag{22.7}$$

Because the coefficient B_1 is independent of the size distribution of the inclusions (see Section 19.1.1), $\zeta_2^{(0)}$ has the same property. The same procedure can be used to get the higher-order terms in (22.2) (Torquato 1985b), but of course this requires information about B_2, B_3, etc.

Since we know the first-order coefficient B_1 for a variety of inclusion shapes, we can compute $\zeta_2^{(0)}$ for these shapes via (22.7). For example, using formula (19.30) for a dilute array of randomly oriented ellipsoids in d dimensions, we have that

$$\zeta_2^{(0)} = \frac{\left[d\sum_{k=1}^{d}(A_k^*)^2\right] - 1}{d-1}, \qquad \text{(ellipsoids)}, \tag{22.8}$$

where the A_k^* are the depolarization factors. Using the results of Section 17.1.2 and (22.8) gives the ensuing limiting three-dimensional cases:

$$\zeta_2^{(0)} = 0 \qquad \text{(spheres)}, \tag{22.9}$$

$$\zeta_2^{(0)} = \frac{1}{4} \qquad \text{(needles)}, \tag{22.10}$$

$$\zeta_2^{(0)} = 1 \qquad \text{(disks)}. \tag{22.11}$$

Similarly, limiting cases of (22.8) in two dimensions are

$$\zeta_2^{(0)} = 0 \qquad \text{(disks)}, \tag{22.12}$$

$$\zeta_2^{(0)} = 1 \qquad \text{(needles)}. \tag{22.13}$$

One can also compute $\zeta_2^{(0)}$ for nonellipsoidal shapes. For example, it follows from the work of Helsing (1994b) on randomly stacked cubes that

$$\zeta_2^{(0)} = 0.11882 \qquad \text{(cubes)}. \tag{22.14}$$

Using the results of Hetherington and Thorpe (1992) for the coefficient B_1 of *regular* polygonal inclusions in the plane, we obtain

$$\zeta_2^{(0)} = 0.20428 \qquad \text{(triangles)}, \tag{22.15}$$

$$\zeta_2^{(0)} = 0.08079 \qquad \text{(squares)}, \tag{22.16}$$

$$\zeta_2^{(0)} = 0.02301 \qquad \text{(hexagons)}. \tag{22.17}$$

It is seen that as the number of sides increases, $\zeta_2^{(0)}$ approaches zero, the circular-cell result, as expected.

ζ_2 for Symmetric-Cell Materials

We now demonstrate that the results above for the leading-order term $\zeta_2^{(0)}$ can be used to obtain ζ_2 for *symmetric-cell* materials (Chapter 8) for *arbitrary* values of ϕ_2. Use of the general expression (8.35) for the three-point probability function $S_3^{(2)} \equiv \hat{S}_3$ of a symmetric-cell material and definition (20.63) immediately yields that

$$\zeta_2 = \phi_2 + (\phi_1 - \phi_2)g, \tag{22.18}$$

where

$$g = I\left[W_3^{(1)}\right], \tag{22.19}$$

$W_3^{(1)}$ is the three-point single-cell weight given by (8.17) or (8.19) with $n = 3$, and I is the integral operator defined by (20.64). The first term ϕ_2 in (22.18) arises after application of the properties (20.68) and (20.69). Since $W_3^{(1)}$ is independent of volume fraction, so is the parameter g. Thus, we can obtain g for different cell shapes by comparing (22.18) to lowest order in ϕ_2 to the leading-order (constant) term in the density expansion (22.2), which yields

$$g = \zeta_2^{(0)}. \tag{22.20}$$

Therefore, g has exactly the same properties as $\zeta_2^{(0)}$, i.e., it depends only on the cell shape (independent of cell size distribution and ϕ_2) and lies in the closed interval $[0, 1]$. The shape parameter g is related to but different from the shape parameter G introduced by Miller (1969) for $d = 3$ and by Beran and Silnutzer (1971) for $d = 2$ [see also Milton (1982)]; here we have $g = (9G - 1)/2$ for $d = 3$ and $g = 4G - 1$ for $d = 2$. We prefer to use the shape parameter g, since it is directly related to the single-cell weight $W_3^{(1)}$ and is exactly given by the leading term in the density expansion of ζ_2.

Observe that for *any* symmetric-cell material at $\phi_1 = \phi_2 = 1/2$, expression (22.18) yields the especially simple result that

$$\zeta_2 = \frac{1}{2}. \tag{22.21}$$

This result applies to any material possessing *phase-inversion symmetry* when $\phi_1 = \phi_2 = 1/2$ (Section 2.2.2).

Consider an isotropic symmetric-cell material made up of d-dimensional ellipsoidal cells. Such space-filling cells must be randomly oriented and possess a size distribution down to the infinitesimally small. Using (22.8), we see that for d-dimensional ellipsoidal cells,

$$g = \frac{\left[d \sum_{k=1}^{d} (A_k^*)^2\right] - 1}{d - 1} \qquad \text{(ellipsoidal cells).} \tag{22.22}$$

Thus, for d-dimensional spherical cells, $\zeta_2 = \phi_2$; for three-dimensional needle-like cells, $\zeta_2 = \phi_1/4 + 3\phi_2/4$; for three-dimensional disk-like cells or two-dimensional needle-like cells, $\zeta_2 = \phi_1$. For symmetric-cell materials with cubical cells ($d = 3$) (Helsing 1994b),

$$\zeta_2 = 0.11882\phi_1 + 0.88118\phi_2 \qquad \text{(cubical cells).} \tag{22.23}$$

This result applies to the three-dimensional random checkerboard. Moreover, using the two-dimensional results above for the regular polygonal inclusions, we have

$$\zeta_2 = 0.20428\phi_1 + 0.79572\phi_2 \qquad \text{(triangular cells),} \tag{22.24}$$

$$\zeta_2 = 0.08079\phi_1 + 0.91921\phi_2 \qquad \text{(square cells),} \tag{22.25}$$

$$\zeta_2 = 0.02301\phi_1 + 0.97699\phi_2 \qquad \text{(hexagonal cells).} \tag{22.26}$$

The result for squares applies to the two-dimensional random checkerboard. We note that five-point bounds on σ_e have been computed for the special case of symmetric-cell materials (Elsayed and McCoy 1973, Elsayed 1974).

ζ_2 for Dense Dispersions

The first comprehensive calculation of ζ_2 for a random model other than the symmetric-cell material was given by Torquato and Stell (Torquato 1980, Torquato and Stell 1983b, Torquato and Stell 1985b) for distributions of identical overlapping spheres ($d = 3$). Analogous two-dimensional calculations were made by Torquato and Beasley (1986a) and by Joslin and Stell (1986b). These results, among others, are summarized in Tables 22.1 and 22.2 and Figure 22.1. The calculation of ζ_2 for overlapping particles

Table 22.1 Three-point parameter ζ_2, defined by (20.67), versus particle volume fraction ϕ_2 for various random arrays of spheres ($d = 3$): symmetric-cell material (SCM) with spherical cells (Miller 1969, Milton 1982), identical overlapping spheres (Torquato et al. 1985), identical hard spheres (Miller and Torquato 1990), and polydisperse hard spheres calculated from (22.33) (Thovert et al. 1990). The unbracketed and bracketed values given for identical hard spheres are from simulation data (Miller and Torquato 1990) and relation (22.30), respectively.

	Three-point parameter ζ_2				
ϕ_2	SCM with spherical cells	Identical overlapping spheres	Identical hard spheres		Polydisperse hard spheres
0.0	0.0	0.0	0.0	(0.0)	0.0
0.1	0.1	0.056	0.020	(0.021)	0.05
0.2	0.2	0.114	0.041	(0.040)	0.10
0.3	0.3	0.171	0.060	(0.059)	0.15
0.4	0.4	0.230	0.077	(0.077)	0.20
0.5	0.5	0.290	0.094	(0.094)	0.25
0.55	0.55	0.320	0.110	(0.102)	0.275
0.6	0.6	0.351	0.134	(0.110)	0.30
0.7	0.7	0.415			
0.8	0.8	0.483			
0.9	0.9	0.558			
0.95	0.95	0.604			
0.99	0.99	0.658			

is straightforward by virtue of the simplicity of S_3 [cf. (5.11)] and the use of efficient Gaussian quadrature techniques to evaluate the threefold integrals (20.66) and (20.67). Joslin and Stell (1986a) and Stell and Rikvold (1987) showed that ζ_2, for overlapping cylinders or spheres, is *insensitive* to polydispersivity effects. For overlapping spheres, the first term in the volume-fraction expansion of ζ_2 is a good approximation to ζ_2 over almost the whole range of ϕ_2 (Berryman 1985b, Torquato 1985b). This is also true for $d = 2$ (Torquato and Beasley 1986a, Joslin and Stell 1986b). For $d = 3$, for example, Torquato (1985b) found that to first order in ϕ_2, $\zeta_2 = 0.5615\phi_2$ (accurate to the number of significant figures indicated).

The evaluation of ζ_2 for random impenetrable (hard) particles, on the other hand, is considerably more complex than for overlapping particles. For isotropic arrays of hard particles, the use of the series expression (5.42) for $S_3^{(2)}$ (consisting of multidimensional integrals involving the two- and three-body functions g_2 and g_3) in the threefold integrals (20.66) and (20.67) results in m-fold integrals with m as large as $3(d+1)$. A general

Table 22.2 Three-point parameter ζ_2, defined by (20.66), versus fiber volume fraction ϕ_2 for the transverse plane ($d = 2$) of various random arrays of unidirectionally aligned infinitely long circular cylinders: symmetric-cell material (SCM) with cylindrical cells (Beran and Silnutzer 1971, Milton 1982), identical overlapping cylinders (Torquato and Beasley 1986a), identical hard cylinders computed from (22.31) (Torquato and Lado 1988a), and polydisperse hard cylinders (Miller and Torquato 1991).

	Three-point parameter ζ_2			
ϕ_2	SCM with cylindrical cells	Identical overlapping cylinders	Identical hard cylinders	Polydisperse hard cylinders
0.0	0.0	0.0	0.0	0.0
0.1	0.1	0.062	0.033	0.05
0.2	0.2	0.123	0.064	0.10
0.3	0.3	0.186	0.095	0.15
0.4	0.4	0.249	0.124	0.20
0.5	0.5	0.312	0.152	0.25
0.6	0.6	0.377	0.179	0.30
0.7	0.7	0.444	0.205	0.35
0.8	0.8	0.514		
0.9	0.9	0.590		
0.95	0.95	0.635		
0.99	0.99	0.687		

technique has been developed by Lado and Torquato (1986) and Torquato and Lado (1988a) to simplify significantly these multidimensional integrals so that they could be evaluated either analytically or reduced to at most a manageable triple quadrature. The basic idea behind this procedure is to expand certain angle-dependent terms of the integrands in orthogonal polynomials and exploit the orthogonality properties of the appropriate basis set. For either $d = 2$ or $d = 3$, ζ_2 can be expressed in the simplified general form

$$\zeta_2 = \frac{\phi_2}{\phi_1}\int_0^\infty dx\, g_2(x) f_\zeta(x) + \frac{\phi_2^2}{\phi_1}\int_0^\infty dz \int_0^\infty dy \int_{|z-y|}^{z+y} dx\, g_3(x,y,z) h_\zeta(x,y,z). \qquad (22.27)$$

Here $f_\zeta(x)$ and $h_\zeta(x,y,z)$ are volume-fraction-independent functions that depend on the space dimension d; they were found explicitly for identical hard spheres ($d = 3$) by Lado and Torquato (1986) and for identical hard (circular) disks ($d = 2$) by Torquato and Lado (1988a). The nontrivial volume-fraction dependence of ζ_2 enters through the functions g_2 and g_3.

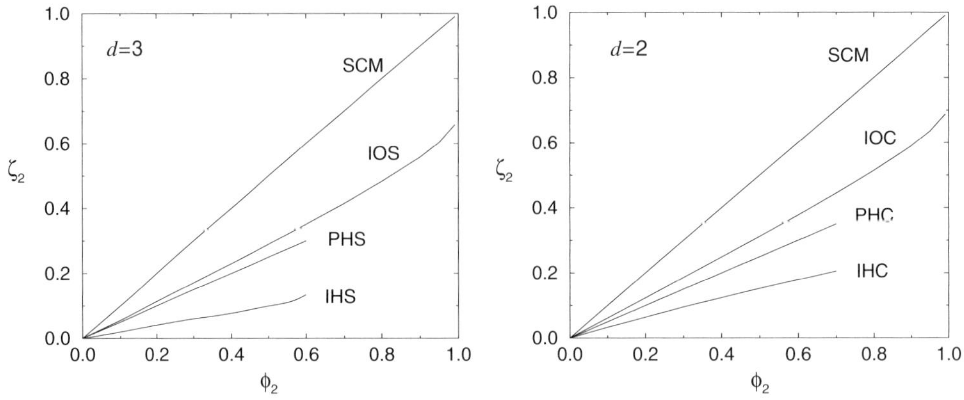

Figure 22.1 Three-point parameter ζ_2 versus inclusion volume fraction ϕ_2 for various random arrays of spheres and aligned, infinitely long circular cylinders (transverse plane). Left panel: Spheres ($d = 3$) as obtained from Table 22.1; symmetric-cell material (SCM) with spherical cells; identical overlapping spheres (IOS); polydisperse hard spheres (PHS); and identical hard spheres (IHS) (simulation data). Right panel: Circular cylinders ($d = 2$) as obtained from Table 22.2; symmetric-cell material (SCM) with cylindrical cells; identical overlapping cylinders (IOC); polydisperse hard cylinders (PHC); and identical hard cylinders (IHC).

Felderhof (1982) used a different procedure to find the expression for $d = 3$ and computed ζ_2 through third order in the sphere volume fraction ϕ_2 in the superposition approximation [cf. (3.41)] for an isotropic *equilibrium* distribution of hard spheres. [Henceforth, we will use the phrase "*equilibrium hard spheres* (*disks*)" and its variants to mean systems along the disordered stable liquid branch or the disordered metastable extension of the liquid branch, as discussed in Section 3.3.] Beasley and Torquato (1986) calculated ζ_2 exactly (up to the indicated number of significant figures) through third order in ϕ_2 for this model using (22.27) and the diagrammatic expansions of g_2 and g_3 discussed in Section 3.2, and obtained

$$\zeta_2 = 0.21068\phi_2 - 0.04693\phi_2^2 + 0.00247\phi_2^3 + \mathcal{O}(\phi_2^4). \tag{22.28}$$

Torquato and Lado (1986) evaluated ζ_2 from (22.27) to all orders in ϕ_2 up to $\phi_2 = 0.6$ (i.e., up to about 94% of the maximally-random-jammed value) using the Verlet-Weis expression for g_2 (Section 3.3.2) and the superposition approximation for g_3. They found approximately linear behavior for $0 \le \phi_2 \le 0.4$. Subsequently, Miller and Torquato (1990) carried out Monte Carlo computer simulations to determine ζ_2 accurately for equilibrium distributions of identical impenetrable (hard) spheres for $0 \le \phi_2 \le 0.6$. These authors actually simulated Torquato's three-point cluster bounds (21.47), which, as noted in Section 22.1.2, are identical to the three-point bounds (21.33) for this model but are easier to determine from simulations. The simulation results for ζ_2 have been compared to the superposition approximations of ζ_2, and to the linear and quadratic formulas

$$\zeta_2 = 0.21068\phi_2, \tag{22.29}$$

and

$$\zeta_2 = 0.21068\phi_2 - 0.04693\phi_2^2, \tag{22.30}$$

respectively, which are simply (22.28) truncated after one and two terms. Remarkably, the quadratic formula (22.30), exact to second order in ϕ_2, follows the simulation data very closely up to and including $\phi_2 = 0.54$, a value along the metastable extension (see Table 3.1). This indicates that cubic and higher-order terms are negligibly small. Note that the linear term of (22.30) is actually the dominant, since the linear formula (22.29), although not as accurate as the quadratic formula (22.30), is a good approximation to the data for $\phi_2 \leq 0.5$. For $0.54 \leq \phi_2 \leq 0.6$, the linear formula (22.29) is the most accurate, with the quadratic formula (22.30) being the next-most-accurate calculation. The full density-dependent superposition approximation (Torquato and Lado 1986) is a good approximation up to $\phi_2 = 0.5$ but increasingly overestimates ζ_2 as ϕ_2 increases above this value.

The accuracy of the quadratic formula (22.30) implies that ζ_2 only incorporates essentially up to *three-body* effects to *lowest order* in ϕ_2, even at high volume fractions: a result consistent with the fact that ζ_2 contains little information about the intrinsically many-body phenomenon of percolation. This property holds for a very general class of realistic statistically isotropic two- and three-dimensional distributions of disks and spheres, respectively, with a polydispersivity in size and an arbitrary degree of penetrability. This is a practically useful conclusion, since the exact calculation of ζ_2 through second order in ϕ_2 for distributions of impenetrable particles is much easier than the corresponding full-density-dependent calculation, which necessarily involves the use of some approximation for g_3, the validity of which is usually questionable at high densities.

Torquato and Lado (1988a) evaluated ζ_2 for isotropic equilibrium distributions of identical hard circular disks from (22.27) using the Percus-Yevick approximation for g_2 (Section 3.3.2) and the superposition approximation (3.41) for g_3. Of course, this models the transverse plane ($d = 2$) of a transversely isotropic fiber-reinforced material composed of equilibrium distributions of *unidirectionally* aligned, infinitely long identical circular hard cylinders in a matrix. They also computed ζ_2 for this two-dimensional model exactly through second order in ϕ_2:

$$\zeta_2 = \frac{\phi_2}{3} - 0.05707\phi_2^2. \tag{22.31}$$

In light of the discussion above, (22.31) should be accurate for the range $0 \leq \phi_2 \leq 0.7$, $\phi_2 = 0.7$ corresponding to about 84% of the putative maximally-random-jammed value for disks ($\phi_{2c} \approx 0.83$). This conclusion has been confirmed by the accurate simulations of Greengard and Helsing (1995), who evaluated ζ_2 for $0.5 \leq \phi_2 \leq 0.7$. At the volume fractions $\phi_2 = 0.5, 0.6, 0.7$ reported in Table 22.2, relation (22.31) is within four percent of the simulation data. The superposition approximation results for ζ_2 are roughly linear for $0 \leq \phi_2 \leq 0.5$ but increasingly overestimate ζ_2 for $\phi_2 \geq 0.5$. Thus, the superposition approximation results are now superseded by relation (22.31) for this model.

Smith and Torquato (1989) computed ζ_2 for random arrays of identical hard circular disks in the cherry-pit model as a function of the impenetrability index λ. Unlike most previous studies, this work considered the *nonequilibrium* random sequential addition process (Section 3.4). Their results are not included in Table 22.2 or Figure 22.1, however. For fixed ϕ_2, the effect of increasing λ is to decrease ζ_2, as expected.

Thovert et al. (1990) computed ζ_2 exactly through first order in ϕ_2 for equilibrium impenetrable (hard) spheres ($d = 3$) with a polydispersivity in size. For the case of a suspension with two different and widely separated particle sizes, they obtained

$$\zeta_2 = 0.35534\phi_2. \tag{22.32}$$

For the instance of a polydisperse suspension containing n different ($n \to \infty$) and widely separated particle sizes, they showed

$$\zeta_2 = 0.5\phi_2. \tag{22.33}$$

This last microstructure yields the largest effect due to polydispersivity. Note that the bidisperse result (22.32) lies exactly midway between the monodisperse result (22.29) and the polydisperse result (22.33). Miller and Torquato (1991) carried out analogous two-dimensional calculations and found that (22.33) also applies to the corresponding polydisperse hard-disk system. Thus, the effect of polydispersivity is again to increase ζ_2.

It is seen that for the wide variety of two- and three-dimensional spheres described in Tables 22.1 and 22.2 and Figure 22.1, ζ_2 is bounded from above by the value for the symmetric-cell material, i.e.,

$$\zeta_2 \leq \phi_2. \tag{22.34}$$

Thus, ζ_2 lies in the smaller closed interval $[0, \phi_2]$ for this class of sphere dispersions that is commonly encountered in practice. [The result (22.34) cannot be true in general. A counterexample is the Hashin–Shtrikman coated-spheres geometry (Section 16.1) in which each concentric shell is itself composed of a packing of much smaller spheres; as the sphere volume fraction within each concentric shell approaches unity, $\zeta_2 \to 1$.] For media in which ζ_2 lies in the interval $[0, \phi_2]$, we can eliminate ζ_2 in favor of the volume fraction ϕ_2 in the three-point upper bound (21.37) to get the following upper bound on the effective conductivity, which involves only phase volume fractions:

$$\sigma_e \leq \langle\sigma\rangle - \frac{\phi_1\phi_2(\sigma_2 - \sigma_1)^2}{\langle\tilde{\sigma}\rangle + (d-1)\langle\sigma\rangle}. \tag{22.35}$$

The bound (22.35) improves upon the Hashin–Shtrikman bound (21.22), since it uses information about ζ_2. For example, when $d = 3$, $\sigma_1 = 1$, and $\sigma_2 = 100$, the Hashin–Shtrikman upper bound gives 40.7, whereas (22.35) gives 34.3.

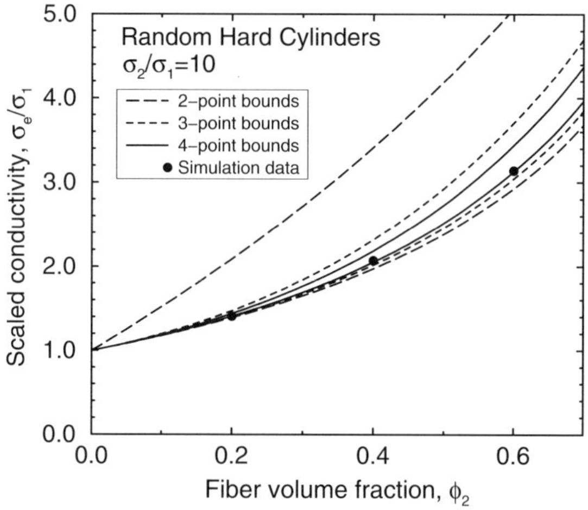

Figure 22.2 Bounds on the scaled effective transverse conductivity σ_e/σ_1 versus ϕ_2 at $\sigma_2/\sigma_1 = 10$ for random arrays of aligned infinitely long identical hard circular cylinders in equilibrium. Three-point bounds (21.33) and four-point bounds (21.42) for $d = 2$ were evaluated by Torquato and Lado (1988a), and simulation data are from Kim and Torquato (1990b). Two-point bounds (21.20) are included.

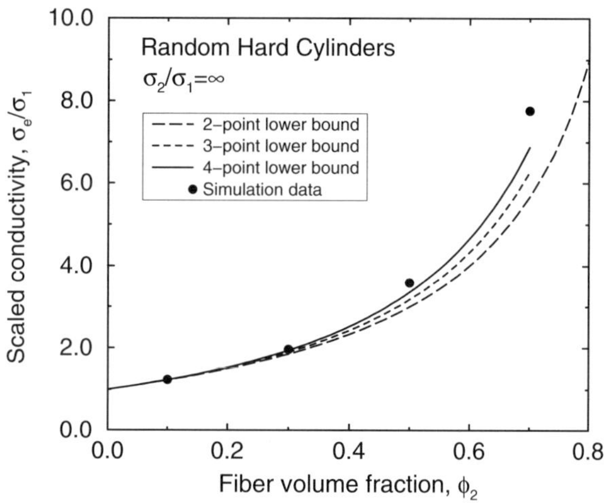

Figure 22.3 As in Figure 22.2 with $\sigma_2/\sigma_1 = \infty$. Simulation datum at $\phi_2 = 0.7$ is from Cheng and Greengard (1997). Upper bounds diverge to infinity. Four-point lower bound, however, gives a relatively sharp estimate of σ_e/σ_1.

Bounds for Dispersions

Three- and four-point bounds on effective transverse conductivity σ_e are depicted in Figures 22.2 and 22.3 for equilibrium arrays of aligned identical hard cylinders ($d = 2$) at $\sigma_2/\sigma_1 = 10$ and $\sigma_2/\sigma_1 = \infty$, respectively, as evaluated by Torquato and Lado (1988a). (Recall from Section 16.1.4 that such a material is fully specified by two effective parameters: the transverse conductivity $\sigma_e \equiv \sigma_e^T$ and longitudinal conductivity $\sigma_e^L = \sigma_1\phi_1 + \sigma_2\phi_2$.) Included in these figures are the corresponding two-point Hashin–Shtrikman bounds (21.20) as well as Brownian-motion simulation data for the effective transverse conductivity σ_e due to Kim and Torquato (1990b). For the case $\sigma_2/\sigma_1 = 10$, the three- and four-point bounds provide significant improvement over the two-point bounds (especially the upper bound); the four-point bounds are very sharp.

As the contrast ratio σ_2/σ_1 increases, the bounds widen, as expected, taking into account the possibility of formation of large conducting clusters in the system, and in the extreme case of superconducting inclusions ($\sigma_2/\sigma_1 = \infty$), all of the upper bounds tend to infinity. However, for the infinite-contrast case, the four-point lower bound provides a very accurate estimate of σ_e for virtually all volume fractions (see Figure 22.3). This is true because the four-point lower bound corresponds to composites (see Figure 21.1) that are topologically similar to equilibrium hard-disk systems. The superconducting disks cannot form clusters for the considered range of volume fractions; i.e., the conducting phase is disconnected. Indeed, for random equilibrium hard-sphere models in *any space dimension*, interparticle contacts occur only at the *maximally random jammed* density (see Section 3.3). These results confirm the remarks made in Section 21.6 about the utility of bounds.

Bounds that improve upon two-point bounds on σ_e of three-dimensional equilibrium hard-sphere systems behave qualitatively the same as the aforementioned two-dimensional results. Figure 22.4 compares two- and three-point bounds for this three-dimensional model with superconducting inclusions ($\sigma_2/\sigma_1 = \infty$) (Miller and Torquato 1990) to corresponding Brownian-motion simulation data of Kim and Torquato (1991) and the three-point approximation (20.83). The best lower bound provides a good estimate of σ_e, and the three-point approximation is quite accurate.

Predicting the effective conductivity of systems of overlapping particles is more challenging, since cluster formation is nontrivial. Figure 22.5 compares four-point bounds for identical overlapping disks (cylinders) (Torquato and Beasley 1986a) that are highly conducting relative to the matrix ($\sigma_2/\sigma_1 = 1000$) to the simulation data of Helsing (1998). Recall that the particle phase percolates at the threshold $\phi_{2c} \approx 0.68$. For the range $0 \leq \phi_2 \leq 0.4$, the lower bound provides a good estimate of the data, implying that the characteristic cluster size Λ_2 of phase 2 (Section 21.6) is relatively small. As ϕ_2 increases from 0.4 to 0.9, the data cross over from the lower to upper bound, corresponding to Λ_2 increasing and becoming macroscopically large once the threshold is crossed. For the range $0.9 \leq \phi_2 \leq 1$, the upper bound provides a good estimate of the data, implying that the characteristic cluster size of phase 1, Λ_1, is relatively small. The behavior of the four-point bounds is consistent with the remarks of Section 21.6.

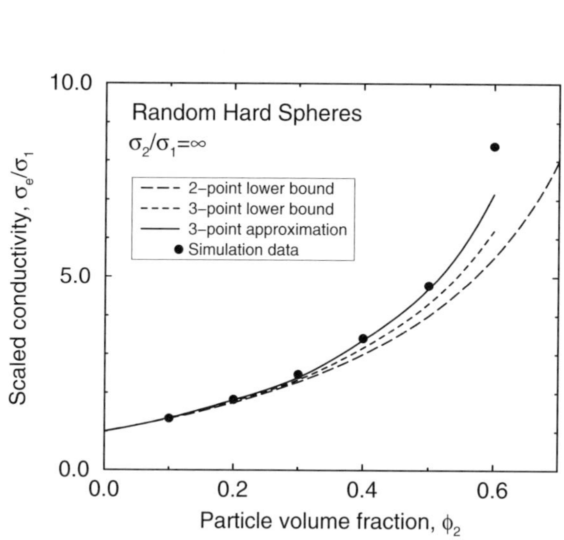

Figure 22.4 Bounds on σ_e/σ_1 versus ϕ_2 at $\sigma_2/\sigma_1 = \infty$ for random arrays of identical hard spheres ($d = 3$) in equilibrium. Three-point bound (21.37) was evaluated by Miller and Torquato (1990), and simulation data are from Kim and Torquato (1991). Two-point bounds (21.20) and the three-point approximation (20.83) are included.

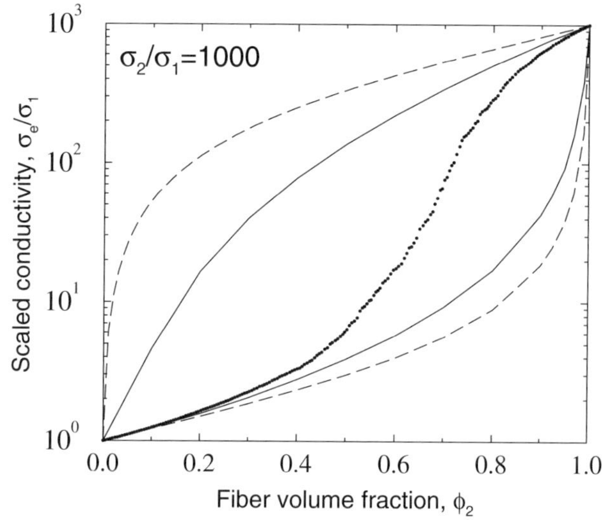

Figure 22.5 Bounds on σ_e/σ_1 versus ϕ_2 at $\sigma_2/\sigma_1 = 1000$ for random arrays of aligned identical overlapping circular cylinders (disks) compared to simulation data (dots) of Helsing (1998). Four-point bounds (21.42) (solid lines) for $d = 2$ were evaluated by Torquato and Beasley (1986a). Two-point bounds (21.20) (dashed lines) are included.

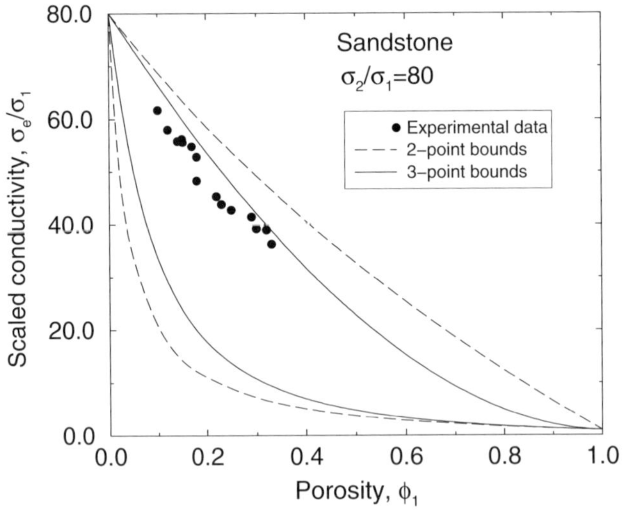

Figure 22.6 Comparison of measured values of σ_e/σ_1 for air-saturated sandstones (Sugawara and Yoshizawa 1962) to the two-point Hashin–Shtrikman bounds (21.20) and three-point bounds (21.35) and (21.37) for $d = 3$ for a system of identical overlapping spheres (phase 2) as a function of porosity ϕ_1. Here $\sigma_2/\sigma_1 = 80$, where phase 2 is the solid phase.

Figure 22.6 compares two- and three-point bounds on σ_e for the three-dimensional overlapping-sphere model (Section 5.1) to measurements Sugawara and Yoshizawa (1962) made on the thermal conductivity of air-saturated sandstones with $\sigma_2/\sigma_1 = 80$. The three-point upper bound gives an accurate estimate of σ_e, since the conducting sandstone phase (modeled by overlapping spheres) is above its percolation threshold ($\phi_{2c} \approx 0.29$). Indeed, both phases of the sandstones and the overlapping-sphere model (Section 10.6) are connected (bicontinuous) for the range of the data.

Let us now consider perfectly insulating ($\sigma_2 = 0$) three-dimensional overlapping spheres in a solid matrix with conductivity $\sigma_1 = 1$. Figure 22.7 compares upper bounds on σ_e [the two-point upper bound (21.22) with the phases interchanged and three-point upper bound obtained from (21.37)] to the simulation data of Kim and Torquato (1992b). [Note that the bound (21.37) becomes an *upper* bound when $\sigma_2 \leq \sigma_1$.] The three-point upper bound and three-point approximation (20.83) (not shown) provide relatively good estimates of σ_e for this system, which is bicontinuous for the wide range of volume fractions $0.29 \leq \phi_2 \leq 0.97$ (Section 10.6).

Figure 22.8 compares three-point lower bounds (21.37) on σ_e for a monodispersion and polydispersion [corresponding to (22.33)] of hard spheres for the case $\sigma_2/\sigma_1 = \infty$. Included in the figure is the three-point approximation (20.83) for the polydispersion. The effect of polydispersivity is to increase σ_e.

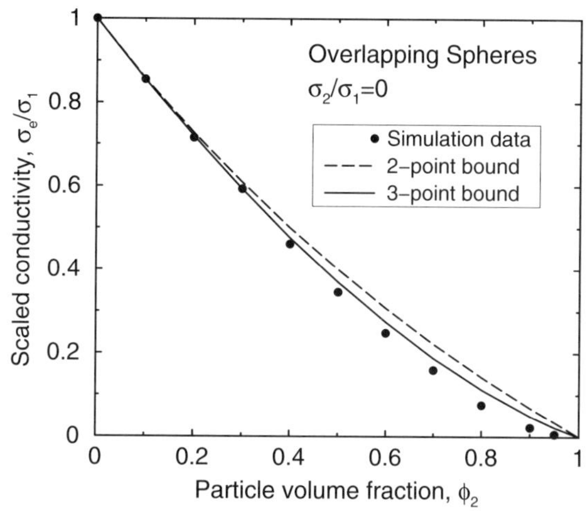

Figure 22.7 Upper bounds on σ_e/σ_1 versus ϕ_2 for a system of insulating identical overlapping spheres ($d = 3$). Simulation data are from Kim and Torquato (1992b).

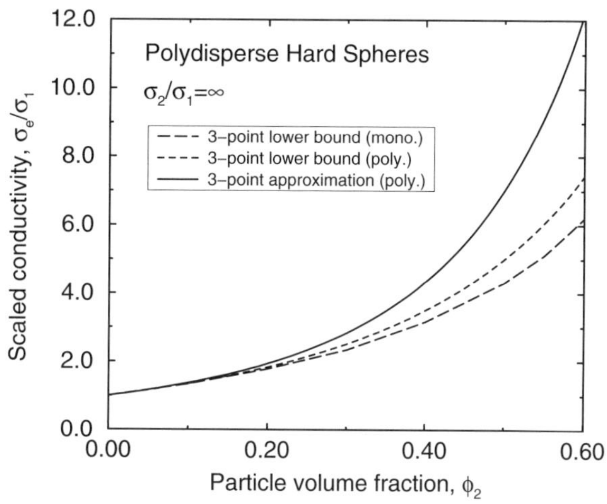

Figure 22.8 Three-point lower bounds (21.37) versus ϕ_2 for a system of superconducting monodisperse and polydisperse spheres ($d = 3$) in equilibrium. Included is the accurate three-point approximation (20.83) for polydisperse spheres.

Bounds for Other Isotropic Media

Three-point contrast bounds on σ_e have been calculated for other isotropic microstructures. Helsing (1993) has computed them for FCC arrays of cubes in a matrix. Roberts and Teubner (1995) have evaluated three-point bounds for certain level cuts of Gaussian random fields, which turn out to be appreciably different from the corresponding bounds for overlapping spheres. Roberts and Knackstedt (1996) have calculated three-point bounds for models derived from two level cuts of Gaussian fields as well as overlapping hollow spheres. Jeulin and Le Coënt (1996) have evaluated three- and four-point bounds for a variety of "random-grain" models in three and two dimensions, respectively. Hyun and Torquato (2000) have computed σ_e as well as four-point bounds on σ_e for two-dimensional periodic arrays of triangular and of hexagonal holes (inclusions) over the entire range of volume fractions.

The first experimental evaluation of ζ_2 was carried out painstakingly by Corson (1974c, 1974d), who determined it by first measuring $S_3^{(2)}$ from photographs of cross-sections of binary mixtures of metals. He used this information to compute the three-point Beran bounds on σ_e for these composites (Corson 1974d). Coker et al. (1996) have evaluated the three-point upper bound on the effective conductivity of the bicontinuous Fontainebleau sandstone shown in Figure 12.14 by extracting $S_3^{(2)}$ and, thus, the parameter ζ_2 from the three-dimensional image of the sample using the techniques discussed in Chapter 12.

Macroscopically Anisotropic Media

Consider computing the two-point anisotropic bounds (21.24) on $\boldsymbol{\sigma}_e$ for three-dimensional arrays of identical aligned inclusions possessing axial symmetry (e.g., circular cylinders and spheroids). When the symmetry axis is aligned in the x_3-direction, the key two-point tensor $\boldsymbol{a}_2$ arising in the bounds (21.24), after combining (20.38), (20.44), and (20.50), is given by

$$\mathbf{a}_2 = -\phi_1\phi_2 \begin{bmatrix} Q & 0 & 0 \\ 0 & Q & 0 \\ 0 & 0 & 1-2Q \end{bmatrix}, \tag{22.36}$$

where

$$Q = \frac{1}{3} - \lim_{\epsilon\to 0} \frac{1}{2\phi_1\phi_2} \int_\epsilon^\infty \frac{dx}{x} \int_0^\pi d(\cos\theta) P_2(\cos\theta)[S_2(\boldsymbol{x}) - \phi_1^2], \tag{22.37}$$

and S_2 is the two-point probability function for phase 1 (matrix).

It has been shown (Willis 1977, Sen and Torquato 1989, Torquato and Lado 1991) that for aligned spheroidal inclusions, Q depends only on the inclusion shape (not on the volume fraction or spatial distribution of the inclusions) and is given explicitly by (17.30) and (17.31) in the prolate and oblate cases, respectively. Torquato and Lado (1991) showed that these results apply to *possibly overlapping* spheroids by employing in (22.37) the scaling relation (7.23) for S_2; this enables one to map results for possibly overlapping spheres ($b/a = 1$) into equivalent results for possibly overlapping spheroids (see Sections 7.2 and 7.3). More generally, for oriented ellipsoids with the structure

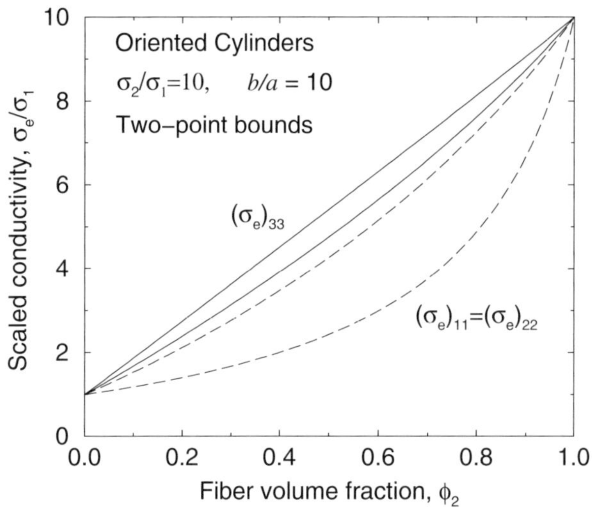

Figure 22.9 Two-point bounds on the diagonal components of the scaled conductivity versus ϕ_2 for a composite consisting of conducting, slender overlapping cylindrical inclusions (Torquato and Sen 1990). Solid and dashed lines are for $(\sigma_e)_{33}$ and $(\sigma_e)_{11} = (\sigma_e)_{22}$, respectively.

of an idealized nematic liquid crystal (Section 7.1), we noted in Section 21.1.2 that $\boldsymbol{a}_2 = -\phi_1\phi_2\boldsymbol{A}^*$, where $\boldsymbol{A}^*$ is the depolarization tensor (17.25) for an ellipsoid. Finally, we observe that the two-point bounds (21.24) yield the exact conductivity tensor (16.10) for a first-rank laminate in the case of aligned disks, in which case $Q = 0$ in (22.36).

Torquato and Sen (1990) have evaluated (22.37), and thus the two-point bounds (21.24) on $\boldsymbol{\sigma}_e$, for a distribution of oriented overlapping circular cylinders (of length $2b$ and diameter $2a$) in a matrix. Figure 22.9 shows the bounds on the three diagonal components of the effective conductivity $(\sigma_e)_{ii}$ (i = 1, 2, and 3) for conducting slender rods. Not surprisingly, the bounds on the effective conductivity in the axial direction $(\sigma_e)_{33}$ are narrower (approaching the arithmetic mean) than the bounds on the effective conductivity in the transverse direction.

Quintanilla and Torquato (1996b) have calculated the two-point anisotropic bounds (21.24) on $\boldsymbol{\sigma}_e$ for two-dimensional, macroscopically anisotropic second-rank laminates and studied how the separation of length scales affects the effective conductivity tensor.

22.1.2 Cluster Bounds

Interestingly, Beasley and Torquato (1986) showed that for the instance of totally impenetrable spheres, the cluster bounds (21.47) are *identical* to the three-point Beran bounds (21.33). The cluster bounds (21.47) are of a functional form that makes them easier to calculate from Monte Carlo computer simulations.

For fully penetrable spheres, the cluster bounds on the effective conductivity are not as sharp as the best three-point contrast-type bounds, but they are simple and analytical (DeVera and Strieder 1977, Torquato 1986b). For the special case of fully penetrable insulating spheres ($\sigma_2 = 0$), the bounds are even simpler and imperceptibly different from the best three-point contrast bounds. In this latter instance, the lower bounds are trivially zero, while the upper bounds reduce to

$$\frac{\sigma_e}{\sigma_1} \leq \frac{\phi_1}{1+\eta/2}, \tag{22.38}$$

where $\eta = -\ln \phi_1$ is the reduced density.

22.1.3 Security-Spheres Bounds

Recall that unlike the aforementioned conventional bounds, the security-spheres upper bound (21.49), at arbitrary ϕ_2, may remain finite for superconducting spheres in a matrix; and the security-spheres lower bound (21.50) may remain nonzero for perfectly insulating spheres in a matrix. Torquato and Rubinstein (1991) have evaluated these bounds for equilibrium arrays of spheres in which each sphere of conductivity σ_2 and diameter D is coated with a thin layer of matrix material of conductivity σ_1 so that the diameter of each composite sphere is D_0. The coating is required in order to ensure that the inner spheres cannot touch. As far as the structure is concerned, one is actually interested in random hard spheres of diameter $D_0 \geq D$ and volume fraction $\hat{\phi}_2 = (D_0/D)^3\phi_2$, where ϕ_2 is the actual sphere volume fraction. Thus, we can use the expression (5.97) for the nearest-neighbor function $H_P(x)$ by evaluating it at the dimensionless distance $(D_0/D)x$ and volume fraction $\hat{\phi}_2$.

The security-spheres bounds are plotted in Figure 22.10 for several values of the dimensionless coating thickness $\epsilon \equiv (D_0 - D)/D$. The bounds become sharper as ϵ is increased, as expected. For these random arrays, the security-sphere bounds (21.49) and (21.50) are generally appreciably sharper than the bounds of Bruno (1991) for $\epsilon = 0.001$ and $\epsilon = 0.01$. For $\epsilon = 0.1$, Bruno's bounds are slightly sharper. Moreover, they are somewhat sharper than the security-sphere bounds (21.49) and (21.50) when the spheres are well separated from one another (e.g., regular lattices).

For cubic lattices of spheres in d dimensions, the nearest-neighbor probability density function H_P is given by (21.51), and the security-spheres bounds (21.49) and (21.50) are particularly simple:

$$\frac{\sigma_e}{\sigma_1} \leq 1 + d\frac{b^d}{b^d-1}\phi_2 \qquad \text{if} \quad \sigma_2/\sigma_1 = \infty, \tag{22.39}$$

$$\frac{\sigma_e}{\sigma_1} \geq \left(1 + \frac{d}{d-1}\frac{b^d}{b^d-1}\phi_2\right)^{-1} \qquad \text{if} \quad \sigma_2/\sigma_1 = 0. \tag{22.40}$$

Here b is the minimum interparticle spacing (in units of the diameter) and is related to the volume fraction via

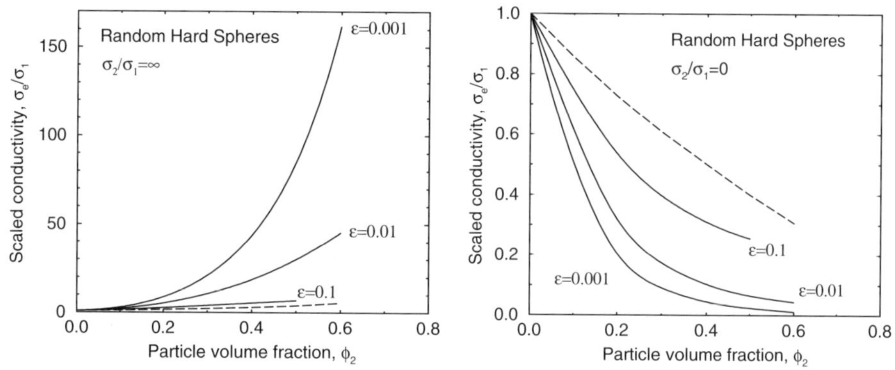

Figure 22.10 Security-spheres bounds on σ_e/σ_1 for random arrays of three-dimensional spheres in equilibrium for several values of dimensionless coating thickness ϵ (Torquato and Rubinstein 1991). Left panel: Upper bounds (21.49) (solid lines) for superconducting spheres. Dashed line is Hashin–Shtrikman lower bound. Right panel: Lower bounds (21.50) (solid lines) for perfectly insulating spheres. Dashed line is Hashin–Shtrikman upper bound.

$$b^d = c(d)\phi_2^{-1}, \tag{22.41}$$

where $c(d)$ is a constant that depends on the dimension d and type of lattice. For example, for a d-dimensional simple cubic lattice

$$c(d) = \frac{\pi^{d/2}}{2^d\Gamma(1+d/2)}. \tag{22.42}$$

The bounds (22.39) and (22.40) are exact through first order in ϕ_2 (see Sections 16.1.5 and 19.1.1) and are relatively sharp up to moderately large volume fractions for both $d = 2$ and $d = 3$ (see Section 16.1.5). Finally, we note that the bounds (22.39) and (22.40) are exact through all orders in ϕ_2 for both $d = 1$ and $d \to \infty$ (see Theorem 20.1).

22.2 Elastic Moduli

We discuss below specific calculations of contrast, cluster, and security-spheres bounds on the effective elastic moduli of heterogeneous materials. Results are reported for model microstructures as well as real materials.

22.2.1 Contrast Bounds

As demonstrated in Section 21.2.2, the key microstructural parameter that determines three-point contrast bounds on the effective bulk modulus K_e of macroscopically isotropic two-phase media is $\zeta_2 = 1 - \zeta_1$; see lower bound (21.87) and upper bound

(21.93) for $d = 2$, and bounds (21.86) for $d = 3$. Corresponding bounds on the effective shear modulus G_e depend on both ζ_2 and $\eta_2 = 1-\eta_1$; see lower bound (21.95) and upper bound (21.100) for $d = 2$, and bounds (21.94) with (21.101) and (21.102) for $d = 3$. The parameter η_2 is defined by (20.168) for any d, and for $d = 2$ and $d = 3$ simplifies according to (20.170) and (20.171), respectively.

In what follows, we describe primarily the evaluation of η_2 and the associated bounds on the effective elastic moduli for isotropic microstructures. The calculation of contrast bounds on the stiffness tensor for anisotropic microstructures is mentioned.

η_2 for Dilute Dispersions

Torquato, Lado and Smith (1987) have used the low-density expansion of the effective moduli to obtain the low-density expansions of the parameters ζ_2 and η_2 for dispersions of inclusions (phase 2) in a matrix (phase 1). In the latter instance, we look to find the coefficients $\eta_2^{(i)}$ in the expression

$$\eta_2 = \eta_2^{(0)} + \eta_2^{(1)}\phi_2 + \cdots . \tag{22.43}$$

In complete analogy with the procedure outlined in Section 22.1.1 for ζ_2, this is accomplished by expanding the coefficients of the cluster expansion (Section 19.2) in powers of the differences in the phase properties $[(K_2 - K_1)$ and $(G_2 - G_1)]$ and then comparing to the weak-contrast expansions (20.181) and (20.182) through lowest order in ϕ_2. (Note that the results for ζ_2 given in Section 22.1.1 also apply to the effective bulk modulus, provided that the composite is elastically isotropic or possesses cubic symmetry.) The low-density expansions of the parameter η_2 found by Torquato et al. (1987) for dispersions of spheres can be generalized to dispersions of randomly oriented ellipsoids using the cluster expansion for the effective shear modulus for the latter system. For example, using the expansion (20.182) for G_e for randomly oriented ellipsoids (involving the shear part of the tensor $\boldsymbol{T}$), we obtain the following limiting three-dimensional cases for the leading order term $\eta_2^{(0)}$ in the density expansion of η_2:

$$\eta_2^{(0)} = 0 \qquad \text{(spheres)}, \tag{22.44}$$

$$\eta_2^{(0)} = \frac{1}{4} \qquad \text{(needles)}, \tag{22.45}$$

$$\eta_2^{(0)} = 1 \qquad \text{(disks)}. \tag{22.46}$$

The analogous limits in two dimensions are given by

$$\eta_2^{(0)} = 0 \qquad \text{(disks)}, \tag{22.47}$$

$$\eta_2^{(0)} = 1 \qquad \text{(needles)}. \tag{22.48}$$

η_2 for Symmetric-Cell Materials

The leading-order term $\eta_2^{(0)}$ determines η_2 for symmetric-cell materials (Chapter 8) for *all values of* ϕ_2. Use of the general expression (8.35) for the three-point probability function of a symmetric-cell material and definition (20.168) immediately yields that

$$\eta_2 = \phi_2 + (\phi_1 - \phi_2)h, \tag{22.49}$$

where

$$h = J\left[W_3^{(1)}\right], \tag{22.50}$$

$W_3^{(1)}$ is the three-point single-cell weight given by (8.17) or (8.19) with $n = 3$, and J is the integral operator defined by (20.166). The first term ϕ_2 in (22.49) arises after use of the properties (20.172) and (20.173). Because $W_3^{(1)}$ is independent of volume fraction, so is the parameter h. Thus, we can evaluate h for different cell shapes by comparing (22.49) to lowest order in ϕ_2 to the leading-order (constant) term in the density expansion (22.43), which yields

$$h = \eta_2^{(0)}. \tag{22.51}$$

Therefore, h has exactly the same properties as $\eta_2^{(0)}$; i.e., it depends only on the cell shape (independent of the cell size distribution and ϕ_2) and lies in the closed interval $[0, 1]$. The shape parameter h is related to but different from the shape parameters E, G, and H derived previously for $d = 3$ (Silnutzer 1972, Milton 1982) and for $d = 2$ (Beran and Silnutzer 1971, Milton 1982); here we have $h = [4(5E - 1) - (9G - 1)]/6$ for $d = 3$ and $h = (8H - 3) - 4(4G - 1)$ for $d = 2$.

For *any* symmetric-cell material at $\phi_1 = \phi_2 = 1/2$, relations (22.18) and (22.49) give

$$\eta_2 = \zeta_2 = \frac{1}{2}, \tag{22.52}$$

which applies to any material with phase-inversion symmetry at $\phi_1 = \phi_2 = 1/2$ (Section 2.2.2). Use of the above results for $\eta_2^{(0)}$ for limiting ellipsoidal shapes, (22.49), and (22.51) enables us to find results for η_2 for symmetric-cell materials with cells of such shape. Thus, for d-dimensional spherical cells, $\eta_2 = \zeta_2 = \phi_2$; for three-dimensional needle-like cells, $\eta_2 = \zeta_2 = \phi_1/4 + 3\phi_2/4$; for three-dimensional disk-like cells or two-dimensional needle-like cells, $\eta_2 = \zeta_2 = \phi_1$.

η_2 for Dense Dispersions

The first thorough evaluation of η_2 for a random model other than the symmetric-cell material was obtained for distributions of identical overlapping spheres independently by Torquato et al. (1985) and by Berryman (1985b). Analogous two-dimensional calculations were obtained independently by Torquato and Beasley (1986b) and by Joslin and Stell (1986b). These results, among others, are summarized in Tables 22.3 and 22.4 and Figure 22.11. The threefold integrals were computed in the same way as ζ_2 (Section 22.1.1). Notice that to a very good approximation, η_2 is determined by the linear term in ϕ_2 for all volume fractions. Joslin and Stell (1986a) demonstrated that η_2 (as ζ_2) is insensitive to polydispersivity effects for the special case of overlapping cylinders. As in the case of ζ_2, the first term in the volume-fraction expansion of η_2 is a good approximation to η_2 over almost the whole range of ϕ_2 (Section 22.1.1). For example, for $d = 3$, Torquato et al. (1987) found that to first order in ϕ_2, $\eta_2 = 0.7468\phi_2$.

Table 22.3 Three-point parameter η_2, defined by (20.171), versus particle volume fraction ϕ_2 for various random arrays of spheres ($d = 3$): symmetric-cell material (SCM) with spherical cells (Miller 1969, Milton 1982), identical overlapping spheres (Torquato et al. 1985, Berryman 1985b), identical hard spheres calculated from (22.54) (Torquato et al. 1987), and the polydispersion of hard spheres calculated from (22.57) (Thovert et al. 1990).

	Three-point parameter η_2			
ϕ_2	SCM with spherical cells	Identical overlapping spheres	Identical hard spheres	Polydisperse hard spheres
0.0	0.0	0.0	0.0	0.0
0.1	0.1	0.075	0.048	0.05
0.2	0.2	0.149	0.097	0.10
0.3	0.3	0.224	0.145	0.15
0.4	0.4	0.295	0.193	0.20
0.5	0.5	0.367	0.241	0.25
0.6	0.6	0.439	0.290	0.30
0.7	0.7	0.512		
0.8	0.8	0.583		
0.9	0.9	0.658		
0.95	0.95	0.710		
0.99	0.99	0.742		

The calculation of η_2 for impenetrable (hard) particles is much more challenging. The multidimensional integrals that result from using the expression (5.42) for $S_3^{(2)}$ in the threefold integrals (20.170) and (20.171) have been considerably simplified (Sen, Lado and Torquato 1987b) by using the aforementioned procedure employed to get relation (22.27) for ζ_2. For either $d = 2$ or $d = 3$, η_2 can be expressed in the simplified general form

$$\eta_2 = \frac{\phi_2}{\phi_1} \int_0^\infty dx\, g_2(x) f_\eta(x) + \frac{\phi_2^2}{\phi_1} \int_0^\infty dz \int_0^\infty dy \int_{|z-y|}^{z+y} dx\, g_3(x, y, z) h_\eta(x, y, z), \tag{22.53}$$

which has the same form as relation (22.27) for ζ_2. Here $f_\eta(x)$ and $h_\eta(x, y, z)$ are volume-fraction independent functions that depend on the dimension d; they were found explicitly for identical hard spheres ($d = 3$) by Sen et al. (1987b) and for identical hard disks ($d = 2$) by Torquato and Lado (1988b). The nontrivial volume-fraction dependence of η_2 resides in the two- and three-body functions g_2 and g_3 discussed in Chapter 3. Torquato et al. (1987) evaluated η_2 exactly (up to the number of significant figures indicated) through first order in ϕ_2 for *equilibrium* interpenetrable spheres, and

Table 22.4 Three-point parameter η_2, defined by (20.170), versus fiber volume fraction ϕ_2 for the transverse plane of various random arrays of unidirectionally aligned, infinitely long circular cylinders ($d = 2$): symmetric-cell material (SCM) with cylindrical cells (Beran and Silnutzer 1971, Milton 1982), identical overlapping cylinders (Torquato and Beasley 1986b), identical hard cylinders calculated from (22.55) (Torquato and Lado 1988b), and polydisperse hard cylinders (Miller and Torquato 1991).

	Three-point parameter η_2			
ϕ_2	SCM with cylindrical cells	Identical overlapping cylinders	Identical hard cylinders	Polydisperse hard cylinders
0.0	0.0	0.0	0.0	0.0
0.1	0.1	0.084	0.070	0.05
0.2	0.2	0.167	0.140	0.10
0.3	0.3	0.250	0.211	0.15
0.4	0.4	0.331	0.283	0.20
0.5	0.5	0.411	0.356	0.25
0.6	0.6	0.490	0.430	0.30
0.7	0.7	0.568	0.505	0.35
0.8	0.8	0.644		
0.9	0.9	0.720		
0.95	0.95	0.760		
0.99	0.99	0.801		

in the totally-impenetrable-sphere limit they obtained

$$\eta_2 = 0.48274\phi_2. \tag{22.54}$$

This is easily arrived at by using the infinitely-dilute-limit expression for g_2 [step-function contribution of (3.52)] in (22.53). Relation (22.54) should serve as an accurate estimate of η_2 for equilibrium impenetrable spheres for $0 \le \phi_2 \le 0.6$ for reasons similar to those given in Section 22.1.1 concerning the corresponding expansion for ζ_2. As in the case of ζ_2, the use of the Verlet-Weis expression for g_2 (Section 3.3.2) and the superposition approximation (3.41) for g_3 in (22.53) results in an η_2 that is approximately linear for $0 \le \phi_2 \le 0.4$, but for $\phi_2 > 0.4$ increasingly overestimates η_2 (Sen et al. 1987a). Again the superposition result (Sen et al. 1987a) is now superseded by (22.54), which is generally more accurate for this model.

Torquato and Lado (1988b) computed η_2 for two-dimensional equilibrium distributions of impenetrable disks (cylinders). They found that for $d = 2$, η_2 through second order in ϕ_2 is exactly given by

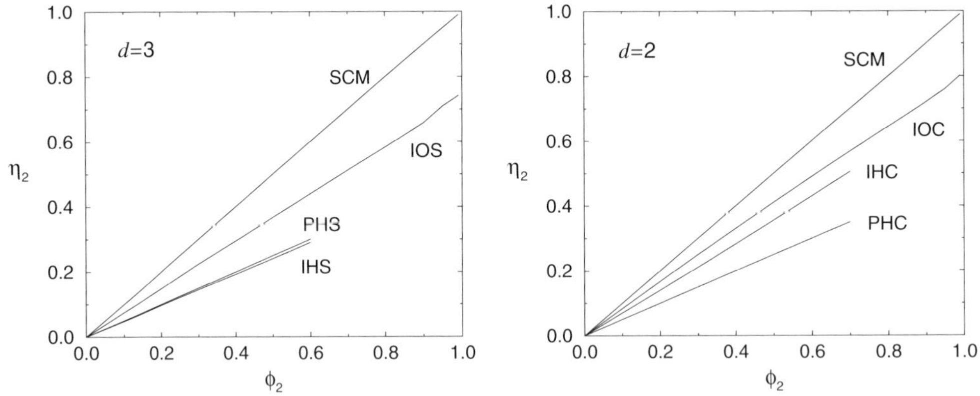

Figure 22.11 Three-point parameter η_2 versus particle volume fraction ϕ_2 for various random distributions of spheres and aligned, infinitely long cylinders (transverse plane). Left panel: Spheres ($d = 3$) as obtained from Table 22.3; symmetric-cell material (SCM) with spherical cells; identical overlapping spheres (IOS); polydisperse hard spheres (PHS); and identical hard spheres (IHS). Right panel: Circular cylinders ($d = 2$) as obtained from Table 22.4; symmetric-cell material (SCM) with circular cylindrical cells; identical overlapping cylinders (IOC); polydisperse hard cylinders (PHC); and identical hard cylinders (IHC).

$$\eta_2 = \frac{56}{81}\phi_2 + 0.0428\phi_2^2. \tag{22.55}$$

Thovert et al. (1990) calculated η_2 exactly through first order in ϕ_2 for equilibrium impenetrable spheres ($d = 3$) with a polydispersivity in size. For a suspension with two different and widely separated particle sizes, they obtained

$$\eta_2 = 0.49137\phi_2. \tag{22.56}$$

For the case of a polydisperse suspension containing n different ($n \to \infty$) and widely separated particle sizes, they obtained

$$\eta_2 = 0.5\phi_2. \tag{22.57}$$

As in the case of ζ_2, the bidisperse result (22.56) lies exactly midway between the monodisperse result (22.54) and the polydisperse result (22.57), which corresponds to the geometry that yields the largest polydispersivity effects. The effect of polydispersivity is to increase η_2, but this is considerably smaller than the effect on ζ_2. Miller and Torquato (1991) have found that (22.57) also applies to the corresponding polydisperse impenetrable-disk (cylinder) geometry. Thus, the effect of *polydispersivity* here (unlike that for $d = 3$) is to *decrease* η_2, albeit by a small amount.

It is noteworthy that for all of the above-reported results for distributions of d-dimensional spheres (symmetric cell materials, overlapping and nonoverlapping spheres with a size distribution, etc.), the inequality

$$\zeta_2 \le \eta_2 \le \phi_2 \tag{22.58}$$

is obeyed (see Tables 22.1, 22.2, 22.3, and 22.4, and Figures 22.1 and 22.11); the upper bound corresponds to the symmetric-cell material. Since many realistic composites fall within this class of materials, the fact that η_2 lies in the smaller interval $[\zeta_2, \phi_2]$ is a useful result. Clearly, these limits cannot apply for arbitrary microstructures. However, when they do apply, one can eliminate η_2 from the three-point bounds in favor of ζ_2 or ϕ_2 as we did to get (22.35). Milton and Phan-Thien (1982) showed that for general three-dimensional media, η_2 lies in the smaller interval $[5\zeta_2/21, (16+5\zeta_2)/21]$.

Bounds for Dispersions

Consider the evaluation of improved bounds on the effective moduli of a transversely isotropic fiber-reinforced material consisting of an equilibrium distribution of *unidirectionally* oriented impenetrable glass cylinders in an epoxy matrix in which $G_2/G_1 = 22$, $G_1/K_1 = 0.21$, and $G_2/K_2 = 0.46$. Recall that such a material is fully specified by three effective moduli (Section 13.3), which we choose to be the transverse bulk modulus $k_e \equiv k_e^T$, transverse shear modulus $G_e \equiv G_e^T$, and longitudinal (axial) shear modulus $\mu_e \equiv G_e^L$. Figure 22.12 depicts the three-point Silnutzer lower bound (21.87) and three-point Gibiansky–Torquato upper bound (21.93) on k_e/k_1 for the composite. Relation (22.30) for ζ_2 is used again. Included in the figure are two-point bounds. The three-point bounds provide significant improvement over the two-point bounds and are tight enough to yield good estimates of k_e for a wide range of volume fractions. As in the case of the effective conductivity, the three-point lower bound will provide a good estimate of k_e/k_1 for a wide range of ϕ_2. Figure 22.12 also shows the three-point Silnutzer lower bound (21.95) and three-point Gibiansky–Torquato upper bound (21.100) on the scaled transverse shear modulus G_e/G_1. Here relation (22.55) is used for η_2. As noted in Section 16.2.11, results for the effective transverse conductivity σ_e translate immediately into equivalent results for axial shear modulus μ_e by means of the following change of variables: $\sigma_i \to G_i$ and $\sigma_e \to \mu_e$. Thus, (21.33) and (21.42) with $d = 2$ give three- and four-point bounds on μ_e. These results for μ_e are qualitatively the same as those shown in Figure 22.12 and hence are not plotted.

For three-dimensional isotropic equilibrium dispersions of impenetrable spheres, the relative improvement of the three-point bounds on the effective moduli K_e and G_e over two-point bounds is very similar to that in the analogous two-dimensional instances just described. Hence, it suffices here to compare the three-point bounds on G_e, as obtained from (21.94), (21.101), (21.102), and (22.54), to the two-point Hashin–Shtrikman bounds (21.73) and to experimental data for spherical glass beads in an epoxy matrix (Smith 1976). Figure 22.13 shows that the three-point bounds provide significant improvement over the two-point bounds. Moreover, the data lie closer to the three-point lower bound, since the particles do not cluster (see Section 21.6).

Consider now the case of three-dimensional overlapping spheres with vanishing moduli ($K_2 = G_2 = 0$) in a solid matrix with $G_1/K_1 = 0.5$. Recall that this model is bicontinuous for a wide range of volume fractions (i.e., $0.29 \le \phi_2 \le 0.97$), as discussed in Section 10.6. Figure 22.14 compares the two-point upper bound (21.72) (but with phases interchanged) to the three-point upper bound (21.35) on the effective bulk

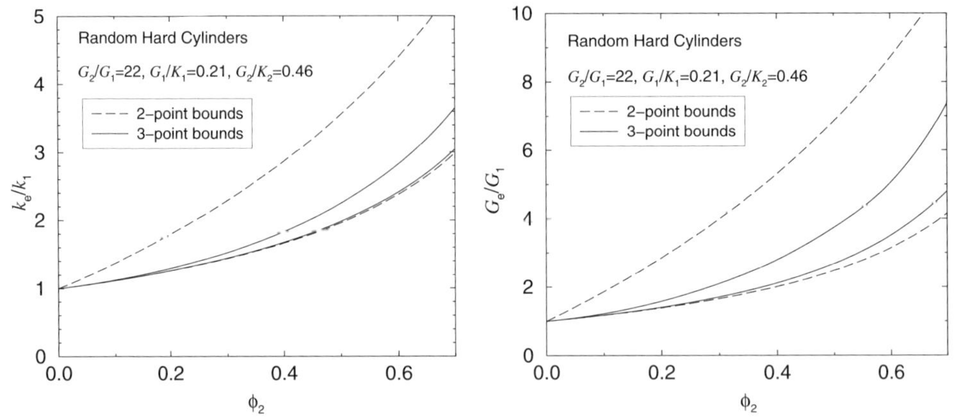

Figure 22.12 Bounds on the scaled effective transverse elastic moduli for a glass–epoxy composite composed of aligned infinitely long nonoverlapping identical circular cylindrical fibers in equilibrium versus fiber volume fraction ϕ_2. Left panel: Bulk modulus from (22.30). Right panel: Shear modulus from (22.55).

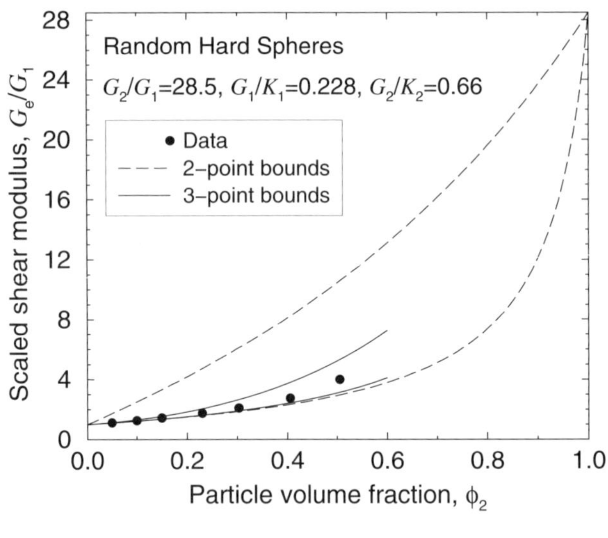

Figure 22.13 Bounds on G_e/G_1 versus ϕ_2 for random equilibrium arrays of identical glass spheres ($d = 3$) in an epoxy matrix. Experimental data are from Smith (1976).

modulus K_e for this model. Included in the figure is the prediction of the three-point approximation (20.184), which, for reasons given in Section 20.2.5, is expected to provide a fairly accurate estimate of K_e. The bounds are qualitatively similar to the corresponding conductivity bounds for perfectly insulating overlapping spheres (see Figure 22.7)

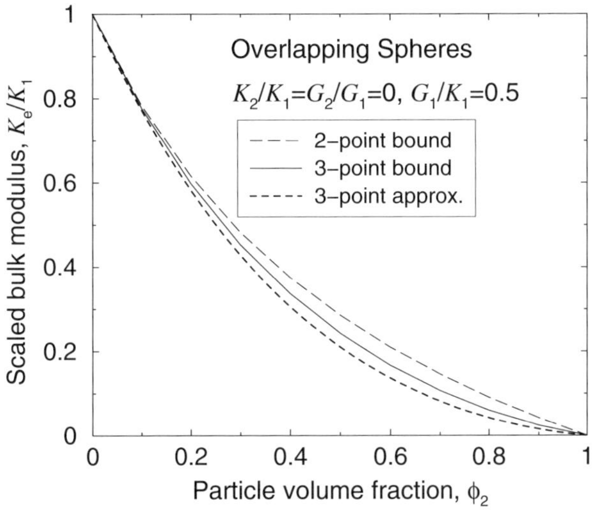

Figure 22.14 Upper bounds on K_e/K_1 versus ϕ_2 for a system of identical overlapping spheres ($d = 3$) with vanishing moduli. The three-point approximation (20.184) is included.

and hence we expect that the three-point upper bound is a relatively sharp bound on the effective bulk modulus.

We note in passing that Davis (1991) used the evaluations of ζ_2 and η_2 for impenetrable spheres (given in Tables 22.1 and 22.3) and the three-point bounds to estimate the effective elastic moduli of metal–matrix composites (silicon carbide particles in aluminum). The theory yielded excellent agreement with experimental measurements of the moduli.

Thovert et al. (1990) have shown that for equilibrium impenetrable spheres with a size distribution, the effect of polydispersivity on three-point bounds on K_e is quantitatively similar to the one on three-point bounds on the effective conductivity σ_e (see Figure 22.8). However, they found the effect of polydispersivity on the effective shear modulus to be considerably smaller.

Bounds for Other Isotropic Media

Three-point bounds on the effective elastic moduli have been determined for other isotropic microstructures. Christiansson and Helsing (1995) have computed them for a class of fiber-reinforced materials consisting of certain hexagonal concentric arrangements of cylindrical shells. This study has shed light on the relationship between the parameters ζ_2 and η_2. The bounds provided sharp estimates of the effective moduli. Hyun and Torquato (2000) have evaluated the effective moduli as well as four-point bounds on them for two-dimensional periodic arrays of triangular and of hexagonal holes (inclusions) over the entire range of volume fractions.

Corson (1974c) calculated the three-point Beran-Molyneux bounds on K_e and three-point McCoy bounds on G_e for binary mixtures of metals, the first such experimental evaluations. Torquato et al. (1999a) have evaluated three-point bounds on the effective moduli of bicontinuous ceramic–metal composites of the type shown in Figure 1.2 by extracting the parameters ζ_2 and η_2 from planar images of the samples. The bounds provided sharp estimates of the effective moduli.

Macroscopically Anisotropic Media

Willis's two-point bounds (21.82) on the effective stiffness tensor $\boldsymbol{C}_e$ are valid for anisotropic multiphase composites. In the special case of two-phase composites in which one phase possesses ellipsoidal symmetry, the key two-point tensor $\boldsymbol{P}^{(0)}$ is related to Eshelby's tensor $\boldsymbol{S}_0$ for an ellipsoid in a matrix with stiffness tensor $\boldsymbol{C}_0$ (see Section 17.2.2) via $\boldsymbol{P}^{(0)} = \boldsymbol{S}_0 : \boldsymbol{C}_0^{-1}$. Weng (1992) evaluated the bounds for oriented ellipsoids in a matrix when the two phases are isotropic. The ellipsoids have a nematic-liquid-crystal structure (Section 7.1). As in the corresponding anisotropic conductivity bounds described in Section 22.1.1, these bounds depend only on the volume fraction and inclusion shape. Note that these bounds, or, equivalently, the bounds (21.79), yield the exact stiffness tensor (16.58) for a first-rank transversely isotropic laminate in the case of aligned disks, in which case Eshelby's tensor is given by (17.98).

22.2.2 Cluster Bounds

Quintanilla and Torquato (1995) showed that their three-point cluster bounds on the effective bulk modulus K_e become identical to the three-point contrast bounds (21.86) when the spheres are totally impenetrable. However, for this geometry the Quintanilla–Torquato bounds on the effective shear modulus G_e are not as sharp as the three-point bounds specified by (21.95) and (21.96) with relations (21.101) and (21.102). For fully penetrable spheres, the Quintanilla–Torquato bounds on the effective moduli are not as sharp as the best three-point contrast-type bounds, but they are simple and analytical. For the special case of fully penetrable spherical cavities ($K_2 = G_2 = 0$), the bounds are even simpler and imperceptibly different from the best three-point contrast bounds. In this latter instance, the lower bounds are trivially zero, while the upper bounds become

$$\frac{K_e}{K_1} \le \frac{\phi_1}{1 + \dfrac{3G_1}{4K_1}\eta}, \qquad \frac{G_e}{G_1} \le \frac{\phi_1}{1 + \dfrac{6K_1 + 12G_1}{9K_1 + 8G_1}\eta}, \tag{22.59}$$

where $\eta = -\ln \phi_1$ is the reduced density.

22.2.3 Security-Spheres Bounds

The security-sphere bounds (21.104) and (21.105) on the effective bulk modulus (Keller et al. 1967) behave qualitatively the same as the corresponding bounds on the effective conductivity (see Figure 22.10). Bruno and Leo (1993) found security-spheres-type bounds on the effective moduli using the complex-variable method. As in the conductivity problem for random arrays of spheres, the security-sphere bounds (21.104) and (21.105) can be much sharper than the Bruno–Leo bounds.

In contrast to the conductivity problem, the Bruno-Leo bounds can be weaker than the security-sphere bounds (21.104) and (21.105) for well-spaced arrays of spheres (e.g., regular lattices). For cubic lattices of spheres in d dimensions, the nearest-neighbor probability density function H_P is given by (21.51), and the security-spheres bounds (21.104) and (21.105) on K_e for extreme phase contrasts are particularly simple:

$$\frac{K_e}{K_1} \leq 1 + \frac{K_1 + 2(d-1)G_1/d}{K_1} \frac{b^d}{b^d - 1} \phi_2 \qquad \text{if} \quad K_2/K_1 = \infty, \tag{22.60}$$

$$\frac{K_e}{K_1} \geq \left[1 + \frac{dK_1 + 2(d-1)G_1}{2(d-1)G_1} \frac{b^d}{b^d - 1} \phi_2 \right]^{-1} \qquad \text{if} \quad K_2/K_1 = 0, \tag{22.61}$$

where b is given by (22.41). The bounds (22.60) and (22.61) are exact through first order in ϕ_2 (see Sections 16.2.4 and 19.2.1) and are relatively sharp up to moderately large volume fractions for both $d = 2$ and $d = 3$ (see Section 16.2.4). Finally, we note that the bounds (22.60) and (22.61) are exact through all orders in ϕ_2 for $d = 1$.

22.3 Trapping Constant

We describe below evaluations of interfacial-surface, void, cluster, security-spheres, and pore-size bounds on the trapping constant $\gamma \equiv (\mathcal{D}\tau\phi_1)^{-1}$ or mean survival time τ. [Other definitions for γ exist that differ by factors involving either the porosity ϕ_1, the diffusion coefficient $\mathcal{D}$, or both; see Richards and Torquato (1987) and Lee, Kim, Miller and Torquato (1989) for further explanations.] Results are given for model microstructures as well as real heterogeneous media.

22.3.1 Interfacial-Surface Lower Bound

The two-point interfacial-surface lower bound (21.114) has been evaluated for *equilibrium* arrays of identical three-dimensional spherical traps in the two extremes of the cherry-pit model: $\lambda = 0$, corresponding to fully overlapping spheres, and $\lambda = 1$, corresponding to nonoverlapping spheres by Torquato (1986d); see also Richards and Torquato (1987). The relevant correlation functions are relatively easy to evaluate for both models (see Sections 5.1 and 5.2). For any λ, the lower bound (21.114) produces the exact Smoluchowski dilute-limit result (19.89), i.e., $\gamma_s = 3\phi_2/R^2$ for spherical traps of radius R in the limit that $\phi_2 \to 0$.

Figure 22.15 compares the lower bound (21.114) for $d = 3$ for perfectly absorbing nonoverlapping spherical traps to a theoretical expression due to Richards (1987) and computer simulations (Lee et al. 1989). It is seen that the bound is relatively sharp. Note also that Richards's theory violates the bound at $\phi_2 \approx 0.5$. Figure 22.16 compares

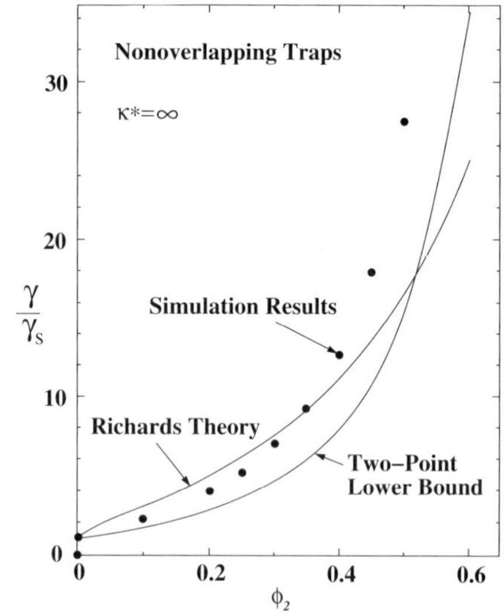

Figure 22.15 Scaled trapping constant γ/γ_s versus trap volume fraction ϕ_2 for equilibrium arrays of three-dimensional nonoverlapping spherical traps of radius R. Comparison of two-point interfacial lower bound (21.114) for $d = 3$, Richards's theory (Richards 1987), and computer simulations (Lee et al. 1989). Here $\gamma_s = 3\phi_2/R^2$ and $\kappa^* = \kappa R/\mathcal{D} = \infty$ (perfect absorption).

simulation data for the *mean survival time* $\tau = (\mathcal{D}\gamma\phi_1)^{-1}$ for perfectly absorbing overlapping spherical traps to the two-point interfacial upper bound on τ, obtained from (21.114) for $d = 3$, and the corresponding pore-size lower bound (21.132) evaluated below [cf. (22.77)]. Note that the bounds are relatively tight; the interfacial upper bound on τ (lower bound on γ) is of course very sharp at low ϕ_2 (data not shown in this range), whereas the pore-size lower bound becomes sharp at high ϕ_2.

The two-point bound (21.114) has been computed for dilute concentrations of identical three-dimensional spheres in the equilibrium cherry-pit model by Rubinstein and Torquato (1988) through order ϕ_2^2. Dividing this bound by γ_s gives

$$\frac{\gamma}{\gamma_s} \geq 1 + \left(\frac{15}{8} + \frac{25}{8}\lambda\right)\phi_2 + \mathcal{O}(\phi_2^2). \tag{22.62}$$

The correction to the Smoluchowski dilute-limit result γ_s accounts for interactions between pairs of traps. As the impenetrability parameter λ increases, the second-order coefficient and hence the lower bound increases, as expected, since the surface area available for reaction increases.

The two-point interfacial surface bound has also been computed for spherical traps with a polydispersivity in size (Miller and Torquato 1989). It was found that the trapping constant γ for the polydisperse case can increase or decrease, relative to the monodisperse case, depending upon whether the *relative* interfacial surface area increases or decreases. Given et al. (1990a) also evaluated bound (21.114) for polydisperse traps.

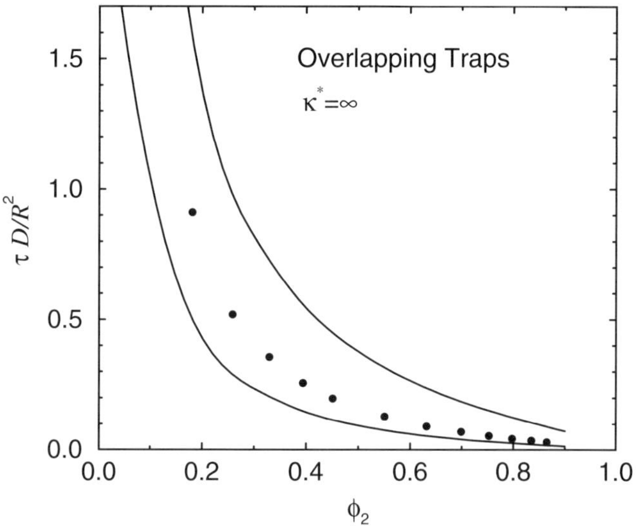

Figure 22.16 Scaled mean survival time $\tau\mathcal{D}/R^2$ versus trap volume fraction ϕ_2 for an array of three-dimensional fully overlapping spherical traps of radius R. Upper bound is two-point interfacial bound (21.114) and the lower bound is pore-size bound (22.77) with $\kappa = \infty$. Black circles are simulation data (Lee et al. 1989) and $\kappa^* = \kappa R/D = \infty$ (perfectly absorbing traps).

22.3.2 Void Lower Bound

For the statistically anisotropic model of oriented, possibly overlapping, spheroidal traps of length $2b$ and maximum diameter $2a$ at number density ρ (i.e., nematic-liquid-crystal structure; see Section 7.1), the two-point void lower bound (21.117) on γ $(d = 3)$ has been determined by Torquato and Lado (1991). Using the expression (7.20) for the void two-point probability function $S_2 \equiv F_{vv}$, they found that

$$\gamma \geq \frac{\phi_2^2 f(b/a)}{4a^2 K_o}, \tag{22.63}$$

where

$$f(b/a) = \begin{cases} \dfrac{2\chi_b}{\ln\left(\dfrac{1+\chi_b}{1-\chi_b}\right)}, & b/a \geq 1, \\ \dfrac{\chi_a}{\tan^{-1}\chi_a}, & b/a \leq 1, \end{cases} \tag{22.64}$$

is a shape-dependent function for which

$$\chi_a^2 = -\chi_b^2 = (a^2/b^2) - 1. \tag{22.65}$$

Also,

$$\phi_2 = \rho\frac{4\pi}{3}a^2 b \tag{22.66}$$

is the spheroid volume fraction, and

$$K_o = \int_0^\infty x[S_2(x) - \phi_1^2]dx \tag{22.67}$$

is a density-dependent parameter such that $S_2(x)$ is the void two-point probability function for isotropic distributions of *spheres* of radius $a = b = R$, with $x = r/(2R)$ a dimensionless distance.

Given the evaluation of the void bound

$$\gamma \geq \frac{\phi_2^2}{4a^2 K_o} \tag{22.68}$$

for spheres with an arbitrary degree of impenetrability, one can obtain corresponding results for oriented prolate and oblate spheroids by normalizing the former with the shape-dependent factor of (22.64). The bound (22.63) depends on the inclusion shape as well as volume fraction ϕ_2 in a complex fashion through the parameter $K_o(\phi_2)$, where ϕ_2 for spheres is set equal to the value of ϕ_2 for spheroids. Since $f(b/a) > 1$ for $b/a < 1$, the void bound for oblate spheroids is always larger than for spheres ($b/a = 1$). Similarly, since $f(b/a) < 1$ for $b/a > 1$, the void bound for prolate spheroids is always smaller than for spheres. Thus, the void bound captures the essential physics of the true behavior of the trapping constant for spheroids relative to spheres (Miller et al. 1991).

Torquato and Rubinstein (1989) have shown that for infinitely dilute ($\phi_2 \to 0$) spherical traps of radius a, $K_o \sim \phi_2/10$, and thus (22.63) yields $\gamma \geq 5\phi_2/(2a^2)$. This implies that unlike two-point interfacial surface and cluster bounds, the void bound does not give the Smoluchowski dilute-limit result. Using the above result in bound (22.63) yields the corresponding result for spheroids at the same number density of spheres:

$$\gamma \geq \frac{5\phi_2}{2a^2} f(b/a) \qquad (\phi_2 \to 0). \tag{22.69}$$

This is to be compared to the exact dilute-limit result for spheroids:

$$\gamma_s = \frac{3\phi_2}{a^2} f(b/a) \qquad (\phi_2 \to 0). \tag{22.70}$$

Thus, the void bound does give the correct shape dependence on the spheroid and, in fact, is correct to within a factor of 5/6 when $\phi_2 \to 0$.

22.3.3 Cluster Lower Bounds

Rubinstein and Torquato (1988) evaluated the three-point cluster bound (21.122) analytically for $d = 3$ for identical spherical traps in the equilibrium cherry-pit model through second order in the volume fraction ϕ_2 and arbitrary values of the impenetrability index λ. Again, dividing this bound by γ_s gives

$$\frac{\gamma}{\gamma_s} \geq 1 + \left[\frac{3}{2} - \lambda^6 + 3\lambda^4 + \frac{9}{8}\lambda^2 - \frac{9}{8}\lambda + \frac{3}{8}(1 + 4\lambda^2)\ln(1 + 2\lambda)\right]\phi_2 + \mathcal{O}(\phi_2^2). \tag{22.71}$$

Comparing (22.71) to the corresponding result (22.62) for the two-point interfacial-surface bound reveals that for $0.6 < \lambda \leq 1$, the former is sharper than the latter, but the converse is true for $0 \leq \lambda \leq 0.6$. The reason why the three-point bound is sharper for large λ and not as sharp for small λ was already discussed in Section 21.3.3.

For fully penetrable traps ($\lambda = 0$), the three-point cluster lower bound has been evaluated analytically for all volume fractions by Rubinstein and Torquato (1988). For this model the correlation functions G_2 and G_3 are trivial, as seen from (5.34). They found for $d = 3$ that

$$\frac{\gamma}{\gamma_s} \geq \frac{-\ln \phi_1}{\phi_1 \phi_2}, \tag{22.72}$$

which again is not as sharp as the interfacial-surface bound (21.114). However, for totally impenetrable spheres ($\lambda = 1$), the three-point cluster bound (21.122) is always sharper than (21.114).

22.3.4 Security-Spheres Upper Bound

Security-spheres *upper bounds* on γ have been computed for simple cubic lattices of identical impenetrable spherical sinks (Rubinstein and Torquato 1988, Torquato and Rubinstein 1989) to all orders in ϕ_2. For example, use of relation (21.51) for the nearest-neighbor probability density function H_P in bound (21.127) yields

$$\frac{\gamma}{\gamma_s} \leq \frac{b}{b-1}\left[1 - \frac{\phi_2 b(1+b)}{2}\right]^{-2}, \tag{22.73}$$

where b is given by (22.41) with (22.42) and $d = 3$. Expanding this result for small volume fractions gives

$$\frac{\gamma}{\gamma_s} \leq 1 + \left[\left(\frac{6}{\pi}\right)^{1/3} + \left(\frac{\pi}{6}\right)^{2/3}\right]\phi_2^{1/3} + \mathcal{O}(\phi_2^{2/3}). \tag{22.74}$$

This upper bound is of the same form as the exact result (16.86), whose coefficient of order $\phi_2^{1/3}$ is approximately 1.7601 and which therefore provides a sharp upper bound. These authors also computed security-spheres upper bounds for random spheres at low trap concentrations.

22.3.5 Pore-Size Upper Bound

For the simple microstructure consisting of nonoverlapping (i.e., disconnected) *three-dimensional* spherical pores of radius R, the pore-size probability density function is given by

$$P(\delta) = \begin{cases} \dfrac{3(R-\delta)^2}{R^3}, & \delta < R, \\[2ex] 0, & \delta > R. \end{cases} \tag{22.75}$$

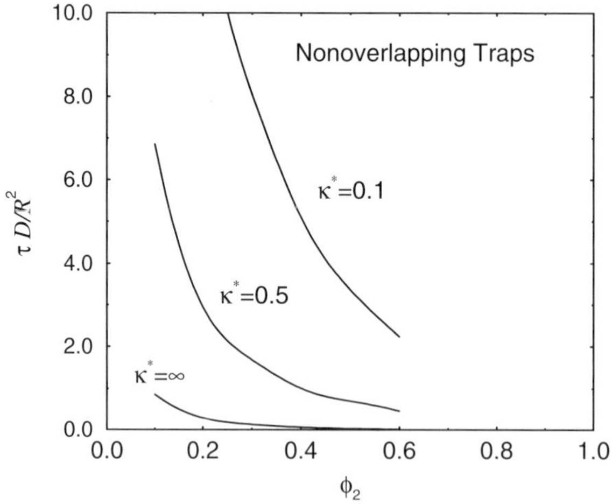

Figure 22.17 Lower bound (21.132) on the dimensionless survival time $\tau\mathcal{D}/R^2$ versus ϕ_2 for an equilibrium array of totally impenetrable spherical traps ($d = 3$) for several values of the dimensionless surface rate constant $\kappa^* = \kappa R/\mathcal{D}$ ($\kappa^* = \infty, 0.5,$ and 0.1).

This expression combined with (2.82) yields the moments as

$$\langle \delta^n \rangle = \frac{6R^n}{(n+1)(n+2)(n+3)}. \tag{22.76}$$

Therefore, for transport inside spherical pores of radius R, inequality (21.132) and relation (22.76) give (Torquato and Avellaneda 1991)

$$\tau \geq \frac{R^2}{16\mathcal{D}} + \frac{R}{3\kappa}. \tag{22.77}$$

Comparison of this result to the exact result

$$\tau = \frac{R^2}{15\mathcal{D}} + \frac{R}{3\kappa}, \tag{22.78}$$

as obtained from (16.83), shows that the bound is very sharp. This is expected, since the mean-square displacement of a Brownian particle (confined to be in a pore region characterized by a single size) is well described by the lower-order moments of $P(\delta)$.

The bounds, of course, will not be as sharp for connected pore regions, especially when there is a wide pore size distribution. Torquato and Avellaneda (1991) computed the pore-size lower bound (21.132) on τ using the pore-size function $P(\delta)$ formula (5.121) for random equilibrium distributions of identical spherical traps in the *cherry-pit model* for various values of the impenetrability parameter λ. Figure 22.16 includes the pore-size bound (21.132) for overlapping traps ($\lambda = 0$) in the diffusion-controlled limit. Figure 22.17 shows the lower bound (21.132) on the survival time versus ϕ_2 for nonoverlapping traps ($\lambda = 1$) for several values of κ.

The pore-size bound (21.132) has been computed for the Fontainebleau sandstone depicted in Figure 12.14 (Coker et al. 1996) and a porous magnetic gel (Rintoul, Torquato, Yeong, Erramilli, Keane, Dabbs and Aksay 1996). Moreover, for a broad class of porous media (including sandstones), a "universal" scaling has been found for the mean survival time τ of three-dimensional media in the diffusion-controlled limit (Torquato and Yeong 1997). The functional form of this approximately universal scaling relation was motivated by the pore-size lower bound on τ (21.132) and is expressible as a simple function of porosity ϕ_1, specific surface s, and mean pore size $\langle\delta\rangle$. Specifically, it is given by

$$\frac{\tau}{\tau_o} = \frac{8}{5}x + \frac{8}{7}x^2, \tag{22.79}$$

where $x = \langle\delta\rangle^2/(\tau_o \mathcal{D})$ and $\tau_o = (3\phi_2)/(\mathcal{D}\phi_1 s^2)$.

We note in passing that the upper bounds (21.133) and (21.134) on the principal relaxation time T_1 for transport interior to nonoverlapping three-dimensional spherical pores of radius a are given by

$$T_1 \geq \frac{a}{3\kappa} + \frac{a^2}{20\mathcal{D}}, \qquad \frac{\kappa a}{\mathcal{D}} \ll 1, \tag{22.80}$$

$$T_1 \geq \frac{a^2}{10\mathcal{D}} + \frac{5a}{32\kappa}, \qquad \frac{\kappa a}{\mathcal{D}} \gg 1, \tag{22.81}$$

where we have used (22.76) and $s = 3/a$. Comapring these bounds to the exact results (13.208) and (13.209) reveals that the bounds provide sharp estimates of T_1 for these simple model microstructures.

22.4 Fluid Permeability

Here we discuss calculations of interfacial-surface, void, cluster, and security-spheres bounds on the fluid permeability k. Results are given for model microstructures as well as real porous media.

22.4.1 Interfacial-Surface Upper Bound

The two-point interfacial-surface upper bound (21.140) for $d = 3$ has been evaluated for random (i.e., *equilibrium*) beds of identical spheres for arbitrary sphere volume fraction ϕ_2 in the two extremes of the cherry-pit model: $\lambda = 0$, corresponding to fully overlapping spheres, and $\lambda = 1$, corresponding to nonoverlapping spheres (Torquato 1986c). The relevant correlation functions are relatively easy to evaluate for both models (see Sections 5.1 and 5.2). The bound has also been computed for spheres with a polydispersivity in size (Torquato and Lu 1990). For any λ, the upper bound (21.140) yields the exact Stokes dilute-limit result (19.103), i.e., $k_s = 2R^2/(9\phi_2)$ for spheres of radius R in the limit that $\phi_2 \to 0$.

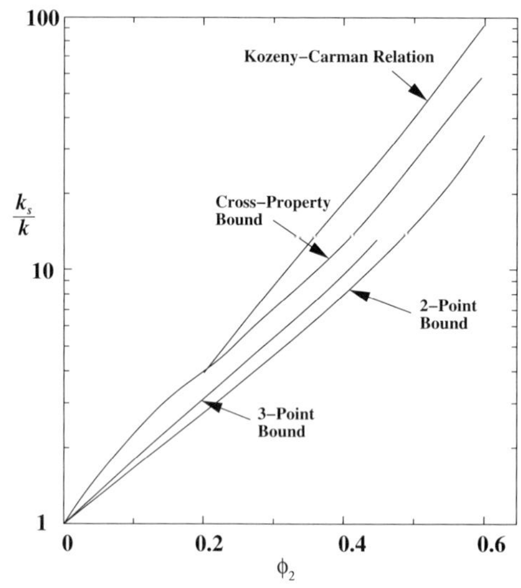

Figure 22.18 Comparison of bounds on the scaled inverse fluid permeability (resistance) k_s/k for equilibrium arrays of identical three-dimensional nonoverlapping spheres of radius R versus sphere volume fraction ϕ_2. Included is two-point interfacial-surface lower bound (21.140), optimized three-point cluster lower bound described below, cross-property bound (23.50), and empirical Kozeny-Carman relation (16.99) with $c = 5$. Here $k_s = 2R^2/(9\phi_2)$.

Figure 22.18 compares, for equilibrium nonoverlapping spheres, the lower bound on k, or equivalently, upper bound on the inverse fluid permeability (resistance) k^{-1}, obtained from (21.140), to the *optimized* three-point cluster lower bound described in Section 22.4.3 and the cross-property bound (23.50) that utilizes information about the mean survival time. Included in the figure is the *empirical* Kozeny-Carman relation (16.99) with $c = 5$. It is seen that the three-point bound is superior to the two-point bound. However, the cross-property bound is the sharpest.

The two-point interfacial-surface upper bound (21.140) also has been computed for dilute concentrations of identical spheres of radius R in the equilibrium cherry-pit model (Torquato and Beasley 1987). This upper bound, for $d = 3$, becomes the following lower bound on the resistance k^{-1}:

$$\frac{k_s}{k} \geq 1 + \left(\frac{15}{8} + \frac{25}{8}\lambda\right)\phi_2 + \mathcal{O}(\phi_2^2). \tag{22.82}$$

Note that this result for the scaled resistance is identical to (22.62) for the scaled trapping constant. Again, as before, the effect of increasing the impenetrability index λ is to increase the scaled resistance.

The interfacial-surface bound has been computed for overlapping spherical grains with a continuous size distribution (Torquato and Lu 1990). These results are summarized in Figure 22.19, where the scaled resistance k_s/k is plotted versus ϕ_2 for the Schulz distribution (6.5). The generalized dilute Stokes result for polydisperse spheres

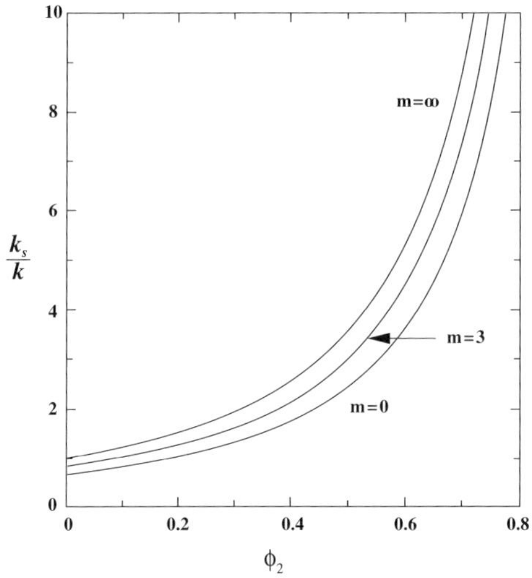

Figure 22.19 Two-point interfacial-surface lower bound (21.140) on the scaled inverse fluid permeability (resistance) k_s/k for polydisperse arrays of three-dimensional overlapping spherical grains versus ϕ_2. The radii are distributed according to the Schulz distribution (6.5). Polydispersivity increases as the parameter m increases (monodisperse case is $m = \infty$).

is given by

$$k_s = \frac{2\langle R^3 \rangle}{9\langle R \rangle \phi_2}. \tag{22.83}$$

Increasing the degree of polydispersivity (i.e., reducing m) decreases the specific surface (relative to the monodisperse limit) and thus decreases the fluid resistance. Interestingly, scaling the bound on the resistance k^{-1} by the square of the specific surface (relative to the monodisperse case) enables all of the curves to collapse, to an excellent approximation, onto the monodisperse result. Thus, given the monodisperse result, one can compute the bound on k^{-1} for any degree of polydispersivity. Note that for any degree of polydispersivity, the two-point bound on k_s/k does not yield the exact result (22.83) in the limit $\phi_2 \to 0$.

It is noteworthy that a scaling ansatz based on the interfacial-surface upper bound (21.140) causes the near collapse of the permeability data of a wide class of porous media, including several types of sandstones, onto a "universal" curve (Martys, Torquato and Bentz 1994b). This relation for k depends on the porosity, specific surface, and the porosity at which the pore space first percolates.

22.4.2 Void Upper Bound

The void upper bound (21.143) on the fluid permeability has been computed by Torquato (1980) for both overlapping and nonoverlapping spheres. As in the case of the void bound on the trapping constant, this bound does not yield the exact result in the infinitely dilute limit. When $\phi_2 \to 0$, it gives $k \leq 4R^2/(15\phi_2)$ for three-dimensional spheres of radius R. The void upper bound (21.143) has also been computed for the Fontainebleau sandstone shown in Figure 12.14 (Coker et al. 1996), but it is not a sharp estimate.

22.4.3 Cluster Upper Bounds

The three-point cluster bound (21.148) on the inverse permeability k^{-1} has been evaluated by Torquato and Beasley (1987) for three-dimensional random arrays of identical spheres in the equilibrium cherry-pit model through second order in ϕ_2. Multiplying this bound by k_s, they obtained

$$\frac{k_s}{k} \geq 1 + \left[\frac{3}{2} - \lambda^6 + \frac{11}{4}\lambda^4 + \frac{5}{6}\lambda^2 - \frac{9}{8}\lambda + \frac{3}{4}(1+3\lambda^2)\ln(1+2\lambda) \right.$$
$$\left. - \frac{\lambda(1+7\lambda)}{16(1+2\lambda)^2} + \frac{\lambda}{16(1+2\lambda)^4}\right]\phi_2 + \mathcal{O}(\phi_2^2). \tag{22.84}$$

Note that bounds (22.71) and (22.84) are not the same for $\lambda > 0$. Comparing the two-point interfacial bound (22.82) to (22.84) reveals that the latter is sharper than the former for $0.52 < \lambda \leq 1$ with the converse holding for $0 \leq \lambda \leq 0.52$. The same qualitative behavior was seen in the trapping problem and occurs here for the same reasons (see Section 22.3.3).

For identical overlapping spheres, the correlation functions G_2 and G_3 are trivial [cf. (5.34)] and the three-point cluster bound (21.148) yields (Torquato and Beasley 1987) the analytical result

$$\frac{k_s}{k} \geq -\frac{\ln\phi_1}{\phi_1\phi_2}. \tag{22.85}$$

This bound was first derived by Weissberg and Prager (1970) using a different procedure. It is not as sharp as the interfacial-surface bound (21.140) for this model (Torquato 1986c) for the reasons described in Section 22.3.3.

The *optimized* three-point cluster bound due to Beasley and Torquato (1989) turns out to be sharper than the bound (21.148) for $0 < \lambda \leq 1$. This optimized bound involves precisely the same multidimensional integrals involving the correlation functions G_2 and G_3 [cf. (21.148)], whose representations in terms of the two- and three-body distribution functions g_2 and g_3 are given by (5.108). The optimized bound was simplified (by expanding certain angle-dependent terms of the integrands in orthogonal polynomials and exploiting the orthogonality properties of the appropriate basis set) and then evaluated for identical impenetrable spheres in the superposition approximation (3.41). This bound is shown in Figure 22.18. The superposition approximation is accurate for $0 \leq \phi_2 \leq 0.45$ but increasingly overestimates k_s/k for $\phi_2 \geq 0.45$.

22.4.4 Security-Spheres Lower Bound

The security-spheres bound (21.154) on k has been computed exactly (to all orders in ϕ_2) for a simple cubic lattice of identical impenetrable spheres and approximately for a dilute random array of such spheres (Rubinstein and Keller 1987, Rubinstein and Torquato 1989). For example, use of relation (21.51) for the nearest-neighbor probability density function H_P in bound (21.154) yields the following lower bound on k for a simple cubic lattice:

$$\frac{k}{k_s} \geq (1-b^5)\left(1-\frac{9}{4}b+\frac{5}{2}b^3-\frac{9}{4}b^5+b^6\right)^{-1}, \tag{22.86}$$

where b is given by (22.41) with (22.42) and $d = 3$. Expanding this result for small volume fractions yields

$$\frac{k}{k_s} \geq 1-\frac{9}{4}\left(\frac{6}{\pi}\right)^{1/3}\phi_2^{1/3}+\mathcal{O}(\phi_2^{2/3}). \tag{22.87}$$

This bound is of the same form as the exact result of (16.100), whose coefficient of order $\phi_2^{1/3}$ is approximately -1.7601.

CHAPTER 23

Cross-Property Relations

An intriguing fundamental as well as practical question in the study of heterogeneous materials is the following: Can different properties of the medium be *rigorously* linked to one another? Such *cross-property* relations become especially useful if one property is more easily measured than another property. Since the effective properties of random media reflect certain morphological information about the medium, one might expect that one could extract useful information about one effective property given an accurate (experimental or theoretical) determination of a different effective property, even when their respective governing equations are uncoupled. Cross-property relations provide a means of ascertaining the possible range of values that different effective properties can possess (i.e., the allowable region in multidimensional property space) and thus have important implications for the design of *multifunctional* composites.

Cross-property relations involving properties with the same or similar governing equations have been established. Prager (1969) found bounds linking the effective thermal conductivity to the effective magnetic permeability (or electrical conductivity) of isotropic two-phase media. Subsequently, Milton (1981a) and Cherkaev and Gibiansky (1992) found optimal bounds of this type. For isotropic two-dimensional two-phase media, Cherkaev and Gibiansky (1993) found cross-property bounds between the effective bulk modulus and effective shear modulus that improve upon the restrictions set forth by the Hashin–Shtrikman bounds (a rectangle in the bulk modulus-shear modulus plane whose sides are determined by the upper and lower bounds on both moduli). Berryman and Milton (1988) obtained corresponding elastic-moduli bounds for isotropic three-dimensional porous media.

Our focus in this chapter will be on cross-property relations between effective properties with different governing partial differential equations. In particular, we will describe rigorous cross-property relations for two types of problems: (1) links between the effective *conductivity* and the effective *elastic moduli* of two-phase isotropic media, and (2) links between the flow properties of porous media, such as the fluid permeability and viscous relaxation times, and effective *diffusion* parameters, such as the mean survival time (or trapping constant), diffusion relaxation times, and effective conductivity. Interesting consequences of both types of cross-property relations, including percolation behavior, will be discussed.

23.1 Conductivity and Elastic Moduli

Can the linear response of a composite to a *mechanical* load be related to the linear response of the same composite to an *electrical (or thermal)* load? Here it will be shown that the effective conductivity and effective elastic moduli are indeed generally related to one another, even though the former is governed by Laplace's equation and the latter by the vector equilibrium equations. Since the determination of the effective conductivity is mathematically equivalent to finding the dielectric constant, magnetic permeability, and diffusion coefficient (see Table 1.1), it follows that the cross-property relations given below connect the elastic moduli to any of these properties as well. We begin with a discussion of elementary bounds and then describe and apply the sharper *translation* bounds.

23.1.1 Elementary Bounds

Milton (1984) showed that for arbitrary d-dimensional isotropic two-phase media, if the phase bulk moduli K_i equal the phase conductivities σ_i, then the effective bulk modulus K_e is bounded from above by the effective conductivity σ_e. This result was trivially extended (Torquato 1992) to the more general situation in which $K_2/K_1 \le \sigma_2/\sigma_1$ and is stated as the following theorem:

Theorem 23.1 *For d-dimensional isotropic two-phase media of arbitrary topology having nonnegative-phase Poisson's ratios ν_i and $K_2/K_1 \le \sigma_2/\sigma_1$, the dimensionless effective bulk modulus K_e/K_1 and dimensionless effective conductivity σ_e/σ_1 are related to one another via the bound*

$$\frac{K_e}{K_1} \le \frac{\sigma_e}{\sigma_1}. \tag{23.1}$$

Proof: The proof follows closely the one given by Milton (1984). We begin with the variational upper bound on the effective bulk modulus:

$$K_e \le \langle \lambda [\mathrm{Tr}\, \hat{\varepsilon}]^2 + 2G\, \hat{\varepsilon} : \hat{\varepsilon} \rangle, \tag{23.2}$$

where $\hat{\varepsilon}$ is a trial strain field such that $\langle \hat{\varepsilon} \rangle = \boldsymbol{I}/d$. This inequality is obtained from the variational upper bound (14.95) and the identity $d\lambda = dK - 2G$ [cf. (13.94)], where λ is the Lamé constant. The essential idea is to use *exact* electric potentials as components of trial displacement fields, i.e., $\hat{\boldsymbol{u}} = \boldsymbol{T}$, so that $\langle\hat{\varepsilon}\rangle = \langle\nabla T\rangle = \boldsymbol{I}/d$. Here T_i are the components of the vector $\boldsymbol{T}$, each of which is an exact electric potential. It follows from the energy representation (14.31) that the scalar effective conductivity is

$$\sigma_e = d\langle\sigma\nabla\boldsymbol{T} : \nabla\boldsymbol{T}\rangle \tag{23.3}$$

and the trial strain is

$$\hat{\varepsilon} = \frac{1}{2}\left[\nabla\boldsymbol{T} + (\nabla\boldsymbol{T})^{\mathrm{T}}\right]. \tag{23.4}$$

Substitution of relation (23.4) into (23.2) and use of the inequalities $[\mathrm{Tr}\,\hat{\varepsilon}]^2 \le d\hat{\varepsilon} : \hat{\varepsilon} \le d\nabla\boldsymbol{T} : \nabla\boldsymbol{T}$ leads to the upper bound

$$K_e \le d\langle K\nabla\boldsymbol{T} : \nabla\boldsymbol{T}\rangle, \tag{23.5}$$

where we have again used the identity $dK = d\lambda + 2G$. In arriving at (23.5), we have assumed that $\lambda_1 \ge 0$ and $\lambda_2 \ge 0$, which is equivalent to assuming that $\nu_1 \ge 0$ and $\nu_2 \ge 0$ (see Section 13.3.2). We can rewrite (23.3) and (23.5) in terms of the indicator function $\mathcal{I}^{(i)}$ for phase i as follows:

$$\begin{aligned}\frac{\sigma_e}{\sigma_1} &= d\langle\mathcal{I}^{(1)}\nabla\boldsymbol{T} : \nabla\boldsymbol{T}\rangle + d\frac{\sigma_2}{\sigma_1}\langle\mathcal{I}^{(2)}\nabla\boldsymbol{T} : \nabla\boldsymbol{T}\rangle,\\ \frac{K_e}{K_1} &\le d\langle\mathcal{I}^{(1)}\nabla\boldsymbol{T} : \nabla\boldsymbol{T}\rangle + d\frac{K_2}{K_1}\langle\mathcal{I}^{(2)}\nabla\boldsymbol{T} : \nabla\boldsymbol{T}\rangle.\end{aligned}$$

Combination of the last two relations, under the assumption that $K_2/K_1 \le \sigma_2/\sigma_1$, proves the theorem. ∎

Using Theorem 23.1, Torquato (1992) was able to derive an upper bound on the effective shear modulus G_e in terms of σ_e and the effective Poisson ratio ν_e or in terms of σ_e only. These results are stated as the following corollary:

Corollary 23.1 *For d-dimensional isotropic two-phase media of arbitrary topology having nonnegative-phase Poisson's ratios ν_i and $K_2/K_1 \le \sigma_2/\sigma_1$, the dimensionless effective shear modulus G_e/K_1 is related to the dimensionless effective conductivity σ_e/σ_1 and effective Poisson's ratio via the bound*

$$\frac{G_e}{K_1} \le \frac{\sigma_e}{\sigma_1}\frac{d[1-\nu_e(d-1)]}{2(1+\nu_e)}. \tag{23.6}$$

A weaker bound on G_e/K_1 just involving σ_e/σ_1 is

$$\frac{G_e}{K_1} \le \frac{d}{2}\frac{\sigma_e}{\sigma_1}, \tag{23.7}$$

which follows immediately from (23.6), provided that $\nu_e \ge 0$.

Proof: The proof of (23.6) follows immediately from Theorem 23.1 and the interrelation $K_e/G_e = 2(1+\nu_e)/\{d[1-\nu_e(d-1)]\}$ [cf. (13.131)]. ∎

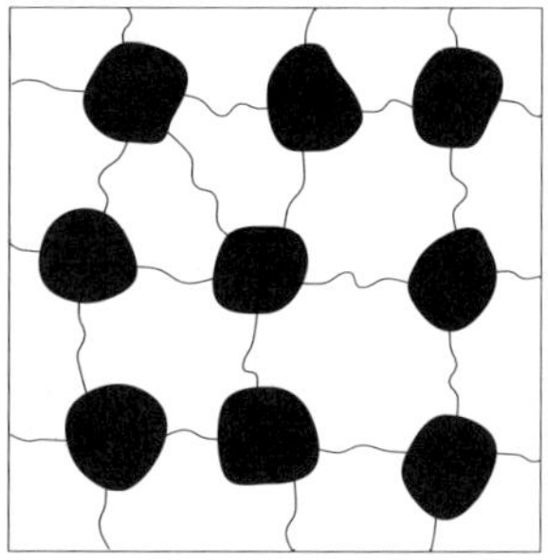

Figure 23.1 A schematic of a compressible but superconducting composite composed of an incompressible and superconducting phase (black) and a compressible finite-conductivity phase (white).

The above cross-property bounds have interesting implications. As a first example, consider a composite in which phase 1 is compressible ($K_1 < \infty$) with finite conductivity ($\sigma_1 < \infty$), and phase 2 is incompressible ($K_2 = \infty$) and superconducting ($\sigma_2 = \infty$). Milton (1984) observed that relation (23.1) implies that the composite cannot be incompressible with finite conductivity, but it does not forbid a composite that is compressible and superconducting. Can one imagine a specific microstructure corresponding to the latter situation? In Figure 23.1 we depict such an example. The thin "wires" of the black superconducting, incompressible phase that connect the larger "blobs" enable the composite to be superconducting as well. However, because the black phase has a finite shear modulus, the thin wires will bend when subjected to hydrostatic loading, rendering the composite compressible.

As a second example, consider phase 2 to be a void phase, i.e., $K_2 = G_2 = \sigma_2 = 0$. Near the connectivity threshold ($\phi_1 \to \phi_{1c}^+$), let us assume that the effective conductivity and elastic moduli have the same thresholds and obey the following scaling laws (Section 9.2.3):

$$\sigma_e \sim (\phi_1 - \phi_{1c})^t, \tag{23.8}$$

$$K_e \sim (\phi_1 - \phi_{1c})^f, \tag{23.9}$$

$$G_e \sim (\phi_1 - \phi_{1c})^g, \tag{23.10}$$

where t, f, and g are the respective critical exponents. In contrast to Chapter 9, where we defined a critical exponent for the Young's modulus only, we more generally define two critical exponents f and g for the bulk and shear moduli. It is often found that $f = g$, implying that the effective Poisson ratio ν_e is a constant at the percolation threshold. Torquato (1992) showed that the above scaling laws in conjunction with relations (23.1) and (23.7) immediately lead to the inequalities

$$f \geq t, \qquad g \geq t, \qquad d \geq 2. \tag{23.11}$$

Thus, the critical exponents for the elastic moduli are generally greater than the critical exponent for the conductivity. Results from theoretical (Feng et al. 1987) and experimental (Smith and Lobb 1979, Benguigui 1986) studies of continuum percolation are consistent with the inequalities (23.11).

23.1.2 Translation Bounds for $d = 2$

The *translation* method (Lurie and Cherkaev 1984, Murat and Tartar 1985, Lurie and Cherkaev 1986, Milton 1990) is a powerful means of obtaining sharp bounds on the effective properties of composites. It is based on bounding the effective properties of a comparison medium whose local properties differ from the original medium by a constant translation tensor that is *quasi-convex*. A description of the translation method is beyond the scope of this book. Gibiansky and Torquato (1993, 1995b) employed the translation method to find the sharpest known bounds on the sets of pairs (σ_e, k_e) and (σ_e, G_e) for two-dimensional two-phase isotropic composites of all possible microstructures at a prescribed or arbitrary volume fraction ϕ_i. These bounds enclose certain regions in the σ_e-k_e and σ_e-G_e planes (see Figure 23.2), of which portions are realizable by certain microstructures and thus are optimal. As before, we will denote the two-dimensional (planar) bulk modulus by k (see Section 13.3).

Before stating the bounds, we must introduce certain definitions and notations. Denote by $F(a_1, a_2, \phi_1, \phi_2, y)$ the function

$$F(a_1, a_2, \phi_1, \phi_2, y) = a_1\phi_1 + a_2\phi_2 - \frac{\phi_1\phi_2(a_1 - a_2)^2}{a_1\phi_2 + a_2\phi_1 + y}, \tag{23.12}$$

which is a scalar version of the inverse Y-transformation (Milton 1991, Cherkaev and Gibiansky 1992). Henceforth, we indicate explicitly only the last argument of F and use the shorthand notation $F_a(y) \equiv F(a_1, a_2, \phi_1, \phi_2, y)$. Now let σ_{1*}, σ_{2*}, k_{1*}, k_{2*} G_{1*}, G_{2*}, G_{3*}, and G_{4*} denote the expressions

$$\sigma_{1*} = F_\sigma(\sigma_1), \qquad \sigma_{2*} = F_\sigma(\sigma_2), \tag{23.13}$$

$$k_{1*} = F_k(G_1), \qquad k_{2*} = F_k(G_2), \tag{23.14}$$

$$G_{1*} = F_G[k_1G_1/(k_1 + 2G_1)], \qquad G_{2*} = F_G[k_2G_2/(k_2 + 2G_2)], \tag{23.15}$$

$$G_{3*} = F_G[k_2G_1/(k_2 + 2G_1)], \qquad G_{4*} = F_G[k_1G_2/(k_1 + 2G_2)]. \tag{23.16}$$

Moreover, let the harmonic average of the phase bulk moduli be defined by

$$k_h = (\phi_1/k_1 + \phi_2/k_2)^{-1} = F_k(0). \tag{23.17}$$

The formulas (23.13)–(23.15) coincide with the Hashin–Shtrikman bounds on the effective conductivity σ_e [cf. (21.20)], effective bulk modulus k_e [cf. (21.70)], and effective shear modulus G_e [cf. (21.73)] for two-dimensional isotropic composites, respectivley. The relations (23.16) coincide with the corresponding Hashin–Shtrikman–Walpole bounds on G_e [cf. (21.78)].

The translation bounds are described by hyperbolas in the σ_e-k_e and σ_e-G_e planes with asymptotes that are parallel to the axes $\sigma_e = 0$, $k_e = 0$ or $\sigma_e = 0$, $G_e = 0$. Every hyperbola in the plane can be defined by three points that it passes through. We denote by

$\text{Hyp}[(x_1,y_1),(x_2,y_2),(x_3,y_3)]$ the segment of the hyperbola that joins the points (x_1,y_1), (x_2,y_2), and when extended passes through the point (x_3,y_3). It may be parametrically described in the x_e-y_e plane as

$$x_e = \langle x \rangle_\alpha - \frac{\alpha(1-\alpha)(x_1-x_2)^2}{x_1(1-\alpha)+x_2\alpha - x_3}, \quad y_e = \langle y \rangle_\alpha - \frac{\alpha(1-\alpha)(y_1-y_2)^2}{y_1(1-\alpha)+y_2\alpha - y_3}, \tag{23.18}$$

where $\langle x \rangle_\alpha = x_1\alpha + x_2(1-\alpha)$, $\langle y \rangle_\alpha = y_1\alpha + y_2(1-\alpha)$, and $\alpha \in [0,1]$.

The translation bounds are stated below as Theorems 23.2 and 23.3 without proofs. Detailed proofs can be found in Gibiansky and Torquato (1995a).

Theorem 23.2 *Bounds on the set of the pairs (σ_e, k_e) for any two-dimensional isotropic composite with a fixed volume fraction $\phi_1 = 1 - \phi_2$ are defined by the following four hyberbolas in the σ_e-k_e plane:*

$$Hyp[(\sigma_{1*},k_{1*}),(\sigma_{2*},k_{2*}),(\sigma_1,k_h)], \qquad Hyp[(\sigma_{1*},k_{1*}),(\sigma_{2*},k_{2*}),(\sigma_2,k_h)],$$
$$Hyp[(\sigma_{1*},k_{1*}),(\sigma_{2*},k_{2*}),(\sigma_1,k_1)], \qquad Hyp[(\sigma_{1*},k_{1*}),(\sigma_{2*},k_{2*}),(\sigma_2,k_2)].$$

The outermost pair of these curves yield the desired bounds.

Theorem 23.3 *Bounds on the set of the pairs (σ_e, G_e) for any two-dimensional isotropic composite with a fixed volume fraction $\phi_1 = 1 - \phi_2$ are defined by the following four hyperbolas in the σ_e-G_e plane:*

$$Hyp[(\sigma_{1*},G_{1*}),(\sigma_{2*},G_{3*}),(\sigma_1,G_1)], \qquad Hyp[(\sigma_{1*},G_{1*}),(\sigma_{2*},G_{3*}),(\sigma_2,G_2)],$$
$$Hyp[(\sigma_{1*},G_{4*}),(\sigma_{2*},G_{2*}),(\sigma_1,G_1)], \qquad Hyp[(\sigma_{1*},G_{4*}),(\sigma_{2*},G_{2*}),(\sigma_2,G_2)],$$

and the segments of the following two vertical straight lines:

$$\sigma_e = \sigma_{1*}, \; G_e \in [G_{1*}, G_{4*}], \qquad \sigma_e = \sigma_{2*}, \; G_e \in [G_{2*}, G_{3*}].$$

The outermost of these curves give us the desired bounds.

Remark:

1. Corresponding cross-property bounds for a composite with *arbitrary* volume fractions can be found as the union of the aforementioned bounds for a fixed volume fraction. The problem of finding this union admits a simple geometrical solution and again is defined by hyperbolas in the conductivity-elastic moduli planes. The reader is referred to Gibiansky and Torquato (1995a) for details. There they also show that the volume-fraction-independent translation bounds are sharper than the bounds (23.1) and (23.7).

The cross-property bounds of Theorems 23.2 and 23.3 are depicted in Figure 23.2 for the following values of the parameters:

$$\sigma_2/\sigma_1 = 20, \quad k_2/k_1 = 20, \quad \nu_1 = \nu_2 = 0.3, \tag{23.19}$$

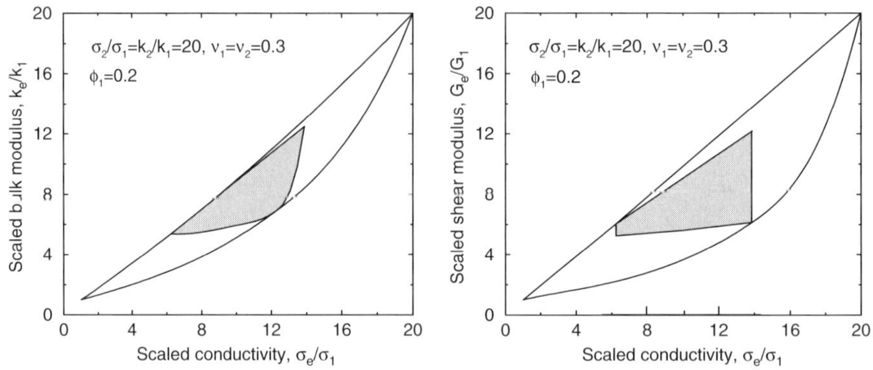

Figure 23.2 Cross-property bounds of Theorems 23.2 and 23.3. The gray regions are for fixed volume fraction ($\phi_1 = 0.2$), and the large lens-shaped regions are for any ϕ_1.

where $\nu_i = (k_i - G_i)/(k_i + G_i)$ is the two-dimensional Poisson ratio of phase i. Figure 23.2 shows that the conductivity-bulk modulus bounds are defined by the large lens-shaped region when the volume fraction is not prescribed and by the smaller gray lens-shaped region when the volume fraction is specified to be $\phi_1 = 1 - \phi_2 = 0.2$. The upper boundary of the gray region is the hyperbola $\text{Hyp}[(\sigma_{1*}, k_{1*}), (\sigma_{2*}, k_{2*}), (\sigma_1, k_1)]$, and it is an *optimal* upper bound, since it corresponds to the doubly coated cylinders (circles) geometry shown in Figure 21.1. The lower boundary of the gray region is the hyperbola $\text{Hyp}[(\sigma_{1*}, G_{1*}), (\sigma_{2*}, G_{3*}), (\sigma_2, G_2)]$, and it is unknown whether there exist any structures that realize it.

For fixed volume fraction, the conductivity-shear modulus pair lie within the gray *curvilinear trapezium* (Figure 23.2). The vertical sides are given by the Hashin–Shtrikman bounds on σ_e. The other two curvilinear sides are two of the hyperbola segments of Theorem 23.3. The two corner points (σ_{1*}, G_{1*}) and (σ_{2*}, G_{2*}) correspond to the laminate composites that realize the Hashin–Shtrikman bounds for elasticity and conductivity (Chapter 21). The other two points (σ_{1*}, G_{4*}) and (σ_{2*}, G_{3*}) correspond to the shear modulus bounds due to Walpole (1966b), but it is not known which structures attain them. The outermost curves of the large lens-shaped region define the bounds for arbitrary volume fractions. Figure 23.2 serves to illustrate that the values of the effective conductivity and effective elastic moduli are constrained to lie in certain regions of the conductivity-elastic moduli planes.

In what follows, we apply the cross-property bounds to some special limiting cases of the phase properties as well as to specific microstructures, including regular and random arrays of circular cylinders.

Equal Phase Moduli

First consider the case of a composite possessing *equal shear moduli* $G_1 = G_2 = G$. This is a trivial instance because neither effective elastic modulus depends on the microstructure (see Section 16.2.6) and therefore neither k_e nor G_e is connected to the effective conductivity of the composite.

In the case of composites with *equal phase bulk moduli* $k_1 = k_2 = k$, the effective bulk modulus $k_e = k$ is independent of the structure, and hence k_e is not connected to the effective conductivity. In the σ_e-G_e plane, the trapezium degenerates into a rectangle bounded by the Hashin–Shtrikman points. Therefore, we again find no connection between the shear modulus and the conductivity.

Perfectly Rigid Superconducting Phase

Assume that phase 2 is *perfectly rigid and superconducting*, i.e. $k_2/k_1 = \infty$, $G_2/G_1 = \infty$, and $\sigma_2/\sigma_1 = \infty$. The hyperbolas in this extreme case degenerate into straight lines, and the bounds of Theorems 23.2 and 23.3 simplify as

$$\sigma_e \geq \sigma_{1*}^{\infty}, \quad k_{1*}^{\infty} \leq k_e \leq k_{1*}^{\infty} + \max\left[\frac{k_1 + G_1}{2\sigma_1}, \frac{2k_2 G_2}{(k_2 + G_2)\sigma_2}\right](\sigma_e - \sigma_{1*}^{\infty}), \tag{23.20}$$

$$\sigma_e \geq \sigma_{1*}^{\infty}, \quad G_{1*}^{\infty} \leq G_e \leq G_{4*}^{\infty} + \max\left[\frac{k_1 + 2G_1}{4\sigma_1}, \frac{k_2 G_2}{(k_2 + G_2)\sigma_2}\right](\sigma_e - \sigma_{1*}^{\infty}), \tag{23.21}$$

where

$$\sigma_{1*}^{\infty} = \frac{1 + \phi_2}{\phi_1}\sigma_1, \qquad k_{1*}^{\infty} = \frac{k_1 + G_1\phi_2}{\phi_1},$$

$$G_{1*}^{\infty} = \frac{k_1 G_1(1 + \phi_2) + 2G_1^2}{(k_1 + 2G_1)\phi_1}, \qquad G_{4*}^{\infty} = \frac{k_1\phi_2 + 2G_1}{2\phi_1}.$$

In contrast to the Hashin–Shtrikman upper bounds on k_e and G_e, the upper bounds (23.20) and (23.21) do not diverge to infinity if σ_e remains finite in this infinite-contrast case. Note that the lower bounds on the elastic moduli are independent of the conductivity and coincide with the corresponding Hashin–Shtrikman lower bounds. Moreover, the upper bounds on k_e and G_e may depend on ratios of the infinite moduli because a very small amount (i.e., volume fraction of order $1/k_2$ or $1/\sigma_2$) of a very rigid conducting material can yield finite effective properties.

Perfectly Insulating Void Phase

Let us now assume that phase 2 is a *void* or *cavity phase*, i.e., $k_2 = G_2 = \sigma_2 = 0$. Here results are presented in the inverse coordinates $(1/\sigma_e, 1/k_e)$ and $(1/\sigma_e, 1/G_e)$. The bounds of Theorems 23.2 and 23.3 simplify as

$$\frac{1}{\sigma_e} \geq \frac{1}{\sigma_{1*}^0}, \quad \frac{1}{k_e} \geq \frac{1}{k_{1*}^0} + \min\left[\frac{(k_1 + G_1)\sigma_1}{2k_1 G_1}, \frac{2\sigma_2}{k_2 + G_2}\right]\left(\frac{1}{\sigma_e} - \frac{1}{\sigma_{1*}^0}\right), \tag{23.22}$$

$$\frac{1}{\sigma_e} \geq \frac{1}{\sigma_{1*}^0}, \quad \frac{1}{G_e} \geq \frac{1}{G_{1*}^0} + \min\left[\frac{(k_1 + G_1)\sigma_1}{k_1 G_1}, \frac{4\sigma_2}{k_2 + 2G_2}\right]\left(\frac{1}{\sigma_e} - \frac{1}{\sigma_{1*}^0}\right), \tag{23.23}$$

where

$$\frac{1}{\sigma^0_{1*}} = \frac{1+\phi_2}{\sigma_1\phi_1}, \quad \frac{1}{k^0_{1*}} = \frac{G_1+k_1\phi_2}{k_1G_1\phi_1}, \quad \frac{1}{G^0_{1*}} = \frac{k_1(1+\phi_2)+2G_1\phi_2}{k_1G_1\phi_1}.$$

Note that the lower bounds are trivially zero, since they correspond to the Hashin–Shtrikman lower bounds.

It is useful to state the corresponding volume-fraction-independent bounds [obtained from (23.22) and (23.23) by eliminating volume fractions]:

$$\frac{1}{\sigma_e} \geq \frac{1}{\sigma_1}, \quad \frac{1}{k_e} \geq \frac{1}{k_1} + \min\left[\frac{(k_1+G_1)\sigma_1}{2k_1G_1}, \frac{2\sigma_2}{k_2+G_2}\right]\left(\frac{1}{\sigma_e} - \frac{1}{\sigma_1}\right), \tag{23.24}$$

$$\frac{1}{\sigma_e} \geq \frac{1}{\sigma_1}, \quad \frac{1}{G_e} \geq \frac{1}{G_1} + \min\left[\frac{(k_1+G_1)\sigma_1}{k_1G_1}, \frac{4\sigma_2}{k_2+2G_2}\right]\left(\frac{1}{\sigma_e} - \frac{1}{\sigma_1}\right). \tag{23.25}$$

These volume-fraction-independent bounds, in conjunction with the percolation scaling laws (23.8)–(23.10), cannot provide improvement upon the critical-exponent bounds (23.11) for composites with a perfectly insulating void phase.

Interestingly, the bounds (23.24) and (23.25) apply as well to *cracked materials* for arbitrary shapes and distributions of cracks. The cracks (phase 2) are taken to have vanishing properties. The reader is referred to Gibiansky and Torquato (1996c) for applications of (23.24) and (23.25) to cracked materials.

Hexagonal Arrays of Cylinders

How sharp are the Gibiansky–Torquato cross-property estimates given an exact determination of one of the effective properties? To examine this question, we consider the case of superconducting perfectly rigid oriented cylinders (phase 2) arranged on the sites of a triangular lattice (hexagonal array) in a matrix such that $k_2/k_1 = \infty$, $G_1/k_1 = G_2/k_2 = 0.4$, and $\sigma_2/\sigma_1 = \infty$. Exact results for the effective conductivity (Perrins et al. 1979) in conjunction with the cross-property relations (23.20) and (23.21) yield bounds on the effective moduli. We make the additional but mild assumption that phase 1 determines the slope of the upper bounds in (23.20) and (23.21), i.e., $(k_1+G_1)/(2\sigma_1) \geq (2k_1G_2)/[(k_2+G_2)\sigma_2]$. Figure 23.3 shows that the cross-property bounds on the effective elastic moduli agree very well with the corresponding simulation data for the effective elastic moduli (Eischen and Torquato 1993). (Note that only the upper bounds on the elastic moduli contain conductivity information.) It is noteworthy that conventional variational upper bounds on the effective properties (such as the Hashin–Shtrikman upper bound) here diverge to infinity, since they do not incorporate information that the perfectly rigid phase is in fact disconnected. In contrast, the cross-property upper bounds use information that the *infinite-contrast phase is disconnected through conductivity data*.

Random Arrays of Cylinders

Conductivity data for *equilibrium* arrays of impenetrable cylinders have been obtained (Kim and Torquato 1990b), but we are not aware of elastic moduli data for the same *random* array. Consider superconducting cylinders ($\sigma_2/\sigma_1 = \infty$) and take $k_2/k_1 = 10$ and $G_1/k_1 = G_2/k_2 = 0.4$. Using the conductivity data in conjunction with Theorem 23.2

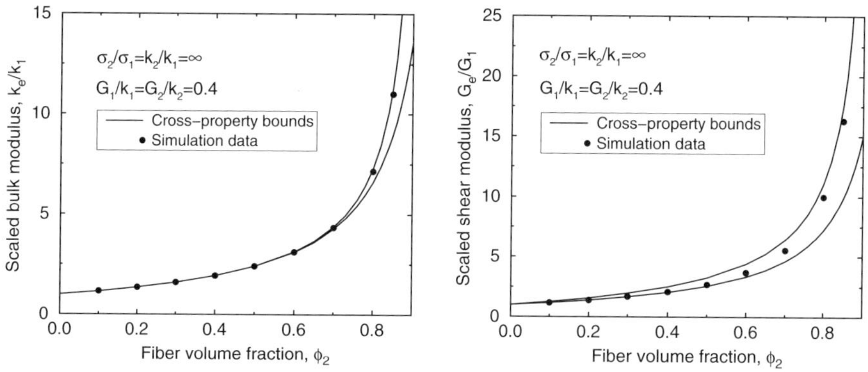

Figure 23.3 Comparison of cross-property bounds (23.20) for k_e and (23.21) for G_e for hexagonal arrays of cylinders to corresponding simulation data (Eischen and Torquato 1993).

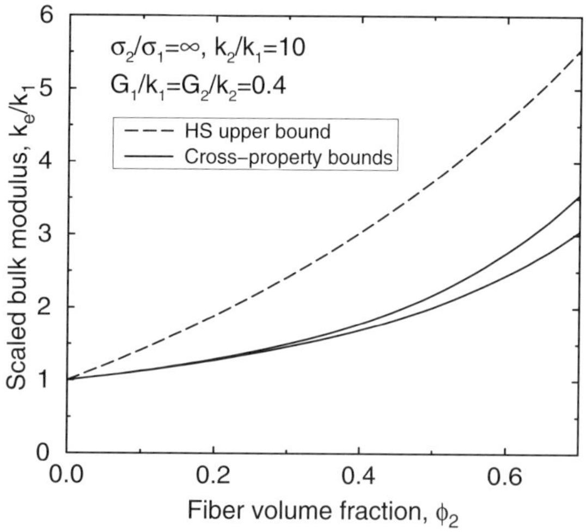

Figure 23.4 Prediction of the cross-property bounds of Theorem 23.2 for k_e for random arrays of impenetrable cylinders using conductivity data (Kim and Torquato 1990b).

enables one to predict the effective bulk modulus. Figure 23.4 shows that the elastic moduli-conductivity bounds on k_e are quite sharp. Moreover, the cross-property *upper bound* provides substantial improvement over the Hashin–Shtrikman *upper bound* on k_e.

23.1.3 Translation Bounds for $d = 3$

Using the translation method, Gibiansky and Torquato (1996b) have derived the sharpest known cross-property bounds linking the effective conductivity σ_e to the effective bulk modulus K_e for three-dimensional two-phase isotropic composites of all possible microstructures at a prescribed or arbitrary volume fraction. Berryman and Milton (1988) have obtained such bounds for the special case of porous media using three-point bounds on the effective conductivity and elastic moduli (Sections 21.1 and 21.2) by eliminating the three-point parameters. Gibiansky and Torquato (1996b) extended the Berryman–Milton procedure to the general two-phase case and showed that the translation bound given below linking σ_e and K_e is sharper. We note, however, that it is highly nontrivial to derive translation bounds for $d = 3$ that connect the effective conductivity to the effective shear modulus. On the other hand, the procedure used by Berryman and Milton (1988) to find bounds on the pair (σ_e, K_e) can also be employed to find bounds on the pair (σ_e, G_e) for general two-phase composites in three dimensions.

Before stating the translation bounds, we first introduce some notations. Let σ_{1*}, σ_{2*}, $\sigma_{1\#}$, $\sigma_{2\#}$, K_{1*}, K_{2*}, σ_a, σ_h, K_a, and K_h denote the expressions

$$\sigma_{1*} = F_\sigma(2\sigma_1), \qquad \sigma_{2*} = F_\sigma(2\sigma_2), \tag{23.26}$$

$$\sigma_{1\#} = F_\sigma(-2\sigma_1), \qquad \sigma_{2\#} = F_\sigma(-2\sigma_2), \tag{23.27}$$

$$K_{1*} = F_K(4G_1/3), \qquad K_{2*} = F_K(4G_2/3), \tag{23.28}$$

$$\sigma_a = \phi_1\sigma_1 + \phi_2\sigma_2 = F_\sigma(\infty), \qquad \sigma_h = \left(\frac{\phi_1}{\sigma_1} + \frac{\phi_2}{\sigma_2}\right)^{-1} = F_\sigma(0), \tag{23.29}$$

$$K_a = \phi_1 K_1 + \phi_2 K_2 = F_K(\infty), \qquad K_h = \left(\frac{\phi_1}{K_1} + \frac{\phi_2}{K_2}\right)^{-1} = F_K(0). \tag{23.30}$$

Here $F_a(y) \equiv F(a_1, a_2, \phi_1, \phi_2, y)$ is given by (23.12). The relations (23.26) and (23.28) coincide with the Hashin–Shtrikman bounds on the effective conductivity σ_e [cf. (21.20)] and effective bulk modulus K_e [cf. (21.70)] of three-dimensional isotropic composites, respectively. The formulas (23.29) and (23.30) coincide with the Wiener bounds on σ_e [cf. (21.13)] and Reuss-Voigt bounds on K_e [cf. (21.63)], respectively. It is unknown whether relations (23.27) have any physical meaning.

The cross-property bounds are again given by segments of hyperbolas in the σ_e-K_e plane with asymptotes that are parallel to the axes $\sigma_e = 0$ and $K_e = 0$. The proof of the theorem below can be found in Gibiansky and Torquato (1996b).

Theorem 23.4 *Bounds on the set of the pairs (σ_e, K_e) for any isotropic composite at a fixed volume fraction $\phi_1 = 1 - \phi_2$ are defined by the segments of the following five hyperbolas in the σ_e-K_e plane:*

$$Hyp[(\sigma_{1*}, K_{1*}), (\sigma_{2*}, K_{2*}), (\sigma_1, K_1)], \qquad Hyp[(\sigma_{1*}, K_{1*}), (\sigma_{2*}, K_{2*}), (\sigma_2, K_2)],$$

$$Hyp[(\sigma_{1*}, K_{1*}), (\sigma_{2*}, K_{2*}), (\sigma_{1\#}, K_h)], \qquad Hyp[(\sigma_{1*}, K_{1*}), (\sigma_{2*}, K_{2*}), (\sigma_{2\#}, K_h)],$$

$$Hyp[(\sigma_{1*}, K_{1*}), (\sigma_{2*}, K_{2*}), (\sigma_a, K_a)].$$

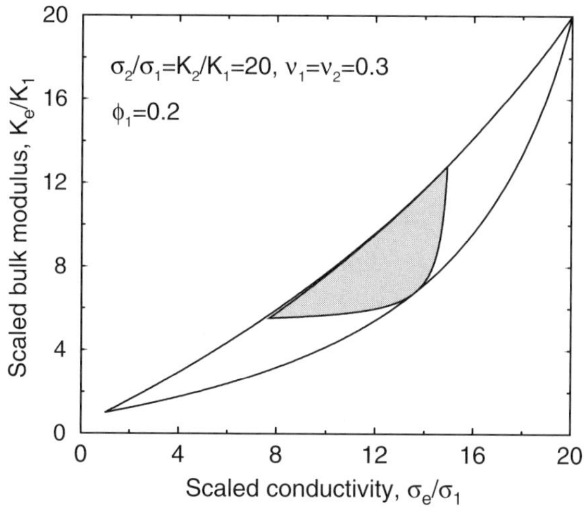

Figure 23.5 Cross-property bounds of Theorem 23.4. The gray lens-shaped region is for fixed volume fraction ($\phi_1 = 0.2$), and the large lens-shaped region is for any ϕ_1.

The outermost pair of these curves give us the desired bounds.

Remark:

1. Theorem 23.4 is not restricted to isotropic composites only but applies as well to anisotropic composites with cubic symmetry.

Depending upon the values of the parameters, any two of the five hyperbolas of Theorem 23.4 can be the outermost pair. Figure 23.5 depicts the conductivity-bulk modulus bounds for the following values of the parameters:

$$\sigma_2/\sigma_1 = 20, \quad K_2/K_1 = 20, \quad \nu_1 = \nu_2 = 0.3, \quad \phi_1 = 0.2. \tag{23.31}$$

It is seen that the conductivity-bulk modulus bounds are defined by the large lens-shaped region when the volume fraction is not prescribed, and by the smaller gray lens-shaped region when the volume fraction is specified to be $\phi_1 = 1 - \phi_2 = 0.2$. The hyperbola Hyp$[(\sigma_{1*}, K_{1*}), (\sigma_{2*}, K_{2*}), (\sigma_1, K_1)]$ is the upper boundary of the gray region. It is an *optimal* upper bound, since it corresponds to the doubly coated spheres geometry (see Section 21.1.2). The lower boundary of the gray region is the hyperbola Hyp$[(\sigma_{1*}, K_{1*}), (\sigma_{2*}, K_{2*}), (\sigma_{2\#}, K_h)]$, and it is unknown whether there exist any structures that realize it. The upper bound of the volume-fraction-independent lens-shaped region is generally sharper than (23.1).

The cross-property bounds are applied below to some special limiting cases of the phase properties as well as to specific microstructures, including regular and random arrays of spheres.

Equal Phase Shear Moduli

The instance of a composite possessing *equal shear moduli* $G_1 = G_2 = G$ is trivial, since neither effective elastic modulus depends on the microstructure (see Section 16.2.6) and therefore neither K_e nor G_e is connected to the effective conductivity of the composite.

Perfectly Rigid Superconducting Phase

Let us assume that *one of the phases is perfectly rigid and superconducting,* i.e., $K_2/K_1 = \infty$, $G_2/G_1 = \infty$, and $\sigma_2/\sigma_1 = \infty$. The hyperbolas of Theorem 23.4 in this extreme case degenerate into straight lines, and the bounds simplify as

$$\sigma_e \geq \sigma_{1*}^{\infty}, \quad K_{1*}^{\infty} \leq K_e \leq K_{1*}^{\infty} + \max\left[\frac{3K_1 + 4G_1}{9\sigma_1}, \frac{6K_2G_2}{(3K_2 + 4G_2)\sigma_2}, \frac{2G_2}{3\sigma_2}\right](\sigma_e - \sigma_{1*}^{\infty}), \tag{23.32}$$

where

$$\sigma_{1*}^{\infty} = \frac{1 + 2\phi_2}{\phi_1}\sigma_1, \quad K_{1*}^{\infty} = \frac{3K_1 + 4G_1\phi_2}{3\phi_1}. \tag{23.33}$$

Unlike the Hashin–Shtrikman and Berryman–Milton upper bounds on K_e, the upper bound (23.32) does not diverge to infinity if σ_e remains finite in this infinite-contrast case. Note that the lower bound on K_e is independent of the conductivity and coincides with the corresponding Hashin–Shtrikman lower bound.

Perfectly Insulating Void Phase

Assume that *one of the phases is composed of voids,* i.e., $K_2/K_1 = 0$, $G_2/G_1 = 0$, and $\sigma_2/\sigma_1 = 0$. It is convenient to present the results in the inverse coordinates, i.e., in the $1/\sigma_e$-$1/K_e$ plane. The bounds of Theorem 23.4 simplify as

$$\frac{1}{\sigma_e} \geq \frac{1}{\sigma_{1*}^0}, \quad \frac{1}{K_e} \geq \frac{1}{K_{1*}^0} + \min\left[\frac{(3K_1 + 4G_1)\sigma_1}{6K_1G_1}, \frac{9\sigma_2}{3K_2 + 4G_2}, \frac{3\sigma_1}{2G_1}\right]\left(\frac{1}{\sigma_e} - \frac{1}{\sigma_{1*}^0}\right), \tag{23.34}$$

where

$$\frac{1}{\sigma_{1*}^0} = \frac{1 + \phi_2}{2\sigma_1\phi_1}, \quad \frac{1}{K_{1*}^0} = \frac{4G_1 + 3K_1\phi_2}{4K_1G_1\phi_1}. \tag{23.35}$$

The lower bounds are equal to zero, since they coincide with the Hashin–Shtrikman lower bounds.

For arbitrary volume fractions, the bounds (23.34) yield

$$\frac{1}{\sigma_e} \geq \frac{1}{\sigma_1}, \quad \frac{1}{K_e} \geq \frac{1}{K_1} + \min\left[\frac{(3K_1 + 4G_1)\sigma_1}{6K_1G_1}, \frac{9\sigma_2}{3K_2 + 4G_2}, \frac{3\sigma_1}{2G_1}\right]\left(\frac{1}{\sigma_e} - \frac{1}{\sigma_1}\right).$$

As in the two-dimensional case, these bounds apply to cracked materials for arbitrary shapes and distributions of cracks (Gibiansky and Torquato 1996c).

Conducting Liquid in Insulating Solid

Consider now the instance when phase 2 is a conducting liquid ($\sigma_2 \neq 0$, $G_2 = 0$) and phase 1 is some insulating solid material ($\sigma_1 = 0, G_1 \neq 0$). In this case, the conductivity-bulk modulus bounds of Theorem 23.4 are restricted by the following three curves:

$$\sigma_e = 0, \qquad K_e \in [K_h, K_{1*}],$$
$$K_e = K_h, \qquad \sigma_e \in [0, \sigma_{2*}],$$
$$\text{Hyp}[(0, K_{1*}), (\sigma_{2*}, K_h), (\sigma_a, K_a)],$$

where σ_{2*}, K_{1*}, K_{2*}, σ_a, K_a, and K_h are given by (23.26), (23.28), (23.29), and (23.30).

Cubic Arrays of Spheres

We can employ the exact results of McPhedran and McKenzie (1978) for the effective conductivity of cubic arrays of spheres and our cross-property relations in order to obtain bounds on the effective bulk modulus. The bounds of Theorem 23.4 are applicable to the bulk modulus of such cubic symmetric composites, although they are not isotropic. The bounds on K_e can then be compared to the exact results of Nunan and Keller (1984) for the effective bulk modulus of cubic lattices. In particular, consider face-centered cubic arrays of superconducting perfectly rigid spherical inclusions (phase 2) in a matrix in which $K_2/K_1 = \infty$, $G_2/G_1 = \infty$, and $\nu_1 = 0.45$. The bounds in this instance are given by the relation (23.32). We make the additional but mild assumption that phase 1 determines the slope of the upper bound in (23.32).

Figure 23.6 summarizes our findings. Note that only the upper bound contains conductivity information. We see that for volume fractions in the range $\phi_2 \leq 0.5$, the cross-property bounds predict the bulk modulus of the composite almost exactly. For higher volume fractions, agreement with the data is still quite good. Typical upper bounds on the effective properties (such as the Hashin–Shtrikman upper bound) diverge to infinity, since they are not able to incorporate the information that the perfectly rigid phase is in fact disconnected. In contrast, the cross-property upper bound uses the important topological information that the *infinite-contrast phase is disconnected* via conductivity information.

Random Arrays of Hard Spheres

Conductivity data for *equilibrium* distributions of mutually impenetrable spheres have been obtained by Kim and Torquato (1991) for several volume fractions and contrast ratios. It is of interest to see how well our cross-property relations predict the elastic moduli in this instance. Let us consider the case of random superconducting spheres ($\sigma_2/\sigma_1 = \infty$) for several volume fractions and take $K_2/K_1 = 10$ and $G_1/K_1 = G_2/K_2 = 0.4$. Note that unlike the previous example, K_2/K_1 is finite. Figure 23.7 shows that the bulk modulus-conductivity bounds of Theorem 23.4 are quite sharp. The cross-property *upper bound* provides large improvement over the Hashin–Shtrikman *upper bound* on K_e.

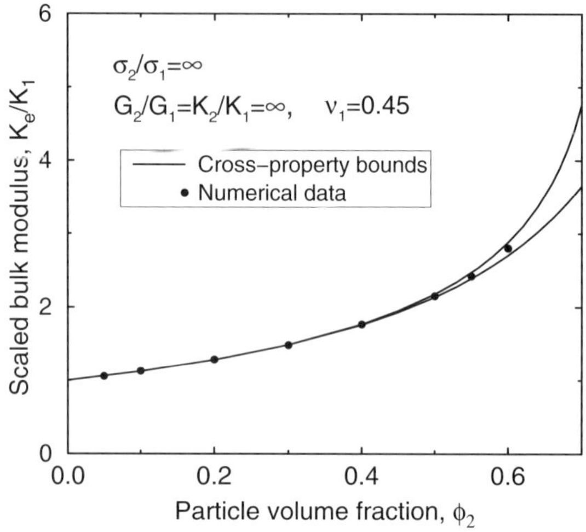

Figure 23.6 Comparison of cross-property bounds (23.32) for K_e for face-centered cubic arrays of spheres to corresponding numerical data (Nunan and Keller 1984).

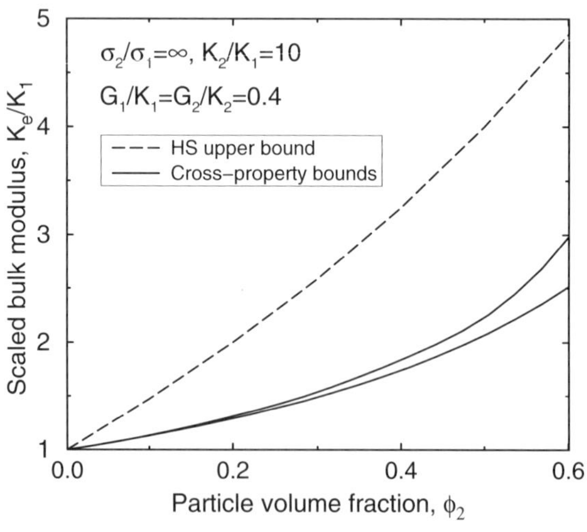

Figure 23.7 Prediction of the cross-property bounds of Theorem 23.4 for K_e for random arrays of impenetrable spheres using conductivity data (Kim and Torquato 1991).

23.2 Flow and Diffusion Parameters

At first glance, it seems surprising that the effective properties characterizing diffusion processes in porous media should be related to the effective properties associated with flow processes. Indeed, in the former case the local governing relations are scalar equations, whereas in the latter instance the local governing relations are generally second-order tensor equations. Nonetheless, there are rigorous links between flow and diffusion parameters, as we now show.

23.2.1 Permeability and Survival Time

We recall that fluid permeability can expressed as

$$\boldsymbol{k} = \langle \boldsymbol{w} \rangle, \tag{23.36}$$

where $\boldsymbol{w}$ is the nonsymmetric second-order tensor that solves the steady-state Stokes equations (13.171)–(13.173) in the fluid phase and is taken to be zero in the solid phase. Similarly, for diffusion-controlled reactions, we have shown that the trapping constant γ obeys

$$\gamma^{-1} = \langle u \rangle. \tag{23.37}$$

Here u is the scalar concentration field that satisfies the steady-state diffusion equation (13.148) in the trap-free region with perfectly absorbing boundary conditions, given by (13.149), and is taken to be zero in the trap phase. Recall that γ is related to the mean survival time τ via the relation $\gamma^{-1} = \mathcal{D}\phi_1\tau$, where $\mathcal{D}$ is the diffusion coefficient and ϕ_1 is the porosity.

Theorem 23.5 *For an ergodic anisotropic porous medium of arbitrary microstructure having a fluid region or a trap-free region $\mathcal{V}_1$ of porosity ϕ_1,*

$$\boldsymbol{k} \le \gamma^{-1}\boldsymbol{I} = \mathcal{D}\phi_1\tau\boldsymbol{I}\ , \tag{23.38}$$

i.e., the symmetric tensor that results from subtracting the fluid permeability tensor $\boldsymbol{k}$ *from the rotationally invariant inverse trapping-constant tensor $\gamma^{-1}\boldsymbol{I}$ is positive semidefinite.*

Proof: We first introduce the auxiliary ergodic and symmetric second order tensor field $\boldsymbol{v}$ defined by

$$\boldsymbol{v} = \boldsymbol{u} - \boldsymbol{w}, \tag{23.39}$$

where

$$\boldsymbol{u} = u\boldsymbol{I} \tag{23.40}$$

is the isotropic tensor formed with the scalar u. Combination of the local equations for u and $\boldsymbol{w}$ and definitions (23.39) and (23.40) yields the governing equations for $\boldsymbol{v}$:

$$\Delta \boldsymbol{v} = -\nabla \boldsymbol{\pi} \qquad \text{in } \mathcal{V}_1, \tag{23.41}$$

$$\nabla \cdot \boldsymbol{v} = \nabla \cdot \boldsymbol{u} \qquad \text{in } \mathcal{V}_1, \tag{23.42}$$

$$\boldsymbol{v} = 0 \qquad \text{on } \partial\mathcal{V}. \tag{23.43}$$

The proof uses the methods of Chapter 14. Specifically, multiplying relation (23.41) from the left by $\boldsymbol{v}$, averaging, and integrating by parts gives

$$\langle \nabla \boldsymbol{v} : \nabla \boldsymbol{v} \rangle = -\langle \boldsymbol{\pi}(\nabla \cdot \boldsymbol{v}) \rangle, \tag{23.44}$$

since the resulting surface integrals vanish. Utilizing the divergence condition of (23.42), we see that this relation becomes

$$\begin{aligned} \langle \nabla \boldsymbol{v} : \nabla \boldsymbol{v} \rangle &= -\langle \boldsymbol{\pi}(\nabla \cdot \boldsymbol{u}) \rangle \\ &= \langle \boldsymbol{u} \cdot \nabla \boldsymbol{\pi} \rangle. \end{aligned} \tag{23.45}$$

The second line of (23.45) follows by integrating the right-hand side of the first line by parts. Multiplying the momentum relation (14.133) of Stokes equations from the left by $\boldsymbol{u}$, averaging, and integrating by parts, gives

$$\begin{aligned} \langle \boldsymbol{u} \cdot \nabla \boldsymbol{\pi} \rangle &= \langle \boldsymbol{u} \rangle - \langle \boldsymbol{w} \rangle \\ &= \gamma^{-1} \boldsymbol{I} - \boldsymbol{k}. \end{aligned} \tag{23.46}$$

The second line of (23.46) follows from the definitions (23.36) and (23.37). The use of (23.39), (23.45), and (23.46) finally gives

$$\langle \boldsymbol{v} \rangle = \langle \nabla \boldsymbol{v} : \nabla \boldsymbol{v} \rangle = \gamma^{-1} \boldsymbol{I} - \boldsymbol{k}, \tag{23.47}$$

which means that $\langle \boldsymbol{v} \rangle$ is positive semidefinite, thus proving the theorem. ∎

Remark:

1. For a finite surface reaction $\kappa < \infty$, the inequality of Theorem 23.5 still holds, albeit in a weaker form than in the diffusion-controlled case ($\kappa = \infty$). It follows that since $\tau(\kappa = \infty) < \tau(\kappa)$, we have from Theorem 23.5 that

$$\boldsymbol{k} \le [\gamma(\kappa)]^{-1} \boldsymbol{I} = \mathcal{D}\phi_1 \tau(\kappa) \boldsymbol{I},$$

   with the equality possibly applying only when $\kappa = \infty$.

How sharp is the bound (23.38) of Theorem 23.5? Are there microstructures that achieve the equality sign of (23.38)? Consider the second query first and assume, without loss of generality, that the coordinate frame is aligned with the principal axes of the medium. The equality sign is achieved whenever a principal component of (23.47) is zero. Thus, it is achieved for transport in *parallel channels* (in the x_3-direction) of constant cross section dispersed throughout a solid or trap region with porosity ϕ_1. For example, for identical channels of arbitrary cross-sectional shape embedded in three dimensions, we recall that relation (16.99) exactly gives

$$k_{33} = \gamma^{-1} = \frac{\phi_1^3}{cs^2}, \tag{23.48}$$

where c is a shape-dependent constant and s is the specific surface. Note that since there is no flow in the other principal directions for this anisotropic geometry, i.e., $k_{11} = k_{22} = 0$, the bound (23.38) is trivially satisfied for these diagonal elements.

Let us now consider the question of the sharpness of the bound (23.38) for media that do not achieve the equality sign. For general microstructures this is a difficult question to answer, since there are relatively few exact results for the permeability tensor and trapping constant for well-defined models. The preponderance of such exact results exists for macroscopically isotropic media in which proposition (23.38) simplifies as

$$k \leq \gamma^{-1} = \mathcal{D}\phi_1 \tau. \tag{23.49}$$

Here k is defined by $\boldsymbol{k} = k\boldsymbol{I}$. Clearly, there are microstructures for which the bound (23.49) is not sharp. For example, for any microstructure with a completely disconnected pore space, k is zero while γ is nonzero, so that $k\gamma = 0$. Less trivially, for any cubic array of narrow tubes in d dimensions, it is easy to see that $k = \gamma^{-1}/d$. For the case of flow and diffusion exterior to isotropic distributions of spheres, the bound (23.49) is substantially sharper. For example, for a dilute bed of three-dimensional spheres or randomly oriented spheroids, $k = 2\gamma^{-1}/3$ (see Sections 19.3.2 and 19.4.2). The d-dimensional generalization of this result for $d \geq 3$ is $k = [(d+1)(d-2)/(d^2-3)]\gamma^{-1}$, and hence the equality of (23.49) is obeyed in the limit $d \to \infty$. Existing analytical results for random distributions of spheres (Childress 1972, Howells 1974, Mattern and Felderhof 1987) and for periodic arrays of spheres (Zick and Homsy 1982, Sangani and Acrivos 1982, Felderhof 1985) demonstrate that bound (23.49) is relatively sharp for low to moderate values of ϕ_2.

For high values of ϕ_2 in beds of particles, bound (23.49) is not sharp, at least in the case of periodic arrays for which we have exact results (Section 16.4.2). However, in the special case of transport through beds of spheres, the following more restrictive bound [closely related to (23.49)] is conjectured to exist:

$$\frac{k}{k_s} \leq \frac{\gamma_s}{\gamma}, \tag{23.50}$$

where γ_s and k_s are the dilute Smoluchowski and Stokes limits, respectively, given by (19.89) and (19.103). A proof of bound (23.50) has yet to be given. The bound (23.50) is sharper than the best available conventional variational bounds on either k or γ for beds of spheres. Thus, given "exact" Brownian-motion simulation data for the trapping constant γ of random porous media composed of spheres with a variable degree of penetrability and a polydispersivity in size (Lee and Torquato 1989, Miller and Torquato 1989, Torquato and Kim 1989), one can obtain sharp upper bounds on the fluid permeability for such models using (23.50). This cross-property bound is compared to variational bounds in Figure 22.18.

Finally, we observe that the trapping constant γ for a dilute bed of spheroids, given exactly by (19.97), provides a relatively sharp upper bound on the fluid permeability tensor $\boldsymbol{k}$ (in the sense of Theorem 23.5) for a dilute bed of aligned spheroids, given exactly by (19.110). Moreover, simulation results for γ for oriented spheroids at high trap volume fractions (Miller et al. 1991) combined with Theorem 23.5 provide upper bounds on the fluid permeability tensor for such anisotropic microstructures.

The relationship between the permeability tensor $\boldsymbol{k}$ and the trapping constant γ is much deeper than anyone previously thought. The fact that Theorem 23.5 exists suggests that techniques used to determine the scalar γ may be used, with some modification, to find the tensor $\boldsymbol{k}$. Indeed, for the aforementioned parallel-channel geometries, the trapping problem is isomorphic to the flow problem (in the direction of the channels). For such microstructures, the mean-square displacement of a *diffusion* tracer before trapping yields not only γ but also k_{33}, and hence the diffusion tracer is equivalent to a "momentum" tracer. The inequality of Theorem 23.5 suggests that a momentum tracer may still exist to yield $\boldsymbol{k}$, but it would not be identical to the diffusion tracer in the trapping problem. The formulation of the permeability problem in terms of a tracer of momentum would represent not only an important breakthrough theoretically, but computationally as well, especially in light of the existence of efficient methods to obtain γ using Brownian-motion simulation techniques (Torquato and Kim 1989, Torquato et al. 1999b).

23.2.2 Permeability, Formation Factor, and Viscous Relaxation Times

Avellaneda and Torquato (1991) showed that the fluid permeability k is related to certain averages of the relaxation times Θ_n and the effective electrical conductivity σ_e or, equivalently, the formation factor $F = \sigma_1/\sigma_e$ (where σ_1 is the conductivity of the fluid). They also proved that k is bounded from above and below in terms of the principal (largest) viscous relaxation time Θ_1 as well as principal diffusion relaxation time T_1. The proofs of the three theorems, given immediately below, follow those given by Avellaneda and Torquato (1991).

Theorem 23.6 *For random porous media of arbitrary topology with scalar permeability k at porosity ϕ_1, the following expression holds:*

$$k = \frac{L^2}{F}, \tag{23.51}$$

where

$$L^2 = \nu \frac{\sum_{n=1}^{\infty} b_n^2 \Theta_n}{\sum_{n=1}^{\infty} b_n^2}. \tag{23.52}$$

Here L is an effective length parameter, F is the formation factor, ν is the kinematic viscosity, Θ_n are the viscous relaxation times, and b_n are the eigenfunction expansion coefficients given by (13.224).

Proof: First we take the Laplace transforms (in time) of the unsteady Stokes equations (13.210)–(13.212) and obtain

$$s\hat{\boldsymbol{v}} = -\nabla(\hat{p}/\rho) + \nu\Delta\hat{\boldsymbol{v}} + v_0\boldsymbol{e}, \tag{23.53}$$

$$\nabla \cdot \hat{\boldsymbol{v}} = 0, \tag{23.54}$$

$$\hat{v} = 0, \tag{23.55}$$

where, for an arbitrary function $f(\boldsymbol{x},t)$,

$$\hat{f}(\boldsymbol{x},s) = \int_0^\infty f(\boldsymbol{x},t)e^{-st}dt. \tag{23.56}$$

Setting $s = 0$ in relations (23.53)–(23.55) yields

$$\nu\Delta\hat{v}(\boldsymbol{x},0) = \nabla[\hat{p}(r,0)/\rho] - v_0\boldsymbol{e}, \tag{23.57}$$

$$\nabla\cdot\,\hat{v}(\boldsymbol{x},0) = 0, \tag{23.58}$$

$$\hat{v}(\boldsymbol{x},0) = 0. \tag{23.59}$$

Letting

$$\boldsymbol{w}(\boldsymbol{x}) = \nu\hat{v}(\boldsymbol{x},0)/v_0, \qquad \pi(\boldsymbol{x}) = \hat{p}(\boldsymbol{x},0)/\rho v_0 \tag{23.60}$$

in (23.57)–(23.59) yields the canonical equations (13.180)–(13.182), which determine the fluid permeability. Thus, the solution of (13.180)–(13.182) can be expressed in terms of the eigenfunctions $\{\Psi_n\}$ that solve (13.214)–(13.216) by Laplace transforming the normal mode expansion (13.213):

$$\frac{\hat{v}(\boldsymbol{x},s)}{v_0} = \sum_{n=1}^\infty b_n\Psi_n(\boldsymbol{x})\frac{\Theta_n}{1+s\Theta_n}. \tag{23.61}$$

Setting $s = 0$ in (23.61) and employing (23.60) gives

$$\boldsymbol{w}(\boldsymbol{x}) = \nu\sum_{n=1}^\infty b_n\Psi_n(\boldsymbol{x})\Theta_n. \tag{23.62}$$

Forming the scalar product of the scaled electric field $\boldsymbol{E}$ [defined by (13.221)] with (23.62), ensemble averaging, and using (13.224)–(13.226) along with the identity

$$\langle\boldsymbol{w}\cdot\boldsymbol{e}\rangle = \langle\boldsymbol{w}\cdot\boldsymbol{E}\rangle$$

yields

$$\begin{aligned} k &= \nu\phi_1\sum_{n=1}^\infty b_n^2\Theta_n \\ &= \frac{\nu}{F}\frac{\sum_{n=1}^\infty b_n^2\Theta_n}{\sum_{n=1}^\infty b_n^2}, \end{aligned}$$

which proves the theorem. ∎

Remarks:

1. Relation (23.51) for k is not a definition but rather is an *exact* result, since the effective length parameter L is directly connected to the *spectrum* of the Stokes operator. Physically, L is the characteristic size of the *dynamically connected part of the pore space* (see Section 13.5).

2. The definition (23.52) of L differs from the one introduced by Avellaneda and Torquato (1991) by a constant factor of 8.
3. For flow within bundles of parallel cylindrical circular tubes of radius a, L is exactly equal to $a/8$ (Avellaneda and Torquato 1991), $F = 1/\phi_1$, and therefore relation (23.51) gives $k = (\phi_1 a^2)/8$, which, as expected, is the exact result given by (16.97).
4. Theorem 23.6 implicitly assumes that the spectrum of the Stokes operator in $\mathcal{V}_1$ is discrete, an assumption that is justified if the pore volume V_1 is finite or if the microstructure is periodic. However, Theorem 23.6 is valid also for arbitrary ergodic porous media, with the series replaced by integrals over the density of states.
5. Theorem 23.6 applies not only to statistically isotropic media but to anisotropic porous media as well, with the obvious modifications.
6. A Laplace-variable-dependent fluid permeability can be defined as

$$k(s) = \nu\langle\hat{\boldsymbol{v}}(\boldsymbol{x}, s)\cdot\boldsymbol{e}\rangle/\upsilon_0. \tag{23.63}$$

   Note that $k(0) \equiv k$ is the usual steady-state or static permeability defined by (13.179). Substitution of (23.61) into the above relation for $k(s)$ gives

$$k(s) = \nu\phi_1 \sum_{n=1}^{\infty} \frac{b_n^2\Theta_n}{1 + s\Theta_n}. \tag{23.64}$$

   If a porous medium saturated with a viscous fluid is subjected to an oscillatory pressure gradient $\nabla p_0(\omega)$ at frequency ω, then the induced averaged velocity $\boldsymbol{U}(\omega)$ will also be oscillatory and proportional to the pressure gradient according to $\boldsymbol{U}(\omega) = -\tilde{k}(\omega)\nabla p_0(\omega)/\mu$. Here $\tilde{k}(\omega)$ is the *dynamic permeability* (Johnson, Koplik and Dashen 1987, Sheng and Zhou 1988) and is related to $k(s)$ according to the relation $\tilde{k}(\omega) = k(s = -i\omega)$.

The calculation of L is as difficult as the calculation of k. Therefore, it is useful to find bounds on L or estimates for it in terms of flow or diffusion properties. In what follows, we describe such results.

Theorem 23.7 *For random isotropic porous media of arbitrary microstructure at porosity ϕ_1, the fluid permeability k is bounded from above according to*

$$k \le \frac{\nu\Theta_1}{F}, \tag{23.65}$$

where ν is the kinematic viscosity, Θ_1 is the principal viscous relaxation time, and F is the formation factor.

Proof: Since the eigenvalues ϵ are positive and $\epsilon_1 \le \epsilon_n (\Theta_1 \ge \Theta_n)$ for $n \ne 1$, it follows that

$$\sum_{n=1}^{\infty} b_n^2\Theta_n \le \sum_{n=1}^{\infty} b_n^2\Theta_1. \tag{23.66}$$

This result in combination with (13.226) and Theorem 23.6 gives the upper bound of Theorem 23.7. ∎

Remark:

1. The upper bound is valid whether the spectrum is discrete or continuous and hence holds for ergodic media.

Theorem 23.8 *For random isotropic porous media of arbitrary microstructure, the fluid permeability k is bounded from above as follows:*

$$k \leq \frac{\mathcal{D}T_1}{F}, \tag{23.67}$$

where $\mathcal{D}$ is the diffusion coefficient of the fluid and T_1 is the principal relaxation time for diffusion among static perfectly absorbing traps.

Proof: This theorem follows immediately from Theorem 23.7 and Theorem 23.9 (discussed below), which states that $\nu\Theta_1 \leq \mathcal{D}T_1$. ∎

Remarks:

1. Assume that the effective conductivity $\sigma_e = \sigma_1/F$ and permeability obey the power laws (9.49) and (9.50), respectively, near the connectivity threshold ($\phi_1 \to \phi_{1c}^+$). Torquato (1992) used these scaling laws and cross-property bound (23.67) to prove that (for $d \geq 2$) the permeability critical exponent e is generally greater than the conductivity critical exponent t, i.e.,

$$e \geq t, \qquad d \geq 2, \tag{23.68}$$

   under the assumption that the percolation thresholds are the same for the two problems. Since the time T_1 does not vanish at the percolation threshold, relation (23.68) immediately follows from the aforementioned relations. Theoretical predictions of these critical exponents for continuum percolation (Feng et al. 1987) are consistent with inequality (23.68); see also Section 9.2.3.
2. The relaxation times are easy to evaluate when transport occurs inside certain simple domains (Avellaneda and Torquato 1991). For example, in the case of unsteady Poiseuille flow interior to a circular cylindrical tube of radius a, the eigenvalues $\epsilon = 1/(\nu\Theta_n)$ are determined from the equation $J_0(a\sqrt{\epsilon}) = 0$, where $J_0(x)$ is the zeroth-order Bessel function. The first zero occurs at $x_0 \approx 2.405$ and thus $\nu\Theta_1 \approx 0.1729a^2$. Therefore, for arrays of parallel tubes at porosity ϕ_1 [$k = (\phi_1 a^2)/8$ and $F = 1/\phi_1$], the upper bound of Theorem 23.7 yields the sharp bound $0.125\phi_1 a^2 \leq 0.1729\phi_1 a^2$. For more general porous media, this suggests that the static permeability k can be reasonably estimated from the principal relaxation time Θ_1 according to Theorem 23.7, provided that the media possess a narrow range of pore sizes. As previously discussed, the flow and diffusion problems are identical for transport interior to parallel tubes and hence $\nu\Theta_n = \mathcal{D}T_n$.
3. Torquato and Kim (1992) have conjectured that for a wide class of realistic porous media, one can replace T_1 in (23.67) with the mean survival time τ, yielding the inequality

$$k \leq \frac{\mathcal{D}\tau}{F}. \tag{23.69}$$

This class includes packed beds of particles, soils, and sandstones. They showed that bound (23.69) can be violated for porous media containing many isolated pores and dead-end regions whose sizes are on the order of or smaller than the exact effective length scale L appearing in (23.51). Relation (23.69) cannot be generally true, since from Theorem 13.3, $T_1 \geq \tau$.

4. There are two noteworthy approximate cross-property relations linking the permeability k to diffusion parameters. Johnson, Koplik and Schwartz (1986) obtained the following very useful three-dimensional approximation:

$$k \approx \frac{\Lambda^2}{8F}, \tag{23.70}$$

where Λ^2 is a dynamically weighted ratio of V_1/S (pore volume to surface area) involving the electric field, i.e.,

$$\frac{2}{\Lambda} = \frac{\int_{\mathcal{S}} |\boldsymbol{E}(\boldsymbol{x})|^2 dS}{\int_{\mathcal{V}_1} |\boldsymbol{E}(\boldsymbol{x})|^2 dV}. \tag{23.71}$$

Here $\mathcal{S}$ and $\mathcal{V}_1$ denote the pore–solid interface and pore phase, respectively. The formula (23.70) provides a good estimate of k for a variety of porous media and is usually superior to the well-known Kozeny–Carman relation (16.99), which just involves the simple length scale V_1/S. Schwartz, Martys, Bentz, Garboczi and Torquato (1993) found that the relation

$$k \approx \frac{\phi_1 \mathcal{D}\tau}{F} \tag{23.72}$$

gives accurate estimates of the permeabilities of realistic models of porous media. Coker et al. (1996) showed that it also provides a good estimate of the permeability of a Fontainebleau sandstone. Note that relation (23.72) is obtained by multiplying the right-hand side of (23.69) by ϕ_1.

23.2.3 Viscous and Diffusion Relaxation Times

Using a classical Rayleigh–Ritz variational principle, Avellaneda and Torquato (1991) proved that the principal viscous and diffusion relaxation times obey the inequality stated immediately below.

Theorem 23.9 *For any porous medium, the following inequality holds between Θ_1 and T_1:*

$$\nu\Theta_1 \leq \mathcal{D}T_1. \tag{23.73}$$

Thus, a measurement of the viscous relaxation time Θ_1 (inversely proportional to the smallest eigenvalue of the Stokes operator) can be used to determine information about the diffusion relaxation time T_1, and vice versa. The relaxation time T_1 can be obtained from NMR relaxation experiments (Wilkinson et al. 1991, Torquato and Kim 1992) and hence can be used to bound the viscous relaxation time Θ_1 from above.

APPENDIX A

Equilibrium Hard-Disk Program

Here we provide a Fortran 77 program to generate equilibrium configurations of equal-sized hard disks via the Metropolis algorithm. It also computes the radial distribution function $g_2(r)$ by averaging over the configurations. The extension of the program to d-dimensional hard spheres is obvious. For simplicity, it does not use a "neighbor list," as described in Chapter 12.

```
PROGRAM MCDISKS
C   ************************************************************
C   ** KEY INPUT PARAMETERS ARE:
C   ** -NUMBER OF PARTICLES: N
C   ** -VOLUME (AREA) FRACTION: FRACT
C   ** -NUMBER OF REALIZATIONS (CONFIGURATIONS): NSTEP
C   ** -NUMBER OF CYCLES BETWEEN DATA SAVES AND
C   **  CALCULATION OF g2(r): ISAVE
C   ** -BIN WIDTH FOR CALCULATION OF g2(r): DELR
C   ************************************************************
C   ** OUTPUT RETURNS:
C   ** -RADIAL DIST. FUNC. (RDF) g2(r) VS. r/D: File="RDF"
C   ** -PARTICLE CONFIGS AT SPECIFIED INTERVALS
C   **  IN CONFIG FILE: File="config"
C   ************************************************************
C   ** PRINCIPAL VARIABLES:
C   ** INTEGER N1            MAXIMUM SIZE OF ARRAYS
C   ** INTEGER N             NUMBER OF PARTICLES
C   ** INTEGER NSTEP           NUMBER OF MONTE CARLO CYCLES
C   ** REAL RX(N),RY(N)        PARTICLE COORDINATES
C   ** REAL DIAM             PARTICLE DIAMETER
C   ** REAL FRACT             DISK AREA FRACTION
C   ** REAL DRMAX              MAX PARTICLE DISPLACEMENT
C   ** REAL RATIO            ACCEPTANCE RATIO
C   ** LOGICAL OVRLAP          TRUE IF PARTICLE I OVERLAPS
C   ** CHARACTER CNFILE          FILE FOR DISK CONFIGURATIONS
C   ************************************************************
C   DECLARING VARIABLES AND COUNTERS
IMPLICIT NONE
INTEGER    N,N1
```

```
PARAMETER ( N1 = 10000)
REAL      RX(N1), RY(N1), GRAT(5000), GRA(5000)
REAL      DRMAX, FRACT, DIAM, RATIO, DELR, SIDE
REAL      RXIOLD, RYIOLD, RXINEW, RYINEW
REAL      RANF, DUMMY, ACM, ACMMVA
REAL      SPACE, RLO, X, Y
INTEGER   STEP, I, NSTEP, IRATIO, IPRINT, ISAVE
INTEGER   HIST(5000), MAXBIN
INTEGER   MNUM, NCONFIG, M, J, UNITFILE
LOGICAL   OVRLAP
CHARACTER CNFILE*80
COMMON / REAL1 / RX, RY, DIAM, DELR, GRAT, GRA
COMMON / INT1 / HIST, MAXBIN
C    **********************************************************
C    ** READ INPUT DATA
WRITE(*,'("**CONSTANT-NVT MC PROGRAM FOR HARD DISKS**")')
WRITE(*,'(/ "NUMBER OF PARTICLES; N=M^2 M=INTEGER")')
WRITE(*,*)'ENTER M'
READ(*,*)MNUM
N=MNUM*MNUM
WRITE(*,'("ENTER THE DISK AREA FRACTION")')
READ(*,*) FRACT
WRITE(*,'("NUMBER OF CONFIGS FOR CALCULATION OF g2(r)")')
READ (*,*) NCONFIG
WRITE(*,'("ENTER NUMBER OF CYCLES BETWEEN DATA SAVES ")')
READ(*,*)ISAVE
C    ** ISAVE = 50 IS A TYPICAL VALUE. SET ISAVE ¿ 50 FOR HIGH FRACT
NSTEP=ISAVE*NCONFIG
C    ** DOES ISAVE CYCLES BETWEEN ANALYSIS FOR g2(r)
WRITE(*,'("ENTER NUMBER OF CYCLES BETWEEN OUTPUT  ")')
READ(*,*)IPRINT
C    ** IPRINT = 50 IS A TYPICAL VALUE
WRITE(*,'("ENTER BIN WIDTH FOR g2(r) IN UNITS OF DIAM.")')
READ(*,*) DELR
WRITE(*,'("ENTER INITIAL MAX. DISPLACEMENT IN UNITS OF DIAM.")')
READ (*,*) DRMAX
C    **  SET DRMAX SO THAT ACCEPTANCE RATIO IS ABOUT 0.5
WRITE(*,'("ENTER INTERVAL FOR UPDATE OF MAX. DISPL.")')
READ(*,*)IRATIO
C    ** IRATIO = 1 IS A TYPICAL VALUE
C    ** OPENING OUTPUT FILE: UNIT=UNITFILE NAME=CNFILE='config'
CNFILE='config'
UNITFILE=10
OPEN(UNIT=UNITFILE,ACCESS='APPEND',FILE=CNFILE)
C    ** WRITE INPUT DATA
WRITE(*,'(" NUMBER OF PARTICLES    =",I10  )') N
WRITE(*,'(" NUMBER OF MC CYCLES    =",I10 )') NSTEP
WRITE(*,'(" OUTPUT FREQUENCY       =",I10  )') IPRINT
WRITE(*,'(" SAVE FREQUENCY         =",I10  )') ISAVE
WRITE(*,'(" RATIO UPDATE FREQUENCY =",I10  )') IRATIO
WRITE(*,'(" DISK AREA FRACTION     =",F10.5 )') FRACT
C    ** COMPUTE DISK DIAMETER FROM DISK AREA FRACTION
DIAM  = (4.0*FRACT/3.14159264/REAL(N))**0.5
DRMAX  = DRMAX * DIAM
SIDE = 1/DIAM
C    ** INTERPARTICLE SEPARATIONS MUST BE ¡ 1/2 BOX LENGTH
MAXBIN = INT((SIDE/2.-1.)/DELR)
C    ** ZERO HISTOGRAM FOR g2(r)
DO I=1,5000
GRAT(I)=0.0
END DO
WRITE(*,'(" NUMBER OF DISKS        =",I10  )' )  N
WRITE(*,'(" DIAM/(BOX LENGTH)      =",F10.5 )' )  DIAM
WRITE(*,'(" DRMAX/(BOX LENGTH)     =",F10.5 )' ) DRMAX
WRITE(*,'(" CONFIGURATION FILE NAME=   ",A    )') CNFILE
C    ** CREATE INITIAL CONFIGURATION: A SQUARE ARRAY
C    ** ORIGIN OF COORDINATE SYSTEM AT THE BOX CENTER
SPACE = 1.0/MNUM
RLO=-1.0*0.5
```

```
RLO=RLO+(.5*SPACE)
M=1
DO I=0,MNUM-1
X=REAL(I)*SPACE
DO J=0,MNUM-1
Y=REAL(J)*SPACE
RX(M)=RLO+X
RY(M)=RLO+Y
M=M+1
END DO
END DO
C    ** ZERO ACCUMULATORS
ACM    = 0.0
ACMMVA = 0.0
WRITE( *, '(//" START OF MARKOV CHAIN    "//)')
WRITE( *, '( "     ACM    RATIO     DRMAX     ")')
C    ************************************************************
C    ** LOOP OVER CYCLES BEGINS
C    ** MC PARTICLE MOVES AND CHECK FOR OVERLAP
C    ************************************************************
DO 100 STEP = 1, NSTEP
C    ** LOOP OVER PARTICLES
DO 99 I = 1, N
RXIOLD = RX(I)
RYIOLD = RY(I)
C    ** MOVE PARTICLE I AND PICKUP THE CENTRAL IMAGE
RXINEW = RXIOLD + (2.0 * RANF( DUMMY) - 1.0 ) *DRMAX
RYINEW = RYIOLD + (2.0 * RANF( DUMMY)- 1.0 ) * DRMAX
*    ** IMPLEMENT PERIODIC BOUNDARY CONDITIONS
RXINEW = RXINEW - ANINT (RXINEW)
RYINEW = RYINEW - ANINT (RYINEW)
C    ** CHECK FOR ACCEPTANCE (NO OVERLAP)
CALL TEST ( RXINEW, RYINEW, I, OVRLAP,N )
IF (.NOT.OVRLAP ) THEN
C    ** ACCEPT MOVE
RX(I) = RXINEW
RY(I) = RYINEW
ACMMVA = ACMMVA + 1.0
ENDIF
ACM = ACM + 1.0
99  CONTINUE
C    ************************************************************
C    ** LOOP OVER PARTICLES COMPLETE
C    ************************************************************
C    ** PERFORM PERIODIC OPERATIONS
C    ** CHANGE MAXIMUM DISPLACEMENT
IF ( MOD ( STEP, IRATIO ) .EQ. 0 ) THEN
RATIO = ACMMVA / REAL ( N * IRATIO )
IF ( RATIO .GT. 0.5 ) THEN
DRMAX = DRMAX * 1.05
ELSE
DRMAX = DRMAX * 0.95
ENDIF
ACMMVA = 0.0
ENDIF
C    ** WRITE OUT RUNTIME INFORMATION
IF ( MOD ( STEP, IPRINT ) .EQ. 0 ) THEN
WRITE(*,'(I8,3F10.4)') INT(ACM/n), RATIO, DRMAX
ENDIF
C    ** WRITE OUT THE CONFIGURATION AT INTERVALS
IF ( MOD ( STEP, ISAVE ) .EQ. 0 ) THEN
CALL RADIAL ( UNITFILE,N,FRACT)
ENDIF
100  CONTINUE
C    ************************************************************
C    ** ENDS THE LOOP OVER CYCLES
C    ************************************************************
C    ** WRITE OUT g2(r) vs. r (IN UNITS OF DIAM)
OPEN(UNIT=17,STATUS='UNKNOWN',FILE='RDF')
```

```
DO 400 I=1,MAXBIN
WRITE(17,395)(1.+(I-0.5)*DELR),GRAT(I)/REAL(NCONFIG)
395    FORMAT('',F7.3,F9.4,F9.4,F9.4)
400  CONTINUE
CLOSE(UNIT=17)
CLOSE(UNIT=UNITFILE)
STOP
END

SUBROUTINE TEST ( RXINEW, RYINEW, I, OVRLAP,N )
C    ************************************************************
C    ** CHECKS FOR OVERLAP OF I WITH ALL OTHER PARTICLES.
C    ** PRINCIPAL VARIABLES:
C    ** INTEGER I                THE PARTICLE OF INTEREST
C    ** INTEGER N                NUMBER OF PARTICLES
C    ** REAL    RXI,RYI          POSITION OF PARTICLE I
C    ** REAL    RX(N),RY(N)      PARTICLE POSITIONS
C    ** REAL    DIAM             PARTICLE DIAMETER
C    ** LOGICAL OVRLAP           TRUE IF PARTICLE I OVERLAPS
C    **
C    ** USAGE:
C    ** CALLED AFTER  TRIAL DISPLACEMENT OF I TO ESTABLISH
C    ** WHETHER THERE IS AN OVERLAP IN THE TRIAL CONFIGURATION.
C    ************************************************************
IMPLICIT NONE
INTEGER   N,N1
PARAMETER ( N1 = 10000 )
REAL      RX(N1), RY(N1),GRA(5000),GRAT(5000)
REAL      RXINEW, RYINEW,R2
REAL      DIAM,DELR,DIAMSQ
INTEGER   I,HIST(5000),MAXBIN
LOGICAL   OVRLAP
REAL      RXIJ, RYIJ
INTEGER   J
COMMON / real1 / RX,RY,DIAM,DELR,GRAT,GRA
COMMON / int1 / HIST,MAXBIN
C    ************************************************************
OVRLAP = .FALSE.
DIAMSQ  = DIAM * DIAM
C    ** LOOPS OVER MOLECULES EXCEPT I
DO 100 J = 1, N
IF ( J .NE. I ) THEN
RXIJ = RXINEW - RX(J)
RYIJ = RYINEW - RY(J)
RXIJ = RXIJ - ANINT ( RXIJ )
RYIJ = RYIJ - ANINT ( RYIJ )
R2 = RXIJ*RXIJ + RYIJ*RYIJ
IF ( R2 .LT. DIAMSQ ) THEN
OVRLAP = .TRUE.
RETURN
ENDIF
ENDIF
100  CONTINUE
RETURN
END

SUBROUTINE RADIAL ( UNITFILE,N,FRACT )
C    ************************************************************
C    ** SUBROUTINE TO WRITE OUT THE CONFIGURATION TO UNIT 10
C    ** AND TABULATE STATISTICS FOR RADIAL DISTRIBUTION FUNCTION
C    ************************************************************
IMPLICIT NONE
INTEGER     N,N1
PARAMETER (  N1 = 10000 )
CHARACTER   CNFILE*(*)
REAL        RX(N1), RY(N1),GRAT(5000),GRA(5000)
REAL        FRACT,DIAM,DELR,SIDE,RXIJ,RYIJ,RIJSQ
REAL        CONST,RLOWER,RUPPER,CIDEAL
INTEGER     CNUNIT,HIST(5000),MAXBIN
```

```
INTEGER     I,IBIN,J, UNITFILE
PARAMETER (  CNUNIT = 10 )
COMMON / real1 / RX,RY,DIAM,DELR,GRAT,GRA
COMMON / int1 / HIST,MAXBIN
C     ************************************************************
SIDE = 1/DIAM
DO I = 1,N
WRITE (UNITFILE,* ) RX(I), RY(I)
END DO
DO IBIN=1,MAXBIN
HIST(IBIN)=0
END DO
C     ** BUILD HISTOGRAM
DO 100 I=1,N-1
DO 99 J = I+1,N
RXIJ = RX(I)-RX(J)
RYIJ = RY(I)-RY(J)
RXIJ = RXIJ - ANINT(RXIJ)
RYIJ = RYIJ - ANINT(RYIJ)
RIJSQ=RXIJ*RXIJ  + RYIJ*RYIJ
RIJSQ=SQRT(RIJSQ)
RIJSQ = (RIJSQ)*SIDE
IBIN = INT((RIJSQ-1.)/DELR) + 1
IF (IBIN.LE.MAXBIN) THEN
HIST(IBIN) = HIST(IBIN) + 2
END IF
99     CONTINUE
100  CONTINUE
CONST = 4.0*FRACT
DO 110 IBIN=1,MAXBIN
RLOWER= REAL(IBIN-1)*DELR+1.0
RUPPER= RLOWER + DELR
CIDEAL = CONST*(RUPPER**2-RLOWER**2)
GRA(IBIN) = REAL(HIST(IBIN))/(REAL(N)*CIDEAL)
GRAT(IBIN)=GRAT(IBIN) + GRA(IBIN)
110  CONTINUE
RETURN
END



REAL FUNCTION RANF ( DUMMY )
C     ************************************************************
C     ** RETURNS A UNIFORM RANDOM VARIATE IN THE RANGE 0 TO 1.
C     ** WARNING!!! GOOD RANDOM NUMBER GENERATORS ARE MACHINE SPECIFIC.
C     ** USE THE ONE RECOMMENDED FOR YOUR MACHINE.
C     ************************************************************
IMPLICIT NONE
INTEGER     L, C, M
PARAMETER ( L = 1029, C = 221591, M = 1048576 )
INTEGER     SEED
REAL        DUMMY
SAVE        SEED
DATA        SEED / 0 /
C     ************************************************************
SEED = MOD ( SEED * L + C, M )
RANF = REAL ( SEED ) / M
RETURN
END
```

APPENDIX B

Interrelations Among Two- and Three-Dimensional Moduli

Consider a d-dimensional linear isotropic homogeneous material with bulk modulus, shear modulus, Young's modulus, and Poisson's ratio denoted by $K^{(d)}$, $G^{(d)}$, $E^{(d)}$, and $\nu^{(d)}$, respectively. These are precisely the moduli contained in the stress–strain relation (13.89) and strain–stress relation (13.90), except that here we explicitly indicate the dependence on dimensionality with the superscript (d).

Our interest here is to connect the two-dimensional (planar) moduli to the three-dimensional moduli by assuming either plane-strain or plane-stress elasticity. For plane-strain elasticity, we take $\epsilon_{11} = \epsilon_{12} = \epsilon_{13} = 0$ in relation (13.90) with $d = 3$. If we compare this simplified three-dimensional expression to relation (13.90) with $d = 2$, we obtain the interrelations

$$E^{(2)} = \frac{E^{(3)}}{[1 - \nu^{(3)}][1 + \nu^{(3)}]}, \qquad \nu^{(2)} = \frac{\nu^{(3)}}{1 - \nu^{(3)}}. \tag{B.1}$$

Similarly, by comparing (13.89) with $d = 3$ under plane-strain conditions and (13.89) with $d = 2$, we obtain

$$K^{(2)} = K^{(3)} + G^{(3)}/3, \qquad G^{(2)} = G^{(3)}. \tag{B.2}$$

For plane-stress elasticity, we take $\sigma_{11} = \sigma_{12} = \sigma_{13} = 0$ in the expression (13.90) with $d = 3$. Comparing this simplified three-dimensional relation to relation (13.90) with $d = 2$ gives

$$E^{(2)} = E^{(3)}, \qquad \nu^{(2)} = \nu^{(3)}. \tag{B.3}$$

Finally, comparing (13.89) with $d = 3$ under plane-stress conditions and (13.89) with $d = 2$ yields

$$K^{(2)} = \frac{9K^{(3)}G^{(3)}}{3K^{(3)} + 4G^{(3)}}, \qquad G^{(2)} = G^{(3)}. \tag{B.4}$$

When distinguishing between the planar and three-dimensional bulk moduli, we denote by $k \equiv K^{(2)}$ the planar bulk modulus (plane-strain or plane-stress) and by $K \equiv K^{(3)}$ the three-dimensional bulk modulus. Note that the interrelations derived above for homogeneous materials do not hold for the effective properties.

References

Aboav, D. A. (1980). The arrangement of cells in a net, *Metallography* **13**, 43–58.

Abramovich, B. and Indelman, P. (1995). Effective permittivity of log-normal isotropic random media, *J. Phys. A: Math. Gen.* **28**, 693–700.

Abramowitz, M. and Stegun, I. A. (1972). *Handbook of Mathematical Functions*, Dover, New York.

Acrivos, A. and Chang, E. (1986). A model for estimating transport quantities in two-phase materials, *Phys. Fluids* **29**, 3–4.

Adler, P. M. (1992). *Porous Media—Geometry and Transports*, Butterworth-Heinemann, Boston.

Aizenman, M. and Barsky, D. J. (1987). Sharpness of the phase transition in percolation models, *Commun. Math. Phys.* **108**, 489–526.

Alberts, B. et al. (1997). *Essential Cell Biology*, Garland Publishing, New York.

Allen, M. P. and Tildesley, D. J. (1987). *Computer Simulation of Liquids*, Oxford University Press, Oxford, England.

Anonymous (1972). What is random packing?, *Nature* **239**, 488–489.

Archie, G. E. (1942). The electrical resistivity log as an aid in determining some reservoir characteristics, *Trans. AIME* **146**, 54–61.

Aurenhammer, F. (1991). Voronoi diagrams—A survey of a fundamental geometric data structure, *ACM Computing Surveys* **23**, 345–405.

Avellaneda, M. (1987a). Iterated homogenization, differential effective medium theory and applications, *Commun. Pure Appl. Math.* **40**, 527–554.

Avellaneda, M. (1987b). Optimal bounds and microgeometries for elastic two-phase composites, *SIAM J. Appl. Math.* **47**, 1216–1228.

Avellaneda, M. and Torquato, S. (1991). Rigorous link between fluid permeability, electrical conductivity, and relaxation times for transport in porous media, *Phys. Fluids A* **3**, 2529–2540.

Avellaneda, M., Cherkaev, A. V., Lurie, K. A. and Milton, G. W. (1988). On the effective conductivity of polycrystals and a three-dimensional phase-interchange inequality, *J. Appl. Phys.* **63**, 4989–5003.

Baer, S. and Lebowitz, J. L. (1964). Convergence of fugacity expansion and bounds on molecular distributions for mixtures, *J. Chem. Phys.* **40**, 3474–3478.

Baker, G. A. (1975). *Essentials of Padé Approximants*, Academic Press, New York.

Balberg, I., Anderson, C. H., Alexander, S. and Wagner, N. (1984). Excluded volume and its relation to the onset of percolation, *Phys. Rev. B* **30**, 3933–3943.

Banavar, J. R. and Schwartz, L. M. (1987). Magnetic resonance as a probe of permeability in porous media, *Phys. Rev. Lett.* **58**, 1411–1414.

Batchelor, G. K. (1959). *The Theory of Homogeneous Turbulence*, Cambridge University Press, Cambridge, England.

Batchelor, G. K. (1972). Sedimentation in a dilute dispersion of spheres, *J. Fluid Mech.* **52**, 245–268.

Batchelor, G. K. (1974). Transport properties of two-phase materials with random structure, *Ann. Rev. Fluid Mech.* **6**, 227–255.

Batchelor, G. K. and O'Brien, R. W. (1977). Thermal or electrical conduction through a granular material, *Proc. R. Soc. Lond. A* **355**, 313–333.

Baus, M. and Colot, J. L. (1987). Thermodynamics and structure of a fluid of hard rods, disks, spheres, or hyperspheres from rescaled virial expansions, *Phys. Rev. A* **36**, 3912–3925.

Baxter, R. J. (1968). Ornstein–Zernike relation for a disordered fluid, *Austr. J. Phys.* **21**, 563–569.

Bayer, B. E. (1964). Relation between granularity and density for a random dot model, *J. Opt. Soc. Am.* **54**, 1485–1490.

Beasley, J. D. and Torquato, S. (1986). Bounds on the conductivity of a suspension of random impenetrable spheres, *J. Appl. Phys.* **60**, 3576–3581.

Beasley, J. D. and Torquato, S. (1989). New bounds on the permeability of a random array of spheres, *Phys. Fluids A* **1**, 199–207.

Bendsøe, M. P. (1995). *Optimization of Structural Topology, Shape and Material*, Springer-Verlag, Berlin.

Bendsøe, M. P. and Kikuchi, N. (1988). Generating optimal topologies in structural design using a homogenization method, *Comput. Methods Appl. Mech. Eng.* **71**, 197–224.

Benguigui, L. (1984). Experimental study of the elastic properties of a percolating system, *Phys. Rev. Lett.* **53**, 2028–2030.

Benguigui, L. (1986). Lattice and continuum percolation transport exponents: Experiments in two-dimensions, *Phys. Rev. B* **34**, 8176–8178.

Benguigui, L. and Ron, P. (1993). Experimental realization of superelasticity near the percolation threshold, *Phys. Rev. Lett.* **70**, 2423–2426.

Bensoussan, A., Lions, J. L. and Papanicolaou, G. (1978). *Asymptotic Analysis for Periodic Structures*, North Holland, Amsterdam.

Bentz, D. P. and Martys, N. S. (1994). Hydraulic radius and transport in reconstructed model three-dimensional porous media, *Transp. Porous Media* **17**, 221–238.

Benveniste, Y. (1986). On the effective thermal conductivity of multiphase composites, *Z. Ang. Math. Phys.* **37**, 696–713.

Benveniste, Y. (1987). A new approach to the application of Mori-Tanaka's theory in composite materials, *Mech. Mater.* **6**, 147–157.

Benveniste, Y. (1996). Thermal expansion of polycrystalline aggregates consisting of elongated crystals and containing cylindrical pores or inclusions, *J. Mech. Phys. Solids* **44**, 137–153.

Benveniste, Y. and Dvorak, G. J. (1992). Uniform fields and universal relations in piezoelectric composites, *J. Mech. Phys. Solids* **40**, 1295–1312.

Beran, M. (1965). Use of the variational approach to determine bounds for the effective permittivity in random media, *Nuovo Cimento* **38**, 771–782.

Beran, M. J. (1968). *Statistical Continuum Theories*, Wiley, New York.

Beran, M. J. and Molyneux, J. (1966). Use of classical variational principles to determine bounds for the effective bulk modulus in heterogeneous media, *Quart. Appl. Math.* **24**, 107–118.

Beran, M. J. and Silnutzer, N. R. (1971). Effective electrical, thermal and magnetic properties of fiber reinforced materials, *J. Composite Mater.* **5**, 246–249.

Berdichevsky, V. L. (1983). *Effective Electrical, Thermal and Magnetic Properties of Fiber-Reinforced Materials*, Nauka, Moscow, Russia.

Berg, H. C. (1983). *Random Walks in Biology*, Princeton University Press, Princeton, New Jersey.

Bergman, D. J. (1978). The dielectric constant of a composite material–A problem in classical physics, *Phys. Rep. C* **43**, 377–407.

Bergman, D. J. (1980). Exactly solvable microscopic geometries and rigorous bounds for the complex dielectric constant of a two-component composite material, *Phys. Rev. Lett.* **44**, 1285–1287.

Bergman, D. J. (1982). Rigorous bounds for the complex dielectric constant of a two-component composite, *Ann. Phys.* **138**, 78–114.

Berk, N. F. (1987). Scattering properties of a model bicontinuous structure with a well defined length scale, *Phys. Rev. Lett.* **58**, 2718–2721.

Berk, N. F. (1991). Scattering properties of the leveled-wave model of random morphologies, *Phys. Rev. A* **44**, 5069–5079.

Berlyand, L. and Wehr, J. (1997). Non-Gaussian limiting behavior of the percolation threshold in a large system, *Commun. Math. Phys.* **185**, 73–92.

Bernal, J. D. (1964). The structure of liquids, *Proc. R. Soc. Lond. A* **280**, 299–322.

Bernal, J. D. (1965). The geometry of the structure of liquids, in *Liquids: structure, properties,solid interactions*, T. J. Hughel (ed.), Elsevier, New York, pp. 25–50.

Berryman, J. G. (1980a). Long-wavelength propagation in composite elastic media. II. Ellipsoidal inclusions, *J. Acoust. Soc. Am.* **68**, 1820–1831.

Berryman, J. G. (1980b). Long-wavelength propagation in composite elastic media. I. Spherical inclusions, *J. Acoust. Soc. Am.* **68**, 1809–1819.

Berryman, J. G. (1983). Random close packing of hard spheres and disks, *Phys. Rev. A* **27**, 1053–1061.

Berryman, J. G. (1985a). Measurement of spatial correlation functions using image processing techniques, *J. Appl. Phys.* **57**, 2374–2384.

Berryman, J. G. (1985b). Variational bounds on elastic constants for the penetrable sphere model, *J. Phys. D.* **18**, 585–597.

Berryman, J. G. (1987). Relationship between specific surface area and spatial correlation functions for anisotropic porous media, *J. Math. Phys.* **28**, 244–245.

Berryman, J. G. (1989). Estimating effective moduli of composites using quantitative image analysis in random media and composites, in *Random Media and Composites*, R. V. Kohn and G. W. Milton (eds.), SIAM, Philadelphia.

Berryman, J. G. and Berge, P. A. (1996). Critique of two explicit schemes for estimating elastic properties of multiphase composites, *Mech. Mater.* **22**, 149–164.

Berryman, J. G. and Blair, S. C. (1986). Use of digital image analysis to estimate fluid permeability of porous materials: Application of two-point correlation functions, *J. Appl. Phys.* **60**, 1930–1938.

Berryman, J. G. and Milton, G. W. (1985). Normalization constraint for variational bounds on fluid permeability, *J. Chem. Phys.* **83**, 754–760.

Berryman, J. G. and Milton, G. W. (1988). Microgeometry of random composites and porous media, *J. Phys. D: Appl. Phys.* **21**, 87–94.

Binney, J. J., Dowrick, N. J., Fisher, A. J. and Newman, M. E. J. (1992). *The Theory of Critical Phenomena: An Introduction to the Renormalization Group*, Oxford University Press, Oxford, England.

Blum, L. and Stell, G. (1979). Polydisperse systems. I. Scattering function for polydisperse fluids of hard or permeable spheres, *J. Chem. Phys.* **71**, 42–46.

Blum, L. and Stell, G. (1980). Polydisperse systems. I. Scattering function for polydisperse fluids of hard or permeable spheres (erratum), *J. Chem. Phys* **72**, 2212.

Blumenfeld, R. and Torquato, S. (1993). Coarse-graining procedure to generate and analyze heterogeneous materials–Theory, *Phys. Rev. E* **48**, 4492–4500.

Bobeth, M. and Diener, G. (1986). Field fluctuations in multicomponent mixtures, *J. Mech. Phys. Solids* **34**, 1–17.

Bochner, S. (1936). *Lectures on Fourier Analysis*, Edwards, Ann Arbor, Michigan.

Boltzmann, L. (1898). *Lectures on Gas Theory*, University of California Press, Berkeley, California. 1964 translation by S. Brush of the original 1898 publication.

Bond, J. R., Kofman, L. and Pogosyan, D. (1996). How filaments of galaxies are woven into the cosmic web, *Nature* **380**, 603–606.

Bonnecaze, R. T. and Brady, J. F. (1991). The effective conductivity of random suspensions of spherical particles, *Proc. R. Soc. Lond. A* **432**, 445–465.

Bonnier, B., Boyer, D. and Viot, P. (1994). Pair correlation function in random sequential adsorption processes, *J. Phys. A: Math. Gen.* **27**, 3671–3682.

Botsis, J., Beldica, C. and Zhao, D. (1994). On strength scaling of composites with long aligned fibres, *Int. J. Fract.* **68**, 375–384.

Boucher, S. (1974). On the effective moduli of isotropic two-phase elastic composites, *J. Composite Mater.* **8**, 82–89.

Brady, J. F. and Durlofsky, L. J. (1988). The sedimentation rate of disordered suspensions, *Phys. Fluids* **31**, 717–727.

Brilliantov, N. V., Andrienko, Y. A., Krapivsky, P. L. and Kurths, J. (1996). Fractal formation and ordering in random sequential adsorption, *Phys. Rev. Lett.* **76**, 4058–4061.

Brinker, C. J. and Scherer, G. W. (1990). *Sol-Gel Science*, Academic Press, New York.

Brinkman, H. C. (1947). A calculation of the viscous force exerted by a flowing fluid on a dense swarm of particles, *Appl. Sci. Res.* **A1**, 27–34.

Broadbent, S. and Hammersley, J. (1957). Percolation processes. I. Crystals and mazes, *Proc. Camb. Phil. Soc.* **53**, 629–641.

Brown, W. F. (1955). Solid mixture permittivities, *J. Chem. Phys.* **23**, 1514–1517.

Brown, W. F. (1974). Properties of two-phase "cell-materials," *J. Math. Phys.* **15**, 1516–1524.

Brownstein, K. R. and Tarr, C. E. (1979). Importance of classical diffusion in NMR studies of water in biological cells, *Phys. Rev. A* **19**, 2446–2453.

Bruggeman, D. (1935). Berechnung verschiedener Physikalischer Konstanten von heterogenen Substanzen, *Ann. Physik (Liepzig)* **24**, 636–679.

Bruno, O. P. (1990). The effective conductivity of an infinitely interchangeable mixture, *Commun. Pure Appl. Math.* **43**, 769–807.

Bruno, O. P. (1991). The effective conductivity of strongly heterogeneous composites, *Proc. R. Soc. Lond. A* **433**, 353–381.

Bruno, O. P. and Leo, P. H. (1993). On the stiffness of materials containing a disordered array of microscopic holes or hard inclusions, *Arch. Rational Mech. Analysis* **121**, 304–338.

Budiansky, B. (1965). On the elastic moduli of some heterogeneous materials, *J. Mech. Phys. Solids* **13**, 223–227.

Bug, A. L. R., Safran, S. A., Grest, G. S. and Webman, I. (1985). Do interactions raise or lower a percolation threshold?, *Phys. Rev. Lett.* **55**, 1896–1899.

Buryachenko, V. A. (2000). Multiparticle effective field and related methods in micromechanics of composite materials, *Appl. Mech. Rev.* **54**, 1–47.

Caflisch, R. E. and Rubinstein, J. (1984). *Lectures in Mathematical Theory of Multi-Phase Flow*, Courant Institute of Mathematical Sciences, New York.

Cahn, J. W. (1965). Phase separation by spinodal decomposition in isotropic systems, *J. Chem. Phys.* **42**, 93–99.

Calef, D. F. and Deutch, J. M. (1983). Diffusion-controlled reactions, *Ann. Rev. Phys. Chem.* **34**, 493–524.

Cancelliere, A., Chang, C., Foti, E., Rothman, D. H. and Succi, S. (1990). The permeability of a random medium—Comparison of simulation with theory, *Phys. Fluids A* **2**, 2085–2088.

Chaikin, P. M. and Lubensky, T. C. (1995). *Principles of Condensed Matter Physics*, Cambridge University Press, New York.

Chandrasekhar, S. (1943). Stochastic problems in physics and astronomy, *Rev. Mod. Phys.* **15**, 1–89.

Chayes, J. T. and Chayes, L. (1986). Bulk transport properties and exponent inequalities for random resistor and flow networks, *Commun. Math. Phys.* **105**, 133–152.

Chen, H. S. and Acrivos, A. (1978a). The solution of the equations of linear elasticity for an infinite region containing two spherical inclusions, *Int. J. Solids Structures* **14**, 331–348.

Chen, H. S. and Acrivos, A. (1978b). The effective elastic moduli of composite materials containing spherical inclusions at non-dilute concentrations, *Int. J. Solids Structures* **14**, 349–364.

Chen, J., Thorpe, M. F. and Davis, L. C. (1995). Elastic properties of rigid fiber-reinforced composites, *J. Appl. Phys.* **77**, 4349–4360.

Cheng, H. W. and Greengard, L. (1997). On the numerical evaluation of electrostatic fields in dense random dispersions of cylinders, *J. Comput. Phys.* **136**, 629–639.

Cheng, H. and Torquato, S. (1997a). Effective conductivity of periodic arrays of spheres with interfacial resistance, *Proc. R. Soc. Lond. A* **453**, 145–161.

Cheng, H. and Torquato, S. (1997b). Effective conductivity of dispersions of spheres with a superconducting interface, *Proc. R. Soc. Lond. A* **453**, 1331–1344.

Cherkaev, A. V. (2000). *Variational Methods for Structural Optimization*, Springer-Verlag, New York.

Cherkaev, A. V. and Gibiansky, L. V. (1992). The exact coupled bounds for effective tensors of electrical and magnetic properties of two-component two-dimensional composites, *Proc. R. Soc. Edinburgh* **122A**, 93–125.

Cherkaev, A. V. and Gibiansky, L. V. (1993). Coupled estimates for the bulk and shear moduli of a two-dimensional isotropic elastic composite, *J. Mech. Phys. Solids* **41**, 937–980.

Cherkaev, A. V., Lurie, K. A. and Milton, G. W. (1992). Invariant properties of the stress in plane elasticity and equivalence classes of composites, *Proc. R. Soc. Lond. A* **438**, 519–529.

Cherradi, N., Kawasaki, A. and Gasik, M. (1994). Worldwide trends in functional gradient materials research and development, *Composites Eng.* **4**, 883–894.

Chiew, Y. C. and Glandt, E. D. (1983). Percolation behavior of permeable and of adhesive spheres, *J. Phys. A: Math. Gen.* **16**, 2599–2608.

Chiew, Y. C. and Glandt, E. D. (1984). Interfacial surface area in dispersions and porous media, *J. Colloid Interface Sci.* **99**, 86–96.

Chiew, Y. C. and Glandt, E. D. (1987). Effective conductivity of dispersions: The effect of resistance at the particle surfaces, *Chem. Eng. Sci.* **42**, 2677–2685.

Chiew, Y. C., Stell, G. and Glandt, E. D. (1985). Clustering and percolation in multicomponent systems of randomly centered and permeable spheres, *J. Chem. Phys.* **83**, 761–767.

Childress, S. (1972). Viscous flow past a random array of spheres, *J. Chem. Phys.* **56**, 2527–2539.

Choy, T. C., Alexopoulos, A. and Thorpe, M. F. (1998). Dielectric function for a material containing hyperspherical inclusions to $\mathcal{O}(c^2)$. I. Multipole expansions, *Proc. R. Soc. Lond. A* **454**, 1973–1992.

Christensen, R. M. (1979). *Mechanics of Composite Materials*, Wiley, New York.

Christensen, R. M. (1993). Effective properties of composite materials containing voids, *Proc. R. Soc. Lond. A* **440**, 461–473; Erratum, **453**, 447 (1997).

Christensen, R. M. (1995). The hierarchy of microstructures for low-density materials, *Z. Ang. Math. Phys.* **46**, S506–S521.

Christensen, R. M. (2000). Mechanics of cellular and other low-density materials, *Int. J. Solids Structures* **37**, 93–104.

Christensen, R. M. and Lo, K. H. (1979). Solutions for effective shear properties in three phase sphere and cylinder models, *J. Mech. Phys. Solids* **27**, 315–330.

Christiansson, H. and Helsing, J. (1995). A numerical study of the ζ_2 and η_2 structural parameters for a class of fiber-reinforced composites, *J. Appl. Phys.* **77**, 4734–4738.

Cichocki, B. and Felderhof, B. U. (1988). Dielectric constant of polarizable, nonpolar fluids and suspensions, *J. Stat. Phys.* **53**, 499–521.

Cinlar, E. (1975). *Introduction to Stochastic Processes*, Prentice Hall, Englewood Cliffs, New Jersey.

Cinlar, E. and Torquato, S. (1995). Exact determination of the two-point cluster function for one-dimensional continuum percolation, *J. Stat. Phys.* **78**, 827–839.

Coker, D. A. and Torquato, S. (1995). Extraction of morphological quantities from a digitized medium, *J. Appl. Phys.* **77**, 6087–6099.

Coker, D. A., Torquato, S. and Dunsmuir, J. H. (1996). Morphology and physical properties of Fontainebleau sandstone via a tomographic analysis, *J. Geophys. Res.* **101**, 17497–17506.

Coniglio, A., De Angelis, U. and Forlani, A. (1977). Pair connectedness and cluster size, *J. Phys. A: Math. Gen.* **10**, 1123–1139.

Conway, J. H. and Sloane, N. J. A. (1993). *Sphere Packings, Lattices and Groups*, Springer-Verlag, New York.

Cooper, D. W. (1988). Random-sequential-packing simulations in three dimensions for spheres, *Phys. Rev. A* **38**, 522–524.

Corson, P. B. (1974a). Correlation functions for predicting properties of heterogeneous materials. I. Experimental measurement of spatial correlation functions in multiphase solids, *J. Appl. Phys.* **45**, 3159–3164.

Corson, P. B. (1974b). Correlation functions for predicting properties of heterogeneous materials. II. Empirical construction of spatial correlation functions for two-phase solids, *J. Appl. Phys.* **45**, 3165–3170.

Corson, P. B. (1974c). Correlation functions for predicting properties of heterogeneous materials. III. Effective elastic moduli of two-phase solids, *J. Appl. Phys.* **45**, 3171–3179.

Corson, P. B. (1974d). Correlation functions for predicting properties of heterogeneous materials. IV. Effective thermal conductivity of two-phase solids, *J. Appl. Phys.* **45**, 3180–3182.

Cramer, H. (1954). *Mathematical Methods of Statistics*, Princeton University Press, Princeton.

Cressie, N. A. C. (1993). *Statistics for Spatial Data*, John Wiley, New York.

Crossley, P. A., Schwartz, L. M. and Banavar, J. R. (1991). Image-based models of porous media—Application to vycor glass and carbonate rocks, *Appl. Phys. Lett.* **59**, 3553–3555.

Cruz, M. E., Ghaddar, C. K. and Patera, A. T. (1995). A variational-bound nip-element method for geometrically stiff problems—Application to thermal composites and porous media, *Proc. R. Soc. Lond. A* **449**, 93–122.

Cule, D. and Torquato, S. (1998). Electric-field distribution in composite media, *Phys. Rev. B* **58**, 11829–11832.

Cule, D. and Torquato, S. (1999). Generating random media from limited microstructural information via stochastic optimization, *J. Appl. Phys.* **86**, 3428–3437.

Davis, L. C. (1991). Third-order bounds on the elastic moduli of metal–matrix composites, *Metall. Trans. A.* **22**, 3065–3067.

Day, A. R., Snyder, K. A., Garboczi, E. J. and Thorpe, M. F. (1992). The elastic moduli of a sheet containing circular holes, *J. Mech. Phys. Solids* **40**, 1031–1051.

Debenedetti, P. G. (1996). *Metastable Liquids: Concepts and Principles*, Princeton University Press, Princeton, New Jersey.

Debye, P. and Bueche, A. M. (1949). Scattering by an inhomogeneous solid, *J. Appl. Phys.* **20**, 518–525.

Debye, P., Anderson, H. R. and Brumberger, H. (1957). Scattering by an inhomogeneous solid. II. The correlation function and its applications, *J. Appl. Phys.* **28**, 679–683.

Dederichs, P. H. and Zeller, R. (1973). Variational treatment of the elastic constants of disordered materials, *Z. Physik A* **259**, 103–116.

Delannay, R. and Le Caer, G. (1994). Topological characteristics of 2D cellular structures generated by fragmentation, *Phys. Rev. Lett.* **73**, 1553–1556.

den Hollander, F. and Weiss, G. H. (1994). Aspects of trapping in transport processes, in *Contemporary Problems in Statistical Physics*, G. H. Weiss (ed.), SIAM, Philadelphia.

DeSimone, T., Stratt, R. M. and Demoulini, S. (1986). Continuum percolation in an interacting system: Exact solution of the Percus–Yevick equation for connectivity in liquids, *Phys. Rev. Lett.* **56**, 1140–1143.

DeVera, A. L. and Strieder, W. (1977). Upper and lower bounds on the thermal conductivity of a random, two-phase material, *J. Phys. Chem.* **81**, 1783–1790.

Diggle, P. J. (1983). *Statistical Analysis of Spatial Point Patterns*, Academic Press, New York.

Dijkstra, M., van Roij, R. and Evans, R. (1999). Direct simulation of the phase behavior of binary hard-sphere mixtures: Test of the depletion potential description, *Phys. Rev. Lett.* **82**, 117–120.

Doi, M. (1976). A new variational approach to the diffusion and the flow problem in porous media, *J. Phys. Soc. Japan* **40**, 567–572.

Domb, C. (1947). The problem of random intervals on a line, *Proc. Camb. Phil. Soc.* **43**, 329–341.

Donsker, M. D. and Varadhan, S. R. S. (1975). Asymptotics for the Wiener sausage, *Commun. Pure Appl. Math.* **28**, 525–565.

Douglas, J. F. and Garboczi, E. J. (1995). Intrinsic viscosity and the polarizability of particles having a wide range of shapes, *Adv. Chem. Phys.* **91**, 85–153.

Drory, A. (1997). Exact solution of a one-dimensional continuum percolation model, *Phys. Rev. E* **55**, 3878–3885.

Dullien, F. A. L. (1979). *Porous Media: Fluid Transport and Pore Structure*, Academic Press, New York.

Durrett, R. (1996). *Probability: Theory and Examples*, Duxbury Press, Belmont, California.

Durrett, R. and Levin, S. A. (1994). Stochastic spatial models—A user's guide to ecological applications, *Phil. Trans. R. Soc. Lond. B* **343**, 329–350.

Dvorak, G. J. (1986). Thermal expansion of elastic-plastic composite materials, *J. Appl. Mech.* **53**, 737–743.

Dvorak, G. J. (1990). On uniform fields in heterogeneous media, *Proc. R. Soc. Lond. A* **431**, 89–110.

Dvorak, G. J. and Benveniste, Y. (1992). On transformation strains and uniform fields in multiphase elastic media, *Proc. R. Soc. Lond. A* **437**, 291–310.

Dykhne, A. M. (1971). Conductivity of a two-dimensional two-phase system, *Soviet Phys. JETP* **32**, 63–65.

Einstein, A. (1906). Eine neue Bestimmung der Moleküldimensionen, *Ann. Phys.* **19**, 289–306.

Eischen, J. W. and Torquato, S. (1993). Determining elastic behavior of composites by the boundary element method, *J. Appl. Phys.* **74**, 159–170.

Elam, W. T., Kerstein, A. R. and Rehr, J. J. (1984). Critical properties of the void percolation problem for spheres, *Phys. Rev. Lett.* **52**, 1516–1519.

Elsayed, M. A. (1974). Bounds for effective thermal, electrical, and magnetic properties of heterogeneous materials using high order statistical information, *J. Math. Phys.* **15**, 2001–2015.

Elsayed, M. A. and McCoy, J. J. (1973). Effect of fiber positioning on the effective physical properties of composite materials, *J. Composite Mater.* **7**, 466–480.

Ersfeld, B. (1997). *Effektive Elastizität einer Suspension von Kugeln*, Ph.D. thesis, RWTH Aachen.

Eshelby, J. D. (1957). The determination of the elastic field of an ellipsoidal inclusion and related problems, *Proc. R. Soc. Lond. A* **241**, 376–396.

Essam, J. W. (1972). Percolation and cluster size, in *Phase Transitions and Critical Phenomena*, C. Domb and M. S. Green (eds.), Vol. 2, Academic Press, London, pp. 197–270.

Evans, J. W. (1993). Random and cooperative sequential adsorption, *Rev. Mod. Phys.* **65**, 1281–1329.

Eyre, D. J. and Milton, G. W. (1999). A fast numerical scheme for computing the response of composites using grid refinement, *Euro. Phys. J.* **6**, 41–47.

Family, F. (1993). Fractal structures and dynamics of cluster growth, in *On Clusters and Clustering*, P. J. Reynolds (ed.), North-Holland, Amsterdam, pp. 323–344.

Fanti, L. A., Glandt, E. D. and Chiew, Y. C. (1988). Cluster volume and surface area in dispersions of penetrable particles or pores, *J. Chem. Phys.* **89**, 1055–1063.

Feder, J. (1980). Random sequential adsorption, *J. Theor. Biol.* **87**, 237–254.

Felderhof, B. U. (1982). Bounds for the effective dielectric constant of a suspension of uniform spheres, *J. Phys. C.* **15**, 3953–3966.

Felderhof, B. U. (1984). Bounds for the complex dielectric constant of a two-phase composite, *Physica A* **126**, 430–442.

Felderhof, B. U. (1985). Wigner solids and diffusion controlled reactions in a regular array of spheres, *Physica A* **130**, 34–56.

Felderhof, B. U. and Deutch, J. M. (1976). Concentration dependence of the rate of diffusion-controlled reactions, *J. Chem. Phys.* **64**, 4551–4558.

Felderhof, B. U., Ford, G. W. and Cohen, E. G. D. (1982). Cluster expansion for the dielectric constant of a polarizable suspension, *J. Stat. Phys.* **28**, 135–164.

Feller, W. (1966). *An Introduction to Probability Theory and Its Applications*, Vol. II, Wiley, New York.

Feng, S., Halperin, B. I. and Sen, P. N. (1987). Transport properties of continuum systems near the percolation threshold, *Phys. Rev. B* **35**, 197–214.

Ferrari, M. and Filipponi, M. (1991). Appraisal of current homogenizing techniques for the elastic response of porous and reinforced glass, *J. Am. Ceram. Soc.* **74**, 229–231.

Finney, J. L. (1970). Random packings and the structure of simple liquids. I. The geometry of random close packing, *Proc. R. Soc. Lond. A.* **319**, 479–493.

Fisher, M. E. (1961). Critical probabilities for cluster size and percolation problems, *J. Math. Phys.* **2**, 620–627.

Fisher, M. E. and Essam, J. W. (1961). Some cluster size and percolation problems, *J. Math. Phys.* **2**, 609–619.

Fishman, R. S., Kurtze, D. A. and Bierwagen, G. P. (1992). The effects of density fluctuations in organic coatings, *J. Appl. Phys.* **72**, 3116–3124.

Flaherty, J. E. and Keller, J. B. (1973). Elastic behavior of composite media, *Comm. Pure Appl. Math.* **26**, 565–580.

Flegler, S. L., Heckman, J. W. and Klomparens, K. L. (1993). *Scanning and Transmission Electron Microscopy: An Introduction*, W. H. Freeman, New York.

Francfort, G. A. and Murat, F. (1986). Homogenization and optimal bounds in linear elasticity, *Arch. Rational Mech. Analysis* **94**, 307–334.

Fredrich, J. T., Menendez, B. and Wong, T. F. (1995). Imaging the pore structure of geomaterials, *Science* **268**, 276–279.

Freed, K. F. and Muthukumar, M. (1978). On the Stokes problem for a suspension of spheres at finite concentrations, *J. Chem. Phys.* **68**, 2088–2096.

Frenkel, D. and Smit, B. (1996). *Understanding Molecular Simulation*, Academic Press, New York.

Frisch, H. L. and Stillinger, F. H. (1963). Contribution to the statistical geometric basis of radiation scattering, *J. Chem. Phys.* **38**, 2200–2207.

Garboczi, E. J. and Day, A. R. (1995). Algorithm for computing the effective linear elastic properties of heterogeneous materials—Three-dimensional results for composites with equal phase Poisson ratios, *J. Mech. Phys. Solids* **43**, 1349–1362.

Garboczi, E. J., Snyder, K. A., Douglas, J. F. and Thorpe, M. F. (1995). Geometrical percolation threshold of overlapping ellipsoids, *Phys. Rev. E* **52**, 819–828.

Gawlinski, E. T. and Redner, S. (1983). Monte-Carlo renormalisation group for continuum percolation with excluded-volume interactions, *J. Phys. A: Math. Gen.* **16**, 1063–1071.

Gawlinski, E. T. and Stanley, H. E. (1981). Continuum percolation in two dimensions: Monte Carlo tests of scaling and universality for non-interacting discs, *J. Phys. A: Math. Gen.* **14**, L291–L299.

Gelhar, L. W. (1993). *Stochastic Subsurface Hydrology*, Prentice Hall, Englewood Cliffs, New Jersey.

Georgiadis, J. G. and Catton, I. (1988). An effective equation governing convective transport in porous media, *J. Heat Transfer* **110**, 635–641.

Ghosh, S. and Mukhopadhyay, S. N. (1993). A material based finite element analysis of heterogeneous media involving Dirichlet tessellations, *Comput. Methods Appl. Mech. Eng.* **104**, 211–247.

Gibiansky, L. V. and Torquato, S. (1993). Link between the conductivity and elastic moduli of composite materials, *Phys. Rev. Lett.* **71**, 2927–2930.

Gibiansky, L. V. and Torquato, S. (1995a). Rigorous link between the conductivity and elastic moduli of fibre-reinforced composite materials, *Phil. Trans. Royal Soc. Lond. A* **353**, 243–278.

Gibiansky, L. V. and Torquato, S. (1995b). Geometrical-parameter bounds on the effective moduli of composites, *J. Mech. Phys. Solids* **43**, 1587–1613.

Gibiansky, L. V. and Torquato, S. (1996a). Phase-interchange relations for the elastic moduli of two-phase composites, *Int. J. Eng. Sci.* **34**, 739–760.

Gibiansky, L. V. and Torquato, S. (1996b). Connection between the conductivity and bulk modulus of isotropic composite materials, *Proc. R. Soc. Lond. A* **452**, 253–283.

Gibiansky, L. V. and Torquato, S. (1996c). Bounds on the effective moduli of cracked materials, *J. Mech. Phys. Solids* **44**, 233–242.

Gibiansky, L. V. and Torquato, S. (1997). Thermal expansion of isotropic multiphase composites and polycrystals, *J. Mech. Phys. Solids* **45**, 1223–1252.

Gibson, L. J. and Ashby, M. F. (1997). *Cellular Solids: Structure and Properties*, second edition, Cambridge University Press, Cambridge, England.

Given, J. A. and Stell, G. (1991). Ornstein–Zernike formalism for reaction rates in random media, *J. Chem. Phys.* **94**, 3060–3069.

Given, J. A., Blawzdziewicz, J. and Stell, G. (1990a). Diffusion-controlled reactions in a polydisperse medium of reactive sinks, *J. Chem. Phys.* **93**, 8156–8170.

Given, J. A., Kim, I. C., Torquato, S. and Stell, G. (1990b). Comparison of analytic and numerical results for the mean cluster density in continuum percolation, *J. Chem. Phys.* **93**, 5128–5139.

Golden, K. (1990). Convexity and exponent inequalities for conduction near percolation, *Phys. Rev. Lett.* **65**, 2923–2926.

Golden, K. M., Ackley, S. F. and Lytle, V. I. (1998). The percolation phase transition in sea ice, *Science* **282**, 2238–2241.

Golden, K. and Papanicolaou G. (1983). Bounds for effective parameters of heterogeneous media by analytic continuation, *Commun. Math. Phys.* **90**, 473–491.

Gonzalez, J. J., Hemmer, P. C. and Hoye, J. S. (1974). Cooperative effects in random sequential polymer reactions, *Chem. Phys.* **3**, 228–238.

Goodier, J. N. (1933). Concentration of stress around spherical and cylindrical inclusions and flaws, *J. Appl. Mech.* **55**, 39–44.

Gotoh, K. and Finney, J. L. (1974). Statistical geometrical approach to random packing density of equal spheres, *Nature* **252**, 202–205.

Grabovsky, Y. and Kohn, R. V. (1995). Microstructures minimizing the energy of a two phase elastic composite in two space dimensions. 2. The Vigdergauz microstructure, *J. Math. Phys. Solids* **43**, 949–972.

Grabovsky, Y., Milton, G. W. and Sage, D. S. (2000). Exact relations for effective tensors of composites: Necessary conditions and sufficient conditions, *Comm. Pure Appl. Math.* **53**, 300–353.

Greengard, L. and Helsing, J. (1995). A numerical study of the ζ_2 parameter for random suspensions of disks, *J. Appl. Phys.* **77**, 2015–2019.

Greengard, L. and Moura, M. (1994). On the numerical evaluation of the electrostatic fields in composite materials, *Acta Numerica* pp. 379–410.

Grimmett, G. (1989). *Percolation*, Springer-Verlag, New York.

Gubernatis, J. E. and Krumhansl, J. A. (1975). Macroscopic engineering properties of polycrystalline materials: Elastic properties, *J. Appl. Phys.* **46**, 1875–1883.

Guinier, A. and Fournet, G. (1955). *Small Angle Scattering of X-Rays*, John Wiley and Sons, New York.

Haan, S. W. and Zwanzig, R. (1977). Series expansions in a continuum percolation problem, *J. Phys. A: Math. Gen.* **10**, 1547–1555.

Haile, J. M., Massobrio, C. and Torquato, S. (1985). Two-point matrix probability function for two-phase random media: Computer simulation results for impenetrable spheres, *J. Chem. Phys.* **83**, 4075–4078.

Hales, T. C. (1998). The Kepler conjecture, preprint.

Hall, P. (1985). On continuum percolation, *Ann. Prob.* **13**, 1250–1266.

Hall, P. (1988). *Introduction to the Theory of Coverage Processes*, Wiley, New York.

Halperin, B. I. and Nelson, D. R. (1978). Theory of two-dimensional melting, *Phys. Rev. Lett.* **41**, 121–124.

Hansen, J. P. and McDonald, I. R. (1986). *Theory of Simple Liquids*, Academic Press, New York.

Hashin, Z. (1962). The elastic moduli of heterogeneous materials, *J. Appl. Mech.* **29**, 143–150, 765–766.

Hashin, Z. (1965a). Viscoelastic behavior of heterogeneous media, *J. Appl. Mech.* **32**, 630–636.

Hashin, Z. (1965b). On elastic behavior of fiber reinforced materials of arbitrary transverse phase geometry, *J. Mech. Phys. Solids* **13**, 119–134.

Hashin, Z. (1970). Theory of composite materials, *Mechanics of Composite Materials*, Pergamon Press, New York.

Hashin, Z. (1983). Analysis of composite materials, *J. Appl. Mech.* **50**, 481–505.

Hashin, Z. (1992). Extremum principles for elastic heterogeneous media with imperfect interfaces and their application to bounding of effective moduli, *J. Mech. Phys. Solids* **40**, 767–781.

Hashin, Z. and Rosen, B. W. (1964). The elastic moduli of fiber-reinforced materials, *J. Appl. Mech.* **31**, 223–232.

Hashin, Z. and Shtrikman, S. (1962). A variational approach to the theory of the effective magnetic permeability of multiphase materials, *J. Appl. Phys.* **33**, 3125–3131.

Hashin, Z. and Shtrikman, S. (1963). A variational approach to the theory of the elastic behaviour of multiphase materials, *J. Mech. Phys. Solids* **11**, 127–140.

Hasimoto, H. (1959). On the periodic fundamental solutions of the Stokes equations and their application to viscous flow past a cubic array of spheres, *J. Fluid Mech.* **5**, 317–328.

Hefland, E., Reiss, H., Frisch, H. L. and Lebowitz, J. L. (1960). Scaled particle theory of fluids, *J. Chem. Phys.* **33**, 1379–1385.

Helsing, J. (1993). Bounds to the conductivity of some two-component composites, *J. Appl. Phys.* **73**, 1240–1245.

Helsing, J. (1994a). Bounds on the shear modulus of composites by interface integral methods, *J. Mech. Phys. Solids* **42**, 1123–1138.

Helsing, J. (1994b). Third-order bounds on the conductivity of a random stacking of cubes, *J. Math. Phys.* **35**, 1688–1692.

Helsing, J. (1998). A high-order accurate algorithm for electrostatics of overlapping disks, *J. Stat. Phys.* **90**, 1461–1473.

Helsing, J., Milton, G. W. and Movchan, A. B. (1997). Duality relations, correspondences and numerical results for planar elastic composites, *J. Mech. Phys. Solids* **45**, 565–590.

Helte, A. (1994). Fourth-order bounds on the effective conductivity for a system of fully penetrable spheres, *Proc. R. Soc. Lond. A* **445**, 247–256.

Henderson, S. I., Mortensen, T. C., Underwood, S. M. and van Megen, W. (1996). Effect of particle size distribution on crystallisation and the glass transition of hard sphere colloids, *Physica A* **233**, 102–116.

Hertz, P. (1909). Über den gegenseitigen durchschnittlichen Abstand von Punkten die mit bekannter mittlerer Dichte im Raume angeordnet sind, *Math. Ann.* **67**, 387–398.

Hetherington, J. H. and Thorpe, M. F. (1992). The conductivity of a sheet containing inclusions with sharp corners, *Proc. R. Soc. Lond. A* **438**, 591–604.

Hilfer, R. (1991). Geometric and dielectric characterization of porous media, *Phys. Rev. B* **44**, 60–75.

Hilfer, R. (1996). Transport and relaxation phenomena in porous media, *Adv. Chem. Phys.* **92**, 299–424.

Hill, R. (1952). The elastic behaviour of a crystalline aggregate, *Proc. Phys. Soc. Lond. A.* **65**, 349–355.

Hill, R. (1963). Elastic properties of reinforced solids: Some theoretical principles, *J. Mech. Phys. Solids* **11**, 357–372.

Hill, R. (1964). Theory of mechanical properties of fiber-strengthened materials. I. Elastic behavior, *J. Mech. Phys. Solids* **12**, 199–212.

Hill, R. (1965a). Theory of mechanical properties of fiber-strengthened materials. III. Self-consistent model, *J. Mech. Phys. Solids* **13**, 189–198.

Hill, R. (1965b). A self-consistent mechanics of composite materials, *J. Mech. Phys. Solids* **13**, 213–222.

Hill, T. (1955). Molecular clusters in imperfect gases, *J. Chem. Phys.* **23**, 617–622.

Hill, T. L. (1987). *Statistical Mechanics: Principles and Selected Applications*, Dover Publications, New York.

Hinch, E. J. (1977). An averaged-equation approach to particle interactions in a fluid suspension, *J. Fluid Mech.* **83**, 695–720.

Ho, F. G. and Strieder, W. (1979). Asymptotic expansion of the porous medium, effective diffusion coefficient in the Knudsen number, *J. Chem. Phys.* **70**, 5635–5639.

Hoover, W. G. and Ree, F. H. (1968). Melting transition and communal entropy for hard spheres, *J. Chem. Phys.* **49**, 3609–3618.

Hori, M. (1973). Statistical theory of effective electrical, thermal, and magnetic properties of random heterogeneous materials. I. Perturbation expansions for the effective permittivity of cell materials, *J. Math. Phys.* **14**, 514–523.

Hoshen, J. and Kopelman, R. (1976). Percolation and cluster distribution. I. Cluster multiple labelling technique and critical concentration algorithm, *Phys. Rev. B* **14**, 3438–3445.

Howells, I. D. (1974). Drag due to the motion of a Newtonian fluid through a sparse random array of small fixed rigid objects, *J. Fluid Mech.* **64**, 449–475.

Howle, L. E., Behringer, R. P. and Georgiadis, J. G. (1993). Visualization of convective fluid flow in a porous medium, *Nature* **362**, 230–232.

Huang, K. (1987). *Statistical Mechanics*, John Wiley, New York.

Hyun, S. and Torquato, S. (2000). Effective elastic and transport properties of regular honeycombs for all densities, *J. Mater. Res.* **15**, 1985–1993.

Hyun, S. and Torquato, S. (2001). Designing composite microstructures with targeted properties, *J. Mater. Res.* **16**, 280–285.

Jackson, J. D. (1990). *Classical Electrodynamics*, Wiley, New York.

Jasiuk, I., Chen, J. and Thorpe, M. F. (1994). Elastic moduli of two-dimensional materials with polygonal and elliptical holes, *Appl. Mech. Rev.* **47**, S18–S28.

Jaster, A. (1998). Orientational order of the two-dimensional hard-disk system, *Europhys. Lett.* **42**, 277–281.

Jeffrey, D. J. (1973). Conduction through a random suspension of spheres, *Proc. R. Soc. Lond. A* **335**, 355–367.

Jeffrey, D. J. (1974). Group expansions for the bulk properties of a statistically homogeneous, random suspensions, *Proc. R. Soc. Lond. A* **338**, 503–516.

Jenkins, J. T. and Hanes, D. M. (1998). Collisional sheet flows of sediment driven by a turbulent fluid, *J. Fluid Mech.* **370**, 29–52.

Jeulin, D. and Le Coënt, A. (1996). Morphological modeling of random composites, in *Continuum Models and Discrete Systems*, K. Z. Markov (ed.), World Scientific, Singapore.

Jeune, B. and Barabe, D. (1998). Interactions between physical and biological constraints in the structure of the inflorescences of the araceae, *Annals Botany* **82**, 577–586.

Jikov, V. V., Kozlov, S. M. and Olenik, O. A. (1994). *Homogenization of Differential Operators and Integral Functionals*, Springer-Verlag, Berlin.

Jodrey, W. S. and Tory, E. M. (1985). Computer simulation of close random packing of equal spheres, *Phys. Rev. A* **32**, 2347–2351.

Johnson, D. L., Koplik, J. and Dashen, R. (1987). Theory of dynamic permeability and tortuosity in fluid-saturated porous media, *J. Fluid Mech.* **176**, 379–402.

Johnson, D. L., Koplik, J. and Schwartz, L. M. (1986). New pore-size parameter characterizing transport in porous media, *Phys. Rev. Lett.* **57**, 2564–2567.

Jones, R. B. and Schmitz, R. (1984). Effective elasticity of a solid suspension of spheres. I. Stress–strain constitutive relation, *Physica A* **125**, 381–397.

Joshi, M. Y. (1974). *A Class of Stochastic Models for Porous Media*, Ph.D. thesis, University of Kansas, Lawrence.

Joslin, C. G. and Stell, G. (1986a). Effective properties of fiber-reinforced composites: Effects of polydispersity in fiber diameter, *J. Appl. Phys.* **60**, 1611–1613.

Joslin, C. G. and Stell, G. (1986b). Bounds on the properties of fiber-reinforced composites, *J. Appl. Phys.* **60**, 1607–1610.

Ju, J. W. and Chen, T. M. (1994). Effective elastic moduli of two-phase composites containing randomly dispersed spherical inhomogeneities, *Acta Mechanica* **103**, 123–144.

Jullien, R., Sadoc, J.-F. and Mosseri, R. (1997). Packing at random in curved space and frustration: A numerical study, *J. Physique I* **7**, 1677–1692.

Kansal, A. R., Torquato, S., Harsh, G. R., Chiocca, E. A. and Deisboeck, T. S. (2000b). Simulated brain tumor growth dynamics using a three-dimensional cellular automaton, *J. Theor. Biol.* **203**, 367–382.

Kansal, A. R., Truskett, T. M. and Torquato, S. (2000a). Nonequilibrium hard-disk packings with controlled orientational order, *J. Chem. Phys.* **113**, 4844–4851.

Kantor, Y. and Bergman, D. J. (1984). Improved rigorous bounds on the effective elastic moduli of a composite material, *J. Mech. Phys. Solids* **32**, 41–62.

Keller, J. B. (1963). Conductivity of a medium containing a dense array of perfectly conducting spheres or cylinders or nonconducting cylinders, *J. Appl. Phys.* **34**, 991–993.

Keller, J. B. (1964). A theorem on the conductivity of a composite medium, *J. Math Phys.* **5**, 548–549.

Keller, J. B., Rubenfeld, L. and Molyneux, J. (1967). Extremum principles for slow viscous flows with applications to suspensions, *J. Fluid Mech.* **30**, 97–125.

Kellogg, O. D. (1953). *Foundations of Potential Theory*, Dover, New York.

Kendall, M. G. and Moran, P. A. P. (1962). *Geometrical Probability*, C. Griffin, London.

Kerstein, A. R. (1983). Equivalence of the void percolation problem for overlapping spheres and a network problem, *J. Phys. A: Math. Gen.* **16**, 3071–3075.

Kerstein, A. R. (1987). Percolation model of polydisperse composite solid propellant combustion, *Combust. Flame* **69**, 95–112.

Kertesz, J. (1981). Percolation of holes between overlapping spheres: Monte Carlo calculation of the critical volume fraction, *J. Phys. Lett.* **42**, L393–L395.

Kertesz, J. and Vicsek, T. (1982). Monte Carlo renormalization group study of the percolation problem of discs with a distribution of radii, *Z. Phys. B* **45**, 345–350.

Kim, I. C. and Torquato, S. (1990a). Monte Carlo calculations of connectedness and mean cluster size for bidispersions of overlapping spheres, *J. Chem. Phys.* **93**, 5998–6002.

Kim, I. C. and Torquato, S. (1990b). Determination of the effective conductivity of heterogeneous media by Brownian motion simulation, *J. Appl. Phys.* **68**, 3892–3903.

Kim, I. C. and Torquato, S. (1991). Effective conductivity of suspensions of hard spheres by Brownian motion simulation, *J. Appl. Phys.* **69**, 2280–2289.

Kim, I. C. and Torquato, S. (1992a). Diffusion of finite-sized Brownian particles in porous media, *J. Chem. Phys.* **96**, 1498–1503.

Kim, I. C. and Torquato, S. (1992b). Effective conductivity of suspensions of overlapping spheres, *J. Appl. Phys.* **71**, 2727–2735.

Kinney, J. H. and Nichols, M. C. (1992). X-ray tomographic microscopy (XTM) using synchrotron radiation, *Ann. Rev. Mater. Sci.* **22**, 121–152.

Kirkpatrick, S., Gelatt, C. D. and Vecchi, M. P. (1983). Optimization by simulated annealing, *Science* **220**, 671–680.

Kirkwood, J. G. and Salsburg, Z. W. (1953). The statistical mechanical theory of molecular distribution functions in liquids, *Discuss. Faraday Society* **15**, 28–34.

Kirste, R. and Porod, G. (1962). Röntgenkleinwinkelstreuung an Kolloiden Systemen. Asymptotisches Verhalten der Streukurven, *Kolloid-Z.* **184**, 1–7.

Kohn, R. V. (1988). Recent progress in the mathematical modelling of composite materials, in *Composite Material Response: Constitutive Relations and Damage Mechanisms*, G. C. Sih et al. (ed.), Elsevier, London.

Kohn, R. V. (1989). Composite materials and structural optimization, in *Smart/Intelligent Materials and Systems*, C. Rogers (ed.), Technomic Press, Lancaster, PA.

Kohn, R. V. and Milton, G. W. (1986). On bounding the effective conductivity of anisotropic composites, in *Homogenization and Effective Moduli of Materials and Media*, J. L. Ericksen et al. (eds.), Springer-Verlag, New York.

Koplik, J., Levine, H. and Zee, A. (1983). Viscosity renormalization in the Brinkman equation, *Phys. Fluids* **26**, 2864–2870.

Kosterlitz, J. M. and Thouless, D. J. (1973). Ordering, metastability and phase transitions in two-dimensional systems, *J. Phys. C* **6**, 1181–1203.

Kratky, K. W. (1978). The area of intersection of n equal circular disks, *J. Phys. A: Math. Gen.* **11**, 1017–1024.

Kratky, K. W. (1981). Intersecting disks (and spheres) and statistical mechanics. I. Mathematical basis, *J. Stat. Phys.* **25**, 619–634.

Kraynik, A. M., Nielsen, M. K., Reinelt, D. A. and Warren, W. E. (1999). Foam micromechanics: Structure and rheology of foams, emulsions and cellular solids, in *Foams and Emulsions*, J.-F. Sadoc and N. Rivier (eds.), Kluwer, New York.

Kroner, E. (1977). Bounds for effective elastic moduli of disordered materials, *J. Mech. Phys. Solids* **25**, 137–155.

Kublanov, L. and Milton, G. W. (1991). Rigorous estimates for the elastic moduli for a two-dimensional two-phase material, unpublished.

Kusy, R. P. (1977). Influence of particle size ratio on the continuity of aggregates, *J. Appl. Phys.* **48**, 5301–5305.

Lado, F. (1968). Equation of state of the hard-disk fluid from approximate integral equations, *J. Chem. Phys.* **49**, 3092–3096.

Lado, F. (1971). Numerical Fourier transforms in one, two, and three dimensions for liquid state calculations, *J. Comput. Phys.* **8**, 417–433.

Lado, F. and Torquato, S. (1986). Effective properties of two-phase disordered composite media. I. Simplification of bounds on the conductivity and bulk modulus of dispersions of impenetrable spheres, *Phys. Rev. B* **33**, 3370–3378.

Lado, F. and Torquato, S. (1990). Two-point probability function for distributions of oriented hard ellipsoids, *J. Chem. Phys.* **93**, 5912–5917.

Lakes, R. (1987). Foam structures with a negative Poisson's ratio, *Science* **235**, 1038–1040.

Lamb, H. (1932). *Hydrodynamics*, Cambridge University Press, Cambridge, England.

Landau, L. D. and Lifshitz, E. M. (1970). *Theory of Elasticity*, Pergamon Press, New York.

Landauer, R. (1952). The electrical resistance of binary metallic mixtures, *J. Appl. Phys.* **23**, 779–784.

Landauer, R. (1978). Electrical conductivity in inhomogeneous media, in *Electrical, Transport and Optical Properties of Inhomogeneous Media*, J. C. Garland and D. B. Tanner (eds.), AIP, New York, pp. 2–43.

Larsen, U. D., Sigmund, O. and Bouwstra, S. (1997). Design and fabrication of compliant micromechanisms and structures with negative Poisson's ratio, *J. Microelectromech. Sys.* **6**, 99–106.

Lebenhaft, J. R. and Kapral, R. (1979). Diffusion-controlled processes among partially absorbing stationary sinks, *J. Stat. Phys.* **20**, 25–56.

Lebowitz, J. L. (1964). Exact solution of generalized Percus–Yevick equation for a mixture of hard spheres, *Phys. Rev.* **133**, A895–A899.

Lebowitz, J. L. and Perram, J. W. (1983). Correlation functions for nematic liquid crystals, *Molecular Phys.* **50**, 1207–1214.

Lee, S. B. (1990). Universality of continuum percolation, *Phys. Rev. B* **42**, 4877–4880.

Lee, S. B. (1993). Connectedness and clustering of 2-phase disordered media for adhesive sphere model, *J. Chem. Phys.* **98**, 8119–8125.

Lee, S. B. and Torquato, S. (1988a). Porosity for the penetrable-concentric-shell model of two-phase disordered media: Computer simulation results, *J. Chem. Phys.* **89**, 3258–3263.

Lee, S. B. and Torquato, S. (1988b). Pair connectedness and mean cluster size for continuum-percolation models: Computer-simulation results, *J. Chem. Phys.* **89**, 6427–6433.

Lee, S. B. and Torquato, S. (1989). Measure of clustering in continuum percolation: Computer-simulation of the two-point cluster function, *J. Chem. Phys.* **91**, 1173–1178.

Lee, S. B. and Torquato, S. (1990). Monte-Carlo study of correlated continuum percolation—Universality and percolation thresholds, *Phys. Rev. A* **41**, 5338–5344.

Lee, S. B., Kim, I. C., Miller, C. A. and Torquato, S. (1989). Random-walk simulation of diffusion-controlled processes among static traps, *Phys. Rev. B* **39**, 11833–11839.

Leutheusser, E. (1986). On the Percus–Yevick equation for a rigid disk fluid, *J. Chem. Phys.* **84**, 1050–1051.

Levin, V. M. (1967). Thermal expansion coefficients of heterogeneous materials, *Mech. Solids* **21**, 9–17.

Levitz, P. (1998). Off-lattice reconstruction of porous media: Critical evaluation, geometrical confinement and molecular transport, *Adv. Colloid Interface Sci.* **77**, 71–106.

Levy, O. and Bergman, D. J. (1994). Critical behavior of the weakly nonlinear conductivity and flicker noise of 2-component composites, *Phys. Rev. B* **50**, 3652–3660.

Lewis, F. T. (1946). The shape of cells as a mathematical problem, *Amer. Scientist* **34**, 359–369.

Li, Y. S. and Duxbury, P. M. (1989). From moduli scaling to breakdown scaling—A moment-spectrum analysis, *Phys. Rev. B* **40**, 4889–4897.

Lifshitz, I. M., Gredeskul, S. A. and Pastur, L. A. (1988). *Introduction to the Theory of Disordered Systems*, Wiley, New York.

Lipton, R. (1988). On the effective elasticity of a two-dimensional homogenised incompressible elastic composite, *Proc. R. Soc. Edinburgh* **110A**, 45–61.

Lipton, R. (1997). Variational methods, bounds, and size effects for composites with highly conducting interface, *J. Mech. Phys. Solids* **45**, 361–384.

Lipton, R. and Vernescu, B. (1996). Composites with imperfect interface, *Proc. R. Soc. Lond. A* **452**, 329–358.

Lorenz, B., Orgzall, I. and Heuer, H. O. (1993). Universality and cluster structures in continuum models of percolation with two different radius distributions, *J. Phys. A: Math. Gen.* **26**, 4711–4722.

Lorenz, C. D. and Ziff, R. M. (1998). Precise determination of the bond percolation thresholds and finite-size scaling corrections for the sc, fcc, and bcc lattices, *Phys. Rev. E* **57**, 230–236.

Lu, B. L. and Torquato, S. (1990a). Local volume fraction fluctuations in heterogeneous media, *J. Chem. Phys.* **93**, 3452–3459.

Lu, B. L. and Torquato, S. (1990b). n-point probability functions for a lattice model of heterogeneous media, *Phys. Rev. B* **42**, 4453–4459.

Lu, B. L. and Torquato, S. (1990c). Photographic granularity—Mathematical formulation and effect of impenetrability of grains, *J. Opt. Soc. Am. A* **7**, 717–724.

Lu, B. L. and Torquato, S. (1991). General formalism to characterize the microstructure of polydispersed random media, *Phys. Rev. A* **43**, 2078–2080.

Lu, B. L. and Torquato, S. (1992a). Lineal-path function for random heterogeneous materials, *Phys. Rev. A* **45**, 922–929.

Lu, B. L. and Torquato, S. (1992b). Nearest-surface distribution functions for polydispersed particle systems, *Phys. Rev. A* **45**, 5530–5544.

Lu, B. L. and Torquato, S. (1992c). Lineal-path function for random heterogeneous materials. II. Effect of polydispersivity, *Phys. Rev. A* **45**, 7292–7301.

Lubachevsky, B. D. and Stillinger, F. H. (1990). Geometric properties of random disk packings, *J. Stat. Phys.* **60**, 561–583.

Lubachevsky, B. D., Stillinger, F. H. and Pinson, E. N. (1991). Disks vs. spheres: Contrasting properties of random packings, *J. Stat. Phys.* **64**, 501–524.

Lurie, K. A. and Cherkaev, A. V. (1984). Exact estimates of the conductivity of composites formed by two isotropically conducting media taken in prescribed proportion, *Proc. R. Soc. Edinburgh* **99A**, 71–87.

Lurie, K. A. and Cherkaev, A. V. (1985). Optimization of properties of multicomponent isotropic composites, *J. Opt. Theor. Appl.* **46**, 571–580.

Lurie, K. A. and Cherkaev, A. V. (1986). Exact estimates of the conductivity of a binary mixture of isotropic materials, *Proc. R. Soc. Edinburgh* **104A**, 21–38.

Mackenzie, J. K. (1962). Sequential filling of a line by intervals placed at random and its application to linear adsorption, *J. Chem. Phys.* **37**, 723–728.

Malliaris, A. and Turner, D. T. (1971). Influence of particle size on the electrical resistivity of compacted mixtures of polymeric and metallic powders, *J. Appl. Phys.* **42**, 614–618.

Manwart, C. and Hilfer, R. (1999). Reconstruction of random media using Monte Carlo methods, *Phys. Rev. E* **59**, 5596–5599.

Manwart, C., Torquato, S. and Hilfer, R. (2000). Stochastic reconstruction of sandstones, *Phys. Rev. E* **62**, 893–899.

Markov, K. Z. (1998). On the cluster bounds for the effective properties of microcracked solids, *J. Mech. Phys. Solids* **46**, 357–388.

Markov, K. Z. (1999). On the correlation functions of two-phase random media and related problems, *Proc. R. Soc. Lond. A* **455**, 1049–1066.

Martys, N. and Garboczi, E. J. (1992). Length scales relating the fluid permeability and electrical conductivity in random two-dimensional model porous media, *Phys. Rev. B* **46**, 6080–6090.

Martys, N., Bentz, D. P. and Garboczi, E. J. (1994a). Computer simulation study of the effective viscosity in Brinkman's equation, *Phys. Fluids* **6**, 1434–1439.

Martys, N., Hagedorn, J. G., Goujon, D. and Devaney, J. E. (2000). Lattice Boltzmann simulations of single and multi-component flow in porous media, in *Mesoscale Modeling: Techniques and Applications*, D. B. Nicolaides and A. Bick (eds.), Marcel Dekker.

Martys, N., Torquato, S. and Bentz, D. P. (1994b). Universal scaling of fluid permeability for sphere packings, *Phys. Rev. E* **50**, 403–408.

Matheron, G. (1967). *Elements pour une Theorie des Milieux Poreux*, Masson, Paris.

Matheron, G. (1975). *Random Sets and Integral Geometry*, Wiley, New York.

Mattern, K. and Felderhof, B. U. (1987). Rate of diffusion-controlled reactions in a random array of spherical sinks, *Physica A* **143**, 1–20.

Mau, S. C. and Huse, D. A. (1999). Stacking entropy of hard-sphere crystals, *Phys. Rev. E* **59**, 4396–4401.

Maxwell, J. C. (1873). *Treatise on Electricity and Magnetism*, Clarendon Press, Oxford.

Mayer, J. E. (1947). Integral equations between distribution functions of molecules, *J. Chem. Phys.* **15**, 187–201.

Mayer, J. E. and Montroll, E. (1941). Molecular distribution, *J. Chem. Phys.* **9**, 2–16.

McCoy, J. J. (1970). On the displacement field in an elastic medium with random variation of material properties, *Recent Advances in Engineering Sciences*, Vol. 5, Gordon and Breach, New York.

McCoy, J. J. (1979). On the calculation of bulk properties of heterogeneous materials, *Q. Appl. Math.* **37**, 137–149.

McKenzie, D. R., McPhedran, R. C. and Derrick, G. H. (1978). The conductivity of lattices of spheres. II. The body centered and face centered cubic lattices, *Proc. R. Soc. Lond. A* **362**, 211–232.

McLachlan, D. S., Blaszkiewicz, M. and Newnham, R. E. (1990). Electrical resistivity of composites, *J. Am. Ceram. Soc.* **73**, 2187–2203.

McLaughlin, R. (1977). A study of the differential scheme for composite materials, *Int. J. Eng. Sci.* **15**, 237–244.

McMillan, B. (1955). History of a problem, *J. Soc. Indust. Appl. Math.* **3**, 119–128.

McNally, J. G. and Cox, E. C. (1989). Spots and stripes: The patterning spectrum in the cellular slime-mold polysphondylium-pallidum, *Development* **105**, 323–333.

McPhedran, R. C. and McKenzie, D. R. (1978). The conductivity of lattices of spheres. I. The simple cubic lattice, *Proc. R. Soc. Lond. A* **359**, 45–63.

McPhedran, R. C. and Milton, G. W. (1981). Bounds and exact theories for the transport properties of inhomogeneous media, *Appl. Phys. A* **26**, 207–220.

Meakin, P. (1988). Models for colloidal aggregation, *Ann. Rev. Phys. Chem.* **39**, 237–267.

Meakin, P. and Jullien, R. (1992). Random-sequential adsorption of disks of different sizes, *Phys. Rev. A* **46**, 2029–2038.

Meester, R. and Roy, R. (1996). *Continuum Percolation*, Cambridge University Press, New York.

Meester, R., Roy, R. and Sarkar, A. (1994). Nonuniversality and continuity of the critical covered volume fraction in continuum percolation, *J. Stat. Phys.* **75**, 123–134.

Mehta, A. and Barker, G. C. (2000). Glassy dynamics in granular compaction, *J. Phys.: Condens. Matter* **12**, 6619–6628.

Mendelson, K. S. (1975). Effective conductivity of two-phase material with cylindrical phase boundaries, *J. Appl. Phys.* **46**, 917–918.

Metcalfe, G., Shinbrot, T., McCarthy, J. J. and Ottino, J. M. (1995). Avalanche mixing of granular solids, *Nature* **374**, 39–41.

Metropolis, N., Rosenbluth, A. W., Rosenbluth, M. N., Teller, A. H. and Teller, E. (1953). Equation of state calculations by fast computing machines, *J. Chem. Phys.* **21**, 1087–1092.

Miller, C. A. and Torquato, S. (1989). Diffusion-controlled reactions among spherical traps: Effect of polydispersity in trap size, *Phys. Rev. B* **40**, 7101–7108.

Miller, C. A. and Torquato, S. (1990). Effective conductivity of hard-sphere dispersions, *J. Appl. Phys.* **68**, 5486–5493.

Miller, C. A. and Torquato, S. (1991). Improved bounds on elastic and transport properties of fiber-reinforced composites: Effect of polydispersivity in fiber radius, *J. Appl. Phys.* **69**, 1948–1955.

Miller, C. A., Kim, I. C. and Torquato, S. (1991). Trapping and flow among random arrays of oriented spheroidal inclusions, *J. Chem. Phys.* **94**, 5592–5598.

Miller, M. N. (1969). Bounds for effective electrical, thermal, and magnetic properties of heterogeneous materials, *J. Math. Phys.* **10**, 1988–2004.

Miloh, T. and Benveniste, Y. (1999). On the effective conductivity of composites with ellipsoidal inhomgeneities and highly conducting interfaces, *Proc. R. Soc. Lond. A* **455**, 2687–2706.

Milton, G. W. (1980). Bounds on the complex dielectric constant of a composite material, *Appl. Phys. Lett.* **37**, 300–302.

Milton, G. W. (1981a). Bounds on the transport and optical properties of a two-component composite material, *J. Appl. Phys.* **52**, 5294–5304.

Milton, G. W. (1981b). Bounds on the electromagnetic, elastic, and other properties of two-component composites, *Phys. Rev. Lett.* **46**, 542–545.

Milton, G. W. (1981c). Bounds on the complex permittivity of a two-component composite material, *J. Appl. Phys.* **52**, 5286–5293.

Milton, G. W. (1981d). Concerning bounds on the transport and mechanical properties of multicomponent composite materials, *Appl. Phys.* **A26**, 125–130.

Milton, G. W. (1982). Bounds on the elastic and transport properties of two-component composites, *J. Mech. Phys. Solids* **30**, 177–191.

Milton, G. W. (1984). Correlation of the electromagnetic and elastic properties of composites and microgeometries corresponding with effective medium approximations, in *Physics and Chemistry of Porous Media*, D. L. Johnson and P. N. Sen (eds.), AIP, New York.

Milton, G. W. (1985). The coherent potential approximation is a realizable effective medium scheme, *Comm. Math. Phys.* **99**, 463–500.

Milton, G. W. (1986). Modelling the properties of composites by laminates, in *Homogenization and Effective Moduli of Materials and Media*, J. L. Ericksen et al. (eds.), Springer-Verlag, New York.

Milton, G. W. (1987). Multicomponent composites, electrical networks and new types of continued fractions. I and II, *Commun. Math. Phys.* **111**, 281–372.

Milton, G. W. (1990). On characterizing the set of possible effective tensors of composites: The variational method and the translation method, *Commun. Pure Appl. Math.* **43**, 63–125.

Milton, G. W. (1991). Modelling the properties of composites by laminates, in *Composite Media and Homogenization Theory*, G. D. Maso and G. F. Dell'Antonio (eds.), Birkhauser, Boston, MA.

Milton, G. W. (1992). Composite materials with Poisson ratios close to -1, *J. Mech. Phys. Solids* **40**, 1105–1137.

Milton, G. W. (2001). *The Theory of Composites*, Cambridge University Press, Cambridge, England.

Milton, G. W. and Golden, K. (1985). Thermal conduction in composites, *18th International Thermal Conductivity Conference*, Plenum Press, New York.

Milton, G. W. and Kohn, R. V. (1988). Variational bounds on the effective moduli of anisotropic composites, *J. Mech. Phys. Solids* **36**, 597–629.

Milton, G. W. and McPhedran, R. C. (1982). A comparison of two methods for deriving bounds on the effective conductivity of composites, *Lecture Notes in Physics* **154**, Springer-Verlag, New York.

Milton, G. W. and Nesi, V. (1999). Optimal G-closure bounds via stability under lamination, *Arch. Rational Mech. Analysis* **150**, 191–207.

Milton, G. W. and Phan-Thien, N. (1982). New bounds on effective elastic moduli of two-component materials, *Proc. R. Soc. Lond. A* **380**, 305–331.

Mohanty, K. K., Ottino, J. M. and Davis, H. T. (1982). Reaction and transport in disordered composite media: An introduction to percolation concepts, *Chem. Eng. Sci.* **37**, 905–924.

Molyneux, J. E. (1970). Effective permittivity of a polycrystalline dielectric, *J. Math. Phys.* **11**, 1172–1184.

Mori, T. and Tanaka, K. (1973). Average stress in matrix and average elastic energy of materials with misfitting inclusions, *Acta Metall.* **21**, 571–574.

Moulinec, H. and Suquet, P. (1998). A numerical method for computing the overall response of nonlinear composites with complex microstructure, *Comput. Meth. Appl. Mech. Eng.* **157**, 69–94.

Mura, T. (1987). *Micromechanics of Defects in Solids*, Martinus Nijhoff, Dordrecht.

Murat, F. and Tartar, L. (1985). Calcul des variations et homogénéisation, in *Les Méthodes de l'Homogénéisation: Théorie et Applications en Physique*, Cours de l'Ecole d'Été d'Analyse Numerique, D. Bergman et al. (eds.), Eyrolles, Paris.

Muthukumar, M. and Cukier, R. I. (1981). Concentration dependence of diffusion-controlled processes among stationary reactive sinks, *J. Stat. Phys.* **26**, 453–469.

Nemat-Nasser, S. and Hori, M. (1993). *Micromechanics: Overall Properties of Heterogeneous Materials*, North-Holland, Amsterdam.

Nesi, V. (1991). Multiphase interchange inequalities, *J. Math. Phys.* **32**, 2263–2275.

Noever, D. A. (1992). Statistical crystallography of surface micelle spacing, *Langmuir* **8**, 1036–1038.

Norris, A. N. (1985). A differential scheme for the effective moduli of composites, *Mech. Mat.* **4**, 1–16.

Norris, A. N. (1989). An examination of the Mori–Tanaka effective medium approximation for multiphase composites, *J. Appl. Mech.* **56**, 83–88.

Norris, A. N., Callegari, A. J. and Sheng, P. (1985). A generalized differential effective medium theory, *J. Mech. Phys. Solids* **33**, 525–543.

Nunan, K. C. and Keller, J. B. (1984). Effective elasticity tensor of a periodic composite, *J. Mech. Phys. Solids* **32**, 259–280.

Nye, J. F. (1957). *Physcial Properties of Crystals*, Oxford University Press, London.

Oberbeck, H. A. (1876). On steady-state flow under consideration of inner friction, *J. Reine. Angew. Math.* **81**, 62–80.

Oden, J. T., Vemaganti, K. and Moes, N. (1999). Hierarchical modeling of heterogeneous solids, *Comput. Methods Appl. Mech. Eng.* **172**, 3–25.

Okabe, A., Boots, B. and Sugihara, K. (1992). *Spatial Tesselations: Concepts and Applications of Voronoi diagrams*, Wiley, New York.

O'Neill, E. L. (1963). *Introduction to Statistical Optics*, Addison–Wesley, Reading, Massachusetts.

Onsager, L. (1931a). Reciprocal relations in irreversible processes. I, *Phys. Rev.* **37**, 405–426.

Onsager, L. (1931b). Reciprocal relations in irreversible processes. II, *Phys. Rev.* **38**, 2265–2279.

Onsager, L. (1944). Crystal statistics. I. A two-dimensional model with an order-disorder transition, *Phys. Rev.* **65**, 117–149.

Onsager, L. (1949). The effects of shape on the interaction of colloidal particles, *Ann. New York Acad. Sci.* **51**, 627–659.

Ornstein, L. S. and Zernike, F. (1914). Accidental deviations of density and opalescence at the critical point of a single substance, *Proc. Akad. Sci. (Amsterdam)* **17**, 793–806.

Ostoja-Starzewski, M. (1993). Micromechanics as a basis of random elastic continuum approximations, *Probab. Eng. Mech.* **8**, 107–114.

Ostoja-Starzewski, M. and Wang, X. (1999). Stochastic finite elements as a bridge between random material microstructure and global response, *Comput. Methods Appl. Mech. Eng.* **168**, 35–49.

Papanicolaou, G. C. and Varadan, S. R. S. (1979). Boundary value problems with rapidly oscillating random coeffecients, *Colloquia Mathematica Societatis J'anos Bolyai* **27**, 835–873.

Paul, B. (1960). Prediction of elastic constant of multiphase materials, *Trans. Metall. Soc. AIME* **218**, 36–41.

Peebles, P. J. E. (1993). *Principles of Physical Cosmology*, Princeton University Press, Princeton.

Penrose, M. D. (1991). On a continuum percolation model, *Adv. Appl. Prob.* **23**, 536–556.

Percus, J. (1964). Pair distribution function in classical statistical mechanics, in *The Equilibrium Theory of Classical Fluids*, H. L. Frisch and J. L. Lebowitz (eds.), Benjamin, New York.

Percus, J. K. and Yevick, G. J. (1958). Analysis of classical statistical mechanics by means of collective coordinates, *Phys. Rev.* **110**, 1–13.

Perrins, W. T., McKenzie, D. R. and McPhedran, R. C. (1979). Transport properties of regular arrays of cylinders, *Proc. R. Soc. Lond. A* **369**, 207–225.

Persson, L. E., Persson, L., Svanstedt, N. and Wyller, J. (1989). An introduction to the homogenization method, *Technical report*, Luleåa, Sweden.

Peterson, J. M. and Hermans, J. J. (1969). The dielectric constants of nonconducting suspensions, *J. Composite Mater.* **3**, 338–354.

Phan-Thien, N. and Milton, G. W. (1982). New bounds on the effective thermal conductivity of n-phase materials, *Proc. R. Soc. Lond. A* **380**, 333–348.

Phani, M. K. and Dhar, D. (1984). Continuum percolation with discs having a distribution of radii, *J. Phys. A: Math. Gen.* **17**, L645–L649.

Pielou, E. C. (1977). *Mathematical Ecology*, Wiley, New York.

Pike, G. E. and Seager, C. H. (1974). Percolation and conductivity: A computer study. I. *Phys. Rev. B* **10**, 1421–1434.

Poladian, L. (1990). *Effective Transport and Optical Properties of Composite Materials*, Ph.D. thesis, University of Sydney, Australia.

Polder, D. and Van Santen, J. H. (1946). The effective permeability of mixtures of solids, *Physica* **12**, 257–271.

Pomeau, Y. (1980). Some asymptotic estimates in the random parking problem, *J. Phys. A: Math. Gen.* **13**, L193–L196.

Ponte Castaneda, P. and Suquet, P. (1998). Nonlinear composites, *Adv. Appl. Mech.* **34**, 171–302.

Portman, J. J. and Wolynes, P. G. (1999). Complementary variational approximations for intermittency and reaction dynamics in fluctuating environments, *J. Phys. Chem. A* **103**, 10602–10610.

Postma, G. W. (1955). Wave propagation in a stratified medium, *Geophys.* **20**, 780–806.

Pouliquen, O., Nicolas, M. and Weidman, P. D. (1997). Crystallization of non-Brownian spheres under horizontal shaking, *Phys. Rev. Lett.* **79**, 3640–3643.

Powell, M. J. D. (1964). The volume internal to three intersecting hard spheres, *Mol. Phys.* **7**, 591–592.

Prager, S. (1961). Viscous flow through porous media, *Phys. Fluids* **4**, 1477–1482.

Prager, S. (1963a). Interphase transfer in stationary two-phase media, *Chem. Eng. Sci.* **18**, 227–231.

Prager, S. (1963b). Diffusion and viscous flow in concentrated suspensions, *Physica* **29**, 129–139.

Prager, S. (1969). Improved variational bounds on some bulk properties of a two-phase random medium, *J. Chem. Phys.* **50**, 4305–4312.

Prall, D. and Lakes, R. S. (1997). Properties of a chiral honeycomb with a Poisson's ratio of -1, *Int. J. Mech. Sci.* **39**, 305–314.

Priestley, M. B. (1981). *Spectral Analysis and Time Series*, Academic Press, New York.

Pugnaloni, L. A., Gianotti, R. D. and Vericat, F. (1997). Comment on "Exact solution of a one-dimensional continuum percolation model," *Phys. Rev. E* **56**, 6206–6207.

Pusey, P. N. and van Megan, W. (1986). Phase behaviour of concentrated suspensions of nearly hard colloidal spheres, *Nature* **320**, 340–342.

Quiblier, J. A. (1984). A new 3-dimensional modeling technique for studying porous media, *J. Colloid Interface Sci.* **98**, 84–102.

Quickenden, T. I. and Tan, G. K. (1974). Random packing in two dimensions and the structure of monolayers, *J. Colloid Interface Sci.* **48**, 382–393.

Quintanilla, J. (1999). Microstructure functions for random media with impenetrable particles, *Phys. Rev. E* **60**, 5788–5794.

Quintanilla, J. and Torquato, S. (1995). New bounds on the elastic moduli of suspensions of spheres, *J. Appl. Phys.* **77**, 4361–4372.

Quintanilla, J. and Torquato, S. (1996a). Lineal measures of clustering in overlapping particle systems, *Phys. Rev. E* **54**, 4027–4036.

Quintanilla, J. and Torquato, S. (1996b). Microstructure and conductivity of hierarchical laminate composites, *Phys. Rev. E* **53**, 4368–4378.

Quintanilla, J. and Torquato, S. (1996c). Clustering properties of d-dimensional overlapping spheres, *Phys. Rev. E* **54**, 5331–5339; Erratum, **56**, 3732 (1997).

Quintanilla, J. and Torquato, S. (1997a). Microstructure functions for a model of statistically inhomogeneous random media, *Phys. Rev. E* **55**, 1558–1565.

Quintanilla, J. and Torquato, S. (1997b). Local volume fraction fluctuations in random media, *J. Chem. Phys.* **106**, 2741–2751.

Quintanilla, J. and Torquato, S. (1999). Local volume fraction fluctuations in periodic hetergeneous media, *J. Chem. Phys.* **110**, 3215–3219.

Quintanilla, J., Torquato, S. and Ziff, R. M. (2000). Efficient measurement of the percolation threshold for fully penetrable discs, *J. Phys. A: Math. Gen.* **33**, L399–L407.

Rahaman, M. N. (1995). *Ceramic Processing and Sintering*, Marcel Dekker, Inc., New York.
Ramshaw, J. D. (1984). Dielectric polarization in random media, *J. Stat. Phys.* **35**, 49–75.
Rayleigh, L. (1892). On the influence of obstacles arranged in a rectangular order upon the properties of medium, *Phil. Mag.* **34**, 481–502.
Reck, R. A. and Prager, S. (1965). Diffusion-controlled quenching at higher quencher concentrations, *J. Chem. Phys.* **42**, 3027–3032.
Reiss, H. and Hammerich, A. D. (1986). Hard spheres: Scaled particle theory and exact relations on the existence and structure of the fluid/solid phase transition, *J. Phys. Chem.* **90**, 6252–6260.
Reiss, H., Frisch, H. L. and Lebowitz, J. L. (1959). Statistical mechanics of rigid spheres, *J. Chem. Phys.* **31**, 369–380.
Reńyi, A. (1963). On a one-dimensional problem concerning random space filling, *Sel. Trans. Math. Stat. Prob.* **4**, 203–218.
Rice, S. O. (1945). Mathematical analysis of random noise, *Bell. Syst. Tech. J.* **24**, 46–156.
Richards, P. M. (1987). Diffusion to nonoverlapping or spatially correlated traps, *Phys. Rev. B* **35**, 248–256.
Richards, P. M. and Torquato, S. (1987). Upper and lower bounds for the rate of diffusion-controlled reactions, *J. Chem. Phys.* **87**, 4612–4614.
Rikvold, P. A. and Stell G. (1985). d-dimensional interprenetrable-sphere models of random two-phase media: Microstructure and an application to chromatography, *J. Colloid Interface Sci.* **108**, 158–173.
Riley, M. R., Muzzio, F. J. and Reyes, S. C. (1998). Calculation of effective diffusivities and reactivities in immobilized cell systems using finite difference methods, *Comput. Chem. Eng.* **22**, 525–534.
Rintoul, M. D. (2000). Precise determination of the void percolation threshold for two distributions of overlapping spheres, *Phys. Rev. E.* **62**, 68–72.
Rintoul, M. D. and Torquato, S. (1996a). Metastability and crystallization in hard-sphere systems, *Phys. Rev. Lett.* **77**, 4198–4201.
Rintoul, M. D. and Torquato, S. (1996b). Computer simulations of dense hard-sphere systems, *J. Chem. Phys.* **105**, 9258–9265; Erratum, **107**, 2698–2698 (1997).
Rintoul, M. D. and Torquato, S. (1997a). Precise determination of the critical threshold and exponents in a three-dimensional continuum percolation model, *J. Phys. A: Math. Gen.* **30**, L585–L592.
Rintoul, M. D. and Torquato, S. (1997b). Reconstruction of the structure of dispersions, *J. Colloid Interface Sci.* **186**, 467–476.
Rintoul, M. D. and Torquato, S. (1998). Hard-sphere statistics along the metastable amorphous branch, *Phys. Rev. E* **58**, 532–537.
Rintoul, M. D., Torquato, S. and Tarjus, G. (1996). Nearest-neighbor statistics in a one-dimensional random sequential adsorption process, *Phys. Rev. E* **53**, 450–457.
Rintoul, M. D., Torquato, S., Yeong, C., Erramilli, S., Keane, D., Dabbs, D. and Aksay, I. (1996). Structure and transport properties of a porous magnetic gel via X-ray microtomography, *Phys. Rev. E* **54**, 2663–2669.
Roach, S. A. (1968). *The Theory of Random Clumping*, Methuen, London.
Roberts, A. P. (1997). Morphology and thermal conductivity of model organic aerogels, *Phys. Rev. E* **55**, R1286–R1289.
Roberts, A. P. and Knackstedt, M. A. (1996). Structure-property correlations in model composite materials, *Phys. Rev. E* **54**, 2313–2328.
Roberts, A. P. and Teubner, M. (1995). Transport properties of heterogeneous materials derived from Gaussian random fields: Bounds and simulation, *Phys. Rev. E* **51**, 4141–4154.

Roberts, A. P. and Torquato, S. (1999). Chord-distribution functions of three-dimensional random media: Approximate first-passage times of Gaussian processes, *Phys. Rev. E* **59**, 4953–4963.

Rodin, G. J. (1993). The overall elastic response of materials containing spherical inhomogeneities, *Int. J. Solids Structures* **30**, 1849–1863.

Rosen, B. W. and Hashin, Z. (1970). Effective thermal expansion coefficients and specific heats of composite materials, *Int. J. Eng. Sci.* **8**, 157–174.

Rowlinson, J. S. (1964). The virial expansion in two dimensions, *Mol. Phys.* **7**, 593–594.

Rubenfeld, L. and Keller, J. B. (1969). Bounds on the elastic moduli of composite media, *SIAM J. Appl. Math.* **17**, 495–510.

Rubinstein, J. and Keller, J. B. (1987). Lower bounds on permeability, *Phys. Fluids* **30**, 2919–2921.

Rubinstein, J. and Torquato, S. (1988). Diffusion-controlled reactions: Mathematical formulation, variational principles, and rigorous bounds, *J. Chem. Phys.* **88**, 6372–6380.

Rubinstein, J. and Torquato, S. (1989). Flow in random porous media: Mathematical formulation, variational principles, and rigorous bounds, *J. Fluid Mech.* **206**, 25–46.

Rushbrooke, G. S. and Scoins, H. I. (1953). On the theory of fluids, *Proc. R. Soc. Lond. A* **216**, 203–218.

Russel, W. B., Saville, D. A. and Schowalter, W. R. (1989). *Colloidal Dispersions*, Cambridge University Press, Cambridge, England.

Rutgers, M. A., Dunsmuir, J. H., Xue, J. Z., Russel, W. B. and Chaikin, P. M. (1996). Measurement of the hard-sphere equation of state using screened charged polystyrene colloids, *Phys. Rev. B* **53**, 5043–5046.

Sahimi, M. (1994). *Applications of Percolation Theory*, Taylor and Francis, London.

Sahimi, M. and Arbabi, S. (1993). Mechanics of disordered solids. II. Percolation on elastic networks with bond-bending forces, *Phys. Rev. B* **47**, 703–712.

Salacuse, J. J. and Stell, G. (1982). Polydisperse systems: Statistical thermodynamics, with applications to several models including hard and permeable spheres, *J. Chem. Phys.* **77**, 3714–3725.

Salsburg, Z. W., Zwanzig, R. W. and Kirkwood, J. G. (1953). Molecular distribution functions in a one-dimensional fluid, *J. Chem. Phys.* **21**, 1098–1107.

Sanchez-Palencia, E. (1980). *Non-homogeneous Media and Vibration Theory*, Springer-Verlag, Berlin.

Sangani, A. S. and Yao, C. (1988). Transport processes in random arrays of cylinders. I. Thermal conduction, *Phys. Fluids* **31**, 2426–2434.

Sangani, A. S. and Acrivos, A. (1982). Slow flow through a periodic array of spheres, *Int. J. Multiphase Flow* **8**, 343–360.

Sapoval, B., Russ, S., Korb, J. P. and Petit, D. (1996). Nuclear relaxation in irregular or fractal pores, *Fractals* **4**, 453–462.

Saslaw, W. C. (2000). *The Distribution of the Galaxies*, Cambridge University Press, New York.

Schapery, R. A. (1968). Thermal expansion coefficients of composite materials based on energy principles, *J. Comp. Mater.* **2**, 380–404.

Scheidegger, A. E. (1974). *The Physics of Flow Through Porous Media*, University of Toronto Press, Toronto, Canada.

Scher, H. and Zallen, R. (1970). Critcal density in percolation processes, *J. Chem. Phys.* **53**, 3759–3761.

Schulgasser, K. (1976a). On a phase interchange relationship for composite materials, *J. Math. Phys.* **17**, 378–381.

Schulgasser, K. (1976b). Relationship between single-crystal and polycrystal electrical conductivity, *J. Appl. Phys.* **47**, 1880–1886.

Schulgasser, K. (1976c). On the conductivity of fiber reinforced materials, *J. Math. Phys.* **17**, 382–387.

Schulgasser, K. (1977). Concerning the effective transverse conductivity of a two-dimensional two-phase material, *Int. J. Heat Mass Transfer* **20**, 1273–1280.

Schulgasser, K. (1987). Thermal expansion of polycrystalline aggregates with texture, *J. Mech. Phys. Solids* **35**, 35–42.

Schulgasser, K. (1992). A reciprocal theorem in two-dimensional heat transfer and its implications, *Int. Comm. Heat Mass Transfer* **19**, 639–649.

Schulz, G. V. (1939). Über die Kinetik der Kettenpolymerisationen, *Z. Physik Chem.* **B43**, 25–46.

Schwartz, L. M. and Banavar, J. R. (1989). Calculation of electrical transport in continuum systems by diffusion simulation, *Physica A* **157**, 230–234.

Schwartz, L. M., Martys, N., Bentz, D. P., Garboczi, E. J. and Torquato, S. (1993). Cross-property relations and permeability estimation in model porous media, *Phys. Rev. E* **48**, 4584–4591.

Scott, G. D. and Kilgour, D. M. (1969). The density of random close packing of spheres, *Brit. J. Appl. Phys.* **2**, 863–866.

Seaton, N. A. and Glandt, E. D. (1986). Spatial correlation functions from computer simulations, *J. Chem. Phys.* **85**, 5262–5268.

Seaton, N. A. and Glandt, E. D. (1987). Aggregation and percolation in a system of adhesive spheres, *J. Chem. Phys.* **86**, 4668–4677.

Sen, A. K. and Torquato, S. (1988). Series expansions for clustering in continuum: Percolation models with interactions, *J. Chem. Phys.* **89**, 3799–3807.

Sen, A. K. and Torquato, S. (1989). Effective conductivity of anisotropic two-phase composite media, *Phys. Rev. B* **39**, 4504–4515.

Sen, A. K., Lado, F. and Torquato, S. (1987a). Bulk properties of composite media: II. Evaluation of bounds on the shear modulus of suspensions of impenetrable spheres, *J. Appl. Phys.* **62**, 4135–4141.

Sen, A. K., Lado, F. and Torquato, S. (1987b). Bulk properties of composite media. I. Simplification of bounds on the shear modulus of suspensions of impenetrable spheres, *J. Appl. Phys.* **62**, 3503–3513.

Sen, P. N., Scala, C. and Cohen, M. H. (1981). A self-similar model for sedimentary rocks with application to the dielectric constant of fused glass beads, *Geophys.* **46**, 781–795.

Serra, J. (1982). *Image Analysis and Mathematical Morphology*, Academic Press, New York.

Sevick, E. M., Monson, P. A. and Ottino, J. M. (1988). Monte Carlo calculations of cluster statistics in continuum models of composite morphology, *J. Chem. Phys.* **88**, 1198–1206.

Sheehan, N. and Torquato, S. (2001). Generating microstructures with specified correlation functions, *J. Appl. Phys.* **89**, 53–60.

Sheng, P. (1995). *Introduction to Wave Scattering, Localization and Mesoscopic Phenomena*, Academic Press, New York.

Sheng, P. and Kohn, R. V. (1982). Geometric effects in continuous-media percolation, *Phys. Rev. B* **26**, 1331–1335.

Sheng, P. and Zhou, M.-Y. (1988). Dynamic permeability in porous media, *Phys. Rev. Lett.* **61**, 1591–1594.

Shepp, L. A. (1967). Covariances of unit processes, *Proceedings of a Working Conference on Stochastic Processes*, Santa Barbara, California.

Shinbrot, T., Alexander, A. and Muzzio, F. J. (1999). Spontaneous chaotic granular mixing, *Nature* **397**, 675–678.

Sigmund, O. (1994). *Design of Material Structures Using Topology Optimization*, Ph.D. thesis, Department of Solid Mechanics, Technical University of Denmark.

Sigmund, O. and Torquato, S. (1996). Composites with extremal thermal expansion coefficients, *Appl. Phys. Lett.* **69**, 3203–3205.

Sigmund, O. and Torquato, S. (1997). Design of materials with extreme thermal expansion using a three-phase topology optimization method, *J. Mech. Phys. Solids* **45**, 1037–1067.

Sigmund, O., Torquato, S. and Aksay, I. A. (1998). On the design of 1-3 piezocomposites using topology optimization, *J. Mater. Res.* **13**, 1038–1048.

Silnutzer, N. R. (1972). *Effective Constants of Statistically Homogeneous Materials*, Ph.D. thesis, University of Pennsylvania, Philadephia.

Smith, J. C. (1976). Experimental values for the elastic constants of a particulate-filled glassy polymer, *J. Res. NBS* **80A**, 45–49.

Smith, L. N. and Lobb, C. J. (1979). Percolation in two-dimensional conductor–insulator networks with controllable anisotropy, *Phys. Rev. B* **20**, 3653–3658.

Smith, P. A. and Torquato, S. (1988). Computer simulation results for the two-point probability function of composite media, *J. Comput. Phys.* **76**, 176–191.

Smith, P. A. and Torquato, S. (1989). Computer simulation results for bounds on the effective conductivity of composite media, *J. Appl. Phys.* **65**, 893–900.

Smoluchowski, M. V. (1917). Versuch einer mathematischen Theorie der Koagulationskinetik kolloider Lösungen, *Z. Phys. Chem.* **92**, 129–168.

Snyder, K. A. (1998). A numerical test of air void spacing equations, *Adv. Cement Based Mater.* **8**, 28–44.

Sokolnikoff, I. S. (1956). *Mathematical Theory of Elasticity*, McGraw-Hill, New York.

Speedy, R. J. (1994). On the reproducibility of glasses, *J. Chem. Phys.* **100**, 6684–6691.

Stanley, H. E. (1971). *Introduction to Phase Transitions and Critical Phenomena*, Oxford University Press, New York.

Stanley, H. E. (1977). Cluster shapes at the percolation threshold: An effective cluster dimensionality and its connection with critical-point exponents, *J. Phys. A: Math. Gen.* **10**, L211–L220.

Stauffer, D. and Aharony, A. (1992). *Introduction to Percolation Theory*, Taylor and Francis, London.

Steinhardt, P. J., Nelson, D. R. and Ronchetti, M. (1983). Bond-orientational order in liquids and glasses, *Phys. Rev. B* **28**, 784–805.

Stell, G. (1964). Cluster expansions for classical systems in equilibrium, in *The Equilibrium Theory of Classical Fluids*, H. L. Frisch and J. L. Lebowitz (eds.), Benjamin, New York.

Stell, G. (1966). Boltzmann's method of evaluating and using molecular distribution functions, *Technical report*, Brooklyn Polytechnic Institute.

Stell, G. (1984). Exact equation for the pair-connectedness function, *J. Phys. A: Math. Gen.* **17**, L855–L858.

Stell, G. (1985). Mayer-Montroll equations (and some variants) through history for fun and profit, in *The Wonderful World of Stochastics : A Tribute to Elliott W. Montroll*, M. F. Shlesinger and G. H. Weiss (eds.), Amsterdam, New York.

Stell, G. and Rikvold, P. A. (1987). Polydispersity in fluids and composites: Some theoretical results, *Chem. Eng. Comm.* **51**, 233–260.

Stokes, G. G. (1851). On the effect of the internal friction of fluids on the motion of pendulums, *Trans. Camb. Phil. Soc.* **9**, 8–106.

Stoyan, D., Kendall, W. S. and Mecke, J. (1995). *Stochastic Geometry and Its Applications*, second edition, Wiley, New York.

Stratton, J. A. (1941). *Electromagnetic Theory*, McGraw-Hill, New York.

Strieder, W. and Aris, R. (1973). *Variational Methods Applied to Problems of Diffusion and Reaction*, Springer-Verlag, New York.

Stroscio, J. A. and Kaiser, W. J. (1993). *Scanning Tunneling Microscopy*, Academic Press, New York.

Suding, P. N. and Ziff, R. M. (1999). Site percolation thresholds for Archimedean lattices, *Phys. Rev. E* **60**, 275–283.

Sugawara, A. and Yoshizawa Y. (1962). An experimental investigation on the thermal conductivity of consolidated porous materials, *J. Appl. Phys.* **33**, 3135–3138.

Swendsen, R. H. (1981). Dynamics of random sequential adsorption, *Phys. Rev. A* **24**, 504–508.

Sykes, M. F. and Essam, J. W. (1964). Exact critical percolation probabilities for site and bond problems in two dimensions, *J. Math. Phys.* **5**, 1117–1127.

Talbot, D. R. S. and Willis, J. R. (1984). The overall sink strength of an inhomogeneous lossy medium. I. Self-consistent estimates, *Mech. Mater.* **3**, 171–181.

Talbot, D. R. S. and Willis, J. R. (1987). Bounds and self-consistent estimates for the overall properties of nonlinear composites, *IMA J. Appl. Math.* **39**, 215–240.

Talbot, J. and Schaaf, P. (1989). Random sequential adsorption of mixtures, *Phys. Rev. A* **40**, 422–427.

Tanemura, M. and Hasegawa, M. (1980). Geometrical models of territory. I. Models for synchronous and asynchronous settlement of territories, *J. Theor. Biol.* **82**, 477–496.

Tao, R. B., Chen, Z. and Sheng, P. (1990). First-principles Fourier approach for the calculation of the effective dielectric constant of periodic composites, *Phys. Rev. B* **41**, 2417–2420.

Tartar, L. (1978). Estimations de coefficients homogénéisés, in *Computing Methods in Applied Sciences and Engineering*, R. Glowinski and J. L. Lions (eds.), Springer-Verlag, New York.

Tartar, L. (1979). Compensated compactness and applications to partial differential equations, in *Nonlinear Analysis and Mechanics*, R. J. Knops (ed.), Pitman, Boston.

Tartar, L. (1980). In *Non-homogeneous Media and Vibration Theory*, Springer-Verlag, Berlin. Appendix.

Tartar, L. (1985). Estimations fines des coefficients homogénéisés, in *Ennio De Giorgi Colloquium*, P. Krée (ed.), Pitman, Boston.

Tassopoulos, M. and Rosner, D. E. (1992). Simulation of vapor diffusion in anisotropic particulate deposits, *Chem. Eng. Sci.* **47**, 421–443.

Taylor, G. I. (1954). The two coefficients of viscosity for an incompressible fluid containing air bubbles, *Proc. R. Soc. Lond. A* **226**, 34–37.

Temam, R. (1979). *Navier–Stokes Equations*, North-Holland, New York.

Teubner, M. (1991). Level surfaces of Gaussian random fields and microemulsions, *Europhys. Lett.* **14**, 403–408.

Thiele, E. (1963). Equation of state for hard spheres, *J. Chem. Phys.* **39**, 474–479.

Thies-Weesie, M. (1995). *Sedimentation and Liquid Permeation of Inorganic Colloids*, Ph.D. thesis, Universiteit Utrecht, Netherlands.

Thompson, A. H., Katz, A. J. and Krohn, C. E. (1987). The microgeometry and transport properties of sedimentary rock, *Adv. Phys.* **36**, 625–694.

Thovert, J. F., Kim, I. C., Torquato, S. and Acrivos, A. (1990). Bounds on the effective properties of polydispersed suspensions of spheres: An evaluation of two relevant morphological parameters, *J. Appl. Phys.* **67**, 6088–6098.

Tirrell, D. A. (1996). Putting a new spin on spider silk, *Science* **271**, 39–40.

Tobochnik, J. and Chapin, P. M. (1988). Monte Carlo simulation of hard spheres near random closest packing using spherical boundary conditions, *J. Chem. Phys.* **88**, 5824–5830.

Tokunaga, T. K. (1985). Porous media gas diffusivity from a free path distribution model, *J. Chem. Phys.* **82**, 5298–5299.

Tolman, R. C. (1979). *Principles of Statistical Mechanics*, Dover Publications, New York.

Torney, D. C. and Goldstein, B. (1987). Rates of diffusion-limited reaction in periodic systems, *J. Stat. Phys.* **49**, 725–750.

Torquato, S. (1980). *Microscopic Approach to Transport in Two-Phase Random Media*, Ph.D. thesis, State University of New York at Stony Brook.

Torquato, S. (1984). Bulk properties of two-phase disordered media. I. Cluster expansion for the effective dielectric constant of dispersions of penetrable spheres, *J. Chem. Phys.* **81**, 5079–5088.

Torquato, S. (1985a). Effective electrical conductivity of two-phase disordered composite media, *J. Appl. Phys.* **58**, 3790–3797.

Torquato, S. (1985b). Bulk properties of two-phase disordered media. II. Effective conductivity of a dilute dispersion of penetrable spheres, *J. Chem. Phys.* **83**, 4776–4785.

Torquato, S. (1986a). Interfacial surface statistics arising in diffusion and flow problems in porous media, *J. Chem. Phys.* **85**, 4622–4628.

Torquato, S. (1986b). Bulk properties of two-phase disordered media. III. New bounds on the effective conductivity of dispersions of penetrable spheres, *J. Chem. Phys.* **84**, 6345–6359.

Torquato, S. (1986c). Microstructure characterization and bulk properties of disordered two-phase media, *J. Stat. Phys.* **45**, 843–873.

Torquato, S. (1986d). Concentration dependence of diffusion-controlled reactions among static reactive sinks, *J. Chem. Phys.* **85**, 7178–7179.

Torquato, S. (1987). Thermal conductivity of disordered heterogeneous media from the microstructure, *Rev. Chem. Eng.* **4**, 151–204.

Torquato, S. (1990). Relationship between permeability and diffusion-controlled trapping constant of porous media, *Phys. Rev. Lett.* **64**, 2644–2646.

Torquato, S. (1991a). Diffusion and reaction among traps: Some theoretical and simulation results, *J. Stat. Phys.* **65**, 1173–1206.

Torquato, S. (1991b). Random heterogeneous media: Microstructure and improved bounds on the effective properties, *Appl. Mech. Rev.* **44**, 37–76.

Torquato, S. (1991c). Trapping of finite-sized Brownian particles in porous media, *J. Chem. Phys.* **95**, 2838–2841.

Torquato, S. (1992). Connection between the morphology and effective properties of heterogeneous materials, in *Macroscopic Behavior of Heterogeneous Materials from the Microstructure*, S. Torquato and D. Krajcinovic (eds.), Vol. 147, Am. Soc. Mech. Eng. - AMD, pp. 53–65.

Torquato, S. (1995a). Nearest-neighbor statistics for packings of hard spheres and disks, *Phys. Rev. E* **51**, 3170–3182.

Torquato, S. (1995b). Mean nearest-neighbor distance in random packings of hard d-dimensional spheres, *Phys. Rev. Lett.* **74**, 2156–2159.

Torquato, S. (1997). Effective stiffness tensor of composite media. I. Exact series expansions, *J. Mech. Phys. Solids* **45**, 1421–1448.

Torquato, S. (1998). Effective stiffness tensor of composite media. II. Applications to isotropic dispersions, *J. Mech. Phys. Solids* **46**, 1411–1440.

Torquato, S. (1999). Exact conditions on physically realizable correlation functions of random media, *J. Chem. Phys.* **111**, 8832–8837.

Torquato, S. (2000a). Modeling of physical properties of composite materials, *Int. J. Solids Structures* **37**, 411–422.

Torquato, S. (2000b). Glass transition: Hard knock for thermodynamics, *Nature* **405**, 521–522.

Torquato, S. and Avellaneda, M. (1991). Diffusion and reaction in heterogeneous media: Pore size distribution, relaxation times, and mean survival time, *J. Chem. Phys.* **95**, 6477–6489.

Torquato, S. and Beasley, J. D. (1986a). Effective properties of fiber-reinforced materials. I. Bounds on the effective thermal conductivity of dispersions of fully penetrable cylinders, *Int. J. Eng. Sci.* **24**, 415–433.

Torquato, S. and Beasley, J. D. (1986b). Effective properties of fiber-reinforced materials. II. Bounds on the effective elastic moduli of dispersions of fully penetrable cylinders, *Int. J. Eng. Sci.* **24**, 435–447.

Torquato, S. and Beasley, J. D. (1987). Bounds on the permeability of a random array of partially penetrable spheres, *Phys. Fluids* **30**, 633–641.

Torquato, S. and Hyun, S. (2001). Effective-medium approximation for composite media: Realizable single-scale dispersions, *J. Appl. Phys.* **89**, 1725–1729.

Torquato, S. and Kim, I. C. (1989). Efficient simulation technique to compute effective properties of heterogeneous media, *Appl. Phys. Lett.* **55**, 1847–1849.

Torquato, S. and Kim, I. C. (1992). Cross-property relations for momentum and diffusional transport in porous media, *J. Appl. Phys.* **72**, 2612–2619.

Torquato, S. and Lado, F. (1985). Characterisation of the microstructure of distributions of rigid rods and discs in a matrix, *J. Phys. A: Math. Gen.* **18**, 141–148.

Torquato, S. and Lado, F. (1986). Effective properties of two-phase disordered composite media. II. Evaluation of bounds on the conductivity and bulk modulus of dispersions of impenetrable spheres, *Phys. Rev. B* **33**, 6428–6435.

Torquato, S. and Lado, F. (1988a). Bounds on the conductivity of a random array of cylinders, *Proc. R. Soc. Lond. A* **417**, 59–80.

Torquato, S. and Lado, F. (1988b). Bounds on the effective transport and elastic properties of a random array of cylindrical fibers in a matrix, *J. Appl. Mech.* **55**, 347–354.

Torquato, S. and Lado, F. (1991). Trapping constant, thermal conductivity, and the microstructure of suspensions of oriented spheroids, *J. Chem. Phys.* **94**, 4453–4462.

Torquato, S. and Lee, S. B. (1990). Computer simulations of nearest-neighbor distribution functions and related quantities for hard-sphere systems, *Physica A* **167**, 361–383.

Torquato, S. and Lu, B. (1990). Rigorous bounds on the fluid permeability: Effect of polydispersivity in grain size, *Phys. Fluids A* **2**, 487–490.

Torquato, S. and Lu, B. (1993). Chord-length distribution function for two-phase random media, *Phys. Rev. E* **47**, 2950–2953.

Torquato, S. and Rintoul, M. D. (1995). Effect of the interface on the properties of composite media, *Phys. Rev. Lett.* **75**, 4067–4070. Erratum vol. 76, 3241 (1996).

Torquato, S. and Rubinstein, J. (1989). Diffusion-controlled reactions. II. Further bounds on the rate constant, *J. Chem. Phys.* **90**, 1644–1647.

Torquato, S. and Rubinstein, J. (1991). Improved bounds on the effective conductivity of high-contrast suspensions, *J. Appl. Phys.* **69**, 7118–7125.

Torquato, S. and Sen, A. K. (1990). Conductivity tensor of anisotropic composite media from the microstructure, *J. Appl. Phys.* **67**, 1145–1155.

Torquato, S. and Stell, G. (1982). Microstructure of two-phase random media. I. The n-point probability functions, *J. Chem. Phys.* **77**, 2071–2077.

Torquato, S. and Stell, G. (1983a). Microstructure of two-phase random media. II. The Mayer–Montroll and Kirkwood–Salsburg hierarchies, *J. Chem. Phys.* **78**, 3262–3272.

Torquato, S. and Stell, G. (1983b). Microstructure of two-phase random media. III. The n-point matrix probability functions for fully penetrable spheres, *J. Chem. Phys.* **79**, 1505–1510.

Torquato, S. and Stell, G. (1984). Microstructure of two-phase random media. IV. Expected surface area of a dispersion of penetrable spheres and its characteristic function, *J. Chem. Phys.* **80**, 878–880.

Torquato, S. and Stell, G. (1985a). Microstructure of two-phase random media. V. The n-point matrix probability functions for impenetrable spheres, *J. Chem. Phys.* **82**, 980–987.

Torquato, S. and Stell, G. (1985b). Bounds on the effective thermal conductivity of a dispersion of fully penetrable spheres, *Int. J. Eng. Sci.* **23**, 375–383.

Torquato, S. and Yeong, C. L. Y. (1997). Universal scaling for diffusion-controlled reactions among traps, *J. Chem. Phys.* **106**, 8814–8820.

Torquato, S., Beasley, J. D. and Chiew, Y. C. (1988). Two-point cluster function for continuum percolation, *J. Chem. Phys.* **88**, 6540–6547.

Torquato, S., Gibiansky, L. V., Silva, M. J. and Gibson, L. J (1998). Effective mechanical and transport properties of cellular solids, *Int. J. Mech. Sci.* **40**, 71–82.

Torquato, S., Kim, I. C. and Cule, D. (1999b). Effective conductivity, dielectric constant, and diffusion coefficient of digitized composite media via first-passage-time equations, *J. Appl. Phys.* **85**, 1560–1571.

Torquato, S., Lado, F. and Smith, P. A. (1987). Bulk properties of two-phase disordered media. IV. Mechanical properties of suspensions of penetrable spheres at nondilute concentrations, *J. Chem. Phys.* **86**, 6388–6393.

Torquato, S., Lu, B. and Rubinstein, J. (1990). Nearest-neighbor distribution functions in many-body sytems, *Phys. Rev. A* **41**, 2059–2075.

Torquato, S., Stell, G. and Beasley, J. (1985). Third-order bounds on the effective bulk and shear moduli of a dispersion of fully penetrable spheres, *Int. J. Eng. Sci.* **23**, 385–392.

Torquato, S., Truskett, T. M. and Debenedetti, P. G. (2000). Is random close packing of spheres well defined?, *Phys. Rev. Lett.* **84**, 2064–2067.

Torquato, S., Yeong, C. L. Y., Rintoul, M. D., Milius, D. L. and Aksay, I. A. (1999a). Elastic properties and structure of interpenetrating boron carbide aluminum multiphase composites, *J. Am. Ceram. Soc.* **82**, 1263–1268.

Truskett, T. M., Torquato, S. and Debenedetti, P. G. (1998). Density fluctuations in many-body systems, *Phys. Rev. E* **58**, 7639–7380.

Truskett, T. M., Torquato, S. and Debenedetti, P. G. (2000). Towards a quantification of disorder in materials: Distinguishing equilibrium and glassy sphere packings, *Phys. Rev. E* **62**, 993–1001.

Underwood, E. E. (1970). *Quantitative Stereology*, Addison–Wesley, Reading, Massachusetts.

Van Den Berg, J. (1981). Percolation theory on pairs of matching lattices, *J. Math. Phys.* **22**, 152–157.

Vanmarcke, E. (1983). *Random Fields: Analysis and Synthesis*, MIT Press, Cambridge, Massachusetts.

Vericat, F., Gianotti, R. D. and Rodriguez, A. E. (1987). Exact pair-connectedness function of a one-dimensional hard rod fluid, *J. Phys. A: Math. Gen.* **20**, 6155–6158.

Verlet, L. (1967). Computer experiments on classical fluids. I. Thermodynamical properties of Lennard-Jones molecules, *Phys. Rev.* **159**, 98–103.

Verlet, L. and Weis, J. J. (1972). Perturbation theory for the thermodynamic properties of simple liquids, *Mol. Phys.* **24**, 1013–1024.

Vezzetti, D. J. (1975). A new derivation of some fluctuation theorems in statistical mechanics, *J. Math. Phys.* **16**, 31–33.

Vigdergauz, S. B. (1989). Regular structures with extremal elastic properties, *Mech. Solids* **24**, 57–63.

Vigdergauz, S. B. (1994). Two-dimensional grained composites of extreme rigidity, *J. Appl. Mech.* **61**, 390–394.

Viot, P., Tarjus, G. and Talbot, J. (1993). Exact solution of a generalizd ballistic-deposition model, *Phys. Rev. E* **48**, 480–488.

Viot, P., Tarjus, G., Ricci, S. M. and Talbot, J. (1992). Random sequential adsorption of anisotropic particles. I. Jamming limit and asymptotic behavior, *J. Chem. Phys.* **97**, 5212–5218.

Visscher W. M. and Bolsterli, M. (1972). Random packing of equal and unequal spheres in two and three dimensions, *Nature* **239**, 504–507.

Walpole, L. J. (1966a). On bounds for the overall elastic moduli of inhomogeneous systems. I, *J. Mech. Phys. Solids* **14**, 151–162.

Walpole, L. J. (1966b). On bounds for the overall elastic moduli of inhomogeneous systems. II, *J. Mech. Phys. Solids* **14**, 289–301.

Watt, J. P., Davies, G. F. and O'Connell, R. J. (1976). The elastic properties of composite materials, *Revs. Geophys. Space Phys.* **14**, 541–563.

Weaire, D. and Hutzler, S. (2000). *The Physics of Foams*, Oxford University Press, Oxford, England.

Weaire, D. and Rivier, N. (1984). Soap, cells and statistics: Random patterns in two dimensions, *Contemp. Phys.* **25**, 59–99.

Weber, H., Marx, D. and Binder, K. (1995). Melting transition in two dimensions: A finite-size scaling analysis of bond-orientational order in hard disks, *Phys. Rev. B* **51**, 14636–14651.

Weissberg, H. L. (1963). Effective diffusion coefficient in porous media, *J. Appl. Phys.* **34**, 2636–2639.

Weissberg, H. L. and Prager, S. (1970). Viscous flow through porous media. III. Upper bounds on the permeability for a simple random geometry, *Phys. Fluids* **13**, 2958–2965.

Wells, D. G. (1991). *The Penguin Dictionary of Curious and Interesting Geometry*, Penguin, London, England.

Weng, G. J. (1984). Some elastic properties of reinforced solids with special reference to isotropic ones containing spherical inclusions, *Int. J. Eng. Sci.* **22**, 845–856.

Weng, G. J. (1992). Explicit evaluation of Willis' bounds with ellipsoidal inclusions, *Int. J. Eng. Sci.* **30**, 83–92.

Wertheim, M. S. (1963). Exact solution of the Percus–Yevick integral equation for hard spheres, *Phys. Rev. Lett.* **10**, 321–323.

Widom, B. (1966). Random sequential addition of hard spheres to a volume, *J. Chem. Phys.* **44**, 3888–3894.

Wiener, O. (1912). Die Theorie des Mischkörpers für das Feld des stationären Strömung, *Abh. Math.- Physichen Klasse Königl. Säcsh. Gesel. Wissen.* **32**, 509–604.

Wilkinson, D. J., Johnson, D. L. and Schwartz, L. M. (1991). Nuclear magnetic relaxation in porous media: The role of the mean lifetime $\tau(\rho, d)$, *Phys. Rev. B* **44**, 4960–4971.

Willis, J. R. (1977). Bounds and self-consistent estimates for the overall properties of anisotropic composites, *J. Mech. Phys. Solids* **25**, 185–202.

Willis, J. R. (1981). Variational and related methods for the overall properties of composites, *Adv. Appl. Mech.* **21**, 1–78.

Winterfeld, P. H., Scriven, L. E. and Davis, H. T. (1981). Percolation and conductivity of random two-dimensional composites, *J. Phys. C.* **14**, 2361–2376.

Witten, T. A. and Sander, L. M. (1981). Diffusion-limited aggregation: A kinetic critical phenomenon, *Phys. Rev. Lett.* **47**, 1400–1403.

Wu, T. T. (1966). The effect of inclusion shape on the elastic moduli of two-phase material, *Int. J. Solids Structures* **2**, 1–8.

Xia, W. and Thorpe, M. F. (1988). Percolation properties of random ellipses, *Phys. Rev. A.* **38**, 2650–2656.

Xu, B., Arias, F., Brittain, S. T., Zhao, X.-M., Gryzbowski, B., Torquato, S. and Whitesides, G. M. (1999). Making negative Poisson's ratio microstructures by soft lithography, *Advanced Materials* **11**, 1186–1189.

Xu, J. and Stell, G. (1988). Pair connectedness and mean cluster size of randomly centered spheres: Beyond the Percus–Yevick approximation, *Technical report*, SUNY Stony Brook. CEAS Report No. 511.

Yau, W. W., Kirkland, J. J. and Bly, D. D. (1979). *Modern Size-Exclusion Liquid Chromotography*, Wiley, New York.

Yeong, C. L. Y. and Torquato, S. (1998a). Reconstructing random media, *Phys. Rev. E* **57**, 495–506.

Yeong, C. L. Y. and Torquato, S. (1998b). Reconstructing random media. II. Three-dimensional media from two-dimensional cuts, *Phys. Rev. E* **58**, 224–233.

Yonezawa, F. and Cohen, M. H. (1983). Granular effective medium approximation, *J. Appl. Phys.* **54**, 2895–2899.

Young, A. P. (1979). Melting and the vector Coulomb gas in two dimensions, *Phys. Rev. B* **19**, 1855–1866.

Zallen, R. (1983). *The Physics of Amorphous Solids*, Wiley, New York.

Zernike, F. and Prins, J. A. (1927). Die Beugung von Röntgenstrahlen in Flüssigkeiten als Effekt der Molekülanordnung, *Z. Phys.* **41**, 184–194.

Zhou, H. X. and Szabo, A. (1996). Theory and simulation of stochastically-gated diffusion-influenced reactions, *J. Phys. Chem.* **100**, 2597–2604.

Zick, A. A. and Homsy, G. M. (1982). Stokes flow through periodic arrays of spheres, *J. Fluid Mech.* **115**, 13–26.

Ziff, R. M. (1977). On the bulk distribution functions and fluctuation theorems, *J. Math. Phys.* **18**, 1825–1831.

Ziff, R. M. and Sapoval, B. (1986). The efficient determination of the percolation threshold by a frontier-generating walk in a gradient, *J. Phys. A: Math. Gen.* **19**, L1169–L1172.

Ziman, J. M. (1979). *Models of Disorder*, Cambridge University Press, New York.

Zinchenko, A. Z. (1994). Algorithm for random close packing of spheres with periodic boundary conditions, *J. Comput. Phys.* **114**, 298–307.

Zoia, G. and Strieder, W. (1997). Transport coefficients in random two-phase media with interfacial resistance, *Phys. Rev. B* **56**, 1249–1253.

Zwanzig, R. (1990). Diffuson-controlled ligand binding to spheres partially covered by receptors: An effective-medium treatment, *Proc. Natl. Acad. Sci. USA* **87**, 5856–5857.

Index

Interdisciplinary Applied Mathematics

Volume 16

Problems in engineering, computational science, and the physical and biological sciences are using increasingly sophisticated mathematical techniques. Thus, the bridge between the mathematical sciences and other disciplines is heavily traveled. The correspondingly increased dialog between the disciplines has led to the establishment of the series: *Interdisciplinary Applied Mathematics*.

The purpose of this series is to meet the current and future needs for the interaction between various science and technology areas on the one hand and mathematics on the other. This is done, firstly, by encouraging the ways that mathematics may be applied in traditional areas, as well as point towards new and innovative areas of applications; and secondly, by encouraging other scientific disciplines to engage in a dialog with mathematicians outlining their problems to both access new methods and suggest innovative developments within mathematics itself.

The series will consist of monographs and high-level texts from researchers working on the interplay between mathematics and other fields of science and technology.